Lecture Notes in Civil Engineering

Volume 88

Lecture Notes in Civil Engineering (LNCE) publishes the latest developments in Civil Engineering - quickly, informally and in top quality. Though original research reported in proceedings and post-proceedings represents the core of LNCE, edited volumes of exceptionally high quality and interest may also be considered for publication. Volumes published in LNCE embrace all aspects and subfields of, as well as new challenges in, Civil Engineering. Topics in the series include:

- Construction and Structural Mechanics
- Building Materials
- Concrete, Steel and Timber Structures
- Geotechnical Engineering
- Earthquake Engineering
- Coastal Engineering
- Ocean and Offshore Engineering; Ships and Floating Structures
- Hydraulics, Hydrology and Water Resources Engineering
- Environmental Engineering and Sustainability
- Structural Health and Monitoring
- Surveying and Geographical Information Systems
- Indoor Environments
- Transportation and Traffic
- Risk Analysis
- Safety and Security

To submit a proposal or request further information, please contact the appropriate Springer Editor:

- Mr. Pierpaolo Riva at pierpaolo.riva@springer.com (Europe and Americas);
- Ms. Swati Meherishi at swati.meherishi@springer.com (Asia - except China, and Australia, New Zealand);
- Dr. Mengchu Huang at mengchu.huang@springer.com (China).

All books in the series now indexed by Scopus and EI Compendex database!

More information about this series at http://www.springer.com/series/15087

Madhavi Latha Gali · P. Raghuveer Rao
Editors

Problematic Soils and Geoenvironmental Concerns

Proceedings of IGC 2018

Springer

Editors
Madhavi Latha Gali
Department of Civil Engineering
Indian Institute of Science
Bengaluru, Karnataka, India

P. Raghuveer Rao
Department of Civil Engineering
Indian Institute of Science
Bengaluru, Karnataka, India

ISSN 2366-2557 ISSN 2366-2565 (electronic)
Lecture Notes in Civil Engineering
ISBN 978-981-15-6236-5 ISBN 978-981-15-6237-2 (eBook)
https://doi.org/10.1007/978-981-15-6237-2

This Springer imprint is published by the registered company Springer Nature Singapore Pte Ltd.
The registered company address is: 152 Beach Road, #21-01/04 Gateway East, Singapore 189721, Singapore

Preface

Indian Geotechnical Conference (IGC 2018) was held at the National Science Complex of the Indian Institute of Science, Bangalore, during 13–15 December 2018. This is the annual conference of the Indian Geotechnical Society (IGS), which was established in the year 1948 with the aim to promote cooperation among the engineers, scientists and practitioners for the advancement and dissemination of knowledge in the field of geotechnical engineering. IGC 2018 was a special event since it coincided with the 70 years celebrations of IGS.

The conference was a grand event with about 700 participants. The conference was inaugurated on 13 December in the presence of President of IGS Prof. G. L. Sivakumar Babu and the Chief Guest Prof. E. C. Shin, Vice-President, Asia, International Society of Soil Mechanics and Geotechnical Engineering (ISSMGE). The conference had 14 keynote lectures and 12 theme lectures presented by eminent academicians and practitioners from different parts of the world. Totally, 313 technical papers under 12 different themes of the conference were presented during the conference in 19 oral presentation sessions and 10 digital display sessions. All the participants of the conference had a common vision of deliberating on current geotechnical engineering research and practice and to strengthen the relationship between scientists, researchers and practising engineers within the fields of geotechnical engineering and to bring focus to problems that are relevant to society's needs and develop solutions. The conference acted as a platform to academicians and field engineers to interact, share knowledge and experiences and identify potential collaborations. The conference also provided opportunity to many young students, researchers and engineers and helped them to get connected to people involved in geotechnical engineering research and practice and national and international groups and technical committees.

All papers submitted to IGC 2018 had undergone a peer-review process and subsequently revised before being accepted. To publish conference proceedings through Springer, selected papers from the conference were grouped into four different volumes, namely Geotechnical Characterization and Modelling, Construction in Geotechnical Engineering, Geohazards and Problematic Soils and Ground Improvement. This book on *Problematic Soils and Geoenvironmental Concerns* has

68 chapters, mainly discussing the challenges associated with certain types of soils and possible remedies and solutions. This book encompasses vast subject areas of geoenvironmental engineering, expansive and collapsible soils, ground improvement and geosynthetics. While some of the chapters on geoenvironmental engineering deal with conventional soil stabilization methods using admixtures, it is quite heartening to see that many chapters discussed trending topics like biochar, erosion hotspots, electrical resistivity tomography and bio-enzymes. Several chapters discussed issues related to remediation of contaminated soils and landfills. The chapters on ground improvement covered stone columns, prefabricated vertical drains, electro-osmosis, soilcrete columns and geosynthetic inclusions to improve strength or drainage properties of soils in different geotechnical applications. Some of the solutions discussed in this book pertain to live projects, providing realistic solutions to deal with problematic soils through soil remediation or ground improvement.

We sincerely thank the Indian Geotechnical Society, especially Prof. G. L. Sivakumar Babu, President, IGS, and Prof. J. T. Shahu, Honorary Secretary, IGS, for their great support in organizing the conference. We also thank the Organizing Committee of IGC 2018; Prof. P. V. Sivapullaiah, Conference Chair; Prof. H. N. Ramesh, Conference Vice-Chair; Dr. C. R. Parthasarathy, Prof. P. Anbazhagan and Prof. K. V. Vijayendra, Organizing Secretaries; and Prof. K. Vijaya Bhaskar Raju, Treasurer, for all their hard work, long working hours spent and responsibility shared in planning and executing various tasks of this outstanding event. The unconditional support extended by the Conference Advisory Committee, Technical Committee, sponsors of the conference, keynote speakers, theme speakers, session chairs, session coordinators, student volunteers, participants, presenters and authors of the technical papers in making the conference a grand success is sincerely appreciated. We thank the entire Springer Team, in particular Swati Meherishi, Rini Christy Xavier Rajasekaran, Muskan Jaiswal and Ashok Kumar, for their hard work and support in bringing out the proceedings of IGC 2018.

Bengaluru, India

Madhavi Latha Gali
P. Raghuveer Rao
(Editors)

Contents

About the Editors

Dr. Madhavi Latha Gali is a Professor in the Department of Civil Engineering, Indian Institute of Science (IISc) Bangalore, India. She completed her Ph.D. from Indian Institute of Technology Madras, and has previously worked as a post-doctoral fellow and assistant professor at IISc and IIT Guwahati respectively. Professor Latha is a member of various professional bodies including IGS, ISSMGE and ISRM, and is the Editor-in-Chief of the Indian Geotechnical Journal, and an Editorial board member in many reputed journals. Her research work focuses on fundamental aspects of soil and ground reinforcement, and she has authored 70 journal articles, 4 book chapters and has developed a web-course on Geotechnical Earthquake Engineering on the NPTEL platform, sponsored by the Ministry of Human Resources Development, Government of India.

Dr. P. Raghuveer Rao is Principal Research Scientist at Department of Civil Engineering and involved in teaching, research and consultancy in the broad area of geotechnical engineering. He has been working with the department since 1989 and has been teaching courses related to subsurface exploration and soil testing, Earth retaining structures, behavior and testing of unsaturated soils, and fundamental of soil behavior for Masters and Doctoral students. His research interests are geotechnical instrumentation, slope stability analysis, numerical modelling, mechanics of unsaturated soils, contaminant transport through soil and reinforced earth structures. He has conducted several field and laboratory tests for design of foundations of different structures like buildings, turbo-generator and water tanks. He has analyzed stability of several embankments, tailing dams and stability of large size surge shafts for a hydropower project through numerical modelling and trial wedge method. He has 21 publications in journals and conference proceedings.

Laboratory Investigations on Geotechnical Properties of Screened Bottom Ash from Two MSW Incineration Plants in Delhi

Garima Gupta, Debanjana Gupta, Manoj Datta, G. V. Ramana, Shashank Bishnoi, and B. J. Alappat

Abstract Municipal solid waste (MSW) incineration has recently started in India with many new MSW waste-to-energy (WtE) plants underway. The process residue of MSW Incineration (MSWI) is primarily bottom ash (BA), which is being dumped to MSW landfills. The present study is an attempt to investigate the geotechnical properties of MSWI BA from two such plants in Delhi with an objective to evaluate its potential for reuse in bulk geotechnical applications. The results have been compared with coal BA (CBA) from a nearby thermal power plant and local sand. It was found that MSWI BA is comparatively coarser but strength properties are in the similar range to that of CBA and local sand. Compaction densities and specific gravity were lower than local soil but much higher than CBA. With new plants emerging all across the country, this study is a starting point for operators to plan the disposal of these residues effectively as well as save the limited land resources.

Keywords Bottom ash · Municipal solid waste · Incineration · Waste to energy · Combustible fraction · Ballistic separators

1 Introduction

The trend followed by most of the European nations and some Asian countries like Japan, China, Taiwan, etc. to incinerate MSW in WtE plants (An et al. 2014; Inkaew et al. 2014; Yu et al. 2013; Chang et al. 2009) is being increasingly adopted in India. Some of these WtE plants have become operational in last few years and many new facilities are being planned under '*Swachh Bharat Mission*' (MNRE 2017). However, no data is available from India on the engineering properties of the MSWI residues which are presently being dumped back to the landfills. Bottom ash (BA), which forms the major proportion (70–80%) of these residues, has a potential for

G. Gupta (✉) · D. Gupta · M. Datta · G. V. Ramana · S. Bishnoi · B. J. Alappat
Department of Civil Engineering, Indian Institute of Technology Delhi, New Delhi, Delhi, India
e-mail: gamma840@gmail.com

M. Latha Gali and R. R. P. (eds.), *Problematic Soils and Geoenvironmental Concerns*, Lecture Notes in Civil Engineering 88,
https://doi.org/10.1007/978-981-15-6237-2_1

bulk reuse in earthworks and road construction, and if the same could be ascertained for the Indian conditions of MSW, the footprint of future landfills can be reduced significantly.

2 Objectives

The present study aims to evaluate reuse potential of MSWI BA for bulk geotechnical applications by comparing its geotechnical properties with BA from a nearby coal-fired thermal power plant and the local soil. Only the fraction passing 4.75 mm which constituted 60–70% of the MSWI BA has been tested for the purpose. The oversized fraction (above 4.75 mm) consisted of glass, ceramic, metal, unburnt organics such as paper, textile, plastic, etc. and gravelly material such as stones, brick bats, sintered material and construction and demolition waste (Gupta et al. 2018).

3 Material and Methods

MSWI BA samples were collected from two MSW WtE plants in Delhi, P1 and P2. The operating temperature in both the plants was in the range of 850–1000 °C. Both the plants were segregating the waste before burning in the furnace. The segregation operations aimed to reduce the moisture content of the waste and separate combustible fraction (e.g., paper, textiles, wood, etc.) from non-combustible fraction (e.g., stones, brick bats, etc.). P1 was using ballistic separators while P2 used trommels to perform segregation operations.

About a ton of BA was sampled winter months, i.e., January and February 2018 for 3–4 days in a two-week period from each plant. It was air-dried for 4–5 days and then sieved through 4.75 mm sieve. The fraction passing 4.75 mm was collected and stored in sealed plastic containers. The reference materials for comparison, i.e., coal bottom ash (CBA) and Yamuna sand (YS) were sampled locally. Grain-size distribution curves, specific gravity, compaction behavior and shear strength properties were studied. Compositional analysis and some physico-chemical properties of MSWI BA have been reported in Gupta et al. (2018).

4 Experimental Study

4.1 Grain-Size Distribution (GSD)

Atleast three set of tests were performed to estimate the GSD of each material as per Indian standards (IS) 2720 Part IV. Table 1 depicts percentage of coarse sand,

Table 1 Comparison of grain-size distribution of MSWI bottom ash with reference materials (wt%)

Size (mm)	Classification	BA P1	BA P2	CBA	YS
4.75–2.00	Coarse sand	11	8	1	0
2.00–0.425	Medium sand	38	33	15	2
0.425–0.075	Fine sand	37	39	67	90
0.075–0	Silt + clay	13	20	16	8

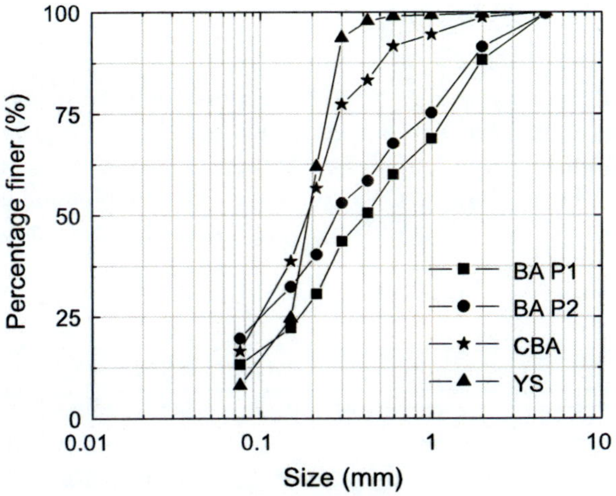

Fig. 1 Grain-size distribution curves of MSWI bottom ash, CBA and YS

medium sand, fine sand and silt plus clay in each of them. Figure 1 shows GSD curve of MSWI BA from WtE plants, P1 and P2, CBA and YS.

4.2 *Specific Gravity*

The specific gravity of all the material was evaluated using density bottle (IS 2720 Part III/Sec 1) as well as Le Chatelier flask (IS 4031 Part XI). Kerosene was used for the experiment to prevent floating of lightweight particles on the surface as well as prevent any reaction with the water. Table 2 shows the range of specific gravities obtained from the test.

Table 2 Comparison of specific gravity values of MSWI bottom ash with reference materials

Material	Specific gravity
BA P1	2.61–2.63
BA P2	2.57–2.60
CBA	2.07–2.11
YS	2.65–2.66

4.3 Compaction Behavior

Compaction behavior was studied by both vibratory compaction (as per IS 2720 Part 14) as well as standard Proctor compaction (as per IS 2720 Part 7) to evaluate the maximum dry density (γ_{dmax}), minimum dry density (γ_{dmin}) and optimum moisture content (OMC). Figure 2 shows OMC versus dry density plot obtained by standard Proctor compaction. The values of γ_{dmax}, γ_{dmin} and OMC from both the tests have been reported in Table 3.

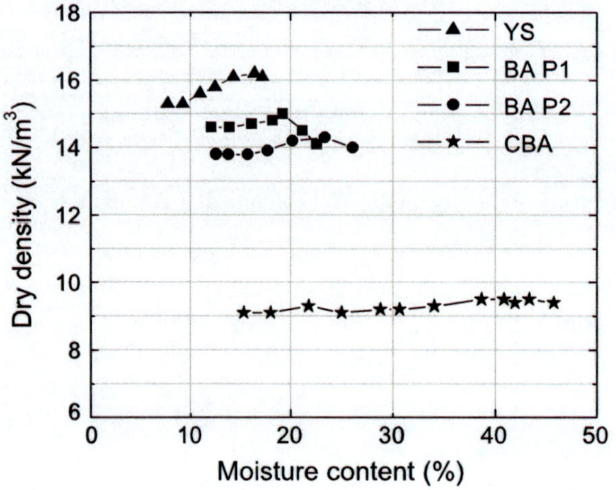

Fig. 2 OMC versus dry density plot

Table 3 Results from vibratory and standard Proctor compaction

Material	Standard proctor		Vibratory	
	OMC (%)	γ_{dmax} (g/cc)	γ_{dmax} (g/cc)	γ_{dmin} (g/cc)
BA P1	19.2	1.51	1.52	1.21
BA P2	23.4	1.43	1.41	1.06
CBA	–	0.95	1.00	0.82
YS	16.4	1.62	1.67	1.37

4.4 Shear Strength Behavior

Consolidated drained (CD) triaxial tests were conducted to study shear strength behavior. The tests were conducted for dense specimens (relative densities of the order of 70% and above) and confining pressures of 100, 200 and 300 kPa. The stress–strain behavior and volumetric change behavior for all the samples are shown in Figs. 3, 4, 5 and 6. Table 4 summarizes values of angle of shearing resistance, φ'.

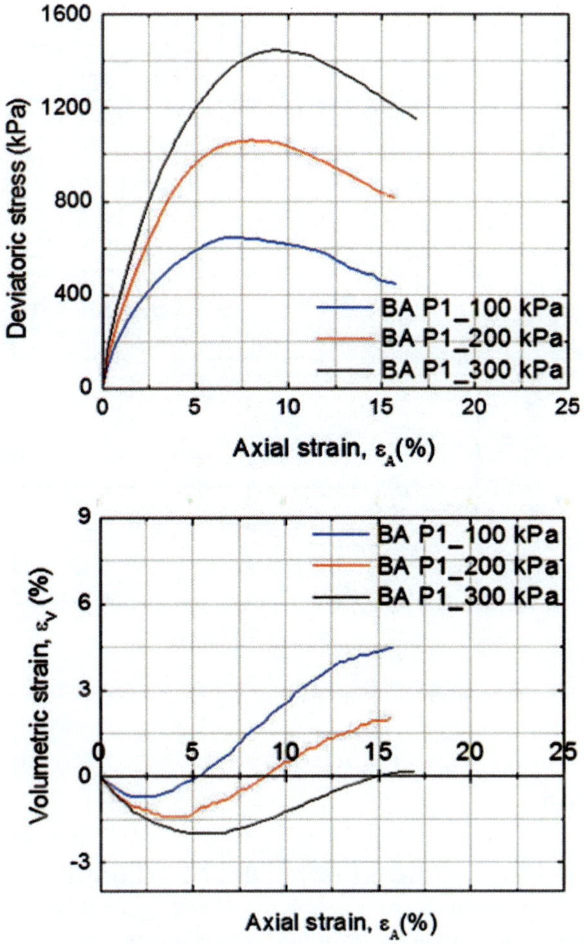

Fig. 3 Stress–strain and volumetric change behavior of MSWI bottom ash from plant P1

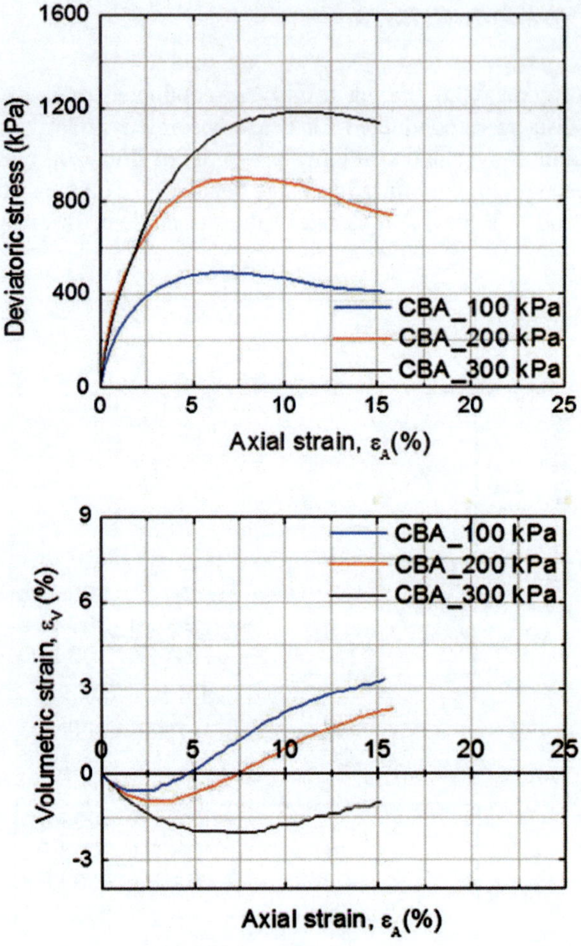

Fig. 4 Stress–strain and volumetric change behavior of bottom ash from coal-fired thermal power plant

5 Discussion

GSD as given in Fig. 1 and Table 1 reveals that MSWI BA from both the plants is comparatively coarser and can be categorized as well-graded sand, whereas CBA is medium-fine sand and YS is primarily fine sand. Percentage fines (passing 75 μm) are highest in MSWI BA from plant P2.

From Table 2, it is observed that specific gravity of MSWI BA is of the order of 2.57–2.63 which falls close to that of Yamuna sand having specific gravity in the range of 2.65–2.66. However, the specific gravity of coal bottom ash is much lower and fall between 2.07–2.11 due to the presence of intra-particle voids. The values obtained are consistent with the values reported in literature (Gupta et al.

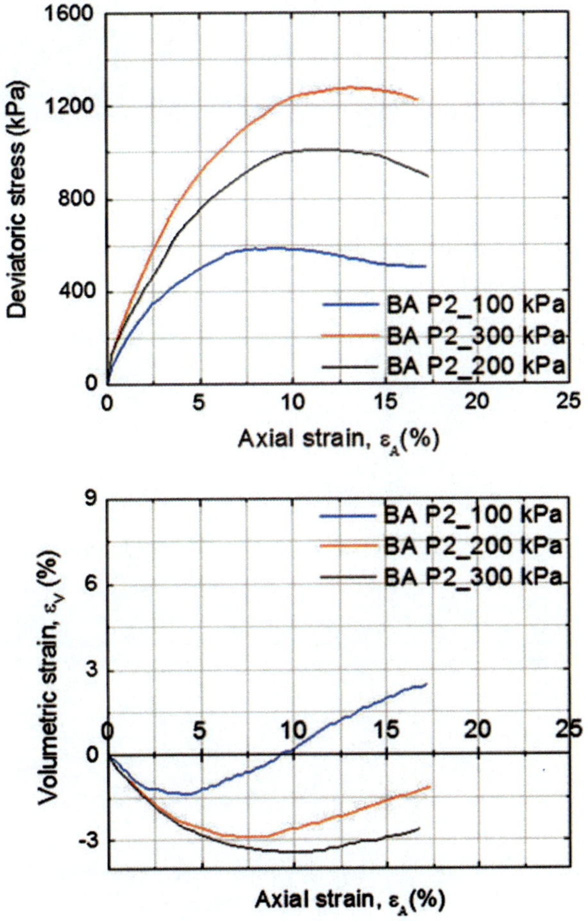

Fig. 5 Stress–strain and volumetric change behavior of MSWI bottom ash from plant P2

2017; Zekkos et al. 2013; Jakka et al. 2010). Presence of unburnt organics (3–6%) in MSWI BA, reported in Gupta et al. (2018), can be the reason associated with its specific gravity values lower than the natural silicates.

Compaction behavior as observed from Fig. 2 suggests that MSWI BA from the two plants has compaction densities slightly lower than YS but much higher than CBA. BA P2 has γ_{dmax} value lower than that of BA P1. Lower compaction densities can be attributed to their lower-specific gravity values. Also, the OMC versus dry density plots are flat for all the samples owing to their granular nature. A minor peak is observed for BA P1, BA P2 and YS unlike CBA which showed no peak. OMC value is highest in BA P2 and lowest in YS which can be attributed to their percentage fines value. Higher the percentage fines, more is the tendency to exhibit higher OMC. From Table 3, it is evident that γ_{dmax} obtained from vibratory compaction is higher than that obtained from Proctor compaction for YS and CBA suggesting that vibratory

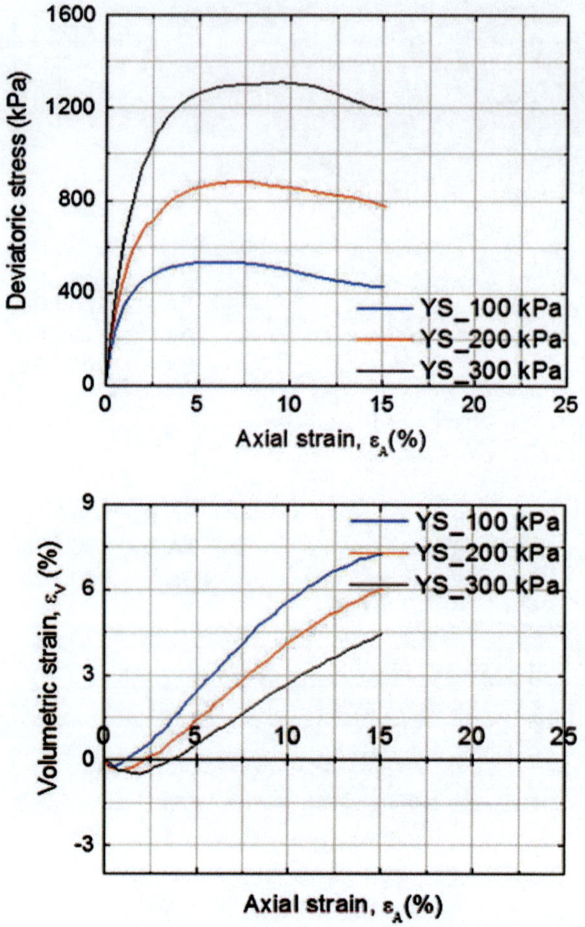

Fig. 6 Stress–strain and volumetric change behavior of Yamuna sand

Table 4 φ' values from CD tests ($c' = 0$)

Size (mm)	φ' (°)
BA P1	42.2
BA P2	38.1
CBA	41.7
YS	41.9

compaction is more suited for these materials to obtain better compaction. However, the values are almost similar for BA P1 and BA P2 indicating the applicability of both the methods for MSWI BA.

The stress–strain behavior (Figs. 3, 4, 5 and 6) and φ' values (Table 4) for all the four samples are comparable. φ' is of the order of 38°–42°. As evident from slopes of stress–strain graph in Figs. 3, 4, 5 and 6, initial elastic modulus is highest in for Yamuna sand and lowest for MSWI BA P1. All four samples initially compress and later dilate as expected for dense soil specimens. However, in comparison to YS, all other samples, i.e., BA P1, BA P2 and CBA, underwent higher compression and lower dilation indicating the possibility of crushing. BA P2 underwent maximum compression and least dilation. It also exhibited lowest φ' value.

6 Conclusions

The results of the study indicate that MSWI BA from WtE plants can possibly be reused as a replacement of natural sand in bulk geotechnical applications such as in embankments, earthfills and road construction. However, more studies are required for assessing leaching of soluble salts and heavy metals, to ascertain its implications on environment before being put to reuse. With new plants emerging all across the country, this study is a starting point for operators to plan the disposal of these residues effectively as well as save the limited land resources.

References

An J, Kim J, Golestani B, Tasneem KM, Al Muhit BA, Nam BH, Behzadan AH (2014) Evaluating the use of waste-to-energy bottom ash as road construction materials. State of Florida Department of Transportation, Contract No.: BDK78-977-20

Chang CY, Wang CF, Mui DT, Cheng MT, Chiang HL (2009) Characteristics of elements in waste ashes from a solid waste incinerator in Taiwan. J Hazard Mater 165(1–3):766–773

Gupta G, Datta M, Ramana GV, Alappat BJ (2017) Feasibility of using MSW incinerator ash in geotechnical applications. In: Indian geotechnical conference, GeoNEst, IIT Guwahati, India

Gupta G, Datta M, Ramana GV, Alappat BJ, Bishnoi S (2018) Feasibility of reuse of bottom ash from MSW waste-to-energy plants in India. In: Zhan L, Chen Y, Bouazza A (eds) The international congress on environmental geotechnics, vol 1. ICEG 2018, Springer, Singapore, pp 344–350

Inkaew K, Saffarzadeh A, Shimaoka T (2014) Characterization of grate sifting deposition ash, unquenched bottom ash and water-quenched bottom ash from mass-burn moving grate waste to energy plant. 土木学会論文集G (環境) 70(7):III_469–III_475

Jakka RS, Ramana GV, Datta M (2010) Shear behaviour of loose and compacted pond ash. Geotech Geol Eng 28(6):763–778

MNRE (2017) Annual report 2016–17. Ministry of New and Renewable Resources, Government of India

Yu J, Sun L, Xiang J, Jin L, Hu S, Su S, Qiu J (2013) Physical and chemical characterization of ashes from a municipal solid waste incinerator in China. Waste Manage Res 31(7):663–673

Zekkos D, Kabalan M, Syal SM, Hambright M, Sahadewa A (2013) Geotechnical characterization of a municipal solid waste incineration ash from a Michigan monofill. Waste Manage 33(6):1442–1450

Stabilization of Old MSW Landfills Using Reinforced Soil

Debanjana Gupta, Manoj Datta, and Bappaditya Manna

Abstract Old municipal solid waste (MSW) unengineered dumps, which have reached considerable height, have sloping sides with inadequate stability. The traditional technique of flattening the slope has to overcome the twin challenges of high cost of moving an enormous volume of waste and availability of area for placing the excavated material. In this study, an attempt has been made to strengthen the slopes of old MSW landfills for heights of 40 m, resting on firm base, by stabilizing them with reinforced soil, placed along the length of the slope and its width restricted to 10 m. Slope stability computations, using limit equilibrium methods, have been made for the original waste slope and that strengthened by the reinforced soil by varying soil strength and geometric properties. The study reveals that the waste slope, stabilized by the use of reinforced soil, can provide a feasible alternative to the removal of large quantity of excess waste.

Keywords Stabilization · Municipal solid waste · Landfill · Reinforced soil · Slope stability · Limit equilibrium method · Berms

1 Introduction

1.1 Background

As the cities are growing at a rapid rate, the landfills which were once used to be at a considerable distance away from the receptors have now become a part of the city itself. Urbanization has also led to the increase in the rate and quantity of waste generation. Unavailability of space to laterally expand the landfills or start the new ones has put immense load on the existing landfill. Consequently, the old dumps are growing vertically with steeper slopes to accommodate more waste. The instability of these side slopes has now become a prime concern associated with old MSW

D. Gupta (✉) · M. Datta · B. Manna
Department of Civil Engineering, Indian Institute of Technology Delhi, New Delhi, Delhi, India
e-mail: gupta.debanjana.06@gmail.com

M. Latha Gali and R. R. P. (eds.), *Problematic Soils and Geoenvironmental Concerns*, Lecture Notes in Civil Engineering 88,
https://doi.org/10.1007/978-981-15-6237-2_2

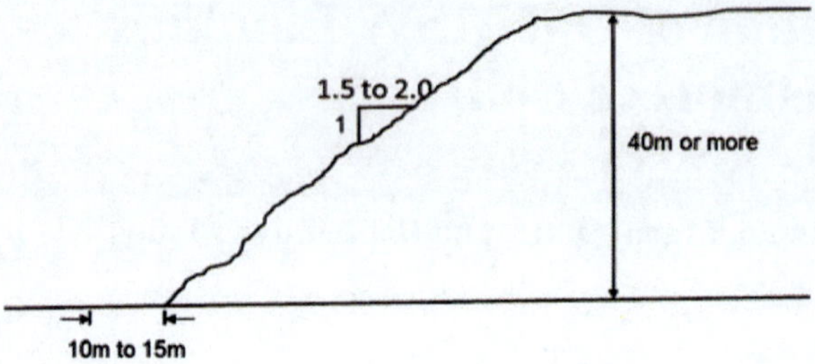

Fig. 1 Old MSW landfill

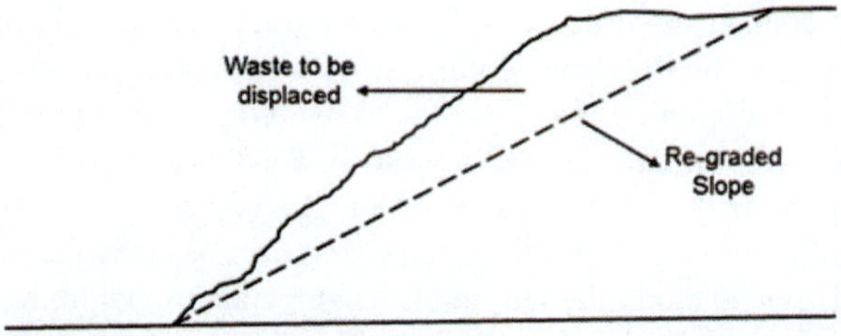

Fig. 2 Traditional method of flattening of slope

landfills. One of the major factors contributing to this instability is the placement of the waste by the 'tipping over' method which leads to steep slopes. The problem becomes grave when the dump reaches greater height (Fig. 1).

1.2 *Re-grading of Slope*

The traditional solution adopted to mitigate this problem is to flatten the slope. However, this solution has certain constraints like (a) cost of excavation and movement of a large volume of the waste (b) availability of area for placing the excavated material (Fig. 2).

1.3 *Stabilization with Berms*

Whenever a small width is available beyond the toe of the landfill, the use of berms (reinforced or unreinforced) for the stabilization of the landfills has been studied extensively. The berms can either be vertical or sloping. Qian and Koerner (2009) in their study credited the use of a reinforced berm as an *attractive alternative* for the expansion of the landfill without increasing the footprint of the existing waste filling boundaries. Basha et al. (2015) also analyzed the prospect of expansion of

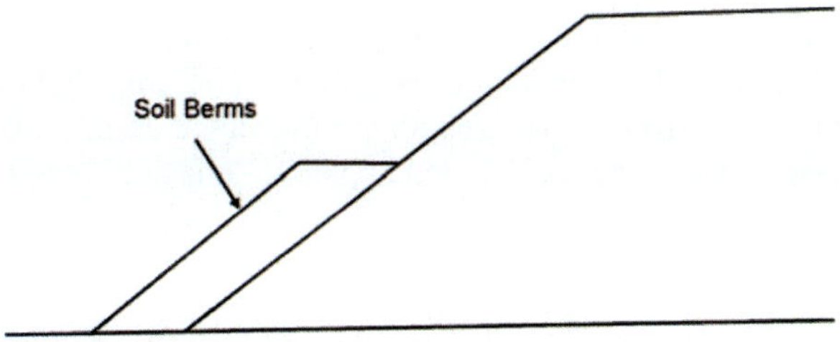

Fig. 3 Provision of berms (unreinforced)

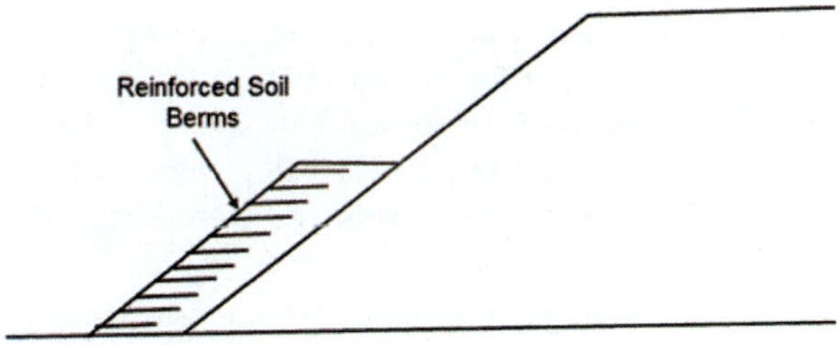

Fig. 4 Provision of berms (reinforced)

the existing landfills vertically by constructing reinforced soil berm and found it to be an economic alternative to the construction of new landfills. In both the papers, translational failure was considered and the slope stability analysis was done using three-part wedge mechanism.

A single berm or multiple berms either continuous or uniquely spaced are also reported to reduce the overall associated risk and harm of a sudden failure of a dumpsite (De Stefano et al. 2016). Koda and Osiński (2015) in a case study for Radiowo landfill, Warsaw reported the construction of berms filled with solid waste, debris and soil residues from a landfill to be the most appropriate slope stability improvement method for the landfill.

In this study, an attempt has been made to strengthen the slope of the old MSW landfills using reinforced soil berm inclined at an angle similar to that of the landfill slope so that they can run parallel along the length. This requires availability of some small space beyond the toe of the landfill. Such a situation is encountered when an inspection road exists at a site at the level of the toe (Figs. 3 and 4).

1.4 Problem Statement

A typical MSW landfill with height of 40 m having steep slopes has been taken up as the problem for analyzing and strengthening the slope stability. Slope angles of 2H:1V and 1.5H:1V are used for the study. A firm base has been taken for analysis, and the water table is considered at great depth. The geotechnical properties for the waste have been taken after Ramaiah et al. (2017). The unit weight is reported as 12 kN/m^3, and the mobilized shear strength parameters at 25 mm displacement are: the angle of shearing resistance as 23° and apparent cohesion intercept as 13 kPa.

2 Objectives

The aforesaid MSW landfill has been analyzed with the following objectives:

(i) Slope stability computations for the original waste slope.
(ii) Slope stability computations for the waste slope along with an unreinforced soil berm by varying the angle of shearing resistance of the soil.
(iii) Slope stability computations for the waste slope along with a reinforced soil berm by varying the strength of geogrids used as reinforcing elements.

3 Methodology

3.1 Method of Analyses

Limit equilibrium method (LEM) has been adopted for the slope stability analysis. Among various methods of slices (MoS) used worldwide, the Morgenstern–Price (M-P) method has been used in this study. A half sine function has been considered to establish a relationship between the interslice shear and normal forces.

Dry and static conditions are considered for the analysis. The acceptable factor of safety (FoS) under such conditions is 1.5, which is the target FoS to be achieved for the problem under consideration.

The software used for the aforesaid analysis is SLOPE/W. A circular failure surface has been assumed throughout the analysis (including part circular and straight line along base).

3.2 Sequence of the Analyses

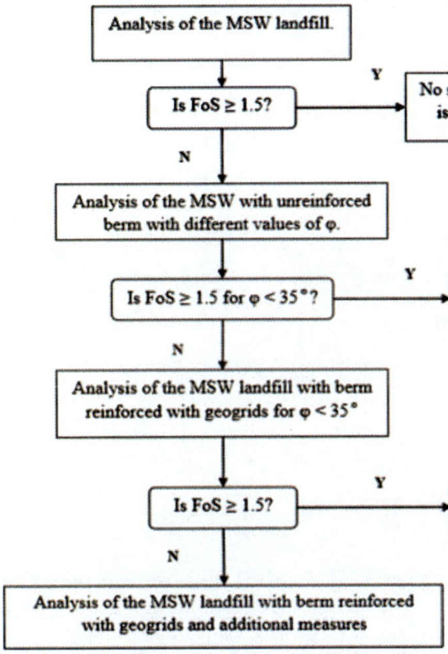

4 Slope Stability Analysis

4.1 Slope Stability Analysis of the Existing MSW Landfill

The analysis of a 40 m high MSW landfill with steep slopes yields the following results (Fig. 5).

Although for the landfill having slope of 2H:1V, the FoS is 1.24, but it is less than the desirable FoS of 1.5. However, for the landfill having slope of 1.5H:1V, the FoS is less than even 1 implying that the slope is inherently unstable.

Hence, in order to provide stability and improving the FoS of the landfills, support to the toe is provided in the form of an unreinforced berm.

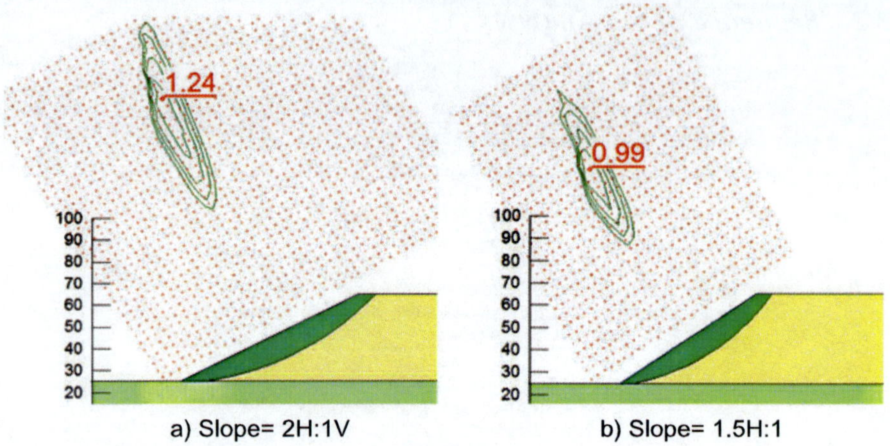

Fig. 5 Failure surface with FoS for the MSW landfill

4.2 *Slope Stability Analysis of MSW Landfill with Unreinforced Berm*

Analyses have been done for the MSW landfill with unreinforced berm made up of cohesionless soil. The unit weight of the soil is taken as 18 kN/m^3, and the angle of shearing resistance (φ) has been varied from 34° to 46°.

The critical failure surfaces obtained in the analyses can be divided into three categories: failure surface passing (a) through the waste mass only above the berm (b) through the waste mass beyond the berm (c) locally through the berm (Fig. 6).

4.2.1 Unreinforced Berm with Height = 10 m and Slope = 2H:1V

For $\varphi \geq 38°$, the failure surface passes through the waste mass only above the berm. Since the properties of the waste remain unaltered, the FoS for all the cases comes out to be the same irrespective of the value of φ. The critical failure surface for $\varphi =$ 34° is local tangential and passes completely through the berm because of the low value of φ (Table 1).

4.2.2 Unreinforced Berm with Height = 20 m and Slope = 2H:1V

The critical failure surface for $\varphi = 34°$ is local tangential and passes completely through the berm because of the low value of φ. However, for $\varphi = 38°$ and $\varphi = 42°$, the failure takes place through the waste mass beyond the unreinforced berm. For φ $= 46°$, the failure surface passes through the waste mass only above the berm and

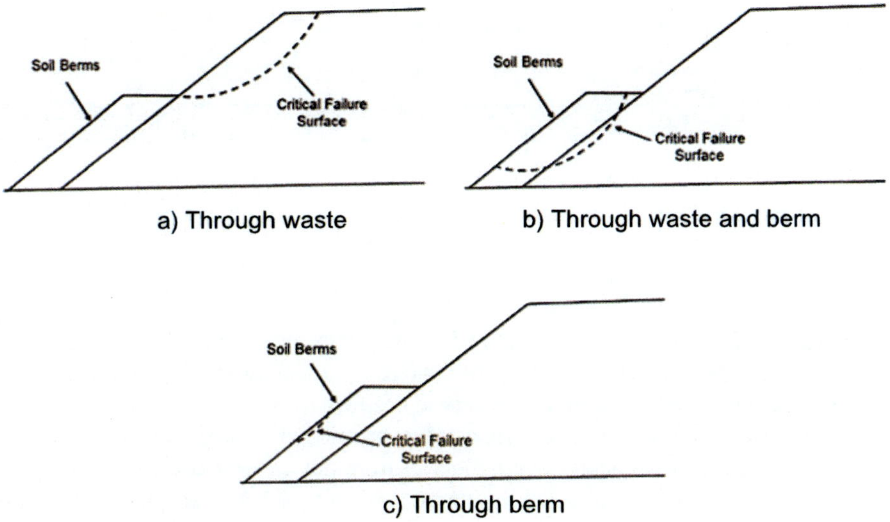

Fig. 6 Typical failure surfaces

Table 1 Factors of safety of landfill with unreinforced berms with height = 10 m and slope = 2H:1V

φ	34°	38°	42°	46°
FoS	1.37	1.38	1.38	1.38

Table 2 Factors of safety of landfill with unreinforced berms with height = 20 m and slope = 2H:1V

φ	34°	38°	42°	46°
FoS	1.35	1.47	1.54	1.56

the FoS is greater than 1.5 implying the said case to be a possible solution to the problem under consideration (Table 2).

4.2.3 Unreinforced Berm with Height = 10 m and Slope = 1.5H:1V

The failure surfaces obtained and their variation with the change of φ are similar to that observed in Sect. 4.2.1. However, the FoS obtained is much less than the desirable value of 1.5 (Table 3).

Table 3 Factors of safety of landfill with unreinforced berms with height = 10 m and slope = 1.5H:1V

φ	34°	38°	42°	46°
FoS	1.04	1.13	1.13	1.13

4.2.4 Unreinforced Berm with Height = 20 m and Slope = 1.5H:1V

The failure surface completely passes through the waste for $\varphi \geq 42°$, and the FoS remains the same for both the cases, i.e., 1.3 as the waste properties remain the same. However, the FoS obtained is less than 1.5 and hence cannot be taken as a solution unlike in the case of 2H:1V. For the other two cases, local tangential failures passing through the unreinforced berm are observed (Table 4).

The stability of the supporting berm is achieved only when higher angle of shearing resistance is used. But locally available soil may not possess such shear strength properties. So, in order to stabilize the supporting soil having lesser values of φ (less than 35°), reinforcement is required.

4.3 Slope Stability Analysis of MSW Landfill with Reinforced Berm

The analyses of reinforced berms with geogrids have been done in this section. The angle of internal friction of the soil (φ) and the interface angle of shearing resistance between the soil and the geogrids (δ) have been taken as 34° and 30°, respectively.

The length of the geogrids has been fixed as 9 m for all the layers, and the spacing between the layers has been kept as 0.8 m. The layout of the geogrids is in accordance with the minimum requirement criteria laid in the manual 'Design of Mechanically Stabilized Earth Walls and Reinforced Soil Slopes—Volume II' developed by FHWA. The product of the reduction factors (ΠRF) is taken as 2. Hence, the ultimate tensile capacity of the geogrids (T_{ult}) is twice the allowable tensile capacity of the geogrids (T_{all}).

As we are trying to make use of a material having $\varphi = 34°$, which when remain unreinforced undergoes local failure, so at first, analyses have been done to make the berm internally stable. So, analyses have been carried out by varying strength of geogrids until a FoS of 1.5 is achieved for the internal stability of the reinforced berm.

Table 4 Factors of safety of landfill with unreinforced berms with height = 20 m and slope = 1.5H:1V

φ	34°	38°	42°	46°
FoS	1.02	1.18	1.3	1.3

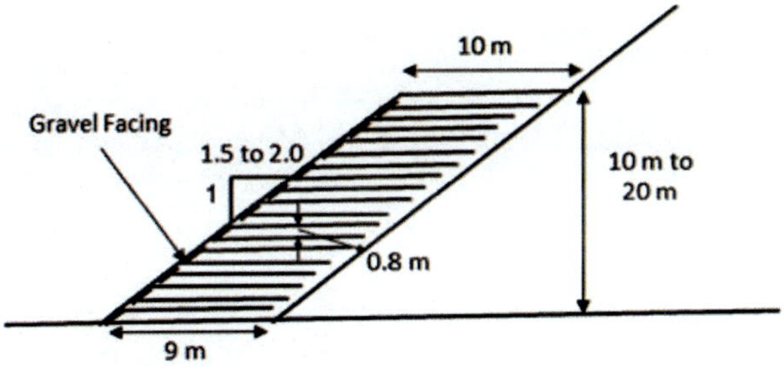

Fig. 7 Reinforcement detail of the berm

It is then followed by the overall stability of the landfill along with the reinforced berm is found out (Fig. 7).

4.3.1 Reinforced Berm with Height = 10 m and Slope = 2H:1V

The analyses for the internal stability of the berm with height 10 m and slope 2H:1V yield the following results (Table 5 and Fig. 8).

Table 5 Factors of safety for internal stability of the reinforced berms for $\varphi = 34°$

T_{all} (kN/m)	0	2
FoS	1.35	1.53

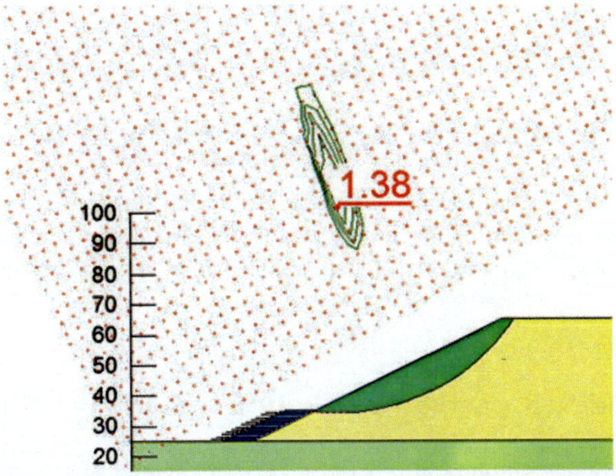

Fig. 8 Failure surface and FoS for berm height = 10 m and slope = 2H:1V for $\varphi = 34°$ (overall stability)

Table 6 Factors of safety for internal stability of the reinforced berms for $\varphi = 34°$

T_{all} (kN/m)	0	2	5
FoS	1.35	1.47	1.53

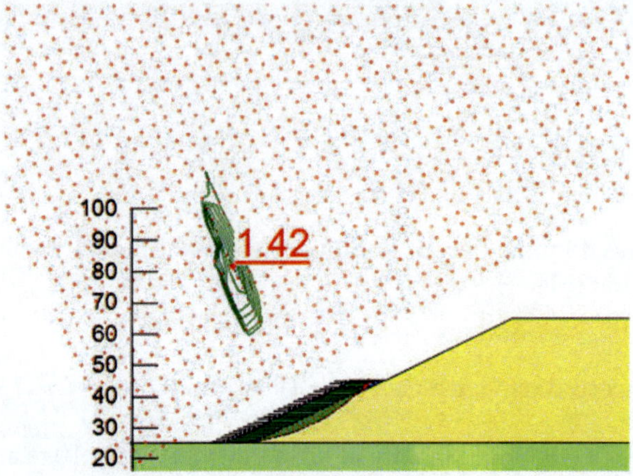

Fig. 9 Failure surface and FoS for berm height = 20 m and slope = 2H:1V for $\varphi = 34°$ (overall stability)

The critical failure surface passes completely through the waste mass which is similar to the unreinforced berm with higher values of φ of soil. The FoS is also the same for both the cases as the waste properties remain unchanged.

4.3.2 Reinforced Berm with Height = 20 m and Slope = 2H:1V

The analyses for the internal stability of the berm with height 20 m and slope 2H:1V yield the following results (Table 6 and Fig. 9).

The critical failure surface passes through the waste mass and the reinforced berm. The FoS is also less than the desirable value. Hence, some additional measures are needed to be taken so that the FoS for the overall stability reaches the target value of 1.5.

4.3.3 Reinforced Berm with Height = 10 m and Slope = 1.5H:1V

The analyses for the internal stability of the berm with height 10 m and slope 1.5H:1V yield the following results (Table 7 and Fig. 10).

The critical failure surface passes through the waste mass. The FoS obtained is same as in the case of unreinforced case with higher value of φ.

Table 7 Factors of safety for internal stability of the reinforced berms for $\varphi = 34°$

T_{all} (kN/m)	0	2	5	10
FoS	1.01	1.22	1.37	1.56

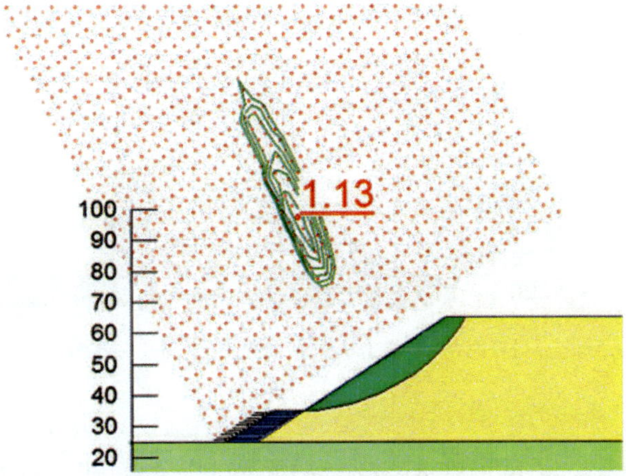

Fig. 10 Failure surface and FoS for berm height = 10 m and slope = 1.5H:1V for $\varphi = 34°$ (overall stability)

Table 8 Factors of safety for internal stability of the reinforced berms for $\varphi = 34°$

T_{all} (kN/m)	0	2	5	10	20	40
FoS	1.01	1.15	1.25	1.32	1.44	1.63

4.3.4 Reinforced Berm with Height = 20 m and Slope = 1.5H:1V

The analyses for the internal stability of the berm with height 20 m and slope 1.5H:1V yield the following results (Table 8 and Fig. 11).

Similar to Sect. 4.3.2, the FoS obtained is less than 1.5, and moreover, the failure surface passes through both the waste mass and the reinforced berm. However, even if the failure surface is made to pass through the waste mass only above the berm, the maximum FoS that can be achieved is 1.3.

4.4 Slope Stability Analysis of MSW Landfill with Reinforced Berm Placed Along with Additional Measures.

For the 20 m high reinforced berm, the overall failure surface is passing through the waste dump beyond the reinforced berm. The geogrids for most of the section also lie

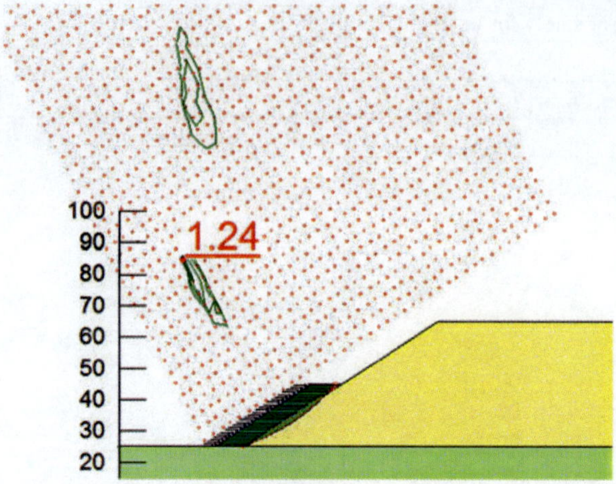

Fig. 11 Failure surface and FoS for berm height = 20 m and slope = 1.5H:1V for $\varphi = 34°$(overall stability)

completely within the failed mass, and as the mobilization of strength for geogrids in this case is not achieved, they no more contribute to the increase in FoS. Hence, any increase in the strength of the geogrids will not help. Further, they cannot be extended into the waste mass due to durability constraints.

One of the ways, in which the problem can be dealt with, is to shift the critical failure surface upwards so that the FoS increases. This can be done either by the construction of a reinforced earth walls at the toe of the berm or by the construction of diaphragm wall within the berm. The solutions are depicted for the 2H:1V slope in Figs. 12 and 13.

For the berm with reinforced earth wall, an additional reinforcement of shorter length is provided at the top of the wall in order to avoid any local failure. For the berm with the diaphragm wall, a few layers of reinforcement at the base of the berm have been curtailed in order to provide the required space for the construction of the wall (Fig. 14).

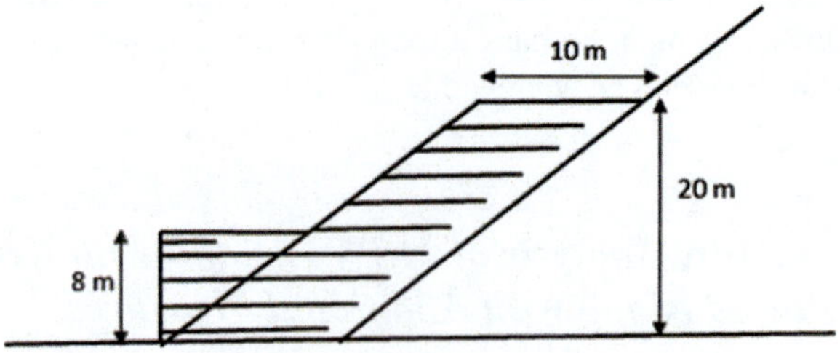

Fig. 12 Reinforced berm with reinforced earth wall

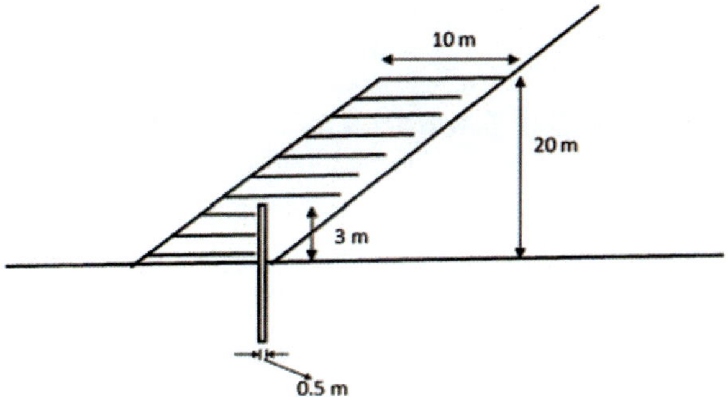

Fig. 13 Reinforced berm with diaphragm wall

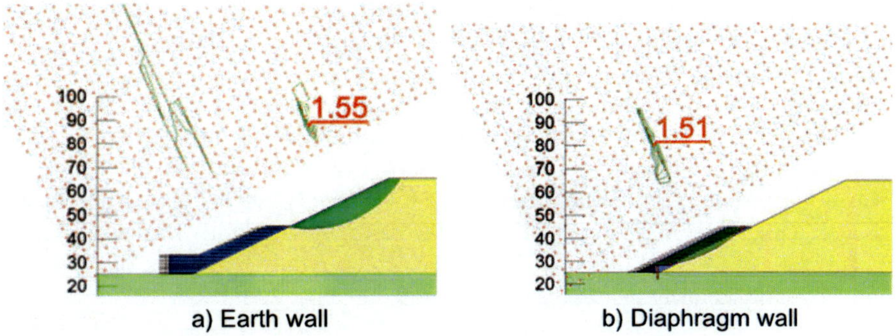

Fig. 14 Failure surface and FoS for berm height = 20 m and slope = 2H:1V for $\varphi = 34°$

For the reinforced earth wall, the failure surface passes through the waste mass and slightly touches the berm; however, the FoS achieved is more than the desired value. For the diaphragm wall, the failure surface although passes through the waste mass beyond the berm, but since the FoS is greater than the target value of 1.5, thus it can be adopted as a solution for the problem under consideration.

However, the analyses for both the cases of an earth wall and diaphragm wall have been done only for the slope stability check. Other necessary internal and external checks are to be done to assess the feasibility of the construction of the walls.

For the landfill having slope of 1.5H:1V, the highest FoS that can be obtained with a 20 m berm is 1.3 (in case of unreinforced berm with φ of the material greater than 42°). The failure surface in this case passes only through the waste mass above the berm. Thus, any enhancement in the berm is not going to change the FoS. Therefore, in order to obtain the target FoS of 1.5, the landfill slope above the berm has to be re-graded to a flatter angle.

5 Results

5.1 *The Existing MSW Landfill*

The slope stability analysis for the existing landfill of height 40 m yields the FoS of 1.24 and 0.99, respectively, for landfill slope of 2H:1V and 1. 5H:1V. As both the FoS are less than the target value of 1.5, some measures are needed to be taken.

5.2 *MSW Landfill with Unreinforced Berm.*

The FoS obtained from the analyses of the landfill along with the unreinforced soil berm with φ of the soil varying from 34° to 46° has been summarized in Tables 9 and 10.

With the provision of unreinforced berm along the length of the landfill, the FoS increases from that in case of the existing landfill. The FoS further increases with increase in the height of the unreinforced berm except for the case when $\varphi = 34°$. This is because the failure obtained in this case is locally tangential to the berm, and as such with the increase in the height of the berm, the instability increases.

However, the desirable FoS of 1.5 can be achieved only for a slope of 2H:1V for high angle of shearing resistance of soil.

5.3 *MSW Landfill with Reinforced Berm*

The FoS obtained from the analyses of the landfill along with the reinforced berm made up of soil having $\varphi = 34°$ which has been summarized in Tables 11 and 12.

Table 9 FoS for landfill with slope 2H:1V and unreinforced berm at an inclination of 2H:1V up to height 10 and 20 m

φ	34°	38°	42°	46°
10 m	1.37	1.38	1.38	1.38
20 m	1.35	1.47	1.54	1.56

Table 10 FoS for landfill with slope 1.5H:1V and unreinforced berm at an inclination of 1.5H:1V up to height 10 and 20 m

φ	34°	38°	42°	46°
10 m	1.04	1.13	1.13	1.13
20 m	1.02	1.18	1.3	1.3

Table 11 FoS for landfill with slope 2H:1V and reinforced berm at an inclination of 2H:1V for $\varphi = 34°$ up to height 10 and 20 m

Berm height	Unreinforced	Reinforced
10 m	1.37	1.38
20 m	1.35	1.42

Table 12 FoS for landfill with slope 1. 5H:1V and reinforced berm placed at an inclination of 1. 5H:1V for $\varphi = 34°$

Berm height	Unreinforced	Reinforced
10 m	1.04	1.13
20 m	1.02	1.24

For 10 m height of the soil placed, with the provision of geogrids as reinforcement, the FoS achieved, by using soil with low angle of shearing resistance, is same as that determined for the unreinforced soil with angle of shearing resistance greater than 35°.

However, for 20 m height of soil placed, same phenomenon is not obtained.

5.4 *MSW Landfill with Reinforced Soil Placed Along the Length of the Landfill with Additional Measures*

A comparative summary of the FoS obtained from the analyses of the existing landfill and that of the landfill with berm of height 20 m made up of soil having $\varphi = 34°$ is given in Table 13.

The target FoS is finally achieved for a landfill having side slope of 2H:1V, after providing a 20 m high reinforced berm made up soil having angle of shearing resistance less than 35° with additional measures.

Table 13 FoS for landfill with reinforced berm up to height 20 m and slope 2H:1V and soil having $\varphi = 34°$

Without berm	With unreinforced berm	With reinforced berm	With reinforced berm and additional measures
1.24	1.35	1.42	1.51–1.55

6 Conclusions

With respect to the results of the slope stability analyses done under dry and static conditions, the following conclusions can be made.

(1) The FoS increases for an MSW landfill with the provision of reinforced berm placed along the length of the landfill inclined at an angle similar to that of the landfill side slope.
(2) The FoS further increases with the increase in the height of the reinforced berm.
(3) For the landfill of height 40 m and slope of 2H:1V, the target FoS of 1.5 can be achieved by (a) placing unreinforced berm with high angle of shearing resistance ($\varphi = 46°$) up to 20 m height (b) by providing reinforced berm for lower value of angle of shearing resistance ($\varphi < 35°$) up to 20 m height along with additional measures.
(4) For the landfill of height 40 m and slope of 1.5H:1V, the maximum FoS that can be obtained with the help of a berm is 1.3. Hence, to achieve the target FoS of 1.5, the landfill slope above the berm needs to be re-graded to a flatter angle.

References

Basha BM, Mahapatra S, Manna B (2015) Optimum dimensions of reinforced soil berm for vertical expansion of municipal solid waste (MSW) landfills. In: Conference information: international foundations congress and equipment expo, San Antonio, Texas, pp 2677–2686

De Stefano M, Gharabaghi B, Clemmer R, Jahanfar MA (2016) Berm design to reduce risks of catastrophic slope failures at solid waste disposal sites. Waste Manage Res 34(11):1117–1125

Federal Highway Administration (FHWA) (2009) Design and construction of mechanically stabilised earth walls and reinforced soil slopes, vol II, FHWA-1 0-024, Washington, DC

Koda E, Osiński P (2015) Application of alternative methods of slope stability improvement on landfills. In: Proceedings of the XVI ECSMGE geotechnical engineering for infrastructure and development, Edinburgh, pp 2717–2722

Qian XD, Koerner RM (2009) Stability analysis when using an engineered berm to increase landfill space. J Geotech Geoenviron Eng 135(8):1082–1091

Ramaiah BJ, Ramana GV, Datta M (2017) Mechanical characterization of municipal solid waste from two waste dumps at Delhi, India. Waste Manage 68:275–291

Effect of Drying and Wetting of Shear Strength of Soil

Naresh Mali, Tarun Semwal, Khushboo Kadian, Manuj Sharma, and K. V. Uday

Abstract Shear strength of soil is a very important property in geotechnical engineering, in order to study the strength of soils. According to classical soil mechanics, the soil is generally assumed to be either fully saturated or completely dry conditions. However, 40% of the natural soil on the earth's surface encountered as an unsaturated/partially saturated condition. The problems faced at the time of designing the retaining structures, considering the soil mass in dry condition, affects in reducing the factor of safety of the slope due to wetting/drying conditions. The present study aims to investigate the loss/gain in the strength of the soils during the wetting and drying process, and it was observed that the strength is higher in drying path than that of wetting path.

Keywords Drying · Wetting · Strength of soil

1 Introduction

Soils are formed due to weathering process from the parent rock. Due to changes in environmental and physical variance, soil formation and their properties change with location and prevailing drainage conditions (Santamarina et al. 2001). These changes in the soil properties are also affected due to wetting and drying conditions caused by environment or imposed conditions (Jayanth et al. 2013). The shear strength of the soil is one of the most important parameters for engineering applications. In most of the engineering applications, the soil was prepared at optimum moisture content and maximum dry density to achieve the desired strength. However, the shear strength of the soil varies due to many prevailing conditions when the soil is subjected to wetting and drying phenomenon (Abu-Hejleh and Znidarcic 1995; Uchaipichat 2010; Nishimura and Fredlund 1999; Tse and Ng 2008).

N. Mali · T. Semwal · K. Kadian · M. Sharma · K. V. Uday (✉)
School of Engineering, Indian Institute of Technology Mandi, Kamand Campus, Mandi, Himachal Pradesh 175005, India
e-mail: uday@iitmandi.ac.in

M. Latha Gali and R. R. P. (eds.), *Problematic Soils and Geoenvironmental Concerns*, Lecture Notes in Civil Engineering 88,
https://doi.org/10.1007/978-981-15-6237-2_3

Soils in nature are subjected to climatic variation and undergo wetting and drying cycles periodically. This can significantly alter the soil hydro-mechanical behaviour and damage the earth retaining structures. Researchers depicted that the shear strength of unsaturated soils is dependent on the dry density and water content of compacted specimens (Nowamooz and Masrouri 2009). Sivakumar et al. (2006) have shown a unique relationship between shear strength, soil suction and dry density for the Kaolin clay. The shear strength of unsaturated soils is affected by water content and the method of compaction and concluded that the specimens compacted dry of optimum exhibits higher strength, brittleness and tendency to swell, whereas on wet of optimum, the soil is found to be more ductile with lower strength and tends to consolidate (Morris et al. 1992; Nishimura and Fredlund 1999; Tse and Ng 2008). Desiccation cracks would also develop on slopes, clay liners and landfill cover due to continuous wetting and drying cycles (Lim and Miller 2004; Tang et al. 2011). Due to the presence of these, cracks will significantly weaken the soil mechanical performance (Velde 2001; Chertkov and Ravina 2002; Chertkov 2002; Albercht and Benson 2001; Tang et al. 2011). Goh et al. (2011) indicated that the shear strength characteristics of soils under wetting and drying cycles are different. The shear strength on the drying path is higher than the wetting path due to the presence of hysteresis effect (Nishimura and Fredlund 1999; Tse and Ng 2008). Earlier studies depicted that the changes in moisture level may lead to losing/gain its strength of the soil. Due to drying, soil leads to cracking, whereas wetting of soil loses it shear strength of the soil, collapses and finally leads to slope failures (Goh et al. 2011; Morris et al. 1992; Nishimura and Fredlund 1999; Tse and Ng 2008).

The above-mentioned studies provided various experimental techniques to investigate the wetting and drying cycle on volume and mechanical behaviour of soil.

The present study has been planned to study the unconfined compression strength of soil samples at different moisture levels. The different water contents are obtained by subjecting the soil samples prepared at optimum moisture content at various conditions.

In what follows, first, the detailed methodologies involved to determine the characterization of the soil used for the study were mentioned. Further, the detailed experimental procedure was mentioned for the preparation of the soil specimen and testing procedure to determine the strength characteristics of the soil specimen subjected to drying and wetting phenomenon. Finally, the results obtained from the prepared soil specimen employing the experimental testing procedure were discussed.

Table 1 Physical properties of soils

Property	Soil 1	Soil 2	Soil 3
Specific gravity	2.68	2.62	2.58
Gravel (%)	0	0	0
Sand (%)	0	0	44
Silt (%)	44	29	37
Clay (%)	56	71	18
Liquid limit, LL (%)	42	381	32
Plastic limit, PL (%)	23	112	23
Plasticity Index, PI (%)	19	268	9
Soil classification*	CI	CH	ML
Maximum dry density (kN/m^3)	17	14	19
OMC (%)	21	29	16
Cohesion (kPa)	176	243	7
Angle of internal friction (°)	0	0	25

*According to IS Classification

2 Materials and Methods

2.1 Soil Description

For the present study, three soil samples were selected and the following tests were conducted such as specific gravity (IS 2720 part III/section 1, 1980), wet sieve and hydrometer analysis (IS 2720, part—IV, 1985), LL and PL (IS 2720, part—V, 1985), standard proctor test (IS 2720, part—VII, 1985) and direct shear tests (IS 2720, part—13, 1986); the results were shown in Table 1.

2.2 Experimental Testing Procedure

The unconfined compression test (UCC) is performed to determine the shear strength characteristics of the soil in accordance with IS 2720-part—10, 1991. The sample preparation for UCC test was prepared by taking 5 kg of soil passing through 4.75-mm sieve in a mould. The sample is taken in a tray, and suitable water content is added, according to optimum moisture content (OMC, %) of that soil and mixed it thoroughly. The sample is compacted by using a 4.89 kg weight of the hammer, and 56 nos. of evenly disturbed blows are given in five equal layers falling from a height of 450 mm. The collar is removed and is trimmed off using the spatula. The sample is extracted using sample extractor, and each sample is taken for testing by maintaining as $L/D = 2$ (approx.,), the desired dimensions of the sample as height and diameter as 76 mm and 38 mm; and these samples are properly sealed till for testing. About 15 nos. of such samples were obtained from each soil. Care is taken to avoid any moisture loss between the testing and subsequent testing of the specimen. The UCC test is conducted for three nos. of the specimen to obtain the strength characteristics of the soil. The average of the two nos. of closer results has been considered as the UCC strength of the soil, and the undrained cohesion values (C_u) have been obtained. However, out of 15 nos. of the specimen, six nos. of the specimen were subjected to the wetting process by maintaining a relative humidity of 95% at the temperature of 25 °C, inside the environmental chamber. Due to the wetting process, the weight of the specimen increases. The weight of the specimen is noted at regular intervals, and after the specimen has reached to an increase of 2% of moisture above OMC, the respective three nos. specimen were tested using UCC apparatus and the desired strength is obtained. Similarly, the same procedure is repeated to obtain the specimen with 4% increase of moisture above OMC. However, remaining six nos. of the specimen were subjected to constant drying at a relative humidity of 35% at a temperature of 25 °C inside the chamber. The drying is continued similar to wetting, and three nos. of each specimen corresponding to a decrease of 2% moisture and 4% moisture are retrieved for consequent testing for their compressive strength.

3 Results and Discussions

3.1 Effect on Strength of the Different Soil (S_1, S_2, S_3) Specimens Prepared at Different Moisture Levels

Figure 1 represents the stress–strain relationship of soil S_1 at different moisture levels, i.e. 17%, 19%, 21%, 23% and 25%, respectively. It has been observed that

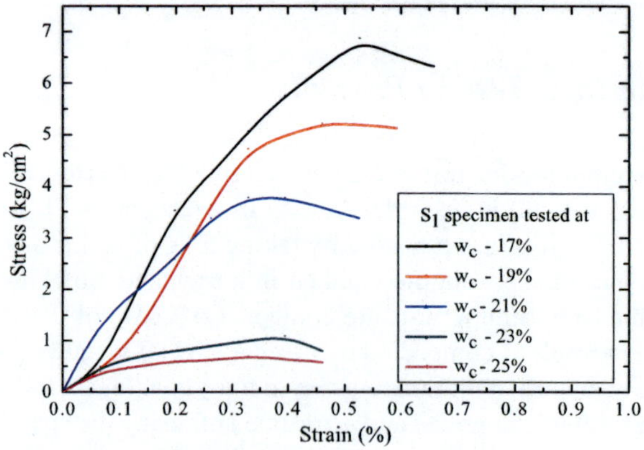

Fig. 1 Stress–strain relationship for soil (S_1) sample

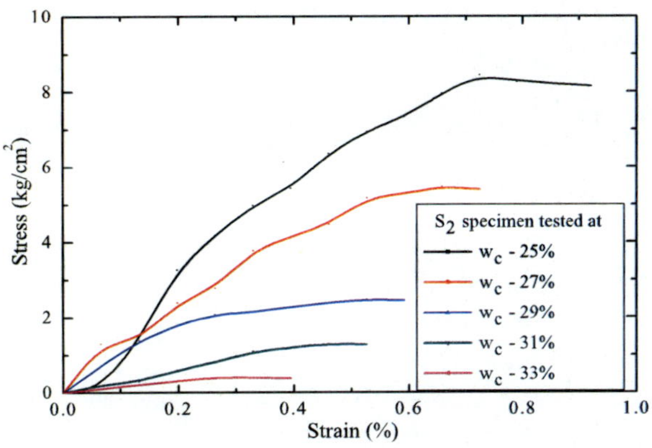

Fig. 2 Stress–strain relationship for soil (S_2) sample

the strength of soil decreases with increases in water content. The strength of soil is higher in drying path than that of wetting path due to decrease in soil volume which leads to increase in suction, thereby affecting strength characteristics of the soil as reported by Uchaipichat (2010), Morris et al. (1992), Nishimura and Fredlund (1999), Tse and Ng (2008). The percentage increase in strain is observed to decrease with the increase in water content.

Effect on strength of the soil (S_2) specimen prepared at different moisture levels

Figure 2 represents the strength of soil S_2 at different moisture levels, i.e. 25%, 27%, 29%, 31% and 33%, respectively. It was observed that the strength of soil decreases with increase in water content. Due to presence of high clay content in the sample and also due to its high Plasticity Index, the strength of the sample is higher in drying path than that of the wetting path as reported by Morris et al. (1992), Nishimura and Fredlund (1999), Tse and Ng (2008).

Effect on strength of the soil (S_3) specimen prepared at different moisture levels

Figure 3 represents the strength of soil S_3 at different moisture levels, i.e. 12%, 14%, 16%, 18% and 20%, respectively. It has been observed that the strength of soil decreases with increase in water content. The strength of the soil is higher in drying path than that of the wetting path as reported by Goh et al. (2011), Nishimura and Fredlund (1999), Tse and Ng (2008). The percentage increase in strain is observed to decrease with the increase in water content.

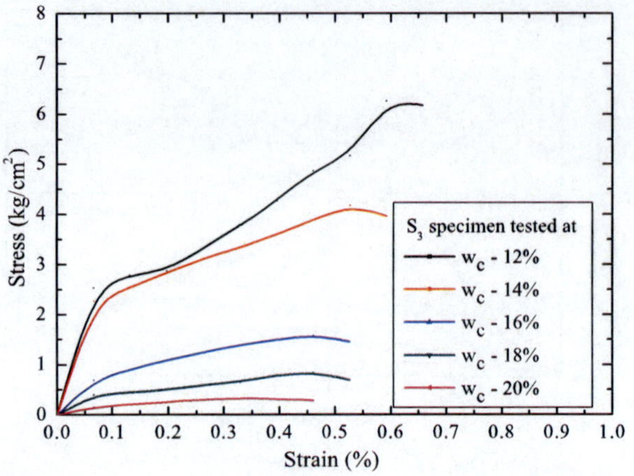

Fig. 3 Stress–strain relationship for soil (S_3) sample

3.2 *Effect of the Undrained Cohesion of Soil Specimens Prepared at Different Moisture Levels*

Figure 4 represents the water characteristic curve (Oberg and Sallfors 1997; Salager et al. 2013) of water content and strength on different soils. The strength of soil S_2 is more than the strength of remaining soils S_1 and S_3 due to more percentage of clay content in S_2 and also due to its high Plasticity Index. The strength of soil S_1 is more compared to the remaining soils, due to more percentage of clay and silt content as compared to remaining soils. The strength of soil sample S_3 is less compared to

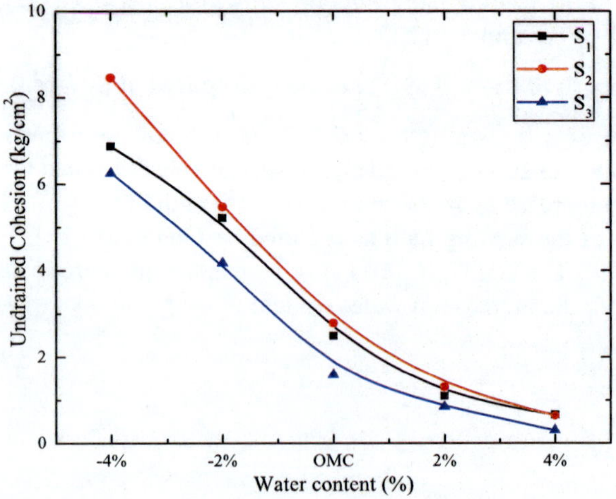

Fig. 4 Effect of water content on the undrained cohesion of soils

remaining soils due to less percentage of clay content and also due to presence of the silt and sand contents in it. Hence, it was observed that the undrained cohesion value is decreasing with the increase in water content.

The strength of the soil is higher in drying path compared to wetting path. The higher strength in the drying path relates to soil–water evaporation process. Therefore, the soil volume is increasing (swelling), for wetting path due to reduction in effective stress, whereas in the drying path, the soil volume reduces due to suction build-up, and therefore, their effect is observed on the strength characteristics of soil. In addition to suction, the strength of the soil also significantly depends on soil structure, i.e. pore sizes, heterogeneous zones and formation of cracks. This is one of the main reasons that why the slopes are prone to soil instability immediately or after rainfall.

4 Conclusions

The present study reported the results of unconfined compression tests for various soil specimens prepared at different moisture levels. The change in the strength of soil has been observed due to wetting and drying effect. It was observed that the strength of soil is found to be higher in drying path compared to wetting paths. The percentage increase in strain was observed to decrease with the increase in water content.

References

Abu-Hejleh AN, Znidarcic D (1995) Desiccation theory for soft cohesive soils. J Geotech Eng ASCE 121(6):493–502
Albrecht BA, Benson CH (2001) Effects of desiccation on compacted natural clay. J Geotech Eng 127(1):67–75
Chertkov VY (2002) Characteristic crack dimension of saturated drying soils. J Sci Res Dev. vol IV, Israel
Chertkov VY, Ravina I (2002) Combined effects of interblock and interaggrate capillary cracks on the hydraulic conductivity of swelling clay soils. Water Resour Res 38(8):1157
Fredlund DG, Rahardjo H (1993) Soil mechanics for unsaturated soils. Wiley, New York
Goh SG, Rahardjo H, Leong EC (2011) Shear strength of unsaturated under drying and wetting conditions. J Geotech Eng ASCE 10(1):594–606
Jayanth S, Iyer K, Singh DN (2013) Continuous determination of drying path SWRC of fine-grained soils. Geomech Geoeng 8(1):28–35
Lim YY, Miller GA (2004) Wetting-induced compression of compacted Oklahoma soils. J Geotech Geoenviron Engi 130(10):1014–1023
Morris PH, Graham J, Williams DJ (1992) Cracking in drying soils. Can Geotech J 29:263–277
Nishimura T, Fredlund DG (1999) Unconfined compressive shear strength of an unsaturated silty soil subjected to high total suction. In: Proceedings of the international symposium on slope stability engineering, IS-SHIKOK, vol 2, pp 757–762

Nowamooz H, Masrouri F (2009) Density-dependent hydro-mechanical behavior of a compacted expansive soil. Eng Geol 106:105–115

Oberg A, Sallfors G (1997) Determination of shear strength parameters of unsaturated silts and sands based on water retention curve. J Geotech Eng 20:40–48

Salager S, Nuth M, Ferrari A, Laloui L (2013) Investigation into water retention behaviour of deformable soils. Can Geotech J 50(2):200–208

Santamarina JC, Klein KA, Fam MA (2001) Soils and waves: particulate materials behavior, characterization and process monitoring. Wiley Publications, New York

Sivakumar V, Tan WC, Murray EJ, Mckinley JD (2006) Wetting, drying and compression characteristics of compacted clay. Géotechnique 56:57–62

Tang CS, Cui YJ, Shi B, Tang AM, Liu C (2011) Desiccation and cracking behavior of clay layer from slurry state under wetting and drying cycles. Geoderma 166:111–118

Tse EYM, Ng CWW (2008) Effects of wetting and drying cycles on unsaturated shear strength. In: Proceedings of 1st European conference on unsaturated soils, CRC Press/Balkema, Leiden, The Netherlands, pp 481–486

Uchaipichat A (2010) Prediction of shear strength for unsaturated soils under drying and wetting process. Electron J Geotech Eng 15:15–25

Velde B (2001) Surface cracking and aggregate formation observed in a Rendzina soils, La Touche (Vienne) France. Geoderma 99:261–276

Yang HP, Wang XZ, Xiao J (2014) Influence of wetting and drying cycles on strength characteristics of Nanjing expansive soils. Chin J Geotech Eng 36(5):951–954

Influence of the Rate of Construction on the Response of PVD Improved Soft Ground

Priyanka Talukdar and Arindam Dey

Abstract The major problem of embankment on soft soil foundations is to overcome excessive settlement, initiating undrained failure of the infrastructure, provided proper ground improvement is not planned. One of the most widely used solution to this problem is to stabilize the soft soil foundations under embankment, by the installation of prefabricated vertical drains (PVD), which is carried out throughout the world. It is observed that the installation of vertical drains highly increases the settlement rate, improves pore water pressure dissipation, and decreases the lateral deformation of the soft clay foundation. The paper presents a numerical model to analyze the response of PVD-reinforced soft soil under embankment loading to the rate of embankment construction by finite element modeling. Apart from predicting the dissipation of excess pore water pressure, lateral displacement and the resulting consolidation settlement with time, the stability factors for different rates of embankment construction have been studied. It is observed that higher stability is manifested by the embankment having a slower rate of construction.

Keywords Embankment · Soft soil · PVD · Finite element modeling · Rate of construction

1 Introduction

The exact estimation of the nature of embankments constructed on soft clay stabilized with vertical drains has posed to be a challenging problem, although significant advancement has been made in the past few years through extensive numerical modeling. Soft soil foundations can cause excessive settlement, initiating undrained failure of the infrastructure if proper ground improvement is not carried

P. Talukdar (✉) · A. Dey
Department of Civil Engineering, Indian Institute of Technology Guwahati, Guwahati, Assam, India
e-mail: priyanka.talukdar1990@gmail.com

A. Dey
e-mail: arindamdeyiitg16@gmail.com

M. Latha Gali and R. R. P. (eds.), *Problematic Soils and Geoenvironmental Concerns*, Lecture Notes in Civil Engineering 88,
https://doi.org/10.1007/978-981-15-6237-2_4

out (Indraratna et al. 1992). Therefore, it is imperative to apply adequate ground improvement techniques to the existing soft soils prior to the construction to prevent unacceptable excessive and differential settlement along with enhancement of the bearing capacity of the foundations. Among various methods of soft soil improvement, the installation of prefabricated vertical drains (PVD) is one of the well-established and effective techniques practiced worldwide (Wang 2009). The other forms of such hydraulic installations commonly functioning as vertical drains include stone columns and sand drains.

As reported by researchers (Guetif et al. 2007), vertical drains, such as stone columns, not only act as reinforcement, possessing greater strength and stiffness in comparison with the surrounding soil, but they also speed up the time-dependent dissipation of excess pore water pressure caused by surcharge loading due to shortening of the drainage path. Various analytical and numerical solutions have already been developed for understanding the load transfer mechanism of soft soil reinforced with stone columns. Among the most significant contributions, the studies by Wang (2009), Alamgir et al. (1996) are noteworthy.

Well-documented classical solution for vertical drains (single-drain analysis) has been carried out (Barron 1948; Hansbo 1981) and has frequently been utilized in settlement prediction, especially along the centerline of the embankment. Due to the fast development in computer capability and increasing popularity of the finite element method in geotechnical engineering, a detailed numerical analysis of the behavior of the soft soil stabilized with multiple vertical drains can now be carried out. The characteristics of vertical drains and their smear effects have been well-documented by Dey (2008). Although single-drain analysis is often sufficient to model the soil behavior along the embankment centerline (symmetric geometry), multidrain analysis is essential to incorporate the effect of changing gravity load along the embankment width to accurately predict settlements and lateral displacements. Limited case studies employing various forms of multidrain analysis have been described in the recent past (Indraratna et al. 1994; Chai et al. 1995; Indraratna and Redana 1997; Anvesh Reddy and Dey 2014). The effect of strain rate on the strength of soft cohesive soils has been extensively investigated since the early work of Terzaghi (1931) followed by Casagrande and Wilson (1951). Experimental results have consistently shown that the loading rate has an effect on the undrained shear strength and that stress–strain relations of natural soft clays are strain rate-dependent (Perloff and Osterberg 1963; Bjerrum 1972; Lo and Morin 1972; Graham et al. 1983; Kulhawy and Mayne 1990). Bjerrum (1973) attributed the time effect to the 'interparticle creep' along the direction of shear due to the presence of the viscous absorbed water between clay particles. The undrained creep during shear in the soil gives rise to increased pore pressures, decreased effective stresses, and decreased strengths; consequently, the undrained shear strength is rate-dependent (Ladd and Foott 1974).

In this study, a multidrain analysis has been carried out to understand the response of an embankment on PVD improved soft soil to the varying rates of construction.

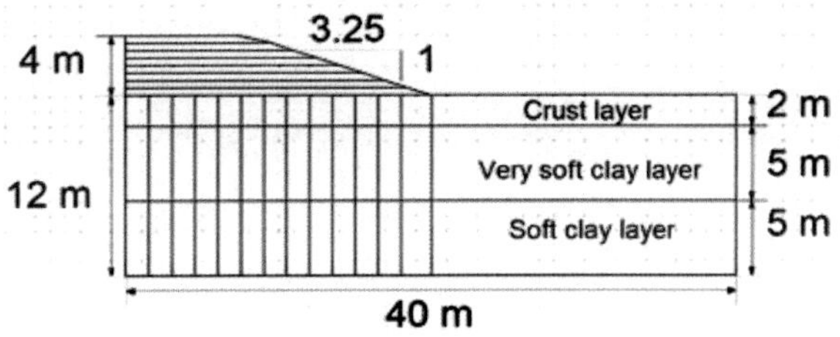

Fig. 1 Model used for the present study

2 Methodology

2.1 Model for the Present Study

In the present work, a simple FE model has been used to understand the effect of the rate of construction of the embankment on the foundation improved by PVDs. The model used in the study, along with its relevant dimensions, is shown in Fig. 1. The embankment is constructed to a height of 4 m with 3.25H:1V side slope. The embankment is constructed by eight lifts, each lift having a height of 1 m. The foundation layer consists three layers; starting from the bottom, it consists of 5 m soft clay layer, 5 m very soft clay layer, and 2 m overconsolidated crust layer.

The horizontal spacing between the vertical drains is 1.5 m, except at the embankment toe where the spacing is 2 m (this was done purely for modeling convenience so that there is a drain at the embankment toe).

The sand blanket is not included in the model as a separate material. The effect of the sand blanket can be modeled by specifying a zero-pressure boundary condition along the ground surface. The physical implication is that there will be no build-up of positive pore pressures at the ground surface. Any water arriving at the ground surface will have the opportunity to disappear through the sandy material. The boundary condition simulates this effect. This is much simpler than trying to include the sand blanket in the model which achieves the same objective.

2.2 Materials

The soft clay and very soft clay are characterized here by using the modified Cam-Clay (MCC) constitutive model available in Sigma/W. It is an ideal constitutive model for this case, since it can account for pore pressure changes arising from mean effective stress and deviatoric stress changes, which is an important feature in soft clay behavior. The soil properties used to represent the stress–strain behavior and strength of the soft clay and very soft clay are shown in Table 1. The stiffness of the soil is controlled by the slopes of the isotropic normal compression line (Lambda) and the unloading-reloading line (Kappa).

Table 1 Soil properties for the soft clay and very soft clay layers

Parameter	Soft clay	Very soft clay
Constitutive model	Modified clam clay	Modified clam clay
Over-consolidation ratio	1.5	1.5
Poisson's ratio (ν)	0.334	0.334
Lambda (λ)	0.5	0.5
Kappa (κ)	0.1	0.1
Initial void ratio	2.7	2.75
Friction angle (φ')	25.4	23.1
Unit weight (γ) (kN/m^3)	15	14
K_{sat} (m/day)	2.5e−005	5.67e−005

Table 2 Soil properties for the crust layer

Parameter	Soft clay
Constitutive model	Linear elastic
Young's modulus E (kPa)	10,000
Poisson's ratio (ν)	0.334
Unit weight (γ) (kN/m^3)	16
K_{sat} (m/day)	0.000248

The crust layer has been assigned a linear elastic constitutive model. This is, in part, used for numerical stability, as beyond the toe of the embankment, the in situ stresses are very small and the ground will tend to heave and go into tension. This can cause numerical convergence problems, since the elastic-plastic model, for example, cannot accommodate tension. Using a linear elastic model avoids this problem. The properties of the crust layer have been shown in Table 2.

In a Sigma/W fully coupled consolidation analysis, it is necessary to define a volumetric water content (VWC) function and a hydraulic conductivity function, even though the soil is saturated and remains saturated during the embankment loading. The volumetric water content function remains unused for saturated conditions, but is nonetheless required. An approximate estimated function is consequently adequate. As with the VWC function, an approximate hydraulic conductivity function is adequate, since only the saturated conductivity (K_{sat}) is used in the analysis. The VWC and the hydraulic conductivity functions for the soft clay, very soft clay, and the crust layers have been shown in Figs. 2 and 3, respectively.

The embankment fill is modeled using a simple linear elastic model with total stress parameters. The material used for the embankment construction is coarse sand and gravel. The pore pressure for such a material can be ignored in the numerical model. This is achieved in Sigma/W by assigning 'total stress material properties' to the fill. Hydraulic properties are consequently not required for the fill. All other relevant material properties are listed in Table 3.

Fig. 2 Volumetric water content functions

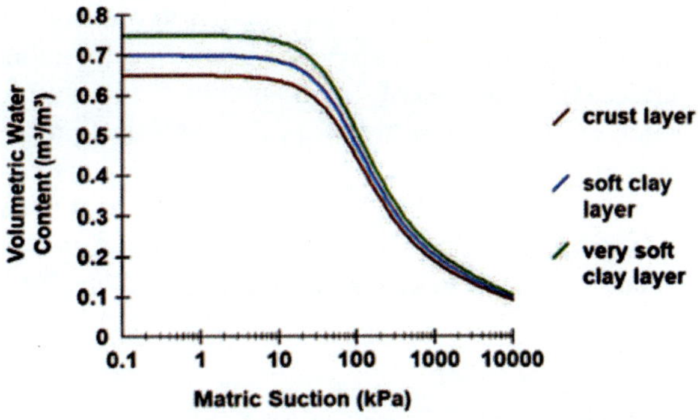

Fig. 3 Hydraulic conductivity functions

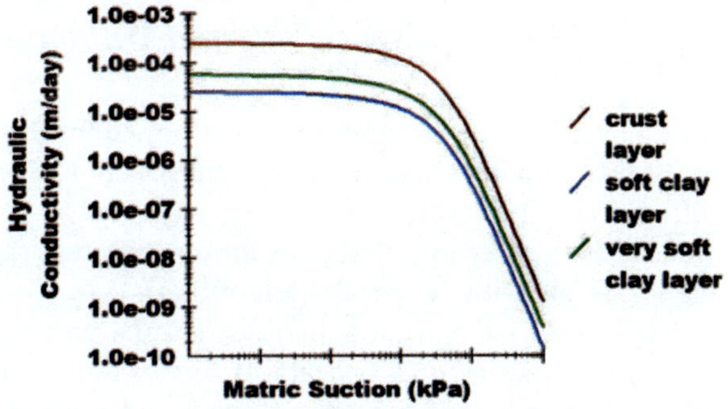

Table 3 Soil properties for the embankment fill

Parameter	Soft clay
Constitutive model	Linear elastic
Young's modulus E (kPa)	10,000
Poisson's ratio (ν)	0.334
Unit weight (γ) (kN/m^3)	18

2.3 Analyses Techniques

An initial in situ analysis has been carried out in Sigma/W to establish the initial ground stresses. The K_o condition in Sigma/W is specified through Poisson's ratio. The specified unit weights are used to apply the self-weight of the material. An initial water table has been specified along the ground surface during the in situ analysis. The subsequent sequential loading process carried out on the foundation, in the form of addition of embankment lifts, which has been modeled stage-wise using the 'coupled stress pore water pressure (PWP) analysis' approach. The embankment was constructed in eight different lifts, each lift having a height of 1 m. Three different rates (0.5, 0.05, and 0.005 m/day) have been used to construct the embankment. A stress-based stability analysis had been performed after the end of construction, with the aid of the Slope/W module in order to obtain the corresponding stability values. The entry–exit specification had been used to define a wide range of circular slip surfaces in the Slope/W. The boundary conditions are, in essence, the driving force which results in the solutions of the numerical problems. In the present simulation, the horizontal displacements along the left and the right sides of the foundation have been fixed, whereas the displacement of the bottom has been fixed in both horizontal and vertical directions. A zero-pressure boundary condition has been used along the ground surface to model the effect of a sand blanket.

3 Results and Discussions

3.1 Stability Analysis (With and Without PVDs)

Stress-based stability analysis was performed for the models to check the variation of the FoS with the different rates of construction. The factor of safety values for the different rates has been shown in Figs. 4a–c and 5a–c. The figures show the stability values for the three different rates for two different cases: without the installation of the PVDs and with the installation of PVDs.

These results demonstrate significant variation in stability of the embankment with the rate of construction. The stability factors for very rapid construction rates, which will lead to undrained behavior of the foundation soil, are very less as compared to the stability values for slow construction rates which permit complete drainage of the foundation soil. The analysis also indicates that installation of the PVDs results in higher stability values.

3.2 Seepage and Deformation Analysis (With PVDs)

The build-up of excess pore water pressure in the foundation at the end of construction of the embankment, near its centerline, for the three different rates has been shown in Fig. 6.

Figure 6 clearly indicates that the rise of excess PWP in the soft soil for rapid rates is higher when compared to the excess PWP values for slow rate of construction. The maximum value of excess PWP for 0.005 m/day is 29.02 kPa, whereas it is 75.4 kPa and 110.7 kPa for the construction rates of 0.05 m/day and 0.5 m/day, respectively.

The development of lateral displacements in soft clay foundations during and after the construction of embankments has been the subject of numerous studies in recent years. The observations of the detrimental effect of lateral displacements on the behavior of adjacent structures have resulted in the need for such studies. For example, in the particular case of piles installed close to or within embankments, lateral displacements have been found to produce bending moments (Heyman

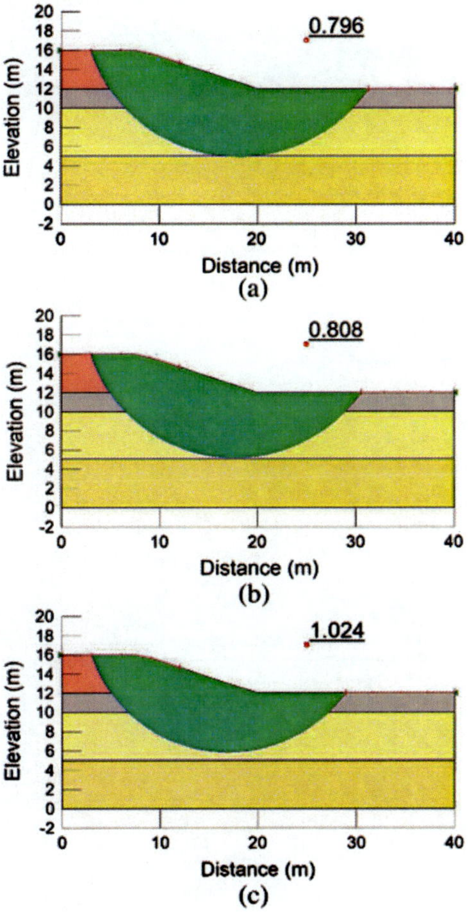

Fig. 4 Stability of embankment on soft soil without PVDs **a** rate 0.5 m/day **b** 0.05 m/day **c** 0.005 m/day

and Boersma 1961), resulting in undesirable movements of piled bridge abutments (Marche and Chapuis 1974), and can even lead to the structural failure of the piles (Marche and Lacroix 1972). On the other hand, there has been a suggestion that lateral displacements can be considered as a good indicator of the stability of embankment foundations (Franx and Boonstrag 1948). The variation of the lateral displacements of the soft soil foundation, at the end of construction time, just below the toe region has been shown in Fig. 7.

Figure 7 indicates that varying the rates of construction did not pronounce much variation in the lateral displacements. The lateral displacements for all the rates give similar trend, and the variation in their magnitudes is not substantial. This can be attributed to the fact that the variation of excess PWP in the foundation for different

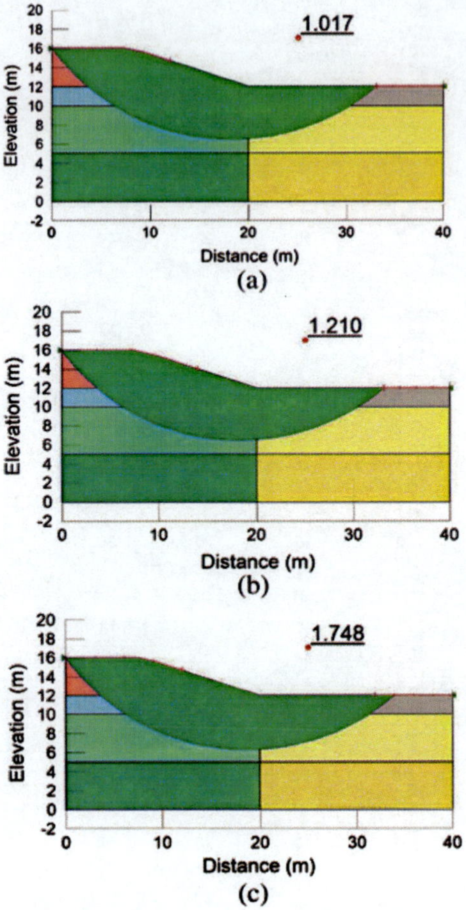

Fig. 5 Stability of embankment on soft soil with PVDs **a** rate 0.5 m/day **b** 0.05 m/day **c** 0.005 m/day

rates of construction is high below the embankment, and it reduces near and beyond the toe region as shown in Fig. 8.

In Fig. 8, the excess PWP was calculated at an elevation of 8 m along the distance from the centerline of the embankment. It is very clearly understood that the excess PWP values below the embankment is very high, and it reduces as it is measured toward the toe. This nature is consistent for all the three different rates of construction. The excess PWP is being negligible near the toe region irrespective of the different rates; these rates will not pronounce any significant variation in the lateral displacement of the soft soil foundation near the toe region.

The ground surface settlement profiles for the three different rates at the end of the construction are shown in Fig. 9. The figure indicates that dissipation of excess PWP at slow rates results in higher settlements below the embankment as compared to the settlements at rapid rates.

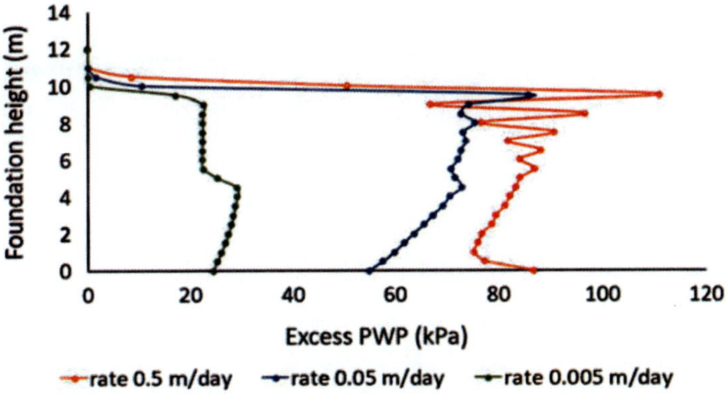

Fig. 6 Variation of excess PWP at the centerline of the embankment for different rates of construction

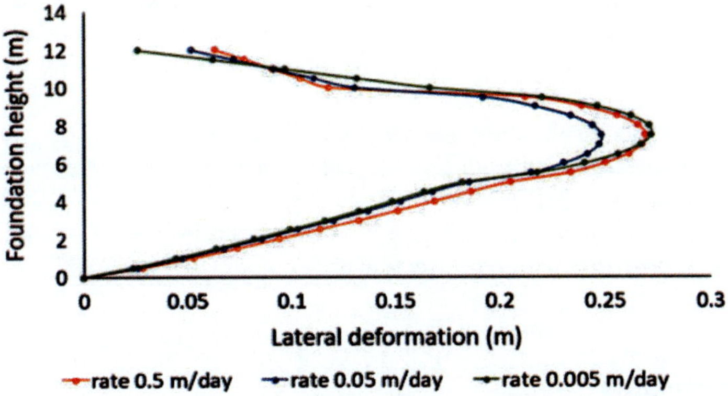

Fig. 7 Variation of lateral displacement below embankment toe for different rates of construction

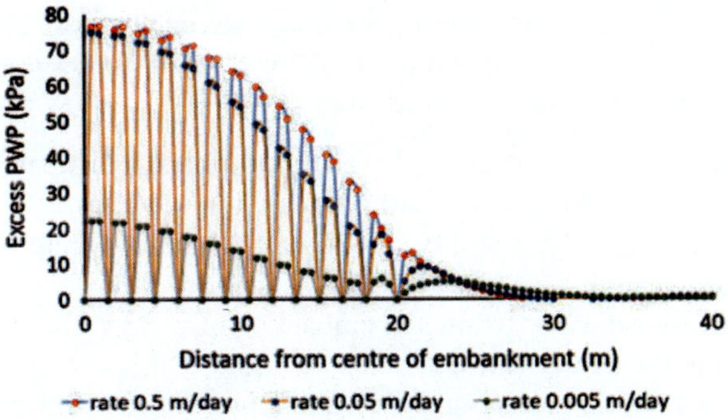

Fig. 8 Variation of excess PWP in the foundation from the center of embankment

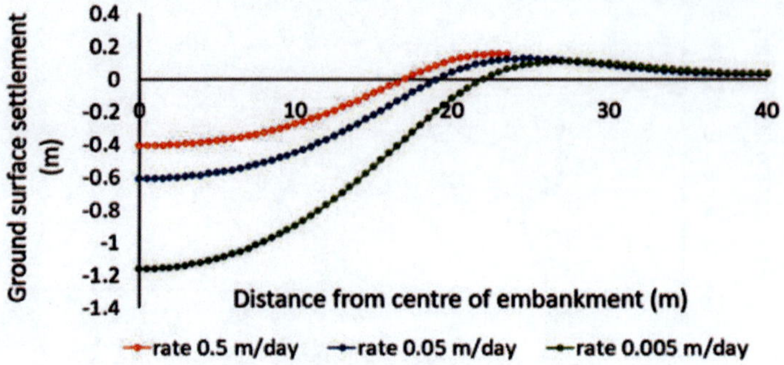

Fig. 9 Ground surface settlement for different rates of construction

In Fig. 9, the maximum settlement of 1.2 m is observed for 0.005 m/day, while the same is noted as 0.6 m and 0.4 m for the construction rates 0.05 m/day and 0.5 m/day, respectively. Beyond the toe region, the in situ stresses are small and the ground has a tendency to go into tension, which is manifested as a heave for all the different rates.

The settlements for the different rates have been presented in Fig. 10a, showing the deformed mesh, along with the vertical displacement contours at the end of construction (8, 80, 800 days).

Figure 10 shows that, the faster rate of construction induces lesser settlement as it results in slower dissipation of excess PWP.

4 Conclusions

The study aimed at the response of the embankment on PVD improved soft soil to the different rates of construction. An embankment on soft soil improved by the installation of PVDs has been simulated for three different construction rates. Following are the conclusions obtained from the analyses:

(a) Installation of PVDs results in higher stability values as compared to the stability values of an embankment on a soft ground without PVDs.
(b) It is observed that the stability of the embankment having a slower rate of construction is high.
(c) The rise of excess PWP below the embankment is very high for rapid rates as compared to the slower rates.
(d) The excess PWP is maximum below the embankment, and it reduces toward the toe.
(e) The rate of construction does not have significant effect on the lateral displacement near the toe region.

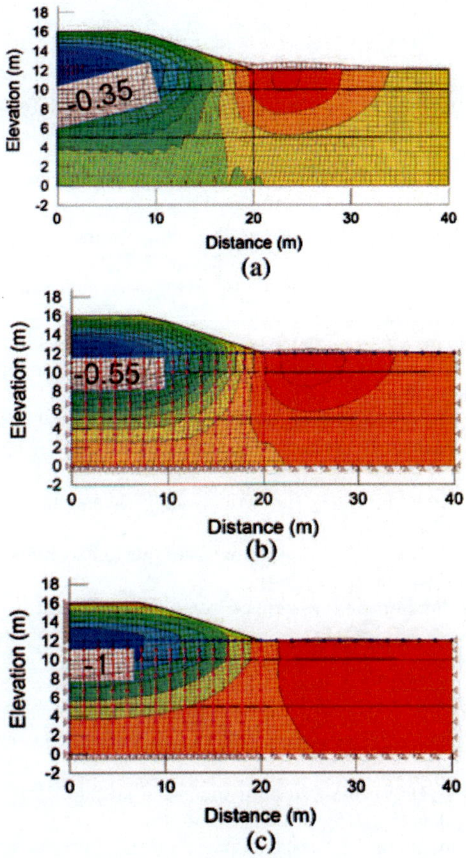

Fig. 10 Observed deformed mesh along with vertical displacement contours for different rates of construction **a** rate 0.5 m/day **b** 0.05 m/day **c** 0.005 m/day

(f) The ground surface settlement at the end of construction is less for rapid rates as it corresponds to the undrained behavior of the foundation soil with slow dissipation of excess PWP.

References

Alamgir M, Miura N, Poorooshasb HB, Madhav MR (1996) Deformational analysis of soft ground reinforced by columnar inclusions. J Comput Geotech 18(4):267–290

Anvesh Reddy M, Dey A (2014) Compensated raft foundation on a preloaded soil improved by vertical drains. GeoShanghai—2014: ground improvement and geosynthetics, Shanghai, China. Geotechnical Special Publication GSP 238, pp 515–525

Barron RA (1948) Consolidation of fine-grained soils by drain wells. Trans Am Soc Civ Eng 113:718–742

Bjerrum L (1972) Embankments on soft ground. In: Proceedings of the ASCE specialty conference on earth and earth-supported structures, vol 2, Purdue University, West Lafayette, Indiana, pp 1–54

Bjerrum L (1973) Problems of soil mechanics and construction on soft clays. In: Proceedings of the 8th international conference on soil mechanics and foundation engineering, vol 3, Moscow, pp 111–159

Casagrande A, Wilson SD (1951) Effect of rate of loading on the strength of clays and shales at constant water content. Geotechnique 2:251–263

Chai JC, Miura N, Sakajo S, Bergado D (1995) Behavior of vertical drain improved subsoil under embankment loading. Soil Found 35(4):49–61

Dey A (2008) Vertical drains and smear effects: a brief review. In: GeoSymposium 2008, national symposium on geoenvironment, geohazards, geosynthetics and ground improvement: events and practices, 4G, New Delhi, India, pp 333–341

Franx C, Boonstrag C (1948) Horizontal pressures on pile foundations. In: Proceedings, 2nd international conference on soil mechanics and foundation engineering, vol I, Rotterdam, The Netherlands, pp 131–135

Graham J, Crooks JHA, Bell AL (1983) Time effects on the stress-strain behaviour of natural soft clays. Geotechnique 33:327–340

Guetif T, Bouassida M, Debats JM (2007) Improved soft clay characteristics due to stone column installation. J Comput Geotech 34(2):104–111

Hansbo S (1981) Consolidation of fine-grained soils by prefabricated drains. In: Publications Committee of the ICSMFE (ed) Proceedings of the 10th international conference on soil mechanics and foundation engineering, Stockholm, vol 3. A.A. Balkema, Rotterdam, The Netherlands, pp 677–682

Heyman L, Boersma L (1961) Bending moments in piles due to lateral earth pressure. In: Proceedings, 5th international conference on soil mechanics and foundation engineering, vol 11, Paris, France, pp 425–429

Indraratna B, Redana IW (1997) Plane strain modeling of smear effects associated with vertical drains. J Geotech Eng ASCE 123(5):474–478

Indraratna B, Balasubramaniam A, Balachandran S (1992) Performance of test embankment constructed to failure on soft clay. J Geotech Eng ASCE 118(1):12–33

Indraratna B, Balasubramaniam AS, Ratnayake P (1994) Performance of embankment stabilized with vertical drains on soft clay. J Geotech Eng ASCE 120(2):257–273

Kulhawy FH, Mayne PW (1990) Manual on estimation soil properties for foundation design. Electric Power Research Institute, Palo Alto, California

Ladd CC, Foott R (1974) New design procedure for stability of soft clays. J Geotech Eng ASCE 100(7):763–786

Lo KY, Morin JP (1972) Strength anisotropy and time effects of two sensitive clays. Can Geotech J 9(3):261–277

Marche R, Chapuis R (1974) Contrôle de la stabilité des remblais par la mesure des déplacements horizontaux. Can Geotech J 11(1):182–201

Marche R, Lacroix Y (1972) Stabilité des culées deponts établies sur des pieux traversant une couche molle. Can Geotech J 9:1–24

Perloff WH, Osterberg JO (1963) The effect of strain rate on the undrained shear strength of cohesive soils. In: Proceedings of the 2nd Pan-American conference on soil mechanics and foundation engineering, vol 1, Rio de Janeiro, Brazil, pp 103–128

Terzaghi K (1931) The static rigidity of plastic clays. J Rheol 2(3):253–262

Wang G (2009) Consolidation of soft soil foundations reinforced by stone columns under time dependent loading. J Geotech Geoenviron Eng 135(12):1922–1931

Influence of Cement Clinker and GGBS on the Strength of Dispersive Soil

Samaptika Mohanty, N. Roy, and S. P. Singh

Abstract The dispersive soil is a highly erodible soil containing a high percentage of exchangeable sodium ions. In the present work, an attempt has been made to enhance the strength of dispersive soil by adding various proportions of ground granulated blast furnace slag (GGBS) (5, 10, and 15%) and cement clinker (5, 10, 15, 20, 25, and 30%), respectively. The combined effect of GGBS and cement clinker on stabilizing dispersive soil is also studied. The soil used in this study is a virgin and identified as a highly dispersive soil by conducting a double hydrometer test. For these purposes, mechanical properties of the various mixes are investigated by the standard Proctor tests, unconfined compressive strength (UCS), and California bearing ratio (CBR). The strength of dispersive soil under UCS is found to increase significantly by adding various proportions of GGBS and cement clinker.

Keywords Dispersive soil · GGBS · California bearing ratio · Unconfined compressive strength

1 Introduction

Dispersive soils are generally clayey and silty soils, which are highly susceptible to erosion and contain a high percentage of exchangeable sodium ions. This type of soil saturated with sodium ions, and it can break down easily and quickly into fine particles and wash away when it is exposed to water. Generally, some fraction is montmorillonite. Dispersive clays have inequity in the electrochemical forces among

S. Mohanty (✉)
Ph.D. Research Scholar, Civil Engineering, NIT, Rourkela, India
e-mail: mohantysamaptika@gmail.com

N. Roy · S. P. Singh
Professor, Civil Engineering, NIT, Rourkela, India
e-mail: nroy@nitrkl.ac.in

S. P. Singh
e-mail: spsingh@nitrkl.ac.in

M. Latha Gali and R. R. P. (eds.), *Problematic Soils and Geoenvironmental Concerns*, Lecture Notes in Civil Engineering 88,
https://doi.org/10.1007/978-981-15-6237-2_5

their particles. This inequity develops the small soil particles in the dispersive soil to be repulsed rather than attracted to each other. Colloidal particles are observed when water flows through dispersive soils. Sinkholes in farms with a top layer of dispersive soils will make the land unusable as the soil is washed out. The plate-shaped particles influence the ease of particle dispersion. There are many potential problems associated with these soils. Dispersive soil has a high swell- shrink problem that means when this type of soil is exposed to the water, it gets severe erosion. These soils show different natures in erosion in comparison to ordinary clay soils. Ouhadi and Goodarzi (2006) studied the fundamental aspects of the interaction between dispersive soil and alum (aluminum sulfate), which are examined and evaluated. Bhuvaneshwari et al. (2007) studied the characteristic of dispersive soil stabilizing with lime and fly ash. Umesha et al. (2009) observed the use of lime and cement to bind the soil clay particles and reduce the dispersivity. Umesha et al. (2011) determined standard tests such as shrinkage limit and unconfined compressive strength tests to quantify the dispersivity of the soils. Abbasi and Nazifi (2013) evaluated and modified the Sherard diagram for the determination of the dispersion potential of clayey soils. Vakili et al. (2013) experimented with a dispersive core soil, which is achieved by mixing with lime and pozzolan, separately or simultaneously. Turkoz et al. (2014) used the magnesium chloride ($MgCl_2$) solution with dispersive soil. The effect of magnesium chloride ($MgCl_2$) solution on the engineering properties of clay soils is evaluated. Goodarzi and Salimi (2015) observed that the potential use and effectiveness of dispersive soil (artificial dispersive soil by adding $NaNO_3$) stabilization using two types of industrial by-product; including granulated blast furnace slag (GBFS) and basic oxygen furnace slag (BOFS) are investigated.

2 Motivation and Objective

To stabilize the dispersive soil for its utilization in civil engineering purposes.

3 Materials

In the present study, the sodium bentonite was used as dispersive soil, and for the mixture material, sodium bentonite GGBS and cement clinker were used. The dispersive soils were collected from Cuttack, Odisha, India, which contains montmorillonite and sodium as the predominant clay mineral. GGBS was collected from Rourkela steel plant, Rourkela, and Odisha, India. The GGBS was used in this experiment powdered in a ball mill. Heating clay and limestone make cement clinker in a rotary kiln at high temperature. It was procured from Shiva cement limited, Rourkela (Odisha). The cement clinker was used in the experiment first crushed in a crusher machine and then fined in a ball mill.

4 Methodology

The dispersive soil, cement clinker, and GGBS were ovens dried. After that, the geotechnical properties like grain size analysis, specific gravity, double hydrometer test, CBR test, compaction characteristics, and UCS test are performed to characterize the strength of dispersive soil. Different percentages of cement clinker and GGBS were mixed with soil to find the optimum strength.

5 Results and Discussion

The experimental work to characterize various combinations of dispersive soil, GGBS, and cement clinker has been done and summarized below. Table 1 shows the physical properties of cement clinker and GGBS, respectively. Whereas Table 2 shows the mix proportions understudies. To enhance the strength of dispersive soil by adding various percentages of ground granulated blast furnace slag (GGBS) and cement clinker. The combined effect of GGBS and cement clinker on stabilizing dispersive soil is also studied.

5.1 Dispersiveness Test

A parallel hydrometer test is then made on a soil sample, but without a chemical dispersing agent, the percentage of dispersion is the ratio of the dry mass of particles smaller than 0.005 mm diameter of the second test to the first expressed as a percentage. Figure 1 shows the double hydrometer test of dispersive soil from which the percentage of dispersion is calculated using the above formula. As per ASTM D4221-99 code, the value under <30% classified as non-dispersive (ND), 30–50%

Table 1 Physical properties of raw materials

Sl. No.	Properties	Dispersive soil	Cement clinker	GGBS
1	Liquid limit (%)	428	33.02	2.8
	Plastic limit	62%	Non-plastic	Non-plastic
	Plasticity index	366%		
2	Specific gravity	2.74	3.24	2.84
3	OMC (%)	34.12	14.36	14.17
	MDD (kN/m^3)	13.98	18.01	16.60
4	Unconfined compressive strength, q_u (kPa)	175.47	110.38	176.19

Table 2 Mix proportion under study

Specimen code	Dispersive soil	Cement clinker	GGBS
S	100	–	–
SS9505	95	5	–
SS9010	90	10	–
SS8515	85	15	–
SC9505	95	–	5
SC9010	90	–	10
SC8515	85	–	15
SC8020	80	–	20
SC7525	75	–	25
SC7030	70	–	30
S4M1	85	10	5
S4M2	80	15	5
S4M3	75	20	5
S4M4	70	25	5
S4M5	85	5	10
S4M6	80	10	10
S4M7	75	15	10
S4M8	70	20	10
S4M9	65	25	10
S4M10	55	30	15

S—dispersive soil, SS—soil-GGBS mix, SC—soil-cement clinker mix, S4M1—soil-GGBS-cement clinker mix

classified as intermediate dispersive (ID), >50% classified as highly dispersive (HD) is shown in Table 2.

5.2 *Grain Size Distribution*

Figure 2 shows the grain size distribution curve of cement clinker and GGBS. Cement clinker used in the present study is found to be coarse-grained and poorly graded. However, the GGBS is grounded to sizes under 75 μ and utilized.

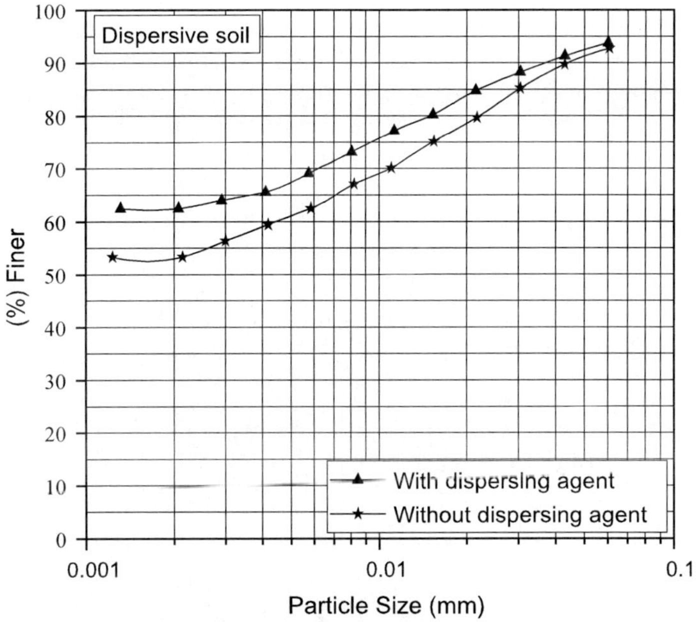

Fig. 1 Artifacts empowered by artificial Intelligence

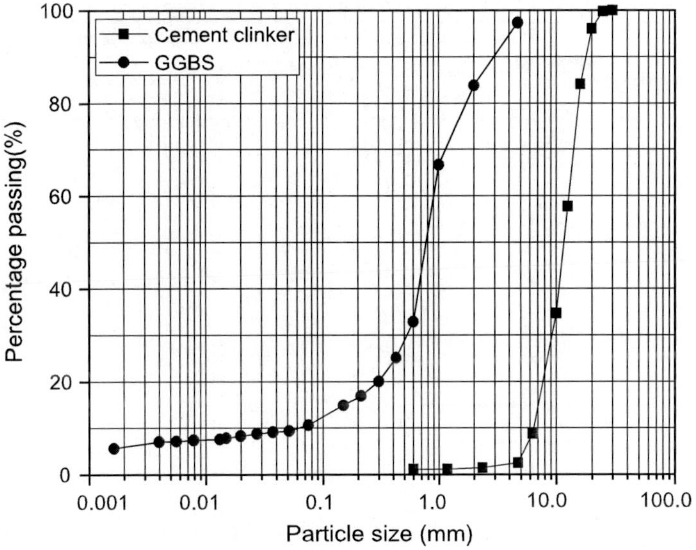

Fig. 2 Grain size analysis curve

5.3 Compaction Characteristics

The standard Proctor compaction tests on dispersive soil and dispersive soil treated with GGBS, cement clinker and GGBS dispersive are determined. Figure 3a shows the MDD of dispersive soil, which is found to be 13.98 kN/m^3 at an OMC of 34.12%. The MDD of GGBS is found to be 16.60 kN/m^3 at an OMC of 14.17%, and the MDD of cement clinker found to be 18.05 kN/m^3 at an OMC of 14.55% (IS 2720-7) (1980). Cement clinker when added to the dispersive soil, there is a decrease in the OMC, while the MDD of the soil increases (Fig. 3b). In addition to GGBS with the dispersive soil, there is a decrease in OMC and an increase in MDD (Fig. 3c). The compaction curve of soil-GGBS-cement clinker mix is shown in Fig. 3d.

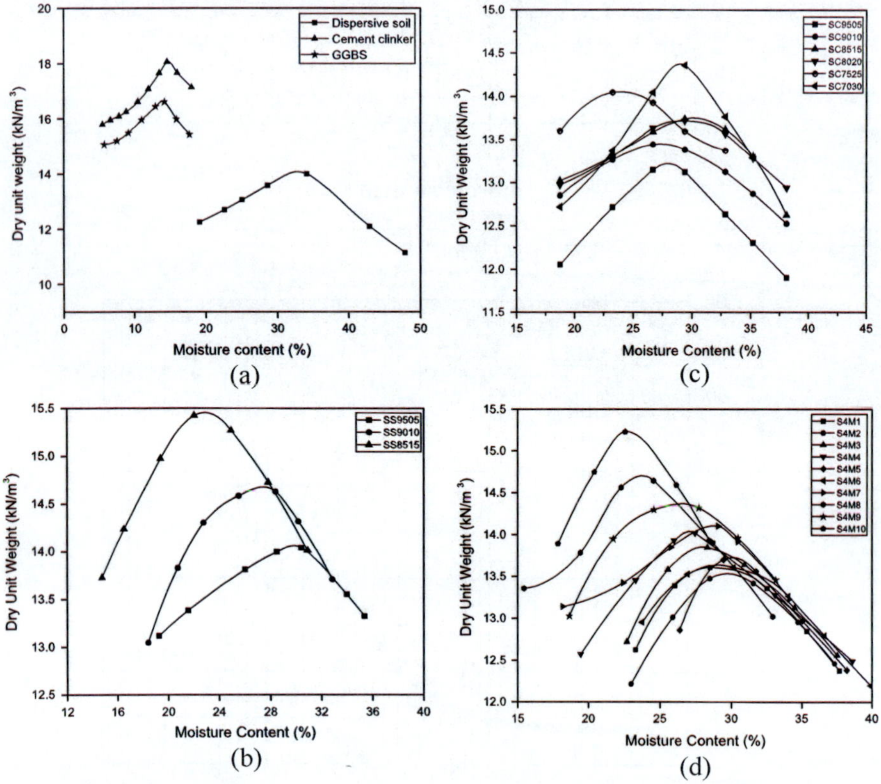

Fig. 3 Compaction curve of **a** raw materials, **b** soil-GGBS mix, **c** soil-cement clinker mix, and **d** soil-GGBS-cement clinker mix

5.4 CBR Test

A series of tests were performed using the CBR method on untreated and treated specimens, which was prepared at the optimum moisture content and maximum dry unit weight. CBR values are closely correlated to bearing capacity and compressive strength of compacted subgrade. Therefore, this test is most suitable to measure the suitability of any compacted subgrade (Indraratna et al. 1994). CBR tests are performed for both soaked and unsoaked condition on the specimens as per ASTM D 2166-16 (2016). Though the CBR value of dispersive soil is low, so before use as subgrade material, the soil can be treated. CBR value for both soaked and unsoaked condition of the dispersive soil, soil-cement clinker mix, soil-GGBS mix, soil-GGBS-cement clinker mix is presented in Fig. 4. Figure 4 (a) shows that the SC7030 mix gives higher CBR value in soaked condition than other percentages soil-cement clinker mix. Figure 4 (b) depicts that the CBR values of soaked CBR are found to

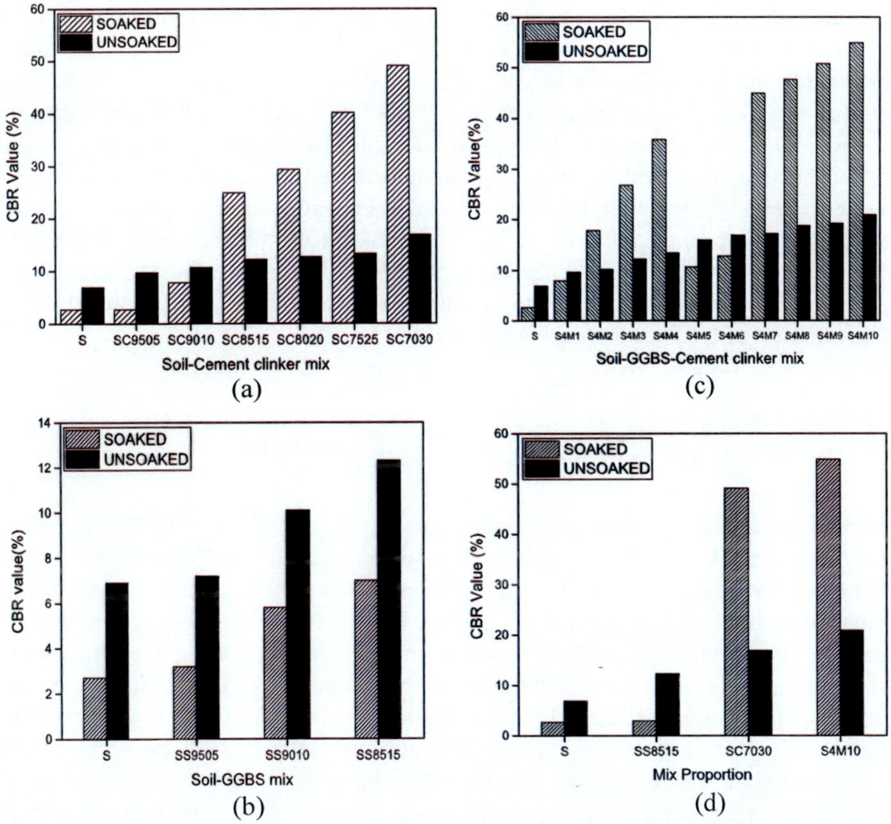

Fig. 4 CBR value versus **a** soil-cement clinker mix, **b** soil-GGBS mix, **c** soil-GGBS-cement clinker mix, and **d** mix design versus CBR value

be less than the unsoaked condition for the soil-GGBS mix. On the other hand, in the case of soil-GGBS-cement clinker mixes the CBR values are higher in soaked conditions as compared to unsoaked conditions except for the S4M1 mix as seen in Fig. 4 (c). Figure 4 (d) reveals that the S4M10 mix gives higher CBR value in the soaked condition as compared to unsoaked condition. CBR values are higher for the hydration process, which leads to the formation of the cementitious compounds and increase the bearing capacity. The improvement in soaked CBR value up to 90% is very stable for subgrade material (Emesiobi 2000). High CBR value reduced the height of embankment, which will reduce the cost of road construction. As per the guidelines of IRC: SP: 72 (2007); a minimum soaked CBR of 20 has been recommended for the subbase layer of low volume road pavements.

5.5 *UCS Test*

The unconfined compressive strength test samples are prepared at the optimum moisture contents determined from the compaction curves. Figure 5 shows the results obtained from UCS analysis of soil mixed with different percentages of cement clinker and GGBS at curing periods 0, 7, 14, 28, 60, and 90 days in normal room temperature. Figure 5a shows that the addition of 30% cement clinker, the strength of soil-cement clinker mix gives higher strength as compared to other percentages of cement clinker. Figure 5b shows the soil-GGBS blend gives higher strength compared to different percentages. Figure 5c shows that S4M10 (55% soil, 15% GGBS, and 30% clinker mix) gives the highest strength (5.93 MPa) and the S4M1 combination gives the lowest (1.90 MPa) strength as compared to other mixtures. Figure 5d shows the variation of UCS versus curing periods.

5.6 *Mineralogical and Morphological Analysis*

Mineralogical and morphological analysis such as X-ray diffraction (XRD) and scanning electron microscope (SEM) tests are also conducted. The X-ray pattern of raw materials (Fig. 6a) shows that the soil in the crystalline phase contains quartz, kaolinite, montmorillonite, and aluminum silicate. It is seen that GGBS contains gehlenite, coesite, akermanite, and melilite found to be an amorphous phase. Cement clinker showed an amorphous and contains calcium silicates, tricalcium aluminate, and tetracalcium aluminoferrite. The cementitious compounds, CSH, CASH, CSHH, CAOCH, and calcite were detected. The stronger reflections from cementitious compounds such as ettringite, CASH, CSH, and calcite peaks were observed in the SC7030 mix. No trace of the ettringite was detected for the mixture of SS8515. GGBS can produce a low quantity of cementitious compounds on its own. For S4M10 mix, CASH, CAOCH, CSH, ettringite, and calcite peaks few new peaks are observed.

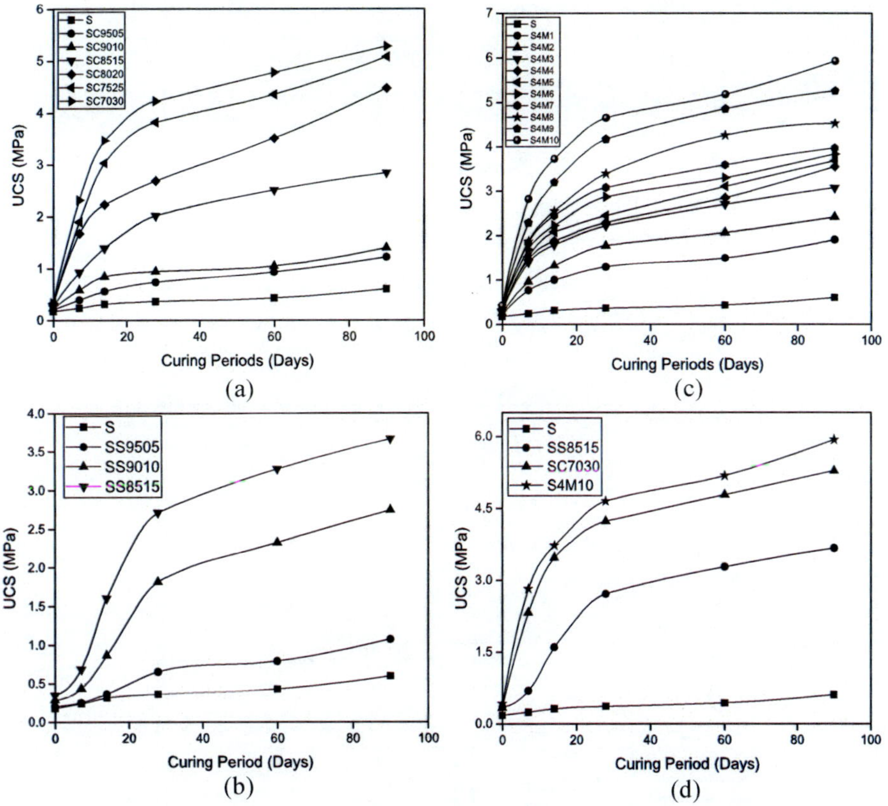

Fig. 5 UCS versus curing period of **a** soil-cement clinker mix, **b** soil-GGBS mix, **c** soil-GGBS-cement clinker mix, and **d** curing periods versus mix proportion

Both untreated and treated soil specimens are cured for 90 days at room temperature and examined in XRD and SEM.

In SEM, analysis has shown that the dispersive soil particles are bonded together with stabilizing materials formed hydrated particles and cementitious compounds because of the reaction between the soil and the stabilizing agents as shown in Fig. 7a, b. From Fig. 7c, it is clearly shown that the formation of hydrated gel, ettringite (needle-like structure) and compacted matrix are observed when the soil mix with GGBS and cement clinker mix. It is important to stress that Ca, Si, Al, and O elements are essential for the formation of CSH and CASH crystals, which are the main cementation products of the stabilized soil.

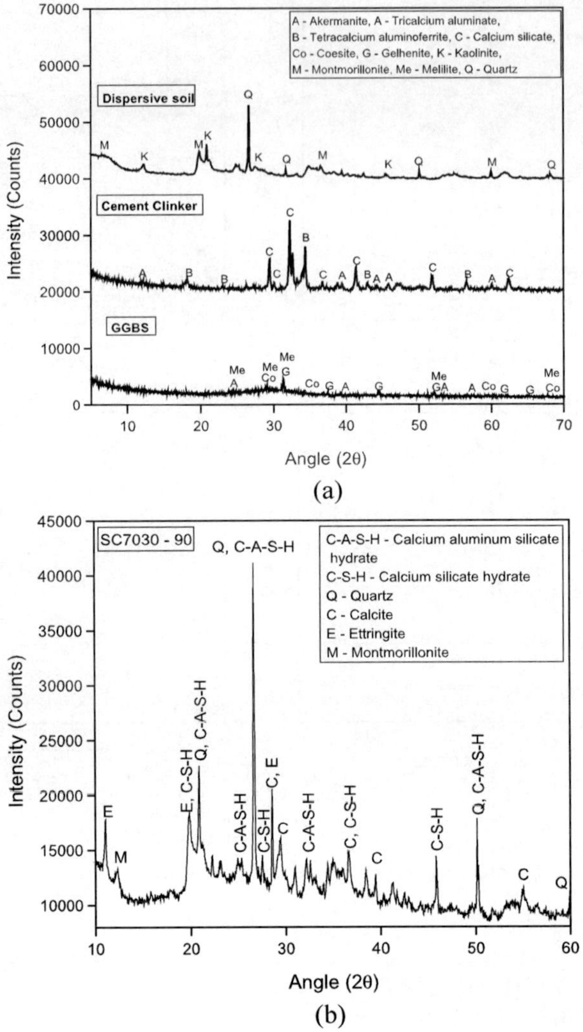

Fig. 6 XRD analysis of **a** raw materials **b** soil-cement clinker mix, **c** soil-GGBS mix, and **d** soil-GGBS-cement clinker mix

6 Conclusions

As observed from the double hydrometer test results, the dispersion percentage of soil found to be 90.66%, which is classified as highly dispersive soil. In XRD analysis, a higher quantity of calcium aluminum silicate hydrate (CASH) and calcium silicate hydrate (CSH) and ettringite, which coated the residues of silt particles and floccules and helped to enhance the mechanical strength. CBR values are higher because the hydration process of cement clinker bound the soil particles and enhanced the

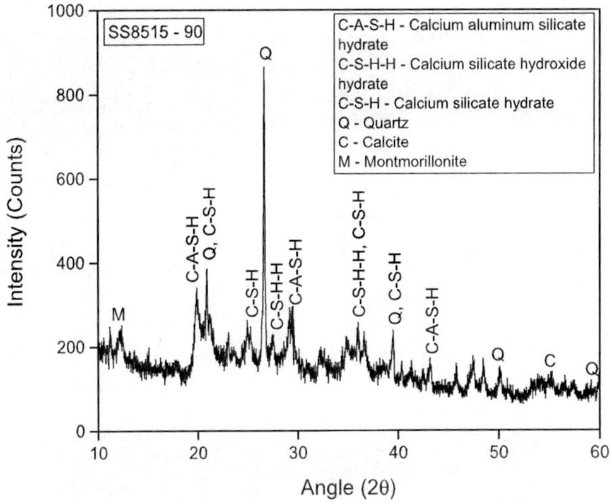

(c)

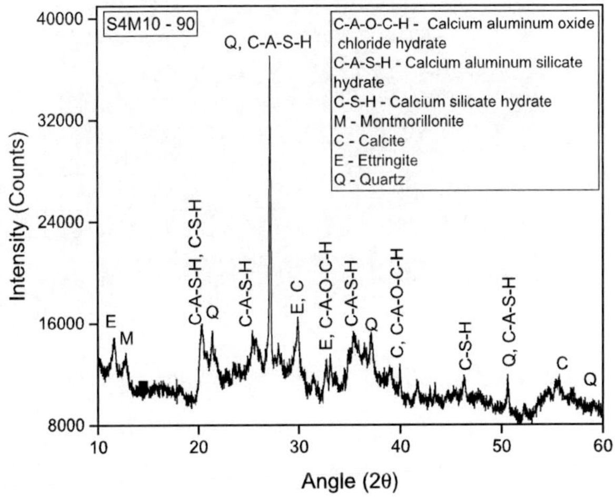

(d)

Fig. 6 (continued)

strength. Therefore, based on the investigation presented above, it is observed that when an additive like cement clinker and GGBS ash are used, the dispersive nature of soil reduces and its strength increases.

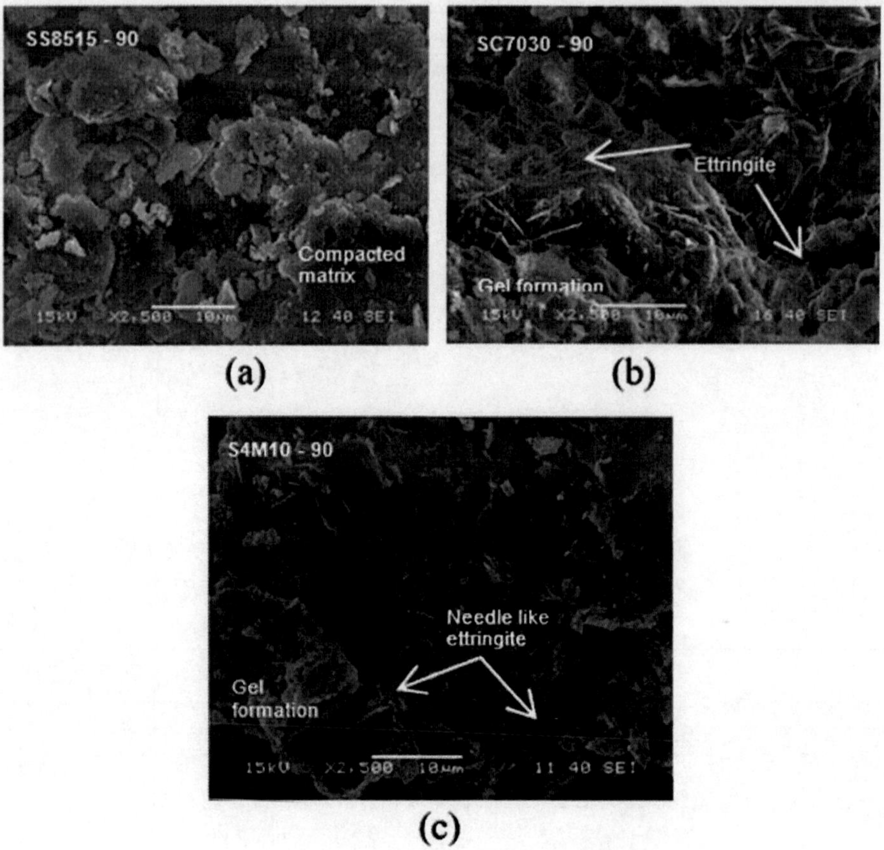

Fig. 7 SEM images of **a** soil-GGBS mix, **b** soil-cement clinker mix, and **c** soil-GGBS-cement clinker mix

References

Abbasi N, Nazifi MH (2013) Assessment and modification of sherard chemical method for evaluation of dispersion potential of soils. Geotech Geol Eng 31:337–346

ASTM D4221-99—Standard test method for dispersive characteristics of clay soil by double hydrometer

Bhuvaneshwari S, Soundra B, Robinson RG, Gandhi SR (2007) Stabilization and microstructural modification of dispersive clayey soils. In: First international conference on soil and rock engineering, Colombo, Sri Lanka, 5–11 Aug

Goodarzi AR, Salimi M (2015) Stabilization treatment of a dispersive clayey soil using granulated blast furnace slag and basic oxygen furnace slag. Appl Clay Sci 108:61–69

IS 2720-Part 7 (1980) Methods of test for soils-determination of water content dry density relationship using light compaction. Bureau of Indian Standards, New Delhi, India

IS 2720-Part 10 (1991) Methods of test for soils-determination of unconfined compressive strength. Bureau of Indian Standards, New Delhi, India

IS 2720-Part 16 (1987) Methods of test for soils-laboratory determination of CBR. Bureau of Indian Standards, New Delhi, India

Ouhadi VR, Goodarzi AR (2006) Assessment of the stability of a dispersive soil treated by alum. Eng Geol 85:91–101

Turkoz M, Savas H, Acaz A, Tosun H (2014) The effect of magnesium chloride solution on the engineering properties of clay soil with expansive and dispersive characteristics. Appl Clay Sci 101:1–9

Umesha TS, Dinesh SV, Sivapullaiah PV (2009) Control of dispersivity of soil using lime and cement materials. Sci Appl 3:8–15

Umesha TS, Dinesh SV, Sivapullaiah PV (2011) Characterization of dispersive Soils. Mater Sci Appl 2:629–633

Vakili AH, Selmat MR, Moayedi H, Amani H (2013) Stabilization of dispersive soils by pozzolan. Forensic Eng ASCE 726–735

List of Conferences

Mohanty S, Roy N, Singh SP (2017a) Strength characteristics of dispersive soil by using industrial by-products. In: International congress and exhibition sustainable civil infrastructures: innovative infrastructure geotechnology. Springer, Egypt, Cham, pp 293–302

Mohanty S, Roy N, Singh SP (2017b) Treatment of dispersive soil by using flyash (FA) and granulated blast furnace slag (GBFS). In: Indian geotechnical conference 2017 GeoNEst, IIT Guwahati, p 312

Permeability Index of Mechanically Biologically Treated Waste and Its Application in Bioreactor Landfills

P. Sughosh, M. R. Pandey, and G. L. Sivakumar Babu

Abstract The hydraulic conductivity of waste is an important parameter that influences the design of leachate recirculation system in a bioreactor landfill. The hydraulic conductivity of waste varies with respect to depth and time in a landfill. These variations can be attributed to the settlement of waste, and the corresponding changes in the void ratio can be related to permeability. In this study, experiments were conducted to find the values of permeability index (C_k, the ratio of the slope of the void ratio—coefficient of permeability relationship) at four different placement densities of waste. The relation between the compression index (C_c) and C_k for waste is discussed. Estimation of permeability using C_k is presented and compared with the reported values in the literature. The effect of initial placement density of waste on the design of injection wells in a typical bioreactor landfill is presented to demonstrate the applicability of C_k and C_c values with the help of HYDRUS-2D simulations.

Keywords Bioreactor landfill · Leachate · Recirculation · Permeability index · Compression index

1 Introduction

The total settlement in a landfill is the sum of immediate, creep, and biodegradation settlements and can be up to around 20–50% of the total depth of landfill (Swati and Joseph 2008; Tchobanoglous et al. 1993). Settlement of waste results in the change in void ratio/density of waste. The variations in the density have corresponding changes in the properties of waste such as the hydraulic conductivity and unsaturated properties of waste. Leachate recirculation system (LRS) is an integral part of the bioreactor

P. Sughosh (✉)
Center for Sustainable Technologies, Indian Institute of Science, Bangalore, India
e-mail: sughosh.p@gmail.com

M. R. Pandey · G. L. Sivakumar Babu
Department of Civil Engineering, Indian Institute of Science, Bangalore, India

M. Latha Gali and R. R. P. (eds.), *Problematic Soils and Geoenvironmental Concerns*, Lecture Notes in Civil Engineering 88,
https://doi.org/10.1007/978-981-15-6237-2_6

landfill, and the hydraulic conductivity of waste is an important parameter used in the design of LRS. The formulas or the modeling tools used to determine the spacing of the leachate recirculation wells are dependent on the hydraulic conductivity of waste (Maler 1998; Tomothy 2016). The hydraulic conductivity is known to vary as a function of density and age of waste (Bleiker et al. 1995; Reddy et al. 2011; Beaven et al. 2010). Therefore, any change in the density of waste, which may be due to the overburden stress (depth) or different placement density as per the design will considerably affect hydraulic conductivity and hence the performance of the landfill. In this paper, studies pertaining to the effects of placement densities of waste and overburden stress in the landfill are studies as they can considerably influence the spacing of the recirculation wells.

Therefore, the aim of the study is to:

- Formulate an experimental program to estimate the hydraulic conductivity of waste corresponding to different depths in a landfill.
- Estimate the value of the permeability index (C_k = ratio of the slope of the void ratio—coefficient of permeability) of waste.
- Study the influence of placement density of waste on the spacing of recirculation wells in a bioreactor landfill.

2 Methodology

Consolidation tests were performed with wastes compacted to different initial densities. During consolidation tests, the hydraulic conductivity of waste was determined under different loading conditions. These values correspond to the hydraulic conductivity of waste in a landfill at different depth for a particular placement density and time. With these results, the permeability index of waste was also estimated. The experimental data was used to run the numerical simulation (HYDRUS-2D) of the landfill recirculation system to study the influence of the placement density on the well spacings.

2.1 Experimental Studies on Compression Index (C_c), Hydraulic Conductivity (k), and Unsaturated Properties of Waste

2.1.1 Compression Index and Hydraulic Conductivity of Waste

One-dimensional consolidation test was carried in an oedometer to determine the compression index of waste. The consolidation experiments were carried out on waste samples at different densities (550, 700, 850, and 1050 kg/m^3). The experiments were carried out in standard consolidation apparatus as per the ASTM D 2435 (2011) standard. The load was applied with standard weights using a lever system, and the deformation was measured with a dial gauge. The waste sample (particle size <4 mm or <10 mm) was filled in a standard consolidation ring (having diameter of 60 mm and height 20 mm) and compacted by the method of static compaction. The sample was subjected to constant vertical stress of 6.25 kPa, and the compression was measured at different time intervals. The hydraulic conductivity was measured by the variable head method once the dial gauge reading was constant for that particular loading. For the next load increment, the normal stress was increased to 12.5 kPa, and compression and hydraulic conductivity were monitored. This procedure was repeated for normal stresses of 25, 50, 100, 200, 400, and 800 kPa.

2.1.2 Estimation of Unsaturated Properties of Waste

In this study, the filter paper method as specified in ASTM D5298 (2016) was followed for the determination of suction of waste. The experimental data points (suction–moisture content) of waste at a particular dry density (800 kg/m^3) were used for curve fitting, and the van Genuchten (VG) parameters were estimated.

2.2 Estimation of Waste Density at Different Depth in the Landfill

The density of the waste increases with the depth of the landfill due to increase in overburden pressure. The total settlement of the landfill was calculated as per the Marques equation (Marques et al. 2003). The corresponding changes in the dry densities at any depth of the landfill were obtained from the settlement of the individual layer. The corresponding changes in the hydraulic conductivity of waste were estimated as per the procedure described in Sects. 2.1.1. The effect of density on unsaturated properties was not considered for the present study. As the effect of biodegradation was not considered, the time frame used in the settlement equation was set to one day.

2.3 Estimation of Hydraulic Conductivity of Waste at Different Depth in the Landfill

At a particular initial placement density of the waste, the permeability index gives the relationship between the k and void ratio/density. The variation of void ratio/density as a function of overburden pressure is given by the compression index (C_c). For a wide variety of natural soils, the ratio of C_k to C_c is in the range of 0.5–2 (Mesri and Rokhsar 1974) but there is no such range of values reported for MSW. As per

Table 1 Compression index and permeability index of waste at different placement density

Bulk density (kg/m^3)	C_c	C_k	C_k/C_c
1050	0.408	0.199	0.487
700	0.507	0.346	0.682
850	0.440	0.297	0.674
550	0.422	0.458	1.085

the methods suggested in Sect. 2.1, both C_k and C_c were determined for waste with initial placement densities (bulk) of 550, 700, 850, and 1050 kg/m^3 and were shown in Table 1. Thus, for any bulk density of waste, the C_k can be determined by knowing the C_c value and from Fig. 4. From the C_k value, the hydraulic conductivity can be obtained for the waste of different dry density present at various depth of landfill.

2.4 *Hydrological Modeling the Leachate Recirculation System (LRS)*

A landfill cross section of 50 m length and 20 m depth was selected to demonstrate the hydraulic performance of LRS. HYDRUS-2D was used for simulation, and landfill depth was subdivided into five equal layers of 4 m each. A vertical well of 17 m deep and 0.3 m diameter having a screen depth of 3 m from the bottom of well was selected. The well dimension selected was similar to that reported in Khire and Mukherjee (2007). The model input mainly consisted of the well dimensions, recirculation rate, period of simulation, material properties, and boundary conditions. Simulation studies on leachate injection at the rate of 5 m^3/day were carried out under continuous operation until steady-state conditions were established. No flux boundary condition (BC) was assumed for top and sidewalls while the bottom was considered as free drainage BC.

3 Results

3.1 *Compression Index of Waste*

From the consolidation experiments, the variation of void ratio with respect to different load increment for four different placement densities is obtained and is shown in Fig. 1.

The compression index of the waste measured was in a narrow range (0.4–0.5, from Table 1) for bulk densities in the range of 550–1050 kg/m^3.

This indicated that the settlement and hence the change in density of waste with an increase in overburden pressure (depth of landfill) would be in a small range for

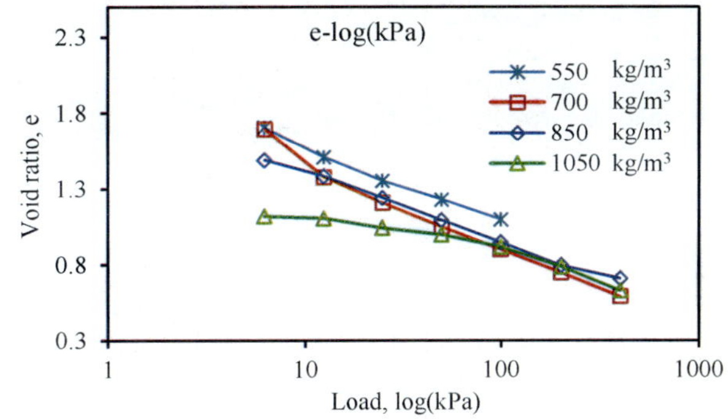

Fig. 1 Variation of void ratio as a function of overburden stress for different initial placement densities of waste

all placement densities. From Fig. 1, it can be observed that for a higher placement density, the change in void ratio is not so significant up to 50 kPa.

3.2 *Estimation of Hydraulic Conductivity*

Generally, the hydraulic conductivity decreases with depth and age of waste. Various laboratory studies on hydraulic conductivity of waste report the range from 1.6×10^{-3} to 3.7×10^{-8} m/s (Powrie and Beaven 1999; Reddy et al. 2009; Stoltz et al. 2010). The variation of void ratio and bulk density with respect to the hydraulic conductivity of waste for four different placement densities are shown in Figs. 2 and 3, respectively. The "k" value decreases with an increase in bulk density and from C_k value of Table 1, we can infer that with an increase in initial placement density the k value decreases (due to increase in the slope of e-log(k) graph). The range of k observed for four different placement densities under increasing overburden pressure is between 1.0×10^{-8} and 1.0×10^{-5} m/s. The results obtained in this study are within the range reported in the literature (Fig. 4).

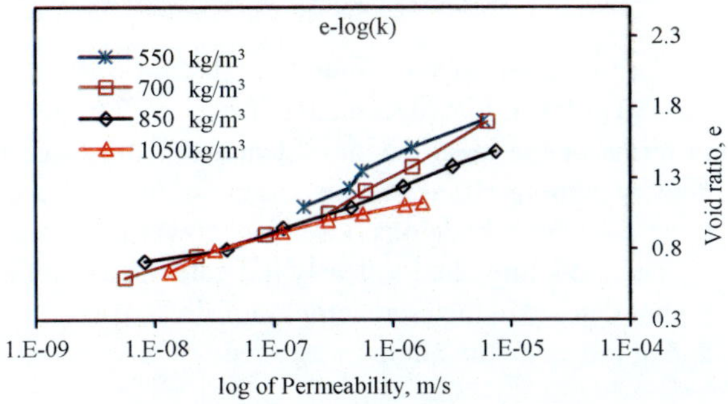

Fig. 2 Variation of void ratio as a function of the hydraulic conductivity of waste for different initial placement densities

Fig. 3 Variation of bulk density as a function of the hydraulic conductivity of waste for different initial placement densities

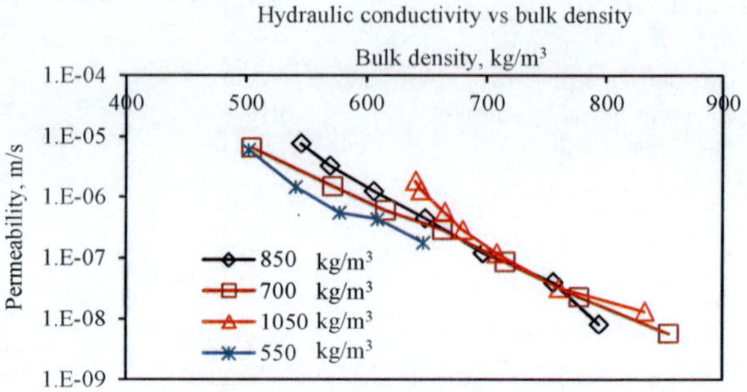

Fig. 4 Relation between the initial placement density of the waste and the ratio of C_k and C_c

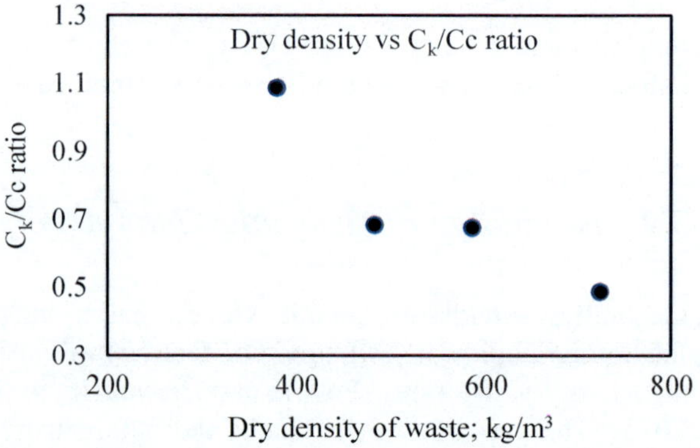

3.3 *Modeling the Effect of Placement Density on the LRS*

The behavior of the LRS with waste having bulk densities of 550 and 850 kg/m^3 was considered for modeling studies. In both cases, the variation of density with the depth of landfill (at every 4 m depth) was calculated as per Sect. 2.3 and was used as the input to the model. For a continuous injection of leachate at 5.5 m^3/day, the saturation lines (SL) for the 550 and 850 kg/m^3 placement densities are as shown in Fig. 5.

For the same leachate injection rate, the wetted width (>90% saturation contour) for 850 kg/m^3 initial placement density is 1.7 times greater than the 550 kg/m^3. Even in terms of the areal extent of leachate saturation achieved, the higher placement density is more efficient. This might be due to the spread of the saturation contours toward the high *k* regions (*k* reduces toward the depth of landfill).

The modeling results clearly indicate the advantage of compacting the waste at a higher density in the bioreactor landfills. It also helps in accommodating more waste and at the same time help in increasing the spacing between the recirculation wells.

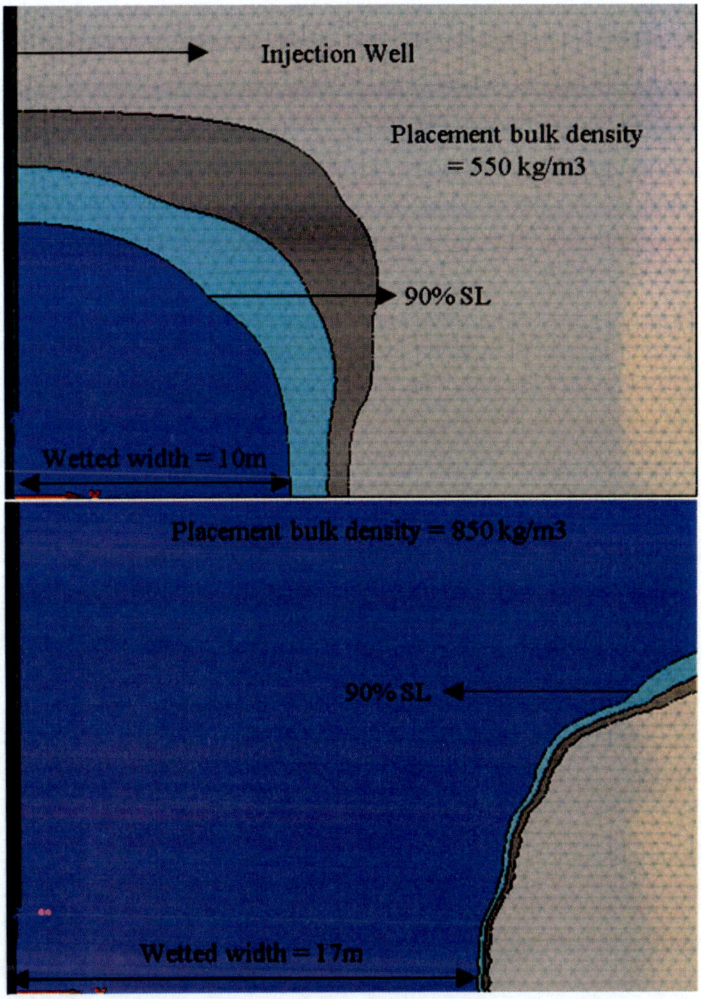

Fig. 5 Distribution of saturation lines in landfills with 550 and 850 kg/m^3 initial placement densities

4 Conclusions and Remark

The paper presents the effect of the initial placement density of waste on the spacing of the leachate recirculation wells in a bioreactor landfill. Consolidation and hydraulic conductivity tests were carried out on the waste samples and the relationship between void ratio, C_c and k were obtained. The permeability index for waste samples was determined in this study which is not reported elsewhere in the literature. The C_c values were in a very narrow range while the C_k values were found to be inversely related to the placement density. The results of numerical simulations show that the saturation lines are dependent on the placement density of the waste in the bioreactor landfill. The study can be applied to space the vertical wells in an optimal way by considering the variation of wetted width with the initial placement density of waste.

References

ASTM D2435/D2435M-11 (2011) Standard test methods for one-dimensional consolidation properties of soils using incremental loading. ASTM International, West Conshohocken, PA

ASTM D5298-16 (2016) Standard test method for measurement of soil potential (suction) using filter paper. ASTM International, West Conshohocken, PA

Beaven RP, Powrie W, Zardava K (2010) Hydraulic properties of MSW. In: Geotechnical characterization, field measurement, and laboratory testing of municipal solid waste, pp 1–43

Bleiker DE, Farquhar G, McBean E (1995) Landfill settlement and the impact on site capacity and refuse hydraulic conductivity. Waste Manage Res 13(6):533–554

Khire MV, Mukherjee M (2007) Leachate injection using vertical wells in bioreactor landfills. Waste Manage 27(9):1233–1247

Maler T (1998) Analysis procedures for design of leachate recirculation systems. Water Qual Int 37–40

Mesri G, Rokhsar A (1974) Theory of consolidation of clays. J Geotech Eng Div 100:889–904

Powrie W, Beaven RP (1999) Hydraulic properties of household waste and implications for landfills. Proc Inst Civ Eng-Geotech Eng 137(4):235–237

Reddy KR, Hettiarachchi H, Parakalla NS, Gangathulasi J, Bogner JE (2009) Geotechnical properties of fresh municipal solid waste at Orchard Hills Landfill, USA. Waste Manage 29(2):952–959

Reddy KR, Hettiarachchi H, Gangathulasi J, Bogner JE (2011) Geotechnical properties of municipal solid waste at different phases of biodegradation. Waste Manage 31(11):2275–2286

Stoltz G, Gourc JP, Oxarango L (2010) Liquid and gas permeabilities of unsaturated municipal solid waste under compression. J Contam Hydrol 118(1–2):27–42

Swati M, Joseph K (2008) Settlement analysis of fresh and partially stabilised municipal solid waste in simulated controlled dumps and bioreactor landfills. Waste Manage 28(8):1355–1363

Tchobanoglous G, Theisen H, Vigil SA, Alaniz VM (1993) Integrated solid waste management: engineering principles and management issues, vol 949. McGraw-Hill, New York

Tomothy GT (2016) Sustainable practices for landfill design and operation. Springer, New York

Effect of Ethanol on Compressibility Swelling and Permeability Characteristics of Bentonite–Sand Mixtures

Tribenee Saikia, Binu Sharma, and Safi Kamal Rahman

Abstract Sand–bentonite mixtures are used to construct clay liner in engineered landfills. Clay-lined landfills when exposed to organic wastes produce leachate which affects the properties of the clay liner extensively, and it may become unreliable in being a permanent barrier to the wastes. This work is to investigate the behaviour of different bentonite–sand mixtures through one-dimensional consolidation using ethanol–water solution mixed in the ratio of 20:80 by volume, as the permeating fluid. The same series of experiments were performed with pure water also as the permeating liquid to get a good comparison of the change in properties. The percentage swell, swelling pressures and permeability were high in the mix with pure water solution. Compressibility was found to be high in the mix with ethanol–water solution. Good correlation exists between per cent swell and swelling pressure in the mix using ethanol–water solution and in the mix using water.

Keywords Bentonite–sand mixtures · Ethanol · Permeability · Swelling · Landfill liners

1 Introduction

Sand–bentonite mixtures are used to construct clay liner in engineered landfills. The clay liner has to be impermeable and should have the strength to carry the load of the wastes disposed upon it. Bentonite having very low permeability serves well as one of the liner materials, and to prevent the development of cracks on desiccation, sand should be added (Mollins et al. 1996). The permeability of sand reduces after addition of bentonite to the samples even as low as 2% by weight (Otoko and Otoko 2014). Swelling is another important property which affects the stability of the clay liner and has to be taken into consideration while analysing the parameters for selection of the liner material. The swelling pressure and swelling potential decrease with the increasing sand content irrespective of initial compaction condition. Soil

T. Saikia (✉) · B. Sharma · S. K. Rahman
Assam Engineering College, Guwahati, Assam, India
e-mail: tribeneesaikia@gmail.com

M. Latha Gali and R. R. P. (eds.), *Problematic Soils and Geoenvironmental Concerns*, Lecture Notes in Civil Engineering 88,
https://doi.org/10.1007/978-981-15-6237-2_7

with fine sand exhibited relatively higher swelling pressure and swelling potential (Srikanth and Mishra 2015). They also found that the hydraulic conductivity of the sand–bentonite mixtures was very much dependent on the particle size of sand.

Clay-lined landfills are exposed to organic wastes which when coming in contact with water, produce various organic liquids or leachate. These affect the properties of the clay liner extensively and it may become unreliable in being a permanent barrier to the wastes. According to Singh et al. (2003), montmorillonite swells by sorbing as much as 8% mass of organic solvent. The ability of clays to swell in hydrocarbon solvents is because of the ability of the solvent to disrupt the attractive van der Waals forces operating between the adjacent organic cations overcoming any enhanced electric double-layer repulsion expected to form because of the lower dielectric constant of organic solvents compared to water as given by Slade and Gates (2004). According to Estabragh et al. (2014), the compression index of soil increases with the increase in the concentration of ethanol.

This work is to investigate the behaviour of different bentonite–sand mixtures like swelling, compressibility and permeability, through one-dimensional consolidation at very low unit weight, using ethanol–water solution as the permeating fluid. These behaviours are then compared with the values obtained with pure water as a pore fluid.

2 Materials and Methodology

One-dimensional consolidation tests have been performed in six bentonite–sand mixtures. The bentonite–sand mixture was mixed in the following proportion by percentage of dry weight as B:S = 100:0, 90:10, 80:20, 70:30, 60:40 and 50:50 (B: bentonite, S: sand). Dry bentonite–sand mixtures were placed initially in the consolidation cell at their loosest dry state and then allowed to saturate. The tests are performed to understand the change in the properties of bentonite–sand mixtures on addition of ethanol–water solution as the pore fluid.

2.1 Materials Used for Study

Commercially available bentonite and dry sand were used for the study. The sand used in the study was classified according to the Unified Soil Classification System (USCS) and was found to be poorly graded sand (SP). The X-ray diffraction spectra for the bentonite show that it is predominantly a montmorillonite with some amount of sand, i.e. the quartz mineral. The liquid limits and the plastic limits of the mixtures were determined according to the code IS 2720 (Part 5) (1985), the values of which are shown in Table 1. Figures 1 and 2 show the variation of liquid limit and plastic limit values, respectively, with the various bentonite and sand mixtures.

Table 1 Properties of bentonite–sand mixtures

Bentonite:Sand	Liquid limit (%)	Plastic limit (%)	Plasticity index (%)
100:0	195	35	160
90:10	165	32.5	132.5
80:20	140	30	110
70:30	123	27	96
60:40	110	24	86
50:50	100	20	80

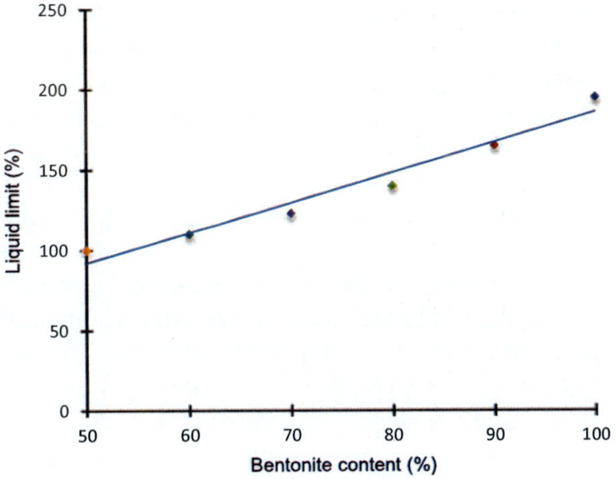

Fig. 1 Variation of liquid limit with bentonite content

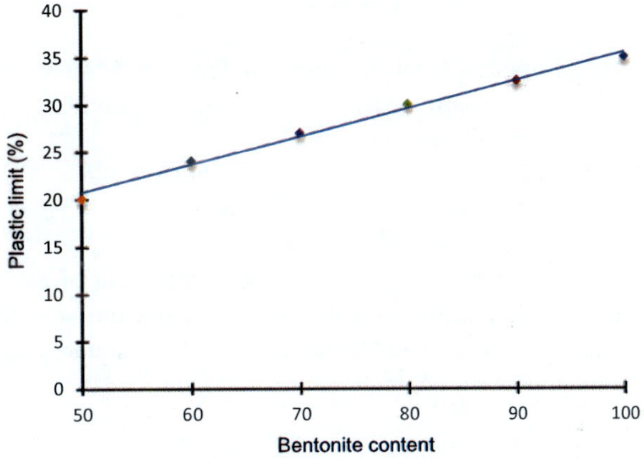

Fig. 2 Variation of plastic limit with bentonite content

2.2 The Consolidation Test Procedure

This research work was carried out in a conventional fixed ring consolidation cell in laboratory to measure the swelling, compressibility and permeability behaviour of the various bentonite–sand mixtures. The mixture was placed in the cutter loosely up to a height of 2/3rd of the cutter, i.e. 13.33 mm. Consolidation test was conducted following the procedure given in the code IS: 2720 (Part 15) (1965). The consolidometer was assembled with the dry soil specimen and porous stones at top and bottom of the specimen, providing a filter paper between the soil specimen and the porous stones. Silicone grease was applied to the inner surface of the ring to eliminate the friction between the ring and the soil sample.

A seating load of 5 kN/m^2 was placed on the loading hanger, and the initial reading of the dial gauge was noted. The sample was then allowed to saturate fully. The first test series was conducted with pure water as the permeating fluid. The second test series was conducted with ethanol–water solution (20% ethanol by volume). As a result, after saturation, the sample started swelling. The dial gauge reading was taken after every 24 h and total swelling percentage was plotted with respect to time. After full swelling, the dial gauge reading was constant and the value was recorded. The swollen sample was then subjected to small amounts of consolidation pressure until the swelling pressure was obtained which is indicated by the returning of the dial gauge to the initial reading before swelling. The double incremental load was then applied up to 640 kN/m^2 as per the code of practice.

2.3 Permeability Calculations

The permeability of the soil samples with different pore fluids was calculated theoretically from the coefficient of consolidation, C_v, values obtained after each stress increment from Eq. (1).

$$k = C_v \cdot m_v \cdot \gamma_w \tag{1}$$

In the above equation, k is the coefficient of permeability, m_v is the coefficient of volume compressibility and γ_w is the unit weight of water. The coefficient of consolidation was determined by Taylor's square root of time fitting method.

3 Results and Discussion

The properties of a clay soil like porosity, swelling and permeability are affected by the interaction of the clay particles with the permeating liquid. This interaction is dependent on the fluid properties such as dielectric constant, ionic strength and surface tension. Determination of dielectric constant is not done in this study. It is seen from literature that the dielectric constant of ethanol is lower than pure water; so it is assumed that the ethanol solution used in this experiment also has lower dielectric constant than water. The dielectric constant, however, is influenced by the concentration of the organic compound present in the solution. The solution used in this study has only 20% ethanol by volume with water which indicates that there will not be much decrease in the value of dielectric constant from that of pure water.

3.1 Swelling Behaviour

Swelling occurs when the soil particles absorb the solvent and increases in volume. Free swell was allowed to occur in the samples saturated with water and with ethanol–water solution separately. The swelling in the soil mixtures is due to the montmorillonite content in the bentonite. Figures 3 and 4 represent the variation of swelling

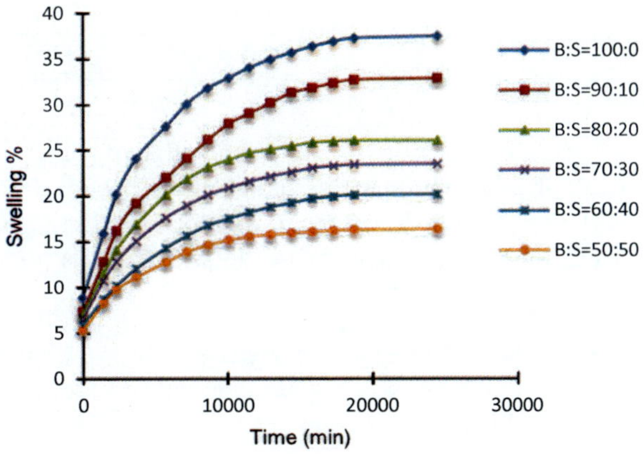

Fig. 3 Swelling % versus time with water as pore fluid

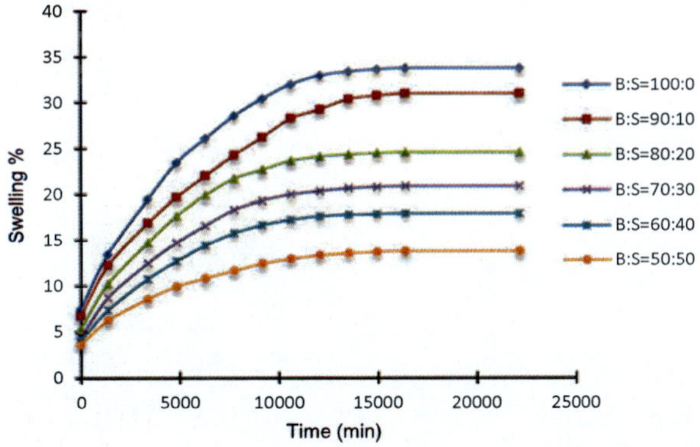

Fig. 4 Swelling % versus time with ethanol–water solution as the pore fluid

Table 2 Variation of swelling pressure for samples with water as the pore fluid

Bentonite:Sand	Total swelling (%)	Swelling pressure (kN/m^2)
100:0	37.73	210
90:10	32.85	180
80:20	26.03	165
70:30	23.48	145
60:40	20.18	120
50:50	16.35	100

Table 3 Variation of swelling pressure for samples with ethanol–water as the pore fluid

Bentonite:Sand	Total swelling (%)	Swelling pressure (kN/m^2)
100:0	33.76	200
90:10	30.91	170
80:20	24.53	160
70:30	20.86	130
60:40	17.93	100
50:50	13.73	80

percentage with time for water and ethanol–water solution as pore fluids, respectively. It is seen that maximum swelling occurred in pure bentonite and it reduces with the addition of sand to it. The total swelling percentage and swelling pressure for all the samples are given in Tables 2 and 3 for pore fluid water and ethanol–water solution, respectively.

Comparison of Swelling Percentage The swelling versus time with pure water and ethanol–water solution has been plotted together for two of the sample mixtures as shown in Figs. 5 and 6.

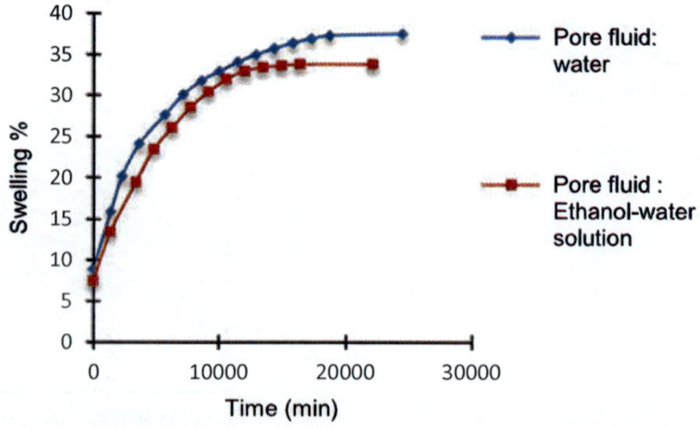

Fig. 5 Swelling % versus time for bentonite:sand = 100:0

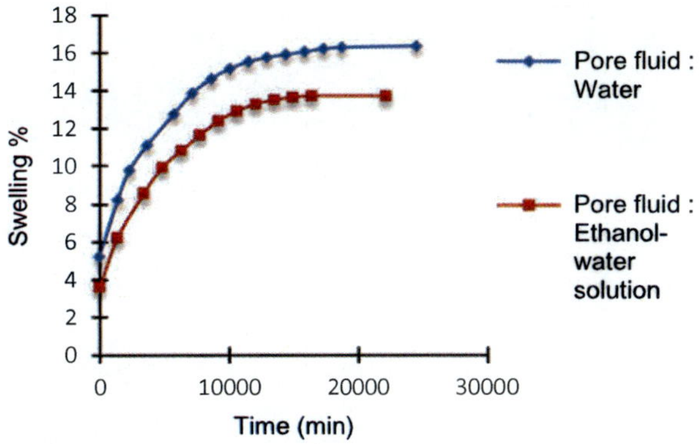

Fig. 6 Swelling % versus time for bentonite:sand = 50:50

From the figures, it can be seen that for the same bentonite-to-sand ratio, the swelling percentage is higher for pure water as pore fluid. The ability of clays to swell is supposed to be related to the ability of the solvent to disrupt the attractive van der Waals forces operating between the adjacent organic cations. But due to lower dielectric constant of ethanol–water solution than water, its presence in the pore fluid to some extends retards the swelling of the clay. Due to this, differences in the maximum swelling percentages have been seen in both the pore fluid solutions.

Comparison of Swelling Pressure Figure 7 shows that with the increase in sand content in the mixtures, the swelling pressure decreases. Another observation is that for the same bentonite-to-sand ratio, the swelling pressure is higher for the sample saturated with pure water which is evident from the fact that the dielectric constant for ethanol is lesser than water which does not help in the expansion of the diffused double layer. Good correlation exists between per cent swell and swelling pressure in the mix using ethanol–water solution and in the mix using water, the correlation coefficient in both the cases being 0.99.

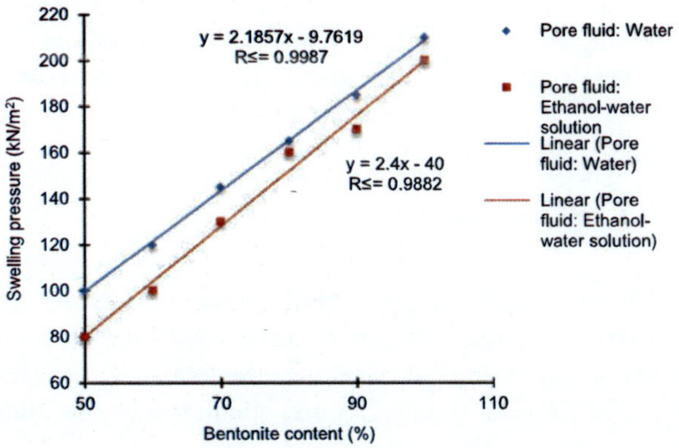

Fig. 7 Variation of swelling pressure with bentonite content

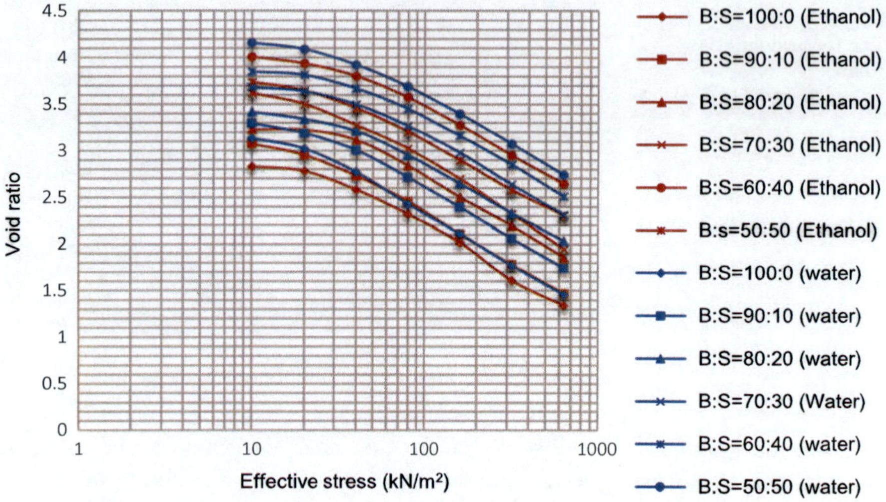

Fig. 8 Void ratio versus log of effective stress

3.2 Consolidation Test Results

The void ratio value decreases as the applied stress increases. Figure 8 shows that with the increase in sand percentage in the samples, the void ratio of the mixture slightly increases. It has been observed that the pure bentonite sample (bentonite:sand = 100:0) has the lowest void ratio at 640 kN/m^2 effective stress value. From Fig. 8, it can be seen that for the same bentonite-to-sand ratio of the mixture, at a given effective stress, the void was higher for the sample with pure water as the pore fluid. It is because of the fact that the bentonite–sand mixtures swelled more when ethanol was not present in the solution.

The compression index, as shown in Fig. 9, decreased with the addition of sand to the samples. But for the same ratio of bentonite and sand, the compression index was found to be higher for the sample with ethanol–water solution as the pore fluid. This is in agreement with Estabragh et al. (2014), where it was stated that the compression index of soil increases with the increase in concentration of ethanol.

3.3 Permeability

The coefficient of permeability increases as the quantity of sand in the sample mixtures increases but it decreases with the increasing effective stress. The coefficients of permeability versus effective stress are shown in Figs. 10 and 11, respectively, for pore fluid water and ethanol–water solution. The permeability values are

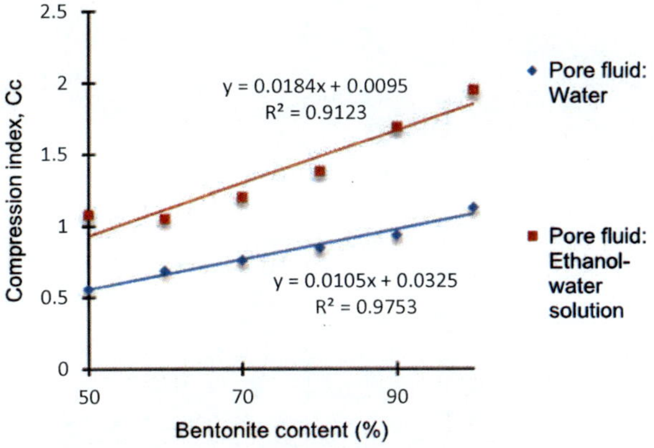

Fig. 9 Variation of compression index with bentonite content

Fig. 10 Theoretical permeability of samples with water as pore fluid

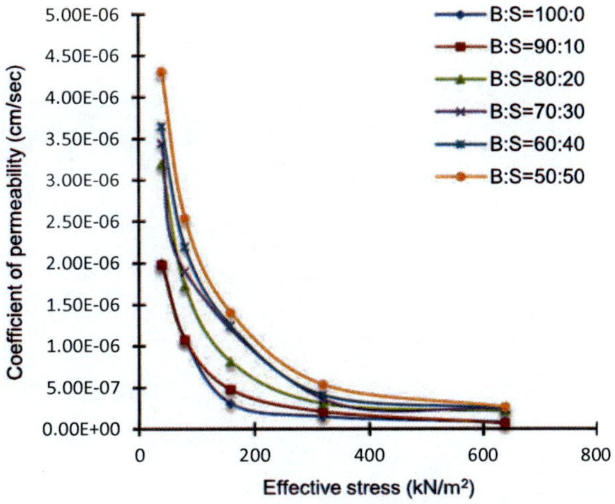

less than 5×10^{-7} cm/s and become insignificant as the effective stress crosses 320 kN/m^2.

Figures 12 and 13 show the void ratio versus log of coefficient of permeability for pore fluids water and ethanol–water solution, respectively. The log of coefficient of permeability was found to vary linearly with void ratio over the full range of pressure increments to which the samples were subjected.

Comparison of Permeability Figures 14 and 15 compare the coefficient of permeability with pore fluid water and ethanol–water solution for bentonite:sand = 100:0 and 50:50, respectively. It was observed that the samples with ethanol–water solution as pore fluid showed lower values of the coefficient of permeability than with pure water for the same bentonite:sand ratio. It can be explained by the fact that the samples swell more with pure water as the pore fluid. The presence of ethanol in the

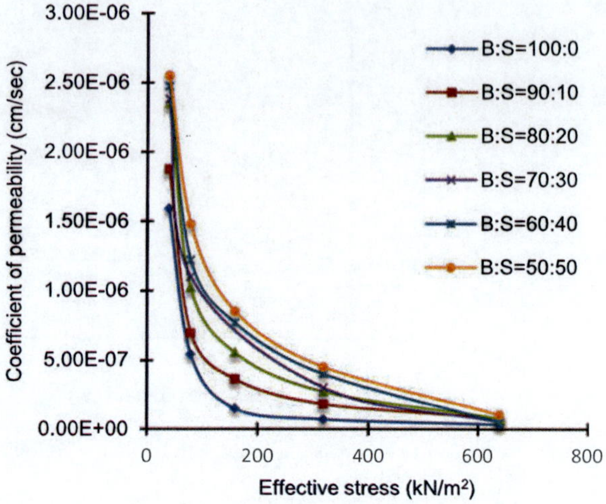

Fig. 11 Theoretical permeability of samples with ethanol–water solution as pore fluid

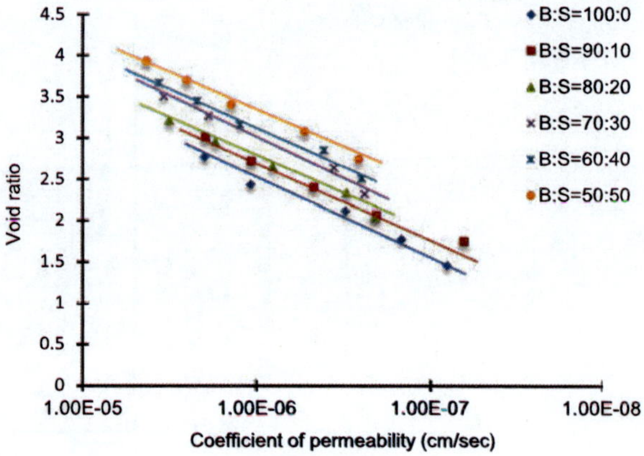

Fig. 12 Void ratio versus log of coefficient of permeability with water as pore fluid

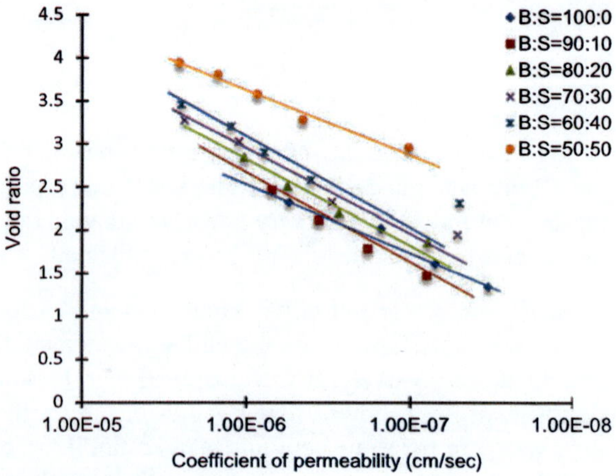

Fig. 13 Void ratio versus log of coefficient of permeability with water as pore fluid

Fig. 14 Void ratio versus log of coefficient of permeability for bentonite:sand = 100:0

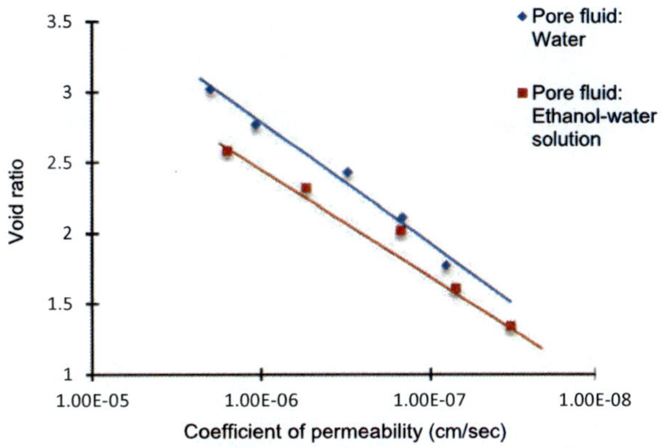

Fig. 15 Void ratio versus log of coefficient of permeability for bentonite:sand = 50:50

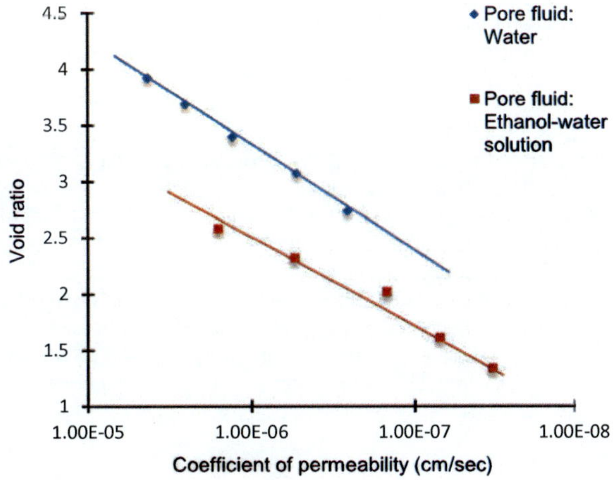

pore fluid reduces the swelling of a particular soil sample, which in turn reduces the void ratio. This, in turn, affects the permeability also.

4 Conclusion

The Atterberg limits decreased with the addition of sand to bentonite and the values were directly proportional to the percentage of bentonite present in the samples tested. From the swelling test, it was found that the swelling percentage was highest for the sample with 100% bentonite irrespective of the pore fluid used. It was also seen that the total swelling percentages were higher for all the samples with pure water as pore fluid as compared to ethanol–water solution. For the same proportion of bentonite and sand, the sample saturated with pure water showed a higher value of swelling pressure than with ethanol–water solution. The swelling pressure decreased with the addition of sand to bentonite for both the conditions. The compression index decreased with the increase in sand content and it had higher values for the samples with ethanol–water solution as the pore fluid. The coefficient of permeability of the samples increased with the increasing sand content. The coefficient of permeability of the samples with ethanol–water solution was low in comparison to the ones with pure water.

References

Estabragh AR, Beytolahpour I, Moradi M, Javadi AA (2014) Consolidation behavior of two fine-grained soils contaminated by glycerol and ethanol. Eng Geol 178:102–108

IS: 2720 (Part 5) (1985) Method of test for soils: determination of liquid and plastic limit. Bureau of Indian Standards, New Delhi, India

IS: 2720 (Part 15) (1965) Methods of test for soils: determination of consolidation properties. Bureau of Indian Standards, New Delhi, India

Mollins LH, Stewart DI, Cousins TW (1996) Predicting the properties of bentonite–sand mixtures. Clay Miner 31:243–252

Otoko RG, Otoko GU (2014) The permeability of ocean sand with bentonite. Int J Eng Technol Res 2(1):01–06. ISSN: 2327-0349

Singh N, Megharaj M, Gates WP, Churchman GJ, Andersonm J, Kookana RS, Naidu R, Chen Z, Slade PG, Sethunathan N (2003) Bioavailability of an organophosphorus pesticide, fenamiphos, sorbed on an organoclay. J Agric Food Chem 51:2653–2658

Slade PG, Gates WP (2004) Permeability of an organo-modified bentonite to ethanol-water solutions. Clays Clay Miner 52(2):192–203

Srikanth V, Mishra AK (2015) A laboratory study on geotechnical characteristics of sand–bentonite mixtures and the role of particle size of sand. Int J Geosynth Ground Eng 2, 3

Characterization of Heavy Metals from Coal Gangue

Mohammed Ashfaq, M. Heera Lal, and Arif Ali Baig Moghal

Abstract Coal gangue, a residue obtained during the coal mining process, accounts for 10–15% of raw coal produced. Usually, this coal gangue is transported and stacked loosely in the nearby areas. Long term piling up of this mine waste can volatilize large amounts of potentially toxic heavy metals which have the ability to infiltrate surrounding ecosystems. In the present study, coal gangue from an open cast mining area in Bhupalpally in Telangana state, India, has been analyzed for its physical and chemical characteristics. Further, column leaching test was performed to evaluate the leachability of selected heavy metal ions simulating field conditions using deionized double distilled water. The column tests proved that the selected heavy metal ions from coal gangue can be easily mobilized.

Keywords Characterization · Heavy metals · Leaching

1 Introduction

Mining of coal is often associated by generation of huge volumes of natural rock masses, popularly known as 'Coal gangue' which usually accompany the carboniferous beds. Coal gangue is characterized by great diversity in grain size and petrographic composition and amounts to more than 20% of the total solid waste generated during the mining (Keefer and Sajwan 1993; Plewa and Mysłek 2001; Malhotra and Mehta 2002; Asokan et al. 2005). Among the various solid wastes generated by different industries, it is identified as the largest with relatively higher impact on the environment (Wu et al. 2017).

Among the Coal Combustion Residues, a lot of study has been focused on coal and fly ash. Coal ash has been successfully utilized for many years in various applications (Yao et al. 2015) and fly ash has been used as fill material, liner material for waste containment, to stabilize toxic and heavy metal diced soils, as subbase material, as

M. Ashfaq · M. Heera Lal · A. A. B. Moghal (✉)
Department of Civil Engineering, NIT Warangal, Warangal 506004, India
e-mail: baig@nitw.ac.in

M. Latha Gali and R. R. P. (eds.), *Problematic Soils and Geoenvironmental Concerns*, Lecture Notes in Civil Engineering 88,
https://doi.org/10.1007/978-981-15-6237-2_8

soil additives to increase crop productivity (Moghal and Sivapullaiah 2011, 2012, 2013; Moghal 2017). But, relatively lesser attention has been given to coal gangue.

The current avenues of coal gangue utilization include re-combustion for power generation, in the production of fertilizers, bricks, concrete, and cement (Wu et al. 2017). Though the rate of coal gangue has spiked in the recent times, its huge volume of generation with relatively higher rate of generation necessitates the identification of alternative utilization approach (Ashfaq et al. 2019). Previous researchers have established the application of coal gangue in retrospective backfilling of abandoned mine sites as a sustainable solution to reduce the problem of subsidence and disposal of solid waste. Wu et al. (2017) and Jabłonska et al. (2017) have identified the potential of coal gangue as an adsorbent and landfill liner material. However, new applications of coal gangue are still sought. In this context, in the present study, physical and chemical analyses of coal gangue and presence of heavy metal have been evaluated. Also, column leaching test on coal gangue has been performed to assess the leaching potential of selected heavy metals from coal gangue.

2 Materials and Methodology

Coal gangue has been procured from Kakatiya Coal mines, Bhupalpally, Telangana. Post procurement, to avoid possible alteration in the chemical and physical properties, the samples were sealed in a zip pouches. Air-dried samples were grounded using jaw crusher and subsequently were allowed to pass through 1 mm mesh sieve to maintain consistency and uniformity of the tested samples.

Acid digestion technique was used to evaluate the initial concentrations of selected trace metal concentration. A liquid-to-solid ratio of 20 was maintained during the leaching tests performed in the column apparatus. The physical properties studied include natural moisture content, specific gravity, pH, total dissolved solids (TDS) and electrical conductivity (TDS). All the tests were performed in accordance with their respective ASTM codes.

3 Results and Discussion

3.1 Physical Properties

Moisture Content The natural moisture content of coal gangue calculated after oven drying the sample was in the range of 2–5%. This variation in value might be due to the heterogeneity of coal gangue.

Specific Gravity (Gs) The average specific gravity value of coal gangue was found to be 2.50. The lower value of specific gravity may be due to varying iron content

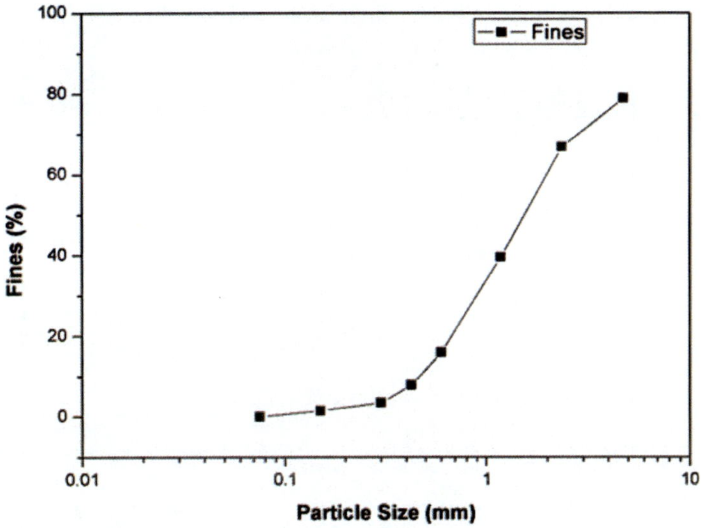

Fig. 1 Grain distribution curve

and presence of entrapped air. This lower 'Gs' leads to lower unit weight and thus coal gangue finds application as a backfill material.

Grain Size Distribution The crushed portion of coal gangue was uniformly mixed to make a representative sample of 1 kg for performing sieve analysis test. The particle size distribution curve of the samples used in the study is presented in Fig. 1. Based on the Unified Classification system (USC), the coal gangue samples are comparable to silty sands (SM).

3.2 *Chemical Properties*

pH The pH was determined at varying liquid-to-solid ratios (4–100) and is presented in Fig. 2. Prior to test, the pH meter was calibrated using three different buffer solutions with pH of 4, 7 and 9.2. From the results, it is evident that the natural coal gangue has a neutral pH of about 7. With the increase in L/S ratio, coal gangue exhibited acidic pH with proportionate increase in acidity with the liquid content. The linear increase in acidity with the L/S ratio can be attributed to the attack of H^+ ions on the mineral phases of the coal gangue.

EC and TDS The instrument used for measuring pH was used for measuring EC and TDS. From results presented in Fig. 3, it was observed that EC and TDS exhibited an identical behavior with the change in L/S ratio. It was noted that with the increasing L/S ratio, both EC and TDS linearly decreased. Similar to the observations made for

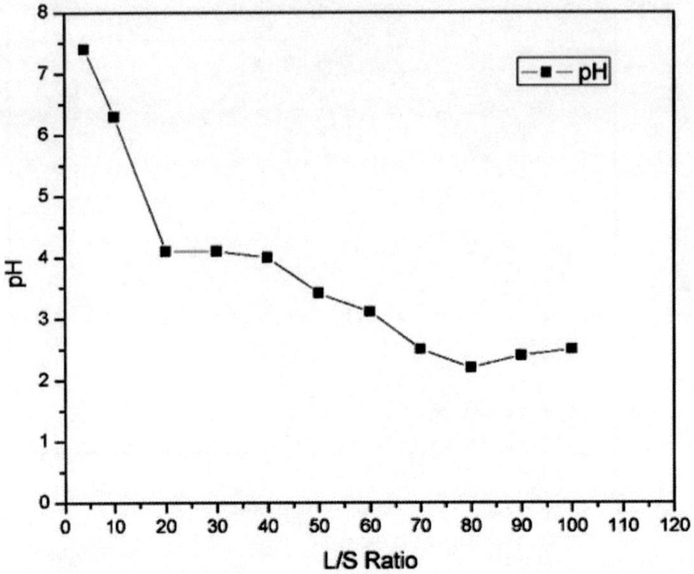

Fig. 2 pH versus L/S ratio

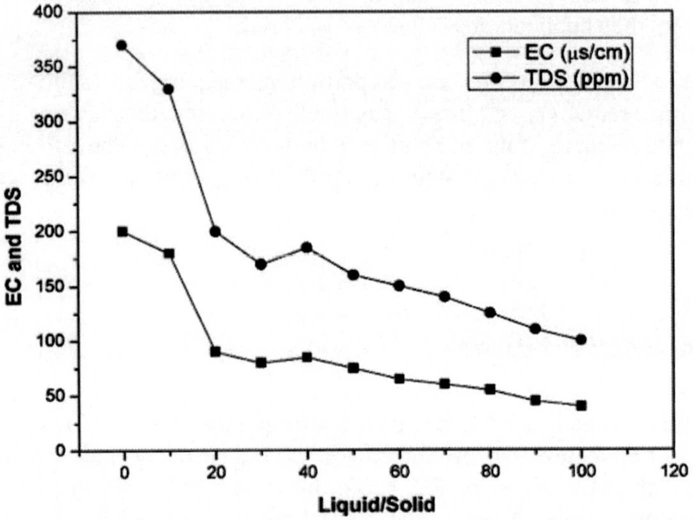

Fig. 3 EC and TDS versus L/S ratio

Table 1 Leached concentrations of heavy metals from coal gangue

Heavy metals	Cd	Cr	As	Se	Ba	Pb
Leachate (mg/L)	0.8	47.5	17.8	3.7	120	12.5
US EPA limit (mg/L)	1.0	5.0	5.0	1.0	100	5.0

pH, the highest fall in EC and TDS was noted at L/S ratio of 20. The sudden fall can be related to the abrupt decrease in the pH of the leachate. Due to lack of EC and TDS data for the coal gangue from other regions in India, validation of the obtained results could not be done.

3.3 *Leaching Behavior of Heavy Metals*

The leached concentrations of selected heavy metals at liquid-to-solid ratio of 20 are presented in Table 1. The concentration of leachate is compared with US EPA regulatory to evaluate the level of toxicity of the selected heavy metals. From the results presented in Table 1, it is evident that except Cadmium (Cd), all other selected heavy metals were found to be more than regulatory limits prescribed by US EPA.

Further it can be observed that heavy metals have the potential to mobilize from coal gangue, subsequently contaminating surrounding surface and ground-water bodies. The observations made are in consistency with previous studies by Cao et al. (2016) and Wang et al. (2016).

4 Conclusions

In the present study, the physical and chemical characteristics of coal gangue were examined. Also, the leaching behavior of heavy metals from coal gangue was studied under column leaching experiments. The following conclusions were drawn from the study.

- The moisture content and specific gravity of coal gangue were found to be lesser than general value of soils and other combustion residues.
- Gradation distribution curve of coal gangue is analogous to that of silty sand.
- Both electrical conductivity values and total dissolved solids content decreased consistently with increase in L/S ratio.
- pH of coal gangue leachate extracted relying on double distilled water reduces with increase in L/S ratio.
- Barring cadmium (Cd), all other elements showed higher mobilization levels from coal gangue when deionised distilled water was used as extractant.

Acknowledgements The authors are indebted to Singareni Collieries Company limited authorities for help rendered by permitting the procurement of coal gangue from Kakatiya coal mines, Bhupalpally.

References

Ashfaq M, Heeralal M, Moghal AAB, Murthy VR (2019) Carbon footprint analysis of coal gangue in geotechnical engineering applications. Indian Geotech J. https://doi.org/10.1007/s40098-019-00389-z

Asokan P, Saxena M, Asolekar SR (2005) Coal combustion residue environmental implications and recycling potentials. Resour Conserv Recycl 43(3):239–262

Cao Z, Cao Y, Dong H, Zhang J, Sun C (2016) Effect of calcination condition on the microstructure and pozzolanic activity of calcined coal gangue. Int J Miner Process 146:23–28

Jabłonska B, Kityk AV, Busch M, Huber P (2017) The structural and surface properties of natural and modified coal gangue. J Environ Manage 190:80–90

Keefer RF, Sajwan K (1993) Trace element in coal and coal combustion residues. Advances in trace substances. Research. Lewis Publishers, CRC Press, Florida

Malhotra VM, Mehta PK (2002) High-performance high-volume fly ash concrete. Marquardt Printing Ltd., Ottawa

Moghal AAB (2017) A state-of-the-art review on the role of fly ashes in geotechnical and geoenvironmental applications. J Mater Civ Eng 29(8):11

Moghal AAB, Sivapullaiah PV (2011) Effect of pozzolanic reactivity on compressibility characteristics of stabilised low lime fly ashes. Geotech Geol Eng 29(5):665–673

Moghal AAB, Sivapullaiah PV (2012) Retention characteristics of Cu^{2+}, Pb^{2+} and Zn^{2+} from aqueous solutions by two types of low lime fly ashes. Toxicol Environ Chem 94(10):1941–1953

Moghal AAB, Sivapullaiah PV (2013) Role of lime leachability on the geotechnical behaviour of fly ashes. Int J Geotech Eng 6(1):43–51

Plewa F, Mysłek Z (2001) Industrial waste management in underground mining technologies

Wang JM, Qin Q, Hu SJ, Wu KN (2016) A concrete material with waste coal gangue and fly ash used for farmland drainage in high groundwater level areas. J Cleaner Prod 112(1):631–638

Wu H, Wen Q, Hu L, Gong M, Tang Z (2017) Feasibility study on the application of coal gangue as landfill liner material. Waste Manage 63:161–171

Yao ZT, Ji XS, Sarker PK, Tang JH, Ge LQ, Xia MS, Xi YQ (2015) A comprehensive review on the applications of coal fly ash. Earth Sci Rev 141:105–121

Critical Review for Utilization of Blast Furnace Slag in Geotechnical Application

Bhavin G. Buddhdev and Ketan L. Timani

Abstract In the present state, the utilization of various waste materials in a different geotechnical application is increase manifold. All the waste materials possess different characteristics depending upon its raw materials and processing techniques. Utilization of these waste materials needs greater concern of its properties and compatibility with parent material. Out of many waste materials available, blast furnace slag is one of them. Blast furnace slags are by-products of metallurgical processes. It is glassy material, vesicular textures, typically with sand-to-gravel-size particles can be converted into powder form as well as particular desire size particles. In this paper, critical review has been made for the utilization of this blast furnace slag in various geotechnical applications depending upon the situation in weak and problematic soil condition. The review of this nature will lead the path in which various experimental program can be perform to find out the solution of different on-site geotechnical problems. The blast furnace slag has versatile characteristics by which it can compatible with all types of soil and provide appropriate solution for any soil stabilization process. Based on the review findings, some suggestion can be made for various geotechnical problems by utilizing this underutilized material, especially in Indian context. This paper ultimately become source for one of the interesting areas of research in which utilization of blast furnace slag in geotechnical application will have technical advantage as well as beneficial in protection and perseverance of natural resources.

Keywords Blast furnace slag · Geotechnical application · Waste materials · Problematic soil

B. G. Buddhdev (✉) · K. L. Timani
Applied Mechanics Department, Vishwakarma Government Engineering College, Chandkheda, Ahmedabad, Gujarat, India
e-mail: bhavin_buddhdev@hotmail.com

M. Latha Gali and R. R. P. (eds.), *Problematic Soils and Geoenvironmental Concerns*, Lecture Notes in Civil Engineering 88,
https://doi.org/10.1007/978-981-15-6237-2_9

1 Introduction

In the current era, research into new and innovative uses of waste materials is continually advancing. Scope of utilization of waste materials can be decided based on factors like availability, technical suitability, environment impact and economic benefits. At present, along with series of waste materials production, most of the countries including India are producing millions of tons of blast furnace slag (BFS) which is the by-product of steel industries. Traditionally unutilized BFS is stockpiled in the steel plants, and eventually land filled at slag disposal sites. The current methods of stockpiling and land filling are not sustainable, so disposal of BFS has become a significant concern both to slag processor companies and to environmental agencies in the last decades. In the literature, very limited information on the engineering properties of BFS was available. Research that focuses on engineering properties of blast furnace slag was scarce. Due to this, in comparison with other recyclable materials, such as fly ash, bottom-ash, silica fume and tire shreds, BFS was underutilized. BFS is the by-product of metallurgical operations, typically containing gangue from the metal ore, flux material and unburned fuel constituents. As a product of calcinated flux stone, alumina and silica phases present in iron ore, the four major oxide phases present in BFS are CaO, SiO_2, Al_2O_3, and MgO. These oxides account for approximately 95 percent of the BFS composition (Hammerling 1999). The physical properties of BFS are largely controlled by various methods by which how it is cools and solidifies. The color of BFS coarse aggregate usually varies from light to dark gray, depending on chemical composition. The void structure of the ACBFS heavily influences the physical properties, including the bulk specific gravity and the absorption (Lewis 1982). Based on the chemical and physical properties, BFS can be converted into any desire size fraction to make it suitable for various geotechnical applications. Utilization of BFS in geotechnical application till now is mainly for soil improvement of expansive soil like black cotton soil as well as in road construction as aggregate in flexible and rigid pavements. Some of the cases, BFS is utilized as improvement in sub-base materials for road construction has been noted. Still utilization of BFS based on site-specific condition as well as application in foundation of various structures is missing in literature. Based on the availability of BFS in national context, the proper guidelines at the production unit as well as collection and convergence in required form is not established as compared with global context. Current Indian guideline for utilization of BFS in various civil engineering applications is only limited to use BFS as aggregates and cement production. Guidelines and regulation pertaining to storage, stock keeping, treatment, processes and distribution as per the requirement are not available for utilization of BFS in India as compared with global scenario as most of country producing BFS have well-defined guideline and regulation. Looking from literature, BFS has a potential physical and chemical characteristic which enables this material to utilize in various problematic geotechnical applications. Based on the needs of the projects, experimental program can be carried out to meet with appropriate solution by using BFS. Experimental results based on the requirement are helpful in framing some guidelines and policies to encourage utilization of this material, especially in Indian context. The waste material utilization in the improvement of soil properties is of great interest since this would lead to cost effectiveness as well as eco-friendliness of the construction. This study is an attempt to understand the effectiveness of utilization of BFS in certain area which is undiscovered till date.

2 Literature Review

2.1 General

Many researchers had put their sincere and elaborate efforts to use the BFS produced from steel processing units in different civil engineering applications. Research papers and related studies were reviewed. Some journal papers were also reviewed. Many of them were about BFS was utilized for geotechnical applications like earth-fill, embankment, sub-base and base course for pavement, etc. Provisions in the international codes for related topic were also reviewed as only few provisions were available in the Indian context. The following paragraph illustrates the initial development in the utilization of BFS as well as the latest scenario for different geotechnical applications.

2.2 Historical Background and International Scenario

According to Hendrik (2002) slag has been used for construction purposes, especially road metal since Roman times. The dramatic expansion in Europe of iron and steel production associated with the industrial Revolution led to increase in slag production. By the early nineteenth century, slag output was rapidly outpacing consumed and there was growth of unattractive slag heaps on valuable industrial land nearby the industrial zones. By the mid-nineteenth century, research had demonstrated a number of new uses for slag, particularly as an aggregate for concrete and as a cementations material in its own right. Consumption remained relatively modest until the twentieth century, when a major new use in asphalt blend for road paving was developed and became popular in step with demand for smooth roads by the growing automobile owning public. This use, together with the rapid growth of concrete usage worldwide, led to the consumption of most existing slag heaps and a current consumption roughly apace with new slag production. Research is ongoing to expand and refine the uses of slag. Slag is properly recognized as a valuable coproduce of iron and steelmaking and not as a waste product.

BFS is obtained during the manufacture of iron and steel and possesses inherent hydrated properties. It was utilized for making different types of construction materials (www.Academicjournals.org/ijps/pdf/.../Safiuddin%20et%20al.pdf). According to Emery (1980), loose dry unit weight values for palletized BFS range from 8.2 to 10.4 kN/m^3. BFS is glassy material, typically with sand-to-gravel-size particles. Most of the studies in the literature focus mainly on the chemical composition and mineralogy of BFS to assess its cementitious properties rather than its mechanical properties. As per Noureldin and McDaniel (1990) reported on some of the engineering properties of blast furnace slag.

Research that focuses on engineering properties of BFS was very limited. The case histories available in the literature detailing some of the unsuccessful attempts of using this slag seem to have decreased the confidence level in promoting its utilization for various applications. Therefore, in comparison with other recyclable materials, this slag was underutilized.

Ghionna et al. (1996) studied the possibility of using slag as structural fill material in landfill embankments. A trial embankment was constructed with slag. The study indicated that diluting slag with inert materials, such as gravel and sand, can reduce the swelling potential. Plate load tests performed on the embankment showed satisfactory elastic modulus values (as high as 55 MPa).

Wild et al. (1998) improved the unconfined compressive strength of sulfate containing soils by stabilizing them with lime and GGBS.

According to Lewis (2003), air-cooled blast furnace slag is an ideal material for use in base and sub-base courses. Inherent properties of the slag provide advantages in engineering performance and in economy. It provides outstanding durability and weight savings of 10–20% over natural aggregate materials in the same applications. Slag has been used in base courses since the time of the Roman Empire, with many examples of such use found in England. Base course construction with slag in the USA dates back to the 1860s. Gradations of slag base and sub-base aggregates used have covered all types of applications.

Poh et al. (2006) showed that there is potential in utilizing BOF slag fines in stabilization of fine-grained soils when activators are used.

Skid resistance was a measure of the minimum force at which a tire prevented from rotation and slides on the pavement surface. Development of sufficient skid resistance is an important requirement of road safety. In this regard, BFS is a favorable aggregate for binding asphalt applications as BFS aggregates are angular and have very rough surface texture. Therefore, pavement surfaces incorporating BFS had shown superior skid characteristics than asphalt surfaces incorporating natural aggregates (Asi 2007).

BFS were used as sub-base materials for pavement construction, and promising results were found by Houben et al. (2010). In most of the research works, the combination of different waste materials like BFS, fly ash, steel slag, etc., was taken to know different properties (Kim and Lee 2012; Guo and Shi 2012).

According to Irem and Prezzi (2009) studies evaluating the geotechnical properties of BFS are very limited. This may be due to the undesirable properties of BFS, that is, its volume instability and high specific gravity. However, BFS possess other very favorable properties like self-cementation, high friction angle, etc. Based on the final report on use of slag in subgrade application, procedure for conducting experiment on BFS mostly following ASTM standard and procedure was used.

The maximum and minimum dry density tests were performed on BFS samples according to ASTM D4253 and ASTM D4254, respectively. Samples of fresh and aged BFS were characterized through a series of laboratory tests (specific gravity, grain-size analysis, X-ray diffraction, compaction, maximum and minimum density, large-scale direct shear, consolidated drained triaxial and long-term swelling tests). The effects of gradation on the engineering properties of both fresh and aged BFS samples were investigated. The corrosion potential of BFS samples was evaluated based on the electrical resistivity and pH test results. The report highlighted that due to the chemical vulnerability of BFS, testing on BFS can be done for fresh samples of BFS as well as aged samples after one year duration.

In UK, "waste protocols project (2007) by Environment Agency" on BFS was conducted. The information would have enabled the Environment Agency to re-consider its view of the material status of BFS; the Environment Agency was (and is) of the opinion that BFS is a waste, whereas, in the view of the industry, BFS is a by-product, which provided certain conditions are satisfied, is not a waste.

BFS manufactured in the UK consistently meets the quality specifications contained within the suite of European product standards relating to aggregates and GGBFS like BS EN 13242, BS EN 13285, BS EN 13043, BS EN 14227-2, BS EN 12620 and BS EN 15167. This wide range of standards gives confidence in the quality of BFS material, making it acceptable to the market, and thereby assuring certainty of use.

The European standards organization, CEN, has finalized the following standards related to the BFS for its various applications:

1. EN 13242:2002/AC: 2004—Aggregates for unbound and hydraulically bound mixtures for use in civil engineering work and road construction.
2. EN 13285:2003—Unbound mixtures and specification.
3. EN 13043:2002/AC: 2004—Aggregates for bituminous mixtures and surface treatments for roads, airfields and other trafficked areas.
4. EN 12620:2002/AC: 2004—Aggregates for concrete.

These standards refer to the use of BFS as an aggregate in unbound, hydraulically bound, asphalt and concrete mixtures, respectively. BFS is used widely throughout the UK as a construction aggregate and complies fully with the physical and chemical quality requirements detailed in EN 13242. This standard supersedes a variety of aggregate standards including British Standard BS 1047 and recognizes BFS as a manufactured aggregate suitable for use alongside aggregates from natural and recycled sources.

3 National Scenario

This section covers the information related to the production of BFS in India as well as consumption details in respective area. It includes the research undertake by researcher of Indian origin and current research and development take place in recent past.

In last 5 years, an enhancement of over 18% for blast furnace slag and basic oxygen furnace slag has been achieved (GOI 2013–2014). By continuing thrust on solid waste consumption at the integrated iron and steel plants, total solid waste utilization has increased from 77% in the 2007–2008 to 86%; an increase of 9% over the last 5 years (GOI 2013–2014). "As per IMY, August 2017, Indian Minerals Yearbook 2016," the total production of BFS in India during the year 2015–2016 is in range of approximately 2500–2700 metric tons.

Physical properties and chemical composition of blast furnace slag for typical Indian slag samples are narrated in Tables 1 and 2, respectively.

Table 1 Physical properties of Indian blast furnace slag

S. No.	Properties	Test method	Values
1	Aggregate impact value	IS:2386 (IV)	17–25%
2	Los Angeles abrasion value	IS:2386 (IV)	28–32%
3	Flakiness index	IS:2386 (I)	12%
4	Elongation index	IS:2386 (I)	9%
5	Water absorption	IS:2386 (III)	1.5–3%
6	Specific gravity (kg/cm^3)	IS:2386 (III)	2650
7	Bulk density (kg/cm^3)	IS:2386 (III)	1800

Source https://pmgsy.nic.in/wastematerials.pdf

Table 2 Chemical composition of typical Indian slag samples

S. No.	Constituents	Values
1	Loss on ignition	0.65
2	SiO_2	33.41
3	Fe_2O_3	0.89
4	Al_2O_3	20.05
5	CaO	34.24
6	MgO	8.86
7	Na_2O	0.16
8	K_2O	0.82
9	SO_3	NIL
10	TiO_2	NIL

https://www.cpcb.nic.in

In India, a huge amount of natural resources like soil and aggregates are being utilized for development of road projects like National Highway Development Program (NHDP), Pradhan Mantri Gram Sadak Yojana (PMGSY) and Chief Minister Gram Sadak Yojana (CMGSY) programs. This leads to resource exploitation of naturally available materials. BFS, as a substitute of store aggregate/chips, has been acceded by the Indian Road Congress (IRC) and Bureau of Indian Standards (BIS) for road construction (https://www.ieindia.org).

It was also concluded in an investigation that slag may be utilized in the building of sub-grade and embankments. Granulated blast furnace slag (GBFS) should be able to use as a partial replacement of unmodified aggregate up to 20–30% in the construction of granular sub-base layers, also maximum un-soaked California bearing ratio (CBR) value was increased by 40.78% when 20% replacement with GBFS, whereas the 4-day soaked CBR value was increased by 46.60% (Subrahmanyam et al. 2014).

In the roadway, BFS was used as a means of ground stabilization offers a degree of stabilization equivalent to that of the traditional method of using rock aggregate. BFS which provides the following advantages when it used as a coarse aggregate for sub-base (Sen and Mishra 2010):

1. Slag that has been water quenched tends to have a lowered wear resistance and soundness;
2. For most sub-base applications in which above two properties are critical, air cooled as opposed to water quenched, slag is used. In order to meet most state coarse aggregate specifications, most often air-cooled slag is crushed to a ¾ in. particle size or less, once properly sized, these by-products can serve as suitable substitutes for native coarse aggregate in this application; and
3. The sections of roadway in which blast furnace slag was used as a means of providing soft ground stabilization provided a degree of stabilization equivalent to that of the traditional method of using rock aggregate.

Maximum dry density of soil increases while plasticity characteristics gradually decreases with increase in slag content, and thus the CBR value of soil increases and, therefore, increases soil strength (Biradar et al. 2014; Rao et al. 2014; Singh and Ali 2014). Slag content in natural soil increases its workability by reducing its liquid limit and thus its plasticity (Rao et al. 2014). It was recommended (Chaubey and Ali Jawaid 2016) for natural soil with 25% slag as an optimum stabilization ratio for soil and can be used for sub-grade as well as in construction of low permeability liners with the addition of any additives such as bentonite.

According to Rajan (2014) the iron and steel slags are non-metallic and does not have hazardous substances. Slag is an alternative construction material with superior environmentally friendly qualities and better product features (Kumar and Kumar 2015). Porosity and permeability of soil can be reduced by iron and steel slag (Karthik and Doraikkannan 2015; Khan and Shinde 2013).

Yadu and Tripathi (2013) studied the effect of granulated blast furnace slag in the engineering behavior of stabilized soil. The results show that inclusion of GGBS increases the strength of soft soils. Similarly, the significant improvement has been observed for un-soaked and soaked CBR value of soils.

According to Chenna (2014–2015), the challenges faced by the Indian steel industry to manage the wastes are multiple which are as follows:

1. No Eco-system for Utilization: Every waste management system should have a generator, facilitator and consumer for utilization of the waste products. Such an eco-system does not exist in India.
2. Lack of Incentives: Incentives for setting up waste-based manufacturing units and their utilization are absent.
3. No Mandatory Norms: Absence of laws mandating use of wastes in appropriate sector, limits their utilization and therefore production of waste-based products.
4. Technical Limitations: A number of metallurgical sludge/dust have high zinc, chlorides, etc., which render the impractical for recycling.
5. Lack of Standards: Indian standards do not exist for a number of steel plant waste-based products. For example, use of steel slag for road sub-grades, coarse and fine aggregates in concrete, blended cements, etc., does not exist.
6. Form of Wastes: Complexities in respect to the form of the wastes exist. Whereas some of the wastes are purely metallurgical in nature and some are siliceous in nature. Therefore, this necessitates application of diverse technology options for treatment of the wastes.
7. Limited Technology Deployment: The restraints mentioned above have led to low penetration of waste technologies in India. An assessment of waste technologies deployment in India can be made by the matrix. Figure 1 shows the level of application against complexity. From the matrix and the information provided by 14 Indian steel plants, it is seen that most of the steel plants are practicing low application to low complexity solutions or at the most low application to medium complexity solutions. Thus, the deployment of waste management technologies is minimal in Indian steel plants.

These challenges have resulted in mere utilization of only 30–35% of the wastes generated. Therefore, the task on hand is shifting gear to manufacture more medium and high application waste products and utilizing them by harnessing the opportunity provided by the potential growth in infrastructure development in India by 2020 to 2025.

"Commonwealth Scientific and Industrial Research Organization (CSIRO)" carried out investigations and suggested commercially interesting applications of slag. The applications include (i) base course and top course to asphalt roads, (ii) anti-skid surfacing for roads on accident-prone intersections, (iii) low-strength concrete for footpaths, (iv) controlled low-strength fill for backfill required for trench stabilization and (v) concrete sub-base for rigid pavements. Environmental Scientists and Toxicologists completed an industry wise "Human Health and Ecological Risk Assessment" (HHERA) which demonstrated that iron and steel slag poses no meaningful threat to human health or environment when used in a variety of residential, agriculture, industrial and construction application. The chances of uptake of the metals in the slag matrix by human, other animals or plants are remote. The study carried out by an independent nationally renowned chemical laboratory has demonstrated that blast furnace do not pose any threat to human or plant life, further revealed

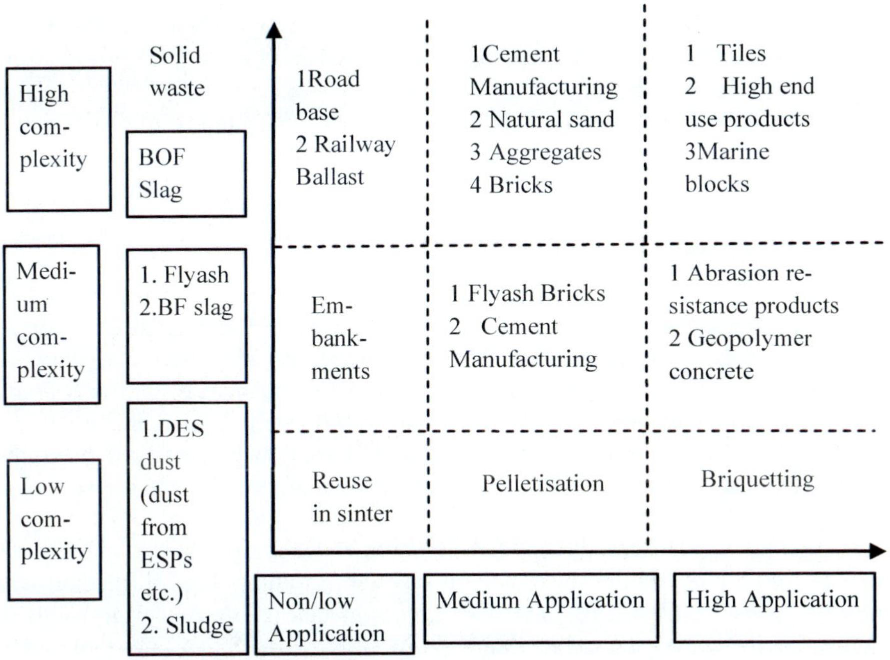

Fig. 1 Matrix of solid wastes application against complexity

that the use of slag has very positive environmental benefits in terms of protection (IMY 2017).

Salient Point from literature review:

Following points were important from literature review regarding the application of BFS in geotechnical engineering

1. Mostly GGBFS have been used for red soil, black cotton soil and expansive soil for soil improvement.
2. GGBFS have been used with lime to reduce the expansive tendencies of lime stabilized sulfate bearing clay soils
3. Addition of GGBFS increases CBR, unconfined compressive strength, dry density and decreases liquid limit and plastic limit of soil.
4. Mostly GGBFS is used with pozzolanic materials fly ash, lime, etc., with cohesive types of soil.
5. Liquid limit of GGBFS can be found out experimentally but plastic limit and plasticity index value cannot evaluate as it is non-plastic materials. Wide variation is observed in the value of liquid limit depending upon the source of GGBFS.
6. Particle size distribution of soil and GGBFS needs to be studied for compatibility of soil-GGBFS mix.

7. OMC and MDD value is decreases with increase in GGBFS percentage in mix. The optimum value of increase in MDD with GGBFS is up to 20–30% in some cases but OMC value is always decreases.
8. Unconfined compressive strength increases up to 40% addition of GGBFS from 10% and after increases in GGBFS decreases the UCS.

4 Recent Update

In order to achieve full utilization of BFS, cast house slag granulation plants (CHSGPs) will be installed where the facility is unavailable. New blast furnaces have been commissioned with CHSGPs to enhance granulation of BFS at RSP and ISP. SAIL is also making an initiative for creation of databank on melting behavior of various steelmaking slags of SAIL Plants at RDCIS, Ranchi. Further, to maximize utilization of the granulated BFS, slag-based JV cement plants have been set up at Bhilai and Bokaro (IMY 2017).

Slag atomizing technology (SAT) of atomizing molten slag is an innovative development of slag utilization. It is a multi-function, multi-application method with a great managerial approach to resolve environmental and technical problems by converting the slag into precious slag ball (PS balls). The PS balls are stable spinel structure that induces superior physical as well as chemical characteristics and free of any pollution issues (https://www.meconlimited.co.in).

"Bureau of Indian Standard (BIS)" published IS 383:2016 Coarse and Fine Aggregates for Concrete-Specification and IS 16714:2018 Ground Granulated Blast Furnace Slag For Use In Cement, Mortar and Concrete-Specification recently in the area of utilization BFS.

5 Conclusions

Based on critical review on utilization of BFS in geotechnical application, it can be understand that wide range of opportunity exists for this waste material. The use of BFS in various civil engineering application and construction was from many decades but with the advancement in production techniques of BFS, characteristics and production capacity changes lot over a period of time. Looking to the current shortage of availability of quality natural materials, BFS has a potential for its utilization in geotechnical application. There is almost need to strengthen the guidelines and policies with respect to Indian context as very fewer information is available. Much research work related to utilization of BFS with cohesive soil had been done mostly in expansive soil mixed with BFS in powder form (GGBFS). Research in area of cohesionless soils is very less by using different size fraction of BFS. Initiative have been taken in form of establishment of BFS processing plants at few places as well as BIS actively work on publication of standard in this allied area. Based on this review, some area will be open for experimental work which ultimately results in fruitful research outcome. This experimental work also provides the feasibility information regarding utilization of BFS in certain area of geotechnical application. Latest guidelines and policies published by the concern authority are mainly related to utilization of BFS in cement and concrete production. Therefore, research in area of utilization of BFS in geotechnical application will be helpful in framing guidelines and policies in Indian context.

References

Asi IM (2007) Evaluating skid resistance of different asphalt concrete mixes. Build Environ 42(1):325–329

Biradar KB, Kumar AU, Satyanarayana PVV (2014) Influence of steel slag and fly ash on strength properties of clayey soil: a comparative study. Int J Eng Trends Technol (IJETT) 14(2)

Chaubey S, Ali Jawaid SM (2016) Soil stabilization using steel slag. Global J Res Anal 5(1)

Chenna K (2014–15) Potential infrastructural growth in India by 2025 vis-a-vis utilizations of waste—a sustainable material. Metall Mater Eng Div Board Annu Tech 1:2014–2015

Emery JJ (1980) Palletized lightweight slag aggregate. In: Proceedings of concrete international, Concrete Society

Ghionna V, Pedroni S, Tenani P, Veggi S (1996) Geotechnical investigation on steel slags mixtures for landfills embankments construction. In: Proceedings of the second international conference on environmental geotechnics, Osaka, Japan, 5–8 Nov, Balkema, Rotterdam, pp 709–714

GOI (2013–2014) Government of India. Outcome Budget of Ministry of Steel

Guo X, Shi H (2012) Utilization of steel slag powder as a combined admixture with ground granulated blast furnace slag (GGBFS) in cement-based materials. J Mater Civ Eng. https://doi.org/10.1061/(ASCE)MT.1943.5533.0000760

Hammerling DM (1999) Calcium sulfide as blast furnace slag used as concrete aggregate. M.S. thesis, Michigan Technological University, Houghton, MI

Hendrik G (2002) Slag-iron and steel. U.S. Geological Survey Minerals Yearbook

Houben L, Akbarnejad S, Molenaar A (2010) Performance of pavements with blast furnace base courses. In: Paving materials and pavement analysis, pp 476–483

https://steel.gov.in/Outcome%20Budget%20(2013-14)/outcome%20budget.pdf

https://pmgsy.nic.in/wastematerials.pdf

https://www.cpcb.nic.in

https://www.ieindia.org/PDF_IMAGES/CouncilData/MMDB_VOL1.pdf?Event_Id=1210. Accessed on 30 June 2016

https://www.meconlimited.co.in/writereaddata/MIST_2016/sesn/tech_4/3.pdf. Accessed on 24 June 2016

https://www.academicjournals.org/ijps/pdf/.../Safiuddin%20et%20al.pdf

https://www.ijmer.com, vol 5(1), Jan 2015

IMY (2017, Aug) Slag-iron and steel. Indian Minerals Yearbook 2016 (Part—II: Metals & alloys), 55th edn

Irem ZY, Prezzi M (2009) Final report on use of steel slag in subgrade applications, FHWA/IN/JTRP-2009/32, Oct 2009

IS 383:2016 (2016) Coarse and fine aggregates for concrete—specification

IS 16714:2018 (2018) Ground granulated blast furnace slag for use in cement, mortar & concrete—specification

Karthik D, Doraikkannan J (2015) Experimental investigation of silica fume and steel slag in concrete. IJMER. ISSN: 2249-6645

Khan R, Shinde SB (2013) Effect of unprocessed steel slag on the strength of concrete when used as fine aggregate. Int J Civ Eng Technol 4(2):231–239

Kim H, Lee H (2012) Effects of high volume of fly ash, blast furnace slag, and bottom ash on flow characteristics, density, and compressive strengths of high-strength mortar. J Mater Civ Eng. https://doi.org/10.1061/(ASCE)MT.1943-5533.0000624

Kumar PR, Kumar P (2015) Use of blast furnace slag as an alternative of natural sand in mortar and concrete. Int J Innov Res Sci Eng Technol 4(2)

Lewis D (2003) Construction of air-cooled blast furnace slag base courses. National Slag Association, MF 183-4

Lewis DW (1982) Properties and uses of iron and steel slags. National Slag Association

Noureldin AS, McDaniel RS (1990) Evaluation of surface mixtures of steel slag and asphalt. Transportation Research Record 1269, Transportation Research Board, National Research Council, Washington, D.C., pp 133–149

Poh HY, Ghataora SG, Ghazireh N (2006) Soil stabilization using basic oxygen steel slag fines. J Mater Civ Eng ASCE 18(2):229–240

Rajan MS (2014) Study on strength properties of concrete by partially replacement of sand by steel slag. Int J Eng Technol Sci IJETS™ 1(6). ISSN (P): 2349-3968, ISSN (O): 2349-3976

Rao K et al (2014) A laboratory study on the effect of steel slag for improving the properties of marine clay for foundation beds. Int J Sci Eng Res 5(7)

Sen T, Mishra U (2010) Usage of industrial waste products in village road construction. Int J Environ Sci Dev 1(2). ISSN: 2010-0264

Singh P, Ali SMJ (2014) Strength properties of blast furnace slag mixed with alluvial soil. GJESR Res Pap 1(9)

Subrahmanyam KV, Arun Kumar U, Satyanarayana PVV (2014) A comparative study on utilization of waste materials in GSB layer. SSRG Int J Civ Eng (SSRG-IJCE) 1(3)

Waste Protocols Project (2007) A technical report on the manufacturing of blast furnace slag and material status in the UK by Environment Agency August

Wild S, Kinuthia JM, Jones GI, Higgins DD (1998) Effects of partial substitution of lime with ground granulated blast furnace slag on the strength properties of lime stabilized sulphate bearing clay soils. Eng Geol 51(1):37–53

Yadu L, Tripathi RK (2013) Effects of granulated blast furnace slag in the engineering behavior of stabilized soil. Procedia Eng 51:125–131

Effect of Inorganic Salt Solutions on the Hydraulic Conductivity and Diffusion Characteristics of Compacted Clay

Partha Das and T. V. Bharat

Abstract Clayey soils rich in montmorillonite content are increasingly used as liner material for various landfill facilities because of their very low saturated hydraulic conductivity, high sorption potential and better self-sealing capacity. However, the performance assessment of such a facility requires long-term hydraulic conductivity and diffusion studies of the clayey soil under the influence of different pore fluids. In this paper, an extensive study was carried out to understand the effect of various inorganic salt solutions such as lithium chloride, sodium chloride and potassium chloride on the montmorillonite rich bentonite clay by performing long-term hydraulic conductivity tests and transient through-diffusion tests. A software package named CONTRADIS was utilized in the present study which was able to perform inverse analysis for estimating the diffusion parameters from the experimental data. The experimental study suggests that the electrolyte concentration changes the hydraulic conductivity and diffusion coefficient significantly. The hydrated cationic radius also has considerable effect on the equilibrium hydraulic conductivity of the soil.

Keywords Through-diffusion · Long-term hydraulic conductivity · CONTRADIS

1 Introduction

For restricting the migration of the contaminants to the surrounding environment, bentonite soils are used as potential barrier material. Bentonites barriers are expected to have saturated hydraulic conductivity less than 10^{-9} m/s throughout the life of a landfill for effectively reducing the contaminant flow through them. However, the hydraulic conductivity of such barrier materials was found to increase significantly when permeated with inorganic salt solutions (Lee and Shackelford 2005). The hydraulic conductivity of fine grained soils like bentonites is greatly influenced by the minerals present in the soil, the composition of exchangeable cations and the cations present in the pore fluid (Rao and Mathew 1995). Hence, a study on long-term

P. Das (✉) · T. V. Bharat
IIT Guwahati, Guwahati, India
e-mail: partha.das@iitg.ac.in

M. Latha Gali and R. R. P. (eds.), *Problematic Soils and Geoenvironmental Concerns*, Lecture Notes in Civil Engineering 88,
https://doi.org/10.1007/978-981-15-6237-2_10

effect on the clay barriers under the influence of different inorganic salt solutions becomes very important (Shackelford et al. 2000; Kolstad et al. 2004).

Also, flow through fine grained soils is mostly governed by the diffusion mechanism, as the hydraulic conductivity is low due to which the advective transport is negligible (Bouazza and Van Impe 1997; Shackelford et al. 1989). In such a condition, the estimation of the model parameters like the effective diffusion coefficient and retardation factors becomes very essential for the design of the liner facilities. Various laboratory techniques are adopted to estimate the model parameter (Barone et al. 1992; Shackelford and Daniel 1991; Van Loon and Jakob 2005; Robin et al. 1987). Amongst all the technique, through-diffusion technique is mostly preferred as is it a non-destructive testing methodology (Bharat 2009). The laboratory estimated diffusion coefficient was found to vary considerably when salt solutions were allowed to diffuse through compacted soil over a longer period of time (Das and Bharat 2017). The influence of the size of the cations and the valence of the cations was also found to be significant. Most of the previous studies were concerned with the size and valence effect cations present in the pore fluid; however, an extensive study on the influencing factors on both hydraulic and mass transport parameter, when permeated with varied electrolyte solutions, remains to be completed.

In this paper, a detailed experimental study on the influence of different pore fluids having a wide range of concentrations on the compacted bentonite soil was presented. Long-term hydraulic conductivity tests were conducted using falling head technique by permeating the compacted bentonite soil with different salt solutions like lithium chloride, sodium chloride and potassium chloride. Also, through-diffusion tests were conducted to understand the variation in the mass transport parameters for the pore fluids in which the hydraulic conductivity variation was significant. A Java-based application package named CONTRADIS was utilized to obtain the model parameters from the experimental concentration profiles, by performing inverse analysis (Das et al. 2016).

2 Materials and Methods

In the present work, bentonite soil procured from Barmer district in Rajasthan was used. The index and surface properties of the studied bentonite are given in Table 1.

2.1 *Long-Term Hydraulic Conductivity Tests*

Air dried soil sample was compacted in a hydraulic conductivity cell fabricated from acrylic solid tube. The internal diameter of the cell was 2.4 cm, and the thickness was kept to be 1 cm. The soil was statically compacted at a bulk density of 1.4 g/cc. After compaction, the soil sample was rigidly held between two porous stones on either end by fixing the caps supported by threading on the outer side of the cell.

Table 1 Index and surface properties of the studied bentonite (Bharat and Das 2017)

Property	Value
Liquid limit	393
Plastic limit	50
Shrinkage limit	18
Specific gravity	2.77
Montmorillonite content (%)	55
Specific surface area (m^2/g)	495
Cation exchange capacity (meq/100 g)	71.7

Hydraulic conductivity tests were conducted by falling head technique in accordance with ASTM D5856. A head of 8.75 m was provided to saturate the soil sample, and saturation was ensured when the fall in head became steady. Saturation was also cross checked by monitoring the weight of the soil sample with time, a constant weight of the soil sample when achieved indicated complete saturation. Hydraulic conductivity tests were commenced after saturation was completed, by permeating the soil sample with different salt solutions having different concentrations. Care was taken to remove the distilled water after saturation process and then filling of the burettes with the pore fluid of required concentrations. The fall in head was regularly monitored until equilibrium hydraulic conductivity was reached.

The test was terminated when the hydraulic conductivity values become constant over a period of 3–4 days. Three series of hydraulic conductivity tests using different electrolytes were conducted to ascertain the dependency of equilibrium hydraulic conductivity on the size of ions and electrolyte concentration. The series consisted of 0.01, 0.1, 0.5 and 1 molar aqueous solutions of lithium chloride (LiCl), sodium chloride (NaCl) and potassium chloride (KCl). All the electrolytes were analytical/laboratory grade, procured either from Merck (Germany) or Spectrochem (India) and supplied by a local vendor. The tests were not conducted for concentration less than 0.01 M as any concentration below this does not influence the exchange process and the exchangeable cations on the surface dominate the clay behavior (Rao and Mathew 1995).

The hydraulic conductivity test result showed that the equilibrium Ks value increased more than the limiting value when the soil sample was permeated with KCl solutions even at lower concentrations. Hence, diffusion tests were conducted with only KCl solution as the pore fluid.

2.2 *Laboratory Through-Diffusion Test*

Through-diffusion test setups were also fabricated from acrylic tubes having internal diameter 2.4 and 1 cm thickness. The air dried soil sample was compacted statically at the same density as was maintained for the long-term hydraulic conductivity

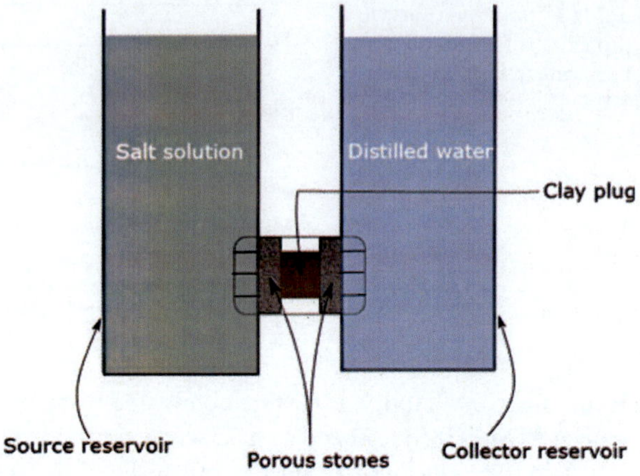

Fig. 1 Illustration of laboratory through-diffusion setup

tests. The diffusion cell was connected on either ends to two reservoirs (source and collector) as shown in Fig. 1. Both the reservoirs were then filled with distilled water to saturate the soil sample. After saturation was ensured by constant weight method as explained earlier, diffusion tests were commenced. The solution in the source reservoir was replaced with salt solution of known concentration, and the water in the collector reservoir was removed and filled with fresh distilled water. During the diffusion process, the solution in both the reservoirs was stirred to avoid deposition of any salts at the bottom. At frequent interval of time, samples were collected from both the reservoirs and analyzed for concentration using flame photometer. Diffusion tests were conducted using KCl solution as the pore fluid having same concentration range that was used in the hydraulic conductivity tests.

3 Theoretical Background for Diffusion

The governing differential equation that describes the transport of chemical species through saturated soil is given as

$$\frac{\partial c}{\partial t} = \frac{D_e}{R_d} \frac{\partial^2 c}{\partial x^2} \tag{1}$$

where c is the salt concentration, t is the time of diffusion, D_e is the effective diffusion coefficient, x is the distance along the length of the soil sample, and R_d is the retardation factor which describes the sorption potential of a soil. The retardation factor for linear sorption case is given as

$$R_d = 1 + \frac{\rho K_d}{n} \tag{2}$$

where K_d is the distribution coefficient, ρ is the dry bulk density and n is the porosity of the clay plug. Equation 3 gives the initial condition, and the boundary conditions of the diffusion experiment are given by Eqs. 4 and 5 (Rowe and Booker 1985):

$$c(0 < x < L, t = 0) = 0 \tag{3}$$

$$c(x = 0, t) = c_0 + \frac{nD_e}{H_s}\int_0^t \left(\frac{\partial c}{\partial x}\right)_{x=0} \mathrm{d}t \tag{4}$$

$$c(x = L, t) = -\frac{nD_e}{H_c}\int_0^t \left(\frac{\partial c}{\partial x}\right)_{x=L} \mathrm{d}t \tag{5}$$

Equation 4 describes the boundary condition at the source reservoir, and Eq. 5 describes the boundary condition at the collector reservoir. Here, c_0 is the initial concentration of the contaminant species at time $t = 0$; H_s and H_c are the equivalent heights of source and collector reservoirs, respectively.

The analytical solution to the governing differential equation (not shown here) considering the above initial and boundary conditions was given by Bharat (2013) using Laplace transformation. This analytical solution was utilized to estimate the model parameters which are the effective diffusion coefficient and retardation factor by inverse problem. The inverse problem was solved by minimizing the error between the experimental concentration profile and the theoretical data. The minimization was performed by using particle swarm optimization technique (Bharat et al. 2012). A Java-based application suite was developed to perform such inverse analysis problem. The relative concentration profiles obtained experimentally were given as input in the software package along with the density and equivalent heights of the reservoir. The software was then capable of estimating the optimized value of the model parameters from the input data.

4 Results and Discussions

4.1 Long-Term Hydraulic Conductivity Test Results

The variation of the equilibrium hydraulic conductivity with time for all the salt solutions at different concentrations was shown (Figs. 2, 3 and 4). It can be observed from the figures that the hydraulic conductivity increases with time for all the concentration, maximum being observed for a concentration of 1 M. This behavior is governed by the diffused double layer thickness (DDL). As the concentration of a particular pore fluid increases, the inter-particle attractive force increases between the clay platelets, and as a result, the DDL thickness is suppressed. With the supression of the

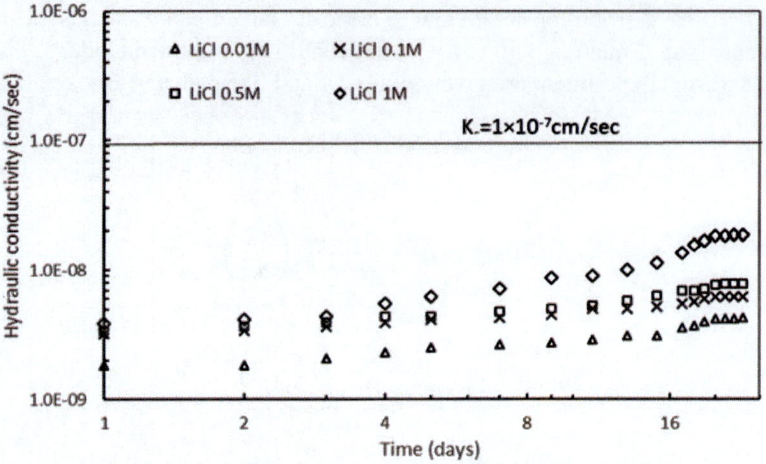

Fig. 2 Temporal variation of hydraulic conductivity of the bentonite with LiCl as the permeant fluid

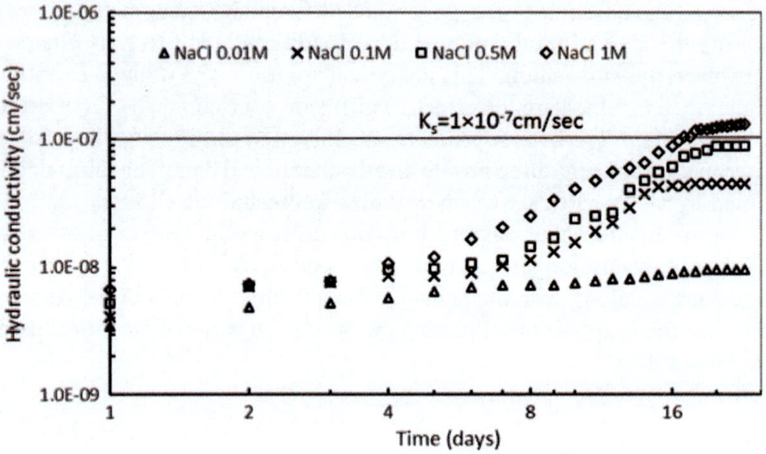

Fig. 3 Temporal variation of hydraulic conductivity of the bentonite with NaCl as the permeant fluid

DDL thickness, inter-particle forces increases which leads to the formation of flocculated structure which in turn results in significant increase of *Ks*. The equilibrium hydraulic conductivity of the soil exposed to LiCl of 0.01 M was 4.27×10^{-9} cm/s, whereas for the same soil, when exposed to LiCl solution of 1 M concentration, the equilibrium hydraulic conductivity increases to 1.88×10^{-8} cm/s, an increase of 4.4 times. Another important aspect that can be brought out from Figs. 2, 3 and 4 is that at 0.01 M concentration, the equilibrium hydraulic conductivity in the presence of Li^+ and Na^+ is of the same order; however, the conductivity changes by at least

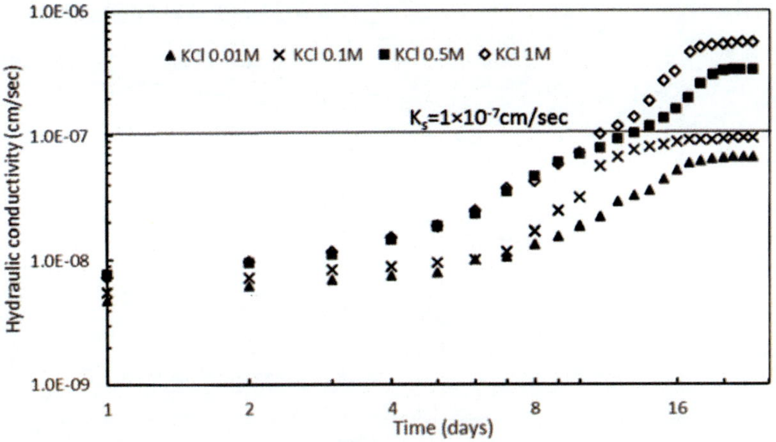

Fig. 4 Temporal variation of hydraulic conductivity of the bentonite with KCl as the permeant fluid

one order of magnitude at a concentration of 1 M. This can be attributed to the fact that at higher concentrations, more exchangeable cations are present which are able to replace the cations in the clay surface, and thus, it compresses the DDL layer reflecting an increase in the equilibrium hydraulic conductivity.

With Li^+ as the diffusing ion, even at higher concentration of 1 M, the limiting value of hydraulic conductivity for liners is not crossed. However, at higher concentration for both sodium and potassium ions, when present in the pore fluid, the limiting value is exceeded. The hydrated cationic radius of K^+ and Na^+ being less than Li^+ ion, the hydraulic conductivity increase is very significant as the DDL is suppressed more compared to the lithium ion providing more flow path for the fluid to migrate and attain higher equilibrium hydraulic conductivity. This particular behavior is not properly understood by the Gouy–Chapman theory, which considers the ions to be point charge and the influence of the size of the cations is neglected. It is, however, qualitatively understood that any decrease in the hydrated ionic radius of the same valence suppresses the double layer thickness (Rao and Mathew 1995).

The influence of the hydrated cationic radius on the equilibrium hydraulic conductivity is shown in Fig. 5. It can be concluded from this figure that with the decrease in the cationic radius of the exchangeable ions, the equilibrium hydraulic conductivity increases. The increase was significant at a concentration of 1 M where the equilibrium hydraulic conductivity in the presence of Li^+ ion was 1.88×10^{-8} cm/s, and in the presence of K^+ ion, it was observed to be 5.58×10^{-7} cm/s, an increase of almost 29 times.

The lowest increase, however, was observed for a concentration of 0.01 M. So for an increase in concentration leads to an increase in the hydraulic conductivity in the presence of all the three salt solutions, and also, any decrease in the cationic radius causes increase in the hydraulic conductivity. This is due to the suppression of the DDL thickness in both the cases which in turn increases the effective void size that

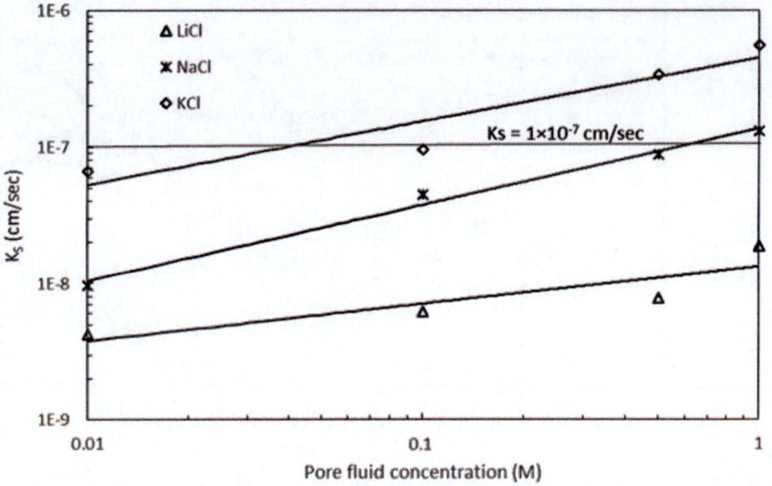

Fig. 5 Variation of equilibrium conductivity of the bentonite with salt concentration for different salts

is changed from micropores to macropores (Rao and Mathew 1995) for which there is a significant increase in the equilibrium hydraulic conductivity. The increase in cationic sizes, therefore, is favorable to the performance of the landfill liner.

4.2 Through-Diffusion Test Results

More detrimental consequences were observed in terms of equilibrium hydraulic conductivity, when KCl was present in the pore fluid. Under such a condition, it becomes very important to understand the migration rate of potassium ions corresponding to different concentrations. Hence, diffusion tests were conducted only with KCl as the diffusing chemical species having the same concentration range as was used in the hydraulic conductivity tests. The experimental concentration profiles for both source and collector reservoirs, with time for compacted soil subjected to KCl solutions of different concentration, are shown in Fig. 6. The effect of concentration can be qualitatively understood from this figure. It can be seen here that when the concentration of KCl solution was 0.01 M, the relative concentration of the source reservoir was around 0.65 after a period of two months. However, at 1 M KCl solution, the relative concentration reached 0.5, indicating a very fast migration rate at this concentration. This is in coherence with what was observed from the hydraulic conductivity test results.

In order to quantitatively understand the influence of different concentrations of KCl electrolyte solutions, experimental plots were utilized to obtain the model parameters which are the effective diffusion coefficient and the retardation factors.

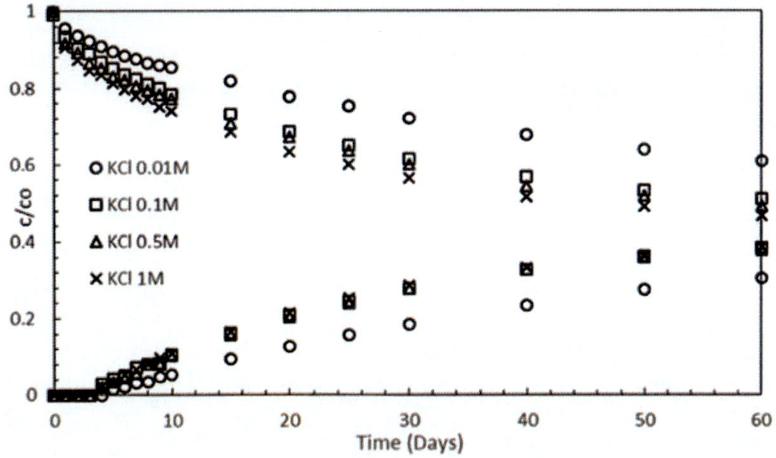

Fig. 6 Temporal data of measured salt concentration in source and collector reservoirs for KCl

The Java application CONTRADIS uses the analytical solution of the governing differential diffusion equation to obtain the inverse analysis (theoretical) plots for an optimized value of the model parameters. Figures 7, 8, 9 and 10 give a comparison of the experimental concentration profile with the theoretical results. It is observed that the theoretical data fits very well with the experimental results.

The effective diffusion coefficient, D_e, in the case of potassium ion (K^+) is found to have increased with the increase in concentration (Figs. 7, 8, 9 and 10). At a concentration of 1 M, the effective diffusion coefficient increases to one order of magnitude (Fig. 10) compared to the case when the concentration was 0.01 M (Fig. 7). The

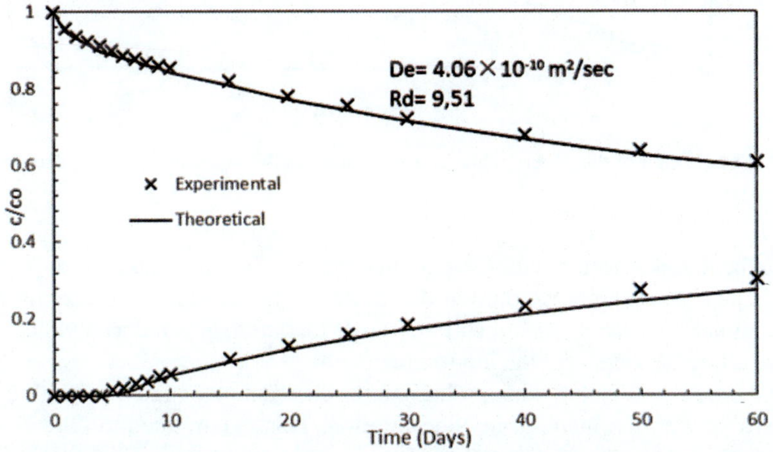

Fig. 7 Comparison of the experimental concentration profile with the theoretical data for 0.01 M KCl

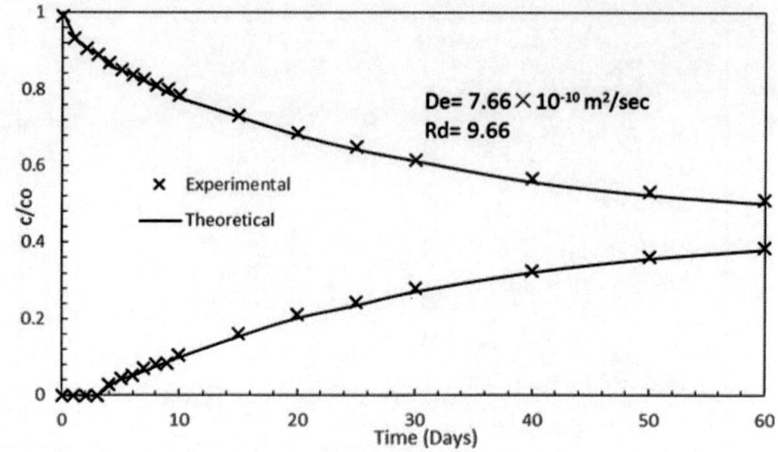

Fig. 8 Comparison of the experimental concentration profile with the theoretical data for 0.1 M KCl

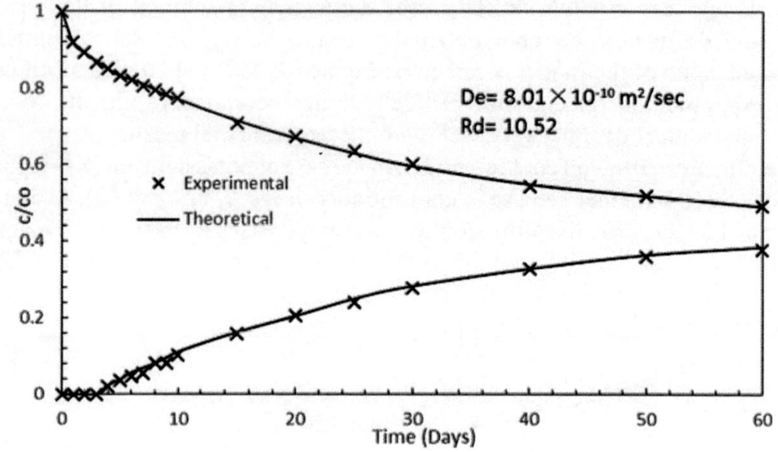

Fig. 9 Comparison of the experimental concentration profile with the theoretical data for 0.5 M KCl

reason behind this is that with increase in concentration more number of potassium ions is present in the bulk solution that replaces the exchangeable cations present in the mineral surfaces. As a result, the DDL thickness gets reduced which in turn allows a free flow path for the ion to move.

The retardation factor, which relates to the sorption potential of the soil, is found to increase with the increase in concentration. This is attributed to the fact that at higher concentrations, more number of potassium ions which are present in the bulk solutions are sorbed onto the mineral surfaces to effectively satisfy the charge density. Table 2 shows the comparison of hydraulic conductivity and diffusion coefficient of

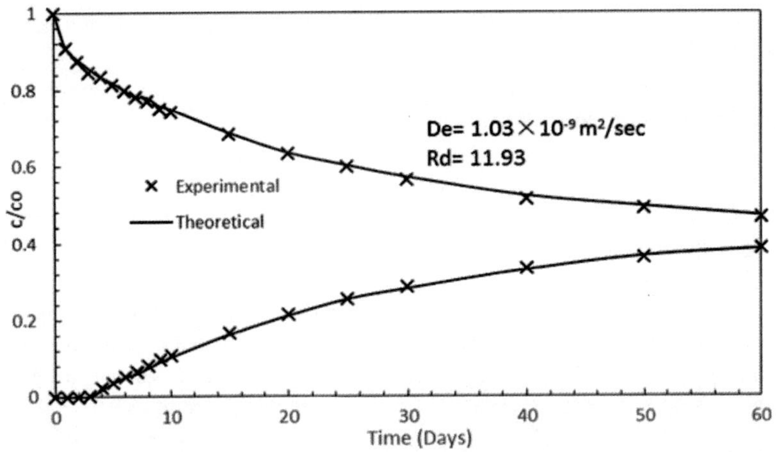

Fig. 10 Comparison of the experimental concentration profile with the theoretical data for 1 M KCl

Table 2 Comparison of hydraulic conductivity and diffusion coefficient with KCl as pore fluid

Salt used	Conc. (M)	Keq (m/s)	D_e (m^2/s)
KCl	0.01	6.69×10^{-10}	4.06×10^{-10}
	0.1	9.48×10^{-10}	7.66×10^{-10}
	0.5	3.38×10^{-9}	8.01×10^{-10}
	1	5.60×10^{-9}	1.03×10^{-9}

soil when it is permeated with KCl solution of varied concentrations. As discussed earlier when the concentration of the pore fluid increases, it leads to the suppression of the DDL thickness resulting in the formation of a pervious flocculated structure. As a result, there is a significant increase in the equilibrium hydraulic conductivity. The effective diffusion coefficient also increases considerably at this stage as the mobility of the ion through the soil pores increases.

5 Summary and Conclusions

Long-term hydraulic conductivity tests were conducted by permeating a saturated soil with different salt solutions having a wide range of concentrations. The cations present in the salt solutions are found to have significant impact on the equilibrium hydraulic conductivity values. The hydraulic conductivity increases with the increase in the pore fluid due to the suppression of the DDL thickness. The limiting value of the hydraulic conductivity for liner design was crossed when sodium ions are present in the pore fluids at concentrations greater than 0.5 M. When potassium ions are present in the pore, the limiting value is exceeded even at a lower concentration of 0.1 M.

The size of the diffusing ion also plays a significant role in changing the equilibrium hydraulic conductivity value. The hydrated cationic radius of lithium ion being large compared to sodium and potassium ion, the migration rate is found to be significantly low when lithium is present in the pore fluid. Bigger size cations like lithium is found to have hydraulic conductivity value much lesser than the limiting value for liner design and as such they do not pose any threat to the liners in the long run.

Laboratory through-diffusion tests were conducted only for KCl as the diffusing chemical species. Only KCl solution was selected for diffusion testing because even at lower concentrations potassium ions were able to exceed the limiting value of hydraulic conductivity for the liner design, and hence, it had more detrimental impact compared to the other pore fluids. It was observed that the migration rate of the potassium ions increased rapidly as the concentration was increased. This was reflected by the effective diffusion coefficient value that was obtained by inverse analysis through CONTRADIS. The effective diffusion coefficient, when potassium was present in the electrolyte solution, increased to one order of magnitude when the concentration increased to 1 M. Both equilibrium hydraulic conductivity and effective diffusion coefficient increased with concentration of KCl, due to the presence of small DDL thickness all around the clay platelets resulting in free flow paths for the ions to diffuse.

References

ASTM D5856-15 (2015) Standard test method for measurement of hydraulic conductivity of porous material using a rigid-wall, compaction-mold permeameter. ASTM International, West Conshohocken, PA, 2015

Barone FS, Rowe RK, Quigley RM (1992) A laboratory estimation of diffusion and adsorption coefficients for several volatile organics in a natural clayey soil. J Contam Hydrol 10:225–250

Bharat TV (2009) Metaheuristics for parameter estimation in contaminant transport modeling through soils. In: Proceedings of 4th international young geotechnical engineers conference, Alexandria (Egypt), 2009

Bharat TV (2013) Analytical model for 1-D contaminant diffusion through clay barriers. Environ Geotech 1(4):210–221

Bharat TV, Das DS (2017) Physicochemical approach for analyzing equilibrium volume of clay sediments in salt solutions. Appl Clay Sci 136:164–175

Bharat TV, Sivapullaiah PV, Allam MM (2012) Robust solver based on modified particle swarm optimization for improved solution of diffusion transport through containment facilities. Expert Syst Appl 39(12):10812–10820

Bouazza A, Van Impe WF (1997) Liner design for waste disposal sites. Environ Geol 35(1)

Das P, Bharat TV (2017) Effect of counter-ions on the diffusion characteristics of a compacted bentonite. Indian Geotech J. https://doi.org/10.1007/s40098-017-0241-y

Das P, Manparvesh SR and Bharat TV (2016) Salt diffusion in compacted plastic clay: experimental and theoretical evaluation. In: International conference on soil and environment, ICSE 2016, Bangalore

Kolstad D, Benson C, Edil T (2004) Hydraulic conductivity and swell of non-prehydrated GCLs permeated with multi-species inorganic solutions. J Geotech Geoenviron Eng 130:1236–1249
Lee JM, Shackelford CD (2005) Impact of bentonite quality on hydraulic conductivity of geosynthetic clay liners. J Geotech Geoenviron Eng 131:4–77
Rao NS, Mathew PK (1995) Effects of exchangeable cations on hydraulic conductivity of a marine clay. Clays Clay Miner 43(4):433–437
Robin MJL, Gillham RW, Oscarson DW (1987) Diffusion of strontium and chloride in compacted clay-based materials. J Soil Sci Soc Am 51:1102–1108
Rowe RK, Booker JR (1985) 1-D pollutant migration in soils of finite depth. J Geotech Eng 111(4):479–549
Shackelford CD, Daniel DE (1991) Diffusion in saturated soil. I. Results for compacted clay. J Geotech Eng 117(3):485–506
Shackelford C, Daniel DE, Liljestrand HM (1989) Diffusion of inorganic chemical species in compacted clay soil. J Contam Hydrol 4(3):241–273
Shackelford CD, Benson CH, Katsumi T, Edil TB, Lin L (2000) Evaluation of the hydraulic conductivity of GCLs permeated with non-standard liquids. Geotext Geomembr 18:133–161
Van Loon LR, Jakob A (2005) Evidence for a second transport porosity for the diffusion of tritiated water (HTO) in a sedimentary rock (Opalinus clay-OPA): application of through- and out-diffusion techniques. Transp Porous Media 61:193–214

Use of Kota Stone Powder to Improve Engineering Properties of Black Cotton Soil

Dayanand Tak, Jitendra Kumar Sharma, and K. S. Grover

Abstract Kota city is having major problem of disposing of Kota stone slurry powder. It is a by-product which is obtained during cutting, grinding and polishing operations of locally available Kota stone in stone industries. One of the innovative ground improvement techniques can be practiced by using Kota stone slurry powder (KSSP) as stabilizing agent or admixture. This technique may simultaneously offer the other advantages such as space saving, environmental sustainability and material availability. In the present study, Kota stone slurry powder is used as a stabilizing material for black cotton soil. The cone penetration, shrinkage limit, differential free swell, standard proctor test, UCS test and swell pressure test are conducted on soil and mix specimens to predict and ascertain the behaviour of mix. Results of these experiments conclude that engineering properties of the black cotton soil like shrinkage limit, maximum dry density (MDD), optimum moisture content (OMC) and UCS are improved by mixing Kota stone slurry powder depending on its proportion. With the percentage increase of Kota stone slurry powder, swelling characteristics of expansive soil are also improved such as decrement in the swelling percentage and swelling pressure. UCS decreases beyond 20% of mix.

Keywords Black cotton soil · Kota stone slurry powder (KSSP) · Unconfined compressive strength (UCS) · Swelling index · Standard proctor test · Swell pressure

D. Tak (✉)
Civil Engineering Department, Government Polytechnic Kota, Kota, India
e-mail: dayanandtak83@gmail.com

J. K. Sharma · K. S. Grover
Civil Engineering Department, Rajasthan Technical University, University Department, Kota, India
e-mail: jksharma@rtu.ac.in

K. S. Grover
e-mail: kdsgrover@gmail.com

M. Latha Gali and R. R. P. (eds.), *Problematic Soils and Geoenvironmental Concerns*, Lecture Notes in Civil Engineering 88,
https://doi.org/10.1007/978-981-15-6237-2_11

1 Introduction

Kota stone is generally used for cladding of the walls, decorating the floors and many other ways. It is having oil resistant, non-water absorbent, non-slippery and excellent stone resolvability properties. It is available in the form of slabs and tiles. Kota stone deposits spread in some areas of Kota and some part of Jhalawar District in Rajasthan. Fine slurry is produced during cutting, grinding and polishing of the Kota stones in stone industries. Every year approximately 3–3.25 lakh metric ton of stone slurry is produced and disposed of into local convenient places. This influences about 5 to 10 hectares useful land every year. This waste causes general environmental and economical drawback and health problems. Some of them are enlisted as loss in soil fertility due to increase of alkalinity in soil, contamination of underground water, health problem due to slurry dust suspended particles in air, stone slurry clusters or heaps, etc. Kota stone slurry is a semi-liquid substance consisting of particles originated from the sawing and polishing process, and water is used to cool and lubricate the sawing and polishing machines. Kota stone slurry powder can be used as a filler or stabilizing material and as a pozzolanic material. In this paper, through various laboratory experiments, results are presented for a possible use of KSSP as stabilizing or filler material. Singh et al. (2018) worked on replacement of marble slurry by stabilizing black cotton soil with Kota stone slurry.

Kota stone slurry powder is obtained by drying of Kota stone slurry. Kota stone is basically a flaggy lime stone. This slurry is having the characteristics which are very useful for improving the properties of expansive soil and helps in escalating the solid waste dumping problem.

Black cotton soil is in general highly expansive soil, and the variation of seasonal moisture content is responsible for substantial distress to the structures that are built over the soils. In black cotton soil, when water content decreases, the soil gets shrink, particles get rearranged, and movement takes place. Differential settlement may occur during swelling or shrinkage of soil. Alternate swelling and shrinkage in black cotton soil lead to the development of cracks in the structure. Therefore, the treatment of black cotton soil is required to avoid such a situation.

Stabilization is a technique to improve the properties of natural soil by using chemical or physical means. The aim of this study is to improve the geotechnical properties of black cotton soil economically. To determine the improvement in geotechnical properties of black cotton soil, the standard proctor test, shrinkage limit, swell index, swell pressure and unconfined compressive strength (UCS) tests are conducted with different percentage of Kota stone slurry powder in laboratory.

Many research works have been introduced worldwide in the direction of utilizing marble dust/ash/lime/cement/lime containing material to stabilize expansive soil for enhancing the characteristics of black cotton soil or expansive soils. Cokca (2001) founded that use of fly ash to improve engineering properties of an expansive soil. Basha et al. (2005) performed work on stabilization of expansive soil with rice husk and cement. From the test results, it is observed that the plasticity of soil reduces, but other engineering properties improve at optimum dosage of rice husk and cement.

Koonnamas et al. (2006) Conducted tests to analyse the effect on volume change of expansive soil with recycled ash with fibre. Palaniappan and Stalin (2009) concluded that marble dust powder is effective waste material for stabilization of expansive soil. It improves the engineering properties of the expansive soil. Modak et al. (2012) completed their work to improvement in black cotton soil by using admixtures. Vishvakarma and Singh (2013) studied the utilization of marble with expansive soil to enhance the soil properties.

Singh and Yadav (2014) studied the effect of marble dust on index proper ties of black cotton soil on different of water content. They mixed 0–40% marble dust in black cotton soil and concluded that the differential free swell, plastic limit and liquid limit of black cotton soil are decreasing by increasing the percentage of marble dust. It is also observed that the shrinkage limit of black cotton soil is increasing. Gupta and Sharma (2014) observed by tests the influence of marble dust, fly ash and Beas sand on sub grade characteristics of expansive soil. The series of test conducted in laboratory on fly ash, sand stabilized black cotton soil which further blended with 0–20% marble dust and concluded that the 15% marble dust is sufficient to increase the CBR-soaked value up to 200% approximately.

Sabat (2014) performed experiments to determine the improvement of expansive soil by mixing marble powder with rice husk ash. Muthukumar et al. (2015) completed work to improve the engineering properties of expansive soil. It also helps to disposal problem of marble powder and improves the ecological system.

Malik and Priyadarshee (2017) did work based on compaction and swelling behaviour of black cotton soil mixed with non-cementitious materials. A series of proctor tests and one-dimensional swell test were conducted to study the effect of the compaction and swelling behaviour. It was observed that swelling pressure, OMC were increasing, but MDD remained same.

The main objective of this study is to utilize the waste material Kota stone slurry powder to improve the geotechnical properties of black cotton soil. A laboratory investigation was conducted on black cotton soil and Kota stone slurry powder mixed in varied percentage. The test results showed a significant improvement in consistency, compaction, UCS and swell properties of black cotton soil.

2 Material Used

2.1 Kota Stone Slurry Powder

Locally available Kota stone slurry is used in mix which is produced from cutting, polishing in stone industries.

2.2 *Black Cotton Soil*

The black cotton soil is found mostly in the central and western parts and covers approximately 20% of the total land area of India. Black cotton soil is inorganic clays of medium to high compressibility and forms a major soil group in India. Black cotton soil is characterized by high shrinkage and swelling properties. The samples of black cotton soil were taken from a field in Amrit Nagar, Bohrkhera, district Kota in Rajasthan state for this study.

3 Experimental Program, Results and Discussions (As Per Is Codes Mentioned)

Objective: To study the behaviour of expansive soil using Kota stone slurry powder by stabilization technique, the percentage of mixing Kota stone slurry powder is 0–20% at interval of 5%.

The following geotechnical properties were studied by using various percentage of Kota stone slurry powder.

- Atterberg limits (liquid limit, plastic limit), Shrinkage limit [IS 2720 (Part 5)].
- Free well index [IS 2720 (Part 40)].
- Optimum moisture content, maximum dry density by standard proctor test [IS 2720 (Part 7)].
- Specific gravity and swell pressure [IS 2720 (Part 41)].

3.1 *Results of Black Cotton Soil and Kota Stone Slurry*

Laboratory test results of black cotton soil and Kota stone slurry powder are presented in Table 1.

From the test results, it is observed that the black cotton soil and KSSP are having 55% and 32.8% liquid limit, respectively. The plastic limit of soil is 28.62%, but Kota stone slurry powder is found as non-plastic material. The maximum dry density of black cotton soil and Kota stone slurry powder is determined as 1.67 gm/cm^3 and 1.69 gm/cm^3, respectively, at optimum moisture content of 19% and 20.70%.

3.2 *Specific Gravity*

The specific gravity is determined by density metre (50 ml) in laboratory. The specific gravity of black cotton soil and mix specimen is calculated, and the result of mix specimen is given in Table 2.

Table 1 Characteristics of black cotton soil and Kota stone slurry powder

Properties	Black cotton soil	Kota stone slurry powder
Liquid limit	55%	32.8
Plastic limit	28.62%	NP
Swell index	50%	–
Shrinkage limit	7.82%	22.49%
Specific gravity	2.57	2.63
MDD	1.67 gm/cm^3	1.69 gm/cm^3
OM C	19%	20.70%
UCS	2.80 kg/cm^2	–
Swell pressure	1.89 kg/cm^2	–
Volumetric shrinkage	120%	16.88%

Table 2 Specific gravity of specimen

Particulars	Sp. gravity
Black cotton soil	2.57
Kota stone slurry powder	2.63
Black cotton soil + 5% KSSP	2.59
Black cotton soil + 10% KSSP	2.60
Black cotton soil + 15% KSSP	2.608
Black cotton soil + 20% KSSP	2.61
Black cotton soil + 25% KSSP	2.615

Specific gravity of black cotton soil is 2.57, and specific gravity of Kota stone slurry powder is 2.63. By the adding of Kota stone slurry powder, it is observed that the specific gravity of mix specimen increases with increasing the percentage of KSSP in soil. Percentage increase in specific gravity of mix is up to 1.6% with increase of Kota stone slurry powder to 20%.

3.3 Differential Free Swell (DFS)

$$\text{DFS} = (V - Vk)/Vk$$

The DFS test is also performed in laboratory, and it is observed that the black cotton soil is highly expansive soil in nature. If KSSP percentage is varied from 5 to 20%, the differential free swell is continuously decreasing. The test results are presented in Table 3.

The DFS value for black cotton soil is 50%, while KSSP is non-swelling in nature. The DFS value of black cotton soil continuously decreases with increasing the percentage of Kota stone slurry powder as shown in Table 3. The maximum

Table 3 Swelling index of specimen

Particulars	Swell index (%)	Percentage decrease (%)
Black cotton soil	50	–
Kota stone slurry powder	NS	–
Black cotton soil + 5% KSSP	45	10
Black cotton soil + 10% KSSP	40	20
Black cotton soil + 15% KSSP	40	20
Black cotton soil + 20% KSSP	35	30
Black cotton soil + 25% KSSP	30	40

decrease in DFS value is determined for 25% mix specimen, which is decreased by 40% from the DFS value of black cotton soil.

3.4 *Plastic Limit*

From the results, it is observed that when the mixing percentage of KSSP increases, the plastic limit decreases continuously. With the increase of mix percentage from 5 to 20%, the plastic limit decreases from 26.21 to 22.51%. Corresponding percentage decrements in plastic limit are 1.54% to 15.4%, respectively, in comparison with black cotton soil. The test results are given in Table 4.

Table 4 Plastic limits of specimen

Particulars	Plastic limit (%)	Per. decr. (%)
Black cotton soil	26.62	–
Kota stone slurry powder	NP	–
Black cotton soil + 5% KSSP	26.21	1.54
Black cotton soil + 10% KSSP	25.82	3.0
Black cotton soil + 15% KSSP	25.14	5.55
Black cotton soil + 20% KSSP	22.51	15.40
Black cotton soil + 25% KSSP	19.86	25.39

Table 5 Liquid limit of specimen

Descriptions	Liquid limit in %	Percentage decreasing (%)
Black cotton soil	55	–
Kota stone slurry powder	32.80	–
Black cotton soil + 5% KSSP	54	1.82
Black cotton soil + 10% KSSP	53.22	3.23
Black cotton soil + 15% KSSP	52.11	5.25
Black cotton soil + 20% KSSP	48.86	11.16
Black cotton soil + 25% KSSP	47.21	14.16

3.5 Liquid Limit Test

The liquid limit of black cotton soil, Kota stone slurry powder and mix specimen of Kota stone slurry is determined by cone penetration test. The liquid limit of black cotton soil and Kota stone slurry powder is obtained 55% and 32.80%, respectively. With the mix percentage of KSSP increases 5% to 25%, the percentage decrease in liquid limit is from 1.82% to 14.16%, respectively. The test results of the liquid limit are given in Table 5.

3.6 Volumetric Shrinkage

Shrinkage limit is the water content where further loss of moisture will not result in any more volume change. The volumetric shrinkage is determined for black cotton soil, Kota stone slurry and mix specimen. The volumetric shrinkage is defined as the change in volume expressed as a percentage of the dry volume when water content is reduced from a given value of the shrinkage limit. The volumetric shrinkage of black cotton soil, Kota stone slurry is determined 7.52% and 16.88%, respectively. By adding and increasing the Kota stone slurry powder in soil, a continuous increment in shrinkage limit and a continuous decrement in volumetric shrinkage are observed. The test results of shrinkage limit and volumetric shrinkage are shown in Table 6.

Table 6 Shrinkage limit and volumetric shrinkage of specimens

Descriptions	Shrinkage limit	Per. increase in SL %	Volumetric shrinkage in %
Black cotton soil	7.82	–	86.70
Kota stone slurry powder	22.49	–	16.88
Black cotton soil + 5% KSSP	8.04	6.91	97.61
Black cotton soil + 10% KSSP	8.54	13.56	90.74
Black cotton soil + 15% KSSP	9.36	24.47	86.0
Black cotton soil + 20% KSSP	10.26	36.46	74.2
Black cotton soil + 25% KSSP	11.28	50.00	60.13

3.7 Proctor Test

The standard proctor test is performed to determine the optimum moisture content (OMC) and maximum dry density (MDD) of black cotton soil, Kota stone slurry and mix specimen. The MDD of black cotton soil and Kota stone slurry powder is found to be 1.67 and 1.685 gm/cm^3 for corresponding OMC of 19.4 and 20.7%. From the mix specimen results, it is observed that the maximum dry density continuously increases with increasing the percentage of Kota stone slurry powder in black cotton soil. The test results of proctor test are given in Table 7.

Maximum dry density (MDD) is slightly increased by 2.7% with increasing of Kota stone slurry powder up to 20% on 18.01% OMC. The graphical presentation of proctor test for black cotton soil and mix specimen of Kota stone slurry is shown in Figs. 1 and 2.

3.8 UCS Test

The unconfined compressive strength (q_u) is the load per unit area at which the cylindrical specimen of a cohesive soil is in compression. The test result of UCS is given in Table 8 (Fig. 3).

Table 7 Proctor test for specimen

Descriptions	Optimum moisture content in %	Maximum dry density in gm/cm^3
Black cotton soil	19.4	1.67
Kota stone slurry powder	20.7	1.69
Black cotton soil + 5% KSSP	19.3	1.676
Black cotton soil + 10% KSSP	19.05	1.679
Black cotton soil + 15% KSSP	18.26	1.702
Black cotton soil + 20% KSSP	18.2	1.707
Black cotton soil + 25% KSSP	18.01	1.712

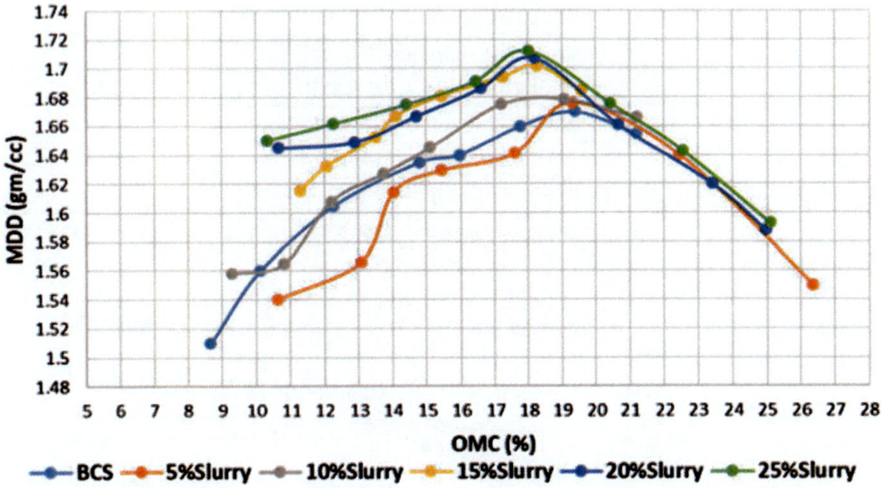

Fig. 1 Standard proctor test results for BCS with different proportions of slurry

The UCS value of black cotton soil is obtained as 2.80 kg/cm^2, and it is increased up to 4.11 kg/cm^2 when 20% KSSP is mixed in black cotton soil. The percentage increase in UCS value is up to 46.21% for the 20% mixing of Kota stone slurry powder. The percentage increase in the UCS value is shown in Fig. 4. It is observed that the UCS value decreases beyond 20% mixing of slurry powder. Hence, it is advised to use slurry as a mixing material only up to 20%.

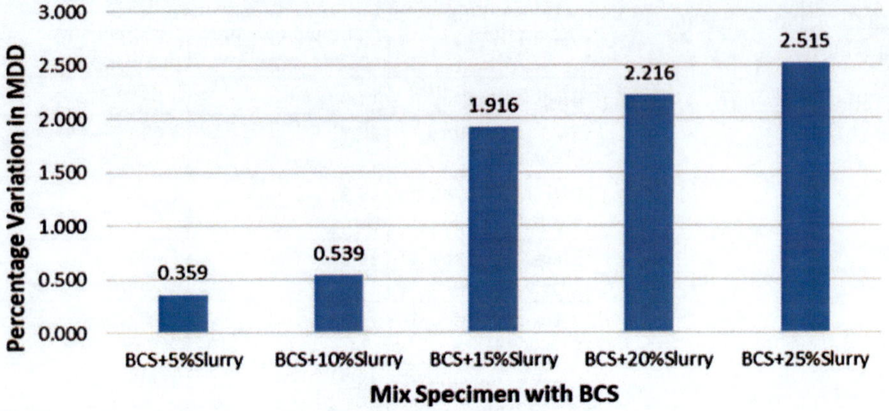

Fig. 2 Percentage variation in MDD of mix specimen

Table 8 UCS Test for Specimen

Descriptions	UCS (kg/cm^2)	Per. variation
Black cotton soil	2.80	–
Kota stone slurry powder	–	–
Black cotton soil + 5% KSSP	2.91	8.86
Black cotton soil + 10% KSSP	3.10	17.83
Black cotton soil + 15% KSSP	3.55	32.34
Black cotton soil + 20% KSSP	4.11	46.21
Black cotton soil + 25% KSSP	3.18	13.13

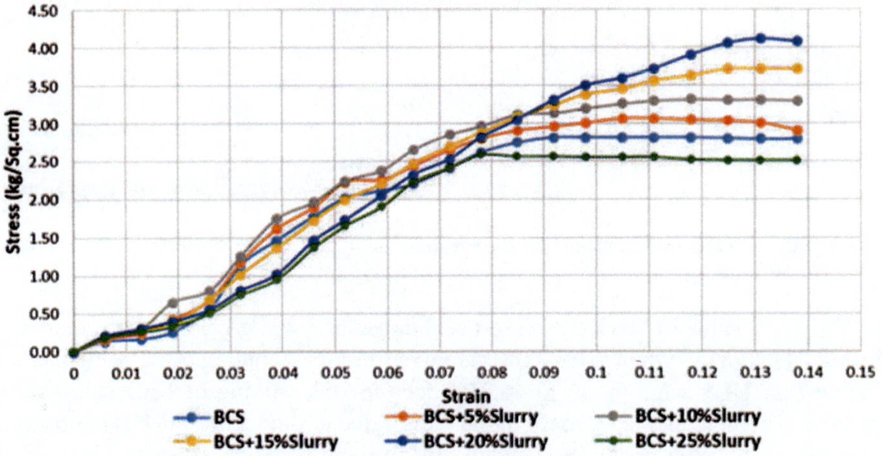

Fig. 3 UCS test results for BCS with different proportions of slurry

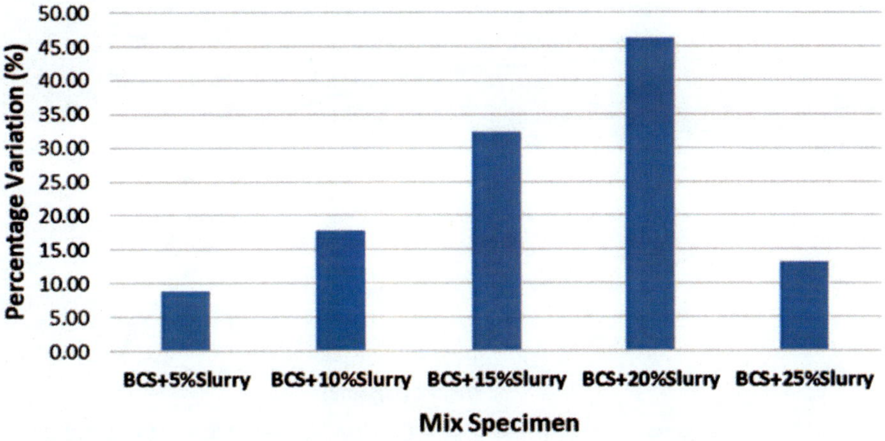

Fig. 4 Percentage variation in UCS for mix specimen

3.9 Swell Pressure Test

The swelling pressure tests are performed for black cotton soil and mix specimen of Kota stone slurry powder. The test results of the specimens are given in Table 9, and graphical presentation of results is shown in Fig. 5.

From Figs. 5 and 6, it is established that with the increasing percentage of Kota stone slurry powder, the value of swelling pressure is decreasing. The percentage decrements in swelling pressure of black cotton soil are 14.29 and 42.12% with the corresponding percentage of mixing of Kota stone slurry powder of 5 and 20%.

Table 9 Swelling pressure for specimen

Descriptions	Total swelling	Per. swell	Swell pressure (kg/cm^2)
Black cotton soil	0.82	4.1	1.89
Kota stone slurry powder	–	–	–
Black cotton soil + 5% KSSP	0.65	3.25	1.62
Black cotton soil + 10% KSSP	0.65	3.25	1.50
Black cotton soil + 15% KSSP	0.58	2.90	1.21
Black cotton soil + 20% KSSP	0.55	2.75	1.09
Black cotton soil + 25% KSSP	0.50	2.45	1.08

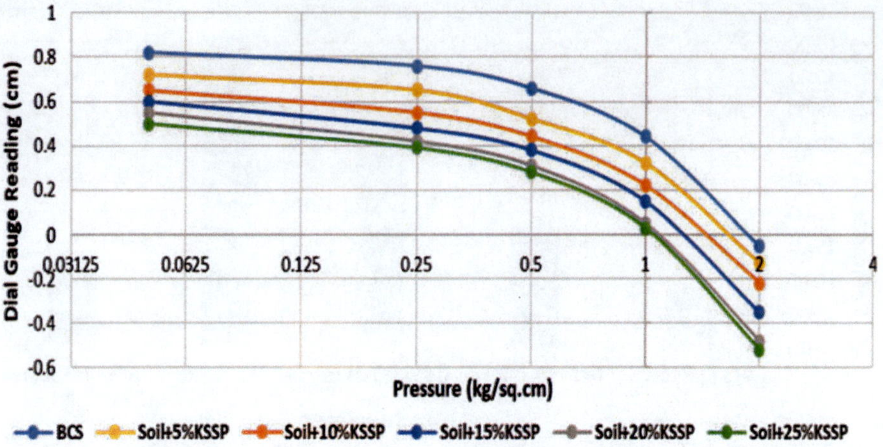

Fig. 5 Dial gauge reading with swell pressure of black cotton soil with different proportions of slurry

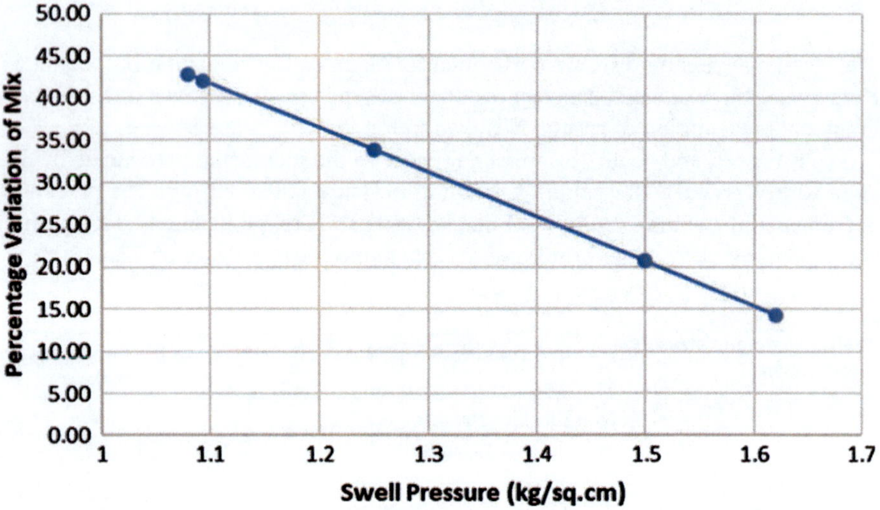

Fig. 6 Percentage variation of mix v/s swell pressure

4 Conclusion

Based on the experimental investigations, the conclusions are obtained:

- It is observed that when the Kota stone slurry powder is mixed with black cotton soil, the specific gravity continuously increases with its increasing percentage, but effect is marginal.

- The liquid limit and plastic limit of black cotton soil are determined 55 and 28.62%. The Kota stone slurry powder is non-plastic material. When the percentage of mixing of Kota stone slurry powder is increased, the plastic limit decreases from 26.62 to 22.51%. The liquid limit of black cotton soil also decreases with increasing the percentage of Kota stone slurry powder.
- The black cotton soil is having shrinkage limit of 7.52% with volumetric shrinkage of 120%. The shrinkage limit of black cotton soil increases with the increase in mix percentage of KSSP and the reason for increment in shrinkage limit is KSSP particles which are finer than black cotton soil particles. Therefore the KSSP particles behave as filler material. The volumetric shrinkage decreases with increase in the percentage of KSSP in black cotton soil.
- The maximum dry density of black cotton soil is determined 1.67 gm/cm^3 at 19.4% OMC. The maximum dry density of Kota stone slurry powder is also determined 1.69 gm/cm^3 at 20.7 OMC. It is observed that when the percentage of Kota stone slurry powder increases in black cotton soil, the maximum dry density increases, and optimum moisture content decreases. The reason of increasing maximum dry density of black cotton soil after adding KSSP is the Kota stone slurry powder particles work as filler and also bonding material.
- From the swell index, it is observed that the black cotton soil is highly expansive soil, and Kota stone slurry powder is non-swelling materials. From the mix specimens of Kota stone slurry powder, the swell index of black cotton soil decreases from 50 to 35% with increasing the percentage of Kota stone slurry powder.
- The unconfined compressive strength of the black cotton soil is 2.80 kg/cm^2. It is concluded that the Kota stone slurry powder works as filler material, which increases the UCS value of black cotton soil by adding Kota stone slurry powder. It is also observed that the unconfined compressive strength increases with increasing percentage of Kota stone slurry powder in black cotton soil up to 20%, and beyond 20%, UCS decreases with increasing of mix.
- The swelling pressure of black cotton soil is determined 1.89 kg/cm^2. It is explained that Kota stone slurry powder is a non-plastic material and it decreases the expansive behaviour of black cotton soil. Due to this reason, the swelling pressure of black cotton soil decreases with increasing percentage of Kota stone slurry powder.

References

Basha EA, Hashim R, Mahmud HB, Muntohar AS (2005) Stabilization of residual soil with rice husk ash and cement. Constr Build Mater 19(6):448–453

Cokca E (2001) Use of fly ashes for the stabilization of an expansive soil. J Geotech Geoenviron Eng 17:73–77. https://doi.org/10.1007/BF02838423

Gupta C, Sharma RK (2014) Influence of marble dust, fly ash and beas sand on sub grade characteristics of expansive soil. IOSR J Mech Civ Eng (IOSR-JMCE) 13–18. e-ISSN: 2278-1684, p-ISSN: 2320-334X

Indian Standard Code: IS 1498–1970, Classification and identification of soils for general engineering purposes (first revision)

Indian Standard Code: IS 2720 (Part 10) (1973) Determination of unconfined compressive strength (first revision)

Indian Standard Code: IS 2720 (Part 40) (1977) Determination of free swell index of soils

Indian Standard Code: IS 2720 (Part 41) (1977) Determination of swelling pressure of soils

Indian Standard Code: IS 2720 (Part 5) (1985) Determination of liquid limit and plastic limit (second revision)

Indian Standard Code: IS 2720 (Part 7) (1980) Determination of water content, dry density relation using light compaction (second revision).

Koonnamas P, Anand J, Puppala PE, Vanapalli SK, Inyang H (2006) Enhancing the volume change properties of sulphate rich expansive soils. https://doi.org/10.1061/(ASCE)0899-156(2006)18:2(295)

Malik V and Priyadarshee A (2017) Effect of rice husk ash stone dust on compaction and swell effect of expansive soil. Int J Geotech Eng. ISSN 1938-6362. https://doi.org/10.1080/19386362

Muthukumar M et al (2015) Utilization of waste marble is used to enhance the engineering properties of expansive soil. It also help to reduces disposal problem of marble powder and preserve the ecological system. Int J Eng Trends Tech 22(11). ISSN 231-5381

Palaniappan KA, Stalin VK (2009) Utility effect of solid waste in problematic soil. Int J Eng Res Ind Appl 2(1):313–321

Modak PR, Prakash B et al (2012) Stabilization of black cotton soil using admixture. Int J Eng Innov Technol 1(5)

Sabat AK (2014) A review of literature on stabilization of expansive soil using solid waste. https://doi.org/10.10610(ASCE)MT,1943-5533001153

Singh P S, Yadav RK (2014) Effect of marble dust on index properties of black cotton soil. Indian J Eng Res Sci Technol 3(3). ISSN: 2319-5991

Singh V et al (2018) Study paper on stabilization of black cotton soil with Kota stone sludge replacement of marble slurry. Natl J Multidiscip Res Dev 3(1):1053–1056 (ISSN 2455–9040)

Vishvakarma A, Singh RR (2013) utilisation of marble powder to enhance the soil properties and protect environment. J Environ Res Dev 7(4A):1479–1483

Amelioration of Expansive Clay Using Recycled Bassanite

E. Krishnaiah, D. Nishanth Kiran, and G. Kalyan Kumar

Abstract Solid waste management is a serious setback over the world. Therefore, reduction, reuse and recycling of waste have become major issue in recent days. The use of recycled bassanite in ground improvement projects is initiated recently in Japan to eliminate the huge quantities of gypsum wastes. The use of recycled bassanite has a positive effect on the environment and economy. It has challenges like release of fluorine more than the standard limits results in contaminate soil. This paper represents the effect of bassanite on clay stabilized soil by taking in consideration of the compressive strength and release of fluorine. Test results showed that the addition of bassanite had a significant effect on the improvement of compressive strength by increasing the amount of bassanite. The release of fluorine increases with the amount of bassanite in soil mixture, and it had a negative effect on the improvement of strength and consuming the amount of admixture. Recycled bassanite produced from gypsum wastes had a potential to be used as a stabilizer material for expansive clay soil and meets the standards of environment.

Keywords Recycled bassanite · Black cotton soil · Soil stabilization · Unconfined compressive strength

1 Introduction

Expansive soils have a peculiar problem of swelling and shrinkage due to moisture content variations. These swell-shrink movements cause considerable distress to the buildings and pavements either through heave or through settlement depending on the applied stress level and the soil swelling pressure. Hence, design and construction of civil engineering structures with expansive soils are a challenging task for geotechnical engineers around the world. There are many conventional mechanical treatments available for control of these problems. These include prewetting, soil replacement with compaction and moisture control, surcharge loading and thermal

E. Krishnaiah (✉) · D. Nishanth Kiran · G. Kalyan Kumar
Department of Civil Engineering, NIT Warangal, Warangal, Telangana, India
e-mail: erukalkrishna@gmail.com

M. Latha Gali and R. R. P. (eds.), *Problematic Soils and Geoenvironmental Concerns*, Lecture Notes in Civil Engineering 88,
https://doi.org/10.1007/978-981-15-6237-2_12

methods. However, these methods have their own limitations with regards to their effectiveness and costs.

New methods are still being researched to reduce the swell-shrink potential of expansive soils. Stabilization of expansive soils with various additives including lime, cement, fly ash and calcium chloride has achieved some success. In general, cement-stabilized soils are prone to high temperature cracking, whereas strength gain through lime stabilization is moderate. Furthermore, both stabilizers can be leached out from soils by water with passing time. Moreover, with cyclic wetting and drying, the beneficial effect of lime stabilization is partially lost on lime treated soils due to partial breakdown of cementation bonds and reduction in dry unit weight and moisture content. However, in this study, recycled bassanite is used to improve the properties of expansive soil. Increasing the use of waste and recycled materials in earthwork projects has created the necessity for a better understanding of the durability and strength performance of these materials against weathering conditions. In general, there are several researchers who have examined the utilization of different types of waste and recycled materials as a stabilizing agent to enhance the strength of weak soil.

The disposal of gypsum waste plasterboards in landfill sites presents many environmental challenges because of the increase in Fluorine solubility above the acceptable limits. As a result, the disposal of gypsum wastes at traditional landfill sites is prohibited. The wastes must be disposed in controlled landfill sites, which lead to increased disposal costs (Kamei and Horai 2008; Ahmed et al. 2011a). The most attention is to protect the environment by recycling gypsum waste plasterboards and reusing the produced recycled bassanite as an additive material for ground improvement. This application meets the challenges of our society by producing useful material from wastes, reducing the huge quantities of gypsum waste produced every year, minimizing the operation of landfill sites and the costs of disposal. In the present study, recycled bassanite is used as a stabilizer agent for improvement of clay soil. Additionally, this study will evaluate the use of furnace slag cement as solidification agent to improve the environmental properties of clay soil treated with recycled bassanite.

2 Methodology

The materials used in the present work are black cotton soil, recycled bassanite and furnace slag cement, which were collected from a site near NIT Warangal campus. Geotechnical characterization was done based on I.S. 2720, and Scanning Electron Microscope (SEM) test was performed on the stabilized soil. The soil passing through 425 μ sieve has a plasticity index (PI) of 68% and a liquid limit (LL) of 98%. According to IS Soil Classification, the soil is classified as a clayey soil with high plasticity (CH). Particle size analysis was performed on the soil by following the standard methods, and the result was shown in Fig. 1. Cement was having more amount of fines. The Standard Proctor compaction tests were also performed on

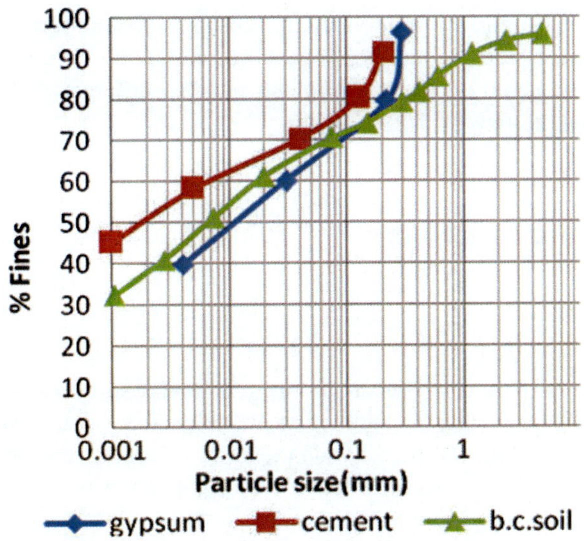

Fig. 1 Particle size distribution curve of the soil, gypsum and cement

the soil to determine the maximum dry density ($\gamma_{d\max}$) and the optimum moisture content (OMC) which are 1.58 g/cc and 21.6%, respectively.

The recycled bassanite used in the present study was crushed and then screened to remove any solid or contaminated materials, such as paper, wood, fibers, paints and stones. The crushed gypsum waste was heated under temperatures ranging between 140 and 160 °C for a certain time to remove three quarters of the water molecules, and then, recycled bassanite was produced as presented in the following equation.

$$CaSO_4.2H_2O \xrightarrow{140^\circ-160^\circ} CaSO_4.1/2H_2O + 3/2H_2O$$

3 Sample Preparation and Testing

The recycled gypsum and tested soil were mixed in dry state to achieve homogeneous and isotropic properties for the mixture. The water content was determined for each sample before and after mixing process to know the effect of adding recycled gypsum on the water content. The unit weight for untreated and treated soil samples was determined by placing the sample in a cylindrical steel container with a known volume. To determine the optimum content of the recycled gypsum–soil (*B*/*S*), pH meter test was performed. Furnace slag cement was mixed with *B*/*S* ratio to prevent the solubility of recycled gypsum.

Cylindrical specimens were modeled for compressive strength. Oil was used to lubricate the inner sides of the molds to ensure no friction occurs during sample extraction. Subsequently, samples were subjected to different curing times 7, 14, 28 days at temperature (21° ± 1 °C).

4 Results and Discussions

4.1 H Data

Different contents of recycled bassanite mixed with clay soil pH values were shown in Fig. 2, and it appears that the additive of pure recycled gypsum (without cement) to clay soil has an insignificant effect on the value of pH. A very limited change in the PH value in clay soil observed when recycled gypsum was introduced. Since the measured pH values were found to be between 7 and 8 which are the neutral values for environmental regulations. The pH value for the mixture of 6% gypsum–soil is 7.57, and further, there is no much increase in pH value. When cement was introduced to tested soil, the pH value increased rapidly, and this related to the cement being alkaline material.

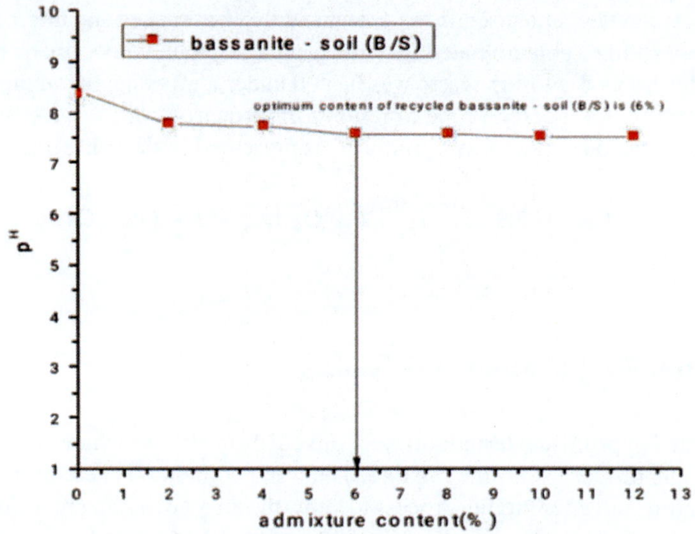

Fig. 2 Effect of recycled bassanite–cement content on pH values of stabilized

4.2 Unit Weight

The effect of recycled gypsum content on the unit weight clay soil was examined. The values of unit weight for different samples treated with recycled gypsum were plotted against the contents of recycled gypsum and are presented in Fig. 3.

The increase in the content of recycled gypsum is associated with a significant increase in the unit weight for different contents of recycled gypsum 0, 4, 6 and 8% of B/S ratio. It is attributed to the tendency of the calcium component of recycled gypsum to encourage the soil particles to flocculate. Flocculation of soil particles was observed because the clay particles carry negative charges and attract calcium, which has two positive charges (Ca^{++}). The recycled bassanite is a hemi-hydrate calcium sulfate ($CaSO_4.1/2\ H_2O$), can absorb water from clay soil and was able to recover three quarters of the water which was missed during the heating process, to become hydrate calcium sulfate ($CaSO_4.2H_2O$). The absorption of water from the particles of clay soil is associated with the decrease of voids between the soil particles. The volume for the clay–gypsum mixture decreased, and unit weight increased. Increase in content of recycled bassanite treated with clay soil, unit weight is increased, and water content is decreased vice versa.

Cement is introduced in the gypsum–soil mixture, unit weight is increased when compared to the soil was treated with recycled gypsum alone (Fig. 4). Three different furnace slag cement contents ranging from 1 to 3% cement–soil (*C*/*S*) are used for the reduction of fluorine content. The maximum unit weight of the soil is 17.6 kN/m^3 when treated with 2.25% of *C*/*S* and 6% of B/S ratios which is more when compared to soil treated with *B*/*S* ratio of 6% only (17.01 kN/m^3).

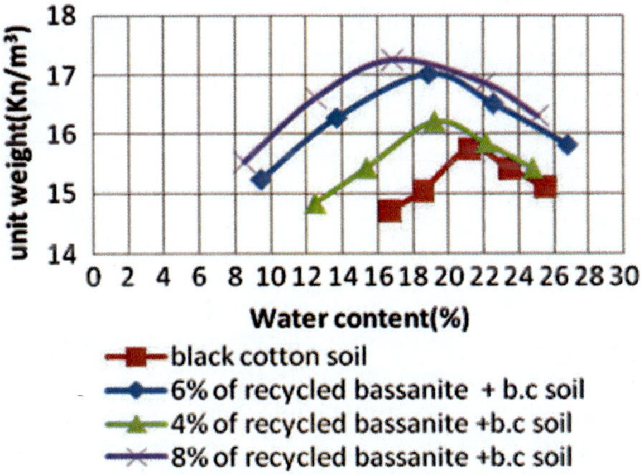

Fig. 3 Effect of recycled gypsum content on unit weight of stabilized soil specimens

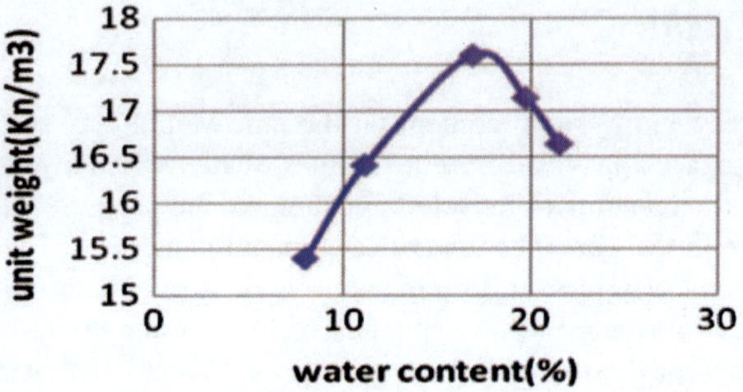

Fig. 4 Effect of cement content on unit weight of clay–gypsum mixture

4.3 Unconfined Compressive Strength

Ultimate compressive strength against the different content of recycled gypsum is shown in Fig. 5. The compressive strength of 6% recycled gypsum content increases with the increase of curing period. Gypsum contains a significant amount of calcium sulfate. The attraction between the soils particles and calcium improves the bonding between the soil particles and strengthens the soil. The strength of stabilized soil–gypsum increases with the curing period. The strength of black cotton soil without any admixture is 60 kPa. The curing period has a significant effect on the strength of stabilized clay–gypsum up to the 7 days curing and it is found to be 134.05 kPa.

Effect of cement content on the strength of clay–gypsum mixture was shown in Fig. 6 for different curing periods. From Figs. 5 and 6 noticed that the compressive strength of 6% recycled gypsum content for curing period of 28 days is 300 kPa, whereas 2.25% of cement (*C*/*S*) with 6% *B*/*S* mixture at 7 day curing time is 800 kPa. It has been shown that 28 days of curing for soil–gypsum treated with cement are adequate for the reaction between the soil and stabilizing agents used to be completed.

5 Conclusion

The main contribution of this work was to show the potential application of recycled bassanite (bassanite), and the following conclusions can be drawn:

The treatment of clay soil with bassanite significantly improves its unit weight and strength performance. Both unit weight and compressive strength of clay–gypsum mixture increased with increase in bassanite content. The soil–gypsum mixture exhibited brittle behavior during unconfined compression, and strain failure decreased slightly with addition of bassanite. It does not have any adverse effect on the pH; the average for the measured value was found to be 8 to 7 neutral value.

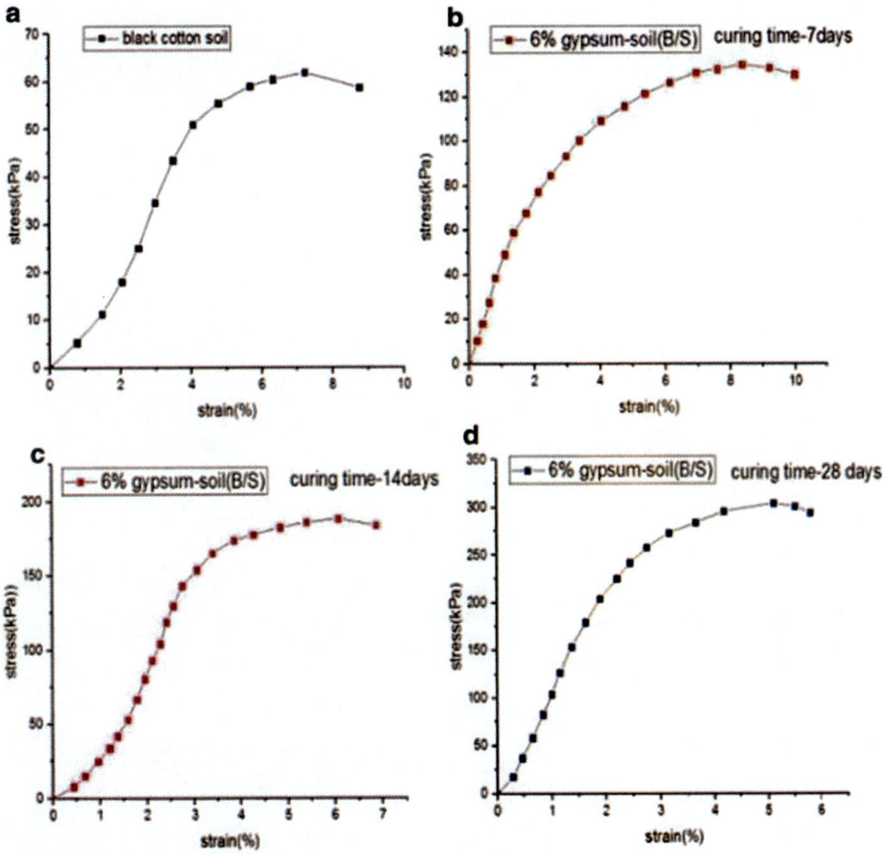

Fig. 5 Effect of recycled gypsum content on ultimate compressive strength, **a** black cotton soil, **b** 7 days curing time, **c** 14 days curing time, **d** 28 days curing time

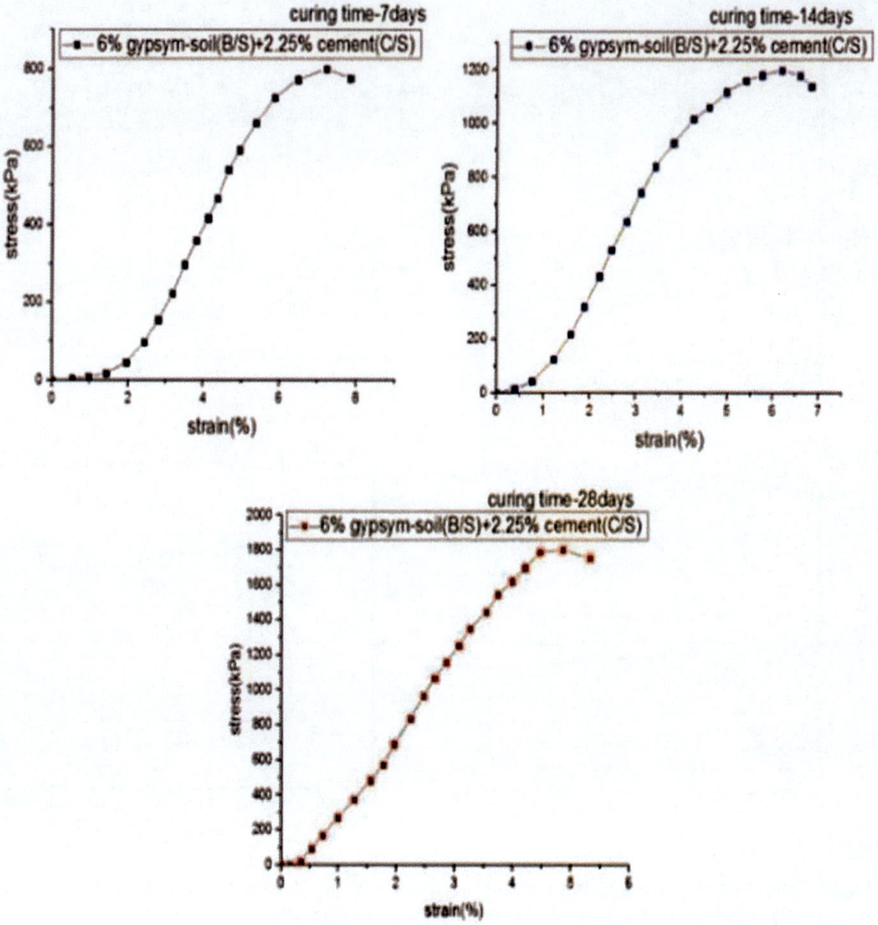

Fig. 6 Relationship between compressive strength and different curing times (days) with different admixtures

References

Ahmed A (2015) Compressive strength and microstructure of soft clay soil stabilized with recycled bassanite. Appl Clay Sci 104:27–35

Ahmed A, Ugai K, Kamei T (2011a) Investigation of recycled gypsum in conjunction with waste plastic trays for ground improvement. Constr Build Mater 25(1):208–217

Kamei T, Horai H (2008) Development of solidification technology for fluorine contaminated bassanite using Portland blast-furnace (B) cement. Jpn Geotech J 4(1):91–98

Influence of Biochar on Geotechnical Properties of Clayey Soil: From the Perspective of Landfill Caps and Bioengineered Slopes

P. V. Divya, Ankit Garg, and K. P. Ananthakrishnan

Abstract The present study investigated the influence of biochar on various geotechnical properties of clayey soil. Biochar, a carbon-rich product, was amended with the soil at different percentages such as 5, 10 and 15%. Laboratory tests were conducted to evaluate the feasibility of utilization of biochar in enhancing the integrity of soil slopes with an emphasis on landfill covers and bioengineered slopes. Scanning electron microscopy (SEM) tests were conducted to analyse micro-porous structure of biochar and soil-biochar mix. The SEM micrographs indicated the porous structure of the biochar and a good interaction with the clay particles. It was concluded from the present study that the biochar amended soil can have steeper and stable slopes compared to soil without biochar amendments. The compressibility of the soil was also decreased with biochar amendment.

Keywords Biochar · Bioengineered soil slopes · Soil stabilization

1 Introduction

Biochar is a carbon-rich solid product and it can be produced from several types of biomass. Biochar finds its application in carbon sequestration (Lehmann and Joseph 2015), improving agricultural productivity (Das et al. 2016), soil water retention (Bordoloi et al. 2018), contaminant remediation and in preventing landfill gas emissions (Yargicoglu and Reddy 2017), etc. Soil fertility, microbial activity and water holding capacity can be enhanced along with the reduction in leaching of nutrients with the amendment of biochar to the soil. Thus, biochar has the potential to be used

P. V. Divya (✉) · K. P. Ananthakrishnan
Department of Civil Engineering, Indian Institute of Technology Palakkad, Palakkad, Kerala, India
e-mail: divyapv@iitpkd.ac.in

A. Garg
Department of Civil and Environmental Engineering, Shantou University, Shantou, Guangdong, China
e-mail: ankit@stu.edu.cn

M. Latha Gali and R. R. P. (eds.), *Problematic Soils and Geoenvironmental Concerns*, Lecture Notes in Civil Engineering 88,
https://doi.org/10.1007/978-981-15-6237-2_13

in bioengineered slopes. Biochar has a highly porous structure and it was reported that the surface area per gram of biochar range from ten to hundreds of square metres (Lehmann and Joseph 2015).

Biochar amended soil can also be used in landfill covers to reduce landfill gas emission (Reddy et al. 2015). A landfill cover is usually composed of several layers, each with a specific function. The final cover system must enhance surface drainage, vegetation, minimize infiltration and control the release of the landfill gases. In case of bioengineered slopes, the design of surface cover should minimize water infiltration and also maximize shear strength of soils (Bordoloi et al. 2018). In both the applications, vegetation growth should be enhanced in long term. This is exactly, where properties of biochar must be optimized to consider vegetation growth in compacted engineered soil in long term. A preliminary study was conducted to study the influence of biochar on geotechnical properties of compacted clayey soil. Apart from basic index properties, shear strength as well as factor of safety of slope with and without biochar amendment is also investigated.

2 Materials Used in the Present Study

2.1 Biochar

The biochar used in the present study was obtained from pyrolysis of hard wood of *Prosopis juliflora*. The pyrolysis temperature is around 500 °C. Details of pyrolysis method are mentioned in Bordoloi et al. (2018). Based on the elemental analysis, % carbon (C) in biochar is around 72%, whereas nitrogen (N) is around 0.19%. The C/N ratio is around 382. The cation exchange capacity (CEC) of biochar is around 18 cmol/kg. Figure 2 shows the scanning electron micrographs (SEM) of the biochar. As shown in Fig. 1, biochar has got a highly porous structure.

The structure, however, is slightly different from that reported by Bordoloi et al. (2018). In Bordoloi et al. (2018), SEM images show biochar to be closer to platy structure and was more randomly arranged. However, in this study, the structure of biochar follows a floral pattern. This difference in structure might be due to the difference in source (waste), from where the biochar is produced. It indicates that the source, from where biochar is produced as well as pyrolysis conditions could highly influence the micro-structure and properties of the biochar. The biochar received was crushed into smaller sizes. The grain size distribution of biochar used as a soil amendment is shown in Fig. 2. It can be observed that the biochar used in the present study is predominantly of sand-sized particles with very less percentage fines.

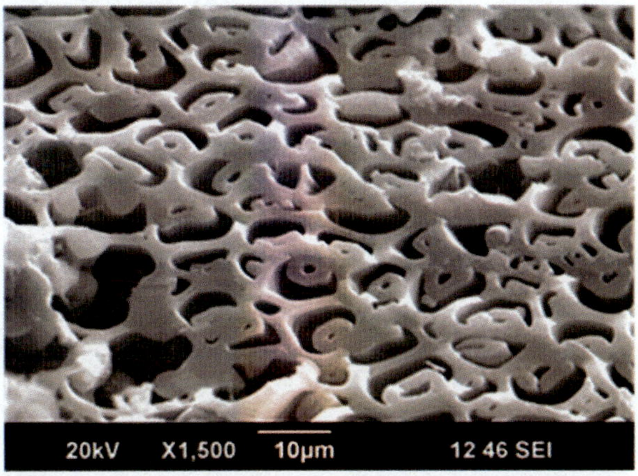

Fig. 1 Scanning electron micrograph (SEM) of the biochar

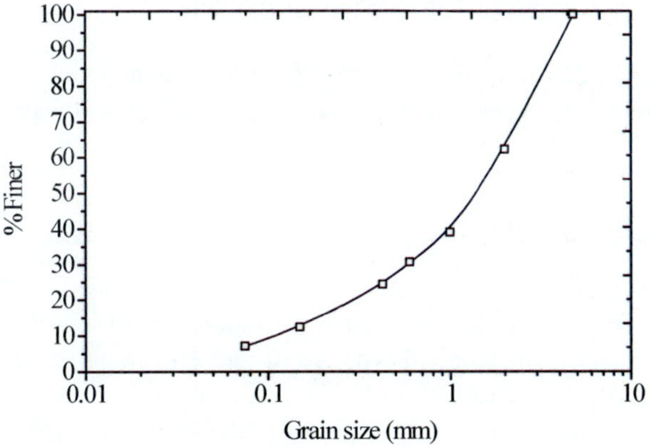

Fig. 2 Grain size distribution of biochar used in the present study

2.2 *Soil*

Soil used in the present study was collected from a site near IIT Palakkad, Kerala. The soil is classified as CI by the Indian Standard Soil Classification System. Figure 3 shows the grain size distribution of the soil used in the present study.

The liquid limit and plasticity index were 48% and 25%, respectively. The maximum dry density and optimum moisture content (OMC) as obtained from standard Proctor compaction test were found to be 16.66 kN/m^3 and 18.18%, respectively. The shear strength parameters are obtained from direct shear tests corresponding to maximum dry density and OMC of the soil. The cohesion intercept and angle of friction were 9.29 kN/m^2 and 25°, respectively.

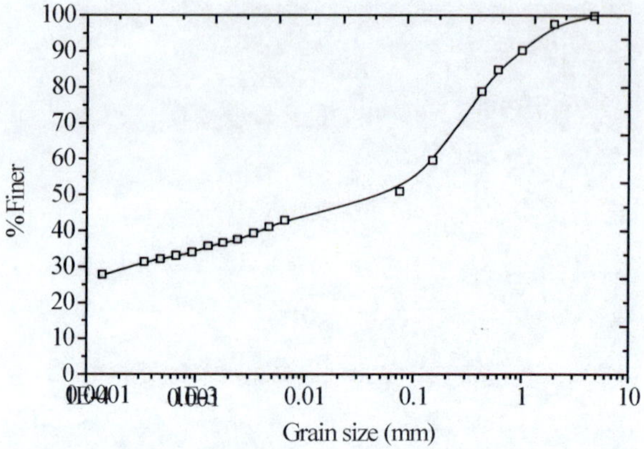

Fig. 3 Grain size distribution of the soil

3 Experimental Investigations

Biochar was amended with the soil at different percentages; 5, 10 and 15%. The influence of biochar on various geotechnical properties of soil was studied.

3.1 Compaction Characteristics

The authors performed the standard Proctor compaction test on soil with and without biochar amendment. Soil blended with 5, 10 and 15% biochar yielded maximum dry unit weight of 16.3, 15.9, 14.9 kN/m^3, respectively, and OMC of around 19%; compared to a maximum dry unit weight of 16.66 kN/m^3 and OMC of 18.18% for soil without any amendments. There was a reduction in the maximum dry density of the soil with biochar amendment as shown in Fig. 4. This could be due to the porous structure of the biochar as shown in Fig. 2.

3.2 Plasticity Characteristics

In order to study the influence of biochar on the plasticity characteristics of the soil, Atterberg limits tests were conducted on soil with and without biochar amendment. The variation of liquid limit, plastic limit and plasticity index is shown in Fig. 5. It can be found that the liquid limit of biochar first decreases slightly and then increases with biochar amendment. The liquid limit is found to slightly increase from 49 to 54% with an increase in biochar from 0 to 15%. This is also likely due to porous structure of biochar, which can retain water.

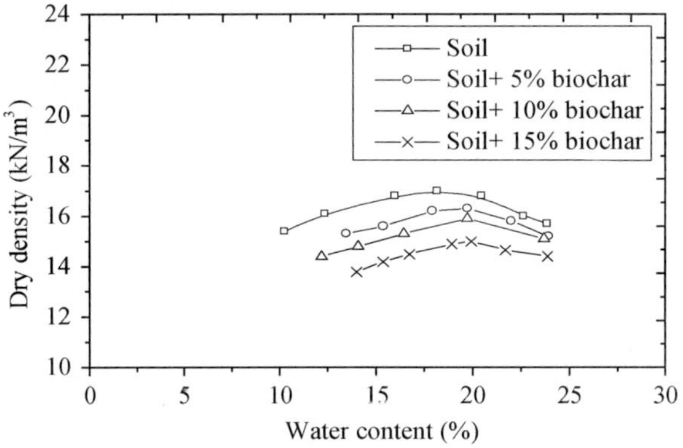

Fig. 4 Compaction curve for soil with and without biochar

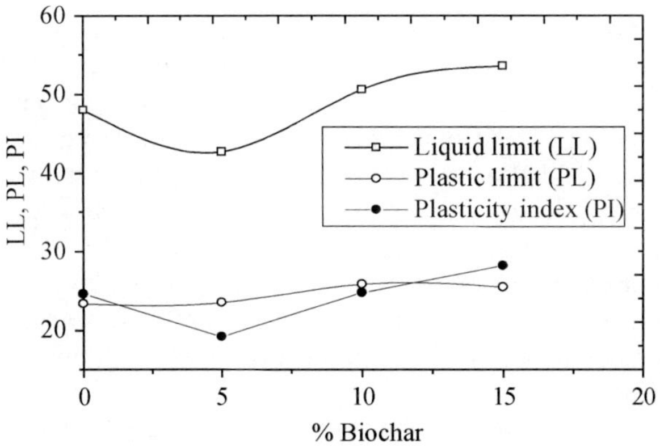

Fig. 5 Variation of liquid limit, plastic limit and plasticity index with % biochar

Plasticity index is also found to increase slightly with increase in biochar content. It indicates that biochar is able to increase the potential of soil to yield. This has significance in performance of final cover in slopes and landfill, which undergoes longer durations of drying conditions.

3.3 Compressibility Characteristics

The compressibility characteristics of the soil with and without biochar were studied in conventional oedometer apparatus. The tests were conducted in accordance with IS 2720 (part XV)-1986. The tests were conducted for different normal stresses of 12.5, 25 and 50 kN/m^2. The soil samples with and without biochar were moist compacted at maximum dry density and optimum moisture content as obtained from standard

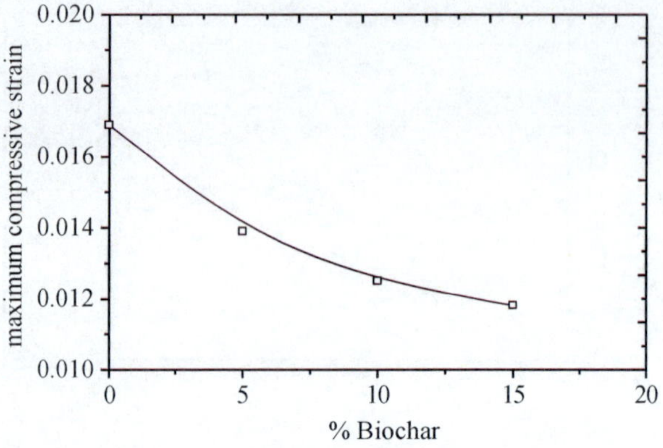

Fig. 6 Variation of maximum compressive strain with % biochar

Proctor compaction tests. Each stress increment was maintained and settlements were recorded for duration of 24 h till further settlement with time is negligible.

The stress range selected was low considering the fact that the range of overburden pressure expected on biochar amended soils in landfill covers and bioengineered slopes are less than 50 kN/m^2. However, though not covered in the present paper, it would be interesting to study the compressibility characteristics of biochar amended soil at higher pressure range also.

The maximum value of compressive strain was calculated from the value of maximum settlement or compression obtained from time versus settlement data. Figure 6 shows the variation of maximum compressive strain with percentage of biochar amendment to the soil.

There is a decreasing trend in the compressibility and settlement characteristics as the percentage of biochar amendment increased. As the percentage of the biochar was increased to 15%, there was 1.4 times decrease in the value of compressive strain experienced by the soil for the selected pressure range. The decreasing trend of compressive strain indicates the contribution of the inter-particles (soil particle-biochar particle) resistive force.

3.4 Shear Strength Characteristics

Shear strength characteristics of the soil with and without biochar were studied by conducting a series of direct shear tests with various percentages of biochar amendments. The tests were conducted in accordance with IS 2720 (part XIII) and the horizontal shear force and shear displacements were noted. The normal stress versus shear stress at failure is plotted in Fig. 7.

Shear strength parameters, cohesion intercept and angle of internal friction were obtained from Fig. 7. The variation of shear strength parameters of soil with different

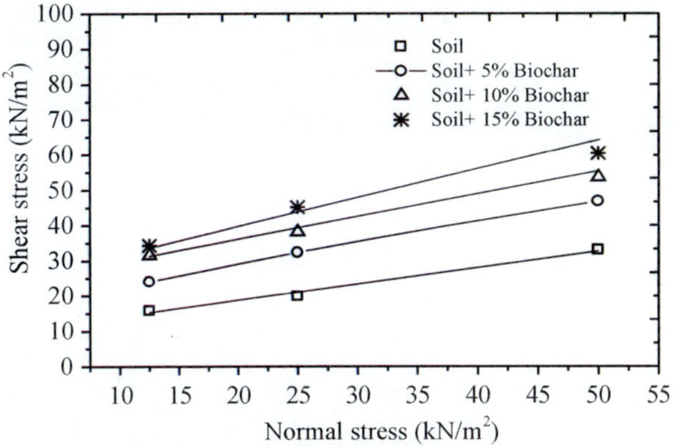

Fig. 7 Shear stress versus normal stress with % biochar

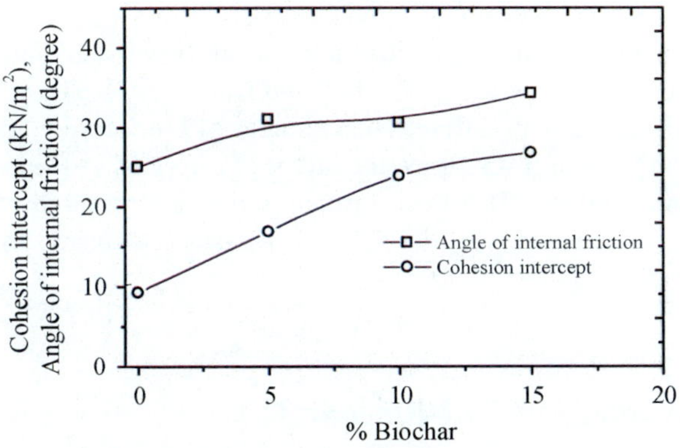

Fig. 8 Variation of shear strength parameters with % biochar

percentage of biochar is plotted in Fig. 8. As can be seen from Figs. 7 and 8, there is an improvement in the shear strength characteristics of the soil with the biochar amendment.

3.5 *Simplified Stability Analysis of Biochar Amended Soil Slopes*

A simplified stability analysis was conducted in order to ensure the slope stability of biochar amended soil slopes. Since the shear strength of the soil was increased by biochar amendment it is expected that slope stability of landfill covers and bioengineered slopes can be enhanced by biochar amendment. Infinite slope stability analysis was conducted assuming a landfill cover slope of 1V:3H which assumes failure

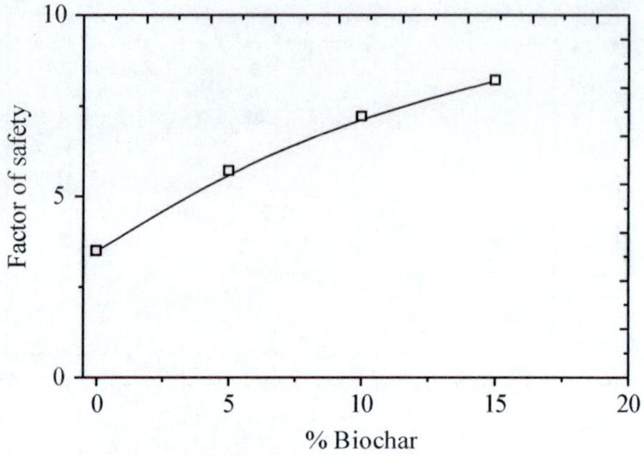

Fig. 9 Variation of factor of safety of soil slopes with % biochar

surface occurs parallel to the slope. The factor of safety was calculated from the ratio of resisting stress to driving stress (Gulhati and Datta 2005). The variation of factor of safety of soil slopes with different percentages of biochar amendments is shown in Fig. 9. As can be seen from Fig. 9, the factor of safety of soil slope was improved with biochar amendments. However, further studies are required for analysing the stability of slopes, especially at the onset of transient seepage through the slopes (Garg et al. 2015).

3.6 *SEM Analysis of Soil-Biochar Mix*

In order to understand the micro-mechanical interaction between the biochar and the soil particles and also to understand the particle morphology, it was also decided to conduct SEM analysis in the present study. As can be observed from Figs. 10a and 10b, there is good interaction between clay particles and the biochar. This indicates the biochar properties would have stronger dominance on soil properties.

Further, as observed from the figures, there are three types of pores in soil-biochar mix, i.e., intra-soil pores (soil-soil), intra-biochar pores (biochar-biochar) and inter-soil-biochar pores (soil-biochar). It is the relative proportion of these pores that would overall determine the geotechnical behaviour of soil-biochar mix. It is expected that with the increasing proportion of biochar, the intra-biochar pores would dominate more and can affect geotechnical behaviour of soil.

a. At a magnification of X500

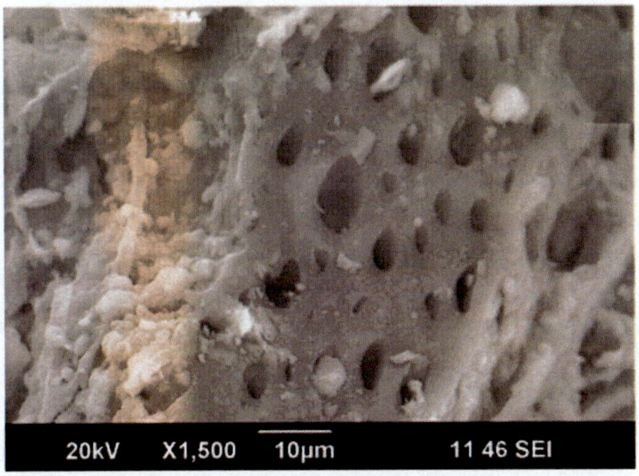

b. At a magnification of X1,500

Fig. 10 SEM images of soil-biochar mix

4 Conclusions

From the present study, it was concluded that the compressibility of the soil was decreased with biochar amendment. As the percentage of the biochar was increased, there was 1.4 times decrease in the value of compressive strain experienced by the soil for the selected pressure range. There was a reduction in the maximum dry density of the soil with biochar amendment, though there was no significant variation in the value of optimum moisture content. As the percentage of biochar amendment in the soil was increased from 5–15%, there was a slight increase of plasticity index (about 1.5 times) and liquid limit (1.3 times) of the clayey soil selected. The shear strength parameters of the soil were improved with the biochar amendments. There was a maximum increase of about 3 times in the value of cohesion intercept and 1.4 times increase in the value of angle of internal friction with biochar amendments. The present study indicates that biochar amended soil can have steeper and stable slopes compared to soil without biochar amendments. The SEM micrographs indicated the porous structure of the biochar and a good interaction with the clay particles.

The stress range selected in the present study was low; considering the fact that the range of overburden pressure expected on biochar amended soils in landfill covers and bioengineered slopes are less than 50 kN/m^2. However, it would be interesting and further works are warranted to study the compressibility and shear strength characteristics of biochar amended soil at higher pressure range and also at fully saturated conditions.

References

Bordoloi S, Garg A, Sreedeep S, Lin P, Mei G (2018) Investigation of cracking and water availability of soil-biochar composite synthesized from invasive weed water hyacinth. Bioresour Technol 263:665–677

Das O, Sarmah AK, Bhattacharyya D (2016) Biocomposites from waste derived biochars: mechanical, thermal, chemical, and morphological properties. Waste Manage 49:560–570

Garg A, Coo JL, Ng CWW (2015) Field study on influence of root characteristics on soil suction distribution in slopes vegetative with *Cynodon dactylon* and *Schefflera heptaphylla*. Earth Surf Proc Land 40(12):1631–1643

Gulhati SK, Datta M (2005) Geotechnical engineering. Tata McGraw-Hill, New Delhi

Lehmann J, Joseph S (eds) (2015) Biochar for environmental management: science, technology and implementation, 2nd edn. Routledge, Abingdon

Reddy KR, Yaghoubi P, Yukselen-Aksoy Y (2015) Effects of biochar amendment on geotechnical properties of landfill cover soil. Waste Manage 33(6):524–532

Yargicoglu EN, Reddy KR (2017) Microbial abundance and activity in biochar amended landfill cover soils: evidence from large-scale column and field experiments. J Environ Eng 143(9):04017058

Remediation of Lead Contaminated Soil Using Olivine

Linu Elizabeth Peter and M. K. Sayida

Abstract Amidst the numerous techniques of remediation researched, the immobilization technique achieved through solidification/stabilization appears to be most effective due to its ability to entrap the waste within a solid cementitious matrix and its cost effectiveness. Due to growing environmental concerns, there is a need to replace conventional stabilization binders with more efficient, environment-friendly stabilizer. In this study, artificially lead contaminated Kaolinite is treated using different concentrations of Olivine, a natural mineral. The effectiveness of the treatment is assessed through unconfined compressive strength test and column leachate test. The optimum concentration of Olivine for remediation is determined. The experimental results show that olivine can be effectively used for the treatment of lead contaminated soil.

Keywords Remediation · Solidification · Olivine · Lead contamination

1 Introduction

Improper dumping of wastes and poor management of industrial discharges in the recent years have led to severe soil contamination. Due to their tendency to bioaccumulate, heavy metal contaminants persist in the soil without undergoing degradation. In addition to being toxic to the environment and public health, they also affect the mechanical behaviour of soils. Lead is a highly toxic, non-biodegradable element found in the environment. Anthropogenic activities like mining and smelting operations, combustion and incineration processes, automobile emissions have lead to excessive lead levels in the air and soil.

In situ immobilization achieved by stabilization/solidification has been found to be an effective remediation technique for treating contaminated soils. Stabilization/solidification processes retain the contaminants and convert them into less toxic

L. E. Peter (✉) · M. K. Sayida
Department of Civil Engineering, College of Engineering Trivandrum, Thiruvananthapuram, Kerala, India
e-mail: lizpete25@gmail.com

M. Latha Gali and R. R. P. (eds.), *Problematic Soils and Geoenvironmental Concerns*, Lecture Notes in Civil Engineering 88,
https://doi.org/10.1007/978-981-15-6237-2_14

or mobile forms by physical or chemical means. Most processes use cement-based binders. However, due to growing environmental concerns on the overuse and over dependance on cement leading to high energy consumptions and high carbon footprint, there is a need to replace cement-based binders. Recently, phosphate-based binders were developed which proved to be effective for lead immobilization. Excess phosphate additions affect the soils chemical structure. Studies conducted by Jin et al. (2016) and Suzuki et al. (2013) showed the effectiveness of MgO-based binders in treating heavy metal contaminated soils. Commercial MgO in addition to being expensive is also a cause of concern due to its carbon footprint.

In this study, Olivine, a natural mineral found on the Earth's subsurface, was chosen as the binder for lead immobilization. Olivine is a magnesium iron silicate comprising of 49–50% MgO. MgO has been effectively used for contaminated soil remediation (Garcia et al. 2004). Currently, there are no known publications on the use of Olivine for the treatment of heavy metal contaminated soil. Olivine if found effective in lead immobilization can replace commercial MgO and phosphate-based binders and will prove to be an economical, environmental-friendly alternative.

2 Experimental Program

In this study, artificially contaminated Kaolinite was prepared at two lead contamination levels, moderate and high. The lead contaminated Kaolinite was then treated at varying percentages of Olivine. The effectiveness of the stabilization/solidification treatment is evaluated based on unconfined compressive strength test and column leachate test.

2.1 Materials

The Kaolinite used for the study was purchased from English India Clays Ltd. Table 1 shows the properties of Kaolinite. The soil is classified as CI as per IS classification.

The Olivine used for the study was purchased from Industrial Minerals and Refractories, Salem. Table 2 gives the properties of Olivine as provided by the manufacturer.

2.2 Preparation of Artificially Contaminated Soil

Lead nitrate salt of analytical reagent grade was used for preparing lead solutions used for artificially contaminating the soil. The soil was artificially contaminated with predetermined concentrations of lead nitrate to prepare moderately contaminated soil with lead concentration of 700 mg/kg and highly contaminated soil of 1500 mg/kg

Table 1 Properties of Kaolinite

Soil property	Value
Specific gravity	2.52
Liquid limit (%)	50
Plastic limit (%)	26
Shrinkage limit (%)	21
Plasticity Index (%)	24
Max. dry density (g/cc)	1.54
Optimum moisture content (%)	27
Unconfined compressive strength (kN/m^2)	140

Table 2 Properties of olivine

Property	Value
Chemical composition (%)	
MgO	49
SiO_2	41
Fe_2O_3	9
Al_2O_3	0.5–2.0
CaO	0.2
Melting point	1600 °C
Free silica content	<0.1%
Bulk density (g/cc)	3.2–3.4

lead concentration represented as M. Pb and H. Pb, respectively. The maximum permissible lead level in the soil is 500 mg/kg as per US EPA criteria. Soil samples were left undisturbed for 30 days. These were then dried and powdered before use.

2.3 *Standard Proctor Test*

Olivine treatment of the contaminated clay involved mixing the clay with Olivine at varying percentages of 15, 20, 25, 30 and 35%. The standard Proctor test was conducted in accordance with IS 2720 (Part 7) to determine maximum dry density and optimum moisture content.

2.4 *Unconfined Compressive Strength Test*

Samples were prepared at the optimum moisture content and maximum dry density of each percentage of Olivine in soil and thoroughly mixed to achieve a uniform blend. The samples were prepared in cylindrical moulds of 38 mm diameter and 76 mm height and cured for 7, 14 and 28 days before testing. The test was conducted as per IS 2720 (Part 10).

2.5 Column Leach Test

In soil environments, both the environmental impact of hazardous materials and the effectiveness of *S/S* treatment technologies are evaluated based on the leaching of the concerned pollutants. In the present study, a flow-through column leach test setup was developed to evaluate the long-term effectiveness of the proposed Olivine-based S/S technology to immobilize heavy metals in soils. The column leach test was conducted in accordance with US EPA (2013) standards. Columns of 8 cm diameter were filled with samples at optimum moisture content and 97% of maximum dry density obtained from Proctor test. The samples were filled to a height of 20 cm from the base. The base of the column was provided with a leachate collection mechanism.

2.6 Alternate Wetting–Drying Test

The resistance of a stabilized/solidified soil to degradation caused by external environmental stresses is evaluated through durability tests. These environmental stresses are simulated by subjecting the S/S treated solid to aggressive physical weathering conditions in the form of wetting, drying, freezing and thawing processes. In this study, alternate wetting and drying test was used to test the durability of the treated soil.

The test was conducted in accordance with IS 4332 (Part IV). Samples for durability test were prepared in the UCC mould at optimum moisture content and maximum dry density. Samples were submerged in potable water at room temperature for 5 h. They were then placed in an oven and dried at 70° ± 3 °C for 42 h. Thus, each cycle includes one wetting and one drying with a total duration of 48 h. After the ith drying or wetting cycle, the mass loss (ML) of the specimen is calculated by Eq. 1.

$$\mathrm{ML}(\%) = (m_o - m_i)/m_o \times 100 \quad (1)$$

Here, m_o is initial mass of sample, and m_i is mass after ith cycle. The same procedure is repeated for the remaining samples till the failure occurred or up to 12 cycles.

3 Results and Discussions

3.1 Compaction Characteristics

Figures 1 and 2 show the variation in maximum dry density and optimum moisture content at varying Olivine content for moderate and high lead contamination, respectively.

For moderately contaminated soils, the dry density was found to increase till 25% Olivine after which it reduced. Optimum moisture content reduced with increase in Olivine content up to 25% after which it reduced. For highly contaminated soils,

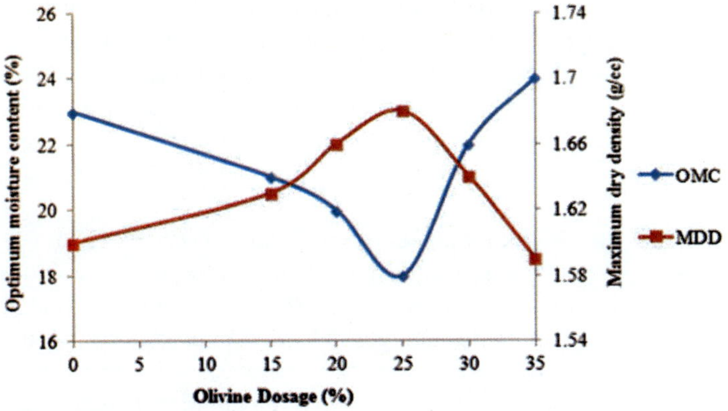

Fig. 1 Variation in maximum dry density and optimum moisture content with varying binder dosage for moderately Pb contaminated soil

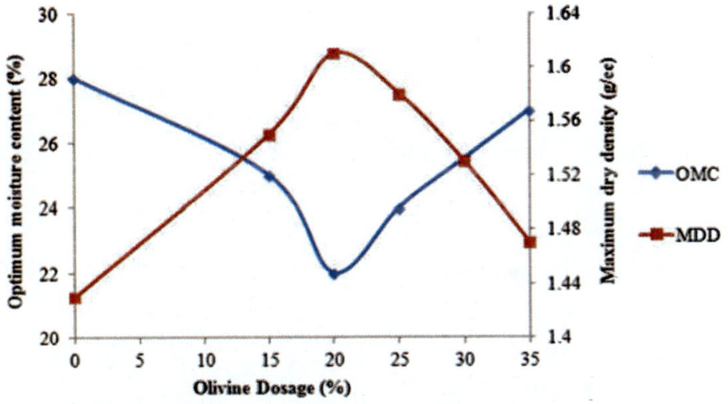

Fig. 2 Variation in maximum dry density and optimum moisture content with varying binder dosage for highly Pb contaminated soil

maximum dry density increased up to 20% and reduced thereafter. Optimum moisture content decreased up to 20% Olivine content and increased thereafter.

3.2 *Unconfined Compressive Strength Results*

Figures 3 and 4 show the stress strain curves of Olivine treated soil without curing for moderately and highly lead contaminated samples, respectively.

Figures 5 and 6 show the stress–strain curves of samples at 28 days of curing for moderately and highly lead contaminated samples, respectively. The unconfined compressive strength test results show that strength of the moderately lead contaminated soil improved with increase in Olivine content up to 25% beyond which it reduced. After 28 days of curing, a three times increase in strength was observed. The compressive strength increased from 171 kN/m^2 for untreated lead contaminated soil to 680 kN/m^2 for 25% Olivine treated samples. The UCC test results of highly Pb contaminated soils show a 3.45 times increase in strength for soil treated at 20% Olivine. The compressive strength increased from 85 to 380 kN/m^2 at 28 days of curing.

Figure 7 shows the variation of 28 day strength at varying binder dosages. The optimum binder content for moderately contaminated soil was found to be 25% and that for highly contaminated soil was 20%. This indicates that increase in concentration of Pb enhances the strength of hydration products formed.

Figure 8 shows the effect of curing on the compressive strength of 25% Olivine treated M. Pb samples and 20% Olivine treated H. Pb samples. The strength was found to significantly increase with curing days. The increment in strength upon Olivine

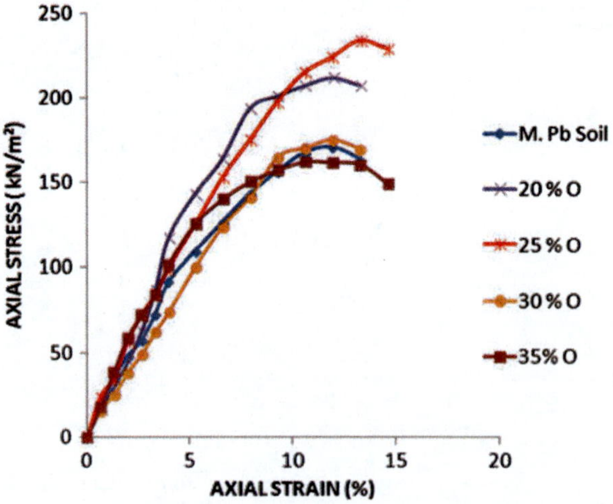

Fig. 3 Stress–strain curves of moderately Pb contaminated samples without curing

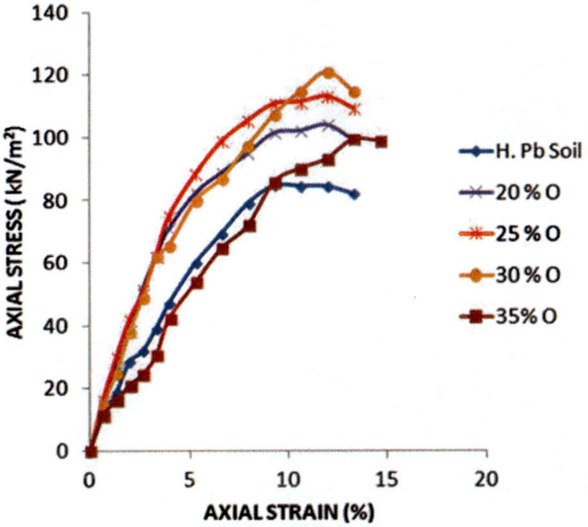

Fig. 4 Stress–strain curves of highly Pb contaminated samples without curing

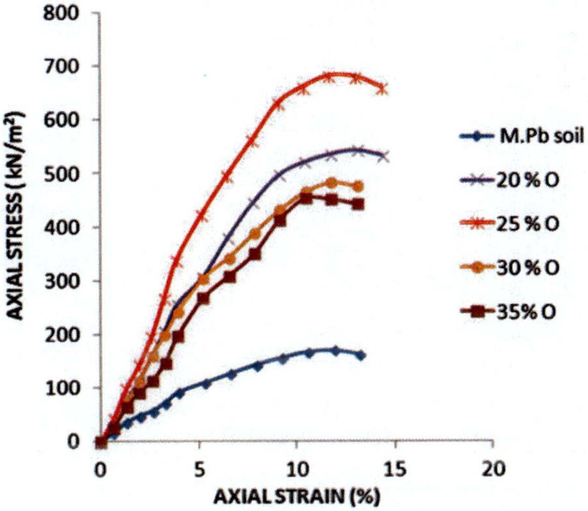

Fig. 5 Stress–strain curves of moderately Pb contaminated soils at 28 days curing

treatment can be attributed to the formation of $Mg(OH)_2$, also called Brucite, the main hydration product of Olivine. Brucite has two sheets of hydroxide ions in hexagonal close packing, with a sheet of Mg atoms between them attached to form a layered structure (Jin and Al-Tabbaa 2014). Hence, the heavy metals get encapsulated within the layers or get adsorbed on the surface of Brucite.

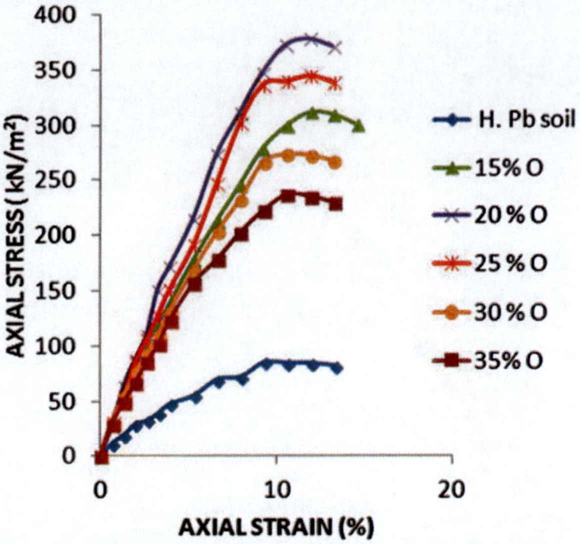

Fig. 6 Stress–strain curves of highly Pb contaminated soils at 28 days curing

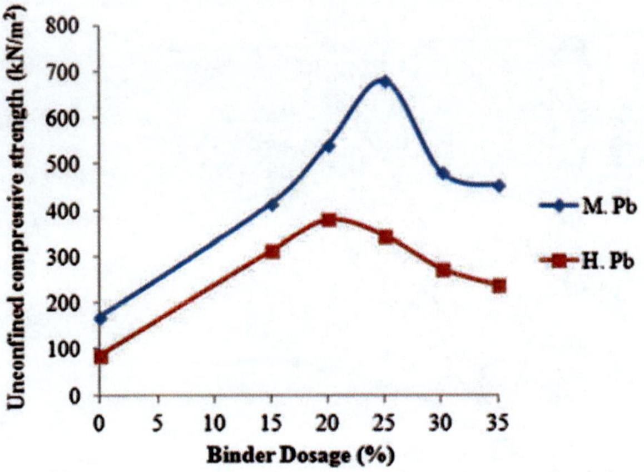

Fig. 7 Variation in 28 day strength of moderate and highly Pb contaminated soil with binder dosage

The results obtained showed that with respect to the regulatory limit for unconfined compressive strength of stabilized/solidified soils specified by US EPA which is 350 kN/m^2, soils at both contaminant levels satisfied the criteria.

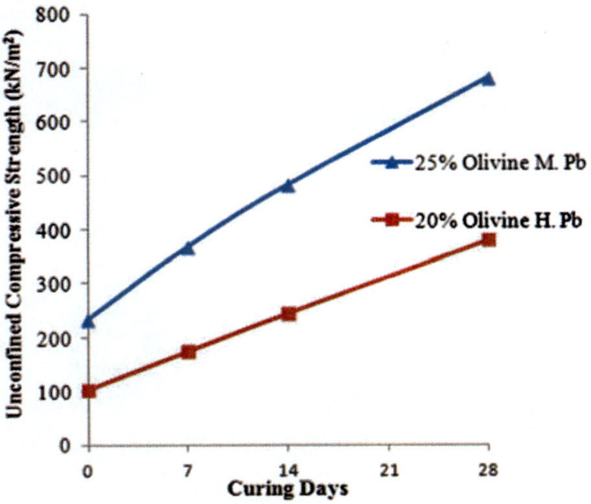

Fig. 8 Effect of curing on compressive strength

3.3 *Leachate Analysis*

The leachate analysis was conducted only for samples that showed maximum strength improvement. The leachate collected from the column leach test setup was analysed using atomic adsorption spectrometer. Figure 9 shows the heavy metal concentration in leachate of untreated and treated sample at optimum Olivine content for both contamination levels.

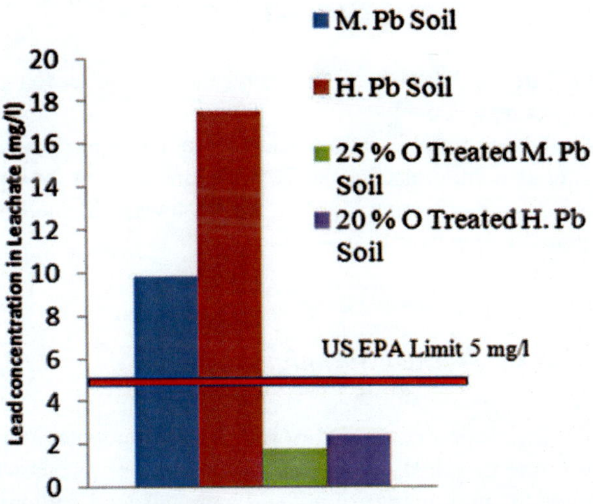

Fig. 9 Concentration of lead in leachate

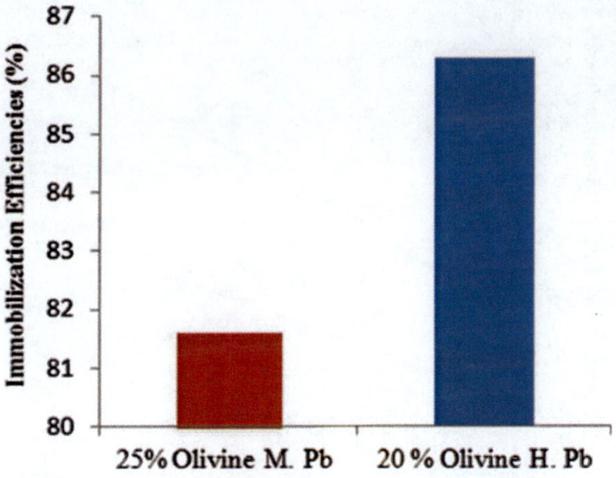

Fig. 10 Immobilization efficiencies achieved by S/S remediation

The lead concentration in the leachate for both moderate and high contaminated soils after treatment was found to be lesser than the US EPA limit of 5 mg/l.

3.4 *Immobilization Efficiency*

The immobilization efficiency (E) of the optimum binder dosage for each heavy metal can be evaluated using Eq. 2.

$$E(\%) = [(C_o - C_i)/C_o] \times 100 \quad (2)$$

where C_o and C_i are the heavy metal concentration in the leachate from untreated and treated samples, respectively.

Figure 10 shows the immobilization efficiency of samples treated at 25% Olivine content for moderate contamination and 20% Olivine content for higher contamination. An immobilization efficiency of 82% and 86% was obtained for moderate and high contamination, respectively.

3.5 *Durability of Treated Samples*

The test results of the wet–dry durability in terms of percentage of mass loss after each cycle are plotted against number of cycles. Figure 11 shows the mass loss percentage for lead contaminated samples. Untreated lead contaminated soils did not withstand the test beyond the second and third cycle for moderate and high level

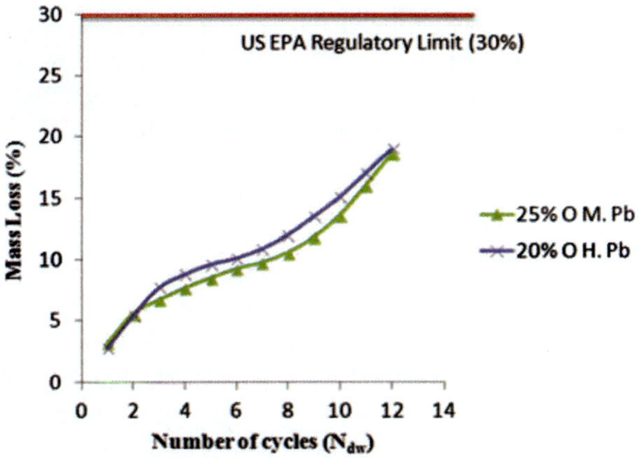

Fig. 11 Percentage mass loss for treated samples

of contamination, respectively. Moderately lead contaminated samples at optimum binder dosage showed a mass loss of 16%, and highly lead contaminated samples at optimum binder dosage showed a mass loss of 17%. Thus, the mass loss percent at both levels of contamination was below the US EPA regulatory limit of 30%.

4 Mechanism of Immobilization

As the main constituent of Olivine is MgO, based on available literature on remediation of contaminated soils using MgO-based binder and hydration mechanisms of Olivine, the mechanism of solidification and immobilization can be explained.

The entrapment of heavy metal in the stabilized soil matrix can be attributed to three distinct mechanisms. First, the high pH conditions induced in the soil upon the addition of binder result in the formation of insoluble heavy metal hydroxides. This makes it difficult for heavy metal species to find their way into ground water. The second possible mechanism is the incorporation of heavy metal ions into the crystal structures of the cementitious compounds formed upon stabilization. These consequently are not readily dissolved in water, thus not showing up into the leachate chemical composition. The third mechanism is surface adsorption or physical entrapment of the heavy metal species through sorption onto particle surfaces. Binders that rely on the third mechanism alone for immobilization are less effective (Suzuki et al. 2013).

Lead being the most common among heavy metal contaminants, a large number of studies have been conducted on its mechanism of immobilization. Based on these, it can be said that, in the alkaline environment provided by the Olivine, lead precipitates to lead hydroxide making it difficult for the lead to leach out.

Contrary to what is observed for cement-based binders, increase in lead contamination led to increased immobilization efficiencies. Excess presence of lead is usually found to retard the hydration process of binders generally used (Wang et al. 2015). In this study, higher contamination level required lesser Olivine for maximum strength attainment. This indicates that excess lead activates the hydration process of Olivine. Detailed investigations into the microstructure of the hydration products formed are necessary to explain this observation.

Initially, addition of binder creates an alkaline environment that causes leaching out of Si and Al from the soil, forming an alumina-silicate-hydrate gel. As the hydration of MgO progresses, strength improvement occurs. At the latter part of the curing period, the presence of lead hydroxides causes the Mg ions to provide additional nucleation sites for precipitation. This results in the formation of magnesium silicate hydrate (M–S–H) gel. The formation of A–S–H and M–S–H and Brucite ($Mg(OH)_2$) is responsible for the strength gain (Chiang et al. 2014).

The strength reduction at higher Olivine dosages can be attributed to the excess Olivine that was not mobilized in the chemical reactions which consequently occupied space within the sample and therefore reducing bond in the mix.

5 Conclusions

A number of previous studies showed that the menace of heavy metal contamination of soil and ground water resources is growing at an alarming rate. This study focused on assessing the potential of a natural mineral, Olivine for use in the stabilization/solidification of lead contaminated soils. The following conclusions can be drawn from the study.

1. Olivine, being a rich source of MgO, has the potential to remediate lead contaminated soils.
2. The optimum binder dosage was found to be around 20–25% Olivine depending on the level of contamination in the soil.
3. Strength of soils improved with Olivine addition, showing a three times increment for moderate lead contamination and 3.45 times increase for high lead contamination.
4. Contrary to the retardation effect of excess lead seen on conventional binders, in this study, excess lead was found to activate the hydration process. This necessitates detailed investigations into the microstructure of hydration products formed.
5. The concentration of lead in the leachate was below the US EPA maximum permissible limit of 5 mg/l for both levels of contamination.
6. Immobilization efficiency of 82% for moderate contamination, and 86% for higher contamination was achieved.
7. The mass loss percentage obtained from the durability test was below the US EPA limit of 30% at both levels of contamination.

Overall, the study showed that Olivine is a promising new binder that can be used for remediation of heavy metal contaminated soil. Extensive research into the hydration mechanisms of Olivine and the effect of excess heavy metals will help develop an optimum binder with better immobilization efficiency. It is imperative to adopt economical and flexible remediation techniques and binder materials that cause minimal environmental impact and at the same time ensure that the reuse potential of the treated soils are not adversely affected due to changes in its engineering properties. Utilization of Olivine is one such promising remediation technique that needs to be extensively explored.

References

Chiang WS, Ferraro G, Fratini E (2014) Multiscale structure of calcium and magnesium–silicate–hydrate gels. J Mater Chem 32:91–98

Garcia MA, Chimenos JA, Fernandez AI, Miralles L, Segarra M, Espiell F (2004) Low-grade MgO used to stabilize heavy metal in highly contaminated soils. Chemosphere 56:481–491

IS: 2720 (Part 10) (1991) Indian standard methods of tests for soils-part 10-determination of unconfined compressive strength. Bureau of Indian Standards, New Delhi

IS: 4332 (Part 4) (2010) Indian standard methods of test for stabilized soils-part 1V-wetting and drying, and freezing and thawing tests. Bureau of Indian Standards, New Delhi

Jin F, Al-Tabbaa A (2014) Evaluation of novel reactive MgO activated slag binder for the immobilisation of lead and zinc. Chemosphere 117:285–294

Jin F, Wang F, Al-Tabbaa A (2016) Three-year performance of in-situ solidified/stabilised soil using novel MgO-bearing binders. Chemosphere 144:681–688

Suzuki T, Akira N, Masakazu N, Hideki N, Fujii H, Tasaka Y (2013) Lead immobilization in artificially contaminated kaolinite using magnesium oxide-based materials: immobilization mechanisms and long-term evaluation. Chem Eng J 232:380–387. ELSEVIER

US EPA (1999) Solidification/stabilization resource Guide. EPA/542B-99-002, Office of Solid Waste and Emergency Response

Wang F, Jin F, Al-Tabbaa A (2015) The performance of blended conventional and novel binders in the in-situ stabilisation/solidification of a contaminated site soil. J Hazardous Mater 285:46–52. ELSEVIER

Release of Dark Colored Leachate from Mined Aged Municipal Solid Waste from Landfills

Mohit Somani, Manoj Datta, G. V. Ramana, and T. R. Sreekrishnan

Abstract The present study focuses on the release of dark colored leachate from soil-like material (SLM) reclaimed from aged municipal solid waste (MSW) at three dumps of India (located at Delhi, Hyderabad and Kadapa). If the material is to be used in filling low-lying areas or in embankments or subgrade, the leaching of colored liquid can cause coloration of the surrounding water bodies and ground water. Local soil was used as a base material for comparison of release of color. Mined soil-like material from MSW releases dark and objectionable color of leachate. The intensity of color of leachate from mined SLM is found to range between 380 and 400 Hazen, 460 and 480 Hazen and 900 and 1000 Hazen in the samples from Delhi, Kadapa and Hyderabad landfill, respectively. In contrast, the intensity of color in water released from local soil varies between 25 and 30 Hazen. The study concludes that small amount of mined SLM requires large quantity of water (70–100 times) for washing the material before the intensity of color reduces to acceptable level, thereby demonstrating the potential to significantly impact nearby ground water wells. Thermal treatment is observed to be reducing the color to acceptable level before reuse.

Keywords Aged municipal solid waste · Colored leachate · Landfill mining · Soil-like material

1 Introduction

Municipal solid waste (MSW) dumps/landfills having large quantities of accumulated garbage (also termed as aged waste or legacy waste) have become a serious threat to society, and their rehabilitation in an environmentally sustainable way is one of

M. Somani (✉) · M. Datta · G. V. Ramana
Department of Civil Engineering, Indian Institute of Technology Delhi, New Delhi, India
e-mail: msomani02@gmail.com

T. R. Sreekrishnan
Department of Biochemical Engineering and Biotechnology, Indian Institute of Technology Delhi, New Delhi, India

M. Latha Gali and R. R. P. (eds.), *Problematic Soils and Geoenvironmental Concerns*, Lecture Notes in Civil Engineering 88,
https://doi.org/10.1007/978-981-15-6237-2_15

the major challenges faced by urban local bodies in developing countries like India (Somani et al. 2017, 2018, 2019a). Landfill mining is one of the possible alternatives for reducing the deposited old waste and reclaiming the dumps for other purposes. It is an approach whereby solid waste that has previously been landfilled is excavated and processed. The waste which has been deposited for years and decades can be mined for reuse, recovery and recycling (Somani et al. 2019b).

Aged waste may be defined as the MSW that has been disposed of in an unlined dumpsites more than 10 years ago (Oettle et al. 2010). Several such dumpsites in India are more than 20–30 m high and 10–30 years old. Soil-like material (SLM) in the present study has been defined as the fraction of aged MSW passing 4.75 mm as per standard of practice from geotechnical perspectives. From the grain size distribution, it was found that 55–60% of aged MSW lies below 4.75 mm. The high soil-like content in the aged MSW is likely the result of daily covering soil, street sweepings, drain silt and construction and demolition (C&D) waste along with the humification of organic matter in the fresh MSW (Parrodi et al. 2017). Critical aspects before re-using aged MSW in earthworks are identified as follows: organic content; total dissolved salts; release of dark colored leachate; heavy metals. In this study, attention is focused on "release of dark colored leachate" from the aged MSW (Somani et al. 2019b).

This study deals with the aged MSW collected from three dumpsites of India located at Delhi, Hyderabad and Kadapa. The present study is aimed at assessing the release of dark colored leachate before using the fine fraction of mined material as earthfill. The aspect of release of dark colored leachate from mined soil-like material has not been reported in the literature. However, if the material is to be used in filling low-lying areas or in embankments or subgrade, the leaching of colored liquid can cause coloration of the surrounding water bodies and ground water. It may have aesthetic issues.

2 Materials and Methods

2.1 Site Description and Landfill Mining Details

The material used in this study was collected from the old dumpsites located in Delhi, Hyderabad and Kadapa. All these sites were developed before current regulations (MoEF 2000, 2016) were in place. Table 1 presents the general information about all the three dump sites.

The samples were collected and segregated between December 2016 and February 2017 from all the three landfills. Samples were taken from a depth of 4–5 m from Delhi and Hyderabad dump sites as shown in Fig. 1. However, at Kadapa dumpsite, capping was underway. So, samples were collected from stockpiles of mined and segregated waste. From Delhi landfill, samples were collected from three locations, and from Hyderabad landfills, samples were collected from two locations. All the

Table 1 General information about all the landfills

Landfill	Started in	Area (acre)	Height (m)	Quantity of waste (MT)
Delhi landfill	1994	40	65–70	6 million
Hyderabad landfill	1999	350	N.A	N.A
Kadapa landfill	1965	10	10–12	0.2 million

Fig. 1 Sample excavation from Delhi landfill

collected samples were screened at the site and then sealed in separated pre-cleaned polythene bags and stored in cooling cabinets.

Figure 1 shows the excavation of aged waste from Delhi landfill. After excavation, waste is further processed through the set of sieves available at the site. Similar process was followed for Hyderabad landfill. The screening of waste performed at the Delhi landfill site is shown in Fig. 2. The sieves used in the present study are 80, 50, 20 and 4.75 mm. From on-site grain size distribution, it was observed that 40–50% fall below 4.75 mm.

2.2 *Laboratory Characterization for Color Study*

The present study is focused on the mined material finer than 4.75 mm. The methodology adopted to analyze the color of water leached out from soil-like material is as follows:

- 10 g of oven dried material was diluted in 100 ml of water.
- It was shaken for 30 min at 200 rpm in a mechanical rotary shaker.

Fig. 2 Screening of waste at Delhi landfill

- Then, it was allowed to settle down for 24 h.
- After 24 h, the solution was filtered through Whatman filter paper no. 42.
- Filtrate was further centrifuged for 8 min @ 8000 rpm.
- The intensity of color of the filtrate was measured by Lovibond Tintometer.

3 Results and Discussions

3.1 Release of Dark Colored Leachate

To measure the intensity of colored leachate from aged MSW, the SLM was first heated at 60 °C till constant weight, and then, the test was performed as per the procedure mentioned in Sect. 2.2. The intensity of color on Hazen scale was found as shown in Table 2.

Figure 3 presents the color of leachate released by the SLM from aged MSW. It can be observed that the color released by SLM in comparison with the color released by local soil is dark brown. The color of the water leached out from local soil appears to be clean and transparent, whereas the color released by reclaimed material is darker and objectionable. The intensity of the leached color to some extent can be correlated with the organic content of the SLM.

The organic content was analyzed by loss on ignition at 550 °C and is presented in Table 3.

From Table 3, it can be observed that the intensity of color depends upon the organic content to some extent. The organic content in local soil was found as least

Table 2 Intensity of color released from SLM collected from all the landfills

S. No.	Name of landfill	Intensity of color (Hazen scale)
1	Delhi landfill (location 1)	375 ± 7
2	Delhi landfill (location 2)	712 ± 7
3	Delhi landfill (location 3)	288 ± 10
4	Hyderabad landfill (location 1)	925 ± 25
5	Hyderabad landfill (location 2)	740 ± 30
6	Kadapa landfill	475 ± 10
7	Local soil (Delhi silt)	25 ± 3
8	Ultra-pure water	2 ± 1

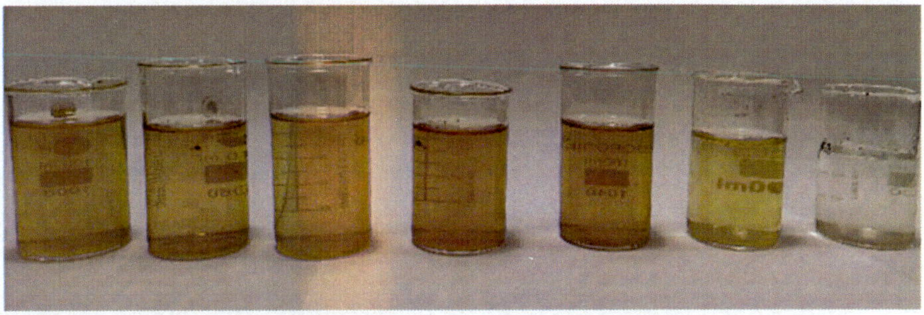

Fig. 3 Color of leachate released by SLM (L to R: Delhi location 1, Delhi location 2, Delhi location 3, Hyderabad location 1, Hyderabad location 2, Kadapa, Delhi silt)

Table 3 Organic content

S. No.	Name of landfill	Organic content (%)
1	Delhi landfill (location 1)	8.80 ± 0.4
2	Delhi landfill (location 2)	4.07 ± 0.08
3	Delhi landfill (location 3)	7.30 ± 0.07
4	Hyderabad landfill (location 1)	15.26 ± 0.28
5	Hyderabad landfill (location 2)	8.52 ± 0.23
6	Kadapa landfill	6.55 ± 0.16
7	Local soil (Delhi silt)	1.12 ± 0.03

with respect to the soil derived from aged MSW, and hence, the intensity of color was also found as very low, i.e., 25 Hazen, whereas the intensity of color was found as 700–900 Hazen in SLM from Hyderabad landfill which is having high organic content of about 15%. However, in SLM, from location 2 of Delhi landfill, organic content was found less in comparison with SLM from other landfills, but still, intensity of

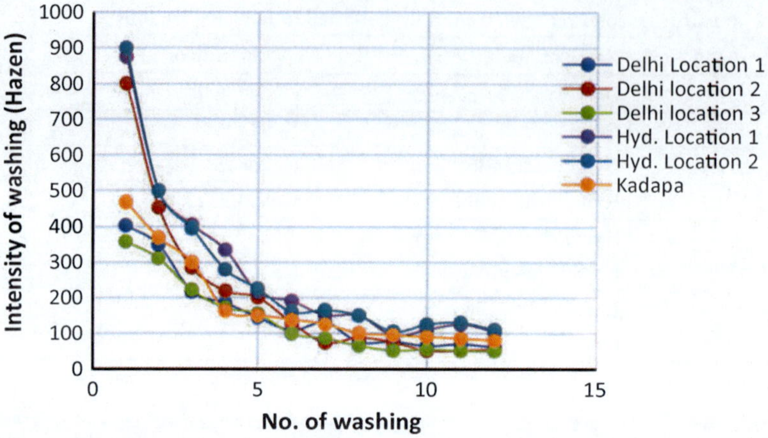

Fig. 4 Change in the color of water after repeated wash

color was found higher. Therefore, it can also be inferred that the release of color also depends on the type of organic content.

3.2 Influence of Infiltrating Pore Water on the Intensity of Color

To analyze the influence of infiltrating pore water on the intensity of release of color, the SLM was washed several times with distilled water, and the change in the color of the leached water is observed as shown after every washing in Fig. 4. The operation of washing was performed for 12 times. During the first washing, the intensity of color ranged as 350–500 in case of Delhi landfill, while after 5th and 6th wash, it reduces to 150 Hazen there after slowly reduces and becomes constant.

From Fig. 3, it can be observed that large quantity of water is required to bring down the intensity of color within the lower limit. Table 4 shows the amount of water required to wash 1 g of soil-like material to reduce the intensity of color of leached water.

3.3 Influence of Thermal Treatment on the Intensity of Color

To measure the influence of heating on the intensity of colored leachate released from aged MSW, the SLM was heated to different temperatures for 2 h duration starting from 60 °C at regular interval till the leachate become colorless. It was found that after heating SLM to 500 °C, the leachate becomes colorless. Release of

Table 4 Water required to wash 1 g SLM to reduce intensity of color

Name of landfill	Quantity of water required to reduce color of water (ml)	
	<100 Hazen	<50 Hazen
Delhi landfill (location 1)	72	115
Delhi landfill (location 2)	70	120
Delhi landfill (location 3)	58	105
Hyderabad landfill (location 1)	105	Constant after 10th wash
Hyderabad landfill (location 2)	98	
Kadapa landfill	55	85
Local soil (Delhi silt)	Already found as 25 Hazen	

colored leachate was also observed without heating SLM and also after air drying at room temperature for 3 days. The intensity of color in the leachate released by SLM without heating/air drying was observed less in comparison with the leachate released by SLM with heating/air drying as shown in Table 5. The intensity of color on Hazen scale was found as shown in Fig. 5.

Figure 5 shows the intensity of color in the SLM collected from all the landfills, and it can be seen that after heating the material at 500 °C, if water is percolated through the SLM, then the color is negligible. It was found that after the initial round

Table 5 Comparison of intensity of color with and without heating

Particular	Intensity of color (Hazen unit)		
	Delhi location 1	Delhi location 2	Delhi location 3
As received	**120**	**307**	**105**
Air drying	223	502	202
After heating at 60 °C	227	753	202

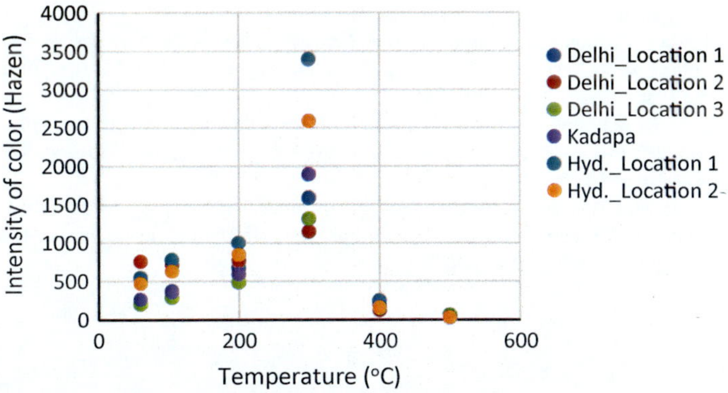

Fig. 5 Variation of color after thermal treatment

of heating, the intensity of color increases and peaks around 300 °C; the release of color was maximum at this temperature.

4 Conclusions

The present study assessed the concerns associated with the release of dark colored leachate from soil-like material mined from dumps which has not been reported in the literature so far. Results reveal that small amount of soil can cause large amount of pore fluid (70–100 times) throughput to become colored, thereby having significant impact especially on ground water wells. Organic content present in SLM was found to be the main cause giving rise to the color. Effectiveness/reduction of color by washing was not observed to be a very effective method because even after 10 washings (10:1 liquid: solids), some residual color remained. Heating of material at 500 °C was found to be more effective in elimination of color. It reduced the color from 380–1000 PCU to 25–72 PCU.

References

MoEF (2000). Municipal solid waste (Handling and Management) Rules, 2000. Government of India. Available at: https://www.moef.nic.in/legis/hsm/mswmhr.html. Accessed 15 Dec 2017

MoEF (2016) Municipal solid waste management rules, 2016. Government of India. Available https://www.moef.nic.in/sites/default/files/SWM%202016_0.pdf. Accessed 15 Dec 2017

Oettle NK, Matasovic N, Kavazanjian E, Rad N, Conkle C (2010) Characterization and placement of municipal solid waste as engineered fill. Global Waste Management Symposium, San Antonio

Parrodi JC, Hollen D, Pomberger R (2017) Characterization of fine fractions from landfill mining: a review of previous landfill mining investigations. In: Proceeding of sixteenth international waste management and landfill symposium, 2–6 Oct, Sardinia, Italy

Somani M, Datta M, Sreekrishnan TR, Ramana GV (2017) Critical aspects relating to re-use of aged municipal solid waste for geotechnical purposes. In: Indian geotechnical conference GeoNEst. IIT Guwahati, India, p 108

Somani M, Datta M, Ramana GV, Sreekrishnan TR (2018) Investigations on fine fraction of aged municipal solid waste recovered through landfill mining: case study of three dumpsites from India. Waste Manage Res 36(8):744–755

Somani M, Datta M, Ramana GV, Sreekrishnan TR (2019) Leachate characteristics of aged soil-like material from MSW dumps: sustainability of landfill mining. ASCE J Hazardous Toxic Radioact Waste 23(4):04019014

Somani M, Datta M, Gupta SK, Sreekrishnan TR, Ramana GV (2019) Comprehensive assessment of the leachate quality and its pollution potential from six municipal waste dumpsites of India. Bioresour Technol Rep 6(1):198–206

Erosion Hotspots and the Drivers of Erosion Along the Part of West Bengal Coast, India

Anindita Nath, Bappaditya Koley, Subhajit Saraswati, Kaushik Bandyopadhyay, and Bidhan Chandra Ray

Abstract Recent years, the rate of coastal erosion is considerably increased in India due to human interference and natural drivers. The coastline of the Western part of West Bengal is severely eroded. Months from June to October, as the monsoon wind affects the direction and magnitude of the waves affecting the coastline with intense erosion hotspots making zone vulnerable for the coastal community. Transportation gets open during monsoon induces erosion. Frontal beaches seaward of seawalls dissolves bringing wave breakers closer to seawalls. Abrasion at the base of seawalls accelerates slumping. In the monsoon season, overtopping of the landward of the seawall also causes high waves breaking very near to the seawalls. Many places in alongshore have been observed seawalls abruptly ended which is indicate the 'end erosion hotspot'. The paper tries to access the drivers and processes conducting to erosion hotspots.

Keywords Abrasion · Erosion hotspot · Seawall · Vulnerable

A. Nath (✉) · S. Saraswati · K. Bandyopadhyay
Department of Construction Engineering, Jadavpur University, Kolkata 700106, India
e-mail: aninditan286@gmail.com

S. Saraswati
e-mail: subhajitsaraswati@gmail.com

K. Bandyopadhyay
e-mail: drkaushik99@gmail.com

B. Koley
School of Oceanographic Studies, Jadavpur University, Kolkata 700032, India
e-mail: bappadityakoley2012@gmail.com

B. C. Ray
Department of Chemistry, Jadavpur University, Kolkata 700032, India
e-mail: drbidhanray@gmail.com

M. Latha Gali and R. R. P. (eds.), *Problematic Soils and Geoenvironmental Concerns*, Lecture Notes in Civil Engineering 88,
https://doi.org/10.1007/978-981-15-6237-2_16

1 Introduction

Coastal erosion is a massive threat faced by all countries having a coastal zone. Coastal erosion is a cause due to various natural and anthropogenic factors. Coastal erosion along Digha–Mandarmani coastal stretch which is south-western part of West Bengal, India, has been studied in recent investigation based on direct field investigation and structural interpretations. The present investigation takes a look into distinctly several coastal segments with varied morphological signature and different types of human induce interventions. The efforts to protect coastal beaches from massive coastal erosion through different construction were initiated along coastal region of West Bengal. Various structures like seawall, geo tube, damp, etc., were constructed from more than 50 years ago to protect the coast due to coastal erosion. The process of coastal protection measures has been implemented along the shoreline successfully in some areas. On the other hand, the protection method has collapsed which increases as the rate of erosion and erosion hotspots. End erosion and scouring have been reported, mainly in areas adjoining seawalls (Thomas et al. 2013). The erosion hotspots are activated day to day by constructing some unplanned coastal protection. From intense field investigation, erosion hotspots are found in south-western part of West Bengal. The concentration of present study is to elucidate all the etiological and helpful factors for coastal erosion along south-west part of West Bengal.

2 Selected Study Region

The 534.27 km^2 coastlines of south-west part of West Bengal from Udaipur to Kanai Chatta area, are selected for the present study (Fig. 1). The study area is associated with coastal erosion, alongshore breakwater, overtopping of seawall, etc., which are the factors of increasing coastal vulnerability and erosion hotspots (Table 1). The study area experiences strong longshore current from south-west to north-east direction during monsoon season and weak long shore current from north-east to south-west direction during the winter season; the reason behind the seasonal differences is wind direction (Jana and Bhattacharya 2012). The studied coastline is dominated by high-energy macro-tidal environment (Tidal range 5.5–5.8 m in GPS, RRI). The chosen studied coastal beaches are flat with low gradient and slightly concave upward undulating and composed of medium to fine-grained sand.

3 Material and Method

Geomorphological set-up of every erosion hotspots is also evaluated through direct field observations. Effect of breakwaters and seawall on the coast were observed during field survey. Important information (like rainfall data, tidal data and wind

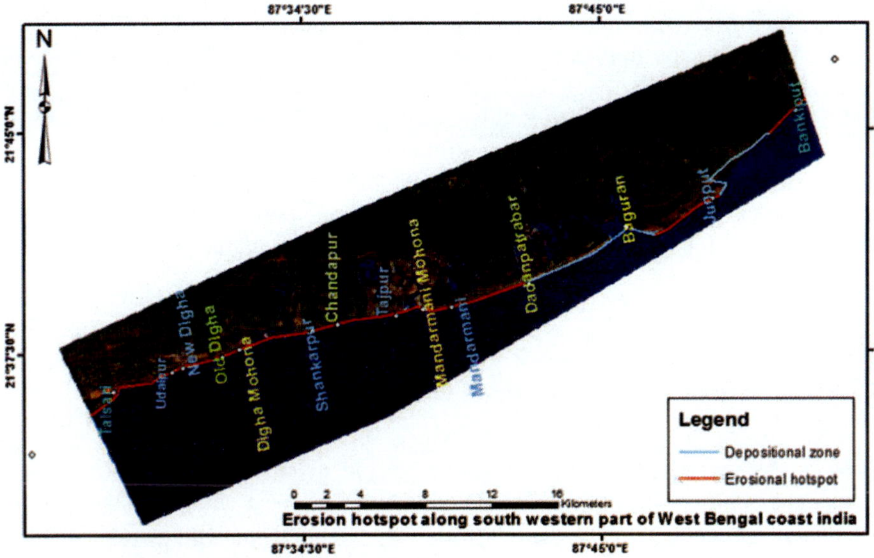

Fig. 1 Erosion hotspot along West Bengal coast, India

Table 1 Erosion with different artificial and coastal morphology

Location	Affected area	Remarks
Udaipur	Erosion and nearshore break water	Heavy inlet erosion towards south
Jatranala	Sever erosion	Seawall overtopping, sand dunes and vegetation washed away
Digha Mohona	Active erosion	Sand bund constructed massive erosion near bund wall
Shankarpur	Under active erosion	Slumping seawall, under water inundation, collapsing shore preventing method seawall end erosion
Chandpur	Successive erosion	Overtopping seawall, sand dunes properly washed out, water logged into village
Tajpur	Active erosion	Sea beach erosion, vegetation washed out, seawall end erosion
Mandarmani	Severe erosion towards Balusai Canal and sapua canal Mohona area	Active beach erosion no seawall observed
Mandarmani Mohona	Massive erosion	No sand bund, eroded maximum area
Dadanpatrabar	Erosion	No seawall observed
Junput	Severe erosion	No seawall, water inundation

Source Field observation

velocity direction data) on the nature and characteristics of coastal erosion has been obtained through site in charge of River Research Institute (RRI) office at Digha. Local information also collected from interaction with local communities who shared traditional knowledge. The entire study area was visited, and detailed sketch has been prepared to understand the role of natural and man-made structures for protecting coastal erosion. The study also investigates the present and earlier shoreline position and impact of different structure near shoreline through local and previous information. Different signatures prove the changes of shoreline and the changing scenarios of coastal erosion and sedimentation and these were recorded.

4 Outcome of the Field Observation

Field investigation has given a hint that erosion occurred at down drift side of breakwaters and seawall. Seawall end points are other locations of erosion. Erosion protection measures unplanned design or inappropriate implementations are also locations of erosion. Vital erosion spots along Digha–Mandarmani coast are Digha, Shankarpur, Tajpur, Digha Mohana, Mandarmani Mohana, Udaipur, Talsari, Dadanpatrabar and Junput. In selected study area, inlet areas are more vulnerable than other coastal area. These inlet areas (Talsari Mohana, Digha Mohana, Mandarmani Mohana) are under active erosion hotspots.

5 Significant Drivers of Erosion

In coastal area, many features are helpful to induce erosion hotspots. The study reveals that south-west part of West Bengal is mainly eroded by some physical phenomena like rainfall, high velocity wave, lack of sand transportation and artificial phenomena such as seawall, growth of urbanization and hotel construction near shoreline.

6 Discussions

The sea beach at Digha has been eroding for a quite long time. A seawall was constructed over a length of 3464 m in stages (1973–1982) to protect the township of Digha. The seawall extends between Jatranulla on the west to Seahawk hotel on the east. But over the last few years, erosion of dune started beyond the Eastern end of the existing sea wall. Swash and backwash of water level are unavoidable system with high velocity in high tidal nature (Adegoke 2014). If the protection structure is constructed near shore, then the slope angle of structure is very steep towards the sea. High velocity wave can create a maximum high pressure upon the structure like seawall, due to the continuous frictional activity, seawall started to collapse and make an erosion hotspot. To avoid the slope failure and reduce erosion hotspots, brick block or pitching on the slope as it is providing a smoother or easier surface for the waves which travel more heights than they do on earthen surface beside, these provide same frictional resistance which could be balanced by outward dripping pressure. Coastal morphology also plays a vital role to increase coastal erosion hotspots.

6.1 Induced Erosion and Shore Protecting Structure

In Digha–Mandarmani coastal stretch, shore protection methods were applied in many times. In 1975, a local seawall (red bolder, sand and cement) was constructed at Jatranulla to Seahawk that had collapsed due to high tide wave (6.385 m in GPS) which was working well from 1975 to 1996. After the downfall of that seawall, this was constructed again with red bolder and new concrete materials. The boulder had no sufficient resistance against frontal wave energy. The existing seawall was designed as a rubble mound wall having a crest level of 18 ft (5.49 m) GTS and slope of 6:1 for upper portion and 12:1 for lower part. Small boulder of laterite was used as armour stones each weighing 70–80 kg which could not resist the wall. The seawall was designed using some parameters:

Significant wave height (Hs)
Wave period (Ts)
Maximum wave height
Maximum high water level MHWL

Here, max wave height was considered 1.6 m or 5.25 ft (data collected from RRI).

The entire area between Digha Mohana and hotel Seahawk was washed out and overtopping of the seawall (1998), during this time water logged into Shankarpur, Chandpur, Tajpur village area (high tide line 6.055 m in GPS). Due to high tidal wave, geo tube was also washed out (5.510 m in GPS). In 1975–2001 was the time under successive coastal erosion and formation of active erosion hotspots. A guard wall was constructed without a proper plan in Shankarpur bus stand to Chandpur with clay material.

Laterite boulders were moved up and rolled by wave action to upper level on the slopes. Due to abrasion from wave action, the boulder size was reduced. Because of the steepening of slope, the run up would be more and this might be a cause for overtopping. The seawall is under deterioration (2018) (Plate 1). In some places of Digha Shankar, seawall observed some deficiency of proper filtration method. Graded filter is not working properly. The motion of the water pushes and pulls the rolling particles and carries salt and suspended particles to crest and trough. In Digha Shankarpur, beach material consists of very sand, silt and clay with low permeability. Wave water percolated between the block fault areas due to high velocity wave. During return flow of water after overtopping, the sand beneath the wall was likely to be washed out seaward causing loss of support of wall (Plate 2). This internal erosion occurred because of inadequate functioning of the filter layer beneath the wall. As a result, it accelerates the gap between the laterite boulders. Underwater inundation is a major problem of Digha, Shankarpur and Mandarmani coastal area (Plate 3).

On the other hand, a seawall is constructed in Jatranulla along Orissa border. In this area, overtopping of seawall is observed in high tide wave (Plate 4). High tide wave can easily overtop the seawall because crest of seawall is not higher than high tide wave in some areas. The distance between toe and crest of seawall is smaller

Plate 1 Collapsing seawall at Shankarpur

Plate 2 No repair work of seawall

Plate 3 Water inundation and under cutting

Plate 4 Wave overtopping of seawall

according to wave energy so wave can overtop seawall easily. To avoid this situation, seawall crest level is to be made higher calculating the maximum wave height by using the Eq. 1 by Stevenson and Molitor:

$$H = 0.17\sqrt{U \times F} + 2.5 - \sqrt[4]{F} \quad (1)$$

where U—wind speed in hour, F—fetch in miles, H—wave height in feet.

From Shankarpur to Tajpur beach, seawall has abruptly ended (Plate 5) and 'end erosion hotspots' were noticed. Wave overtopping and severe wave attack are occurring in Digha and New Digha coast. Mandarmani Mohana area is under successive erosion hotspots. Being an open spot in a seawall stretch, fishing gaps at Digha Mohana and Digha experience a pressure gradient towards the gap which causes pushing of wave/swash into the gap accelerating erosion along Tajpur Shankarpur Mandarmani is due to the down drift effect of breakwater. Coastal road, houses and cultivated land are highly affected along these areas (Plate 6). In south west

Plate 5 Abruptly end of seawall at Shankarpur

Plate 6 Water over flow at land

part of West Bengal coast, the shore protection structures are observed in many places (Noujas and Thomas 2015). Frontal beaches seaward of seawall has disappeared bringing wave breakers closer to seawalls. Abrasion at the base of seawall increase slumping. In some places seawalls have to abruptly end of alongshore which cause 'End erosion hotspots' (Plate 7). Few problem of seawall design, seawall was collapsing in some areas and the certain 'end' of the seawall near Shankarpur was the reason of increasing erosion extremely (Plate 8). Major erosion and hotspots are observed at inlet area of Mandarmani and Digha Mohana. Casuarina vegetation is eroded and washed away by high tidal wave (Plate 9), and the result is improper shape of seawalls which itself weakens the protection measure making them highly vulnerable to erosion (Plates 7 and 8).

Plate 7 End erosion hotspots

Plate 8 Failure of seawall at Shankarpur

Plate 9 Vegetation washed away by wave

6.2 Seasonal Erosion Due to Monsoon Rainfall

Monsoon current is the major responsible force increasing seasonal erosion. Seasonal erosion is observed throughout the sandy coasts along south-western part of West Bengal coast during south-west monsoon. During the month of August and September, one foot of sand deposit has been removing in single tide. Such scour, when too much, may dislodge the apron block sand, then toe wall and subsequently the revetment supported by it may be damaged. The gaps between the concrete or brick blocks in the upper berm were considered as weak point in so much as it helped an easy access for the wave suddenly coming in contact with horizontal berm, became unstable and collapsed on the upper berms a consequent of large quantity of water passing under the revetment (https://shodhganga.inflibnet.ac.in/bitstream/10603/162873/9/09_chapter%201.pdf). But in winter, there is a steady accretion of sand. Depending on monsoon wind direction and speed, the wave's heights are usually higher and drastic in nature (Table 2). In monsoon month mainly in August and September when storm combines with spring tide, it reaches a height of 6 ft above the wall crest. Especially on this time, the weight of armour stone needs more weight than present size to maintain the pressure of existing seawall from wave abrasion.

Table 2 Seasonal measurement of wave parameter in Digha (based on RRI, West Bengal)

Station	Season	Wave velocity (m/s)	Wave length (m)	Wave height (m)	Wave steepness	Wave energy	Breaker type
Digha	January–February	5–6/s	15–22	0.7–0.9	0.05–0.04	Medium	Plunging
Digha	August–September	6–8/s	30–40	0.8–1.5	0.03–0.04	High	Plunging

Source River Research Institute (Mohanpur, West Bengal)

7 Conclusion

Erosion hotspots along south-west part of West Bengal of India are identified from intense field survey. It is observed that the erosion hotspots are mostly dependent on coastal morphology and different coastal structures. This part of coastal area, seawall is a main coastal protecting structure. For the design of this structure, it is important to collect the proper data regarding wave height, wave period, direction, angle and mostly the material on which these structures are to be constructed. Selection of proper area to construct seawall and study of the beach profile and water level are also important aspects. In this coastal area, seawall has been noticed with improper and unplanned methodology. Other important criteria to construct seawall with toe protection method prevent waves from scouring and undercutting the beach. In studied area, some beaches may also be introduced toe failure, and this failure progresses throughout the entire structure (Plate 10). Wave overtopping is a major reason to increase erosion because high tide line is not properly estimated to construct a seawall in West Bengal, India. In the present investigation, seawalls on both sides are abruptly terminated without proper placement and severe wave attacks are observed (Plate 5). The area in the lee of the structure would experience maximum inundation when the water flowing back to the beach would erode the maximum in situ soil (Plate 11) which maintain the stability of seawall on both sides of the opening (Central Marine Fisheries Research Institute 1984). Rubble mound seawalls are the commonly used measure for preventing coastal erosion in India (Plate 12). Shankarpur beach experiences the 'end erosion hotspots' which is the most vulnerable area. Location selection for seawall is a major problem in some areas. If Digha shore protecting method would be constructed according to Coastal Regulation Zone (CRZ) notification and estimated the water level properly, then overtopping problem becomes reduced. Tidal inlets and adjoining areas are also location of coastal erosion. Many times, the lee side slope and crest are gradually

Plate 10 Failure of toe protection method

Plate 11 In situ soil eroded

Plate 12 Rubble mound seawall

damaged due to continual overtopping. It is possible to prevent these damages and upliftment of the performance of structure due to regular interval maintenance (Plate 2). Maintenance of damaged seawall is not properly planned based on appropriate designs since most of their works are executed on piece meal basis as an emergency measure.

Acknowledgements We would like to thank Amiya Bera, Deputy Director (Structure), Irrigation and Waterways Department, Government of West Bengal, for sharing his valuable Ideas. We are conveying our sincere thanks to River Research Institute (especially Dr. Bibhas Chandra Barman) for data collection.

References

Adegoke PB (2014) An empirical study of flood wave impact pressures to determine the effeteness of new seawall designs using a dam-break approach, pp 110–180

A report Coast of West Bengal: an introduction, Chapter 1. https://shodhganga.inflibnet.ac.in/bitstream/10603/162873/9/09_chapter%201.pdf

Central Marine Fisheries Research Institute (1984) Mudbanks of Kerala coast. Bulletin 31:74

Jana A, Bhattacharya AK (2012) Assessment of coastal erosion vulnerability around Midnapur–Balasore coast, Eastern India using integrated remote sensing and GIS techniques. J Indian Soc Remote Sens. https://doi.org/10.1007/s12524-012-0251-2

Noujas V, Thomas KV (2015) Erosion hotspots along Southwest Coast of India. Aquat Procedia 4:548–555. https://doi.org/10.1016/j.aqpro.2015.02.071

Thomas KV, Kurian NP, Shahul Hameed TS, Sheela Nair, Reji Srinivas L (2013) Shoreline management plan for selected location along Kerala coast. Report submitted to ICMAM Project Directorate, MoES. Centre for Earth Science Studies, Thiruvananthapuram, pp 308

Use of Electrical Resistivity Tomography in Predicting Groundwater Contamination Due to Non-engineered Landfill

Debaprakash Parida, Arindam Saha, and Ashim Kanti Dey

Abstract One of the major causes of groundwater contamination in urban areas is the flow of leachate from non-engineered landfills. Use of conventional test wells to predict the contamination is a cumbersome process. In this pilot study, a small-scale landfill having dimensions 4 m × 2.4 m × 0.6 m height was prepared to determine the efficiency of electrical resistivity tomography (ERT) in predicting groundwater contamination due to uncontrolled landfill. ERT survey was performed for five months, and the variations in groundwater contamination were observed. Water samples from the site were analysed for contaminant concentrations. The resistivity of the soil was found to be decreased considerably from around 200 to 20 Ω-m. This low resistivity zone ensured the presence of leachate flow followed by contamination of groundwater. The increased depth of low resistive zone below landfill evidenced the extended depth of groundwater contamination. Charts were prepared by correlating the contaminant concentrations in groundwater with the resistivity values. This chart may help in predicting the groundwater contamination and contaminant concentration using ERT, without making any borehole at a site. The study has evidenced the potential of ERT in predicting groundwater contamination.

Keywords Electrical resistivity tomography (ERT) · Leachate · Groundwater contamination · Non-engineered landfill

1 Introduction

Contamination of groundwater due to the leachate flow is a serious concern regarding the preservation of fresh drinkable groundwater in today's scenario. The presence of abandoned or existing non-engineered landfills poses threat to the fresh groundwater below the same land. Landfills are the facilities for the containment of wastes to safeguard the surrounding environment. Poorly maintained or improperly designed landfills impose adverse environmental effects on its immediate environment (Biswas

D. Parida (✉) · A. Saha · A. K. Dey
National Institute of Technology Silchar, Silchar, Assam, India
e-mail: debaprakash.dp@gmail.com

M. Latha Gali and R. R. P. (eds.), *Problematic Soils and Geoenvironmental Concerns*, Lecture Notes in Civil Engineering 88,
https://doi.org/10.1007/978-981-15-6237-2_17

et al. 2010). Many underground water sources are found to be severely affected by the uncontrolled dump sites. When rainwater percolates through the dump sites, it extracts toxic chemicals from the wastes. This toxic liquid that emanates from the dump sites is called leachate. Groundwater sources are more prone to contamination as the leachate percolates through the soil below the dump site. It is very difficult and time consuming to predict such contamination using only conventional methods. Moreover, conventional methods fail to provide full information on contamination of the whole site. Tracing of groundwater contamination is found to be more convenient with the use of geophysical techniques. Electrical resistivity tomography (ERT) method is more convenient than the other available geophysical methods to delineate the leachate plume emanating from the dump site (Lopes et al. 2012).

ERT is an advanced near-surface geophysical technique evolved to solve the complex field problems. This is a non-destructive method which does not need trenches, borehole (BH) and other destructive methods to serve the intended purpose. ERT survey needs a resistivity recording device and electrodes to determine the subsurface data including physical and geological characteristics (Abdul Nassir et al. 2000). Though this method is not very new, its extensive use in investigating the emerging environmental concerns needs to be explored to a greater extent. The depth of investigation can be increased by increasing the spacing between the electrodes along the resistivity profiles. A typical current flow pattern and potential distribution in a resistivity survey are shown in Fig. 1.

Geoelectrical resistivity method is based on the electrical property of the geomaterial to be surveyed. The knowledge of resistivity of various geomaterial is important in finding the composition of the earth material, moisture content, conductivity, etc. Resistivity imaging survey has played an important role in the field of applied geophysics for about last three decades. This technique was developed by Conrad Schlumberger in 1912, who performed the very first geological resistivity imaging experiment at Normandy. In 1915, Frank Wenner developed a similar idea in USA which offered a greater evolution to the field of applied geophysics in solving the difficult geological and geotechnical field problems. The theory of geoelectrical imaging is widely applied in the field of mining, hydrogeology, archaeological and engineering investigations. Knowledge of this technique has also proved to be very much effective in environmental and geotechnical investigation. Nowadays, geoelectrical imaging survey is mostly in use in various fields of civil engineering due to its interdisciplinary applications in subsurface investigation. Geophysicist

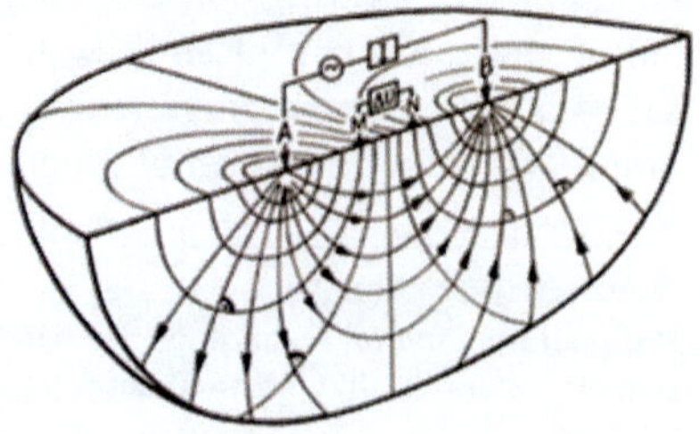

Fig. 1 Current flow pattern and potential distribution in subsurface

and geotechnical engineers have put their effort in implementing the geophysical technique as an easy, economic and cost-effective alternative to the conventional geotechnical methods being applied to the field of complex geotechnical, geological, environmental, geoenvironmental and hydro-geological problems.

The different types of array used during electrical resistivity tomography survey are Wenner array, dipole–dipole array, Wenner–Schlumberger array, pole–dipole array, pole–pole array. In comparison with Wenner array, the Wenner–Schlumberger array can better detect the horizontal subsurface resistivity variations. This array is suitable to characterize both horizontal and vertical structures as it is moderately sensitive to both types of structures. Wenner–Schlumberger array is the best choice between the dipole–dipole array and Wenner array where geological structures in both horizontal and vertical directions are expected in subsurface. In this research work, Wenner–Schlumberger array has been used as leachate will flow in both vertical and lateral directions.

This resistivity imaging method is not susceptible to the telluric noise which has a potential to affect the measured data observed at the site. This geophysical method is very much effective in observing both vertical and lateral variations in the subsoil properties. Electrical resistivity technique offers greater field productivity among other near-surface geophysical technique. The subsurface can be visualized in both 2D and 3D systems by the use of electrical resistivity imaging technique (Aizebeokhai 2010). Greater quality and reliability in subsurface resistivity data can be obtained through this technique. This technique measures the apparent resistivity data of the subsurface which is then processed, and inversion is carried out to get the true results. The observed resistivity of the subsurface material at the site is the apparent resistivity value. True resistivity values are obtained by the inversion of the apparent resistivity values using inversion software such as RES2DINV and RES3DINV in 2D and 3D resistivity survey, respectively.

This paper summarizes an extensive study on the effectiveness and applicability of the electrical resistivity technique in detecting the flow of contaminant at a contaminated site caused by a landfill by performing in situ investigations and laboratory experiments of collected groundwater from the observatory boreholes. The objective of this study is to correlate the resistivity values with contaminated groundwater parameters below a non-engineered landfill site.

2 Site Selection and Characterization

A municipal solid waste (MSW) dump was prepared within the NIT Silchar campus at a location 24.75 °N latitude and 92.79 °E longitude. From the meteorological data, it is observed that the average precipitation per annum is about 300 cm at the study area. The site was so chosen that the water table at that site is at a shallow depth from the ground surface. Flow of leachate from the modelled dump to the underlying groundwater was observed within a shorter time span as the groundwater table was at a shallow depth from the ground surface. For the selection of the site, resistivity survey was carried out at three possible places in the campus after a preliminary investigation. The observed resistivity profiles at the sites, before selecting the final one, are presented in Figs. 2, 3 and 4. Among the possible sites available, the site 3 corresponding to Fig. 4 was finally selected as per the requirement of the experiment. There was no evidence of groundwater table at sites 1 and 2 for which these two were rejected for the proposed experimental work. Moreover, these two sites were filled soil where large amount of gravel and broken brick pieces were present. The selected site was at the foot of a stepped slope near a lake as shown in Fig. 5. The hydro-geological features at the site are presented in Fig. 6. The gravimetric flow of groundwater was along the slope of the natural ground surface towards the lake which was evidenced from the piezometric analysis of depth of groundwater in the observation wells at the site.

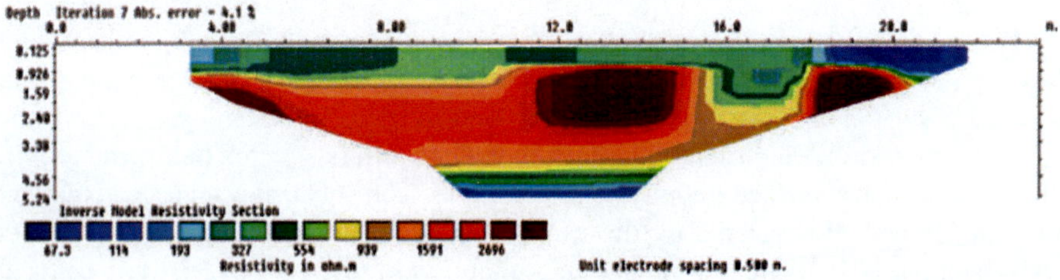

Fig. 2 Resistivity profile at site 1

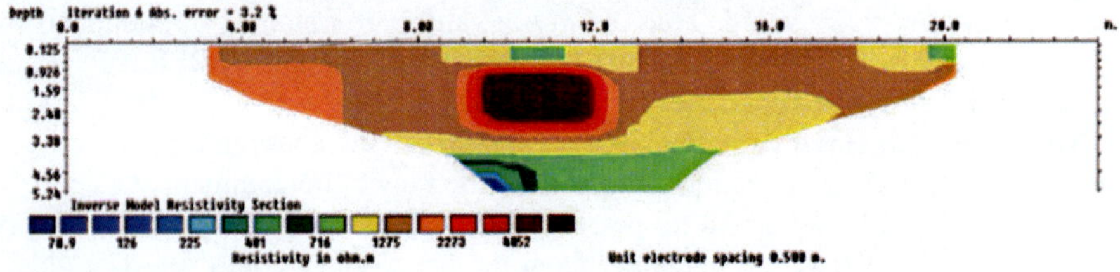

Fig. 3 Resistivity profile at site 2

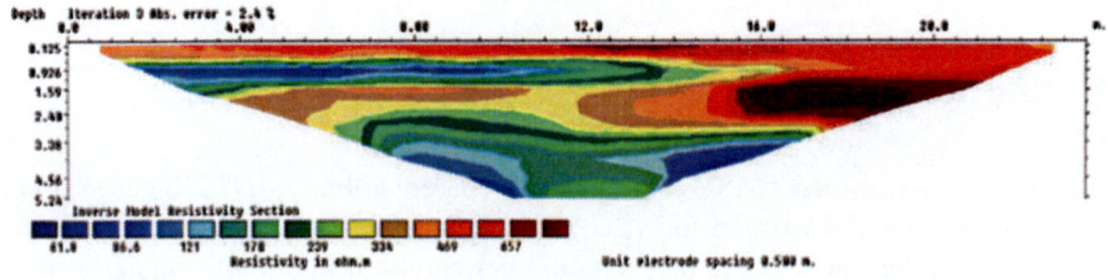

Fig. 4 Resistivity profile at site 3

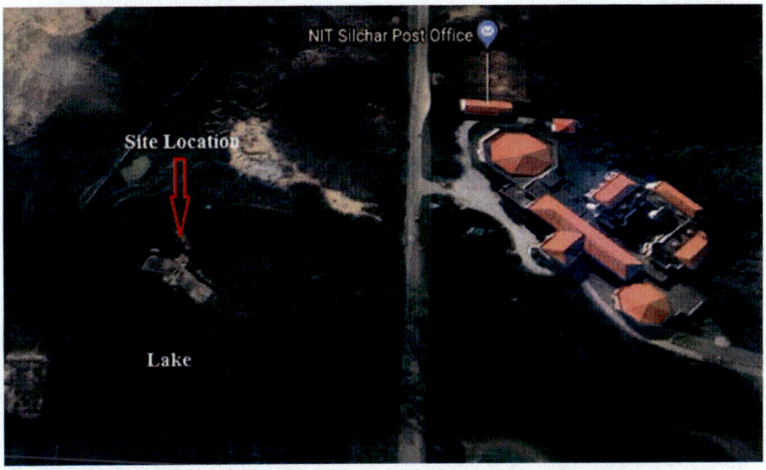

Fig. 5 Satellite image of the selected site

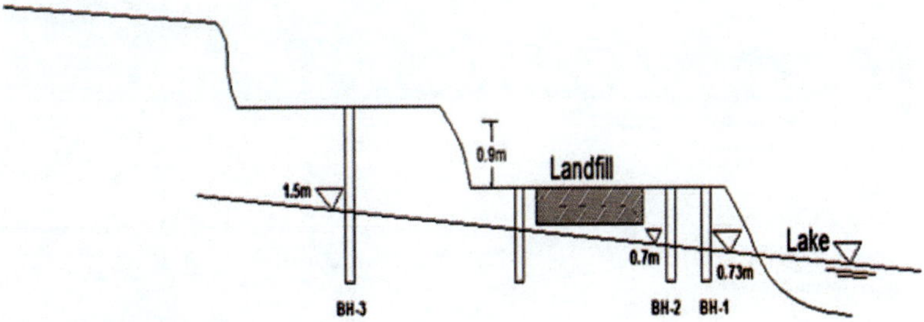

Fig. 6 Hydro-geological features at selected site

3 Materials and Methods

3.1 Landfill Modelling and Waste Characterization

The MSW dump was prepared in November 2017. The depth of groundwater table was at 0.7 m from the ground surface. It was observed that silty clay soil was present at the site having coefficient of hydraulic conductivity of 3.174×10^{-6} cm/sec.

A trench was prepared with size 4 m × 2.4 m × 0.6 m. The trench was made nearer to a lake. The prepared trench and the modelled MSW dump are shown in Figs. 7 and 8, respectively.

The trench was filled with discarded paper, food waste, cow dung, disposable plastics which were used to replicate the original MSW dumps. Cow dung was used as a waste material as it contains some minerals which cause groundwater contamination. Food waste attributes 40% of the total municipal solid waste in a

Fig. 7 Trench prepared to accommodate the wastes

Fig. 8 **Model landfill at the site**

landfill for which these were considered as the governing reasons for groundwater contamination at a landfill site. The model landfill was kept open to allow the rainfall and run-off to pass through wastes in the landfill followed by production of leachate from the model landfill. In the absence of rainfall, the landfill was watered on regular basis so as to produce leachate out of the landfill. It was observed that the top surface of the filled waste has been settled for an average depth of 16 cm which is due to the result of decomposition of organic waste in the dump.

3.2 Data Acquisition and Processing

Two ERT profiles on the downstream side, one on the upstream side and one on the landfill itself, were marked to conduct the resistivity survey along these profiles throughout the research period. The two downstream profiles were at a perpendicular distance of 0.5 and 1.5 m from the landfill, respectively, for profiles 2 and 3. A total of 48 electrodes were used at a spacing of 0.5 m resulting in a 23.5-m-long profile. Two- and three-dimensional resistivity tests were performed at every month to predict the groundwater contamination and visualize the flow of leachate from the model landfill. Positions of the ERT profiles near the model landfill are shown in Fig. 9.

Through the arrangement, it was intended to measure the lateral as well as vertical spreading of the leachate below the model landfill. Resistivity survey was conducted throughout the research period along the predefined profiles from November 2017 to March 2018 at a regular interval. Accuracy of the electrode positions on the predefined profiles was ensured by inserting bamboo sticks inside the holes when no experiment was carried. Square grid pattern was used to arrange the electrodes for three-dimensional resistivity investigation surrounding the landfill. A spacing of 1.2 m was kept along X direction and that of along Y direction was kept 1.5 m. Syscal Junior-II was used as a multichannel resistivity measuring for the acquisition of data at the location. For optimization of input voltage, the instrument was equipped with automatic adjustment of output current value to result in a best measurement quality.

A minimum of 3 stackings up to a maximum of 6 stackings were set with 99.5% confidence limit to get improved data quality during the experiments. For data acquisition, save energy mode was enabled with 800Vpp as maximum usable voltage. For the experiment, a total of three types of array configurations were used, e.g. Wenner array, dipole–dipole array and Wenner–Schlumberger array.

Fig. 9 Model landfill at the site

The collected water samples were analysed for physicochemical parameters in laboratory. Efforts were made to correlate the observed physicochemical parameters of the contaminated groundwater with the true resistivity values. It was intended to determine the quality of the contaminated groundwater below and around the landfill by conducting resistivity survey without collecting the contaminated samples from the site for laboratory analysis. Three observatory wells, two on downstream side and the other on upstream side, were dug to monitor the contamination of groundwater and for collection of samples. The downstream boreholes were at 0.4 and 1.4 m away from the landfill, whereas the upstream borehole was at 1.2 m away from the landfill. To ensure the collection of the contaminated groundwater samples from the boreholes, the depth of observation wells was made 1.2 m from the ground surface. Resistivity imaging survey was carried out once in every month along predefined positions, and water samples were collected from the boreholes each time after the resistivity survey. As the natural slope of groundwater was towards the downstream side of the dump resulting in flow of the contaminated water towards the profiles 1 and 2 at the downstream side of the dump, the governing resistivity profiles are presented and analysed in the result and discussion section. The water quality parameters measured in the laboratory include alkalinity, pH, turbidity, sulphate, chloride, nitrate, iron, hardness, dissolved oxygen.

4 Results and Discussion

4.1 Resistivity Survey

Two-dimensional resistivity profiles were taken along the landfill at three different places. All the 2D ERT profiles at the site evidenced the depth of groundwater table at a shallow depth which was the main reason in choosing this site for the intended research work. ERT profiles 1 and 2 were at the downstream side of the landfill which was intended to capture the lateral flow of the leachate from the landfill towards the nearby lake. From the survey, it was known that the flow of groundwater was towards the profile 1 which was near a lake at the study site. It was observed from the profiles 1, 2 and 3 that the subsurface of the site is composed of clay, silt and sand. Analysis of the resistivity profiles indicated that the topsoil up to about 0.6-m depth was dominated by saturated clayey soil mixed with some portion of silt and the lower depth was dominated by the sandy soil which was later confirmed from the laboratory analysis of the soil samples taken from different depths. Generally, clayey soils have resistivity value within 100 Ω-m. The resistivity values below 100 Ω-m in the profiles 1, 2 and 3 were inferred as the presence of saturated clayey soil at the site. From the profile 3, the high resistivity values near the surface and at the left of the profile were inferred as the presence of red colour sandy clay. The resistivity values of the saturated clay at profiles 1, 2 and 3 were measured before and after dumping the waste in the trench for a comparative analysis. It was also intended to observe the effect of leachate flow on the resistivity values those obtained in these profiles. Profile 3 witnessed the heterogeneity of the subsurface soil at the site.

A 3D ERT survey was also carried out around the proposed trench area to get a better idea on the stratification of the site along the depth. The presence of saturated clayey soil as well as the groundwater table was also observed in the 3D survey. This 3D survey was also intended to observe its efficiency in capturing the leachate flow after dumping the waste into the trench.

The two-dimensional resistivity profiles were taken at an interval of one month, and the water samples were also collected at the same time so as to facilitate the development of correlation chart between the resistivity values and the water quality parameters. The resistivity values from each of the profiles were interpreted and are discussed in this section. Resistivity profiles along the predefined locations on each month and the water quality parameters representing each of the resistivity profiles on respective months are also presented in this section. Later, the water quality parameters and the resistivity values for the corresponding profile for corresponding months are correlated and presented in the following sections.

Resistivity and Water Quality Data for Profile 1. Resistivity survey was carried out for a period of five months from November 2017 to March 2018, and water quality parameters were also determined for each of the investigated months for profile 1. The resistivity profiles for each of the investigated months are presented in Figs. 10, 11, 12, 13 and 14, and the test results for water quality parameters are shown in Table 1.

After one month of dumping and watering the waste in the landfill, resistivity profile 1 which is 1.5 m away from the landfill at downstream side of landfill showed reduced value of resistivity from about 80 to 70 Ω-m which was suspected as a result of leachate intrusion. As there was no significant decrease in resistivity value in the subsoil, therefore it was suspected as very negligible intrusion of the leachate

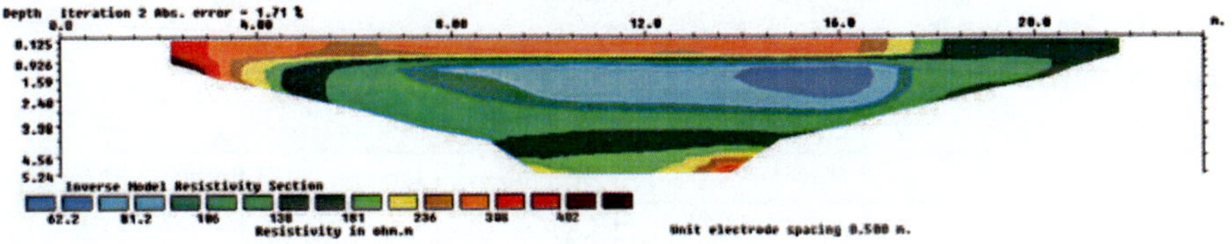

Fig. 10 Resistivity profile before filling the waste

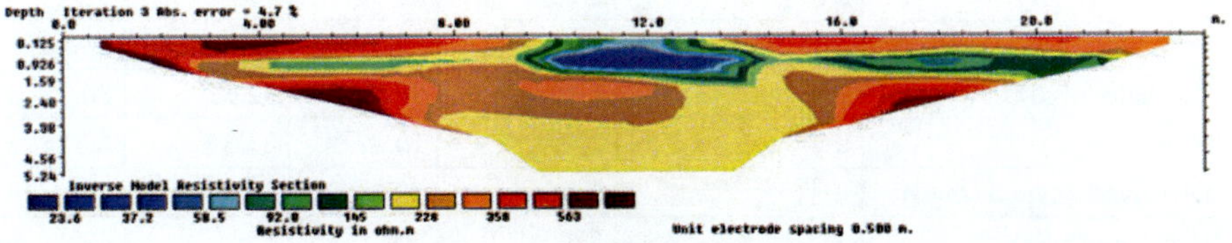

Fig. 11 Resistivity profile after one month of filling

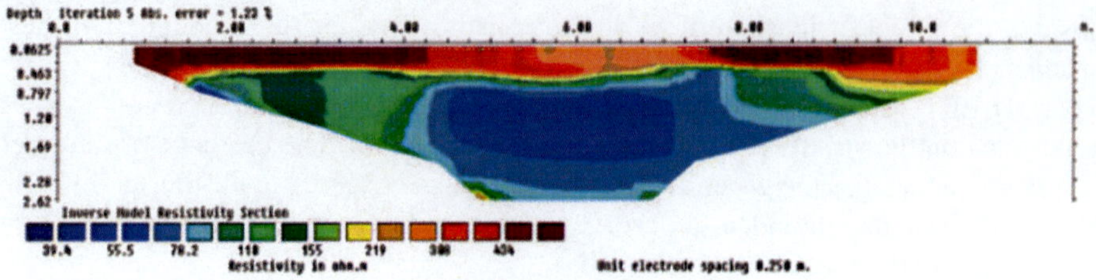

Fig. 12 Resistivity profile after two months of filling

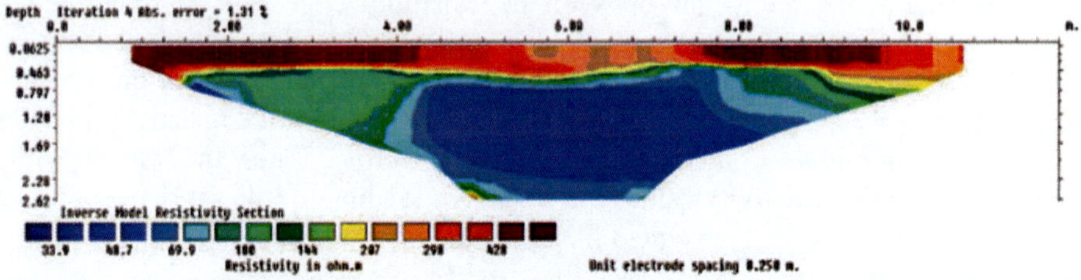

Fig. 13 Resistivity profile after three months of filling

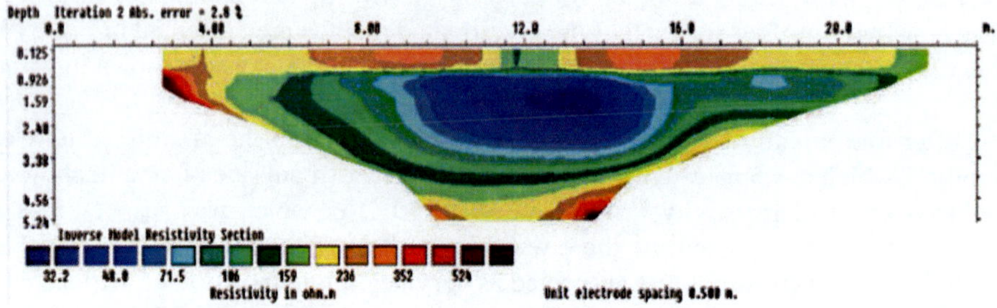

Fig. 14 Resistivity profile after four months of filling

Table 1 Physicochemical analyses of water samples for ERT Profile 1

Parameters	Observed value				
	November	December	January	February	March
pH	6	5.8	5.1	5.3	5.5
Turbidity (NTU)	0.90	1.20	2.10	2.30	2.40
Alkalinity (mg/l)	10	40	30	26	22
Chloride (mg/l)	8.01	20.02	16.52	11.01	9.51
Iron (mg/l)	0.042	0.056	0.085	0.071	0.056
Sulphate (mg/l)	0	5.25	38.73	32.67	30.22
Nitrate (mg/l)	0	2.42	10.17	8.93	8.72
Dissolved oxygen (mg/l)	2	3.2	2.56	2.36	1.99
Hardness (mg/l)	38	52	40	54	42

transported through the saturated pores of clayey soil in the subsurface. The physicochemical analysis of the water sample, collected near this profile, witnessed very negligible increase in the contaminant concentration. The small reduction in the resistivity value below profile 1 was therefore discovered as a result of negligibly increased contaminant concentration in the groundwater. This change in resistivity value informed about the sensitivity of the resistivity survey in capturing the changes in contaminant concentration in the groundwater.

To detect the contaminated zone very minutely, the electrode spacing in the second month has been reduced to 0.25 m. The maximum depth of investigation is up to 2.62 m. ERT profile 1 conducted during the month of December has shown a drastic reduction in resistivity values. The resistivity of the contaminated zone is ranging between 50 and 55 Ω-m, much lower than the resistivity before contamination. Physicochemical analysis has also shown increase in contaminant concentration. The width of contaminated zone is around 1.5 m, and the contaminated zone was about from depth 0.8 to 1.7 m. In third month, the resistivity value of the contaminated portion value decreased from the second month and was in the range of 40–45 Ω-m. The wide contaminated zone was 3.13 m, and contaminated zone started from the depth of 0.84 to 2.28 m. There was some precipitation during the fourth month, which affected the changes in contaminant concentration in groundwater along with the leachate. It was observed that the contaminant concentration somehow reduced in the groundwater due to the precipitation. The resistivity survey also witnessed some similar effect due to precipitation. Most of the resistivity data were observed to follow the changes in the concentration of the physicochemical parameters of the groundwater. The area of the contaminated zone increases largely. There is rise in water table following the precipitation. The contamination zone started from a depth of 0.6 m and spread up to a depth of 3.1 m. The maximum depth of flow of leachate increased at subsequent months as predicted from electrical resistivity tomography. The variation of depth of groundwater table noted from observatory well 1 is also tabulated below. The variation of maximum depth of flow and the width of flow are presented in Table 2.

Table 2 Groundwater table and resistivity details for ERT Profile 1

Time	Depth of natural groundwater level (m)	Resistivity range of the contaminated zone (Ω-m)	Width of contaminated zone (m)	Maximum depth of contaminated zone (m)
Before contamination	0.7	70–75	NA	NA
1st month	0.75	55–60	1.5	1.2
2nd month	0.78	50–55	2.5	1.7
3rd month	0.8	40–45	3.5	2.3
4th month	0.65	45–50	5.5	3.1

Resistivity and Water Quality Data for Profile 2. Along profile 2, the resistivity survey was also carried out which is nearer to the landfill and the corresponding water quality parameters were also determined in the laboratory. Two-dimensional resistivity profiles along the profile 2 are presented in Figs. 15, 16, 17 and 18, and the test results for water quality parameters are presented in Table 3.

The physicochemical analysis of water samples witnessed more fluctuations in borehole 2 which was nearer to the landfill rather in the borehole which was away from the landfill. The parameters which were more affected by the contamination in borehole 2 include pH, acidity, alkalinity, chlorides, sulphate, nitrate, turbidity, hardness and iron. The pH value indicated the acidic nature of the groundwater where the value was reduced from 6.0 before contamination to nearer 5.5 after contamination. The parameter which was most influenced by the contamination was iron concentration. Iron concentration was increased from 0.042 mg/l before contamination to 0.959 mg/l after contamination, which informed about the sensitivity of iron contamination in the groundwater at the investigated site. Even within one month of waste filling, iron concentration exceeded its standard value of 0.300 mg/l recommended by the Indian standard code of practice for drinking water. Similarly, sulphate concentration was also more pronounced after the leachate being introduced in the groundwater.

After the precipitation during March 2018, changes occurred in contaminant concentration in groundwater along with the leachate. It was observed that the contaminant concentration somehow reduced in the groundwater due to the precipitation. The resistivity survey also witnessed some similar effect due to precipitation. Most of the resistivity data were observed to follow the changes in the concentration of the physicochemical parameters of the groundwater. Resistivity survey also showed very less flow of leachate below the profile which was representative for the

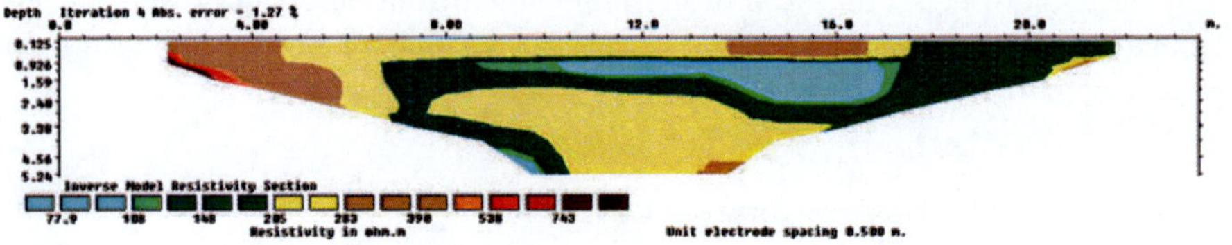

Fig. 15 Resistivity profile before filling the waste

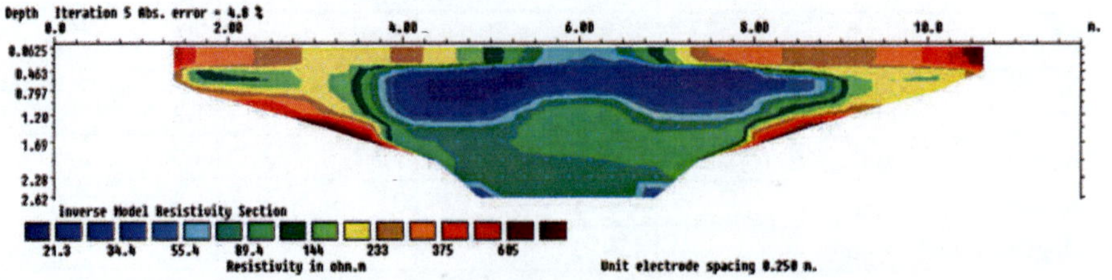

Fig. 16 Resistivity profile after two months of filling

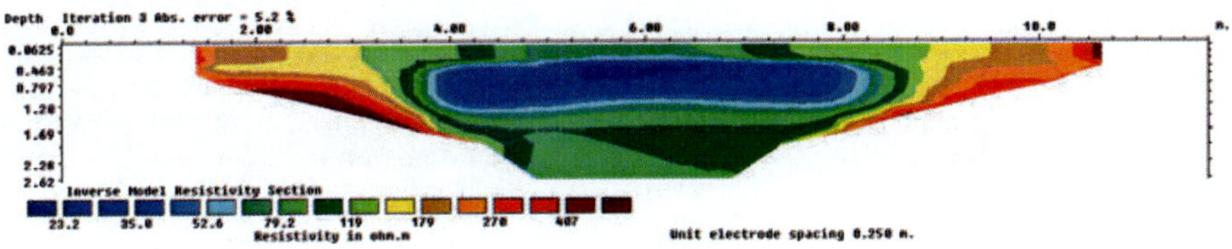

Fig. 17 Resistivity profile after three months of filling

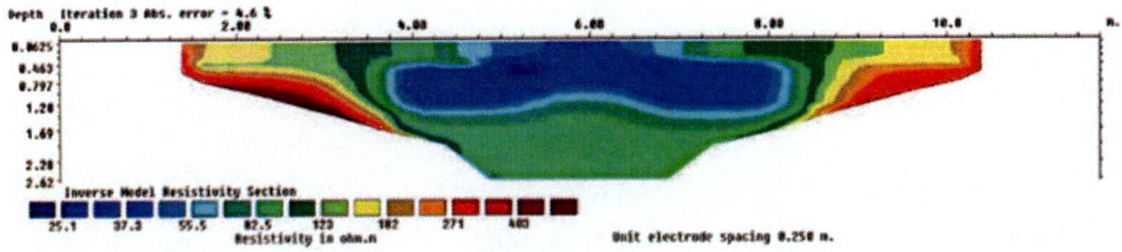

Fig. 18 Resistivity profile after four months of filling

Table 3 Physicochemical analyses of water samples for ERT Profile 2

Parameters	Observed values				
	November	December	January	February	March
pH	6.00	5.80	4.10	4.50	4.70
Turbidity (NTU)	0.90	2.20	5.70	8.80	7.40
Alkalinity (mg/l)	10.00	160.00	378.00	200.00	160.00
Chloride (mg/l)	8.01	93.10	77.59	68.58	52.56
Iron (mg/l)	0.042	0.776	0.776	0.987	0.959
Sulphate (mg/l)	11.23	20.86	160.00	167.87	150.98
Nitrate (mg/l)	3.06	12.50	42.60	45.90	42.65
Dissolved oxygen (mg/l)	3.153	2.956	2.365	2.266	2.069
Hardness (mg/l)	38.00	46.00	44.00	100.00	70.00

borehole 1 data. From the laboratory analysis of the groundwater, it was observed that the water sample collected from borehole 2, 0.5 m away from the landfill, was more contaminated than that of in borehole 1. Higher values of physicochemical parameters indicated degraded quality of groundwater in borehole 2. The variations of depth of groundwater table noted from observatory well 2 were also measured. The variation of maximum depth of flow and the width of flow are presented in Table 4.

Three-Dimensional Resistivity Survey. Two three-dimensional resistivity profiles were performed, one before the construction of landfill and another after one month of construction of landfill. The three-dimensional resistivity profile clearly showed the position of waste filled trench within 0.6-m depth. The waste has a resistivity

Table 4 Groundwater table and resistivity details for ERT Profile 2

Time	Depth of natural groundwater level (m)	Resistivity range of the contaminated zone (Ω-m)	Width of contaminated zone (m)	Maximum depth of contaminated zone (m)
Before contamination	0.70	75–80	NA	NA
1st month	0.72	40–50	4	0.9
2nd month	0.75	30–40	4.1	1
3rd month	0.77	20–30	4.2	1.2
4th month	0.64	30–35	4.5	1.6

of 20–25 Ω-m. The flow path of leachate was clearly visible which was indicated by the reduced resistivity value within the depth of 0.6–1.29 m in Fig. 19. In this resistivity profile, it was also seen that the flow of leachate was towards the lake which was present at the downstream of the landfill. A little flow of leachate was observed beyond the depth of 0.6 m after one month of construction of landfill.

Three-dimensional resistivity data showed the resistivity values in three horizontal layers along the depth at the site. The resistivity data reflected to a great extent a similar type of soil and groundwater flow zones as in case of two-dimensional resistivity profiles at the same site. The groundwater zones were interpreted to be

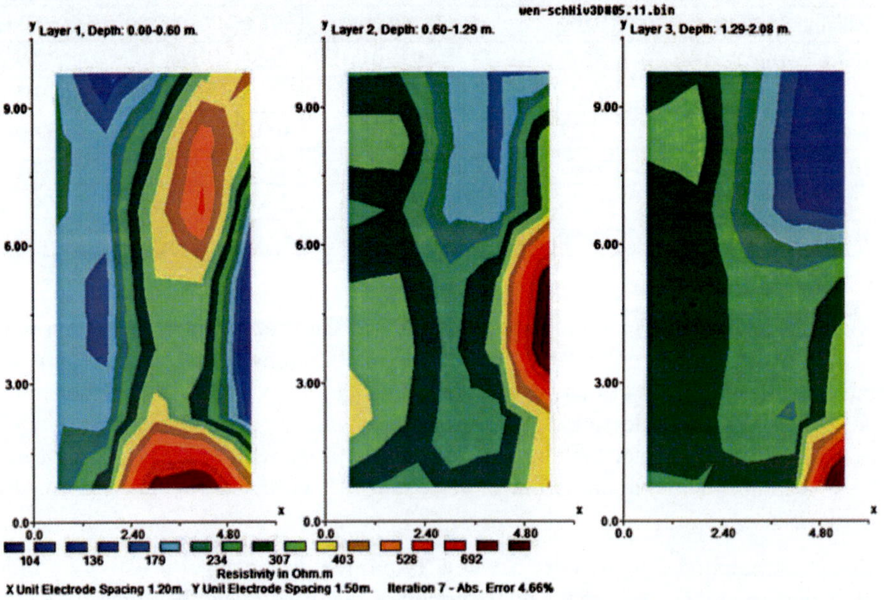

Fig. 19 3D resistivity profile before contamination

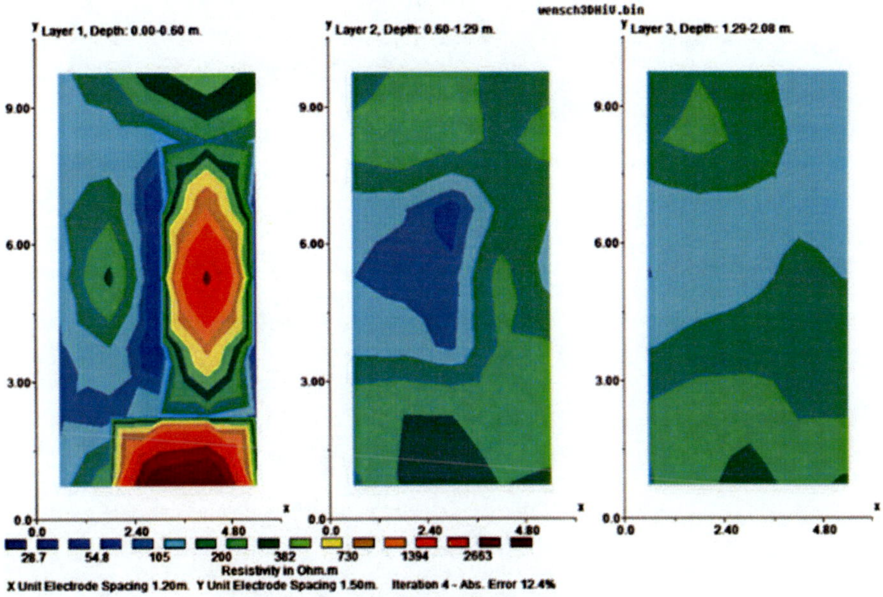

Fig. 20 3D resistivity profile after one month of filling

below 0.6-m depth which the two-dimensional resistivity data have shown earlier. The slope of the groundwater flow was found towards the lake, and the same data were observed during the investigation using piezometric analysis. From this profile, it was observed that a high resistive zone conforming the presence of gravelly area was present at the site which was also evidenced during the boring of the test wells (Fig. 20).

The three-dimensional resistivity profile after one month of contamination clearly showed the flow of leachate from the landfill to its downstream side towards the lake present at the site. The low resistive zone at the middle of the profile in second layer of the resistivity profiles is interpreted as the landfill, and the low resistivity zone surrounding the landfill was observed as the flow of low resistive liquid known as leachate originated from the landfill. The absence of low resistivity zone in the third layer of the resistivity profiles indicated that the flow of leachate from the landfill is limited to a depth of about 1.3 m in vertical direction. But the contaminated zone in the third layer was observed to be extended in horizontal lateral directions. Some high resistivity zones were also observed at the site from this resistivity profiles in case of the earlier resistivity profile.

Correlation of Resistivity Data with Water Quality Parameters. Efforts were made to analyse the physicochemical parameters of each water sample collected from the boreholes over a period of five months and to correlate each parameter with the resistivity values corresponding to each observed month. Correlation charts are being prepared with electrical resistivity and the water quality parameters including pH,

Fig. 21 Correlation of resistivity with pH values

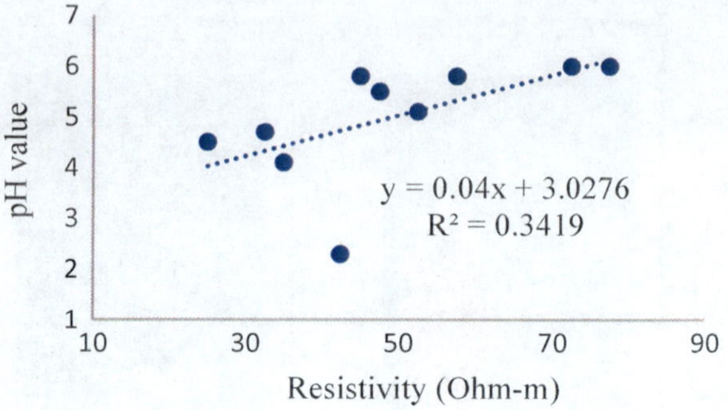

Fig. 22 Correlation of resistivity with chloride

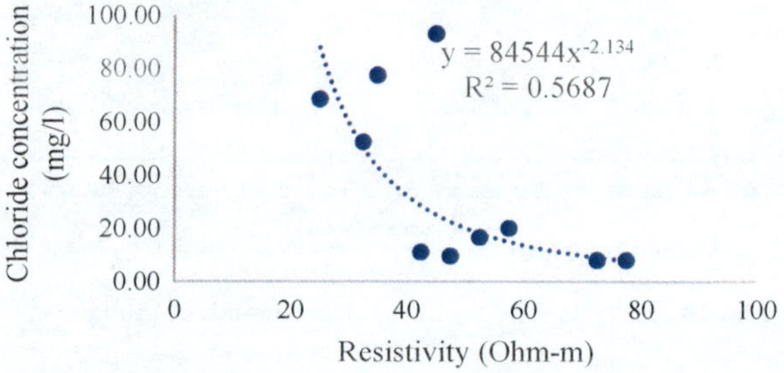

Fig. 23 Correlation of resistivity with iron

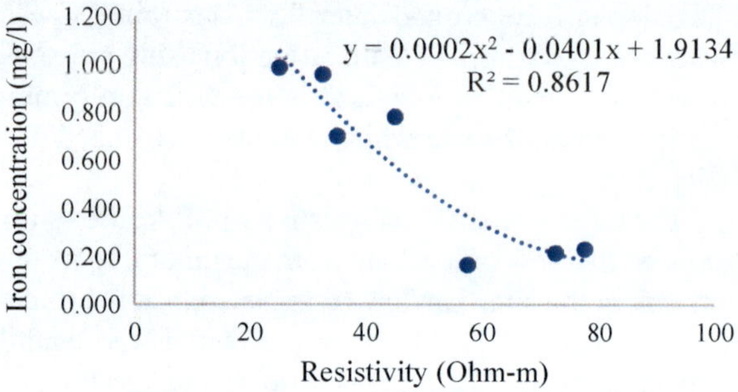

chloride concentration, iron concentration, nitrate concentration, sulphate concentration and hardness concentration. The correlations of resistivity value with the water quality parameters are presented in Figs. 21, 22, 23, 24, 25 and 26.

5 Conclusions

Prediction and remediation of contaminated groundwater below a non-engineered landfill were the prime objectives of this research. Conventional method of solving the

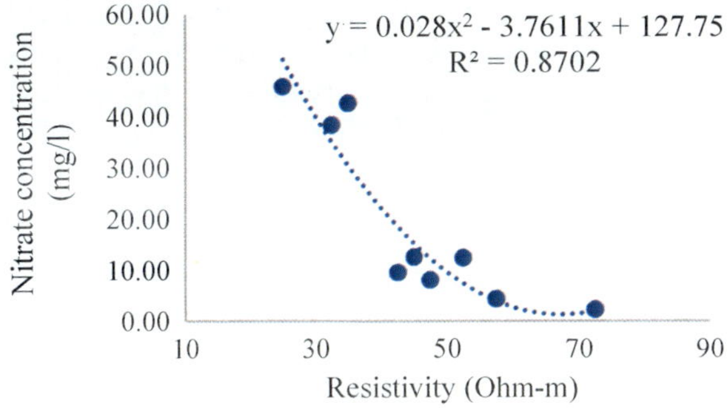

Fig. 24 Correlation of resistivity with nitrate

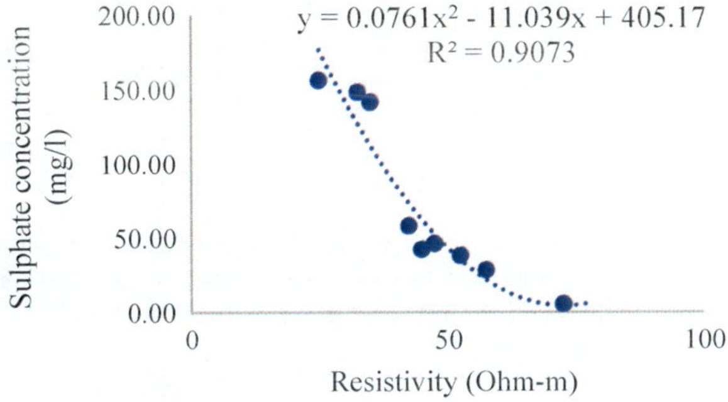

Fig. 25 Correlation of resistivity with sulphate

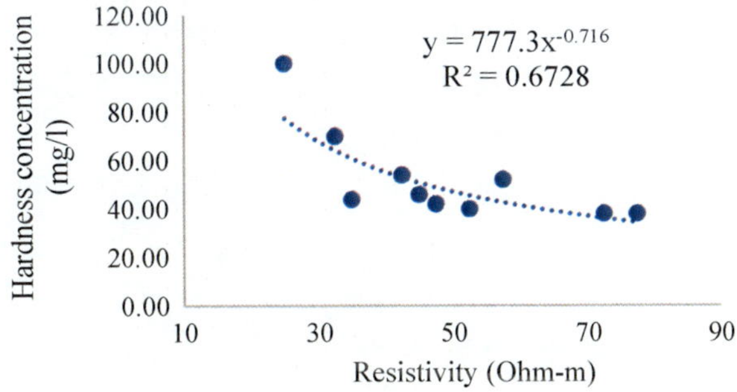

Fig. 26 Correlation of resistivity with hardness

concerned issues being found inefficient, the author has applied advanced geophysical techniques in solving the concerned issues of groundwater contamination. For this research work, electrical resistivity tomography was used to predict the groundwater contamination and to delineate the leachate plume below a landfill. The following conclusions can be drawn from this research work:

1. Resistivity survey accurately reflects the changes in contaminant concentrations in groundwater contaminated by non-engineered landfills.

2. The regression coefficient value of 0.87, 0.90 and 0.86 resulted from the correlation of resistivity value, respectively, with nitrate, sulphate and iron concentrations indicates the accuracy of the resistivity survey in determining the contaminant concentrations in polluted groundwater at a landfill site.
3. Non-engineered landfills have a great potential in contaminating the immediate environment within a very short period of time if no remedial measure is adopted.

References

Abdul Nassir SS, Loke MH, Lee CY, Nawawi MNM (2000) Salt-water intrusion mapping by geoelectrical imaging surveys. Geophys Prospect 48(4):647–661. https://doi.org/10.1046/j.1365-2478.2000.00209.x

Aizebeokhai AP (2010) 2D and 3D geoelectrical resistivity imaging: theory and field design. Sci Res Essays 5(23):3592–3605. https://doi.org/10.1007/978-3-030-01572-5_57

Biswas AK, Kumar S, Babu SS, Bhattacharyya JK, Chakrabarti T (2010) Studies on environmental quality in and around municipal solid waste dumpsite. Resour Conserv Recycl 55(2):129–134. https://doi.org/10.1016/j.resconrec.2010.08.003

Lopes DD, Silva SMCP, Fernandes F, Teixeira RS, Celligoi A, Dall LH (2012) Geophysical technique and groundwater monitoring to detect leachate contamination in the surrounding area of a landfill-Londrina (PR-Brazil). J Environ Manage 113:481–487. https://doi.org/10.1016/j.jenvman.2012.05.028

Effect of Filament Type and Biochemical Composition of Lignocellulose Fiber in Vegetation Growth in Early Plant Establishment Period

Rojimul Hussain, Sanandam Bordoloi, Vinay Kumar Gadi, Ankit Garg, K. Ravi, and S. Sreedeep

Abstract Early plant establishment period for bioengineered structure is important to assess its soil erosion protection and green restoration capability. The use of lignocellulose-based fibers has been recently explored for enhancing vegetation growth. Moreover, the soil–lignocellulose fiber composite has been reported to enhance the water retention and reduce erosion potential. However, the effects of fiber type, i.e., filament type and biochemical composition has not been explored. The objective of the study aims to explore these effects on early plant establishment period by incorporating two contrasting lignocellulose fibers–coir and water hyacinth. Soil column of bare soil and soil–fiber composite compacted at 0.9MDD were prepared and instrumented for measurement of suction and moisture content. An indigenous grass species *Axonopus Compressus* was transplanted and the growth parameter was monitored for a period of 60 days under greenhouse condition. The results indicated that monofilament fiber such as WH is better suited as a lignocellulose fiber for grass growth in compacted soil based on its water retention, easy root propagation and resistance to desiccation cracking.

Keywords Lignocellulose fibre · Vegetation growth · Biochemical composition · Coir fibres · Water hyacinth fibres · Natural fibres

1 Introduction

Vegetation is often used for erosion protection and stabilization of bioengineered structure, which includes slopes, embankment, landfill cover and many other green infrastructures. The performance of these structures mainly relies on the growth

R. Hussain (✉) · S. Bordoloi · V. K. Gadi · K. Ravi · S. Sreedeep
Department of Civil Engineering, Indian Institute of Technology Guwahati, Guwahati, Assam 781039, India
e-mail: rojimul93@gmail.com

A. Garg
Department of Civil and Environmental Engineering, Shantou University, Shantou, Guangdong, China

M. Latha Gali and R. R. P. (eds.), *Problematic Soils and Geoenvironmental Concerns*, Lecture Notes in Civil Engineering 88,
https://doi.org/10.1007/978-981-15-6237-2_18

of vegetation. Early establishment of plants in these structures is important for assessing soil erosion protection and green restoration capability. Many a time researchers used different organic materials or supplement to improve vegetation growth. Recently, the use of sustainable lignocellulose-based fiber materials has been explored for enhancing vegetation growth. However, the efficacy of different fiber type in vegetation growth or early plants establishment has rarely been studied.

The lignocellulose-based fiber which is referred as natural fiber mainly comprises of three biochemical compositions namely cellulose, hemicellulose, and lignin. The characteristics of these fibers are primarily governed by the biochemical composition. In the recent past, the application of these fibers in geotechnical engineering has gained popularity in the form of fiber and geotextile-reinforced soil (Bordoloi et al. 2016). The application of natural fibers extracted from coir and water hyacinth (WH) in soil improved soil shear strength significantly (Hejazi et al. 2012 and Bordoloi et al. 2018a). Soil desiccation crack, which is harmful to the integrity of many bioengineered structures reduced significantly. While soil water retention capacity improved significantly upon inclusion of WH fiber (Bordoloi et al. 2018c); therefore, the application of lignocellulose based fiber in soil has multiple beneficial features that may potentially be effective for vegetation growth or early establishment of plants.

The objective of the present study is to explore the effect of filament type and biochemical composition of lignocellulose fiber in vegetation growth in early plant establishment period. To meet the objective, experimental column studies were carried out by compacting soil with two contrasting natural fibers such as coir and WH having different filament (single filament for WH and multiple filaments for coir) and biochemical composition. The choice of WH fiber lies on the easily available and waste management since, WH is known among the world's most invasive weeds (Malik et al. 2007) which invades water bodies of most tropical to lower temperate countries and multiplies at a very high rate (17.5 metric tons/hectare/day), thus costing millions of dollars to control it (Simberloff et al. 1997 and Shoeb and Singh 2000).

2 Materials and Method

2.1 Soil

Soil used for the experimental study was collected from a hill side located inside the campus of Indian institute of technology, Guwahati, India. The presence of dead plant roots, stones, and any other impurities in the soil has been removed manually. Finally, the soil has been characterized for different indexed properties as per ASTM (D1557 (2015), D2487-11 (2011), D422-63 (2007), D4318 (2010), E1755-01 (2007)) code recommendation. The indexed properties of the soil tested have been listed in Table 1.

Table 1 Soil index properties

Soil property	Value
Specific gravity	2.63
Particle size distribution (%)	
Coarse sand (4.75–2 mm)	0
Medium sand (2–0.425 mm)	7
Fine sand (0.425–0.075 mm)	20
Silt (0.075–0.002 mm)	50
Clay (<0.002 mm)	23
Atterberg limits (%)	
Liquid limit	41
Plastic limit	23
Shrinkage limit	19
Compaction properties	
Optimum moisture content (%)	17
Maximum dry density (kg/m^3)	1720
USCS classification	ML (Low plastic inorganic silt)

2.2 Fiber

Two natural fibers extracted from water hyacinth (*Eichhornia Crassipes)* and coir (*Cocos Nucifera*) has been used in this study. Figure 1 shows the pictorial view of the fibers used. Coir is a multifilament fiber, whereas water hyacinth (WH) is single filament fiber. These are lignocellulosic materials primarily constitute of cellulose, hemicellulose, and lignin. The physiochemical properties of these fibers mainly depend on biochemical composition. Tensile strength of the fiber is mainly controlled by the cellulose content (Hearle and Morton 2008), while lignin content is responsible for resistance against bio-degradation. Hemicellulose binds the cellulose polymers, absorbs moisture, and provides secondary strength (Spiegelberg 1966). Therefore, the characterization of biochemical composition of these fibers is essential for its practical application as a reinforcing material. The physio-biochemical properties of fibers used in this study have been characterized and listed in Table 2.

Figure 2 shows the field emission scanning electron microscope (FE-SEM) images of WH and coir, which represent the surface morphology of the given fibers at two different magnifications (100 and 500X). It is clearly visible from the figure that the filament of coir is much finer than that of WH fiber. In other word, the surface of WH fiber is relatively coarser than that of coir fiber.

Fig. 1 Pictorial view of fibers used in the study

Table 2 Properties of fibers used in this study

Physical properties			Biochemical composition			
	Breaking tensile strength (MPa)	Elongation at break (%)	Cellulose (%)	Hemi-cellulose (%)	Lignin (%)	Ash (%)
WH	313 ± 8	14 ± 1	46 ± 2.6	21 ± 2	11 ± 0.9	11 ± 1.5
Coir	152 ± 11	25 ± 2.2	42 ± 1.45	13.5 ± 1.8	39 ± 2.5	2.6 ± 0.5
Procedure	IS-1670-(1991)		Jenkins (1930)	TAPPI Test Methods (1996)	Goering and Van Soest (1970)	ASTM E1755-01

2.3 Grass Species

Axonopus Compressus, a grass species commonly adopted in green infrastructure application (Ng et al. 2014; Avilés-Nova et al. 2008) for their abundant availability, drought resistance characteristics, and high root strength has been selected for this study. Figure 3 shows the typical shoot and root component of the selected grass species. Roots are fibrous in nature with their tensile strength of around 50–75 MPa

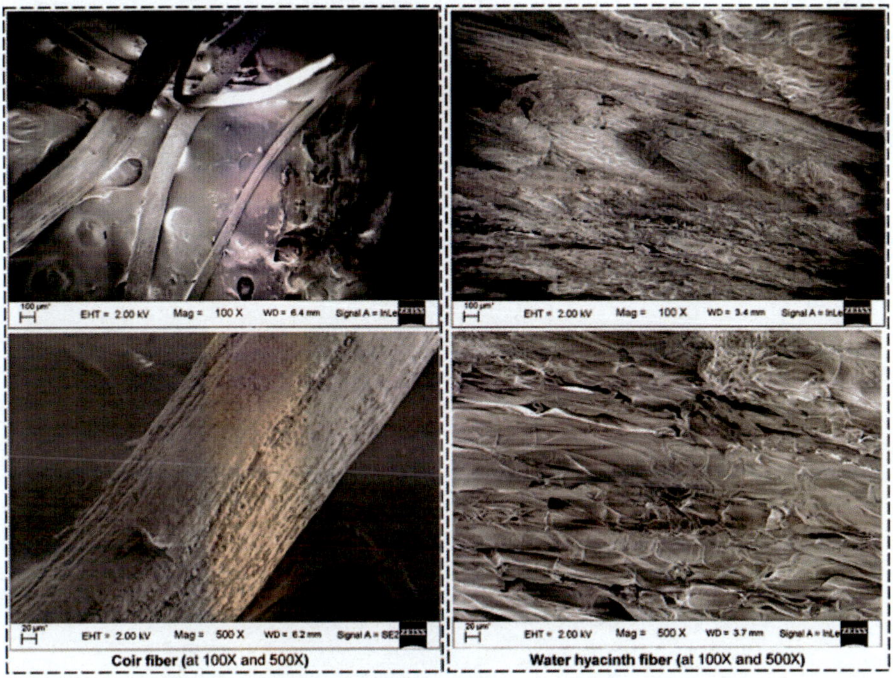

Fig. 2 FE-SEM images of coir and water hyacinth fiber

Fig. 3 Overview of typical shoot and root of the selected grass species (*Axonopus Compressus*)

(Zhang et al. 2014; Noorasyikin and Zainab 2016 and Zhong et al. 2016). Shoot length varies from 5 to 15 cm (Bordoloi et al. 2018c). It spreads by creeping stolon's and underground rhizomes to form close-knit (Skerman and Riveros 1990).

2.4 Test setup and procedure

Compacted soil columns were prepared by compacting bare soil (without fiber) and soil–fiber composites in a PVC cylindrical mold of size 250 mm diameter and 250 mm length with perforated base plate for water drainage. Soil was compacted at a density of 0.9MDD and 15% (w/w) initial moisture content up to a height of 170 mm from the base of the mold as shown in Fig. 4. The selection of compaction condition was based upon the compaction criteria of a soil embankment (Li et al. 2016). Thereafter, the instruments such as MPS-6 (measurement range, 9–10^5 kPa) sensor for measuring soil suction and EC-5 sensor for measuring volumetric water content (θ) were installed diagonally at a depth of 40 mm from the soil surface as shown in Fig. 4. The locations of the sensors were selected to capture the effect of vegetation roots on soil moisture dynamics (Bordoloi et al. 2018b). Finally, all the soil columns were putted in a green house where the temperature and relative humidity were maintained at 25 ± 1 °C and 60 ± 3%, respectively.

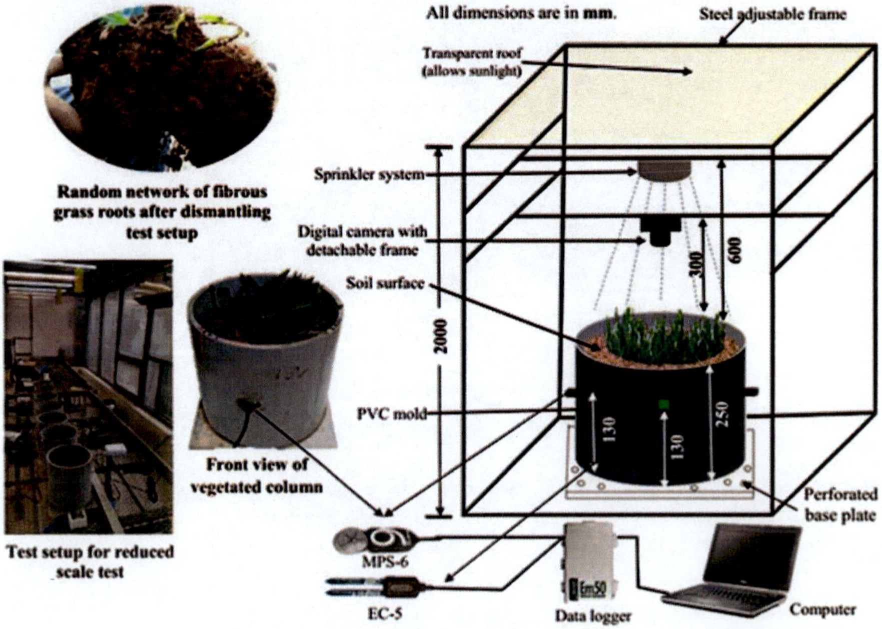

Fig. 4 Schematic diagram of setup for conducting reduced-scale test on vegetated soil

To grow the vegetation, rhizomes of the grass species were collected from the nearby grassland and transplanted on the soil surface by digging small holes. All the soil columns were irrigated frequently by a controlled flow (mariotte bottle) shower system (Fig. 4) to maintain field capacity, which is important for growth and development of roots (Leung et al. 2015). The sensors were then connected to a EM-50 data logger for continuous (15-min interval) monitoring of suction and moisture content. The images of grass surface were captured regularly at one-day interval using a commercial DSLR Camera (Canon EOS 600D) with lens range (18–55 mm), horizontal/vertical resolution of 72 dpi mounted on a steel adjustable frame (Fig. 4). Images of soil surface for crack measurement were captured after cutting off all the grasses from the soil surface at the end of each drying cycle. Three replicates for each soil and soil–fiber composites have been tested.

2.5 Determination of Grass Density (GD) and CIF

In the present study, the growth of vegetation quantified in terms of grass density (GD), which is defined as the ratio of green grass surface area to the soil surface area is considered. The GD was determined from the captured images by using an open-source public domain image-processing software ImageJ (Rasband 2011) as explained in Fig. 5. The captured images were cropped to remove boundary shrinkage and the corresponding total surface area (in terms of pixel) is calculated. Thereafter, the cropped images were threshold by RGB color space and the corresponding pixels

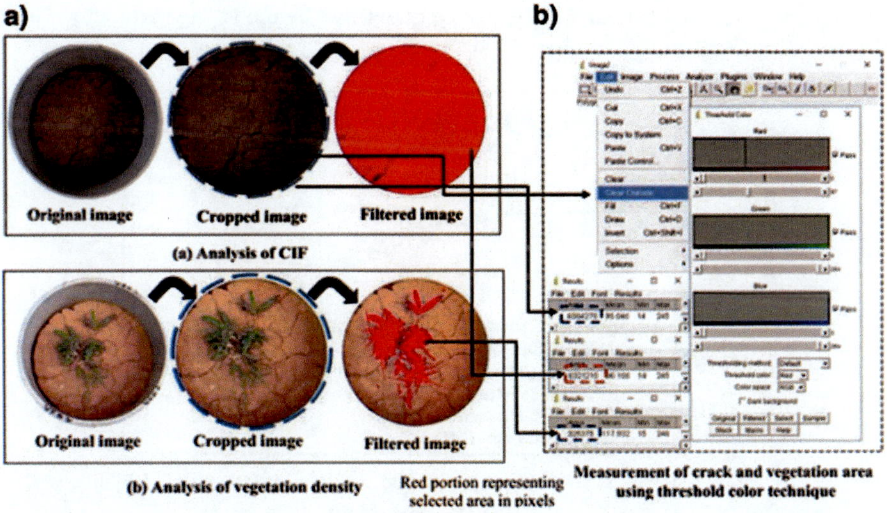

Fig. 5 Determination of **a** CIF and **b** GD by image analysis using threshold color technique

occupied were by the green grass calculated. The accuracy (±1.9%) of the analysis was verified through the calibration of pixel area of images into original area.

Crack developed after each drying cycle was quantified by a factor called crack intensity factor (CIF), which is widely adopted by the researchers for its simplicity (Li et al. 2016 and Bordoloi et al. 2018b). It is defined as the ratio of total surface crack area to the total soil surface area considered. Similar to GD, the determination of CIF from the captured images has been systematically explained in Fig. 5. Here, the cropped images were threshold and the corresponding pixel area occupied by the soil matrix has been calculated. Thereafter, the crack area was calculated by deducting the soil matrix area from the total soil surface area.

3 Results and Discussion

3.1 *Effect of Fiber Type on Soil Water Retention*

Figure 6 shows the SWRCs of different soil–fiber composites as well as bare soil for fifth-drying cycle under vegetated condition. The measured suction and moisture content data obtained from the data logger were fitted to van Genuchten (1980) model using a commercially available software called RETC. The fitting parameters for the SWRCs are presented in Table 3.

It can be observed from the Fig. 6 that the addition of fiber (both coir and WH) improved the soil water retention with respect to bare soil. Moreover, the improvement was higher in soil with WH fiber and followed by coir fiber amended soil. This change in water retention or SWRC can be quantified by the fitting parameters listed

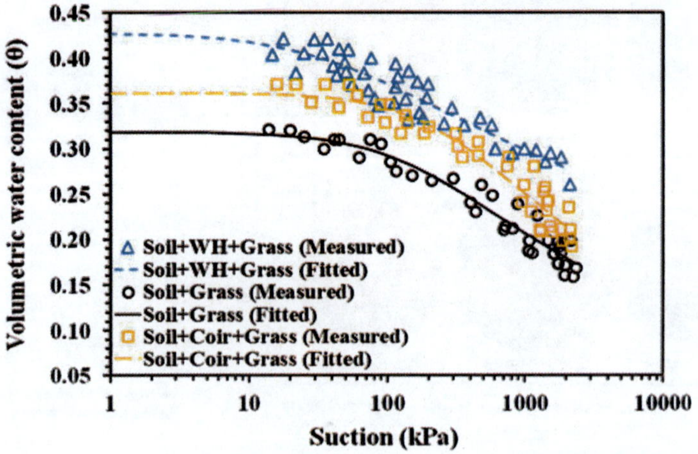

Fig. 6 Soil water retention characteristics (SWRC) for bare soil, soil–WH fiber composite and soil–coir fiber composite

Table 3 A summary of fitting coefficients for SWRC using van Genuchten (1980) model

	θ_s [m^3/m^3]	θ_r [m^3/m^3]	α [kPa^{-1}]	n [-]	m [-]
Soil + grass	0.318	0	0.0066	1.2	0.180
Soil + grass + WH	0.427	0	0.0310	1.1	0.094
Soil + grass + coir	0.361	0	0.0035	1.2	0.207

in Table 3. The saturated volumetric water content (θ_s) is observed to be very high in WH-fibered soil (42.7%) as compared to coir-fibered soil (36.1%) and bare soil (31.8%). The increase of θ_s upon fiber inclusion indicates an increase in porosity or void ratio in soil–fiber composites that hold more water. This can be justified by the hydrophilic nature of the fibers that can be observed from the surface morphology and inherent biochemical composition. It can be seen from the FE-SEM images (Fig. 2) that the surface of both the fibers stacked with fine pores that is conducive to absorb moisture (Methacanon et al. 2010).In addition, WH has high hemicellulose content as compared to coir which is a bio-polymer with available hydroxyl groups that attracts water molecules (Rowell and Stout 2002) and stores on the fiber surface. As a consequence, volume or porosity increases.

The parameter 'α' (inverse of which called air entry value AEV, defined as a particular suction value at which air starts entrapping into soil) is another important parameter for defining SWRC. The 'α' parameter is observed to be very high (i.e., lower AEV) in WH-fibered soil as compared to coir-fibered and bare soil. The lower AEV indicates an early start of desaturation or desaturation at low suction in WH-fibered soil. This is may be attributed to the early escape of excess water absorbed and stored on the WH-fibered surface by the hemicellulose content, i.e., water presence in the soil–fiber interface. The parameter 'n', i.e., the slope of the curve represents the rate of desaturation of the soil under continuous drying. The parameter 'n' is observed to be low in WH-fibered soil (1.1) as compared to coir-fibered soil (1.2) and bare soil (1.2). However, no variation in coir-fibered soil is observed with respect to bare soil. The low 'n' value in WH-fibered soil indicates water retention for longer duration after AEV, i.e., slow desaturation will occur in WH-fibered soil as compared to coir-fibered and bare soil.

3.2 *Effect of Fiber Type on Vegetation Growth*

Figure 7 highlights the vegetation growth in terms of grass density (GD) with time. After transplantation, all the soil columns were continuously irrigated for 18 days for growing grass from rhizomes and it is marked as initial stage. Thereafter, all the soil columns were subjected to five drying-wetting cycles as shown in Fig. 7. It can be observed from the fig that the GD increased with time and the rate of increment was higher after third-drying cycle. Among the three soil–fiber combinations tested, WH-fibered soil showcased higher GD in entire monitoring period and it was followed

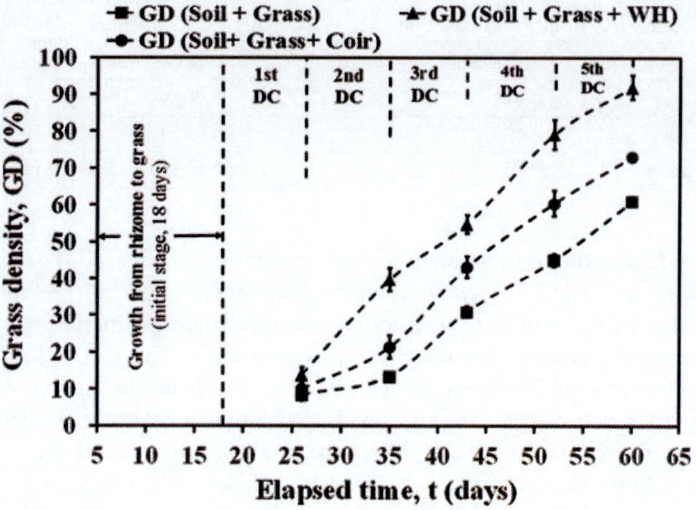

Fig. 7 Variation of grass density (GD) during the monitoring period for bare soil, soil–WH fiber composite and soil–coir fiber composite

by coir-fibered soil. The higher GD in fibered soil (both WH and coir) as compared to bare soil was mainly attributed to the higher water retention of fibered soil over bare soil as seen from Fig. 6.

In addition, the lignocellulosic compound (biodegradable) of these fibers degrades with time while amended in soil and supplies nutrient to the plant for growth that might be the potential reason for the higher GD in fibered soil. Furthermore, the higher GD in WH-fibered soil over coir-fibered soil was mainly due to the higher water retention capacity of WH-fibered soil as compared to coir-fibered soil (Fig. 6). Critically, it can also be observed from Fig. 7 that the growth of 60% GD took 60 days for bare soil, 52 days for coir-fibered soil, and 45 days for WH-fibered soil. Therefore, the inclusion of WH fiber in soil highly reduced early plant establishment period with respect to the bare soil as well as coir-fibered soil.

3.3 *Effect of Fiber Type on Soil CIF*

Figure 8 depicts the soil crack variation in terms of crack intensity factor (CIF) with time. It has been observed from the figure that in all soil–fiber combination tested, the CIF increased with time and reached peak at third-drying cycle beyond which CIF remained almost constant. The constant CIF after third-drying cycle in all the cases may be attributed to the higher growth of vegetation roots that resist crack (Zhou et al. 2009) through mobilization of its tensile strength against tensile stress developed in soil by interface friction between soil and root surface (bridging effect, i.e., root tightly hold soil). Moreover, it may be the reason that the peak CIF observed

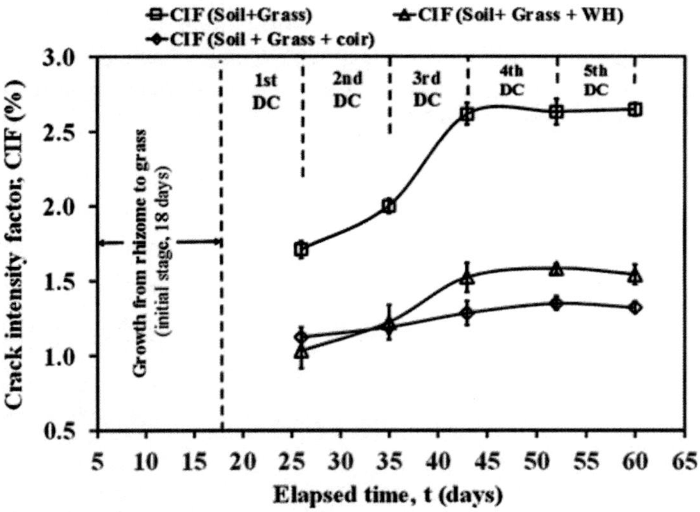

Fig. 8 Variation of CIF during the monitoring period for bare soil, soil–WH fiber composite and soil–coir fiber composite

was the maximum crack potential for the soil–fiber combination considered (i.e., soil type and compaction condition) beyond which further increase of tensile stress (due to heterogeneity in generated soil suction) in soil caused no effect.

Moreover, the inclusion of fiber in soil highly reduced the crack potential as compared to the bare soil. Lowest CIF was observed for coir-fibered soil and it was followed by WH-fibered soil. The lower CIF in fibered soil over bare soil mainly attributed to the utilization of fiber tensile strength against tensile stress developed in soil through interface friction between soil and fiber surface (bridging effect) that reduced crack potential. In addition, the higher water retention in fibered soil may also be the reason for low CIF since the development of crack is mainly related to the soil moisture content (moisture loss). Higher water retention (high moisture content) in fibered soil leads to low crack potential (CIF). Furthermore, the lower CIF in coir-fibered soil as compared to WH-fibered soil may be attributed to the multifilament nature of coir fiber since the specific surface area in multifilament coir fiber is much higher than the single filament WH fiber that offered more interface friction (bridging) to reduce crack (CIF).

4 Conclusion

In the present study, the efficacy of two contrasting natural fibers, namely water hyacinth and coir fiber in promoting vegetation growth in early plant establishment period has been investigated. For experimentation, compacted soil columns were prepared for both bare and fibered (WH and coir) soil and instrumented for monitoring various parameters. A grass species (*Axonopus Compressus)* were grown accordingly. The experimental results revealed that the inclusion of natural fibers in soil significantly improves soil water retention and vegetation growth whereas significantly reduced crack potential as compared to bare soil. More conveniently, the addition of single filament WH fiber in soil is observed to be more effective than a multifilament coir fiber in terms of water retention and vegetation growth. Therefore, the application of WH fiber in green infrastructure would be beneficial for water retention in drought condition and promoting vegetation growth.

References

ASTM D1557 (2015) Standard test methods for laboratory compaction characteristics of soil using modified Effort (56,000 ft-lbf/ft3 (2,700 kN-m/m^3)). Annual Book of Standards 4

ASTM D2487-11 (2011) Standard practice for classification of soils for engineering purpose (unified soil classification system). www.astm.org

ASTM D422–63 (2007) Standard test method for particle-size analysis of soils. ASTM International, West Conshohocken, p 1520

ASTM D4318 (2010) Standard test methods for liquid limit, plastic limit, and plasticity index of soils. ASTM International, West Conshohocken, PA, 2003. https://doi.org/10.1520/D4318-10

ASTM E1755-01 (2007) Standard method for the determination of ash in biomass. Annual Book of ASTM Standards 11(5), ASTM International, Philadelphia, PA. www.astm.org

Avilés-Nova F, Espinoza-Ortega A, Castelán-Ortega OA, Arriaga-Jordán CM (2008) Sheep performance under intensive continuous grazing of native grasslands of *Paspalumnotatum* and *Axonopus compressus* in the subtropical region of the highlands of central Mexico. Trop Anim Health Pro 40(7):509–515

Bordoloi S, Garg A, Sreedeep S (2016) Potential of uncultivated, harmful and abundant weed as a natural geo-reinforcement material. Adv Eng Mater 5(1):276–288

Bordoloi S, Kashyap V, Garg A, Sreedeep S, Wei L, Andriyas S (2018a) Measurement of mechanical characteristics of fiber from a novel invasive weed: a comprehensive comparison with fibers from agricultural crops. Measurement 113:62–70

Bordoloi S, Hussain R, Gadi VK, Bora H, Sahoo L, Karangat R, Sreedeep S (2018b) Monitoring soil cracking and plant parameters for a mixed grass species. Géotech Lett 8(1):49–55

Bordoloi S, Gadi VK, Hussain R, Sahoo L, Garg A, Sreedeep S, Poulsen TG (2018c) Influence of fiber from waste weed *Eichhornia Crassipes* on water retention and cracking characteristics of vegetated soils. Géotech Lett, 1–25

Goering HK, Van Soest PJ (1970) Forage fiber analyses (apparatus, reagents, procedures, and some applications). USDA AgrHandb

Hearle JW, Morton WE (2008) Physical properties of textile fibres. Elsevier

Hejazi SM, Sheikhzadeh M, Abtahi SM, Zadhoush AA (2012) Simple review of soil reinforcement by using natural and synthetic fibers. Constr Build Mater 31(30):100–116

IS-1670-1991 (1991) Determination of breaking load and elongation at break of single strand textiles and yarns. Bureau of Indian Standards Publications, New Delhi

Jenkins SH (1930) The determination of cellulose in straws. Biochem J 24(5):1428

Leung AK, Garg A, Ng CWW (2015) Effects of plant roots on soil-water retention and induced suction in vegetated soil. Eng Geol 193:183–197

Li JH, Li L, Chen R, Li DQ (2016) Cracking and vertical preferential flow through landfill clay liners. Eng Geol 206:33–41

Malik A (2007) Environmental challenge vis a vis opportunity: the case of water hyacinth. Environ Int 33(1):122–138

Methacanon P, Weerawatsophon U, Sumransin N, Prahsarn C, Bergado DT (2010) Properties and potential application of the selected natural fibers as limited life geotextiles. Carbohyd Polym 82(4):1090–1096

Ng CWW, Leung AK, Woon KX (2014) Effects of soil density on grass-induced suction distributions in compacted soil subjected to rainfall. Can Geotech J 51:311–321

Noorasyikin MN, Zainab M (2016) A tensile strength of bermuda grass and vetiver grass in terms of root reinforcement ability toward soil slope stabilization. In: IOP conference series: materials science and engineering, vol 136, no 1). IOP Publishing, p 012029

Rasband WS (2011) ImageJ. US National Institutes of Health, Bethesda, 616 Maryland

Rowell RM, Stout HP (2002) 7 Jute and Kenaf

Shoeb F, Singh HJ (2000) Kinetic studies of biogas evolved from water hyacinth. In: Proceedings of Agroenviron

Simberloff D, Schmitz DC, Brown TC (eds) (1997) Strangers in paradise: impact and management of nonindigenous species in Florida. Island press

Skerman PJ, Riveros F (1990) Tropical grasses (No. 23). Food & Agriculture Organisations

Spiegelberg HL (1966) The effect of hemicelluloses on the mechanical properties of individual pulp fibers. Tappi 49(9):388–396

TAPPI Test Methods (1996) TAPPI T222. Om-88 klason lignin

van Genuchten M (1980) A closed-form equation for predicting the hydraulic conductivity of unsaturated soils. Soil Sci Soc Am J 44(5):892–898

Zhang M, Chen FQ, Zhang JX (2014) The temporal dynamics of cynodon dactylon soil-root system in soil conservation and slope reinforcement. Adv Mater Res 838:675–679 (Trans Tech Publications)

Zhong RH, He XB, Bao YH, Tang Q, Gao JZ, Yan DD, Li Y (2016) Estimation of soil reinforcement by the roots of four post-dam prevailing grass species in the riparian zone of Three Gorges Reservoir China. J Mt Sci 13(3):508

Zhou Y, Chen J, Wang X (2009) Research on resistance cracking and enhancement mechanism of plant root in slope protection by vegetation. J Wuhan Uni (Natural Science Edition) 5:023

Suitability of Iron Oxide-Rich Industrial Waste Material in Clay Soil as a Landfill Liner

Rosmy Cheriyan and S. Chandrakaran

Abstract Landfill liner is the most important component in the landfill disposal site to prevent the seepage of leachate to the ground below. This study mainly focused to study the effect of the industrial waste material from Kerala Metals and Minerals Limited (KMML) which is very fine powder with locally available clay as a landfill liner material. The effect of both index and engineering properties of mixtures was analyzed with the addition of iron oxide-rich industrial waste material. A mix containing 75% clay soil, 25% industrial waste and 5% bentonite was found to satisfy all the requirements of a clay liner material. Batch adsorption tests were performed on these mixes to study the lead (Pb) adsorption capacity of mixtures and it is observed that as the iron oxide-rich waste content increased in the natural soil, lead removal efficiency of soil is also increased.

Keywords Landfill liner · Iron oxide waste · Batch adsorption test

1 Introduction

The management and handling of municipal solid waste in India has continued to be a severe problem since its generation rate increases per day. Traditional ways of disposing this kind of waste is by open dumping or landfill disposal. Mere dumping in sites causes many problems that include both environmental and health issues. Compared to municipal solid waste incineration, landfill is more attractive due to its less investment, if a suitable disposal area can be found. A proper engineered landfill sites can reduce these direct problems due to the waste disposal. It is widely adopted due to its low expensive nature. The most important component in landfill sites is the bottom liner which acts as an impermeable barrier. Usually clay soil is widely used as bottom liner due to its very low permeability characteristics and its availability.

R. Cheriyan (✉) · S. Chandrakaran
Department of Civil Engineering, National Institute of Technology, Calicut, Kerala 673601, India
e-mail: rosmycheriyan11@gmail.com

M. Latha Gali and R. R. P. (eds.), *Problematic Soils and Geoenvironmental Concerns*, Lecture Notes in Civil Engineering 88,
https://doi.org/10.1007/978-981-15-6237-2_19

The fine industrial waste material from titanium dioxide pigment plant in KMML caused much environmental problems to the residents nearby. Here, in this paper, tried to minimize the exposure of this fine waste powder by amending with the locally available soil to satisfy the landfill liner requirements. This research is mainly intended to evaluate the effect of the iron oxide-rich waste from titanium dioxide pigment plant in locally available clay to improve the strength of the soil and thereby finding out an innovative method to utilize the waste and an alternative solution for waste disposal. This study also intended to study the effect of iron oxide minerals in clay soils. Mechanisms dealing with iron oxide and clay particles are very less investigated. Efficient disposal of this industrial waste as a landfill liner could be a new innovative material.

2 Materials

2.1 *Soil*

Clay was collected from paddy site in Meppady, Wayanad. The clay was collected from 3–4 m depth and packed in polythene bags to preserve the water content and transported to the laboratory. The soil was air dried, pulverized and passed through 4.75 mm and 425 micron sieves and stored separately in air tight containers. The index and engineering properties of soil were determined as per IS standards. Soil is classified as CH indicating that clay is of highly plastic in nature.

2.2 *Industrial Waste*

Industrial waste is collected from titanium dioxide pigment plant in KMML in bags. The colour of this fine waste powder is dark maroon and having low pH. The chemical composition of the waste provided by the company is given in Table 1. This waste material has higher specific gravity than the soil collected.

2.3 *Bentonite*

Bentonite used for the study is commercial clay (sodium bentonite). It is a representative of typical bentonite available for construction purpose. It is an impure clay mostly consisting of montmorillonite. It was supplied by Associated Rubber Chemicals (Kochi) Pvt Ltd. It is classified as clay of high plasticity (CH).

Table 1 Chemical composition of industrial waste

S. No.	Constituent	%
1	Fe_2O_3	88.60
2	TiO_2	0.800
3	V_2O_5	0.640
4	Al_2O_3	1.320
5	MnO	0.760
6	Cr_2O_3	0.185
7	CaO	0.230
8	MgO	1.480
9	SiO_2	0.040
10	TiCl	2.300
11	Moisture	2.60

3 Methodology

3.1 Compaction Tests

The standard proctor tests were conducted in the laboratory as per IS 2720 part VII for various mixtures.

3.2 Falling Head Permeability Test

Coefficient of permeability was determined using variable head permeability test as per IS 2720 part XVII for all mixes.

3.3 Unconfined Compressive Strength Test

Unconfined compressive strengths of soil mixes were tested by applying axial compressive strain to soil sample as per IS 2720-part X for all mixes.

3.4 Atterberg Limits

It includes the determination of liquid limit, plastic limit and shrinkage limit. Tests were conducted as per IS 2720-part V.

3.5 Free Swell Index Test

Free swell index of soil was determined as per IS 2720 part 40. It is the increase in volume of soil without any external constraint when subjected to submergence in water.

3.6 SEM Analysis

The microscopic analysis of the soil sample before and after the addition of industrial waste is done to visualize the changes that have occurred in the structure of the soil. The model used was Hitachi SU 6600. Samples were oven dried to get rid of moisture content. The A beam of electrons is generated by a suitable source, typically a tungsten filament or a field emission gun. Electron beam is accelerated through high voltage and passed through electromagnetic lenses to produce thin beam that scans the surface. The secondary electron image (SEI) is used mainly to image the soil surfaces and gives a high resolution image. The photograph of these secondary electrons which reflected from the samples can be obtained as rather three-dimensional images.

3.7 Batch Adsorption Test

Soil was air dried and passed through 75 μm sieve. An S/L ratio of 1:20 was adopted based on USEPA TCLP procedure. 5 g of adsorbent soil was taken in 100 ml of lead solution of known concentration. The mixture was agitated in shaking incubator for a period of 24 h at 130 rpm. Then the solution is filtered through Whatsman No:42 filter paper and filtered liquid was analyzed for lead concentration using atomic adsorption spectrophotometer (AAS). Linear and Langmuir adsorption isotherms were prepared from the results.

4 Results and Discussion

4.1 Geotechnical and Physical Tests

A series of tests were conducted to know the properties of soil, industrial waste from KMML and bentonite and are tabulated in Table 2

Table 2 Properties of soil, industrial waste and bentonite

Properties of clay	soil	Industrial waste	Bentonite
Liquid limit (%)	70	49	408
Plastic limit (%)	32	34	81
Shrinkage limit (%)	10	–	–
Plasticity index (%)	38	15	327
Clay (%)	30	24	76
Silt (%)	70	76	24
Specific gravity	2.52	4.2	2.56
Optimum moisture content (%)	30	–	–
Maximum dry density (kN/m^3)	14.4	–	–
Coefficient of permeability(cm/s)	6.86×10^{-7}	–	–
Unconfined compressive strength (kN/m^2)	186	–	–
pH	7.8	3.4	7.2

4.2 Standard Compaction Test

A series of standard proctor compaction tests were conducted on soil with varying percentage of industrial waste like 0%, 10%, 20%, 25% and 30%. It is noted that as the percentage of industrial waste increases, there is a gradual increase in the maximum dry density (MDD) and decrease in the optimum dry density (OMC). According to Yusuf et al. (2016), the reason behind the increase in MDD and corresponding decrease in OMC of the soil selected for study with increase in industrial waste material was due to the low hydroscopic nature of industrial waste material attributable to higher concentration of ferric oxide in it. The results are shown in Figs. 1 and 2.

4.3 Permeability Test

The addition of industrial waste lowers the permeability to a certain level due to high concentration of iron oxide in it. The pH of the most samples falls below 5. Researchers like Yusuf and Tomacz stated their findings in their journals such that at low pH ironoxides will precipitates on the surface of clay minerals and once it formed the coating will remain stable even at high pH. But the values do not fall below 1×10^{-7} cm/s. Hence, an additional 5% of bentonite is added to satisfy the requirements of a liner material. The variation of hydraulic conductivity values with industrial waste and bentonite is shown in Fig. 3.

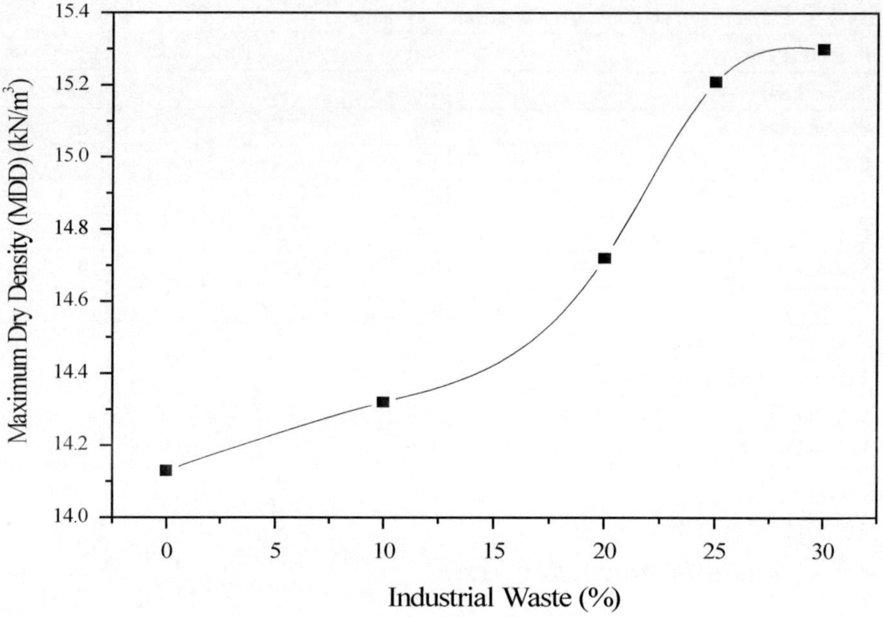

Fig. 1 Variation of MDD with the addition of industrial waste

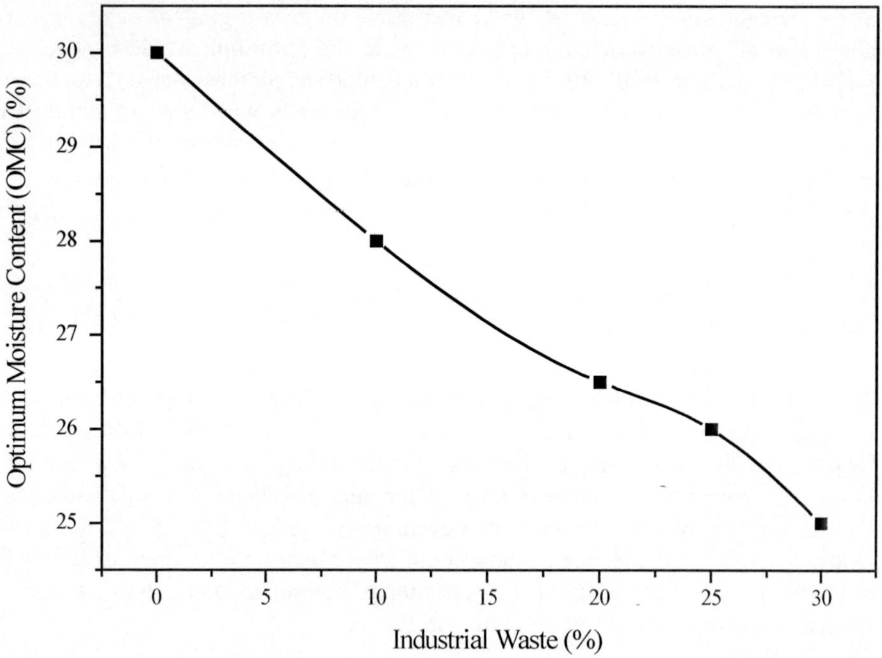

Fig. 2 Variation of OMC with the addition of industrial waste

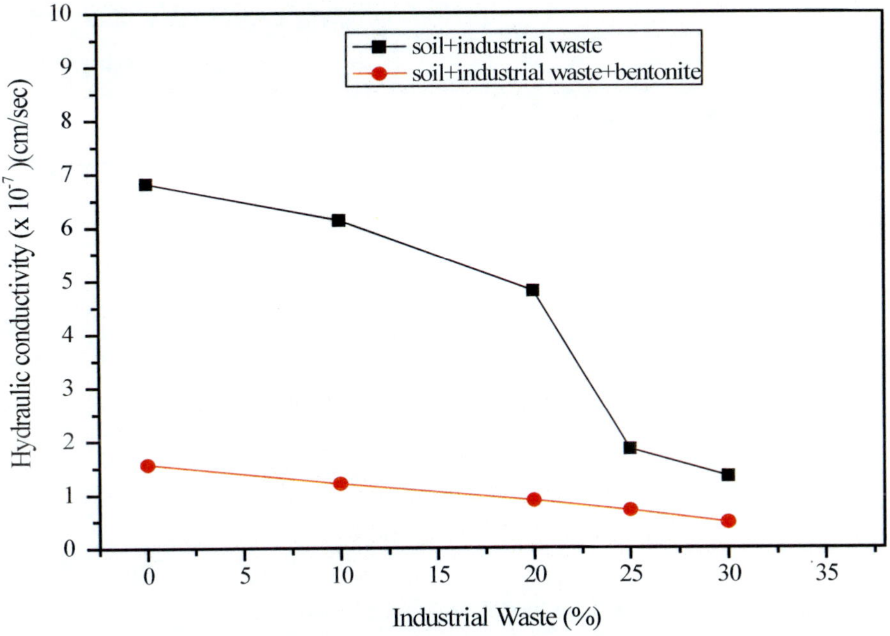

Fig. 3 Hydraulic conductivity values of soil–industrial waste–bentonite mixtures

4.4 *Atterberg Limits*

Both liquid limit and plasticity index decrease with the addition of industrial waste. Addition of industrial waste material produces coarser particles by flocculation and agglomeration. Thus, it requires less water to achieve liquid limit. Plastic limit increases slightly as the industrial waste content increases. Figure 4 shows the trend of these values. The decrease in plasticity indices is true indicator of soil improvement.

4.5 *Unconfined Compressive Strength Test*

Compressive strength for all samples significantly increases with waste material content up to 25%. After that, compressive strength decreased with increase in waste material. Reduction in shear strength is partly attributed to increase in clay size particle that filled the voids between soil particles at their contact points. UCS > 200 kPa is required for the material to perform as a liner material, where iron oxide particle acts as an adhesive agent between clay and increases the number of bonds between clay plates (Tombácz et al. 2004). Thus, strength of composite particle becomes stronger up to limited values (Fig. 5).

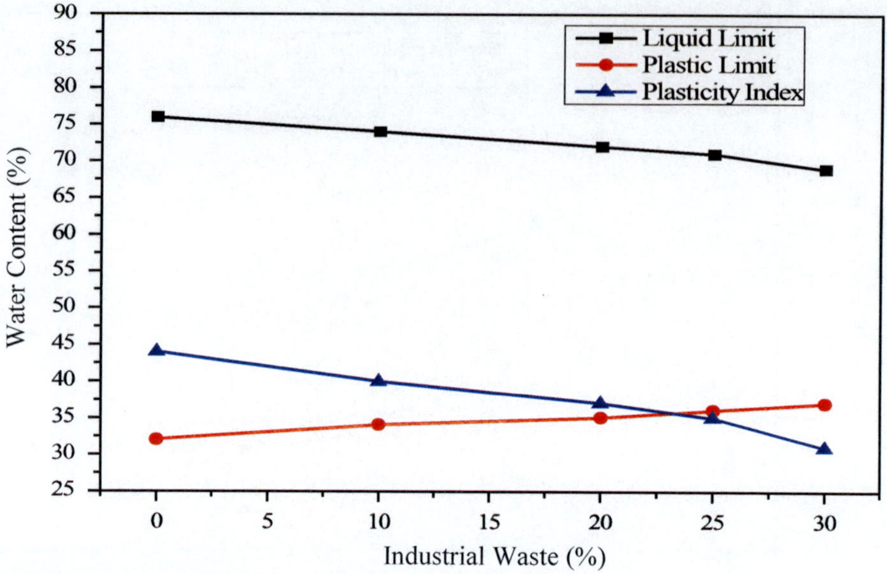

Fig. 4 Variation in plasticity indices with the addition of industrial waste

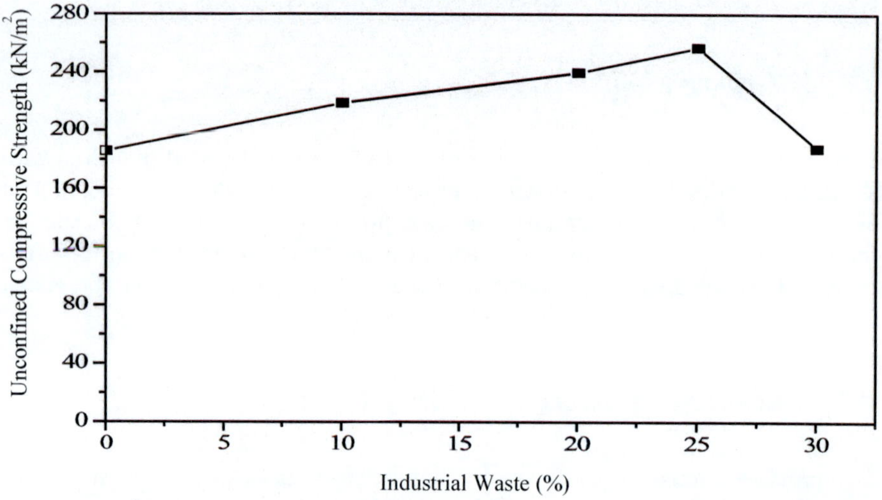

Fig. 5 UCS values with the addition of industrial waste

4.6 *Free Swell Index Test*

As the percentage addition of waste material increases, free swell index value decreases, showing very low degree of expansion. This is due to the presence of

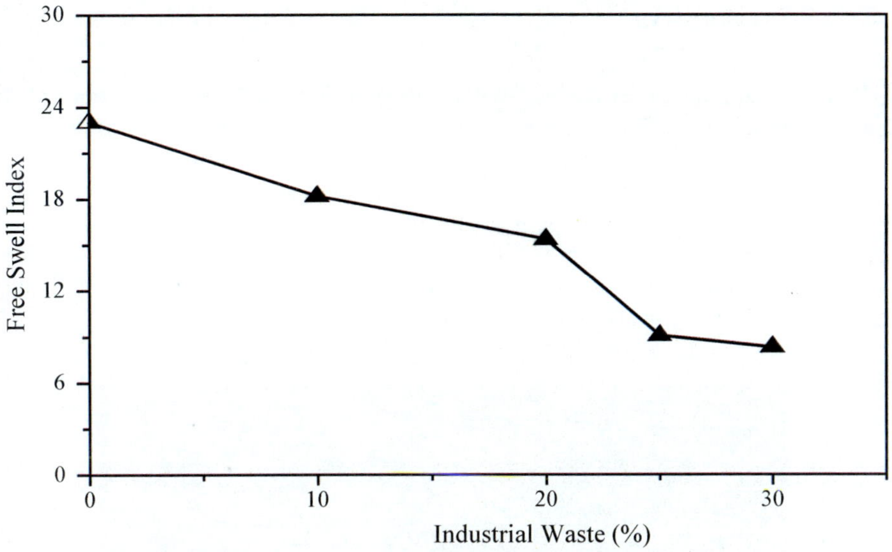

Fig. 6 Free swell index value variation with the addition of industrial waste

iron oxide in it which stabilizes the clay minerals by decreasing critical coagulation concentration, clay dispersion and clay swelling and by increasing microaggregation (Goldberg 1989). Figure 6 shows the variation of free swell index value with the addition of industrial waste.

4.7 Determination of Best Mix

From the studies conducted, the best mix which satisfying all the geotechnical requirements for a liner is found out.

Requirements of landfill liners are:

- Hydraulic conductivity should be less than 1×10^{-7} cm/s.
- It should possess adequate strength against any compressive loads. A minimum strength of 200 kN/m^2 is required.
- Plasticity index, PI > 10 should be there.
- Less swelling property.

From the results, it is found out that soil with 75% soil, 25% industrial waste and 5% bentonite satisfies all these parameters.

4.8 SEM Analysis

All the soil samples were observed under a magnification of 5 microns for uniformity and for comparison. Tests were conducted with natural soil and for optimum liner mixture which contains 25% industrial waste. SEM images were then compared to visualize the changes that caused in its soil morphology.

The SEM image corresponding to the natural soil is shown in Fig. 7 and that of optimum liner mix is shown in Fig. 8. From the figures, it is well explained that iron oxides in the industrial waste get precipitated over the clay particles. It just formed a coating of iron oxide above the clay particles.

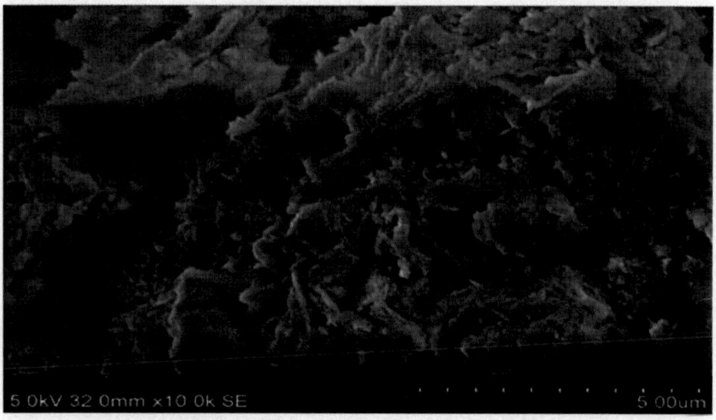

Fig. 7 SEM image of natural soil

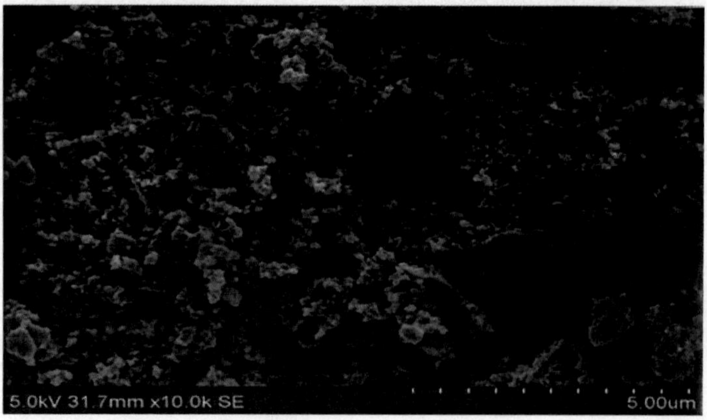

Fig. 8 SEM image of optimum liner mix

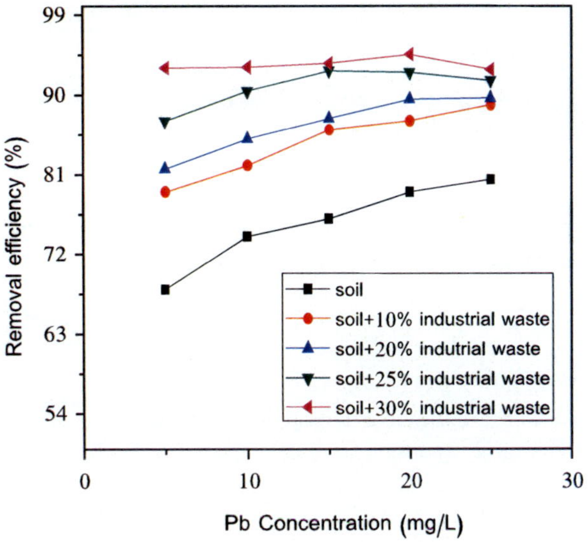

Fig. 9 Removal efficiency of lead from soil for various soil–industrial waste mixtures

4.9 Adsorption Characteristics

Since municipal solid waste contains various heavy metals like Pb, Cr, Cd, Fe, etc. the adsorption capacity of liner mix against these heavy metals needs to be studied.

For the present study, lead metal solution was prepared synthetically for investigating adsorption characteristics of soil–industrial waste mixtures. About 1.615 g of 99% $Pb(NO_3)_2$ is dissolved in distilled water in 1 L volumetric flask up to the mark to obtain 1000 ppm of lead stock solution. Both low and high concentrations of Pb(II) solutions were prepared by diluting with deionized water. Solutions of 5–25 ppm (Pb(II)) were prepared for the analysis.

4.9.1 Batch Adsorption Test

From the test results of batch adsorption test, it is seen that as the amount of iron oxide content in the soil increases in the soil the lead removal efficiency also increases. Figure 9 shows the removal efficiency of soil at different concentration of lead (II).

4.9.2 Adsorption Isotherms

The equilibrium data obtained from the batch adsorption tests were used to get linear and Langmuir adsorption isotherms. After determining concentration of Pb(II) in filtered solution using atomic adsorption spectrophotometer (AAS), the Pb(II) uptake by amended soil mixtures was calculated using the following equation:

$$q = \frac{(C_0 - C_e)\, v}{m} \tag{1}$$

where q is the amount of adsorbate adsorbed (mg adsorbate/g of adsorbent), and C_0 and Ce are the initial and final concentration of adsorbate (mg/L), v is the volume of solution taken (L) and m is the mass of adsorbent taken (g).

There are several adsorption isotherm models. Among them, most commonly used ones are linear isotherm, Langmuir isotherm. The linear adsorption isotherm can be produced by plotting a graph of q versus C_e. The relationship can be expressed as

$$q = K_d C_e \tag{2}$$

where K_d is the distribution coefficient.

The Langmuir equation was developed by Irving Langmuir in 1916, which is commonly written as:

$$q_e = \frac{abC_e}{1 + bC_e} \tag{3}$$

where a is number of moles of solute adsorbed per unit weight of adsorbent in forming a monolayer on the adsorbent surface; b is a constant related to energy of adsorption.

A separation factor or equilibrium parameter (RL) is a dimensionless parameter constant which can be used to predict the affinity between the sorbate and sorbent using Langmuir adsorption isotherm parameters. The expression for R_L is

$$R_L = \frac{1}{1 + bC_i} \tag{4}$$

where b is Langmuir constant, C_i is the initial concentration in mg/L. The value of R_L determines whether Langmuir isotherm is irreversible ($R_L = 0$), linear ($R_L = 1$), favourable ($0 < R_L < 1$) or unfavourable ($R_L > 1$). If R_L value lies between 0 and 1, adsorption is favourable (Table 3).

Table 3 Various parameters of linear isotherms and Langmuir isotherms

Isotherm type	Parameters	Proportion of soil–industrial waste–bentonite				
		95–0–5	85–10–5	75–20–5	70–25–5	65–30–5
Linear	K_d (L/g)	0.10	0.23	0.2268	0.33	0.42
	R^2	0.98	0.95	0.97	0.98	0.93
Langmuir	a (mg/g)	0.264	0.249	0.258	0.25	0.149
	b (L/g)	0.129	0.228	0.26	0.14	0.66
	R^2	0.98	0.99	0.995	0.95	0.8
	R_L	0.606	0.466	0.432	0.33	0.23

Since the regression values, R^2 of linear and Langmuir isotherms are greater values, it can be concluded that adsorption is suitable for both isotherms. The equilibrium parameter, R_L lies between 0 and 1 and it indicates favourable adsorption. Thus, it indicated that adsorption of lead ions on iron oxide amended clay liner is favourable.

Results show that adsorption of lead ions on iron oxide-rich industrial waste obeys both the isotherms. Iron oxide-rich industrial waste is a best adsorbent of Pb ions from soil.

5 Conclusions

- The effect of iron oxide-rich industrial waste on compaction, permeability, unconfined shear strength and plasticity indices and swelling properties was studied.
- The mere addition of industrial waste did not lower the hydraulic conductivity value below 1×10^{-7} cm/s which is mandatory for a liner material.
- With the addition of bentonite (5%), hydraulic conductivity value falls below the required limit and also shows sufficient soil improvement.
- Liquid limit and plasticity indexes were reduced with the addition of iron oxide-rich industrial waste, while plastic limit decreased to a small extent.
- The addition of iron oxide-rich industrial waste decreases the swelling property of natural soil.
- 25% industrial waste and 5% bentonite amended natural soil satisfy the landfill liner requirements. Cracking behaviour of the proposed liner mixture is to be studied for more validity.
- SEM analysis on the samples revealed the effect of iron oxide-rich industrial waste in the soil morphology. The structure become more closely packed with iron oxide coatings that formed over the clay particles.
- Adsorption capacity of natural soil against lead increased with the addition of industrial waste to it. Iron oxide amended clay liner material has great adsorption capacity of lead(II). It can be act as a best adsorbent of Pb ions.

References

Goldberg S (1989) Interaction of aluminum and iron oxides and clay minerals and their effect on soil physical properties: a review. Commun Soil Sci Plant Anal 20(11–12):1181–1207

Humsa TZ, Srivastava RK (2015) Impact of rare earth mining and processing on soil and water environment at Chavara, Kollam, Kerala: a case study. Proc. Earth Planet Sci (11):566–581

Ohtsubo M (1989) Interaction of iron oxides with clays. Clay Sci 7(4):227–242

Singh SK (2010) Sorption and leaching characteristics of heavy metals through clay. IGC-2009

Tombácz E, Libor Z, Illes E, Majzik A, Klumpp E (2004) The role of reactive surface sites and complexation by humic acids in the interaction of clay mineral and iron oxide particles. Org Geochem 35(3):257–267

Umar SEY, Elinwa AU, Matawal DS (2015) Feasibility of using tincal ore waste as barrier material for solid waste containment. IOSR J Mech Civ Eng 2(1):01–10

Yusuf USE, Slim MD, Uchechukwu EA (2016) Hydraulic conductivity of compacted laterite treated with iron ore tailings. Adv Civ Eng

Distribution and Health Risk Assessment of Heavy Metal in Surface Dust in Allahabad Municipality

Pawan Kumar and V. P. Singh

Abstract The environment is getting contaminated due to un-engineered disposal of waste produced by rapid industrialization and urbanization to fulfill the needs of increasing population. Environmental pollutants can harm people's health through a series of complex transport by various exposure pathways. Heavy metals are continuously being dump into the environment which create serious problems to human health. In this study, the soil and dust samples were collected to know their distribution and concentration of various heavy metals like As, Cd, Cr, Cu, Ni, Pb, Zn, etc., and their health risk assessment had been determined. The health risk was assessed using hazard quotient (HQ) and hazard index (HI). The samples were collected from ten different locations including industrial, residential, highways and mixed use in Allahabad, UP (India). The equipment used for heavy metals detection was hand held portable X-ray fluorescence (PXRF) analyzer. The dry and finely powdered soil samples were placed in a plastic container and packed by polypropylene thin film from both sides. Three measurements were performed, and detail descriptions were recorded for each soil samples. The assessment of health risk was analyzed for three exposure pathways: ingestion, dermal contact and inhalation. The main exposure pathway of heavy metals to both children and adults is ingestion. The result shows that the heavy metal concentration is found below hazardous level for adults and at higher level for child below six years. The study will be beneficial for the municipality in terms of non-point source pollution control and management to give the healthy environment to local people.

Keywords Hazard index · Hazard quotient · Health risk assessment · Heavy metals · Contamination

P. Kumar (✉) · V. P. Singh
Motilal Nehru National Institute of Technology Allahabad, Prayagraj, Uttar Pradesh 211004, India
e-mail: rce1652@mnnit.ac.in

V. P. Singh
e-mail: vps15783@mnnit.ac.in

M. Latha Gali and R. R. P. (eds.), *Problematic Soils and Geoenvironmental Concerns*, Lecture Notes in Civil Engineering 88,
https://doi.org/10.1007/978-981-15-6237-2_20

1 Introduction

Exponential industrialization and rapid urbanization have resulted in contamination of soil by metals from anthropogenic sources. The atmospheric deposition of dust and aerosol, vehicle emissions, and various industrial activities are important sources of soil metal pollution (Liu et al. 2016). Soil contamination with hazardous materials is common and burgeoning problems all over the world. The soils around industrial sites are prone to act as sink for metals since metals can be transported long distances and transferred to soil by atmospheric deposition, wastewater and discharge of solid waste containing metals. Major contaminants of soil are heavy metals like Cd, Pb, As, Cu, Cr, Hg, etc. Heavy metals are of great concern due to the large impact on the environment and human health (Ma and Singhirunnusorn 2012). Heavy metal contaminated soil adversely affects the whole ecosystem when these toxic heavy metal migrate into groundwater or are taken up by flora and fauna, which may result in great threat to ecosystems due to translocation and bioaccumulation (Satpathy and Prakash 2014). Whenever soil gets contaminated, it contaminates air and water, consequently our body gets affected by various diseases such as cancer, tuberculosis, chronic lower respiratory diseases as well as many other serious health problems. Toxic metals in urban areas are a subject of great concern, due to their non-biodegradable nature and long resident time. The prolonged presence of the contaminants in the environment and their close proximity to the human population significantly increase the exposure of the urban population to metals via inhalation, ingestion and dermal contact (Suryawanshi et al. 2016). The effect of heavy metals on human health and their cycle is shown in Fig. 1. Cd affects kidney, liver and gastrointestinal tract. Excess Pb causes abdominal pain, confusion, headache, anemia and irritability. Road dust receives a large number of heavy metals inputs from a variety of mobile or stationary sources. In India, soil pollution and road dust become a serious problem with the rapid industrialization and urbanization during last decade. This study had been conducted in Allahabad, where the maximum crowd of people of the world collected to take a bath in the Sangam of three holy rivers known as Triveni of Ganga, Yamuna and Sarasvati. Hence, the urban soil of Allahabad has been intensively affected by human activities and gets contaminated. Therefore, Allahabad, the place of religious faith, art and culture, having high population density and heavy burden of human activities, needs to have such type of assessment.

1.1 Study Area

Allahabad, known as Prayag, is a city located at south east of Uttar Pradesh, India, and it is the administrative headquarters of Allahabad district. It is the most-populous district in UP with its center located at 25° 44′ N and 81° 85′ E. As per census 2011, Allahabad is the seventh most-populous city in the state with an estimated population of 1.21 million in the city with the city's metropolitan area being 70.5 km^2 (27.22 sq.

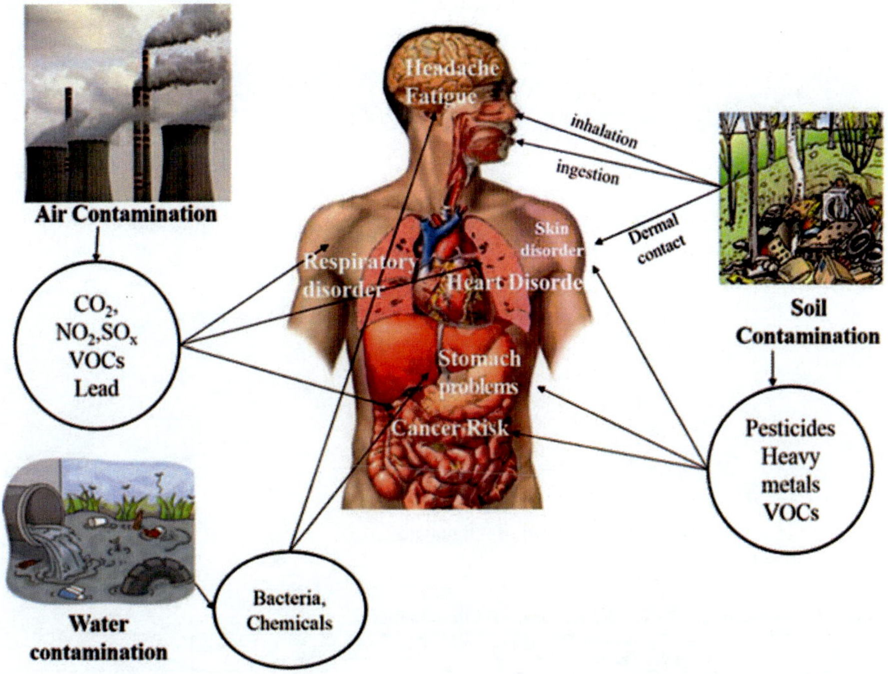

Fig. 1 Contamination source and its effect on human body (Singh 2016)

miles). Overall Allahabad has a very stable and diverse economy comprising various sectors such as State and Central government offices, education and research institutions, real estate, retail, banking, tourism and hospitality, agriculture-based industries, railways, transport and logistics, miscellaneous service sectors and manufacturing. Allahabad is also the head quarter of north central railways.

1.2 Site Investigation and Sample Collection

A simple survey of the site was conducted to identify pollution linkages, not obvious from the desk study. I looked around the site to identify the sources of contamination of soil, surface disturbance, chemical stores, waste storage areas and other adjacent properties. A total of 120 soil and dust samples were collected from ten different places of Allahabad city during November 2017. Sample collection was based on random method in which samples were collected from parks, major streets, educational institutes, commercial areas, industrial areas, etc. (Fig. 2). At each sampling site, about 200 gm of dust and soil samples were collected by sweeping using a polyethylene brush and tray. All the collected samples were stored in a self-locking polyethylene bags, labeled and then transported to the laboratory. Descriptions of

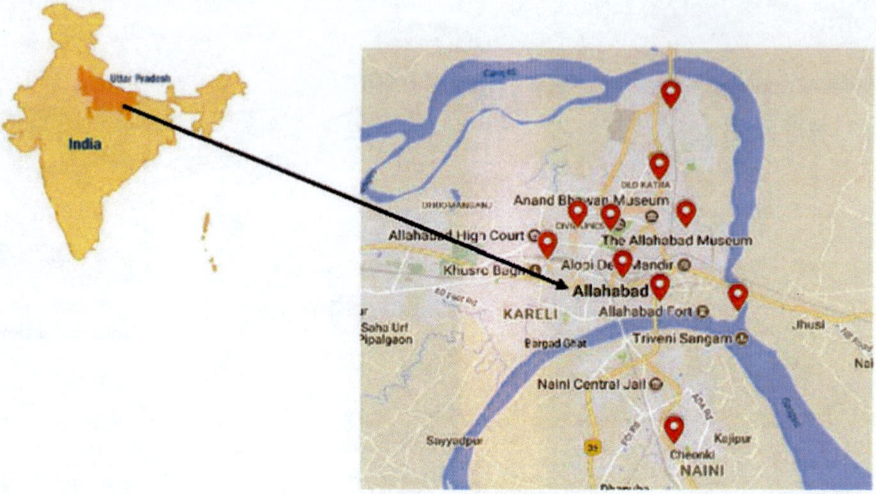

Fig. 2 Map showing sampling location of Allahabad

Table 1 Description of sample location of Allahabad city

S. No.	Location	Latitude	Longitude
SL1	Bank Road Crossing	25° 28′ 13.20″	81° 51′ 43.90″
SL2	Allahpur	25° 26′ 39.28″	81° 52′ 29.65″
SL3	Triveni Sangam	25° 28′ 07.89″	81° 52′ 54.19″
SL4	Minto Park	25° 25′ 55.08″	81° 51′ 44.68″
SL5	Naini (Alstom Company)	25° 23′ 08.92″	81° 52′ 03.83″
SL6	Ram Bagh Railway Station	25° 26′ 22.85″	81° 50′ 55.02″
SL7	Bharat Petroleum Civil Lines	25° 27′ 18.07″	81° 50′ 06.54″
SL8	Allahabad Junction	25° 26′ 40.65″	81° 49′ 27.07″
SL9	Company Bagh	25° 27′ 17.07″	81° 51′ 01.33″
SL10	MNNIT Gym Ground	25° 29′ 41.32″	81° 51′ 53″

sampling locations and Global positioning system (GPS) information are presented in Table 1.

1.3 Sample Preparation and Analysis

Soil samples were dried for at 60 °C for 24 h in the oven. The dry soil sample was finely powdered and then sieved to minus 2 mm mesh size; soil samples were placed in circular white plastic container (height 8.60 mm and inner diameter 10.45 mm)

Fig. 3 Sample preparation

locked by polypropylene thin film from both sides. The average weight of soil sample in each container was 11.40 gm (Fig. 3).

The equipment used for heavy metals assessment was hand held portable X-ray fluorescence (PXRF) analyzer (Elvatech), which is equipped with a 4 W Rh, Au, or Ta anode (per application) X-ray tube and large-area silicon drift detector. An internal instrument standard calibration was performed prior to sample analysis. The calibration of the instrument was done by testing the certified standard soil samples, which 99% match with the testing samples. Three measurements were performed, and a brief description was taken for each soil sample. The time of analysis was fixed at 60 s for each analysis and one sample was analyzed three times after that the average result was shown on the screen of PXRF. The data were recorded in the internal memory of the instrument. All results were transferred from PXRF to computers for further analysis and estimation of results.

2 Heavy Metal Concentrations

The concentrations of heavy metals like Hg, As, Ni, Zn, Cr, Fe, Co, Pb, Cu and Cd in the dust and soil of Allahabad city are estimated. Each heavy metal shows a wide range of values. The mean concentration of these heavy metals was 7.4 mg/kg, 9.2 mg/kg, 38.3 mg/kg, 159.6 mg/kg, 160 mg/kg, 20,996.7 mg/kg, 72.1 mg/kg, 45.7 mg/kg, 49.4 mg/kg, 33.6 mg/kg and 33.6 mg/kg for Hg, As, Ni, Zn, Cr, Fe, Co, Pb, Cu and Cd, respectively. The distribution of these heavy metals had irregular pattern. The concentration of Hg, Co and Cu was higher at Allahpur among ten sampling places, As, Ni and Zn were higher at Allahabad junction, Fe had higher concentration at Bharat petroleum civil lines, Pb at Ram Bagh railway station, Cd at Naini (Alstom company), and Cr was found higher at Sangam area. The concentration of these heavy metals was compared with the background value of the concentration of heavy metals in various standards of different countries and Indian natural soil background value. The source of these heavy metals was reported as of very complex nature which may be associated with transportation, urban industrial and human activities. The industrial processes such as metallurgical, electroplating, chemical and fuel combustion may be main reasons for metal emissions in the environment.

3 Pollution Index (PI)

Pollution indexes of heavy metals were used to assess the degree of metal contamination in the topsoil around the Allahabad municipality. The PI was calculated using the soil environment quality standard of Indian natural soil background values. The PI was defined as follows:

$$PI = \frac{C_n}{S_n} \tag{1}$$

where C_n is the measured concentration of the element n in soils (mg/kg), and S_n is the geochemical background concentration of element n (mg/kg). The degree of heavy metal contamination in the soils can be classified into the following categories: non-pollution ($PI \leq 1$); low-level pollution ($1 < PI \leq 2$); moderate-level pollution ($2 < PI \leq 3$); and high-level pollution ($PI > 3$) (Liu et al. 2016). The concentration of Cu had no effect on human health since $PI < 1$. Cd, Zn and Pb had high level of pollution since they had $PI > 3$ and had adverse effect on human health. The Pb concentration in dusts was attributed to emissions from burning of fossil fuel. Cr and Ni had low level of pollution due to its increase concentration in dust and surface soil (Table 2).

Table 2 Pollution index of heavy metals

Heavy metals	Concentrations (mg/kg)		Pollution index (PI)
	Indian natural soil background	This study	
Cd	0.9	33.6	37.3300
Cr	114	160	1.4035
Cu	56.5	49.4	0.8743
Ni	27.7	38.3	1.3830
Pb	13.1	45.7	3.4885
Zn	22.1	159.4	7.2127

4 Exposure Dose

The model used in this study to calculate the exposure of human to dust metals is based on those developed by Environmental Protection Agency of USA (USEPA). According to the exposure factors handbook (USEPA 2002), the chronic average daily dose (CADD) (mg/kg/day) of a pollutant via ingestion, dermal contact and inhalation as exposure pathways can be estimated using Eqs. (2), (3) and (4).

$$\mathrm{CADD_{ing}} = \frac{c \times R_{\mathrm{ing}} \times CF \times EF \times ED}{BW \times AT} \tag{2}$$

$$\mathrm{CADD_{inh}} = \frac{c \times R_{\mathrm{inh}} \times EF \times ED}{PEF \times BW \times AT} \tag{3}$$

$$\mathrm{CADD_{der}} = \frac{c \times SA \times CF \times SL \times ABS \times EF \times ED}{BW \times AT} \tag{4}$$

where $\mathrm{CADD_{ing}}$ is daily exposure amount of metals through ingestion (mg/kg/day); $\mathrm{CADD_{inh}}$ is daily exposure amount of metals through inhalation (mg/kg/day); $\mathrm{CADD_{der}}$ is daily exposure amount of metals through dermal contact (mg/kg/day). The exposure factors for these models are taken from environmental site assessment guideline (2009), which are soil ingestion rate (R_{ing}), inhalation factor (R_{inh}), exposure duration (ED), exposure frequency (EF), average body weight (BW), average time for carcinogenic and non-carcinogenic risk (AT), conversion factor (CF), surface area of the skin that contact with soil (SA), skin-adherence factor (SF), dermal absorption factor for carcinogenic and non-carcinogenic risk (ABS). The values of these factors are combined and had been taken from the (USEPA 2011; WHO 2002) and actual Indian data which is shown in Table 3.

5 Health Risk Assessment

Sensitivity of individuals is likely to be affected by age, sex, nutritional and pregnancy status and combination of these (IEH 1999). Children may differ from adults in a range of behavioral and physiological parameters like dietary difference. Children can eat much greater quantities of particular foods (e.g., dairy products, soft drinks, soil and some fruit and vegetables) than adults on a body weight basis (William 1999). Gender differences also have to be taken in to consideration when identified potential exposures and pathways because there are anthropometric (e.g., height, weight, body surface area, fat content, muscle mass, etc.) composition differences between males and females. Using USEPA guidelines on toxicity level, we have analyzed our data on the basis of bioavailability method, since we have considered only that amount which is absorbed by central part of human body. We had calculated HQ and HI values which are used to assess human health risk of metal exposure of

Table 3 Factors used for estimation of CADD for carcinogenic and non-carcinogenic risk USEPA (2011) and WHO (2002)

Factor	Unit	Value			
		Adult	Child	For worker	
				Outdoor	Indoor
R_{ing}	mg/day	100	200	100	50
ED	Year	24	6	25	25
EF	day/year	350	350	305	305
BW	Kg	51.5	18.75	51.5	51.5
AT (NCR)	days	8760	2190	25,550	25,550
AT (CR)	days	24,258	24,258	24,258	24,258
CF	Kg/mg	1×10^{-6}	1×10^{-6}	1×10^{-6}	1×10^{-6}
SA	cm^2/event	5700	2800	3300	3300
SF	mg/cm^2	0.07	0.2	0.2	NA
ABS (NCR)	mg/cm^2	0.001	0.001	0.001	0.001
ABS (CR)	mg/cm^2	0.03	0.03	0.03	0.03
R_{inh}	m^3/day	20	20	20	20

soil and dust. In this study, two types of risk assessment have been performed: (1) Non-carcinogenic risk assessment, and (2) carcinogenic risk assessment.

5.1 Non-carcinogenic Risk Assessment

Chronic average daily demand (CADD) has been calculated by using the Eqs. (2), (3) and (4) and factors used for obtaining the value of CADD were taken from USEPA (2011). RfD values are shown in Table 4. After calculating the CADD for the entire exposure pathway, non-carcinogenic hazard quotient (HQ) was calculated by dividing the CADD by reference dose (RfD).

$$\mathrm{HQ}_{ij} = \frac{\mathrm{CADD}_{ij}}{\mathrm{RfD}_{ij}} \tag{5}$$

where i = heavy metal and j = exposure pathway. Reference dose (RfD) is an estimation of maximum permissible risk on human population through daily exposure taking into consideration by sensitivity of group during a lifetime period. HQ ≤ 1 indicates no adverse effect and HQ > 1 indicates adverse health effects according to USEPA (1986). These heavy metals above a certain concentration may pose non-carcinogenic and carcinogenic health risks to human. From hazard quotient, the hazard index had been calculated.

Table 4 Reference dose and cancer slope factor for the heavy metals for different exposure (USEPA 2002)

Metal	CSF	RfD_{ing}	RfD_{inh}	RfD_{der}	References
Cd	6.3	0.001	0.001	0.00001	USEPA (2002); USEPA (1993)
Co	–	0.02	–	–	
Cr (VI)	0.5	0.003	0.0001	0.00006	
Cu	–	0.0371	0.0402	0.0019	
Fe	–	0.7	–	–	
Pb	0.0085	0.0035	0.00352	0.000525	
Zn	–	0.3	0.3	0.06	
Ni	–	0.02	0.0206	0.001	
As	1.5	0.000015	0.0003	0.0003	USEPA (2005)
Hg	–	0.0003	0.00016	0.00016	USEPA (2005)

$$HI = \sum \sum HQ_{ij} \tag{6}$$

$HI \leq 1$ indicates no non-carcinogenic risk to the human health and $HI > 1$ indicates non-carcinogenic risk to the human health (USEPA 2001). Hazard index (HI) of each sample location for adult and child is shown below.

5.2 *Carcinogenic Risk Assessment*

Carcinogenic risk was determined by multiplying lifetime average daily dose (LADD) with cancer slope factor (CSF). It indicates the probability of cancer to the exposed receptor during lifetime period.

$$\text{Cancer Risk} = \text{LADD} \times \text{CSF} \tag{7}$$

Cancer slope factor is shown in Table 4. LADD was calculated using Eq. (2) by replacing average time (AT) with lifetime (70 years × 365 days). Total carcinogenic risk was calculated by the summation of risk induced by all the exposure pathways (Table 5).

$$\text{Total risk} = \text{Risk}_{ing} + \text{Risk}_{inh} + \text{Risk}_{der} \tag{8}$$

According to USEPA (USEPA 2005), the risk $\geq 1 \times 10^{-6}$ indicates carcinogenic risk to the receptor from the exposure pathways. Carcinogenic heavy metals mainly are Pb, Cd, Cr and As and their carcinogenic risk for adult and child are shown in Table 6.

Table 5 Hazard Index (HI) of sample location for adult and child (0–6 years)

S. No.	Sample location	HI (Adult)	HI (Child)
1	Bank Road Crossing	0.42374	2.54162
2	Allahpur	0.57645	3.49176
3	Triveni Sangam	0.66151	4.07258
4	Minto Park	0.33699	2.02750
5	Naini (Alstom Company)	0.56378	3.43857
6	Ram Bagh Railway Station	0.36145	2.16303
7	Bharat Petroleum Civil Lines	0.43038	2.59543
8	Allahabad Junction	0.45214	2.63463
9	Company Bagh	0.32909	2.00114
10	MNNIT Gym ground	0.32181	1.93024
11	Mean value	0.379583	2.68965

6 Conclusions

The heavy metal in the top soil and dust in Allahabad city is investigated deeply in the present study. The average concentration of Hg, As, Ni, Zn, Cr, Fe, Co, Pb, Cu and Cd in sample was 7.4, 9.2, 38.3, 159.4, 160, 20,996.7, 72.1, 45.7, 49.4, and 33.6 mg/kg, respectively, and maximum concentration was 13, 14, 52, 319, 533, 26,550, 127, 101, 122, and 58 mg/kg, respectively. From Table 8, it is clear that the concentration of heavy metals in Allahabad municipality is higher than the background values of the world according to Indian natural soil background value and CNEMC (CNEMC 1990), which indicating that the pollution may come from anthropogenic activities (Table 7).

The Cr, Cu, Zn, Cd and Pb should be paid more attention. Peoples are exposed to pollutants via ingestion inhalation and dermal contact. The main exposure pathway of heavy metal to both adult and children is ingestion. The values of HQ for these pathways in this study decrease in order to ingestion > dermal contact > inhalation. All the values of HQ for single heavy metals are lower than the safe level (<1) except the value of HQ for children via ingestion of Cr at Triveni Sangam which is 1.81575 (>1), needs attentions. The values of HI for adult (mean = 0.3795) are below than the safe level (<1) indicating no health risk for adults while for children HI (mean = 2.68965) is greater than safe level indicating health risk for children from 0 to 6 years old. For every heavy metal, the values of HI for children are higher than those for adults, indicating that children face greater harmful health risk due to dust and top soil of Allahabad municipality. The pollution index of Pb, Zn and Cd is reported as higher than high-level pollution which shows the higher concentration of these heavy metals in dust and soil of Allahabad city. This research will be quite useful for both residents (in taking protective measures) as well as government in alleviating

Table 6 Carcinogenic risk for adult and child for different exposure pathways

Heavy metals	Conc. (mg/kg)	Risk from combined pathways	
		Adult	Child
Bank road crossing			
Pb	35	6.023E-07	3.76059E-06
Cr	139	0.0001449	0.000878532
Cd	29	0.0003809	0.002320409
As	12	3.753E-05	0.000227532
Allahpur			
Pb	67	1.18739E-06	7.19885E-06
Cr	137	0.000142821	0.000865892
Cd	57	0.000748714	0.004560804
As	11	3.44021E-05	0.000208571
Triveni Sangam			
Pb	32	5.6711E-07	3.4383E-06
Cr	533	0.00055565	0.00336876
Cd	21	0.00027584	0.0016803
As	9	2.8147E-05	0.00017065
Minto Park			
Pb	42	7.4434E-07	4.5127E-06
Cr	76	7.9229E-05	0.00048035
Cd	30	0.00039406	0.00240042
As	7	2.1892E-05	0.00013273
Naini (Alstom Company)			
Pb	22	3.899E-07	2.3638E-06
Cr	160	0.0001668	0.00101126
Cd	58	0.0007618	0.00464082
As	10	3.127E-05	0.00018961
Ram Bagh Railway Station			
Pb	101	3.597E-06	1.085E-05
Cr	77	0.0008403	0.0004867
Cd	29	0.0022198	0.0023204
As	7	0.0002176	0.0001327
Bharat Petroleum Civil Lines			
Pb	26	4.61E-07	2.7936E-06

(continued)

Table 6 (continued)

Heavy metals	Conc. (mg/kg)	Risk from combined pathways	
		Adult	Child
Cr	204	0.000213	0.00128936
Cd	23	0.000302	0.00184032
As	10	3.13E-05	0.00018961
Allahabad Junction			
Pb	71	1.2583E-06	7.6286E-06
Cr	89	9.2781E-05	0.00056251
Cd	33	0.00043347	0.00264047
As	14	4.3784E-05	0.00026545
Company Bagh			
Pb	33	5.8484E-07	3.5457E-06
Cr	110	0.00011467	0.00069524
Cd	29	0.00038092	0.00232041
As	5	1.5637E-05	9.4805E-05
MNNIT Gym Ground			
Pb	28	4.9622E-07	3.0085E-06
Cr	75	7.8187E-05	0.00047403
Cd	27	0.00035465	0.00216038
As	7	2.1892E-05	0.00013273

Table 7 Maximum concentration of particular heavy metal sample at location, mean and their standard deviation

Heavy metals	Maximum concentration of particular heavy metals among all sample location		Sample mean ($\bar{x}$)	Sample standard deviation (s)
	Location	Max. Conc (ppm)		
Hg	Allahpur	13	7.4	2.85
As	Allahabad Junction	14	9.2	2.74
Ni	Allahabad Junction	52	38.3	8.2467
Zn	Allahabad Junction	319	159.4	82.79
Cr	Triveni Sangam	533	160	137.737
Fe	Bharat Petroleum Civil Lines	26,550	20,996.7	11,827.98
Co	Allahpur	127	72.1	20.4475

(continued)

Table 7 (continued)

Heavy metals	Maximum concentration of particular heavy metals among all sample location		Sample mean ($\bar{x}$)	Sample standard deviation (s)
	Location	Max. Conc (ppm)		
Pb	Ram Bagh Railway Station	101	45.7	25.5867
Cu	Allahpur	122	49.4	26.916327
Cd	Naini (Alstom Company)	58	33.6	13.06768

Table 8 Comparison of total concentration of heavy metal in dust in different countries and cities (ppm)

Location	Concentrations of metals (ppm)					References
	Cr	Cu	Zn	Cd	Pb	
Hawaii, USA	273	167	434	–	106	Sutherland and Tolosa (2000)
Palermo, Italy	218	98	207	1.1	544	Varrica et al. (2003)
Ottawa, Canada	43.3	65.84	112.5	0.37	39.05	Rasmussen et al. (2001)
Birmingham, UK	–	466.9	534	1.62	48	Charlesworth et al. (2003)
Hong Kong, China	–	110.00	3840.0	–	120	Yeung et al. (2003)
Xi'an, China	167.3	94.98	421.5	–	230.5	Han et al. (2006)
Shanghai, China	159.3	196.8	733.8	1.23	294.9	Shi et al. (2008)
Kuala Lumpur, Malaysia	–	35.5	344	2.9	2466	Ramlan and Badri (1989)
Delhi, India	148.8	191.7	284.5	2.65	120.7	Suryawanshi et al. (2016)
Beijing, China	69.33	72.13	219.20	0.64	201.82	Taner et al. (2008)
Allahabad, India	160	49.4	159.4	33.6	45.7	This study
Indian natural soil background	144	56.5	22.1	0.9	13.1	Kuhad et al. (2008)
China soil guidelines	200	100	250	0.3	300	NEPA (1995)
Canada soil guidelines	64	63	200	10	140	CCME (2007)

heavy metal pollution of urban environment. It would facilitate the decision-makers to manage and/or dispose of contaminated soil dust and minimize health risks on urban inhabitants. In conclusion, it is indicated that the health risk values obtained from this study in dust and top soil of Allahabad municipality are in the negligible range for adults while it is in alarming range for children.

References

CCME (2007) Canadian soil quality guidelines for the protection of environmental and human health. Canadian council of ministers of the environment, Winnipeg

Charlesworth S, Everett M, McCarthy R, Ordonez A, de Miguel E (2003) A comparative study of heavy metal concentration and distribution in deposited street dusts in a large and a small urban area: Birmingham and Coventry, west Midlands, UK. Environment International. 29, pp 563–573

CNEMC (China National Environmental Monitoring Center) (1990) The back grounds of soil environment in china. China Environmental Science Press, Beijing

Han Y, Du P, Cao J, Posmentier ES (2006) Multivariate analysis of heavy metal contamination in urban dusts of Xi'an, central China. Sci Total Environ 355:176–186

IEH (1999) Assessment on the ecological significance of endocrine disruption: effects on reproductive function and consequences for natural population. Assessment A4. MRC institute for environment and health, Leicester, London

Kuhad MS, Malik RS, Singh A, Dahiya IS (2008) Background levels of heavy metals in agricultural soils of Indogangatic plains of Haryana. J Indian Soc Soil Sci 3:700–705

Liu C, Lu L, Huang T, Huang Y, Ding L, Zhao W (2016) The distribution and health risk assessment of metals in soils in the vicinity of industrial sites in Dongguan, China. Int J Environ Res Pub Health 13:832

Ma J, Singhirunnusorn W (2012) Distribution and health risk assessment of heavy metals in surface dusts of Maha Sarakham Municipality. Soc Behav Sci 50:280–293

NEPA (1995) Environmental quality standards for soils. National environmental protection agency of china, Beijing. GB 15618–1995

Ramlan MN, Badri MA (1989) Heavy metals in tropical city street dust and roadside soils: a case of Kuala Lumpur, Malaysia. Environ Tech Lett 10:35–444

Rasmussen PE, Subranmanian KS, Jessiman BJ (2001) A multi-element profile of house dust in relation to exterior dust and soils in the city of Ottawa, Canada. Sci Total Environ 267:125–140

Satpathy D, Vikram Reddy M, Prakash Dhal S (2014) Risk assessment of heavy metals contamination in paddy soil, plants, and grains (*Oryza sativa L.*) at the east coast of India. BioMed Res Int, Hindawi, 11, Article ID 545473

Shi G, Chen Z, Xu S, Zhang J, Wang L, Bi C, Tenj J (2008) Potentially toxic metal concentration of urban soils and roadside dust in shanghai, china. Environ Pollut 156:251–260

Singh VP (2016) Human health risk based assessment of contaminated land and remediation selection appraisal. A thesis, CED, MNNIT Allahabad

Suryawanshi PV, Rajaram BS, Bhanarkar AD, Chalapati Rao CV (2016) Determining heavy metal contamination of road dust in Delhi, India. Atmosfera 29(3):221–234

Sutherland RA, Tolosa CA (2000) Multi element analysis of road dust-deposited sediments in an urban drainage basin, Honolulu, Hawaii. Environ Pollut 110:483–495

Taner PA, Ma HL, Yu PKN (2008) Fingerprinting metals in urban street dust of Beijing, Shanghai and Hong Kong. Environ Sci Technol 42:7111–7117

USEPA (1986) Quality criteria for water. EPA 440/5-86-001. 1 May 1986. Office of Water Regulations and Standards, Washington DC

USEPA (1993) Reference Dose (RfD), description and use in health risk assessments. Background document 1A. Integrated risk information system (IRIS), Washington DC

USEPA (2001) Toxics release inventory executive summary. Office of Water Regulations and Standards, Washington DC

USEPA (2002) Supplemental guidance for developing soil screening levels for superfund sites. OSWER 9355

USEPA (2005) Guidelines for carcinogen risk assessment. US environmental protection agency risk assessment forum, EPA/630/P-03?001F, Washington DC

USEPA (2011) Exposure factors handbook, 2011 edn. National center for environmental assessment, Washington DC

Varrica D, Dongarra G, Sabatino G, Monna F (2003) Inorganic geochemistry of roadway dust from the metropolitan area of Palermo, Italy. Environ Geol 44:222–230

WHO (2002) Water pollutants: biological agents, dissolved chemicals, non-dissolved chemicals, sediments, heat. WHO CEHA, Amman, Jordan

William ER (1999) The built environment and the ecosphere: a global perspective. Build Res Inf 27(4–5):206–220

Yeung ZLL, Kwok RCW, Yu KN (2003) Determination of multi-element profiles of street dust using energy dispersive X-ray florescence (EDXRF). Appl Radiat Isot 58:339–346

Soil Amendment Using Marble Waste for Road Construction

Ankush Kumar Jain, Mrinal Gupta, and Arvind Kumar Jha

Abstract Bulk utilization of marble dust needs to be done to enhance the sustainable construction activities and to embrace safe environment. In the present study, the geotechnical properties of untreated and treated black cotton soil with marble dust are studied in detail to check its potential to improve the properties of soil. The experimental works include particle size analysis, free swell index, specific gravity, compaction characteristics, California bearing ratio (CBR), and pH. The results show that the physical properties of black cotton soil are improved significantly by addition of marble dust. An addition of marble dust leads to an increase in the pH and reduction in the swell index of soil. Further, the CBR values of mixes compacted at their respective optimum water content and maximum dry density increases with addition of marble dust up to 25%. The improvement in the gradation of soil and the formation of cementitious compounds are the key factors to improve the properties of soil with addition of marble dust.

Keywords CBR · Marble dust · Physical properties

1 Introduction

The process of improvement in the properties of soils to meet desired specific criteria by using various methods is generally termed as soil stabilization (Jha and Sivapullaiah 2015). A large portion of India (i.e., 20% area of country) is covered with black cotton soil (a common name of expansive soil). Expansive soil is highly susceptible under moisture variation and possesses a deleterious effect on the structure constructed on it (Wheeler et al. 2003). Hence, expansive soil is considered as a

A. K. Jain · M. Gupta · A. K. Jha (✉)
Manipal University Jaipur, Jaipur, Rajasthan, India
e-mail: erarvindnp@gmail.com

A. K. Jha
Indian Institute of Technology Patna, Bihta, Bihar 801103, India

M. Latha Gali and R. R. P. (eds.), *Problematic Soils and Geoenvironmental Concerns*, Lecture Notes in Civil Engineering 88,
https://doi.org/10.1007/978-981-15-6237-2_21

problematic soil for, and is not used for pavement or, airfield construction (Okagbue and Onyeobi 1999).

Rajasthan state of India produces about 85% of total marble dust of the country which is 90% of the world (Pappu et al. 2006). Many environmental and public issues such as air and water pollution, contamination of surface, and groundwater have been encountered due to production of marble waste (Celik and Sabah 2008).

The metamorphic rock is the parent source of marble formation. Bulk amount of marble slurry/dust generates as by-products during cutting and polishing of marble for commercial use. The byproduct marble slurry/dust is being dumped in an unscientific manner which leads to the reduction in porosity, percolation, and water absorption capacity of soils and thereby, affecting the productivity of agricultural land (Vishwakarma and Rajuput 2013). Several applications of marble waste have been done by construction industries in the manufacture of concrete, bricks, cement, ceramic tiles, acetated calcium carbonate, lime, hollow blocks, tiles, and plastic coating. The utilization of marble waste in construction activities is considered as beneficial in terms of both economic and environmental point of view (Singh et al. 2017). Hence, bulk utilization of marble waste is essential and can be achieved through its possible utilization in highway construction.

It has been reported that the key chemical compositions of marble dust are calcium carbonate ($CaCO_3$), calcium oxide (CaO), magnesium carbonate ($MgCO_3$), and magnesium oxide (MgO). Further, other minor impurities present in marble dust are usually clay minerals, quartz, iron oxide, and graphite (Segadães et al. 2005). The improvement of soils by using chemical stabilizers is due to the ionic reactions between calcium and ions (silica and alumina) in the presence of water (Bell 1996). The ionic reactions cause the formation of cementitious compounds which bind the particles together and thereby, development of compacted matrix. Calcium ion is predominated in marble dust and is considered as an inert material. However, marble dust has several other advantageous properties such as higher shear strength value, low plasticity, granular in nature, and low compressibility. Due to such beneficial properties, it can be utilized to amend the properties of expansive soils for highway construction.

Several researches have been done on the stabilization of problematic soil with the application of waste materials like rick husk, stone quarry dust, GGBS, and fly ash. Still, very few researchers have worked on the utilization of marble dust to stabilize the problematic soil for sustainable and eco-friendly construction. Misra et al. (2009) mentioned that marble slurry waste can be used effectively for subgrade preparation. The plasticity and California bearing ratio (CBR) of red tropical soils are improved with marble dust; however, strength development is not enough to suit as base materials (Okagbue and Onyeobi 1999; Osinubi 2006). Tozsin et al. (2014) suggested that marble waste can be used for soil amendment in neutralizing the acidic behavior. However, relatively very few research has been carried out by considering the factors like environmental, physical, and chemical.

The present work is focused to investigate the potential of marble dust to improve the geotechnical properties of black cotton soil. To accomplish the objectives, the characterizations of soil and marble dust are done by performing various laboratory experimental tests such as Atterberg's limits, particle size analysis, compaction characteristics, CBR, mineralogical, and microstructure. Further, the effects of varying marble dust (0–25%) on the physical and engineering behaviors (i.e., specific gravity, liquid limit, free swell index, pH, compaction characteristics, and CBR) are examined thoroughly.

2 Materials Used and Methodologies Followed

The geotechnical properties of soil and marble dust are presented in Table 1. The soil used in the present work was collected from the depth of 1–1.5 m below the ground surface, Shivdaspura village, Jaipur district, Rajasthan, India. The marble dust is collected from the query site of Kishangarh, Ajmer, Rajasthan. For all experimental purpose, materials passing through IS 425-micron sieve are used. The experimental tests have been done as per Indian Standard (IS) procedures mentioned in Table 1.

The particle size analysis of soils shows the presence of sand-sized particle (4.75–0.075 mm) of 11.00%, silt-sized particle (0.075–0.002 mm) of 13.00%, and clay-sized particle (<0.002 mm) of 76.00%. It is observed that the soil is predominated with clayey size particle whereas marble dust contains a predominant amount of sand size particle (i.e., 96.50%). The specific gravity of soil (i.e., 2.38) is observed to be less than or same as the marble dust (i.e., 2.74). The liquid limit of soil is observed to be higher than marble dust. The free swell index of soil is obtained to be 70%. The pH of soil and marbled dust is observed to be 8.07 and 8.20, respectively.

The optimum water content (OWC) and maximum dry density (MDD) of each specimen have been determined by using mini compaction apparatus developed by

Table 1 Geotechnical properties of soil and marble dust

Property	Soil	Marble dust	Methodologies followed
Sand (4.75–0.075 mm), %	11.00	96.50	(IS)-2720 (Part 4) (1985)
Silt (0.075–0.002 mm), %	13.00	3.50	
Clay (<0.002 mm), %	76.00	–	
Specific gravity	2.38	2.74	IS-2720 (Part 3) (1980)
Liquid limit, %	49.00	17.6	IS-2720 (Part 5) (1985)
Plastic limit, %	29.18	–	IS-2720 (Part 5) (1985)
Plasticity index, %	19.81	–	
Shrinkage limit, %	11.65	–	IS-2720 Part 6 (1972)
Differential free swell index, %	70.00	–	(IS) (2720 Part (40) (1977)
Optimum water content, %	20.80	15.03	Sridharan and Sivapullaiah (2005)
Max. dry density, gm/cm^3	1.62	1.87	
CBR, %	1.62	2.45	IS 2720 Part 16 (1987)
pH value	8.07	8.20	IS-2720 (Part 26) (1987)

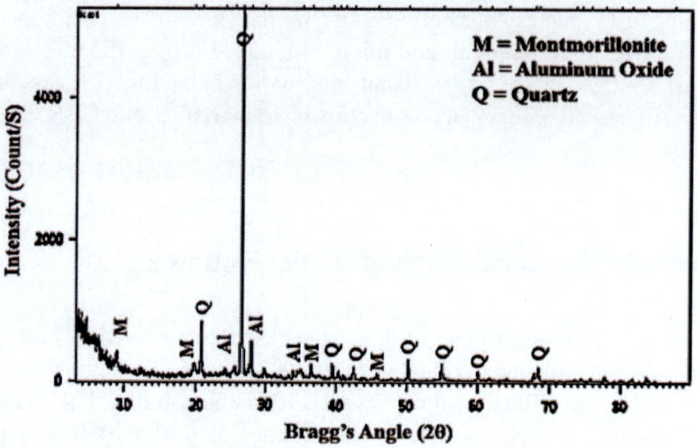

Fig. 1 XRD analysis of black cotton soil

Sridharan and Sivapullaiah (2005). The OWC of soil (i.e., 20.80%) is observed to be higher than the same of marble dust (i.e., 15.03%). However, MDD of soil (i.e., 1.62 g/cm^3) is lower than marble dust (1.87 g/cm^3). The California bearing ratio (CBR) value of marble dust (i.e., 2.45%) is higher than that of soil (i.e., 1.62%).

The mineralogical analysis of samples is determined by using X-ray diffraction (XRD) spectrometer having graphite mono-chromator and Cu-Kα radiation. The sample is scanned from 3° to 90°. The data files of Powder Diffraction Standards (JCPDS 1999) are used to identify the presence of various minerals in thc sample. The XRD analysis of soil confirms the presence of montmorillonite, aluminum oxide, and quartz minerals (Fig. 1).

Field emission scanning electron microscope (FESEM) coupled with energy-dispersive X-ray spectroscopy (EDAX) is performed to examine microstructural and chemical composition of soil. The coating of sample was done by using gold–palladium of with 100 Å thickness. The coating on sample is done in order to avoid charging problem during imaging. Microscopic image of soil (Fig. 2) illustrates several voids with honeycomb networking patterns. The chemical composition analysis of soil revealed the presence of silica (Si) and aluminum (Al) as predominant mineral with minor amount of magnesium (Mg) and sodium (Na) (Fig. 3).

3 Results and Discussion

3.1 Effect of Marble Dust on the Specific Gravity of Soil

Figure 4 shows the effect of varying marble dust content on the specific gravity of black cotton soil.

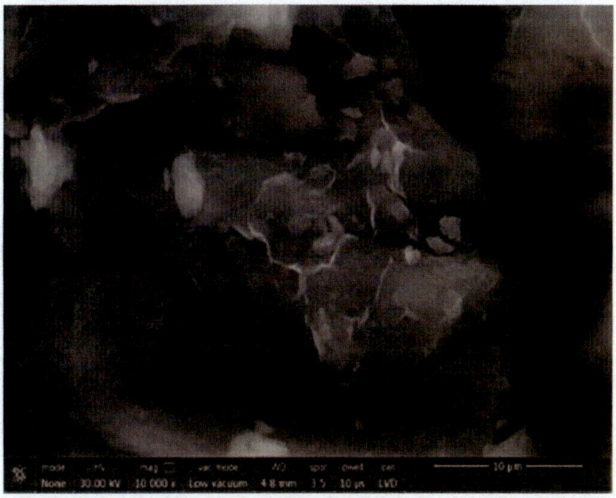

Fig. 2 SEM images of black cotton soil

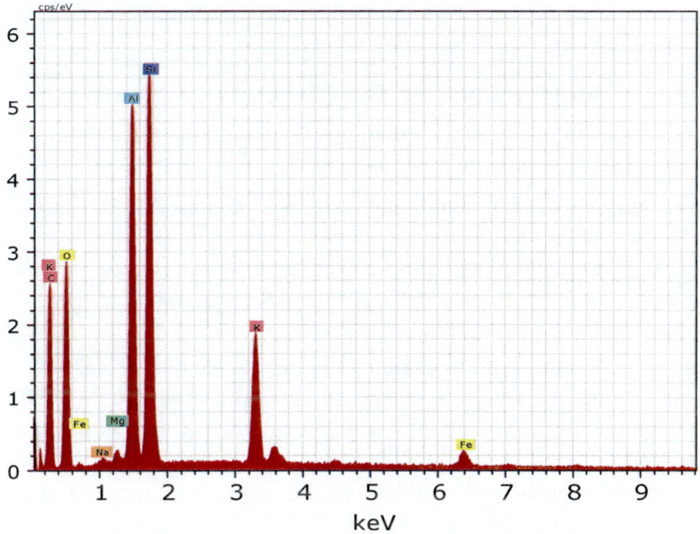

Fig. 3 EDAX analysis of black cotton soil

The investigation shows that the specific gravity increases continuously after replacement of soil with different amount of marble dust up to 20%. The increase in specific gravity with addition of marble dust is attributed to the higher specific gravity value of marble dust (i.e., 2.74) than that of soil (i.e., 2.38).

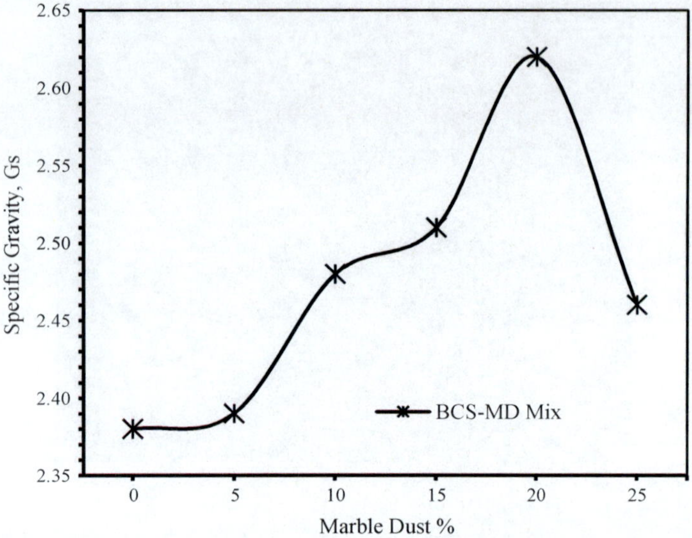

Fig. 4 Variation in specific gravity of black cotton soil (BCS) with addition of marble dust (MD)

3.2 Effect of Marble Dust on the Plasticity Characteristic of Soil

The effect of varying marble dust content up to 25% plasticity of the soil is shown in Fig. 5.

The results show the drastically reduction in the liquid limit of soil with increase amount of marble dust. It is reported that liquid limit of soil is ascribed to the void water present within floccules (Terzaghi and Peck 1967). Hence, the reduction in liquid limit of soil with addition of marble dust may be due to: (i) the aggregation of soil particles at micro-level due to reaction between aluminum and silica present in soil with free calcium present in marble dust and (ii) lower liquid limit value of marble dust compared to soil.

Further, plastic limit of soil reduces with an increase in addition of marble dust and hence, the derivative of liquid limit and plastic limit, i.e., plasticity index also decreases. The decrement in the plastic limit is attributed to the cation exchange process between the cation of the soil and stabilizer (Al-Rawas et al. 2005) and replacement of particles with non-plastic particles of marble dust.

3.3 Effect of Marble Dust on the Shrinkage Limit of Soil

The effect of replacement of soil with varying marble dust on the shrinkage limit is shown in Fig. 6. Shrinkage limit of soil increases with addition of marble dust up to

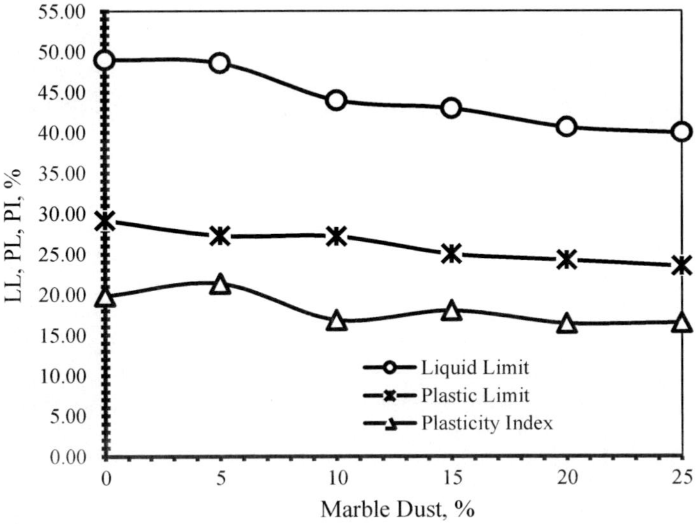

Fig. 5 Variations in liquid limit (LL), plastic limit (PL), and plasticity index (PI) of black cotton soil (BCS) with marble dust

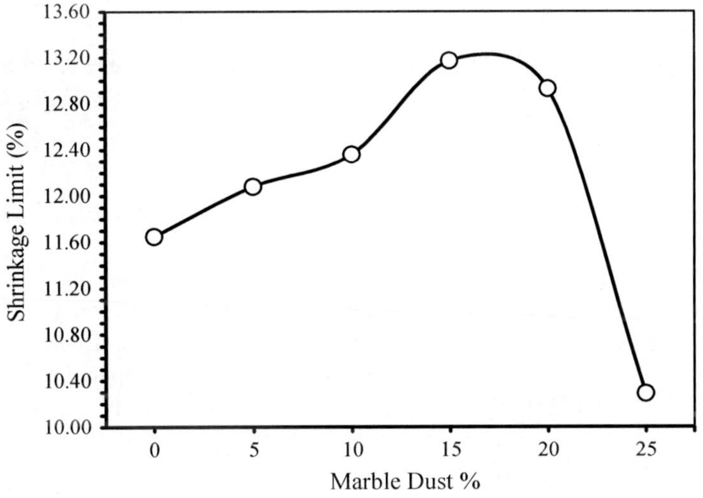

Fig. 6 Variation of shrinkage limit of black cotton soil (BCS) with addition of marble dust

15% and reduces thereafter. The variation in shrinkage limit is due to the change in fabric of soil amended with marble dust.

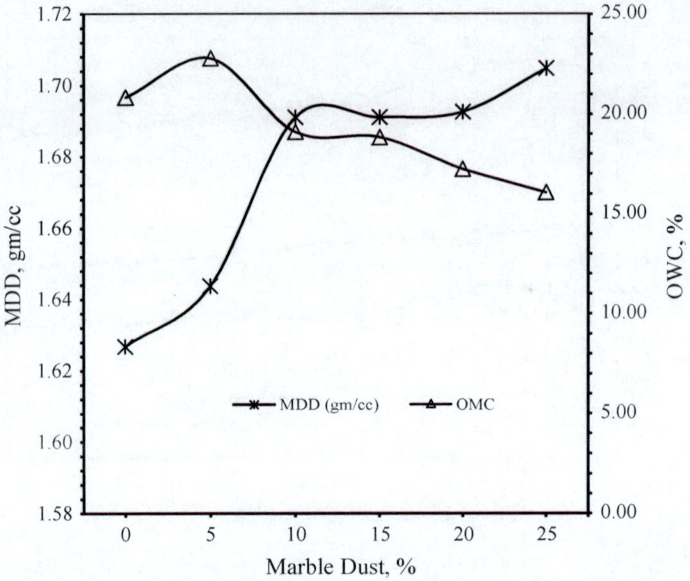

Fig. 7 Variation of compaction characteristics of black cotton soil (BCS) with addition of marble dust

3.4 *Effect of Marble Dust on the Compaction Characteristics (Maximum Dry Density and Optimum Water Content) of Soil*

Influence of marble dust on the maximum dry density (MDD) and optimum water content (OWC) of soil is shown in Fig. 7.

It is observed that addition of marble dust increases the dry density and reduces the optimum water content of soil. The increase in maximum dry density and reduction in OWC of soil with addition of any marble dust content is primarily due to its higher value of maximum dry density and lower value of OMC compared to soil, coarser solid particles, and higher specific gravity of marble dust. Further, this is due to the grain size distribution, which becomes well graded with the addition of quarry dust yielding higher dry density.

3.5 *Effect of Marble Dust on the pH of Soil*

The effect of replacement of soil with varying marble dust on the pH is shown in Fig. 8.

It is observed that pH of soil increases drastically with addition of marble dust up to 15% and reduces thereafter with addition of 20%. The increase in pH value of

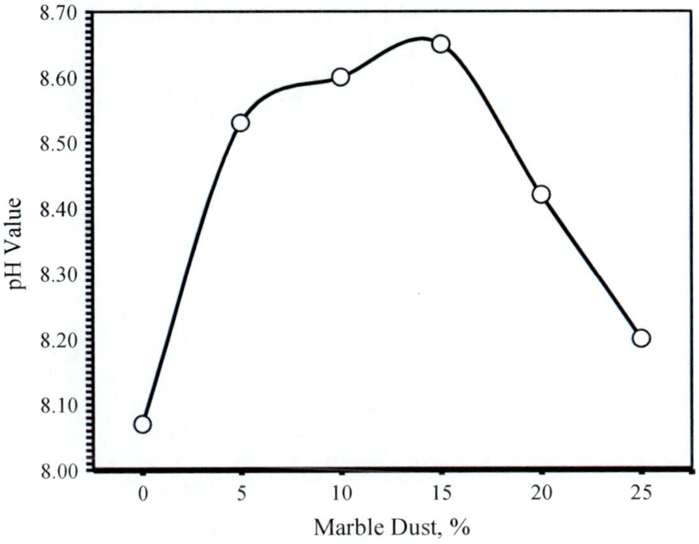

Fig. 8 Variation of pH of soil with addition of marble dust

soil is due to the higher pH value of marble dust (Table 1) and availability of free calcium in the marble dust. The reduction in the pH of soil with 20% marble dust shows the ineffectiveness of free calcium present in the marble dust.

3.6 *Effect of Marble Dust on the Free Swell Index (FSI) of Soil*

Figure 9 shows the reduction in FSI of soil with an increase in the amount of marble dust. The reduction in FSI of soil may be due to the reduction in diffuse double layer which causes the attraction of particles. Further, replacement of swell particle of soil with non-swell particle of marble dust may be the cause of reduction in FSI of soil.

3.7 *Effect of Marble Dust on the California Bearing Ratio (CBR) of Soil*

Influence of marble dust on the California bearing ratio (CBR) of soil is shown in Fig. 10.

The mechanical strength of the material for the highway construction is defined as its CBR value. The investigation shows the higher CBR value of the query dust compared to soil at OWC (Table 1).

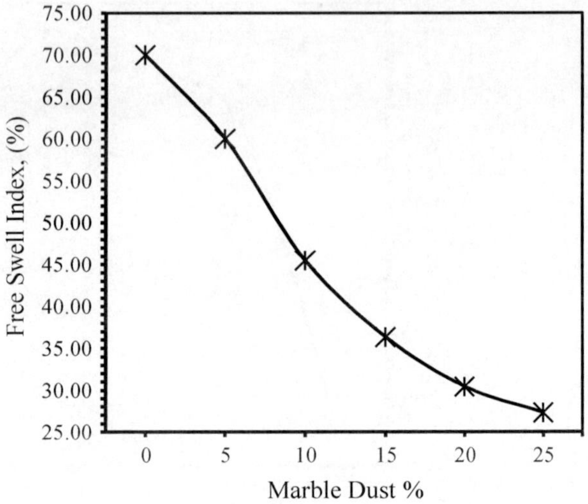

Fig. 9 Variation of FSI (%) of black cotton soil (BCS) with addition of marble dust (MD)

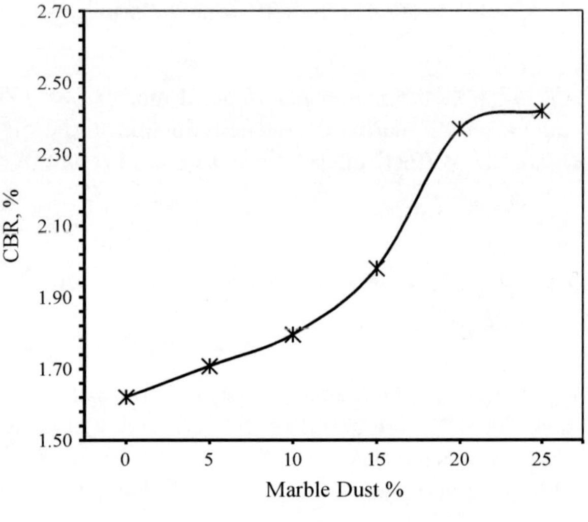

Fig. 10 Variation of CBR of black cotton soil (BCS) with addition of marble dust

CBR values of mix compacted at their respective optimum water content and maximum dry density increase with addition of marble dust up to 20% and show asymptotic behavior thereafter. The increase in CBR value is due to the improvement in angle of shearing resistance of soil–marble dust mixtures. Further, an increase in the maximum dry density with reduction in OMC also leads to the increase in penetration resistance of 50 mm diameter plunger, thereby increasing the CBR values of soil–marble dust mix.

4 Conclusion

The effect of varying amount of marble dust on black cotton soil is investigated in details. The major conclusions that can be drawn from the present study are as follows:

1. The specific gravity of the black cotton soil increases with the increment in the amount of the marble dust up to 20% addition.
2. Plasticity index of soil reduces with a decrease in liquid limit and plastic limit after addition of non-plastic marble dust. The shrinkage limit increases up to 15% addition of marble dust and reduces thereafter.
3. The dry density of black cotton soil increases with reduction in OWC with addition of marble dust. The pH of soil increases with addition of marble dust.
4. Addition of marble dust reduces the free swell index of black cotton soil.
5. CBR of soil increases with the varying amount of marble dust which attributes its utilization for highway construction.

The present study reveals that the marble dust improves the physical as well as engineering behavior of the black cotton soil. Improvement in CBR with the addition of marble dust signifies its utilization for the highway construction. However, CBR of soil–marble dust mixes needs to be done for soaked condition to understand the change in CBR value.

Acknowledgements This study is supported financially by the Research and Development (R&D) division of Manipal University Jaipur (MUJ) under SEED grant [Project Number: MUJ/REGR1435/07]. The authors would like to acknowledge this support.

References

Al-Rawas A, Hago AW, Al-Sarmi H (2005) Effect of lime, cement and sarooj (artificial pozzolan) on the swelling potential of an expansive soil from oman. Build Environ 40:681–687

Bell FG (1996) Lime stabilization of clay minerals and soils. Eng Geol 42:223–237

Bureau of Indian Standards (first revision) IS 2720 (Part 6) (1972) Methods of test for soils: determination of shrinkage factors, New Delhi, India

Bureau of Indian Standards IS 2720 (Part 40) (1977) Methods of test for soils: determination of free swell index of soil, New Delhi, India

Bureau of Indian Standards IS 2720 (Part 3) (1980) Methods of test for soils: determination of specific gravity of soil, New Delhi, India

Bureau of Indian standards IS 2720 (Part 4) (1985) Methods of test for soils: grain size analysis, New Delhi, India

Bureau of Indian Standards (second revision) IS 2720 (Part 5) (1985) Methods of test for soils: determination of liquid limit and plastic limit, New Delhi, India

Bureau of Indian standards IS 2720 (Part 16) (1987) Methods of test for soils: determination of CBR, New Delhi, India

Bureau of Indian Standards IS 2720 (Part 26) (1987) Methods of test for soils: part 26 determination of pH value (second revision), New Delhi, India

Celik MY, Sabah E (2008) Marble deposits and the impact of marble waste on environmental pollution geological and technical characterization of iscehisar (Afyon-Turkey). J Environ Manage 87:106–116
JCPDS (Joint Committee on Powder Diffraction Standards) (1999) Index to the powder diffraction file. International Centre for Diffraction Data, Newtown Square, PA
Jha AK, Sivapullaiah PV (2015) Mechanism of improvement in the strength and volume change behavior of lime stabilized soil. Eng Geol 198:53–64
Misra AK, Mathur R, Rao YV, Singh AP, Goel P (2009) A new technology of marble waste utilization in roads. J Sci Ind Res 69:67–72
Okagbue CO, Onyeobi TUS (1999) Potential of marble dust to stabilise red tropical soils for road construction. Eng Geol 53(3):371–380
Osinubi KJ (2006) Influence of compactive efforts on lime-slag treated tropical black clay. J Mater Civ Eng 175–181
Pappu A, Saxena M, Asolekar SR (2006) Solid wastes generation in India and their recycling potential in building materials. Build Environ 42:2311–2320
Segadães AM, Carvalho MA, Acchar W (2005) Using marble and granite rejects to enhance the processing of clay products. Appl Clay Sci 30:42–52
Singh M, Choudhary K, Srivastava A, Sangwan KS, Bhunia D (2017) A study on environmental and economic impacts of using waste marble powder in concrete. J Build Eng 13:87–95
Sridharan A, Sivapullaiah PV (2005) Mini compaction test apparatus for fine grained soils. Geotech Test J 242:27–35
Terzaghi K, Peck RB (1967) Soil mechanics in engineering practice, 2nd edn. Wiley, New York, USA
Tozsin G, Arol AI, Oztas T, Kalkan E (2014) Using marble wastes as a soil amendment for acidic soil neutralization. J Environ Manage 133:374–377
Vishwakarma A, Rajuput RS (2013) Utilization of marble slurry to enhance soil properties and protect environment. J Environ Res Dept 7(4A):1479–1483
Wheeler SJ, Sharma RS, Buisson MSR (2003) Coupling of hydraulics hysteresis and stress-strain behaviour in unsaturated soils. Geotechnique 53(1):41–54

Effect of Oil Contamination on Geotechnical Properties of Lateritic Soils

M. V. Panchami, J. Bindu, and K. Kannan

Abstract The oil, gas and petroleum industries are considered as a potential source of pollution. Oil contamination has a negative impact on environment. The engineering properties of soil will get adversely affected due to oil contamination making it unsuitable for construction purposes. A good knowledge of change in geotechnical properties of soil due to contamination is required for providing proper recommendations regarding its use in the construction activities. The behaviour of different soils after oil contamination will be different based on the constituents present in it. This paper discusses effect of oil contamination on lateritic soil. Used engine oil was added to soil in the amount of 0%, 2%, 4% and 6% by weight to artificially contaminate the soil, and the variation in properties was identified. The effect of contamination was studied based on the change in consistency limits, compaction characteristics, CBR value and shear strength parameters.

Keywords Oil contamination · Lateritic soil · Consistency limits · Compaction characteristics · CBR · Shear strength

1 Introduction

The environment is getting polluted by many ways. The contamination of soil and water by crude oil is increasing due to oil drilling, leakage from pipes, from mechanical workshops, accident spillage, etc. Oil pollutes water resources, and when it percolates through the soil, it contaminates soil and ground water. Oil contamination harms plants and animals, and thus, the whole food chain will get affected. It spreads as a thin layer on the surface of water that stops oxygen getting to the plants and animals that live in the water. Oil spills also harm air quality. The chemicals in

M. V. Panchami (✉) · J. Bindu
Department of Civil Engineering, College of Engineering Trivandrum, Thiruvananthapuram, Kerala, India
e-mail: sreepanchamimv@gmail.com

K. Kannan
Department of Civil Engineering, Marian Engineering College, Trivandrum, Kerala, India

M. Latha Gali and R. R. P. (eds.), *Problematic Soils and Geoenvironmental Concerns*, Lecture Notes in Civil Engineering 88,
https://doi.org/10.1007/978-981-15-6237-2_22

crude oil are mostly toxic which are found as hydrocarbons. These chemicals cause adverse health effects when inhaled into the body. They are oxidized by oxidants in the atmosphere and form particulate matter which evaporate into the atmosphere. These particulate matters being inhaled can enter the lungs and carry toxic chemicals in to the body.

The various geotechnical properties of soil such as consistency limits, shear strength, CBR value and permeability will get affected due to contamination by oil. This variation in the geotechnical properties will make it unsuitable for construction. The behaviour of different soils due to oil contamination will be different based on the constituents present in them. A good knowledge about change in the properties of soil due to contamination is critical for providing proper recommendations for treatment and use of the same.

The variation in geotechnical properties will depend on the type of soil and the oil by which it is contaminated. The main source of oil is the used engine oil which is a waste product in mechanical workshops, dipots, etc. In this paper, the field of concern is the engine oil contamination effect on laterite soils. Lateritic soils are widespread in tropical areas, and they are highly weathered soils. By conducting suitable tests, the change in the geotechnical properties of lateritic soil due to used engine oil should be found out and evaluated. This is essential in recommending suitable remediation methods for the contaminated soil.

2 Materials

2.1 Soil Sample

The soil selected was lateritic soil, collected near College of Engineering Trivandrum campus. Wet sieve analysis was done for analysing the particle size distribution of soil, and curve was plotted as in Fig. 1. The basic properties were tested and are classified as SC (clayey sands) as per Indian standards. The physical and compaction properties of soil are summarized in Table 1.

2.2 Oil

The oil used here is used engine oil, a waste product from mechanical workshop. The oil is collected from a workshop near to the college campus, Trivandrum. The specific gravity of the oil used is 0.875 g/cc.

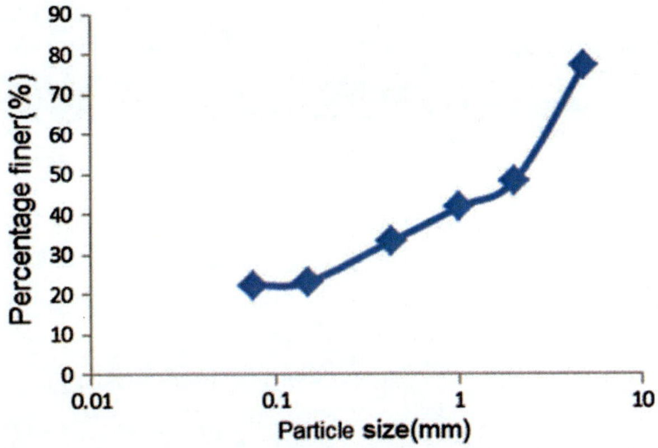

Fig. 1 Particle size distribution of the soil used

Table 1 Characteristics of soil used for the study

Particulars	Value
Specific gravity	2.52
Liquid limit (%)	40
Plastic limit (%)	26
Plasticity index (%)	14
Maximum dry density (g/cc)	1.9

3 Methodology

After particle size distribution, samples were divided into four parts. Degree of contamination is defined as the percentage of volume of oil with respect to dry weight of the soil. Each portion of soil for sampling is mixed thoroughly with oil in the amount 0, 2, 4, 6% by weight of the dry soil samples. The samples were kept for 24 h after mixing it with oil to attain equilibrium. After that, the following tests were carried out.

- Compaction
- Atterberg limits test
- California bearing ratio test
- Unconfined compressive strength test.

3.1 Compaction

Standard Proctor tests were performed on the natural and contaminated samples with 2, 4, 6% as per IS 2720-part 7-1980. It was done to calculate the maximum dry density and optimum moisture content of soil. Here, the pore fluid in this test is a combination of oil and water. So, we have to consider the evaporation of oil also when it is kept in oven for water content determination. Then, this evaporated percentage of oil has to be deducted from the original water content. For that, the following equations can be used.

$$w_{\text{oil}} = \frac{W_{\text{oil}}}{W_{\text{d}}} \tag{1}$$

$$w' = \frac{W_{\text{water}} + W_{\text{oil}}}{W_{\text{d}}} \tag{2}$$

$$w = w' - w_{\text{oil}} \tag{3}$$

where w_{oil} is the oil content, W_{oil} is the weight of oil evaporated, W_{d} is the weight of dry soil, w' is the water and oil content together calculated by weighing the samples before and after placing in oven, w is the actual water content in the soil sample.

During water content determination, the oil percentage that is added will not be completely evaporated by oven drying for 24 h at 105 °C. The rate of oil evaporation decreases with increase in oil content, i.e. 1.43% of oil was evaporated if 2% oil was added to the soil, 1.67% of oil was evaporated if 4% oil was added to the soil, and 1.82% of oil was evaporated if 6% oil was added to the soil.

3.2 *Atterberg Limits Test*

Atterberg limits are characterized by liquid limit, plastic limit and plasticity index. Liquid limit is very necessary in geotechnical engineering for classifying the soil. Liquid limit was found out for natural and contaminated soils using Cassagrande apparatus. The tests were conducted as per IS 2720-part 5-1985. The water content determination was done as in the previous explanation.

3.3 *California Bearing Ratio Test*

California bearing ratio (CBR) tests were conducted for both natural and contaminated soil as per IS 2720-part 16-1987. It was done using a CBR machine. This machine reads both force and penetration of soil compacted at optimum moisture content. Curve was plotted with penetration and load on *X* and *Y* axes, respectively. CBR value was calculated from the graph.

3.4 *Unconfined Compressive Strength Test*

Unconfined compressive strength tests were conducted on natural as well as contaminated soil as per IS 2720-part 10-1986. The tests were conducted at optimum moisture content and maximum dry density.

4 Results and Discussion

4.1 *Compaction*

The results of compaction test are as given in Table 2. The compaction curves for natural and contaminated soils are plotted as shown in Fig. 2. With increase in oil content up to 2%, maximum dry density increased, and optimum moisture content decreased. With further increase in the oil content, dry density and optimum moisture content decreased. The variation in maximum dry density and optimum moisture content with increase in oil percentage is as shown in Figs. 3 and 4, respectively. The decrease in OMC may be due to the coating of crude oil around individual clay particles and disallowing free water from interacting with clay particles. So, amount of water needed by soil to reach its maximum dry density decreased. The decrease in maximum dry density for oil content more than 2% is due to the less packing of

Table 2 Maximum dry density and optimum moisture content at different oil percentages

Oil percentage (%)	Maximum dry density (g/cc)	Optimum moisture content (%)
0	1.9	14.5
2	1.91	12
4	1.87	10.2
6	1.86	6.8

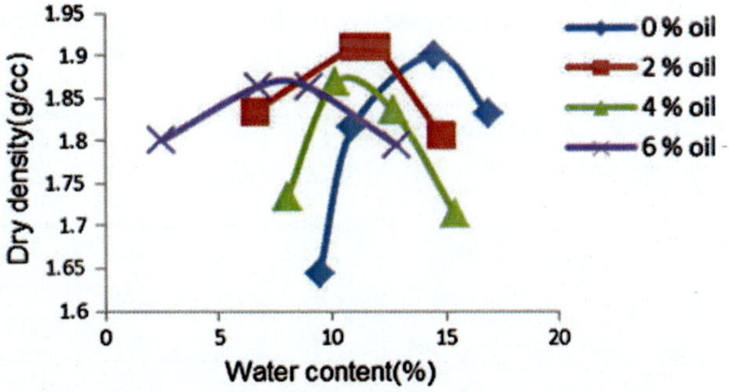

Fig. 2 Compaction curve for soil with different percentages of contamination

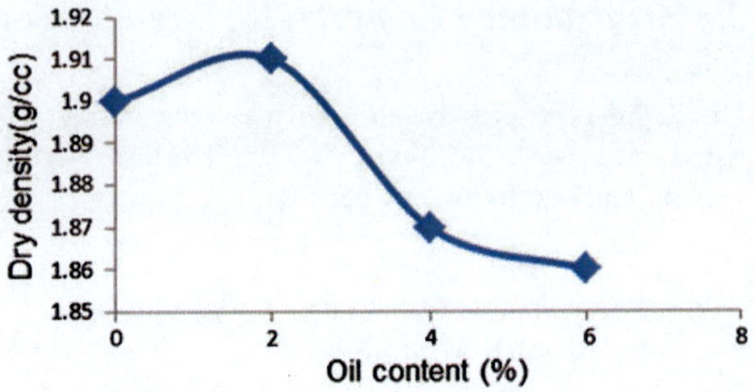

Fig. 3 Variation of dry density with increase in oil percentage

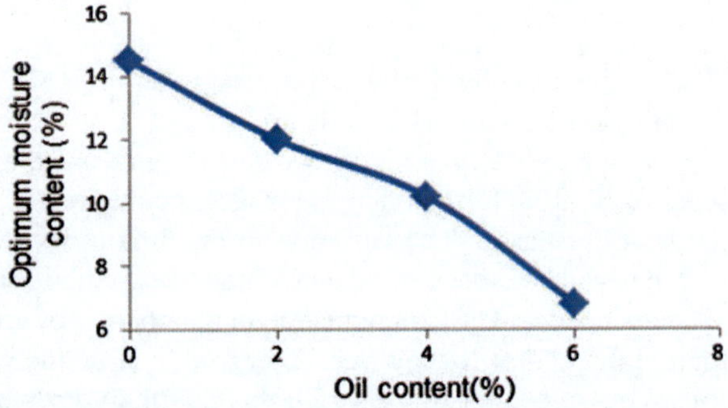

Fig. 4 Variation of optimum moisture content with increase in oil percentage

contaminated soil particles for the same compaction energy used for uncontaminated soil.

4.2 *Atterberg Limits Test*

The liquid limit, plastic limit and plasticity index for samples with different percentages of contamination are as given in Table 3. The variation in Atterberg limits with different oil percentages is as shown in Fig. 5. The liquid limit, plastic limit and plasticity index decreased with increase in oil content. The decrease in liquid limit

Table 3 Atterberg limits at different oil percentages

Oil percentage (%)	Liquid limit (%)	Plastic limit (%)	Plasticity index (%)
0	40	26	14
2	38	24.5	13.5
4	37	24	13
6	35	23	12

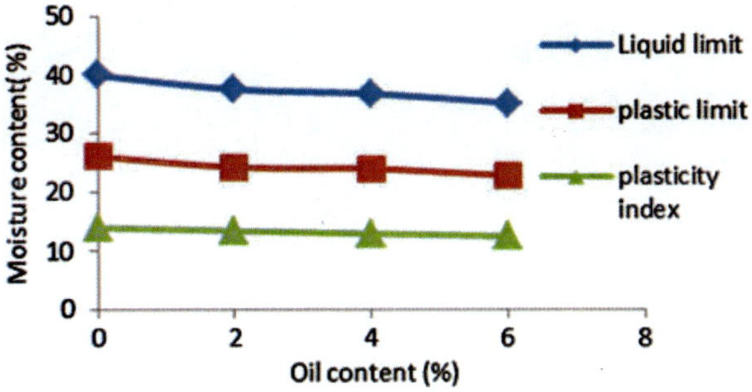

Fig. 5 Variation of Atterberg limits with increase in oil percentage

and plastic limit may be due to the performance of existing non-polar fluids in soil and the decrease in thickness of double layer.

4.3 California Bearing Ratio Test

California bearing ratio test was conducted on uncontaminated and contaminated soil by compacting it at maximum dry density and optimum moisture content. The results are as given in Fig. 6. The CBR values for different oil percentage are as shown in Table 4. The variation in CBR value with increase in oil percentage is as shown in

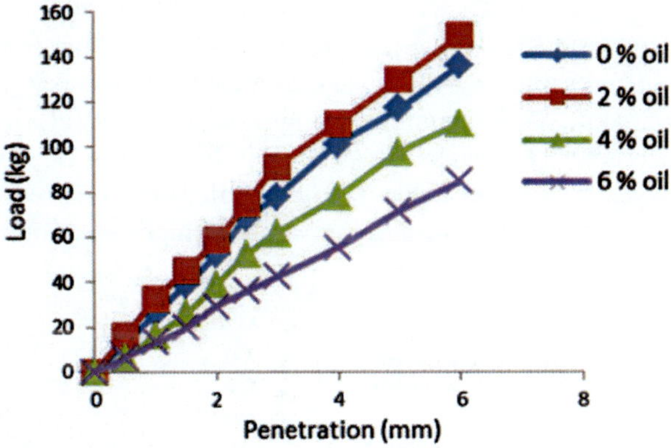

Fig. 6 CBR curve for natural and contaminated soil

Table 4 CBR values for different oil percentages

Oil percentage (%)	CBR value
0	5.69
2	6.32
4	4.77
6	3.47

Fig. 7 Variation of CBR value with different oil percentages

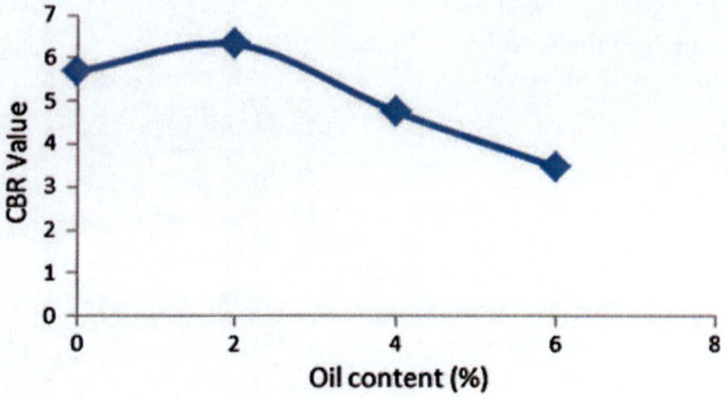

Fig. 7. With increase in oil content up to 2%, CBR value increased, and with further increase in oil percentage, the CBR value decreased. The initial slight increase in CBR value may be due to the clumping together of clay particles facilitated by the crude oil which may cause an increase in inter-particle shearing resistance. Beyond 4% of crude oil, lubrication effect of oil may cause a decrease in shearing resistance of soil and thus a decrease in the CBR value. As the CBR value decreased due to oil contamination, the contaminated soil cannot be used for pavement construction without proper stabilization.

4.4 Unconfined Compressive Strength Test

Unconfined compressive strength tests were conducted on natural as well as contaminated samples by compacting it at maximum dry density and optimum moisture content. The unconfined compressive strength decreased with increase in the oil content. The results are as given in Fig. 8. The UCC value for different oil percentages is given in Table 5. The variation of compressive stress with increase in oil content is also shown in Fig. 9. The compressive strength decreased with increase in the oil content.

Fig. 8 Stress–strain curves of natural and contaminated soils

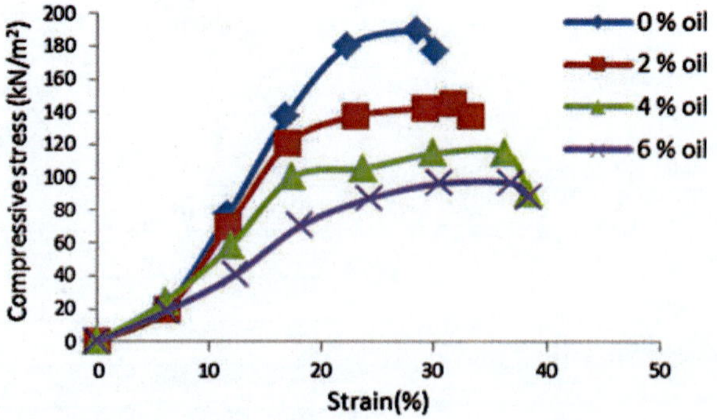

Table 5 Unconfined compressive strength for samples with different oil percentages

Oil percentage (%)	Unconfined compressive strength (kN/m^2)
0	189
2	145
4	116
6	97

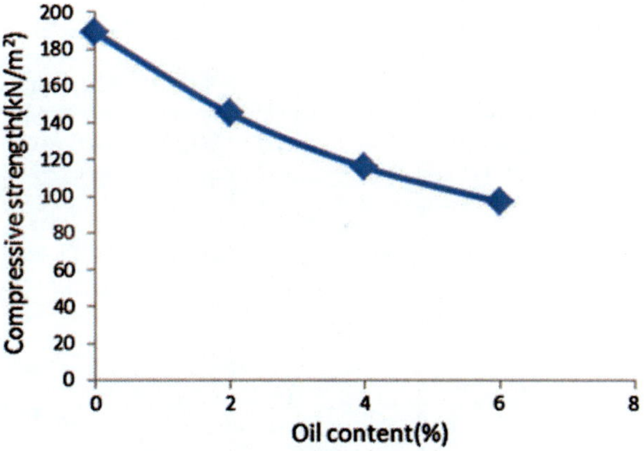

Fig. 9 Variation of unconfined compressive strength with increase in oil content

5 Conclusion

This paper aims at analysing the effect of oil contamination on geotechnical properties of lateritic soil. The variation in properties depends on the type of soil and oil used in the study. The behaviour of different soils due to contamination is different based on the constituents present in it.

Atterberg limits test, compaction, CBR test and UCC tests were conducted in soil samples with varying levels of oil percentage, and the conclusions were as follows:

- Maximum dry density increased up to 2% oil contamination, and with further increase in oil percentage, the value decreased. Optimum moisture content decreased with increase in oil percentage.
- Liquid limit, plastic limit and plasticity index decreased with increase in oil percentage. The decrease in liquid limit and plastic limit may be due to the performance of existing non-polar fluids in soil and the decrease in thickness of double layer.
- CBR value increased up to 2% oil contamination, and with further increase in oil percentage, the CBR value decreased. Thus, the contaminated soil cannot be used for pavement construction without proper stabilization.
- Unconfined compressive strength decreased with increase in oil percentage, and thus, the contaminated soil cannot be used for construction purposes.

Bibliography

Akinwumi II, Diwa D, Obianigwe N (2014) Effects of crude oil contamination on the index properties, strength and permeability of lateritic clay. Int J Appl Sci Eng Res 3(4):816–824

Alhassan HM, Fagge SA (2013) Effects of crude oil low point pour fuel oil and vacuum gas oil contamination on the geotechnical properties sand, clay and laterite soils. Int J Eng Res Appl 3(1):1947–1954

Chaudhary A (2016) Performance evaluation of contaminated cohesive soils. J Civ Environ Eng 6(6):1–4

Iloeje AF, Aniago V (2016) Effect of crude oil on permeability properties of the soil. Int J Trend Sci Res Dev 1(1):39–43

Kermani M, Ebadi T (2012) The effect of oil contamination on the geotechnical properties of fine-grained soils. Soil Sediment Contam Int J 21(5):655–671

Khamehchiyan M, Charkabhi AH, Tajik M (2006) Effects of crude oil contamination on geotechnical properties of clayey and sandy soils. Eng Geol 89(3):220–229

Puri VK (2000) Geotechnical aspects of oil contaminated sands. Soil Sediment Contam 9(4):359–374

Rahman ZA, Hamzah U, Taha MR, Ithnain NS, Ahmad N (2010) Influence of oil contamination on geotechnical properties of basaltic residual soil. Am J Appl Sci 7(7):954–961

A Detailed Geotechnical Investigation on Red Mud and Chemical Analysis of Its Leachate

K. Sarath Chandra and S. Krishnaiah

Abstract Bauxite residue or red mud is the iron-oxide rich waste produced when bauxite ore is processed by the Bayer's process to extract aluminum. Red mud characteristics and production depend on origin, quality and composition of bauxite. Storage and disposal of red mud is the biggest problem faced by the aluminum industries. Red mud residue is associated with being chemically basic, high in heavy metals and low level of naturally occurring radioactive materials. The objective of this study was to characterize red mud which is collected from Hindalco, Belgaum, on the basis of traditional geotechnical methods that may contribute to the behavior of the material and to analyze the chemical constituents of its leachate which helps to understand the leachate characteristics. The characterization of the material included index properties like moisture content, specific gravity, grain size distribution, liquid limit, plastic limit and their indices, engineering properties like shear strength by cohesion intercept and angle of internal friction, compaction characteristics by OMC and MDD and chemical properties by SEM, XRD and TCLP. The findings, in terms of leachate characteristics, showed that red mud has undergone reactions during the leaching process giving leachate with sodium, silica and calcium in highest concentration. Red mud shows acceptable characteristics for potential reuse as a civil engineering material.

Keywords Red mud · Geotechnical investigation · Leachate · Chemical analysis

1 Introduction

As industrialization and urbanization have boomed majorly, so did the negative impacts which puts a debatable question on the safe disposal methods of the industrial waste and one has to look into their negative impacts on the global environment and social life. Out of many elements, aluminum is the third most abundant element on

K. Sarath Chandra (✉) · S. Krishnaiah
Department of Civil Engineering, Jawaharlal Nehru Technological University, Anantapur, Andhra Pradesh, India
e-mail: sarathchandra.k@christuniversity.in

M. Latha Gali and R. R. P. (eds.), *Problematic Soils and Geoenvironmental Concerns*, Lecture Notes in Civil Engineering 88,
https://doi.org/10.1007/978-981-15-6237-2_23

Earth and extraction of this element has drastically increased over time. The increase in extraction is due to the diversity of uses for aluminum in today's society in both industries and products. Red mud is a bauxite residue generated from the aluminum industries during the Bayer's process of extraction of aluminum from the bauxite ore. India alone had a mine generation of 15.4 million tons of bauxite in the year 2014 and expanded to 19 million tons in the year 2015 from only 540 thousand stores. The creation of alumina from 2010 to 2011 by Hindalco was 1.35 million tons. In India, around 4.71 million tons/annum of red mud is delivered which is equivalent to 6.25% of world's generation (Patel 2015).

Due to the presence of large amount of fine particles in red mud, it is always not recommended to dump in an open areas and the high alkalinity (pH > 11) favors leachate and thwarts the sedimentation of red mud if it is placed in storage tanks for a longer period. Incidental fall of a red mud repository dam like Ajka in Hungary leaves an unfavorable effect on the more noteworthy region because of the spreading of red mud particles over a huge territory. In this exploration, an endeavor has been made to control the dispersive idea of red mud and quickening sedimentation by including distinctive added substances (Alam 2018).

On the other side deficiency of hand truck materials, worry over utilization of best soil, force of directions by government associations on the utilization of virgin materials and so on frequently urge the use of mechanical waste (viz. fly fiery remains, red mud, phosphogypsum, waste foundry sand and so on) for the sustainable development of infrastructure (Reddy and Hanumantha Rao 2016). Many researchers worked on converting the different types of industrial and infrastructure waste into an asset in the form of making bricks, tiles, filling materials, etc. (Samal 2013). In this process, red mud also used to make bricks with some additives and as an embankment material as it shows the properties of sand to clay with good compaction characteristics (Panda 2017; Rout 2012).

As such, very limited work has been carried out to establish geotechnical properties of RM waste as filling material in road sub-base or as a backfill material particularly in India. With this in mind, a detailed investigation was made on geotechnical properties of red mud and chemical analysis of its leachate to satisfy it as a good backfill material in both embankments and road sub-bases. To achieve this, a series of sieve analysis, Atterberg's limits, standard proctor tests and unconfined compression tests were performed to confirm it as a satisfying sustainable construction material with the comparison of standard values. An attempt also made to study chemical analysis of its leachate with a possible leachate process and compared with the WHO standards of groundwater to avoid the chemical imbalance of groundwater (Yan et al. 2014).

2 Experimental Investigations

Red mud waste used in this study was collected from the disposal pond of Hindalco Industries Limited, Belgaum, Karnataka. All samples have been collected in dry state having more amount of fines and at a moisture content of 2–3% at room temperature. The specific gravity of sample was determined by using density bottle as per the IS code 2720-3 (1980), and the results are shown in Table 1. For accuracy, average value obtained from tests conducted in triplicate was considered representative, as listed in Table 1. Consistency limits, grain size distribution, pH, compaction tests and CBR tests of RM were determined in accordance with the Indian standard codal provisions IS-2720-PART-5 (1985), IS-2720-PART-4 (1985), IS-2720-PART-26 (1987), IS-2720-PART-7-1980 (1980), IS-2720-PART-16-1979 (1987), respectively. The results obtained are presented in Table 1. Based on coefficient of uniformity and coefficient of curvature of the sieve analysis and consistency limits, the red mud waste was classified as per Indian standard classification system. It has been found that the waste falls under the category of silt of low plasticity (i.e., ML) soil. The optimum moisture content and maximum dry density obtained from the compaction tests indicate the density of the red mud waste which is high at an optimum moisture content to accept this waste as a geotechnical asset. The CBR values of unsoaked and soaked red mud waste are presented in Table 1 that shows the appreciable range of

Table 1 Physical and geotechnical properties of red mud

Property	Values
Specific gravity	2.85
pH	11.04
Liquid limit	39%
Plastic limit	28.5%
Sieve analysis	
Coefficient of uniformity (Cu)	0.67
Coefficient of curvature (Cc)	0.71
IS classification	Silt of low plasticity (i.e., ML)
Compaction test	
Optimum moisture content (OMC)	31.39%
Maximum dry density (MDD)	1.60 kg/m^3
California bearing ratio (CBR)	
Unsoaked	5.28%
Soaked	4.17%
Cohesion	16.82 kPa
Angle of internal friction	22.8°

bearing capacity to use as a subgrade or sub-base material and as backfill material. Shear parameters called cohesion and angle of internal friction were determined as per the standard procedure of unconfined compression test.

In the present study along with the detailed investigation of geotechnical properties of red mud waste, the elements that are present in the red mud waste were identified through SEM, XRF and XRD methods. It is also investigated that if the red mud waste is used in embankments as a backfill material, what are the problems associated with the red mud waste when its leachate comes in contact with the groundwater. Different approaches have been developed to investigate the compatibility of a waste material with the environment. The most general approach involves the application of standardized leaching tests which provide an indication for the possible reuse of the waste material (Brunori 2005). In leaching, soluble constituents are removed from a solid material (such as rock, soil or waste) by a fluid by percolation or diffusion. Therefore, when fill materials come into contact with the leaching liquid (maybe percolating rainwater, surface water, groundwater and liquids present in the fill material), constituents from the solid phase dissolve into the liquid forming leachate.

Table 2 shows the different test methods that are used to collect the leachate in the laboratory. Based on the presence of metals, volatiles, semi-volatiles and PCB's in the red mud waste and the availability of the apparatus and laboratory equipment's, the procedure chosen for the collection of leachate from the red mud waste is toxicity characteristics leaching procedure (TCLP—US EPA test method 1311). The TCLP test involves the extraction of contaminants from a 100 g sample of waste material with an appropriate extraction fluid. A 20:1 liquid-to-solid (L/S) ratio is employed, and the mixture is rotated for 18 ± 2 h at 30 rpm using a rotary agitation apparatus as shown in Fig. 1. The extraction fluid used for the extraction depends on the alkalinity of the waste material. Very alkaline waste materials are leached with a fixed amount of acetic acid without buffering the system (pH 2.88 ± 0.05), while other waste materials are leached with acetic acid buffered at pH 4.93 ± 0.05 with 1 N sodium hydroxide. After rotation, the final pH is measured, and the mixture is filtered. The

Table 2 Procedures to collect the leachate

Sl. No	Test procedures	Extraction fluids	Appropriate for
1	Toxicity characteristic leaching procedure (TCLP) EPA method 1311	Buffered acetic acid, pH 2.88 or 4.93	Metals, semi-volatiles, pesticides, PCBs, volatiles
2	Synthetic precipitation leaching procedure (SPLP) EPA method 1312	Fluid: H_2SO_4 and HNO_3 at pH 4.02	Metals, semi-volatiles, pesticides, PCBs
3	ASTM D3987-85 (ASTM neutral leach)	Reagent water	Semi-volatiles, pesticides, PCBs, cyanide, sulfide, chromium
4	ASTM D5233-92 (ASTM Single Batch)	Buffered acetic acid, pH 2.88 or 4.93	Metals, semi-volatiles, pesticides, PCBs

Fig. 1 Beakers containing mixture with vertical axis rotary machine

filtrate is collected in an appropriate container, and preservative may be added if needed. The filtrate is analyzed for a number of constituents. If these constituent concentrations equal or exceed the concentrations described standards, then waste is characteristically hazardous for toxicity.

Calculation that is used to increase the pH with the addition of more alkaline water in order to achieve the desired chemical proportions to collect the leachate forms the red mud waste.

$$\text{Density of acetic acid} = 1.05\,\text{g/mL}$$
$$\text{pH of acetic acid} = 1.58$$
$$\text{pH of double distilled water} = 6.98$$
$$V = M/\rho.$$
$$\text{For } 1N,\ (60/x)/25\% \quad \text{(per 1000 ml of water) (25\% concentration of acetic acid)}$$
$$= (60/1.05)/25\% = 228.57\ \text{ml in 1000 ml}$$
$$\text{For } 0.5N,\ \text{it is } 114.2 \text{ in 1000 ml}$$
$$\text{Thus, } 114.2\,\text{ml acetic acid in 1000 ml} = X \text{ acetic acid in 100 ml}$$
$$X = (114.2 \times 100)/1000$$
$$= 11.4\,\text{ml acetic acid in 100 ml water.}$$

Table 3 shows the trial and error method to increase pH with the addition of more alkaline water as per the above-mentioned calculations. The alkaline water added from 100 to 600 ml and increased the pH range from 2.15 to 2.88 which is required to collect the leachate from the red mud waste as per the TCPL method. The filtrate

Table 3 Trial and error method to increase pH

Water in ml	pH
100	2.15
120	2.17
140	2.21
160	2.22
180	2.24
220	2.34
300	2.34
320	2.44
520	2.76
600	2.88

is collected with the help of a glass fiber filter in a suitable container. The leachate is then analyzed for the necessary parameters.

3 Results and Discussion

As per the values of sieve analysis and the IS classification system of fine-grained soil, the red mud waste was categorized under the category of silt with low plasticity. Red mud waste has a high specific gravity of 2.85 when compared to normal soil of this category due to the presence of silica, oxides and other compounds, and the high specific gravity indicates better density than the other waste materials. During the processing of the bauxite, it is treated with caustic soda. Therefore, it is highly alkaline and the pH value is 11.04. The liquid limit of the red mud waste is very high; it is observed that it varies with the time of conduction of the test. Plasticity is low compared to the materials of silt and other similar fine aggregates. Standard compaction test depicts that the red mud waste attains appreciable values of dry density at the optimum moisture content. So it can be used as a geotechnical material in many of the construction sites. The CBR values of unsoaked and soaked do not show any huge difference, so the effect of water on the surrounding will not affect much on the bearing capacity of the red mud waste. The cohesion and angle of internal friction give a clear picture of acceptability of this waste into a backfill material in the retaining wall with or without additives. But many researchers suggested that the shear parameters can be increased by using additives.

Further tests like scanning electron microscopy (SEM), powder X-ray diffraction (XRD), X-ray fluorescence (XRF) and chemical analysis of the red mud waste were conducted to understand the chemical constituents of red mud before leachate and after the leachate. The chemical composition of red mud determined by the XRF analyzer is presented in Table 4. Samples were dried to constant weight at a temperature of 110 °C (fully dry, no free water) and ground to fine particles. The main

Table 4 Chemical composition of red mud waste in powder state by XRF analyzer

SN	Component	Result (%)	LLD
1	Fe_2O_3	44.3	0.0016
2	Al_2O_3	18.2	0.0194
3	SiO_2	14.5	0.0057
4	TiO_2	10.5	0.0098
5	Na_2O	9.29	0.295
6	CaO	1.11	0.0117
7	P_2O_5	0.738	0.0059
8	SO_3	0.373	0.0014
9	V_2O_5	0.354	0.0178
10	Cr_2O_3	0.177	0.0044
11	ZrO_2	0.132	<0.0001
12	K_2O	0.105	0.0176
13	CO_2O_3	0.103	0.0021
14	CI	0.0877	0.0015
15	MnO	0.0592	0.0037
16	CeO_2	0.0118	0.0015
17	CuO	0.0113	0.0010
18	SrO	0.0070	0.0002
19	La_2O_3	0.0069	0.0011
20	BaO	0.0058	0.0008

chemical compositions of red mud are Fe_2O_3, Al_2O_3, SiO_2, CaO, Na_2O, TiO, K_2O and MgO. The percentage of these chemical constituents may vary depending upon on the process, property and phase with the type of the bauxite and the alumina and will change over time.

The SEM image of red mud waste is shown in Fig. 2. Combining with the conclusion obtained from the analysis of strength, particle diameter and density, SEM characterization is helpful for the further understanding of the physical performance and the microstructure of red mud. The SEM image red mud waste shows the relatively loose microstructures and high porosities.

Figure 3 explicates the XRD pattern which shows that the main mineral phases of the red mud waste are chantalite ($Na_5Al_3CSi_3O_{15}$), gibbsite ($Al(OH)_3$), muscovite ($KAl_2(FOH)$), calcite ($CaCO_3$) and hematite—Fe_2O_3. It gives a clear identity of the minerals that are present in the red mud waste to observe the minerals which are retained or dissolved in the leachate after allowing the water to pass through the red mud waste.

Finally, leachate was collected from the red mud waste according to the TCLP method of leachate collection. Research shows that the leachate collected in the laboratory is more effective than the leachate collected on site. Based on the literature

Fig. 2 SEM image of red mud waste

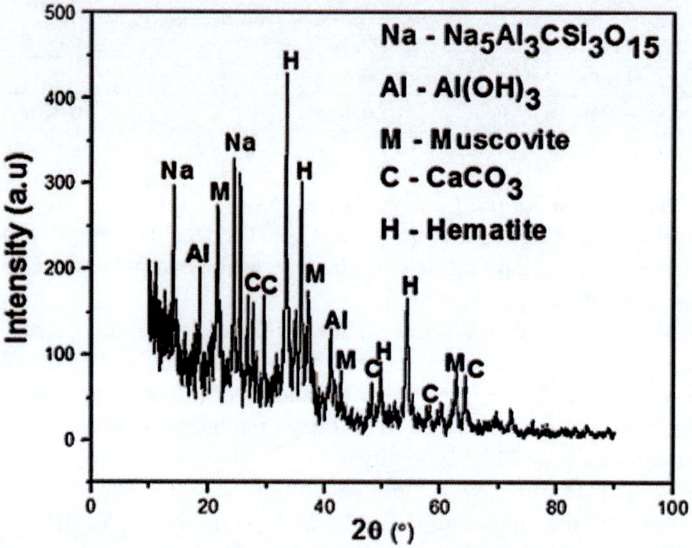

Fig. 3 XRD graph of red mud waste with peak values of minerals

Table 5 Heavy metals present in leachate and comparative limits

Heavy metals	Results obtained (mg/l)	Primary drinking water standard (mg/l)	TCLP hazardous waste limit (mg/l)	WHO standards of drinking water (mg/l)
Arsenic	<0.01	0.05	5.0	0.01
Cadmium	0.04	0.005	1.0	0.003
Chromium	0.76	0.1	5.0	0.05
Lead	0.70	0.015	5.0	0.01
Mercury	<0.001	0.002	0.2	0.001
Copper	0.18	1.3	–	2
Sodium	4.12	20	40	20
Silica	116.09	100	200	100
Calcium	216.8	20–208	–	20–208

study and the level of toxicity, it was decided to analyze the concentration of heavy metals like arsenic, cadmium, chromium, lead, mercury, sodium, calcium, silica and copper in the collected leachate. Table 5 shows the level of heavy metals present in the leachate from the red mud waste, and it is also compared with the primary drinking water standards, TCLP hazardous waste limits and WHO standards of drinking water (2016). It shows the large amount of calcium, silica and sodium that will retain even in leachate.

4 Conclusion

In the present study, extensive tests were conducted on red mud waste samples to obtain a detailed geotechnical properties and chemical analysis of its leachate. From the results, it has been observed that it satisfies all the geotechnical properties which are required to use this waste material as a filling material which can reduce the large amount of waste. It is also noticed the methods to collect the leachate in laboratory. The results obtained from the XRD and XRF shows the range of minerals present in the red mud waste when it is dry. The chemical analysis of the RM leachate after leaching through TCLP procedure shows that there is no significant effect even if this waste enters into the groundwater. However, the presence of heavy metals shows that it is acceptable up to certain concentration only.

References

Alam S (2018) Dispersion and sedimentation characteristics of red mud. J Hazard Toxic Radioactive Waste 22(4):04018025
Brunori C (2005) Reuse of a treated red mud bauxite waste: studies on environmental compatibility. J Hazard Mater 11(1):55–63
IS Code 2720-3-1 (1980) Methods of test for soils. Part 3: determination of specific gravity, section 1: fine grained soils
IS Code 2720-7 (1980) Methods of test for soils. Part 7: determination of water content-dry density relation using light compaction
IS Code 2720-4 (1985) Methods of test for soils. Part 4: grain size analysis
IS Code 2720-5 (1985) Methods of test for soils. Part 5: determination of liquid and plastic limit
IS Code 2720-16 (1987) Methods of test for soils. Part 16: laboratory determination of CBR
IS Code 2720-26 (1987) Method of test for soils. Part 26: determination of pH value
Panda (2017) Characterization of red mud as a structural fill and embankment material using bioremediation. Int Biodeterior Biodegradation 1(19):368–376
Patel (2015) Current status of an industrial waste: red mud an overview. IJLTEMAS 4(8):1–16
Reddy G, Hanumantha Rao B (2016) Evaluation of the compaction characteristics of untreated and treated red mud. In: Geo Chicago, ASCE
Rout S (2012) Utility of red mud as an embankment material. Int J Earth Sci Eng 5(6):1645–1651
Samal S (2013) Proposal for resources, utilization and processes of red mud in India—a review. Int J Miner Process 1(18):43–55
WHO standards of drinking water and ground water limits (2016)
Yan Y, Gao J, Wu J, Li B (2014) Effects of inorganic and organic acids on heavy metals leaching in contaminated sediment. An interdisciplinary response to mine water challenges. China University of Mining and Technology Press, Xuzhou

Engineering Properties of Industrial By-Products-Based Controlled Low-Strength Material

Vinay Kumar Singh and Sarat Kumar Das

Abstract Controlled low-strength material (CLSM) also known by names such as flowable fill or controlled density fill is self-compacted in nature and used in civil engineering works such as backfills, void fills, conduit bedding and many more. These materials have strength lesser than 8.3 MPa. Generally, it consists of Portland cement, fine aggregate, fly ash and water but several other materials such as foundry dust and quarry dust can also be used. In this study, the different engineering properties of CLSM with different mix proportion were evaluated. Mix proportion mainly consists of red mud having 70%, 60% and 50% by weight of total composition along with fly ash of 20%, 30% and 40%, respectively, and phosphogypsum of 10% was studied. Portland cement was used as the binder material in all the composition in the proportion of 10% and 7% of the above composition. For different mix proportions considering flow value as an important parameter firstly percentage of water required for desired flow value (0.15–0.3 m) was evaluated by the trial method and then at those water content along with above mixture, other important engineering properties of CLSM such as compressive strength, bleeding and durability were evaluated. Except for the property of reexcavatibility (strength lesser than 2.1 MPa) for some composition all the other properties of all the composition were within the permissible limit. Overall it can be stated that red mud can be effectively used as one of the materials in CLSM.

Keywords Flowable · Self-compacted · Red mud · Phosphogypsum · Flow value · Bleeding · Durability · Reexcavatibility

V. K. Singh (✉)
Indian Institute of Technology Delhi, Hauz Khas, New Delhi 110016, India
e-mail: shashivinaysingh@gmail.com

S. K. Das
Indian Institute of Technology Dhanbad, Dhanbad, Jharkhand 826004, India

M. Latha Gali and R. R. P. (eds.), *Problematic Soils and Geoenvironmental Concerns*, Lecture Notes in Civil Engineering 88,
https://doi.org/10.1007/978-981-15-6237-2_24

1 Introduction

Controlled low-strength material (CLSM) is also known by different names like self-compacting, self-levelling, flowable fill, controlled density fill, soil-cement slurry, unshrinkable fill and many more. The strength criteria for these materials defined as per American Concrete Institute (ACI) are minimum of 0.3 MPa and maximum of 8.3 MPa but considering reexcavation in future it is limited to 2.1 MPa or less (ACI Committee 229: Controlled Lowstrength Materials (CLSM) 1999).

These materials are different from the low-strength concrete, but the basic principle behind these materials is like that of self-compacting concrete. Therefore, these materials are also called as grout material. These are the self-compacted cementitious material. These materials find their applications in various civil engineering works such as backfill and filling voids where compaction cannot be done such as under-existing buildings and roads, filling abandoned structures beneath the ground and many more. These materials are advantageous than other materials in terms of labour and equipment requirement because of its flow characteristics.

Soil, fly ash, water and cement are the traditional materials of CLSM (ACI Committee 229: Controlled Lowstrength Materials (CLSM) 1999) but now for enhancing the properties admixtures are added. Several researchers (Nataraja and Nalanda 2008; Butalia et al. 2004) studied the potential use of industrial by-products including flue gas desulphurization material as the constituting material. Cementless CLSM, use of quarry dust as a replacement of fine aggregate and use of recycled aggregate are the recent development (Lee et al. 2013; Achtemichuk et al. 2009). Since there is scarcity of fresh water, wastewater can also be used (Al-Harthy et al. 2005) and the quantity of water required is also dependent on the quantity of other materials used (Du et al. 2002). These studies form the scope to explore the use of many other materials in combination with others as a constituent material of CLSM. There are several other industrial by-products which are having some pozzolanic property like rubber, pond ash, bottom ash and granulated blast furnace slag which can be used as a replacement of cement and fine aggregate in the production of CLSM (Wang et al. 2013; Yan et al. 2014; Lini Dev and Robinson 2015; Sheen et al. 2013; Raghavendra and Udayashankar 2014). Fly ash, a well-known pozzolanic waste material, is used in various types of binding a material production. Due to the presence of pozzolanic material in the red mud and phosphogypsum, these materials are also now used in brick production and other construction material.

In the recent study, red mud was used as a replacement for natural sand in self-compacting concrete (Liu and Poon 2016) and (Do and Kim 2016) used red mud as the replacement of cement in the production of CLSM. Phosphogypsum can be used as the replacement of natural sand in the production of concrete and found it as a suitable construction material (Rashad 2017).

The present study deals with the use of red mud, fly ash and phosphogypsum in different combination as constitute material in the production of CLSM by checking the basic engineering properties of CLSM.

2 Materials and Methodology

There is a set of a key property of the CLSM for which different experiments are there by the help of these experiments scope of the material being as a CLSM material is evaluated. For this purpose, first, the geotechnical characterization of the material was done and then the characterization of different mix proportion as a CLSM material experimented.

2.1 Materials Used

Fly ash. The fly ash used in the present study was Class F fly ash and was collected from the location A in India. The fly ash was oven-dried and then the basic geotechnical property was determined. Basic properties of fly ash are discussed in Table 1 and determined as per Indian standard codes (IS: 2720 (Part 3) 1980; IS: 2720 (Part 4) 1985; IS: 2720 (Part 5) 1985; IS: 2720 (Part 7) 1980). From the grain size analysis, the fly ash was found to contain 22% sand-sized particle and 72% silt size particle. The X-ray diffractometer analysis (XRD) of it is shown in Fig. 1. Micromorphology

Table 1 Basic properties of fly ash

Fly ash properties	Value
Specific gravity	2.24
Maximum dry density (kN/m^3)	13.5
Optimum moisture content (%)	24.7

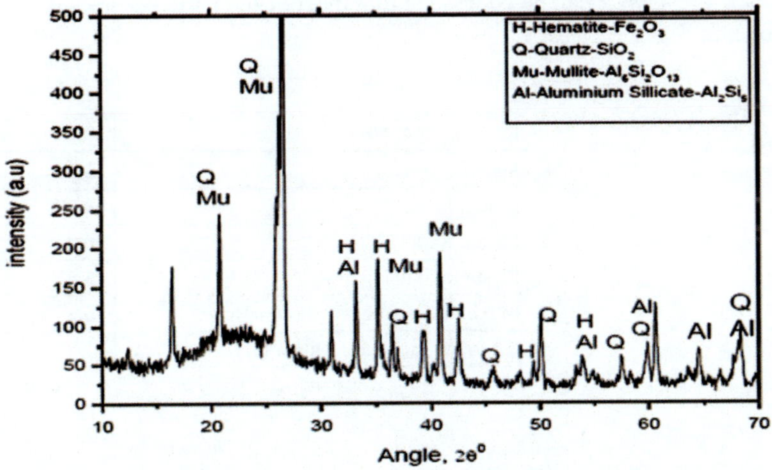

Fig. 1 XRD analysis of fly ash

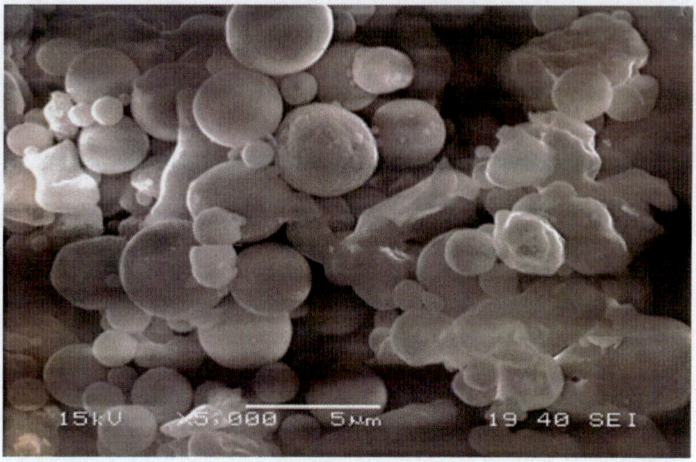

Fig. 2 SEM image of fly ash

study was done by scanning electron microscope (SEM) and from the image of SEM, it can be concluded that most of the particle is circular in shape as shown in Fig. 2.

Red mud. The red mud used in this study was collected from location B in India. The red mud was oven-dried and then the basic geotechnical property was determined. Basic properties of red mud are discussed in Table 2 determined as per Indian standard codes (IS: 2720 (Part 3) 1980; IS: 2720 (Part 4) 1985; IS: 2720 (Part 5) 1985; IS: 2720 (Part 7) 1980). From the grain size analysis, the red mud was found to contain 17% sand-sized particle, 51% silt size particle and 32% clay size particle. Micromorphology study was done by scanning electron microscope (SEM), and from the image of SEM, it can be concluded that most of the particles are flaky in shape as shown in Fig. 3.

Table 2 Basic properties of red mud

Red mud properties	Value
Specific gravity	3.27
Liquid limit (%)	39.89
Plastic limit (%)	36.08
Plasticity index	3.81
Maximum dry density (kN/m^3)	15.2
Optimum moisture content (%)	31.5

Fig. 3 SEM image of red mud

2.2 Experimental Methodology

The present study consists of various experiments for the characterization of different mixtures as a CLSM. In this study, red mud was used in three proportions 70, 60 and 50%, fly ash was used in 20, 30 and 40% while phosphogypsum of 10% in all the mixes. Cement was used in two proportions 7 and 10% of the mix consisting of red mud, fly ash and phosphogypsum, while different water contents of mix containing red mud, fly ash, phosphogypsum and cement were used considering flow as an important parameter. Water content which was able to produce a flow value in between 0.15 and 0.3 m was used and a mixture was obtained and all the basic properties of CLSM were determined. Each mixture is identified by the following notation as explained in Table 3.

The prepared mixture was then characterized by flow value, bleeding, fresh density, unconfined compressive strength (UCS) and durability as a CLSM material.

Determination of flow consistency. Flow consistency is an important parameter of CLSM. Flow consistency is the measure of the spread of the CLSM material. It is determined as per ASTM D6103-04 Standard Test Method for Flow Consistency of Controlled Low-Strength Material (CLSM) (ASTM: D6103-04 2004). In this, a cylindrical mould of length 0.150 m and diameter 0.075 m was taken. Steps followed in this test are listed below.

- The cylindrical mould was placed over non-pervious fibre plate free from any vibration.
- CLSM mixture was poured into the cylindrical mould and was slightly overfilled.
- The surface was strike with a straight edge.

Table 3 Nomenclature of mixes

Red mud (%)	Fly ash (%)	Phosphogypsum (%)	Cement (%)[a]	Water (%)[b]	Nomenclature
70	20	10	10	45	Mix-1
70	20	10	10	50	Mix-2
60	30	10	10	50	Mix-3
60	30	10	10	55	Mix-4
50	40	10	10	55	Mix-5
70	20	10	7	45	Mix-6
70	20	10	7	50	Mix-7
60	30	10	7	50	Mix-8
60	30	10	7	50	Mix-9
50	40	10	7	50	Mix-10

[a]Cement content % is of the mixture containing red mud, fly ash and phosphogypsum
[b]Water content % is of the mixture containing red mud, fly ash, phosphogypsum and cement

- Within 5 s, cylinder was moved in a vertical direction up to a minimum of 0.15 m height.
- Immediately largest measuring diameter was taken in two perpendicular directions.

The materials having a diameter between 0.15 and 0.30 m were acceptable and then with that mix, further characterization was done. The fabricated mould used for the measurement of flow consistency is shown in Fig. 4.

Density. Density plays an important role in replacing or filling any structure or voids. Hence, it is required to know the density of the material while placing as well as the density of material after hardening. The fresh density of CLSM is measured as per ASTM D6023-16 Standard Test Method for Density (Unit Weight), Yield, Cement Content and Air Content (Gravimetric) of Controlled Low-Strength Material (CLSM) (ASTM: D6023-16 2016). It is the measure of the amount of sample in unit volume.

Fig. 4 Cylindrical mould for flow consistency measurement

It is a very important parameter in various applications. For measuring density, a measuring cylinder of 0.001 m^3 was taken. The prepared sample was filled in that cylinder within 5 min of obtaining the CLSM mix and the volume and weight of sample fill in the measuring cylinder were noted. The density for the freshly prepared is given by the ratio of the weight of the sample to the volume of the sample. Hardened density is measured by the sample prepared for UCS by taking the ratio of the weight to volume.

Bleeding. It is the amount of water segregated water after certain duration. As we know that segregation in the concrete leads to the lesser strength of the concrete which is due to the separation of the water from the mix, this process of coming out of the water from the mix is referred as bleeding. Since water content in the CLSM is high, there is a possibility of bleeding so it should be taken proper care. It is measured as per ASTM C940-10a Standard Test Method for Expansion and Bleeding of Freshly Mixed Grouts for Preplaced-Aggregate Concrete in the laboratory (ASTM: C940-10a 2010) according to which it is the volume of water segregated after 3 h to the initial volume taken. For measuring this, 0.001 m^3 measuring cylinder was taken and filled up to the volume of 0.8–0.9 m^3 then covered with a plate so that moisture is not lost and kept it for three hours after that amount of water segregated was collected.

Unconfined compressive strength (UCS). It is an important parameter for evaluating a material as a CLSM material. For the measurement of UCS, mould of dimension 0.07 m height and diameter 0.03 m was made by using a mould of PVC pipe which was as per ASTM D4832-16 Standard Test Method for Preparation and Testing of Controlled Low-Strength Material (CLSM) Test Cylinders (ASTM: D4832-16 2016) and test was done as per IS code (IS: 2720 (Part 10) 1991). Figure 5 shows the UCS mould used in this study. After preparing the sample, it was kept for 7 days and 28 days curing and then it was tested under compression testing machine. Early strength in CLSM is not important because of its highly flowable in nature it requires time to set so that the sample can be extracted properly from the moulds. For CLSM material, 28 days strength is most significant. By seeing the failed sample, it can be concluded that there was a brittle failure in this type of material. Figure 6 shows the failed sample.

California bearing ratio (CBR). It is an important parameter for road construction. So, for evaluating the CLSM material as a fill material in embankments, CBR is done. It is done as per IS: 2720 (Part-16)-1979 (IS: 2720 (Part 16) 1979). Since it is flowable fill, there was no need for compaction. For preventing water to escape out, the CBR mould was pasted to the base plate. The pasted CBR mould is shown in Fig. 7.

Slake durability test. Slake durability test is done for testing the durability of the sample against weathering action of water. It is done as per ASTM D4644-16 Standard Test Method for Slake Durability of Shales and Other Similar Weak Rocks (ASTM: D4644-16 2016). For this test, sample was put into the drums of slake durability and it was rotated in the water at the speed of 20 rpm for 10 min. Then, the

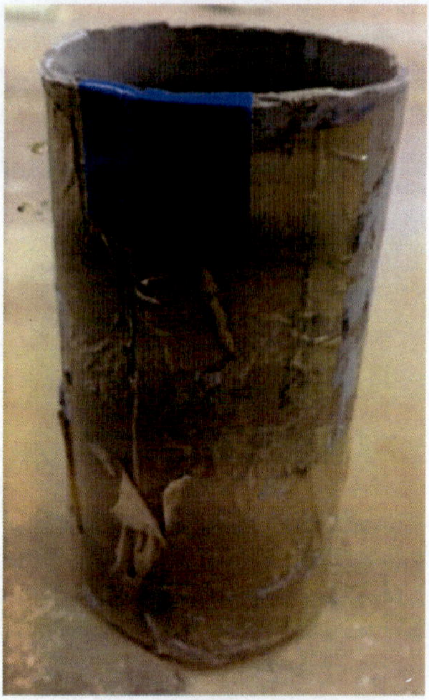

Fig. 5 UCS mould

Fig. 6 Failed UCS sample

Fig. 7 Pasted CBR mould

sample after slaking was collected and kept for oven drying. This procedure was repeated for two cycles and finally, the weight retained percentage was calculated.

Alternate wetting drying test. There are so many places which are subjected to alternate wetting and drying so there is always a need to test the sample for alternate wetting and drying. Due to alternate wetting and drying, there is a change in the volume due to which there is a reduction in the strength of the sample. It is done as per ASTM D559/D559M-15 Standard Test Methods for Wetting and Drying Compacted Soil-Cement Mixtures (ASTM: D559, D559M-15 2015).

3 Results and Discussions

3.1 *Flow Consistency*

Flow consistency was determined for all the mix proportions and the flow values in between 0.15 and 0.3 m were excepted and the rest of the CLSM characterization was done. Table 4 shows the variation of flow value of the different mix proportions. From the results, it can be concluded that with the increase in water content flow is increasing for the same composition while for the different composition the fly ash content is affecting the flow value as more the fly ash more the surface area more the water requirement for better flow consistency. Figures 8 and 9 show the variation of flow value at 10% cement content and 7% cement content, respectively.

Table 4 Flow consistency of the different mix

S. No.	Flow value ($*10^{-3}$ m)
Mix-1	166.5
Mix-2	255
Mix-3	213.8
Mix-4	299.7
Mix-5	246.7
Mix-6	168.3
Mix-7	256.5
Mix-8	215.3
Mix-9	301.7
Mix-10	247.5

Fig. 8 Flow value at 10% cement content

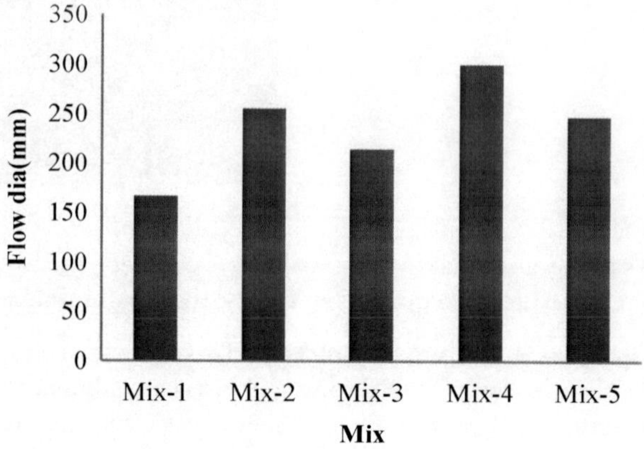

Fig. 9 Flow value at 7% cement content

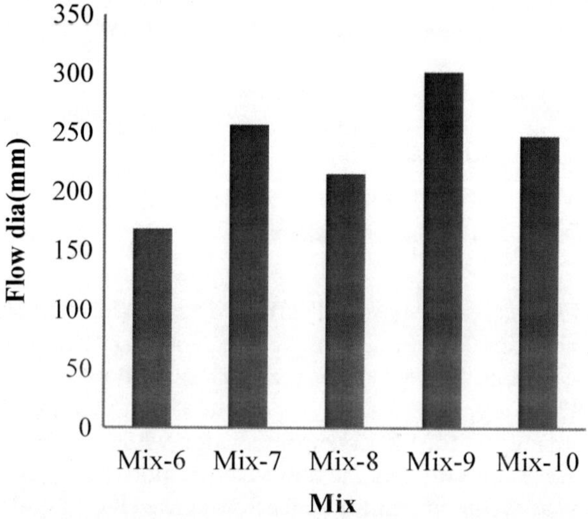

3.2 Density

Fresh density and hardened density at different ages were determined. Fresh density and hardened density of the sample at different ages are given in Table 5. From the results, it can be concluded that with the increase in water content for the same mix proportion the density is decreasing and for different mix proportions the one which contains higher red mud content gives higher density value. Figures 10 and 11 show the variation of fresh density at 10% cement content and 7% cement content, respectively.

Table 5 Densities at different ages

S. No.	Fresh density (kN/m^3)	Hardened density 7 days (kN/m^3)	Hardened density 28 days (kN/m^3)
Mix-1	18.9	18.7	18.2
Mix-2	18.7	18.3	17.8
Mix-3	18.5	17.9	17.7
Mix-4	18.2	17.5	17.1
Mix-5	17.8	17.5	17.5
Mix-6	18.1	18	17.6
Mix-7	17.9	17.7	17.2
Mix-8	16.5	16.4	16.1
Mix-9	15.7	15.3	14.8
Mix-10	15.4	15.2	14.6

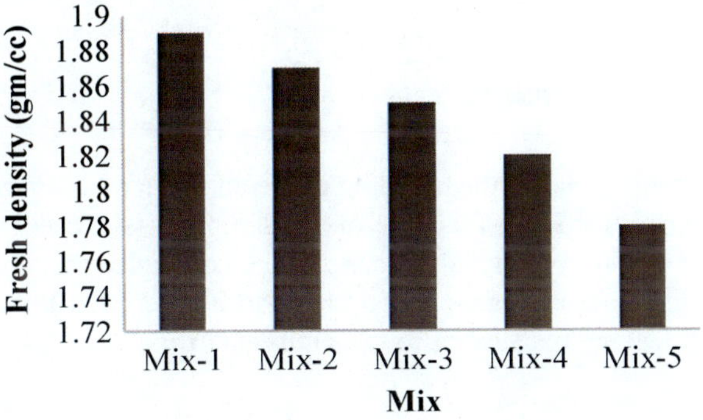

Fig. 10 Fresh density at 10% cement content

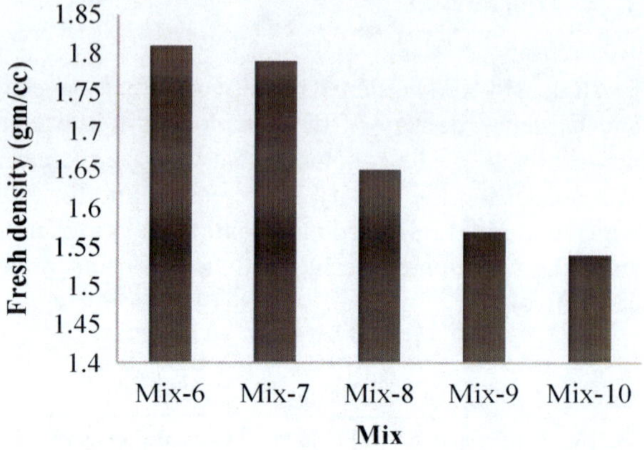

Fig. 11 Fresh density at 7% cement content

Table 6 Bleeding%

S. No.	Bleeding%
Mix-1	0
Mix-2	0
Mix-3	0.05
Mix-4	0.34
Mix-5	0.31
Mix-6	0
Mix-7	0.1
Mix-8	0.15
Mix-9	0.99
Mix-10	0.44

3.3 Bleeding

Table 6 shows the bleeding percentage after 3 h. From the results obtained, it can be concluded that for none of the sample it is exceeding the allowable limit which is 5%. From the result, it can also be concluded that with an increase in water content bleeding increases and it can also be seen that mix having higher RM shows low bleeding because of its plasticity nature.

3.4 UCS

Compression test was done at the age of 7 and 28 days. Table 7 represents the strength of the sample at different ages. Figures 12, 13, 14 and 15 are the stress–strain curve of the UCS sample at 10% and 7% of cement content, respectively, at different ages.

Table 7 UCS of the mix proportions

S. No.	UCS 7 days (*10^6 Pa)	UCS 28 days (*10^6 Pa)
Mix-1	2.71	3.14
Mix-2	1.79	1.82
Mix-3	2.13	2.22
Mix-4	1.36	1.64
Mix-5	1.26	1.42
Mix-6	1.23	1.32
Mix-7	1.11	1.24
Mix-8	1.08	1.16
Mix-9	0.97	1.05
Mix-10	1.01	1.11

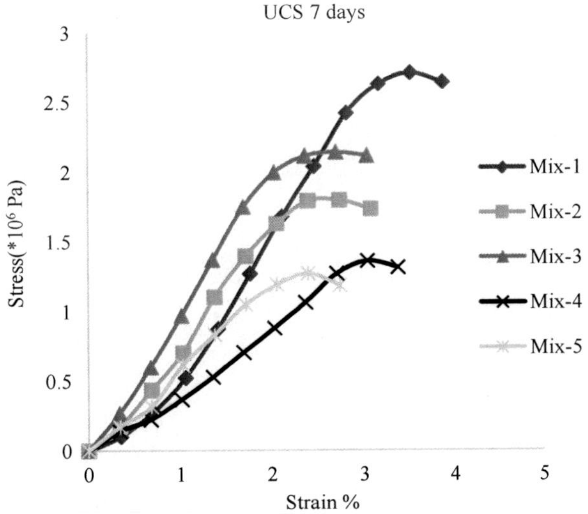

Fig. 12 UCS at 7 days for 10% cement content

From the results, we can see all the compressive strengths are below 8.3 MPa which is required for CLSM. From the results, it can also be concluded that for the mix proportion there is a decrease in the strength value with the increase in water content and it can also be concluded that for different composition at the same water content the compressive strength of that mix containing red mud in more amount is greater than that with the lower red mud content. These results also show that strength also decreases with the decrement of cement content.

Fig. 13 UCS at 7 days for 7% cement content

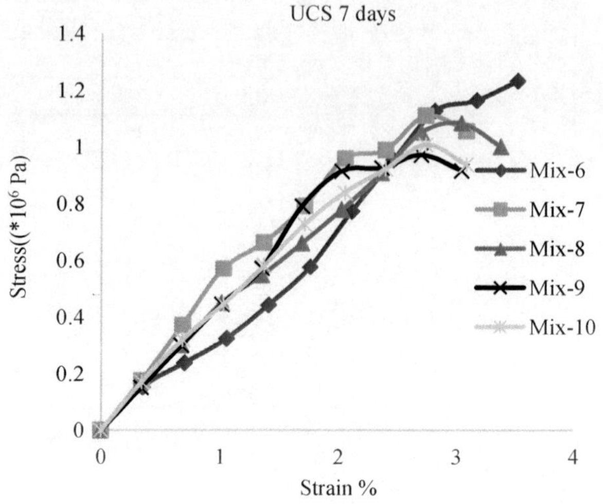

Fig. 14 UCS at 28 days for 10% cement content

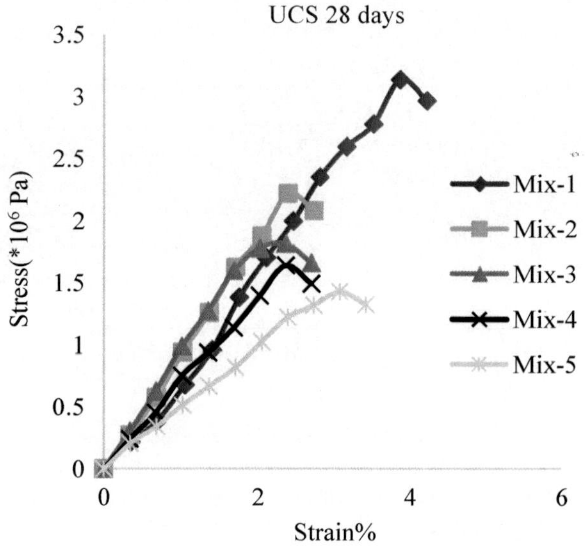

3.5 CBR

Table 8 represents the value obtained after the CBR test after 7 days of curing and then 4 days of soaking. From the results, it can be concluded that the CBR value for all the mix proportion is much more than the required value for the construction work. Form the experimental data, it can be concluded that the CBR value decreases with the increase in water for the same mix proportion while for different mix proportion,

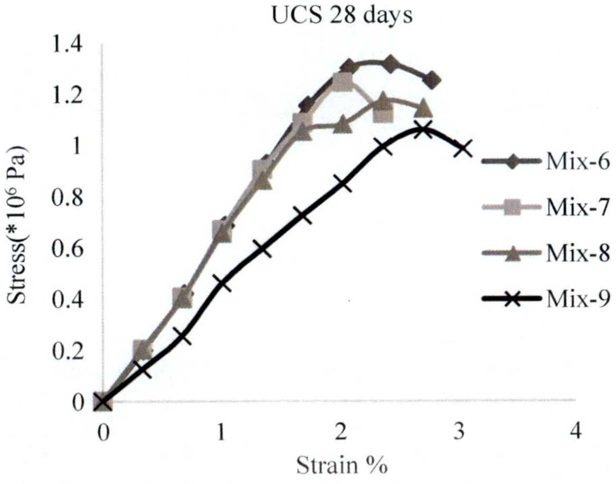

Fig. 15 UCS at 28 days for 7% cement content

Table 8 CBR values

S. No.	CBR%
Mix-1	99.72
Mix-2	93.71
Mix-3	96.81
Mix-4	82.22
Mix-5	83.89
Mix-6	94.64
Mix-7	60.42
Mix-8	43.84
Mix-9	22.99
Mix-10	11.19

it can be concluded that the mix proportion having higher red mud content has a higher value.

3.6 Slake Durability

This test was performed for all the mix proportion but after the first cycle except Mix-1 all were washed away. Table 9 shows the percentage of weight retained after the first and second cycle. Based on results by Gamble's slake durability classification, we can say that these materials are low durable.

Table 9 Slake durability test result

S. No.	%Retained after first cycle	%Retained after second cycle
Mix-1	63.2	50.65

Table 10 Alternate wetting and drying test result

S. No.	UCS (*10^6 Pa)
Mix-1	0.80
Mix-6	0.58

3.7 Alternate Wetting and Drying

The UCS of the sample after 12 cycles was measured. Only Mix-1 and Mix-6 can sustain for 12 cycles and the rest of them failed. Table 10 shows the UCS value of Mix-1 and Mix-6 after 12 alternate wetting and drying cycles and from the results we can say that since all the UCS values are greater than the minimum strength required to be a CLSM material, it can be concluded that these materials are durable to wetting and drying.

4 Conclusion

In this study feasibility of different mix proportions made up of RM, FA, PG, cement, and water were investigated. Experiments were conducted to evaluate the important properties of CLSM with the above material. Based on the experiment conducted, the following major conclusions were drawn:

- The mix containing more amount of fly ash requires more water for attaining higher flow consistency due to the more surface area of it than the red mud.
- The mix containing a higher percentage of red mud is lesser susceptible to bleeding due to its plastic behaviour which is absent in fly ash but for any of the mix proportion bleeding is not found to be outside the acceptable range.
- The density of the freshly prepared sample decreases with the decrease in red mud content as red mud is heavier than fly ash.
- UCS of the prepared mix was decreased with the decrease in red mud content as it can take more load than other materials used and it was also found that the UCS value for the same mix decreases with increase in water content due to the presence of higher capillary pores. However, all the UCS of 28 days was higher than the minimum value for CLSM.
- From the strength perspective, it can be concluded that all the materials can be used as a CLSM material but the places where reexcavatibility is required some material fails.

- The mix with the highest strength was found to be durable to alternate wetting and drying as the UCS value obtained after 12 cycles were more than the minimum value for CLSM.
- From all the conclusion, the major conclusion which can be drawn is that these waste materials can be effectively used as the constitute material of CLSM but before that, we need to look for its environmental aspects.

References

Achtemichuk S, Hubbard J, Sluce R, Shehata MH (2009) The utilization of recycled concrete aggregate to produce controlled low-strength materials without using Portland cement. Cement Concr Compos 31(8):564–569. https://doi.org/10.1016/j.cemconcomp.2008.12.011

ACI Committee 229: Controlled Lowstrength Materials (CLSM) (1999) (ACI 229R-99), Farmington Hills, Michigan: American Concrete Institute, p 15

Al-Harthy AS, Taha R, Abu-Ashour J, Al-Jabri K, Al-Oraimi S (2005) Effect of water quality on the strength of flowable fill mixtures. Cement Concr Compos 27(1):33–39. https://doi.org/10.1016/j.cemconcomp.2004.01.005

ASTM: C940-10a (2010) Standard test method for expansion and bleeding of freshly mixed grouts for preplaced-aggregate concrete in the laboratory. ASTM International, West Conshohocken, PA

ASTM: D4644-16 (2016) Standard test method for slake durability of shales and other similar weak rocks. ASTM International, West Conshohocken, PA

ASTM: D4832-16 (2016) Standard test method for preparation and testing of controlled low strength material (CLSM) test cylinders, ASTM International, West Conshohocken, PA

ASTM: D559/D559M-15 (2015) Standard test methods for wetting and drying compacted soil-cement mixtures. ASTM International, West Conshohocken, PA

ASTM: D6023-16 (2016) Standard test method for density (unit weight), yield, cement content, and air content (gravimetric) of controlled low-strength material (CLSM). ASTM International, West Conshohocken, PA

ASTM: D6103-04 (2004) Standard test method for flow consistency of controlled low strength material (CLSM). ASTM International, West Conshohocken, PA

Butalia TS, Wolfe WE, Zand B, Lee JW (2004) Flowable fill using flue gas desulfurization material. J ASTM Int 1(6):11868. https://doi.org/10.1520/jai11868

Do TM, Kim Y (2016) Engineering properties of controlled low strength material (CLSM) incorporating red mud. Int J Geoeng 7(1). https://doi.org/10.1186/s40703-016-0022-y

Du L, Folliard KJ, Trejo D (2002) Effects of constituent materials and quantities on water demand and compressive strength of controlled low-strength material. J Mater Civ Eng 14(6):485–495. https://doi.org/10.1061/(asce)0899-1561(2002)14:6(485)

IS: 2720 (Part 10) (1991) Methods of test for soils, determination of unconfined compressive strength. Bureau of Indian Standards, New Delhi

IS: 2720 (Part 16) (1979) Methods of test for soils, Determination of California bearing ratio, Bureau of Indian Standards, New Delhi

IS: 2720 (Part 3) (1980) Methods of test for soils, determination of specific gravity, (Reaffirmed 1987). Bureau of Indian Standards, New Delhi

IS: 2720 (Part 4) (1985) Methods of test for soils, determination of grain size analysis. Bureau of Indian Standards, New Delhi

IS: 2720 (Part 5) (1985) Methods of test for soils, determination of liquid and plastic limit. Bureau of Indian Standards, New Delhi

IS: 2720 (Part 7) (1980) Methods of test for soils, determination of water content and dry density using light compaction. Bureau of Indian Standards, New Delhi

Lee NK, Kim HK, Park IS, Lee HK (2013) Alkali-activated, cementless, controlled low-strength materials (CLSM) utilizing industrial by-products. Constr Build Mater 49:738–746. https://doi.org/10.1016/j.conbuildmat.2013.09.002

Lini Dev K, Robinson RG (2015) Pond ash based controlled low strength flowable fills for geotechnical engineering applications. Int J Geosynth Ground Eng 1(4). https://doi.org/10.1007/s40891-015-0035-1.

Liu R-X, Poon C-S (2016) Utilization of red mud derived from bauxite in self-compacting concrete. J Clean Prod 112:384–391. https://doi.org/10.1016/j.jclepro.2015.09.049

Nataraja MC, Nalanda Y (2008) Performance of industrial by-products in controlled low-strength materials (CLSM). Waste Manage 28(7):1168–1181. https://doi.org/10.1016/j.wasman.2007.03.030

Raghavendra T, Udayashankar BC (2014) Flow and strength characteristics of CLSM using ground granulated blast furnace slag. J Mater Civ Eng 26(9):04014050. https://doi.org/10.1061/(asce)mt.1943-5533.0000927

Rashad AM (2017) Phosphogypsum as a construction material. J Clean Prod 166:732–743. https://doi.org/10.1016/j.jclepro.2017.08.049

Sheen Y-N, Zhang L-H, Le D-H (2013) Engineering properties of soil-based controlled low-strength materials as slag partially substitutes to Portland cement. Constr Build Mater 48:822–829. https://doi.org/10.1016/j.conbuildmat.2013.07.046

Wang H-Y, Chen B-T, Yu-Wu Wu (2013) A study of the fresh properties of controlled low-strength rubber lightweight aggregate concrete (CLSRLC). Constr Build Mater 41:526–531. https://doi.org/10.1016/j.conbuildmat.2012.11.113

Yan DYS, Tang IY, Lo IMC (2014) Development of controlled low-strength material derived from beneficial reuse of bottom ash and sediment for green construction. Constr Build Mater 64:201–207. https://doi.org/10.1016/j.conbuildmat.2014.04.087

Influence of the Presence of Zinc on the Behaviour of Bentonite

Saswati Ray, Bismoy Roy Chowdhury, Anil Kumar Mishra, and Ajay Kalamdhad

Abstract In landfill disposal sites, bentonite is used as liner material because of its high contaminant adsorption limit, high swelling capacity and low hydraulic conductivity. However, the effectiveness of liner may reduce due to the chemicals present in the leachate. Swelling and sorption capacity of bentonite may effect in the presence of various chemicals present in the leachate, which in turn reduces the thickness of the diffuse double layer (DDL). Consolidation is an important parameter of liner material, which is essential for settlement calculation. Therefore, it is necessary to study the change in the behaviour of bentonite in the presence of heavy metal. In this investigation, the effects of zinc (Zn^{2+}) of varying concentrations were studied on the behaviour of bentonite. Solutions of zinc 0 (i.e. de-ionized (DI) water), 100 and 1000 ppm concentration were prepared by dissolving salts of zinc nitrate. The results revealed that rate of consolidation and hydraulic conductivity increases with the increase in concentration because of the presence of Zn^{2+} in pore water. However, liquid limit, swelling pressure and swelling potential were decreased. Results also illustrate that at higher concentration, the impact of Zn^{2+} on the behaviour of the bentonite is more significant.

Keywords Bentonite · Heavy metal · Swelling pressure · Hydraulic conductivity · Landfill liner

1 Introduction Section

Excessive exposure of lead can impose potential menace to the ecosystem and can cause an adverse effect on the health of adults and children. Zinc can be severely toxic to the human being if the amount exceeds the permissible limit. Various sources, for example, mining and smelting, sewage water, households, tap water system and industrial wastewater containing zinc are the major sources of lead exposure to the

S. Ray (✉) · B. R. Chowdhury · A. K. Mishra · A. Kalamdhad
Department of Civil Engineering, Indian Institute of Technology Guwahati, Guwahati, Assam 781039, India
e-mail: r.saswati@iitg.ac.in

M. Latha Gali and R. R. P. (eds.), *Problematic Soils and Geoenvironmental Concerns*, Lecture Notes in Civil Engineering 88,
https://doi.org/10.1007/978-981-15-6237-2_25

human being and environment. This exposure can cause toxic symptoms like cholesterol balance, stomach cramp, infertility, vomiting and nausea and can even diminish the function of immune system (Zhang et al. 2012).

Several researchers are attentive on the usage of low-cost adsorbent in order to reduce the processing costs. Clay minerals are economical, readily available and a good substitute for conventional treatment of the leachate as an adsorbent (Sanchez et al. 1999). Clay minerals are used to eliminate the various types of contaminants such as removal of zinc using bentonite as an adsorbent from aqueous solution by the mechanism of adsorption and ion exchange (Mellah and Chegrouche 1997; Kaya and Ören 2005). Other clay minerals like palygorskite clays and raw kaolinite are used as an adsorbent for the removal of various heavy metals from aqueous solution (Potgieter et al. 2006; Yavuz et al. 2003). Leachate contains heavy metal, which can contaminate the surrounding environment and groundwater. In order to prevent contamination by leachate, clay barrier is used in the landfill so that it cannot further migrate. Bentonite is utilized as a liner material due to its low hydraulic conductivity, contaminant adsorption capability and high swelling capacity (Dutta and Mishra 2016). One of the utmost essential properties, which helps in computing the settlement of the clay liner, is the compressibility of liner material (Mishra et al. 2010). Bentonite is an extremely compressible material. Because of the load of the waste piled in the waste disposal site, bentonite is compressed easily. When the surcharge load is applied to a saturated clay, consolidation process starts due to compression, which results in settlement, and the extent is determined after the thorough dissipation of the generated pore water pressure (Yong and Warkentin 1975).

In the past, many studies have been carried out regarding the change in the behaviour of bentonite in the presence of various contaminants. Lo et al. (2004) investigated the movement of lead, zinc and cadmium in the saturated Ottawa sand and bentonite–soil admixture and they concluded that the hydraulic conductivity of bentonite–soil admixture increases considerably when it is permeated with the metal solution. Ouhadi et al. (2006) examined the interaction of the heavy metal ion with bentonite at various pH level. They observed that at high heavy metal concentration and low pH, there is a microstructural change in bentonite. They also concluded that the rheological performance of bentonite is controlled by the osmotic phenomenon. Nakano et al. (2008) conducted the test on three locally available Japanese bentonite and one US bentonite to evaluate the hydraulic conductivity behaviour and lead retention mechanisms, and they observed that carbonate plays the main role at low concentration of lead and precipitate as $PbCO_3$. They also concluded that as a result of higher swelling capacity and montmorillonite content, hydraulic conductivity of the US bentonite is lower in comparison with Japanese bentonite. Du et al. (2015) explored the consequences of different levels of lead contamination on clayey soil/calcium–bentonite backfills. They concluded that with the rise in the lead concentration pH, liquid limit and compression index (C_c) increase and the hydraulic conductivity increases to 50-fold in comparison with clean backfills. Bentonites with different mineralogical compositions were also examined under varying concentration of heavy metals (Dutta and Mishra 2016). It was found that with the rise in the concentration of the metal solution, Atterberg's limit, free swelling, swelling pressure and swelling potential of the soil decreased and the hydraulic conductivity increased. Dutta and Mishra (2017) examined the impact of various heavy metals of varying concentration on the consolidation parameters of bentonite. They concluded that coefficient of consolidation (c_v) increases while, C_c, coefficient of volume changes (m_v) and time to achieve 90% of consolidation (t_{90}) of the bentonites decreases with the rise in heavy metal concentration.

For the usefulness as barrier material in the landfill, hydraulic conductivity and swelling are important parameters of bentonite. Therefore, it is required to know the behaviour of bentonite in the presence of heavy metal in order to predict the fate of heavy metal. The purpose of the present study is to examine the influence of bentonite in the presence of various concentrations of zinc (Zn^{2+}). In the present study, bentonite was evaluated for free swell, liquid limit, swelling pressure, swelling potential and hydraulic conductivity in the presence of various concentrations of Zn^{2+}. The outcome of the study will help engineers in the design of liner systems. The heavy metal studied in the present study was Zn^{2+} as it is very hazardous for human health and environment.

2 Material and Methods

2.1 Bentonite

Bentonite clay used for the investigation was obtained from Barmer district of Rajasthan state of India. The physical and chemical properties of the bentonite are listed in Table 1.

Table 1 Properties of bentonite used in this investigation

Property	Bentonite
Liquid limit (%)	480
Plasticity index (%)	443
Specific gravity	2.73
Free swelling (ml/2 g)	32.5
Clay content (%)	66
Silt content (%)	34
Cation exchange capacity (CEC) (meq/100 g)	40.2
Specific surface area (m^2/g)	396.27
Optimum moisture content (OMC) (%)	33.5
Maximum dry density (MDD) (g/cc)	1.31
pH	8.4

The particle size distribution of the bentonites was analysed by the hydrometer analysis according to ASTM D 422 (2002). The free swelling test for the bentonite was performed. Atterberg's limits were performed according to ASTM D 4318 (2000). The specific surface area (SSA) of the bentonite was analysed by the method described by Cerato and Lutenegger (2002). The method defined by Chapman (1965) was used to analyse the cation exchange capacity (CEC) of the bentonite. ASTM D698 (ASTM 2012) was followed to determine the maximum dry density (MDD) and the optimum moisture content (OMC) of the bentonite.

2.2 *Permeant Liquid*

A major quantity in the leachate contributes to heavy metal. Zinc is chosen for the investigation since it is one of the main contaminant present in leachate and can cause severe effects on environment and human health. Zinc (Zn^{2+}) at concentration of 0 (i.e. de-ionized (DI) water), 100 and 1000 ppm was prepared by dissolving salts of $Zn\,(NO_3)_2$.

2.3 *Determination of Hydraulic Conductivity*

Bentonite was mixed with water at their respective optimum moisture content and kept in the moisture-controlled desiccator for 24 h to achieve the moisture equilibrium. The moisture-equilibrated sample was then statically compacted in oedometer rings of a diameter of 60 mm to a thickness of 15 mm to its MDD. The whole assembly was then placed in the consolidation cell and positioned in the loading frame. The specimens were swamped in the DI water or to the respective metal solutions under the nominal pressure of 4.9 kPa and permitted to swell. The samples were consolidated by increasing the pressure slowly by an increment ratio of 1 (i.e. increased by 4.9, 9.8, 19.6 kPa at each step) to a maximum pressure of 784.8 kPa once the swelling was done. Then, the change in the thickness of soil sample was measured for each pressure increment from the dial gauge readings. The change in the void ratio corresponding to the increase in the overburden pressure was calculated as

$$\Delta e = \frac{\Delta H(1 + e_0)}{H} \tag{1}$$

where ΔH is the change in the thickness of sample due to increase in pressure; H is the initial thickness of the sample; e_0 is the initial void ratio.

From the consolidation test result, a time–settlement curve was obtained at each pressure increment. The coefficient of consolidation (c_v) was obtained using Taylor's square root time ($\sqrt{T}$) method (Taylor 1948). The coefficient of volume change was calculated as

$$m_v = -\frac{\Delta e}{\Delta\sigma(1 + e_0)} \tag{2}$$

where $\Delta\sigma$ is the change in pressure and Δe is the change in void ratio. Coefficient of consolidation (c_v) was determined by the square root of time fitting method given by Taylor (1948) (IS 2720 part XV).

The hydraulic conductivity (k) was calculated by fitting Terzaghi's theory of consolidation (Terzaghi 1943) for various pressure increments using the c_v and m_v as

$$k = c_v m_v \gamma_w \tag{3}$$

where γ_w is the unit weight of the pore fluid.

The compression index (C_c) of a soil is determined to analyse the settlement triggered because of the application of consolidation pressure. Generally, the C_c of a normally consolidated soil is calculated from the gradient of the straight line part of the void ratio–pressure (e-log P) curve as,

$$C_c = \frac{e_i - e_j}{\log \frac{p_i}{p_j}} \tag{4}$$

where e_i and e_j are the corresponding measured void ratio due to change in the consolidation pressure of p_i and p_j at ith and jth steps of loading, respectively.

2.4 Determination of Swelling Potential and Swelling Pressure

A conventional oedometer apparatus was used for the determination of the swelling potential and swelling pressure of compacted bentonite sample. On the compacted bentonite specimen, a sitting load of 4.9 kPa was applied and then was inundated with respective metal concentration and the values of swelling with time were recorded. The measurements were continued until the swelling increment reached negligible values. A standard consolidation test was conducted at this point by applying incremental loads starting with 4.9 kPa and ending with 784.8 kPa. The pressure required to revert the specimen to its initial void ratio was determined as the swelling pressure. It is defined as the value of pressure required to keep the sample at zero swelling after saturating it with saturating fluid. The swelling potential is defined as the percentage swelling of the soil. The minutiae of this method have been described by Sridharan et al. (1986) and Sridharan and Gurtug (2004).

Table 2 Liquid limit and free swelling of bentonite in the presence of zinc

Zinc concentration (ppm)	Bentonite	
	Liquid limit (%)	Free swelling (mL/2 g)
0 (DI water)	480	32.5
100	442	30.2
1000	311	13.6

3 Results and Discussion

3.1 *Effect of Various Concentration of Zn^{2+} on Liquid Limit and Free Swelling*

The effect of various concentration of Zn^{2+} on the liquid limits of the bentonite is listed in Table 2. It shows that with the rise in Zn^{2+} concentration, there is a reduction liquid limit value. With DI water the value of 480% liquid limit decreased to 442% for the Zn^{2+} solution for 100 ppm concentration. With the further rise in Zn^{2+} concentration to 1000 ppm, the liquid limit value reduced to 311%. The percentage reduction in liquid limit was 7.9% and 35.2% for 100 and 1000 ppm, respectively.

The increase in the volume of the bentonite due to the absorbing of water in the absence of external overburden pressure is called free swell. From Table 2, it is seen that with the rise in the concentration of the Zn^{2+} solution from 0 to 100 ppm, the free swelling decreased from 32.5 to 30.2 mL/2 g, though a further rise in the concentration to 1000 ppm, the free swelling of the bentonite reduced from 30.2 to 13.6 mL/2 g. The percentage reduction in free swelling was 7.07 and 58.15% for 100 and 1000 ppm, respectively.

The reduction in free swelling is due to the significant reduction in the DDL thickness because of the increase in metal concentration. The results also indicate that the reduction of liquid limit and free swelling was more prominent at higher concentration.

3.2 *Determination of Swelling Potential and Swelling Pressure*

Table 3 shows that with an increase in metal concentration both swelling potential and swelling pressure decrease.

From Table 2, it is concluded that with the rise in the concentration of the Zn^{2+} solution from 0 to 100 ppm, swelling potential decreased from 34.32 to 23.2%, however, a further rise in the concentration to 1000 ppm, the swelling potential of the bentonite reduced to 16.48%. Similarly, the table also shows that the swelling pressure reduced from 460.9 to 377.6 kPa with the rise in the concentration of the

Table 3 Swelling potential and swelling pressure of bentonite in the presence of zinc

Heavy metal concentration (ppm)	Bentonite	
	Swelling potential (%)	Swelling pressure (kPa)
0 (DI water)	34.3	460.9
100	23.2	377.6
1000	16.5	284.6

Zn^{2+} solution from 0 to 100 ppm. Further rise in the concentration to 1000 ppm, the swelling potential of the bentonite was found to be 284.6 kPa. A maximum decrease in swelling potential and swelling pressure showed with 1000 ppm Zn^{2+} solution with respect to DI water.

3.3 Hydraulic Conductivity

Figure 1 shows that with an increase in Zn^{2+} concentration hydraulic conductivity (k) increases. This is because with an increase in Zn^{2+} concentration, particles come closer to each other and thickness of diffuse double layer decreases, causing an increase in the hydraulic conductivity value. Figure 1 depicts that with the rise in Zn^{2+} concentration from 0 to 100 ppm, the hydraulic conductivity of the bentonite increased slightly, though the hydraulic conductivity increased considerably with the rise in concentration from 100 to 1000 ppm.

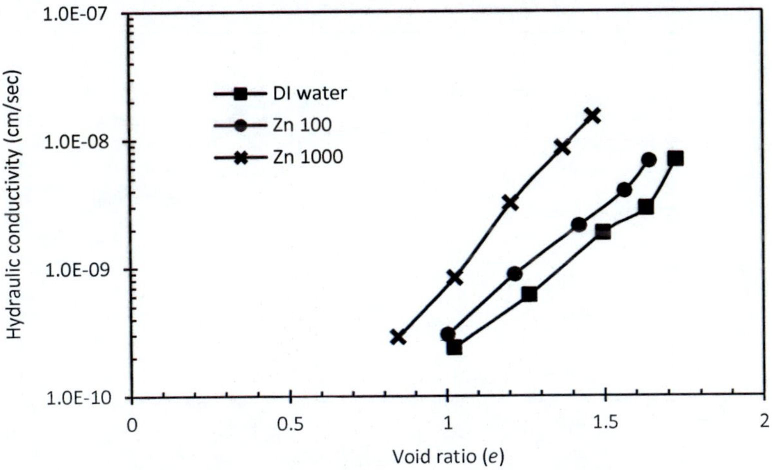

Fig. 1 Hydraulic conductivity values for bentonite at different void ratios

Table 4 Bentonite at a void ratio of 1.1 and 1.4

Concentration	Void ratio	Hydraulic conductivity (cm/s)	Void ratio	Hydraulic conductivity (cm/s)
DI water	1.1	3.01E−10	1.4	8.78E−09
Zn 100	1.1	4.96E−10	1.4	1.90E−09
Zn 1000	1.1	1.31E−09	1.4	1.05E−09

Hydraulic conductivity corresponding to a void ratio of 1.1 and 1.4 was determined from Fig. 1 in order to compare the hydraulic conductivity value of the bentonite at different Zn^{2+} concentrations and void ratio.

It is seen from Table 4 that at higher concentration of Zn^{2+} the increase in hydraulic conductivity is significant. Due to the rise in the concentration of Zn^{2+} from 0 to 100 ppm, the hydraulic conductivity of bentonite was increased from 3.01×10^{-10} to 4.96×10^{-10} cm/s (39.91% increase). With further increase in concentration to 1000 ppm, the hydraulic conductivity increased to 1.31×10^{-10} cm/s (77.02% increase) at a void ratio 1.1.

Similarly, with the increase in void ratio from 1.1 to 1.4, the hydraulic conductivity increased by 61.70% and 19.84% increases % for 100 and 1000 ppm, respectively.

3.4 Void Ratio–Pressure Relationship

Figure 2 shows the void ratio–pressure (*e*-log *P*) relationship for the bentonite at

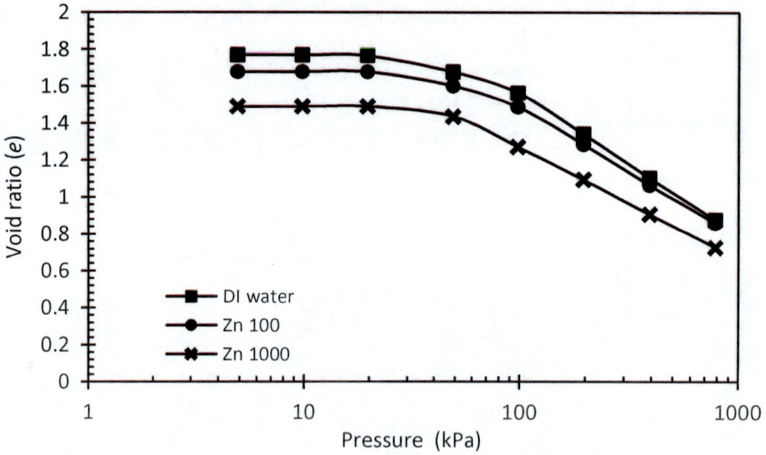

Fig. 2 Void ratio–pressure relationship of bentonite in the presence of heavy metal

different concentrations of Zn^{2+} solution. From Fig. 2, it is shown that the soil specimen permeated with Zn^{2+} solution revealed a lesser value of void ratio in comparison with the specimen permeated with DI water.

As the concentration of Zn^{2+} rises in the pore fluid, more Zn^{2+} ions are adsorbed on the clay surface. Consequently, repulsive force reduces and clay particles come to a closer distance resulting in reduction in DDL thickness.

A comparison between two different Zn^{2+} concentration of 100 and 1000 ppm indicates that at lower metal concentration, the reduction in void ratio was greater as the vertical pressure is applied. This can be attributed to the development of DDL. The interparticle repulsive forces reduce, with the rise in the metal contaminant concentration, causing the compression of the bentonite to a lower void ratio (Bolt 1956; Olson and Mesri 1970; Sridharan and Jayadeva 1982).

3.5 Compression Index (C_c)

A relationship between C_c and different Zn^{2+} concentration in Fig. 3 indicates that with the increase in Zn^{2+} concentration C_c value reduced. With the rise in concentration of Zn^{2+}, particle's orientation turned out to be more flocculated which resists the settlement causing a lower C_c value. The C_c value for the bentonite reduced from 0.76 to 0.70 (7.8% decrease) with an increase in Zn^{2+} concentration from 0 to 100 ppm. With further increase in concentration to 1000 ppm, the C_c value reduced to 0.61 (19.7% decrease).

Figure 4 shows the relationship between C_c and liquid limit. The plot indicates that the C_c value increased from 0.61 to 0.76 with the increase in liquid limit from 311 to 480%.

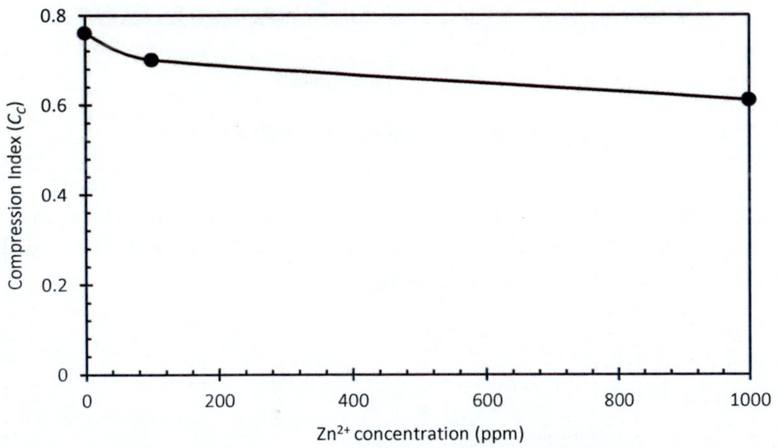

Fig. 3 Effect of Zn^{2+} concentrations on compression index of bentonite

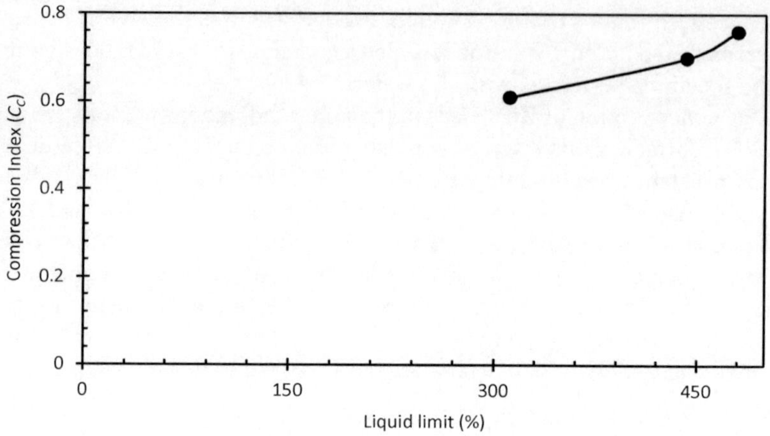

Fig. 4 Plot between the compression index and the liquid limit of the bentonite

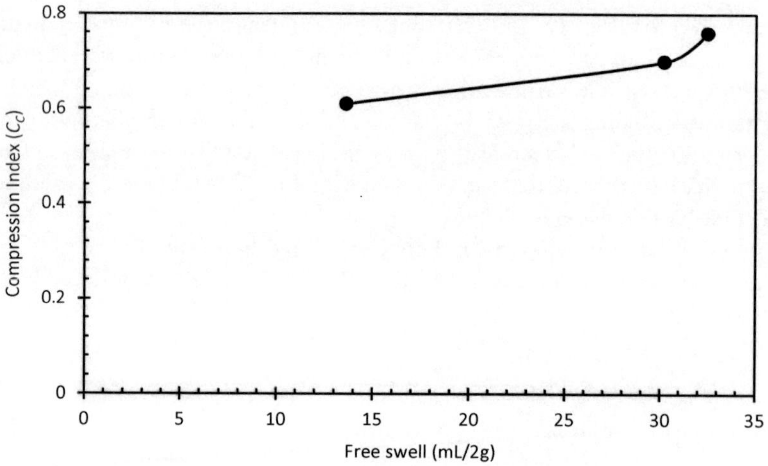

Fig. 5 Plot between the compression index and the free swell of the bentonite

The plot between C_c and free swell in Fig. 5 shows that with the increase in free swell, the C_c for the bentonites increased. With an increase in the free swell from 32.5 to 13.6 mL/2 g, C_c increased from 0.76 to 0.61.

4 Conclusion

This investigation was carried out to study the influence of different concentration of Zn^{2+} on the behaviour of bentonite. Bentonite was studied for their alteration in the liquid limit, free swelling, swelling pressure, swelling potential and hydraulic conductivity caused by the presence of different concentrations of Zn^{2+}. The result revealed that with the increase in Zn^{2+} concentration from 0 to 1000, the free swell, liquid limit, swelling potential and swelling pressure decreased. The reduction in these factors with the rise in concentration is attributed to the decrease in thickness of the diffuse double layer. However, the hydraulic conductivity of the bentonite increases with increase in Zn^{2+} concentration. This study concludes that influence of Zn^{2+} on the behaviour of the bentonite is more significant at higher concentration.

References

ASTM Standard test methods for liquid limit, plastic limit, and plasticity index of soils. D 4318 (2000). American Society for Testing and Materials, Philadelphia

ASTM Standard test method for particle-size analysis of soils. D 422-63 (2002). American Society for Testing and Materials, Philadelphia

ASTM Standard test methods for laboratory compaction characteristics of soil using standard effort, D 698 (2012). American Society for Testing and Materials, Philadelphia

Bolt GH (1956) Physico and chemical analysis of the compressibility of pure clays. Geotechnique 6(2):86–93

Cerato AB, Lutenegger AJ (2002) Determination of surface area of fine-grained soils by the ethylene glycol monoethyl ether (EGME) method. Geotech Test J 25:1–7

Chapman HD (1965) Cation exchange capacity. Methods of soil analysis, part 2 chemical and microbiological properties, 2nd edn. Soil Science Society of America, Madison, pp 891–895

Du YJ, Fan RD, Reddy KR, Liu SY, Yang YL (2015) Impacts of presence of lead contamination in clayey soil–calcium bentonite cutoff wall backfills. Appl Clay Sci 108:111–122

Dutta J, Mishra AK (2016) Influence of the presence of heavy metals on the behaviour of bentonites. Environ Earth Sci 75(11):1–10

Dutta J, Mishra AK (2017) Consolidation behavior of compacted bentonites in the presence of heavy metals. J Hazard Toxic Radioactive Waste 21(3):04017003

Kaya A, Ören AH (2005) Adsorption of zinc from aqueous solutions to bentonite. J Hazard Mater 125(1):183–189

Lo I, Luk A, Yang X (2004) Migration of heavy metals in saturated sand and bentonite/soil admixture. J Environ Eng 130. Special issue: Waste containment barrier materials, pp 906–909

Mellah A, Chegrouche S (1997) The removal of zinc from aqueous solutions by natural bentonite. Water Res 31(3):621–629

Mishra AK, Ohtsubo M, Li LY, Higashi T (2010) Influence of the bentonite on the consolidation behaviour of soil bentonite mixtures. Carbonates Evaporites 25(1):43–49

Nakano A, Li LY, Ohtsubo M, Mishra AK (2008) Lead retention mechanisms and hydraulic conductivity studies of various bentonites for geoenvironment applications. Environ Technol 29:505–514

Olson RE, Mesri G (1970) Mechanisms controlling the compressibility of clay. J Soil Mech Found Div 96(6):1863–1878

Ouhadi VR, Yong RN, Sedighi M (2006) Influence of heavy metal contaminants at variable pH regimes on rheological behaviour of bentonite. Appl Clay Sci 32:217–231

Potgieter JH, Potgieter-Vermaak SS, Kalibantonga PD (2006) Heavy metals removal from solution by palygorskite clay. Miner Eng 19(5):463–470

Sanchez AG, Ayuso EA, Blas O (1999) Sorption of heavy metals from industrial waste water by low-cost mineral silicates. Clay Miner 34(3):469

Sridharan A, Gurtug Y (2004) Swelling behavior of compacted fine grained soils. Eng Geol 72:9–18

Sridharan A, Jayadeva MS (1982) Double layer theory and compressibility of clays. Geotechnique 32(2):133–144
Sridharan A, Rao AS, Sivapullaiah PV (1986) Swelling pressure of clays. ASTM Geotech Test J 9(1):23–33
Taylor DW (1948) Fundamentals of soil mechanics. Wiley, New York
Terzaghi K (1943) Theoretical soil mechanics. Wiley, New York
Yavuz Ö, Altunkaynak Y, Güzel F (2003) Removal of copper, nickel, cobalt and manganese from aqueous solution by kaolinite. Water Res 37(4):948–952
Yong RN, Warkentin BP (1975) Soil properties and behaviour. Elsevier, New York
Zhang X, Yang L, Li Y, Li H, Wang W, Ye B (2012) Impacts of lead/zinc mining and smelting on the environment and human health in China. Environ Monit Assess 184(4):2261–2273

Theoretical Study on Equilibrium Volume of Clay Sediments in Salt Solutions

Dhanesh Sing Das and Tadikonda Venkata Bharat

Abstract Expansive clays exhibit some attractive features such as high swelling capacity and very low hydraulic conductivity, for which it is deemed as the most suitable naturally available material for the design of barrier/buffer material in many geo-environmental engineering practices. Equilibrium sediment volume (ESV) is an important surrogate compatibility test used for assessing the expansiveness of soil. It is also utilized for qualitative understanding of the swelling characteristics of clays, mineral identification, and correlating with different geotechnical properties of clays. A detailed understanding on effect of the influencing parameters on ESV provides useful insights into the fundamental behavior of expansive clays. The main objective of this work was to develop a theoretical model for equilibrium sediment volume of clays in the presence of salt solutions using the concept of diffuse double-layer theory. The ESV showed a linear relationship with the specific surface area of soils.

Keywords Equilibrium sediment volume (ESV) · Specific surface area (SSA) · Diffuse double layer (DDL)

1 Introduction

Expansive clays are being widely considered for potential candidate material for various applications in many geo-environmental practices, e.g., as buffer and backfill material in the high-level nuclear waste repositories (Tripathy et al. 2004); as landfill liner in toxic and municipal solid waste repositories to control the migration of contaminants (Bharat et al. 2009); as shotcrete lining and drilling fluid in tunnel construction; and as reactive barrier in slurry trench cutoff walls (Koch 2002). The ability to resist migration of contaminants due to very low hydraulic conductivity, high-ion adsorption capacity, and high swelling abilities are some of the attractive features for which clays are being recognized as the most suitable naturally available

D. S. Das (✉) · T. V. Bharat
Indian Institute of Technology Guwahati, Guwahati, Assam 780139, India
e-mail: dhaneshsinghdas@gmail.com

M. Latha Gali and R. R. P. (eds.), *Problematic Soils and Geoenvironmental Concerns*, Lecture Notes in Civil Engineering 88,
https://doi.org/10.1007/978-981-15-6237-2_26

candidate for serving the aforementioned roles in various sustainable engineered systems (Schanz and Tripathy 2009).

Equilibrium sediment volume (ESV) is one of the important laboratory index test used for assessing the expansiveness of clays. The test data are useful to understand the swelling characteristics of clays, mineral identification, and correlating with different geotechnical properties of clays. The effect of electrochemical forces on sediment volume of clays is more pronounced as the mechanical pressure acting on the clay particles is zero or negligible. The variation of ESV with pore fluid properties can provide a clear understanding on the characteristics of various physicochemical forces developed in the soil and its relation with volume change behavior of clays. Study on the effect of various pore fluid parameters on the sediment volume behavior of plastic clays is, therefore, very useful in the performance assessment of clays for different geo-environmental applications. A comprehensive understanding on the fundamental behavior of plastic clays can be developed through analyzing the effect of various physicochemical factors on the ESV. Existing literature provides only a qualitative understanding on the ESV behavior of clays based on the experimental study. A detailed theoretical analysis on the various influencing parameters can provide a quantitative understanding on the ESV behavior of expansive clays.

The predicted ESV data were analyzed to study the influence of various pore fluid and soil parameters. Theoretical analysis provided a quantitative understanding on the effect of various electrochemical and soil surface properties on the ESV of soils. The main objective of this work was to develop a theoretical model for equilibrium sediment volume of clays in the presence of salt solutions. The classical Gouy–Chapman diffuse double-layer model was utilized to develop an analytical formulation for estimating the diffuse double-layer thickness of a very low-interacting clay–water–electrolyte system. The model was used to predict the ESV of plastic clays in electrolyte environment.

2 Diffuse Double-Layer Theory

Clay particles, when come in contact with water or electrolyte solutions, diffuse double layer develop around it, following its colloidal nature. The potential distribution in a diffuse double layer around a clay plate, as suggested by Gouy–Chapman model (1911), is illustrated in Fig. 1.

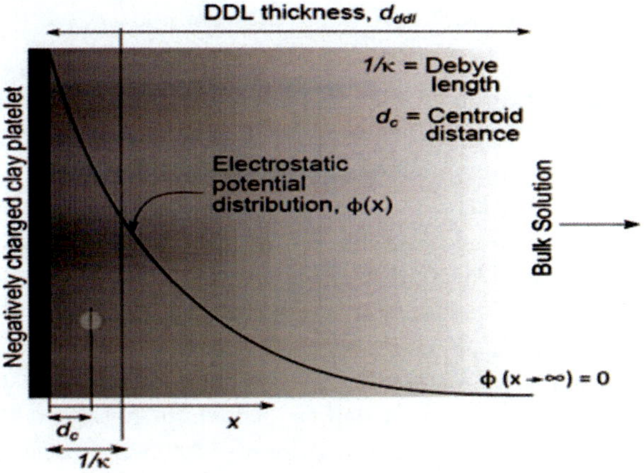

Fig. 1 Illustration of diffuse double layer around a clay platelet (After Bharat and Sridharan 2015b)

2.1 *Non-interacting Clay Platelets*

The electrostatic potential distribution in DDL of a single plate is given by Poisson–Boltzmann equation as shown below:

$$y = 2\ln\left(\frac{\exp(kx) + \exp\left(\frac{z}{4}\right)}{\exp(kx) - \exp\left(\frac{z}{4}\right)}\right) \tag{1}$$

The electric potential near the surface is calculated as,

$$\phi_0 = 0.1725\frac{T}{v}\sinh^{-1}\left(\frac{1256.81 \times \frac{C_{EC}}{S_{SA}}}{\sqrt{n\varepsilon T}}\right) \tag{2}$$

where y is the scaled electrostatic potential $(ve\phi/RT)$, at any distance x, from the surface of the single clay plate, z is the scaled surface potential $(z = ve\phi_0/KT = vF\phi_0/RT)$, $1/\kappa$ is the characteristics length or Debye length, $\kappa = \sqrt{\frac{8\pi ne^2v^2}{\varepsilon KT}}$, e is the elementary electric charge $\left(4.77 \times 10^{-10}\,\text{esu}\right)$, v is the valence of ions, ε is the dielectric constant, n is the molar concentration of the ions in bulk solution (M), K is the Boltzmann constant, F is the Faraday constant, and ϕ is the electric potential at any distance x. C_{EC} and S_{SA} are, respectively, the cation exchange capacity in meq/100 g and the specific surface area in m^2/g of the soil.

2.2 Interacting Clay Platelets

When the distance between two charged plates is such that, the diffuse double layer of both the plates overlap/interact, then the system is considered as interacting system. The resulting potential distribution between the two plates is shown in Fig. 2.

The interaction between diffuse double layer of adjacent plates containing exchangeable cations leads to the existence of osmotic (repulsive) pressures between the plates. The mid-plane potential (ϕ_d) thus developed between the interacting diffuse double layers at equilibrium is the measure of osmotic pressure (P) and the mid-plane distance ($x = d$) between the clay platelets at equilibrium is the measure of void ratio. The void ratio is an easily measurable state variable in soil mechanics. According to Gouy–Chapman model, the relation between void ratio versus osmotic pressure can be obtained using the following relations (Bharat et al. 2013; Bharat and Sridharan 2015a):

$$P = 2nRT(\cosh u - 1) \tag{3}$$

$$\left(\frac{\mathrm{d}y}{\mathrm{d}\xi}\right)_{x=0} = -\frac{C_{\mathrm{EC}}}{S_{\mathrm{SA}}}\left(\frac{2513.6}{\sqrt{\varepsilon n T}}\right) \tag{4}$$

$$\left(\frac{\mathrm{d}y}{\mathrm{d}\xi}\right)_{x=0} = -(2\cosh(z) - 2\cosh(u))^{1/2} \tag{5}$$

$$\kappa t_{\mathrm{DDL}} = -\int_{z}^{u} (2\cosh(y) - 2\cosh(u))^{-1/2}\mathrm{d}y \tag{6}$$

$$e = G_{\mathrm{s}}\gamma_{\mathrm{w}}S_{\mathrm{SA}}t_{\mathrm{DDL}} \tag{7}$$

where P is the consolidation pressure (kPa), R is the gas constant, T is the absolute temperature, u is the scaled midway potential ($ve\phi_d/RT$), ξ is the scaled distance (κx), from the surface of a single clay platelet, G_{s} is the specific gravity of soil particles, e is the void ratio, γ_{w} is the density of water (kg/m^3), t_{DDL} is the midway distance

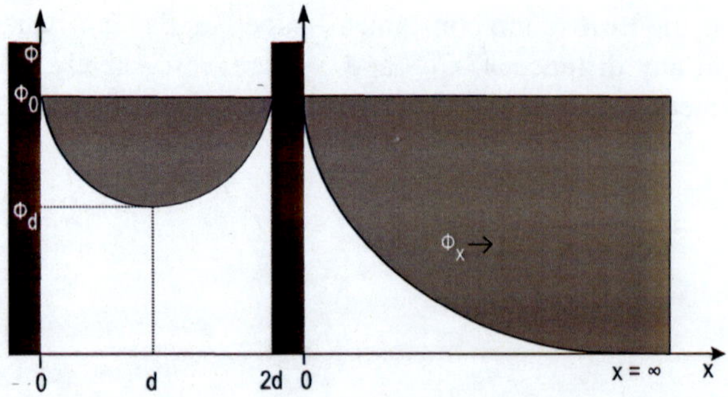

Fig. 2 Illustration of diffuse double layer around interacting and non-interacting plates (After Bharat et al. 2013)

between the particles or DDL thickness, and ϕ_d is the value of the electrostatic potentials at mid-plane.

3 Equilibrium Sediment Volume (ESV)

ESV tests are routinely performed in the laboratory as an index for qualitative assessment of swelling behavior of clay soil. Clay specimen is thoroughly mixed with a given electrolyte solution and brought to equilibrium as illustrated in Fig. 3. During the sedimentation process, clay particles settle under the self-weight with diffuse double developed to the full extent around it and the clay particles are in a parallel arrangement (Fig. 3). Under this condition, the sediment volume is controlled by the diffuse double around the particle as the pressure acting on the clay particles is zero.

Theoretical analysis of the sediment volume at equilibrium was analyzed using the following hypothesis.

3.1 Hypothesis

It was assumed that there was no aggregation of clay particle during settlement as the clay–electrolyte mixture was allowed to settle freely following the mixing procedure.

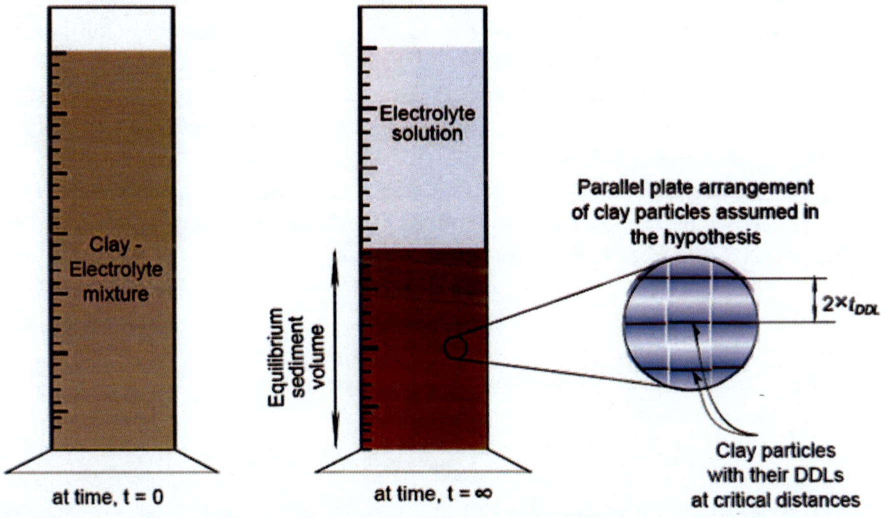

Fig. 3 Illustration of the equilibrium sediment volume test **a** at the beginning and **b** at equilibrium (inset: assumed assumption of parallel clay plate arrangement in the sediment) (After Bharat and Das 2017)

This was ensured by the procedure followed for the preparation of clay–electrolyte mixture. An equilibrium separation distance between the parallel clay platelets is established under the action of both the gravitational force and the surface forces. The downward movement of the clay platelets driven by gravity is counteracted by the electrostatic repulsive forces mobilized between the clay platelets due to the interaction of DDLs. The mobilized electrostatic force keeps on increasing as the clay platelets approach each other and balanced the forces due to gravity at an equilibrium separation distance. To avoid singularity of the integral solution (i.e., Eq. 6), the gravitational forces were considered in the study. At low electrolyte concentration, the electrostatic repulsive forces were predominant as the thickness of DDL was significantly larger. The individual clay platelets settled independently at the bottom of the jar assuming a parallel platelet orientation in presence of such forces (see inset in Fig. 3b for the illustration). At high electrolyte concentration ($n > 1$ M), clay platelets form a flocculated structure as the van der Waals forces become dominant (van Olphen 1977). The van der Waals forces were, however, not considered in this study. The equilibrium sediment volume at different electrolyte concentration was analyzed to understand the effect of the electrolyte concentration on the soil fabric.

In natural soils, the clay platelet size varies in several orders of magnitude. The measured S_{SA} can provide an approximate idea about the average particle size for the bentonite. The DDL thickness around the clay platelets was considered to be independent of the size of the clay platelet, and hence, an uniform DDL thickness was assumed for all the platelet size throughout the bentonite specimen. The S_{SA} showed very negligible or no influence on the DDL thickness as shown in the present results. This is in favor of the above-mentioned assumption. The distance between the platelets was related to the osmotic pressure by solving the DDL-based Eqs. 3 and 6 and using the aforementioned hypothesis. Under equilibrium condition, the osmotic pressure between the platelets is equal to the body forces due to gravity. The body forces can be computed by knowing the clay platelet thickness. The measured equilibrium sediment volume from the experiments can be related to the distance between the platelets as follows:

Consider two parallel clay platelets of area A and thickness t_p, separated by an average distance of $2d$ under the self-weight (combination of clay platelets and the DDLs) as shown in Fig. 4. The volume of the water V_w is

$$V_w = A \times 2t_{DDL}$$

and the volume of soil solids (i.e., clay platelets) is

$$V_s = A \times t_p$$

The void ratio e can be expressed as

$$e = \frac{V_w}{V_s} = \frac{2t_{DDL}}{t_p} \tag{8}$$

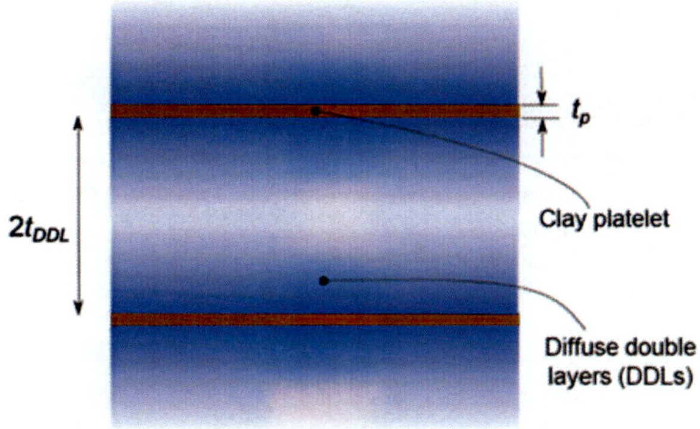

Fig. 4 Illustration of void space between the parallel plate clay system (After Bharat and Das 2017)

Neglecting the contribution of the edges, the specific surface area (surface area per unit mass), S_{SA}, can be written as

$$S_{SA} = \frac{2A}{M_s} = \frac{2}{t_p \times G_s \times \rho_w} \tag{9}$$

where G_s is the specific gravity of the solids and ρ_w is the density of fluid. Substituting for t_p in Eq. 8 using Eq. 9 results

$$e = S_{SA} t_{DDL} \rho_w G_s \tag{10}$$

where t_{DDL} is the thickness of DDL. The equilibrium sediment volume, which is the measured sediment volume per unit mass in the tests, can be related to the void ratio as

$$E_{SV} = (1 + e) \times V_s = \frac{(1 + e)M_s}{G_s \rho_w} = \frac{(1 + S_{SA} t_{DDL} \rho_w G_s)M_s}{G_s \rho_w} \tag{11}$$

where M_s is the dry mass of the bentonite. The DDL thickness, t_{DDL}, is unknown for theoretical estimation of ESV.

3.2 Computation of DDL Thickness

For a very small osmotic pressure at the mid-plane distance, resulting due to the body forces, an analytical formulation was developed. The interaction between the DDLs of the platelets is very small under such forces. Equation 6 can be written in terms of DDL thickness as:

$$\kappa t_{\mathrm{DDL}} = -\int_{z}^{u} (2\cosh(y) - 2\cosh(u))^{-1/2} \mathrm{d}y \tag{12}$$

The mid-plane potential at very low interaction under the body forces is negligible as compared to that of surface potential (i.e., u is negligible). Thus, Eq. 6 was approximated as

$$\kappa t_{\mathrm{DDL}} = -\int_{z}^{y_d} (2\cosh(y) - 2)^{-1/2} \mathrm{d}y \tag{13}$$

where y_d in the upper limit of integration is written as $\cosh^{-1}(1 + P/2nRT)$ (from Eq. 3). The scaled mid-plane potential, however, was considered in the upper limit of the integral to avoid singularity of the solution. The pressure due to body forces can be estimated if the thickness of the clay platelet is known. The mean thickness of the smectite platelets is found to be equal to 10 Å (Nadeau 1985). The particle thickness of up to 40 Å is observed for randomly interstratified minerals. The mean thickness was used in the calculations as a thickness of 40 Å did not provide significant changes in the theoretically estimated ESV values. The weight of an individual platelet due to body force along with the DDLs can be estimated as

$$P = \frac{(t_{\mathrm{p}}\gamma_{\mathrm{s}} + 2t_{\mathrm{DDL}}\gamma_{\mathrm{w}})A}{A} = \gamma_{\mathrm{w}}(t_{\mathrm{p}}G_{\mathrm{s}} + 2t_{\mathrm{DDL}}) \tag{14}$$

where A is the cross-sectional area, γ_{s} is the unit weight of clay platelets, and γ_{w} is the unit weight of pore fluid. Substitution of Eq. 14 in Eq. 13 gives

$$\kappa t_{\mathrm{DDL}} = -\int_{z}^{\cosh^{-1}\left(1+\gamma_{\mathrm{w}}\left(\frac{t_{\mathrm{p}}G_{\mathrm{s}}+2t_{\mathrm{DDL}}}{2nRT}\right)\right)} (2\cosh(y) - 2)^{-1/2} \mathrm{d}y$$

which can alternatively be written as

$$\kappa t_{\mathrm{DDL}} = -\frac{1}{2}\int_{z}^{\cosh^{-1}\left(1+\gamma_{\mathrm{w}}\left(\frac{t_{\mathrm{p}}G_{\mathrm{s}}+2t_{\mathrm{DDL}}}{2nRT}\right)\right)} \cosh(y/2)\, \mathrm{d}y \tag{15}$$

The analytical solution to Eq. 15 was given by:

$$t_{\mathrm{DDL}} = -\frac{1}{\kappa}\left[\ln(\tanh(y/4))\right]_{z}^{\cosh^{-1}\left(1+\gamma_{\mathrm{w}}\left(\frac{t_{\mathrm{p}}G_{\mathrm{s}}+2t_{\mathrm{DDL}}}{2nRT}\right)\right)} \tag{16}$$

and after simplification

$$t_{\mathrm{DDL}} = \frac{1}{\kappa}\left[\begin{array}{l} \ln\left(\tanh\left(\frac{z}{4}\right)\right) - \\ \ln\left(\tanh\left(\frac{1}{4}\cosh^{-1}\left(1 + \gamma_{\mathrm{w}}\left(\frac{t_{\mathrm{p}}G_{\mathrm{s}} + 2t_{\mathrm{DDL}}}{2nRT}\right)\right)\right)\right) \end{array}\right] \tag{17}$$

where $(1/\kappa)$ is expressed in Å and the dimensionless surface potential can be expressed as (Bharat et al. 2013; Bharat and Sridharan 2015a): The scaled surface potential for very low interaction, i.e., $u = 0$ can be obtained as

$$z = \cosh^{-1}\left(1 + 0.5\left[\frac{C_{EC}}{S_{SA}} \times \frac{2513.6}{\sqrt{\varepsilon nT}}\right]^2\right) \tag{18}$$

As the variation of the scaled surface potential with scaled mid-plane potential is negligible at small interactions (i.e., $\kappa t_{\mathrm{DDL}} \gg 1$), Eq. 18 provides an accurate estimation of scaled surface potential. As per Eq. 17, the DDL thickness of around an individual clay platelet in equilibrium sediment volume is dependent on the scaled surface potential, pore fluid parameters, and the DDL thickness (which is unknown). The implicit equation was solved for DDL thickness iteratively. The equation was quickly converged in just two iterations. The accuracy of the proposed analytical solution was verified by comparing with the existing numerical solution (Bharat et al. 2013; Bharat and Sridharan 2015a). The analytical solution was in good agreement with the numerical solution showing a correlation coefficient of $R^2 = 0.999$. The approximation of Eqs. 12, 13 was, therefore, insignificant. The analytical solution was accurate and simple to predict theoretical t_{DDL} by knowing the pore fluid and clay surface parameters. The numerical solution, on the other hand, required the solution of an elliptic integral in each iteration and was tedious to obtain. The proposed analytical solution was then used for theoretical estimation of ESV of bentonites, where only body forces are existing on clay–electrolyte system.

4 Parametric Study

A detailed parametric study was carried out to understand the influence of various pore fluid and soil parameters on the ESV of clays. The ESV was found to decrease with increase in the concentration of the pore fluid (Fig. 5a, c). The tendency of the counterions to diffuse away from the surface reduces as the ion concentration in the bulk solution increases. This results in compression of the diffuse double layer developed around the surface of a clay particle leading to a decrease in the ESV. The valence of the cations in the pore fluids was observed to significantly influence the ESV (Fig. 5a) as the presence of higher valence cations results in a significant reduction in the DDL thickness.

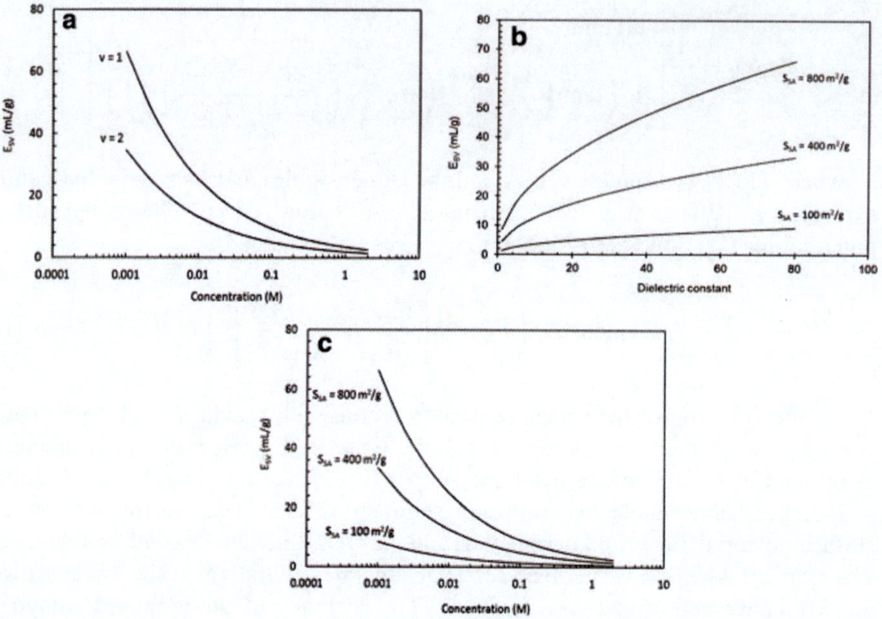

Fig. 5. **a** Variation of ESV with concentration of two pore fluids having two different cationic valences. **b** Variation of ESV of soils having three different S_{SA} with dielectric properties of pore fluid.**c** Variation of ESV of soils having three different S_{SA} with pore fluid concentration

The electric permittivity of the pore fluid has a significant effect on the thickness of diffuse double layer. The ability of the counterions to diffuse away from the surface decreases as the permittivity or the dielectric constant of the pore fluid medium decreases. DDL thickness and therefore consequently the EVR decreased with decrease in the dielectric constant of pore fluid (Fig. 5b).

The theoretical ESV of soils having three different specific surface area in various pore fluids having different dielectric properties is presented in Fig. 5b. The estimated ESV was found to be increasing as the surface area of soils was increased. The influence of dielectric constant was significant for soils having higher surface area, and the effect diminishes as the surface area is reduced. Figure 5c shows the effect of pore fluid concentration on the ESV of soils with different specific surface area. The influence of concentration on the ESV reduces as the surface area of soils is decreased. The cation exchange capacity of soil did not show any influence on ESV; however, the mixed valence in the exchangeable cations has a significant influence on ESV which is evident from the influence of valence of cations.

5 Conclusion

This work provides an analytical model for estimation of the thickness of DDL in a clay–water–electrolyte system at a very low interaction level. The model was utilized to predict the equilibrium volume of clay sediments (ESV) by considering the body forces in the clay–water–electrolyte system. The model incorporates various influencing parameters such as pore fluid concentration, dielectric constant, valence of cations, SSA, and CEC of soil. The parametric study provided a quantitative understanding on the influence of these parameters on the ESV of soils. The specific surface area of soil showed a direct relationship with the ESV.

References

Bharat TV, Das DS (2017) Physicochemical approach for analyzing equilibrium volume of clay sediments in salt solutions. Appl Clay Sci 136:164–175

Bharat TV, Sivapullaiah PV, Allam MM (2009) Swarm intelligence-based solver for parameter estimation of laboratory through-diffusion transport of contaminants. Comput Geotech 36(6):984–992

Bharat TV, Sivapullaiah PV, Allam MM (2013) Novel procedure for the estimation of swelling pressures of compacted bentonites based on diffuse double layer theory. Environ Earth Sci 70:303–314

Bharat TV, Sridharan A (2015a) Prediction of compressibility data for highly plastic clays using diffuse double-layer theory. Clays Clay Miner 63(1):30–42

Bharat TV, Sridharan A (2015b) A critical appraisal of Debye length in clay-electrolyte systems. Clays Clay Miner 63(1):43–50

Koch D (2002) Bentonites as a basic material for technical base liners and site encapsulation cut-off walls. Appl Clay Sci 21(1–2):1–11

Nadeau PH (1985) The physical dimensions of fundamental clay particles. Clay Miner 20(4):499–514

Schanz T, Tripathy S (2009) Swelling pressure of a divalent-rich bentonite: Diffuse double-layer theory revisited. Water Resour Res 45(5)

Tripathy S, Sridharan A, Schanz T (2004) Swelling pressures of compacted bentonites from diffuse double layer theory. Can Geotech J 41:437–450

van Olphen H (1977) An introduction to clay colloid chemistry. Interscience, New York

Influence of Randomly Distributed Waste Tire Fibres on Swelling Behaviour of Expansive Soils

Tejaswani Shukla, Mohit Mistry, Chandresh Solanki, Sanjay Kumar Shukla, and Shruti Shukla

Abstract Reinforcement of soil with fibres in different forms has been a regular practice since early days. Utilizing the waste tire fibres instead of other types of fibres will serve the dual purpose of utilization of waste as well as improving the behaviour of expansive soils. This paper presents the effect of tire fibres on swelling behaviour of expansive soils with a change in size as well as the percentage of fibres. The consolidation tests were performed and swelling potential was identified and compared for different fibre types A ($L/B = 8.95$), B($L/B = 6.387$) and C ($L/B = 4.457$) with varying fibre content as 0.25%, 0.5%, 0.75% and 1.00%. The swelling potential for 12 combinations was computed and compared with that of the virgin soil to calculate the swelling potential ratio (SPR) to find out the optimum percentage and aspect ratio at which the fibres are most effective in reducing the swelling behaviour of the problematic soils.

Keywords Swelling characteristics · Expansive soil · Waste tire fibres · Swelling potential ratio · Aspect ratio

1 Introduction

With the increasing number of vehicles worldwide, a new problem on utilization of waste tires is gaining importance. In this study, an effort to justify using waste tire fibres in soil stabilization is focussed. An increasing trend towards development and industrialization demands on an increase in construction on marine areas, marshy lands and low lands both as a requirement of space as well as a corridor towards development. In India, expansive clayey soils are named as the black cotton soils due to their influential capacity of budding cotton plants. This species of soil covers

T. Shukla (✉) · M. Mistry · C. Solanki · S. Shukla
Department of Applied Mechanics, SVNIT, Surat, India
e-mail: tejaswanishukla23@gmail.com

S. K. Shukla
Discipline of Civil Engineering, School of Engineering, Edith Cowan University, Perth, Australia

M. Latha Gali and R. R. P. (eds.), *Problematic Soils and Geoenvironmental Concerns*, Lecture Notes in Civil Engineering 88,
https://doi.org/10.1007/978-981-15-6237-2_27

around 20% of land area in India. With excessive montmorillonite in its structure, the black cotton soils are a major threat to foundations of structures and leads to various problems in the structure as loss of stability of structure, settlement, swelling, cracking, etc. Another problem associated with waste tires is their decomposition. Due to large volume and non-degradability of rubber, they acquire large areas and lead to low economic benefits. Thus, using waste tires with expansive soils can lead to dual benefits of soil stabilization as well as waste utilization.

Thus, the black cotton soil was used purposely to identify the effects of tire fibres on its swelling behaviour. Many such approaches of improving the expansive soils with fibres have been instigated earlier. Jute amongst the natural fibres and polypropylene amongst the synthetic fibres have gained the highest research attention. Setty and Murthy (1987), Abdi et al. (2008), Viswanadham et al. (2009), and Malekzade and Bisel (2012) conducted research using polypropylene fibres with expansive soils. In all these researches, variation in size and content of polypropylene fibres was made and comparatively the influence was studied. Moreover, with natural fibres, Cyrus and Babu (2005), Al-Akhras et al. (2008) and Babu et al. (2008) conducted research on coir fibres and predicted the utilization of fibres in compacting the swelling effects of expansive soils. In all these studies, a significant improvement in the behaviour of expansive soils is observed with addition of fibres up to a certain limit. Thus, fibre reinforcement is an upcoming solution to swelling behaviour of expansive soils.

Waste fibres have not been much under the research focus unless the recent years where utilization of wastes has been considered an environmental concern. Laskar and Pal (2013) investigated to evaluate the effect of plastic fibre on compaction and consolidation behaviours of reinforced soil. Waste rubber fibres were used to study improvement in swelling behaviour of expansive soils. Trouzine et al. (2012), in Algeria by varying the fibre contents as 10%, 20%, 25%, and 50%. In this study, it is clear that with an increase in scrap rubber tire content, there is an equivalent decrease in liquid limit and plasticity index while an increase is noticed in the plastic limit. Also, the compression index C_c is found to increase with an increasing content of fibres.

Thus, the literature review points out that very little work has been done on investigating the swell-consolidation characteristics of expansive soils reinforced with waste rubber tire fibres. Therefore, this study focuses on extracting the impact of waste tire fibres on swelling potential of expansive soils.

2 Materials

The basic materials used were black cotton soil and waste tire fibres. As Surat is home to black cotton soils, it was a locally available material. All the basic properties of soil were computed. Tire fibres were collected from a local company. These waste tire fibres are obtained by shredding waste tires. By the process of shredding, they were converted into powder forms so that they can be easily mixed with soil as applicable.

Table 1 Physical properties of soil

Soil properties	Values
Specific gravity	2.56
IS classification	CH
Liquid limit (%)	64.4
Plastic limit (%)	30.6
Plasticity index (%)	33.8

An agricultural shredder was locally used to shred the tires. For the purpose of study, the tire fibres were divided into three different types based on their aspect ratio.

Initially, for simplicity, it was divided into three groups using conventional sieve analysis procedure (one retained on 4.75 mm IS sieve, other passing through 4.75 mm IS sieve and retained on 2 mm IS sieve and last passing through 2 mm IS sieve). Later, from each group, some amount of fibres were taken and diameter and length of each fibre were measured with the help of manual scale. Then, the mean values were calculated for the length and width for each type of fibres and the aspect ratio was determined as the representative for each type. Thus, fibres were divided into three types A, B and C with different aspect ratios in order to identify their individual correlation with soil. Specific gravity of fibres was computed to be 1.15 (Tables 1 and 2).

Rubber fibres are usually non-biodegradable as well as highly compressible. Rubber is elastic, and hence, its properties and behaviour as fibres being reinforced into the soil are typically different from that of other natural and synthetic fibres. As compared to other fibres, rubber fibres are thicker in width and shorter in length. Rubber fibres are even available in powder form. In this study, the behaviour was investigated for all the different aspect ratios and different contents of the fibres utilized.

Table 2 Physical properties of tire fibres

Type of fibre	Length (L) (mm)	Breadth (B) (mm)	Aspect ratio (L/B)	Fibre image
A	8.941	1.0	8.941	
B	12.775	2	6.387	
C	18.726	4.20	4.457	

3 Sample Preparation

Due to sticky nature of clayey soils, rubber can be easily bonded with the soil, thus mixing soil with waste tire fibres was easy. Waste tire fibres were randomly sprinkled on soil samples and mixed thoroughly at the optimum moisture content and maximum dry density as computed from the compaction test. Thus, for each type of fibres at 4 different contents of 0.25%, 0.5%, 0.75% and 1.00%, compaction test was performed to analyse the OMC and MDD of waste tire reinforced soils.

Thus, compaction tests for all three types of fibres on all four different contents were performed and with variating fibre length and fibre content, the value of maximum dry density and optimum moisture content was obtained. Table 3 numerates the optimum moisture content and maximum dry density for all 12 types of samples. This data is highly useful to prepare homogenously mixed samples in order to conduct various tests on them. For both strength and compressibility criteria investigation of fibre-reinforced soils, we need the compaction test data in order to identify the optimum mixing moisture content and the maximum dry density of the fibre–soil matrix.

As can be inferred from Table 3, the dry density of fibre-reinforced soils reduces with an increasing length of the fibres as well with an increase in fibre content, that is, more and more fibres we add, lesser is the dry density of the matrix. As specific gravity of fibres is low, more fibres lead to a lighter mixture, thus density keeps on reducing. Short fibres are more adhesive and are easily absorbed in the soil–fibre matrix while long fibres have different planes of roughness, creating more voids thus a less dense mixture leading to reduction in dry density of fibre-reinforced soils. Moreover, with an increase in length of fibres and fibre contents, an effective

Table 3 Sample preparation

Sample no.	Name of sample	Fibre type and content (%)	OMC (%)	MDD (kN/m^3)
1	Virgin soil	–	23.4	15
2	FRSA1	A (0.25)	25	14.72
3	FRSA2	A (0.50)	25	14.68
4	FRSA3	A (0.75)	25.5	14.64
5	FRSA4	A (1.00)	26	14.60
6	FRSB1	B (0.25)	26.5	14.60
7	FRSB2	B (0.50)	27	14.48
8	FRSB3	B (0.75)	27.5	14.40
9	FRSB4	B (1.00)	28	14.28
10	FRSC1	C (0.25)	27.5	14.48
11	FRSC2	C (0.50)	28	14.36
12	FRSC3	C (0.75)	28.5	14.08
13	FRSC4	C (1.00)	29	13.64

increment in the water content of fibre-reinforced soils is observed. As long fibres have rough surface, they absorb certain water, similarly as quantity of fibres increase, the water absorption capacity of fibres enhances, leading to a higher requirement of water, thus water content increases with increase in length and content of fibres for waste tire fibre-reinforced soils.

4 Testing Programme

In order to evaluate the swelling potential of soil, oedometer tests were performed on all 12 samples of fibre-reinforced soils and compared with that of the virgin soil. Swelling potential is an important parameter to classify soils as expansive. High value of swelling potential indicates highly expansive soil. The consolidation sample is prepared in 3 layers at its respective MDD and OMMC in the similar way as compaction samples as prepared. The test was conducted on a standard fixed type odometer with different test samples and a fixed loading schedule. The samples were loaded, reloaded in two cycles and the results were recorded accordingly. The change in sample mass at the end of two cycles of test was recorded, and as a result, the swelling potential was computed (Fig. 1 and Table 4).

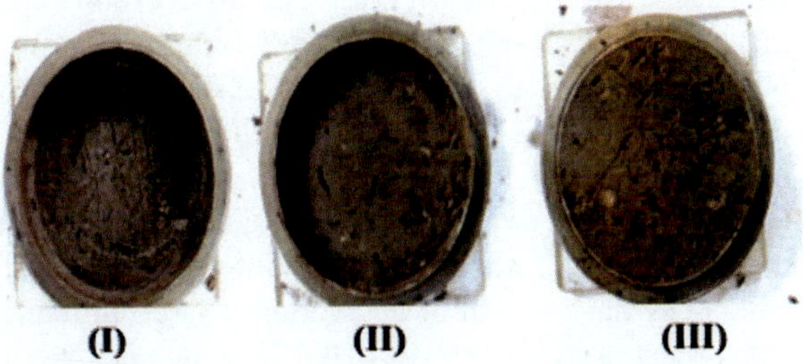

Fig. 1 Stages of sample preparation

Table 4 Loading schedule for consolidation

Loading (kPa)	Unloading (kPa)
5	160
10	40
20	10
40	
80	
160	
320	
640	

$$\text{Swelling potential } (S) = \frac{M_1 - M_2}{M_2} \times 100 \quad (1)$$

where

M_1 Mass of ring + wet soil (before testing)
M_2 Mass of ring + wet soil (after testing).

5 Results and Discussions

Type A fibres are the smaller size of fibres with effective diameter less than 2 mm, width of 1 mm and length as 8.941 mm. The aspect ratio *L/B* of type A fibres is highest as 8.941. Type A fibres being small in size are very effective as they can be easily distributed and occupy the matrix. The drawback with these fibres is their size. Due to small size, these fibres do not develop enhanced adhesion with the soil and have a low contact area, thus lower strength. Still, as a part of matrix, it is quite effective in reducing swelling as well as improve consolidation parameters.

Type B fibres are the fibres with effective diameter between 2 mm and 4.75 mm. This is the medium size of fibres, neither too long, nor too short. They are very effective in developing the bond with the soil. They have length of an average 12.77 mm and width as 2 mm on average. The aspect ratio (*L/B*) is 6.387, which is lower than that of type A fibres. Thus, type B fibres are more effective in reducing swelling potential as compared to type A fibres due to high bonding with soil and high matrix occupation but this improvement is limited to 0.75% fibre content for both types of fibres.

Type C fibres are the fibres with effective diameter greater than 4.75 mm. This is the largest size of fibres. They are not very effective in developing the bond with the soil. They have length of an average 18.726 mm and width as 4.20 mm at average. The aspect ratio (*L*/*B*) owes to 4.457, which is lower than that of type A and B fibres. As this size of fibres is not compatible with consolidation ring size, they do not develop good correlations in consolidation results (Table 5).

Swelling potential is an essential parameter to decide the efficacy of fibre-reinforced soils. For fibre type A, swelling potential reduces with addition of fibres. The reduction is observed up to 0.75% of fibres after which addition of fibres leads

Table 5 Swelling potential of tire fibre-reinforced soil

Fibre content	Type A (%)	Type B (%)	Type C (%)
0	3.14	3.14	3.14
0.25	2.40	0.60	0.69
0.50	2.60	−0.25	0.33
0.75	1.60	0	0.69
1.00	1.69	0.25	1.67

to an increase in swelling. Thus, we infer that there is certain optimum percentage of fibres, which when added to the soil leads it to developing minimum swelling possible with the application of fibres. Here, we try to optimize that fibre content at which swelling in the soil is attained minimum.

Similarly for fibre type B, swelling potential follows the same trend as in fibre type A. It reduces with an increase in fibre content up to 0.5% and increases at 0.75%. Another important observation is the negative value of swelling potential at 0.5% content of fibre type B that imposes shrinkage property. Thus, we infer that with increasing fibre content at a particular size and content of fibres, it may lead to shrinkage of soil. Thus, calculation of exact fibre content at which swelling is minimum without shrinkage is needed to be calculated. Trends in fibres with varying sizes are found common, swelling potential reduces with increasing fibre content, achieving minimum value at certain fibre content and increasing further. Thus, for fibre content B at 0.75%, an ideal result with minimum swelling and no shrinkage was observed. Further by mathematical approach, we determine the optimum concentration of fibres type A and B in soil in order to generate the minimum swelling in the soil.

For fibre type C, swelling potential reduces with increase in fibre content up to 0.50% and increases at 0.75%, further reducing at 1%. As C type fibres are the largest ones and not very compatible with consolidation ring size, they cannot develop good contact with soil with increasing percentages of fibres into soil. Thus, swelling potential starts increasing due to increase in fibre content. As rubber is a highly compressible fibre, higher the content of fibre more is the swelling. For smaller sizes of fibre types A and B, the content in the matrix goes higher by 0.75% while for fibre type C it attains a similar content in the matrix by 0.50%. This leads to enhancement in swelling with increasing size of fibres. Type A fibres being very small in size are less effective while type C fibres being larger in size are less effective. Thus, type B fibres form an ideal choice for reduction in swelling potential of soil along with no shrinkage.

For fibre type A, with the above variation, we infer clearly that with increasing fibre content, swelling potential initially reduces, further increases and then again reduces, but the initial reduction is quite less than the final one. Due to very small size of fibres, type A fibres do not cover up the matrix completely, thus they show irrational behaviour of alternate increase and decrease. Thus, our motive is to compute that optimum content of fibre type A where we attain the minimum value of swelling potential. For this, we go for the mathematical approach. By the method of curve fitting, the above variation can be represented as

$$S_f = 43.953f^4 - 86.803f^3 + 52.876f^2 - 11.487f + S_0 \tag{2}$$

where

S_f swelling potential at respective fibre content, %
f fibre content, %
S_0 swelling potential for virgin soil = 3.14%.

Using mathematical approach, for minimum value of swelling potential, we determine the optimum content of fibre as $f^{S}_{\text{opt},A} = 0.8848\%$. Thus, at this optimum content of fibre type A, we can achieve minimum value for swelling potential as 1.187% (Fig. 2).

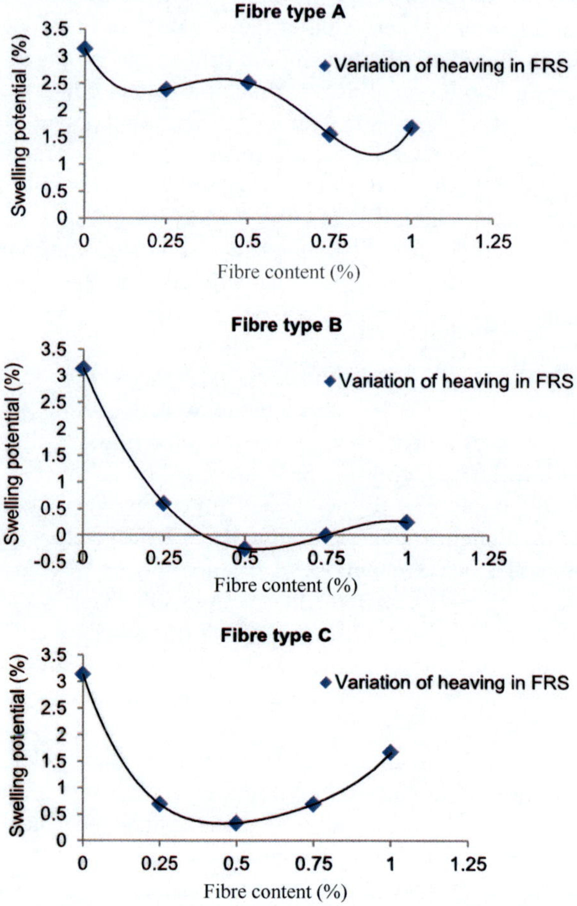

Fig. 2 **a** Variation of swelling potential with fibre content for fibre type A, **b** variation of swelling potential with fibre content for fibre type B and **c** variation of swelling potential with fibre content for fibre type C

For fibre type B, using curve fitting method, the equation for variation of swelling potential is

$$S_f = -5.5631f^4 + 2.0959f^3 + 14.386f^2 - 13.813f + S_0 \quad (3)$$

where

S_f swelling potential at respective fibre content, %
f fibre content, %
S_0 swelling potential for virgin soil = 3.14%.

Using mathematical approach, for minimum value of swelling potential, we determine the optimum content of fibre as $f^S_{\text{opt},B} = 0.5367\%$. Thus, at this optimum content of fibre type B, we can achieve minimum value for swelling potential as −0.263%.

For fibre type C, using curve fitting method, the equation for variation of swelling potential is

$$S_f = 13.826f^4 - 35.497f^3 + 37.347f^2 - 17.142f + S_0 \quad (4)$$

where

S_L Swelling potential at respective fibre length, %
L fibre length in mm,
S_0 Swelling potential for virgin soil = 3.1441%.

Using mathematical approach, for minimum value of swelling potential, we determine the optimum content of fibre as $f^S_{\text{opt},C} = 0.4544\%$. Thus, at this optimum content of fibre type B, we can achieve minimum value for swelling potential as 0.325% (Fig. 3).

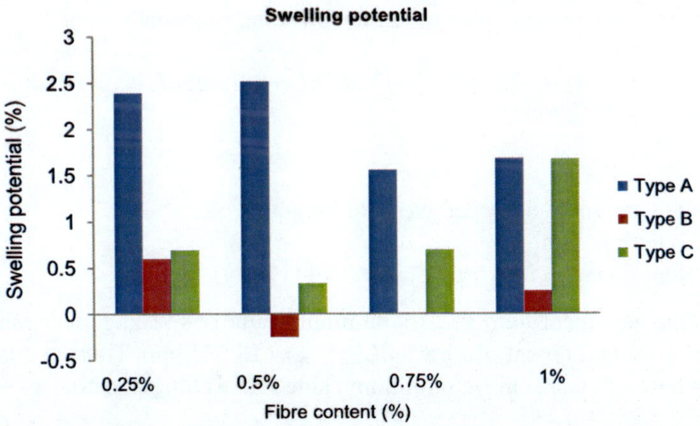

Fig. 3 Variation of swelling potential with fibre content and type

As we can see from the graph values, the value of swelling potential is attained minimum for fibre type B at 0.50%, i.e. −0.25%. The most ideal value obtained is however attained at 0.75% for type B fibres, i.e. 0% which owes to no swelling as well as no shrinkage. Thus, fibre type B at 0.75% leads to minimum swelling potential without any shrinkage, thus it can be the best option in terms of reducing swelling with addition of waste tire fibres.

Similarly, as we compute the optimum content of fibres, we can compute the optimum length of fibres at which the swelling potential minimizes. This optimum length of fibres can be computed at every fibre content varying from 0.25 to 1.00% (Fig. 4).

At 0.25% fibre content, swelling potential reduces with inclusion of fibres, but it is again limited to a certain length of fibres, after which it follows a reduction trend. Again using the mathematical approach we tend to determine the optimum length of fibres at 0.25% of concentration, in order to minimize swelling potential. It is an important parameter in deciding the efficiency of fibre-reinforced soils.

Using curve fitting method, the equation for variation of swelling potential is

$$S_L = 0.0042L^3 - 0.122L^2 + 0.6679L + S_0 \tag{5}$$

where

S_L Swelling potential at respective fibre length, %
L fibre length in mm
S_0 Swelling potential for virgin soil = 3.1441%.

Using mathematical approach, for minimum value of swelling potential, we determine the optimum content of fibre $L^S_{\text{opt},0.25} = 16.065$ mm. Thus, at this optimum length of fibre, we can achieve minimum value for swelling potential as −0.1986%.

Similarly at 0.5% fibre content, swelling potential reduces with inclusion of fibres, but up to certain length of fibres, after which there is a rise in swelling potential. Thus, we optimize the fibre length for minimum swelling potential.

$$S_L = 0.0072L^3 - 0.2075L^2 + 1.2088L + S_0 \tag{6}$$

where

S_L Swelling potential at respective fibre length, %
L fibre length in mm
S_0 Swelling potential for virgin soil = 3.1441%.

Using mathematical approach, for minimum value of swelling potential, we determine the optimum content of fibre as $L^S_{\text{opt},0.50} = 15.633$ mm. Thus, at this optimum length of fibre, we can achieve minimum value for swelling potential as −1.1617%.

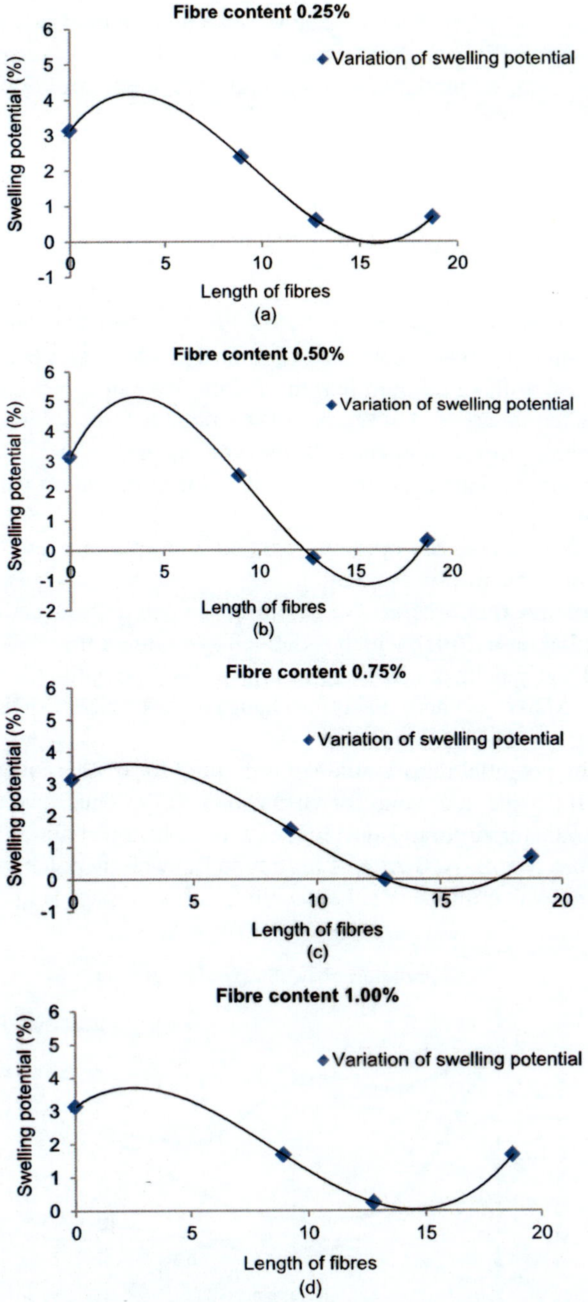

Fig. 4 **a** Variation of swelling potential with fibre content 0.25%, **b** variation of swelling potential with fibre content 0.50%, **c** variation of swelling potential with fibre content 0.75% and **d** variation of swelling potential with fibre content 1.00%

At 0.75% and 1.00% fibre content, again similar behaviour of reduction in swelling potential up to certain limit is observed after which an increase is noticed. Thus, using curve fitting method, we tend to obtain the optimum length for 0.75% and 1.00% as well.

$$S_L = 0.0038L^3 - 0.1007L^2 + 0.4186L + S_0 \text{ (At 0.75\%)} \tag{7}$$

$$S_L = 0.0042L^3 - 0.1084L^2 + 0.4673L + S_0 \text{ (At 1.00\%)} \tag{8}$$

Using mathematical approach, at 0.75% fibre content for minimum value of swelling potential, we determine the optimum content of fibre as $L^S_{\text{opt},0.75} = 15.26$ mm. Thus, at this optimum length of fibre, we can achieve minimum value for swelling potential as −0.4143%. At 1.00% fibre content for minimum value of swelling potential, we determine the optimum content of fibre as $L^S_{\text{opt},1} = 14.679$ mm. Thus, at this optimum length of fibre, we can achieve minimum value for swelling potential as −0.0693%.

Thus, we can compute the optimum content and optimum lengths for fibres in order to minimize the swelling potential.

The effect of fibre reinforcement on swelling behaviour of soil can be expressed as swelling potential ratio (SPR), which is defined as a ratio of the swelling potential of the reinforced soil specimen $(\Delta H/H)_R$ to the swelling potential of the unreinforced soil specimen $(\Delta H/H)_U$, where ΔH is the change in the thickness of the soil specimen (Shukla 2017).

The swelling potential ratio is attained minimum for 0.50% content of fibre type B, i.e. −0.08011 while maximum for virgin soils, i.e. 1. Thus, swelling potential is attained maximum for virgin soil and reduces with inclusion of fibres, being minimum at 0.50% of fibre type B. At 0.75% of fibre type B, swelling potential ratio (SPR) is attained as zero, which implies neither swelling nor shrinkage (Fig. 5).

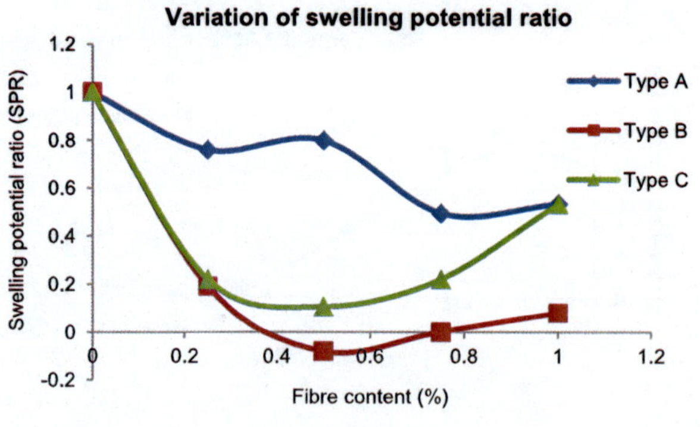

Fig. 5 Variation of swelling potential ratio (SPR) with fibre content

6 Conclusions

In this study, rubber tire fibres were classified on the basis of their size as type A, type B and type C with lengths of fibres as 8.9 mm, 12.8 mm and 18.72 mm, respectively. These fibres were added at four different concentrations to soil at 0.25%, 0.50%, 0.75% and 1.00%. For all fibre types and contents, different samples by randomly mixing fibres with soil at their computed optimum mixing moisture content (OMMC) and maximum dry density (MDD) were prepared and rigorously tested with the consolidation apparatus. Thus, with the study performed, following conclusions can be inferred.

- Black cotton soil observes tremendous reduction in swelling with addition of fibres to it. While swelling potential for virgin soil without fibres is 3.14%, it reduces effectively with addition of fibres. At certain size and content of fibres, instead of swelling, samples start to shrink. The most ideal condition was at 0.75% for fibre type B with 0% swelling. Thus, observed improvement in swelling potential is of 100%.
- Swelling potential values for all types of fibres follow a similar trend. These values reduce on addition of fibres up to 0.75% and increase thereafter. Thus, optimization technique was used to derive the optimum fibre content for all types of fibres with minimum swelling potential values. For fibre type A, $f^S_{\text{opt},A} = 0.88\%$, for fibre type B, $f^S_{\text{opt},B} = 0.54\%$ and for fibre type C, $f^S_{\text{opt},C} = 0.45\%$. Thus, as the fibre size increases, fibre content for minimum value of swelling potential reduces.
- Similarly, at every different content of fibre, we determine the optimum length at which swelling potential is minimum.
- As the length of the fibres increases, the optimum content of fibres reduces and as the fibre content increases, the optimum length of fibres reduces. Thus, lower the fibre content, higher lengths of fibres are needed and as the fibre content increases, we need shorter fibres to reduce the swelling potential. Thus, short fibres at high content or long fibres at low content with good bonding capability are required to minimize the swelling potential of soil.
- For all types of fibres, an equation using curve fitting method was plotted that shows the variation between swelling potential and fibre content (Eqs. 2, 3 and 4). Rubber being a highly compressible material can reduce swelling of soil up to a particular content (0.75% in this case). When added at high concentrations, rubber due to high compressible and rebound nature leads to reduction in swelling potential. When rubber fibres were added at 10–40%, they led to increase in swelling potential.
- Rubber fibres added at low concentrations and aspect ratio lead to high bonding with soil and improvement in swelling characteristics of expansive soils.
- Use of fibre-reinforced soil with waste tire fibres is quite cheap, easily available and applicable.

References

Abdi MR, Ali P, Arjomand MA (2008) Effects of random fibre inclusion on consolidation, hydraulic conductivity, swelling, shrinkage limit and desiccation cracking of clays. Int J Civ Eng 6(4):284–292

Al-Akhras NM, Attom MF, Al-Akhras KM, Malkawi AIH (2008) Influence of fibres on swelling properties of clayey soils. Geosynthetics Int J 15(4):304–309

Babu GLS, Vasuvedan AK, Sayida MK (2008) Use of coir fibres for improving the engineering properties of expansive soils. J Nat Fibres 5(1):61–75

Cyrus S, Babu TJ (2005) Consolidation characteristics of coir reinforced Cochin marine clays. In: Proceedings of '05 Indian geotechnical conference, Ahmedabad, India, pp 125–128

Laskar A, Pal SK (2013) Effect of waste plastic fibres on compaction and consolidation behaviour of reinforced soil. Electron J Geotech Eng (EJGE) 18:1547–1558

Malekzadeh M, Bisel H (2012) Swell and compressibility of fibre reinforced expansive soils. Int J Adv Technol Civ Eng 1(2):42–45.

Setty S, Murthy A (1987) Behaviour of fibre reinforced black cotton soil. In: IGS (90), Bombay, pp 45–90

Shukla SK (2017) Fundamentals of fibre-reinforced soil engineering. Developments in Geotechnical Engineering, Springer Nature Singapore Pvt. Ltd.

Trouzine H, Bekhiti M, Asroun A (2012) Effects of Scrap tyre rubber fibre on swelling behaviour of two clayey soils in Algeria. Geosynthetics Int J 19(2):123–132

Vishwanadham BVS, Phanikumar BR, Mukherjee RV (2009) Effect of Polypropylene tape fibre reinforcement on swelling behaviour of expansive soil. Geosynthetics Int J 16(5):393–401

Influence of Bacteria on Physical Properties of Black Cotton Soil

R. B. Wath and S. S. Pusadkar

Abstract Microbial-induced calcite precipitation (MICP) is a sustainable biological ground improvement technique, which is capable of altering and improving physical properties of black cotton soil. In order to evaluate MICP as a soil ground improvement technique for black cotton soil, four types of microbes were used in this study. These were *Bacillus pasteurii*, *Bacillus subtilis*, *Bacillus megaterium*, and *Pseudomonas*. These microbes in liquid medium were mixed with the soil sample for different percentages (4%, 8%, 12%, 16%, and 20%) by weight of sample. The soil samples were kept for different reaction periods (3 days, 7 days, and 14 days). The physical properties of MICP-treated soil such as liquid limit, plastic limit, shrinkage limit, and swelling pressure were examined. The results show that the plasticity index decreases when upto 20% of microbes were mixed and when microbes were more than 20%, plasticity index increases. The reaction period shows the influence on plasticity index. The shrinkage limit test was performed for soil with 20% bacteria for 3, 7, and 14 days reaction period. The MICP-treated soil shows reduction in shrinkage limit values. Swelling pressure tests were performed for soil with 20% microbes for 7 days reaction period, and swelling pressure was also reduced to 80%. Thus, the use of microbes helps in controlling the swelling and shrinkage characteristics of black cotton soil.

Keywords Black cotton soil · Microbes · MICP · Physical properties

R. B. Wath
Research Scholar, GCOE, Jalgaon, Maharashtra, India
e-mail: wathrb@gmail.com

S. S. Pusadkar (✉)
Professor & HoD, GCOE, Jalgaon, Maharashtra, India
e-mail: ss_pusadkar@yahoo.co.in

M. Latha Gali and R. R. P. (eds.), *Problematic Soils and Geoenvironmental Concerns*, Lecture Notes in Civil Engineering 88,
https://doi.org/10.1007/978-981-15-6237-2_28

1 Introduction

Black cotton soil is causing detrimental effects on buildings, dams, highways, airports, and retaining walls. In India, black cotton soil covers extensive areas in the states of Maharashtra, Karnataka, Andhra Pradesh, Madhya Pradesh, Gujarat, Uttar Pradesh, and Tamil Nadu. The black cotton soils exhibit significant volume change due to change in water content and cause unacceptable differential settlement.

Since last 80 years, extensive research had been carried out all over the world to identify expansive clay, its swelling characteristics, methods of identifying the problems associated with it, and solutions to these problems.

The stabilization techniques for improving the properties of expansive soil and solution to the problems associated with it are many. But most of all these methods are not eco-friendly. To overcome this problem, microbiological process is an uncommon alternative for improvement of properties of soil. Microbiological process is more effective, eco-friendly, and economical than other conventional treatment methods.

In recent years, use of microbial-induced calcite precipitation technique is to alter the engineering properties, and it is gaining attention as a versatile and green method of soil improvement. When a soil is treated using MICP technique, microbial induced calcite bridges adjacent soil particles, cementing soil particles together. The precipitation of calcite between particle–particle also helps in reducing the permeability, compressibility, and increasing soil strength (Dejong et al. 2006; Dhami et al. 2013). Calcite mineralization can occur as a by-product of microbial metabolic activity such as photosynthesis, urea hydrolysis, sulfate reduction, and iron reduction. During these different metabolic processes, the alkalinity or pH of the system increases, favoring the calcite precipitation (Whiffin 2004).

The MICP process regulated mainly by several factors such as (i) concentration of calcium ion, (ii) concentration of dissolved inorganic carbon, (iii) pH, (iv) bacterial cell concentration, (v) availability of nucleation sites, (vi) reactant concentration, and (vii) temperature (Ng et al. 2012, 2014). Temperature has a significant influence on urease activity and the rate of MICP (Van Passen 2009).

The present study was performed to understand feasibility of microbial-induced calcite precipitation on black cotton soil. The problematic soil used was expansive soil, which had high swelling, high compressibility properties, and low bearing capacity. The laboratory tests were conducted to study the influence of microbial-induced calcite precipitate bacteria on swelling, shrinkage limit, and plasticity index with different reaction periods.

Table 1 Physical properties of BC soil

Properties	Value
Free swell index	77%
Specific gravity	2.65
Liquid limit	71%
Plastic limit	32.41%
Plasticity index	39.60%
Shrinkage limit	11%
Maximum dry unit weight	14.6 kN/m^3
Optimum moisture content	22%
Swelling pressure	106 kN/m^2
Soil classification as per IS	CH

2 Materials and Method

2.1 Materials

Black cotton (BC) soil was collected from Buldhana, Maharashtra, for the experimental study. Table 1 summarizes the physical properties of black cotton soil.

2.2 Bacteria

The urease active bacteria used in the current study were *Bacillus pasteurii, Bacillus subtilis, Pseudomonas*, and *Bacillus megaterium.* The bacteria were cultivated in a sterile aerobic batch growth medium consisting of 5 g/L peptic digest of animal tissue, 5 g/L sodium chloride, 1.5 g/L beef extract, and 1.5 g/L yeast extract.

2.3 Methodology

The microbes were mixed with soil sample with different percentages as per given in Table 2. These samples were kept for reaction for the period of 3, 7, and 14 days. After

Table 2 Experimental methodology

Parameters	Characteristics
Microbes in percentage	4, 8, 12, 16, and 20%
Reaction period	3, 7, and 14 days
Tests	Liquid limit, plastic limit, shrinkage limit, swelling pressure

reaction period, the soil samples were used for the tests. The liquid limit and plastic limit tests were performed for all bacteria considered for the study and all percentage of bacteria. The shrinkage limit test and swelling pressure test were conducted for bacteria content of 20%.

3 Result and Discussions

3.1 Liquid Limit

Liquid limit and plastic limit tests were conducted on treated soil sample as per IS 2720 Part 5. The results of liquid limit, plastic limit, and shrinkage limit are discussed below.

The results show that the liquid limit of soil with *B. megaterium* decreases with an increase in the percentages of bacteria and reaction period. Tables 3 and 4 show results of liquid limit for *B. megaterium* and *Pseudomonas.*

The liquid limit of soil treated with *B. megaterium* is decreasing with increasing percentage of bacteria and reaction periods. The reaction period also shows influence of liquid limit with varying percentages of bacteria. For 20% bacterial concentration, liquid limit was reduced by 34% with 3 days reaction period, 41% for 7 days, and 44% for 14 days.

When soil was treated with *Pseudomonas* bacteria, the liquid limit was decreased for all the percentages of and all reaction periods. The liquid limit upto 12% of *Pseudomonas* bacterial concentration shows reduction with 7 days and 14 days reaction period. However, for 16 and 20% bacterial concentration with same reaction periods, the result shows reverse trend.

Table 3 Liquid limit for *Bacillus megaterium*

Reaction period (days)	Liquid limit for different percentage of *Bacillus megaterium*				
	4%	8%	12%	16%	20%
3	65	59.8	55.6	50.2	47.1
7	62	58.8	54.6	47	41.8
14	55	54.4	51.9	44	40

Table 4 Liquid limit for *Pseudomonas*

Reaction period (days)	Liquid limit for different percentage of *Pseudomonas*				
	4%	8%	12%	16%	20%
3	70	65	64.3	61	55
7	57	56	54	60	67
14	55	53	50	62.2	57.5

Table 5 Liquid limit for *Bacillus subtilis*

Reaction period	Percentage of *Bacillus subtilis*				
	4%	8%	12%	16%	20%
3 Days	60	58	55	50	49.5
7 Days	58.5	57	53	49	47
14 Days	57	55	52	48	48

Table 6 Liquid limit for *Sporosarcina pasteurii*

Reaction period	Liquid limit for percentage of *Sporosarcina pasteurii*				
	4%	8%	12%	16%	20%
3 Days	64	59	56	50	48
7 Days	60	56	54	50	48
14 Days	58	54	52	49	46

When *B. subtilis* bacteria with black cotton soil are tested for varying reaction periods, the liquid limit reduces as seen in Table 5. The effect of reaction period with 20% bacterial concentration on liquid limit shows that for 3 days reaction period the liquid limit reduced by 30% of original soil. The variation in liquid limit for 7 and 14 days reaction period is not significant as compared with 3 days reaction period.

The reduction of liquid limit due to precipitation of calcite by *Sporosarcina pasteurii* bacteria can be seen in Table 6. The liquid limit was reduced upto 34% for 20% bacterial concentration for all reaction periods.

3.2 Plastic Limit

The plastic limit of black cotton soil mixed with *B. megaterium* bacteria shows reduction for all percentages and all reaction periods as shown in Table 7. The plastic limit was reduced by 23% for 3 days reaction period, 27% for 7 days, and 31% for 14 days. The increasing reaction period reduces plastic limit.

The *Pseudomonas* bacteria effect on plastic limit of black cotton soil is shown in Table 8. The plastic limit is reduced by 13% and 19% when bacterial concentration

Table 7 Plastic limit for *Bacillus megaterium*

Reaction period (days)	Plastic limit for different percentage of *Bacillus megaterium*				
	4%	8%	12%	16%	20%
3	28.82	28.34	27.69	26.96	25.02
7	28.57	28.15	27.35	24.32	23.6
14	27.14	26.94	25.73	23.88	22.36

Table 8 Plastic limit for *Pseudomonas*

Reaction period (days)	Plastic limit for different percentage of *Pseudomonas*				
	4%	8%	12%	16%	20%
3	31.75	28.17	28.12	29.62	39.40
7	30	29.26	27.7	35.29	23.90
14	29.17	27.69	26.27	32.81	36.20

Table 9 Plastic limit for *Bacillus subtilis*

Reaction period (days)	Plastic limit for different percentage of *Bacillus subtilis*				
	4%	8%	12%	16%	20%
3	38	36.5	36	36	39.5
7	37.5	37	35	36	38
14	37	36	37	36	39

Table 10 Plastic limit for *Sporosarcina pasteurii*

Reaction period (days)	Plastic limit for different percentage of *Sporosarcina pasteurii*				
	4%	8%	12%	16%	20%
3	36.15	31.65	29.78	30	29
7	34	31	30	27.8	26.8
14	33	30	28.5	26	25.5

increases upto 12% with 3 days and 14 days reaction period. The plastic limit was reduced by 26% for 7 days curing periods with 20% bacterial concentration.

The plastic limit of black cotton soil treated with different concentration of *B. subtilis* as shown in Table 9 shows variation with percentage and reaction period. The plastic limit increased with increase in concentration and reaction periods.

Table 10 shows the plastic limit for black cotton soil mixed with *S. pasteurii* bacteria. The plastic limit was reduced by 11, 17, and 21% for 3, 7, and 14 days reaction period with 8–20% bacterial concentration. However, for 4% bacterial concentration, the plastic limit increases as compared to untreated soil.

3.3 Shrinkage Limit Test

Table 11 summarizes the results of shrinkage limit of black cotton soil treated with different bacteria. The shrinkage limit test was conducted on soil sample with 20% bacterial concentration with all reaction periods. The shrinkage limit was observed to be reduced for black cotton soil with all bacteria. 7 days reaction period was found

Table 11 Results of shrinkage limit test

Bacteria	Reaction period (days)	Shrinkage limit (%)
Bacillus megaterium	3	9.75
	7	3.17
	14	4.74
Bacillus pasteurii	3	4.64
	7	3.75
	14	4.33
Pseudomonas	3	7.73
	7	2.5
	14	8.52
Bacillus subtilis	3	2.73
	7	3.81
	14	4.91

effective for MICP treatment as compared to 3 days and 14 days reaction periods except *B. subtilis* bacteria.

3.4 Plasticity Index

The plasticity index was calculated for soil sample treated with various bacteria and reaction period. Figure 1 shows the variation of plasticity index for soil treated with *B. megaterium.* The plasticity index of black cotton soil with *B. megaterium* was reduced with an increase in reaction period. The plasticity index was reduced upto 44.24%, 54%, and 56% for 3 days, 7 days, and 14 days reaction period, respectively.

Figure 2 shows results for plasticity index of different percentages of Pseudomonas with reaction periods. The plasticity index reduced upto 60.6% for 3 days, 42% for

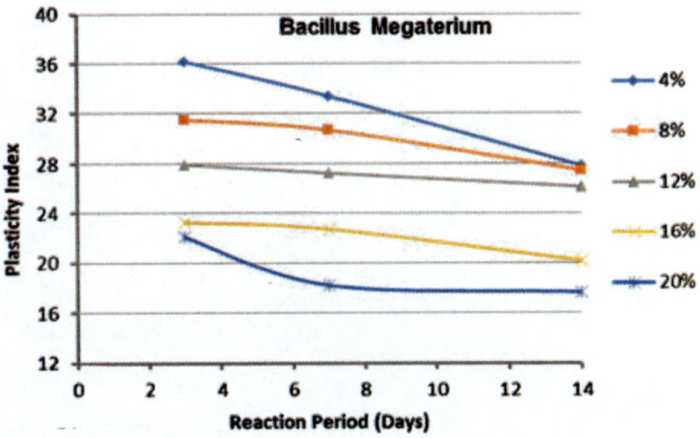

Fig. 1 Plasticity index for *Bacillus megaterium*

Fig. 2 Plasticity index for Pseudomonas

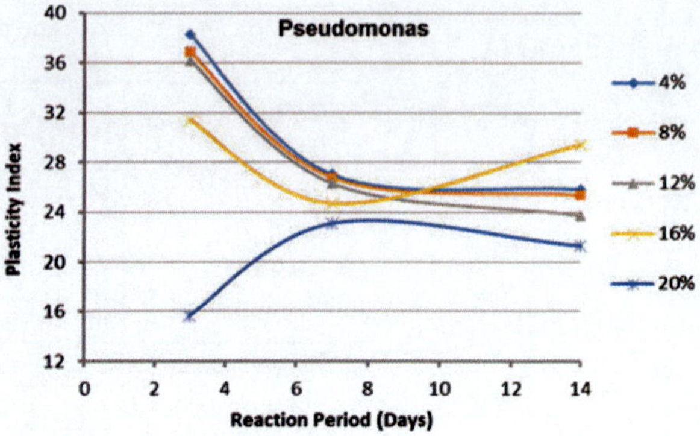

7 days and 46% for 14 days reaction period. The results shown effectiveness of Pseudomonas bacteria in black cotton soil with 3 days reaction period.

Figure 3 shows results of plasticity index for *B. pasteurii* bacteria. Plasticity index reduced by 52% for 20% concentration of *B. pasteurii* with 3 days reaction period and reduced upto 47%, 48% for 7 days, 14 days reaction time with same concentration.

Fig. 3 Plasticity index for *Bacillus pasteurii*

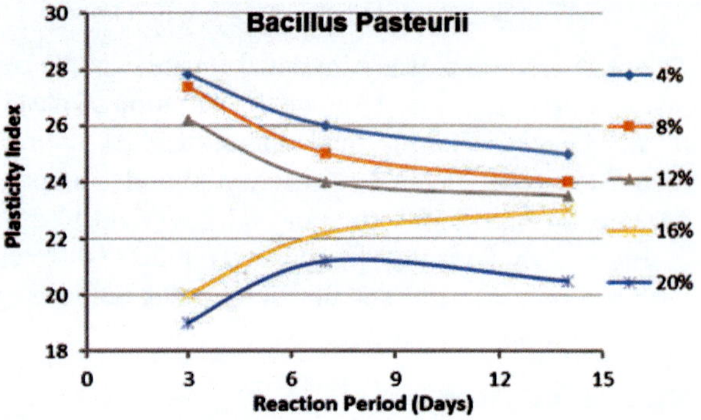

Fig. 4 Plasticity index for *Bacillus subtilis*

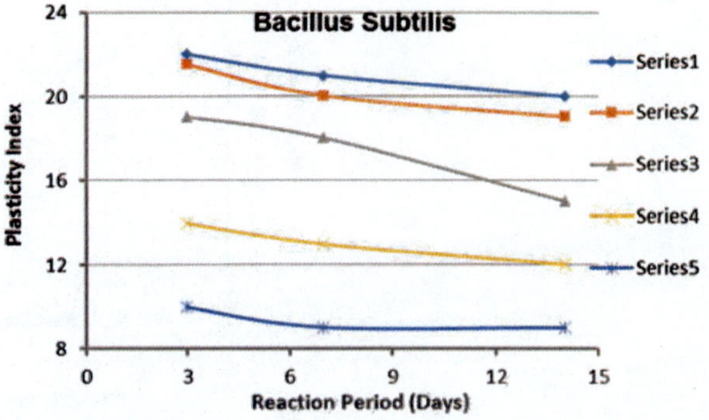

Figure 4 shows the results of plasticity index of soil with *B. subtilis*. Plasticity index show results for all % of bacteria with all reaction period was effective. MICP treatment by *B. subtilis* shows reduction of plasticity index upto 77% for 20% microbial concentration with all reaction period as compared to untreated soil.

3.5 Swelling Pressure Test

The swelling pressure test of the untreated soil and treated soil was conducted as per IS: 2720 Part (XLI)-1977. The soil sample was prepared by mixing 20% of bacteria with soil by weight and keeping it for 7 days for reaction period. After reaction period, the soil was used for the preparation of soil sample for swelling pressure test. The schematic view of swell pressure apparatus used is shown in Fig. 5.

Table 12 summarizes results of swelling pressure test. From the results, it was observed that the MICP treatment is effective to reduce swelling behavior of black cotton soil. From all of these microbes, the swelling pressure was reduced upto 80%.

Fig. 5 Swell pressure test apparatus

Table 12 Results of swelling pressure

Bacteria	Swelling pressure (kN/m^2)
Bacillus megaterium	22.42
Bacillus pasteurii	23.69
Psudomonas	39.24
Bacillus subtilis	25.45

4 Conclusions

The laboratory tests were conducted on black cotton soil with different bacteria, its concentration, and reaction period. The results demonstrate that the use of bacteria helps in controlling swelling and shrinkage characteristic of black cotton soil. The following conclusions can be drawn from the present study.

- Liquid limit reduces for all reaction period and bacterial concentration for all bacteria.
- The liquid limit decreases upto 44%.
- The plastic limit also reduces considerably due to calcite precipitation. However, for few bacteria, it shows an increasing trend.
- The shrinkage limit also reduces with reaction period with 20 percentage of bacterial concentration.
- It can be concluded that MICP treatment is very effective in case of black cotton soil for reducing swelling pressure.

References

Dejong JT, Fritzges MB, Nusslein K (2006) Microbially induced cementation to control sand response to undrained shear. J Geotech Geoenviron Eng 132(11):1381–1392

Dhami NK, Reddy MS, Mukherjee A (2013) Biomineralization of calcium carbonates and their engineering applications: a review. Front Microbiol 4:1–13

Ng W-S, Lee M-L, Hii S-L (2014) Factors affecting improvement in engineering properties of residual soil through microbial-induced calcite precipitation. J Geotech Geoenviron Eng 140(5):1–11

Van Passen LA (2009) Ground improvement by microbial calcite precipitation. Ph.D. dissertation. Delft University of Technology

Ng W-S, Lee M-L, Hii S-L (2012) An overview of the factors affecting microbial-induced calcite precipitation and its potential application is soil improvement. Int Sch Sci Res Innov 6(2):683–689

Whiffin VS (2004) Microbial $CaCO_3$ precipitation for the production of biocement. Ph.D. dissertation, Murdoch University, Perth, Western Australia, Australia

Subgrade Stabilization Using Alkali Activated Binder Treated Jute Geotextile

V. P. Komaravolu, Anasua GuhaRay, and S. K. Tulluri

Abstract Many past research works proved the successful usage of synthetic geotextiles and geogrids as road subgrade reinforcement. However, reinforcing subgrades with natural geotextiles is found to be more economic and eco-friendly. The usage of this is limited by its short degradation time period in soil, and to overcome this, researchers started using treated geotextiles. In this study, a systematic lab investigation has been made to understand the behavior of subgrade strength of roads reinforced with alkali activated binder (AAB) treated jute geotextile. Unreinforced soils, untreated and treated jute reinforced soils are tested for CBR, bearing capacity, and the results showed a considerable increase of CBR, bearing values in treated jute geotextiles. Durability tests such as soil burial tests and tensile strengths of degraded JGT are also carried to study the increase in life expectancy of AAB treated jute geotextile. Alkali activated binder improves life expectancy and mechanical properties of jute, and therefore, treated jute geotextile may be used as alternative material for subgrade soil reinforcement applications.

Keywords Alkali activated binder · Natural geotextile · Geotechnical characterisation · Durability

1 Introduction

Geotextiles have gained a lot of importance in the recent past because of its applications in reinforcing soils, improving drainage, filtration, separation, controlling soil erosion. Presently, man-made geosynthetics are used for this purpose due to their high strength and durability in soil, but the existing research recommends the improvement of soil in an economic and eco-friendly manner by using jute geotextiles (Gupta et al. 2017). Researchers are now showing interest in improving the performance and mechanical properties of natural geotextiles such as jute, coir and flax which have other advantages such as abundant availability, low cost, less abrasiveness, ability

V. P. Komaravolu · A. GuhaRay (✉) · S. K. Tulluri
BITS Pilani Hyderabad Campus, Secunderabad, Telangana 500078, India
e-mail: guharay@hyderabad.bits-pilani.ac.in

M. Latha Gali and R. R. P. (eds.), *Problematic Soils and Geoenvironmental Concerns*, Lecture Notes in Civil Engineering 88,
https://doi.org/10.1007/978-981-15-6237-2_29

to absorb mechanical impact, easy to handle process and environmental friendliness when compared to conventional geosynthetics (Chakraborty et al. 2008).

Jute geotextiles are used for stabilizing bank slopes by reducing the soil erosion. Sanyal and Chakraborty (1994) presented a study in which the slope of Nayachara island, situated in the middle of Hooghly river, was stabilized by bitumen treated jute geotextile. The western bank was protected with a layer of treated jute geotextile covered with riprap of different sizes. After 1.5 years, no disturbance or damage was observed in geotextile protected part, but the tensile strength of geotextile was dropped by 70% due to biological and physical degradation under riprap.

From the past studies, jute is also used to improve subgrade for road construction (Datta 2007) and is found to be more economic, eco-friendly and likely to deform enough to mobilize its full tensile strength and thereby increase the load carrying capacity of subgrade to its full extent (Khan and Rahman 2009). However, the usage of untreated jute is limited by its hydrophilic characteristics and cannot be used in saturated environment for more than 6–9 months. Hence, many chemical treatment procedures are developed to increase life expectancy and to improve mechanical properties of jute geotextile.

Chemical processes such as treatment with copper composition, bitumen, silicate were used to increase the design biodegradability of the JGT to 15 years (Khan and Rahman 2009). According to Wang et al. (2009), raw jute treated with two separate chemical processes, alkali scouring and hydrogen peroxide bleaching reduced the hemicellulose and lignin content of jute, and hence, the breaking tenacity was increased. Acetylation significantly improved the resistance toward biological degradation and reduced water intake but decreased tensile strength (Anderrson and Tillman 1989).

Saha et al. (2012) reported that treatment with a mixture of plant oils in the presence of sodium hydroxide and formaldehyde increased the resistance of jute to biological degradation. A transesterification reaction takes place as the hydroxyl group was replaced by oleic and stearic acid. The reports of Chakraborty et al. (2010) concluded that the tensile strength and elongation at breakage of jute fiber improved by about 41 and 34%, respectively, when treated with alkali (NaOH) and polymer (aqueous emulsion of carboxylated styrene–butadiene co-polymer-based polymer). The degradation of treated jute fiber treated with cement paste becomes constant with time, whereas the untreated jute fibers degraded rapidly with time. Durability of JGT is increased when it is treated with N-vinyl pyrrolidone and ethyl hexyl acrylate in the presence of plasticizers and an initiator (Uddin et al. 1997).

A typical unpaved road cross section consists of a subbase layer, and a base layer placed directly over the subgrade available. So, the bearing capacity is very low especially for soft subgrade soils. In such cases, a geotextile layer is reinforced. According to a case study presented by Rao (2003), a 360 m length of the unpaved road was constructed near Kakinada Port in the Andhra Pradesh state. In this, a single layer of 3 mm untreated jute geotextile was reinforced to increase the subgrade strength of road as the top strata of soil profile consisted of soft clay, underlain by sandy silt soil. Before the construction of road, the CBR values of soil were very low, so a reinforcement of jute layer is provided, and when the road is tested after

30 months of construction, it was found that the unsoaked CBR and soaked CBR were 2.87 and 2.96 times of initial. Even after 7 years of construction, there was no visible damage to road. It was also observed that the water content of that soil decreased with time and dry density of soil increased with time.

From CBR test results conducted on a road section constructed with jute–polypropylene (PP) composite, Basu et al. (2009) concluded that the CBR value of the reinforced road was approximately 80% greater than the unreinforced one after a time period of 18 months. The rutting also decreased by the use of jute–PP composite. The test results of Ravi Shankar and Suresha (2006) indicate that the thickness of aggregate layer can be reduced to half by reinforcing the subgrade. Moreover, the strength gain of relatively soft subgrade soils under cyclic loading is remarkable high, which may lead to substantial economic design.

The present study aims to use alkali activated binder (AAB) treated jute geotextile to increase the bearing capacity and also to reduce the thickness of unpaved road construction, thereby reducing the cost of construction. AAB is produced by the reaction between Class F fly ash and alkali activator solution, which is a mixture of sodium silicate and sodium hydroxide (Kar et al. 2014). The utilization of fly ash, which is a major waste product of thermal power plants, could avoid its disposal as solid waste as well as can improve the properties of jute as a geotextile. The strength gain in AAB is due to silica polymerization (Provis and Van Deventer 2009).

Few research works were done on the properties of alkali activated binder, and as per the reports of Gupta et al. (2018), treating jute geotextile with 0.35 w/s, alkali activated binder does not affect the permeability of JGT adversely, increases tensile strength and elongation by 37.7%, 47.76%, respectively, increases bearing capacity of reinforced sand by 30%. It also increases the axial stiffness of JGT by 70% and decreases friction angle between sand and treated jute by 32.5%. However, the XRD analysis of treated jute showed the formation of AAB layer on jute without any change in jute's mineralogy.

The above results prove that AAB treatment of jute can increase the mechanical properties of jute geotextile by considerable amount without any change in mineralogy of jute and can be used for soil reinforcement effectively. It was also anticipated that after modification with AAB microbial degradation of jute fiber can be either delayed or prevented.

2 Materials

Commercially available Tossa jute (*Corchorus olitorius*) which is available in a roll of 0.91 m (1 yd.) width and 30 m length is used. The sodium hydroxide pellets of 99% pure, industrial grade sodium silicate solution of composition 55.9% water, 29.4% SiO_2 and 14.7% Na_2O and Class F fly ash are used for the preparation of AAB. A model unpaved road is prepared in laboratory for a systematic investigation of bearing capacities using Red loam soil as subgrade material, stone dust as subbase material and fine aggregates as base material.

Table 1 Amount of AAB applied per m^2 of JGT

Water/solid ratio	AAB (kg/m^2)	Fly ash (kg/m^2)	NaOH (kg/m^2)	Sodium silicate (kg/m^2)	Water (kg/m^2)
0.35	3.44	2.18	0.058	0.706	0.498

3 Experimental Setup

3.1 Treatment of Jute Geotextile

Treatment mix consists of fly ash, sodium hydroxide, sodium silicate and the mass ratio is maintained as 400:10.57:129.43, respectively (Kar 2013). The treatment solution was prepared by blending sodium hydroxide, sodium silicate and water accompanied by constant stirring till a clear blend was obtained. The water quantity can be varied to prepare different w/s AAB mixtures. The reaction is allowed to cool down to normal temperature for 24 h. Then, the alkaline solution is mixed with calculated amount of fly ash and is applied to jute geotextile. The quantity of fly ash is calculated from the initial water-to-solids ratios considered. In this study, a water-to-solid ratio (w/s) of 0.35 is considered. The amount of raw materials required for 0.35 w/s AAB to treat one square meter of JGT is given in Table 1.

3.2 Model Pavement

The design of model unpaved road is made as per IRC: SP:77 2008 considering the CBR values of subgrade soil. The model pavement is tested for bearing capacity, and results are attached in Fig. 4.

- Thickness of granular subbase (stone dust) is 230 mm
- Thickness of granular base (fine aggregate) is 250 mm.

4 Experimental Methodology

4.1 CBR Test

CBR is the ratio of force per unit area required to penetrate a soil mass with standard circular piston at the rate of 1.25 mm/min to that required for the corresponding penetration of a standard material. It is developed for evaluating the bearing capacity of subgrade soil for design of roads. So, a series of CBR tests are performed on the unreinforced soil and also on soil reinforced with untreated and treated jute samples of 0.35 w/s ratio to evaluate the change in bearing capacities of reinforced and unreinforced soil. Soaked CBR test is carried out following IS-2720-part 16-1979. Soil is placed in five layers in the mold compacting each layer by 56 blows with a 4.90 kg hammer dropping from a height of 450 mm above the soil. Different tests are conducted by reinforcing untreated, AAB treated jute sample in soil at an effective depth of one-third distance from the top surface of soil (Surendra and Damgir 2011). Then, the sample is allowed to soak for 96 h before testing in CBR test apparatus.

4.2 Tests on Model Unpaved Road

Bearing Capacity Test. A model plate load test is performed on the gravel road model prepared to assess bearing capacities of unreinforced and jute reinforced roads. According to ASTM D1195 (2015), the tank dimension should be five times the width of the foundation to depreciate scaling issues. The cross-sectional area of the steel tank was 120 cm × 91 cm with a height of 91 cm, and a square footing of size 20 cm × 20 cm and 25 mm depth was used. The model gravel road is prepared in it as per the calculated depths with bottom layer as subgrade layer (red loam soil), subbase (stone dust) on it and base layer (fine aggregate) as shown in Fig. 1. Each layer is compacted thoroughly. A hydraulic load cell applies the load on a rectangular footing, and the displacement was measured by the four LVDTs placed at each corner of the plate. Similarly, tests were also carried on the model by reinforcing the subgrade soil with untreated and treated jute geotextile samples placed at a depth of 50 mm from top

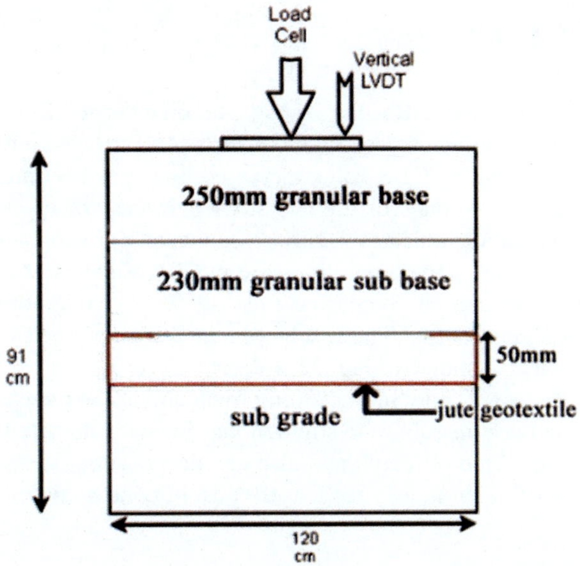

Fig. 1 Design of model pavement with JGT reinforcement in a model plate load test setup

Table 2 Properties of tested soils

Properties	Red soil	Stone dust
Grain size analysis		
Sand %	72.61	95.12
Silt and clay %	27.3	4.88
Specific gravity	2.67	2.50
Maximum dry density (kN/m^3)	18.92	19.41
Optimum moisture content (%)	17.5	10.25
Unsoaked CBR (%)	7.14	25.42
Soaked CBR (%)	4.28	11.8

surface of subgrade layer and wrapped at edges for having a grip. As per IS: 1888 (1982), the strain was maintained at a constant rate of 0.02 mm/min.

5 Results

5.1 Soil Properties

The properties of the subgrade soil (red loam soil) and subbase base soil (stone dust) are presented in Table 2.

5.2 CBR Test

The soaked CBR value of subgrade soil when reinforced with jute geotextile is 6.81%, and when reinforced with 0.35 w/s, AAB treated jute is 14.73%. It is interesting to note that the CBR value increased by about 1.6 times in case of untreated jute and about 3.4 times of the actual CBR in case of alkali activated binder treated jute and which indicates the immense increase of subgrade strength and its stabilization. However, a maximum increase in CBR value can be obtained in ground condition, as soil can be compacted to 94–95% of maximum dry density safely (Khan and Rahman 2009). The results obtained from soaked California bearing ratio (CBR) tests are shown in Fig. 2.

As we know setting strong foundations can increase the life of constructions, the increase in subgrade strength can increase the life of unpaved road, decreases or avoids the rut depth and also can decrease thickness of the pavement construction which enormously reduces the cost of construction.

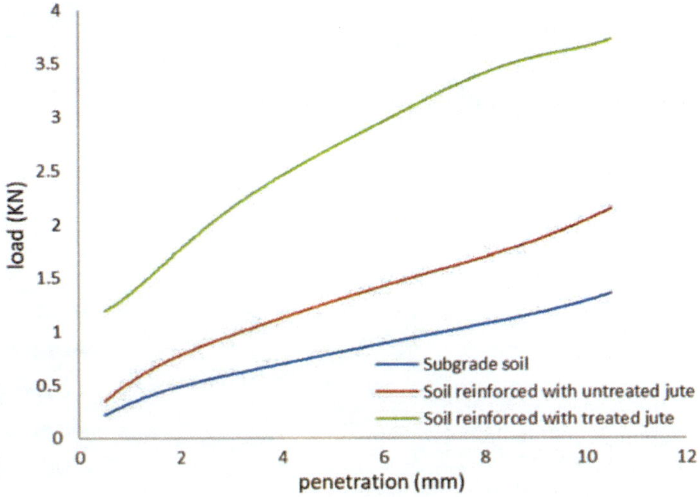

Fig. 2 Relationship between penetration (mm) and load applied (KN), obtained from CBR test

5.3 *Bearing Capacity*

Model plate load tests also revealed that the increase in bearing capacity of unpaved road reinforced with untreated jute is more pronounced in AAB treated JGT. The bearing capacity improvement factor (BC_{IF}) given in Eq. 1 of soil reinforced with untreated jute is 1.22 and that of AAB treated jute reinforced soil is 1.53. This again represents a substantial increase, and the reason for the increase of bearing capacity is due to the high amount of pressure taken by AAB treated jute when compared to that of untreated jute. The bearing capacity improvement factors of different w/s AAB treated jute reinforced in only sand at different depths are given in Fig. 3.

$$BC_{IF} = \frac{\text{bearing capacity of reinforced soil}}{\text{bearing capacity of unreinforced soil}} \tag{1}$$

From the above figure, the increase in bearing capacity of 0.35 w/s AAB treated JGT is about 1.25 times, and also, from the results of Gupta et al. (2018), increase in bearing capacity of AAB treated jute is about 35%, we can conclude that bearing capacity of model unpaved road will be increased by the use of treated jute reinforcement. However, these results are specific to the present test conditions.

Tests conducted by Milligan et al. (1986) indicate that the general mechanism and behavior observed in the model tests are reproduced at large scale as well. However, extrapolation of results from the present tests to the prototype cases can be done through a suitable scaling law (Butterfield 1999; Fakher and Jones 1996). The load-settlement curves of the model unpaved road with unreinforced, reinforced with AAB treated, untreated jute geotextiles are shown in Fig. 4.

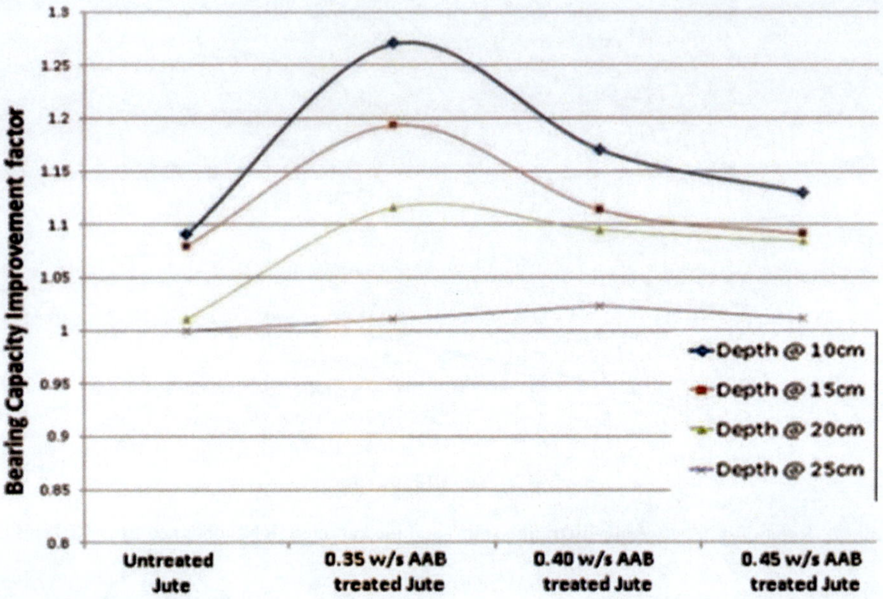

Fig. 3 Bearing capacity improvement factors of different w/s AAB treated jute reinforced at different depths in sand

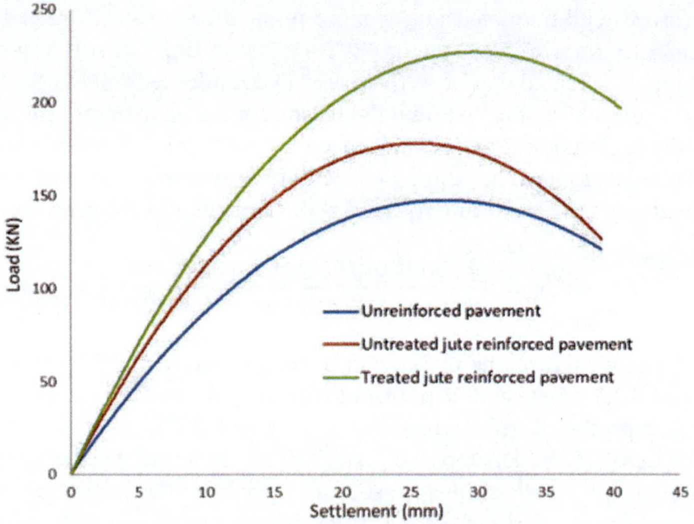

Fig. 4 Bearing capacities of model unpaved road obtained from model plate load test

From Fig. 4, we can observe the increase in load-bearing capacity of soil and also the increase in settlement corresponding to maximum load in case of reinforced soil. The increase is however more in case of AAB treated jute that the untreated jute reinforcement.

The bearing capacity value of soil reinforced with 0.35 w/s AAB treated JGT is found to be 235 kN/m^2 whereas that of untreated jute is found to be 185 kN/m^2 and that of soil without reinforcement is 155 kN/m^2. The bearing capacity increase in treated jute also depends on the water-to-solid ratio of AAB, with the increase or decrease of 0.35 w/s, the bearing capacity of pavement decreases because the layer of AAB coating becomes gradually softer with increasing w/s ratio, thus reducing the flexibility of jute geotextile, and if w/s decreases, the strength of AAB decreases leading to a lower bearing capacity of jute fabric. A w/s ratio of 0.35 is found to be optimum for achieving maximum load-bearing capacity (Gupta et al. 2018).

6 Results

6.1 *Bearing Capacity*

Durability is one of the reasons for the limited usage of natural geotextiles as reinforcements. The degradation period of a natural geotextile is very low when compared to that of any geosynthetics. Durability tests are also conducted to know the behavior and sustainability of AAB treated jute in soil. Soil burial test is done to test the durability by keeping treated and untreated jute samples in sand with a high water content of 12% for a specific time period. The sand was stored in a closed container to maintain the relative humidity and was changed after each week to ensure the uniformity of saturation. The degradation of jute can be observed visually in Fig. 5. Degradation is also measured by narrow strip tensile strength at different time periods, and results are shown in Fig. 6. Surprisingly, the result after one month of exposure showed no degradation in treated jute sample, whereas the untreated jute started to degrade. However, the treated jute will also start degrading after specific period, but there is no much disadvantage because by that time treated JGT provides a self-sustaining subgrade for most of soils, i.e., the gain in strength of subgrade compensates the loss of strength in JGT. This concluded that usage of AAB treated jute is definitely preferable to the untreated jute geotextile as soil reinforcement. The degradation comparison of both the samples of untreated and treated jute is shown in Fig. 5.

6.2 *Tensile Strength Test*

The tensile strength of both treated and untreated geotextile is determined by an MCS universal testing machine (UTM). The width and the gauge length of a tested

Fig. 5 Degradation of untreated jute geotextile (left) when compared to 0.35 w/s AAB treated jute (right) after exposure for 1 month in soil burial test

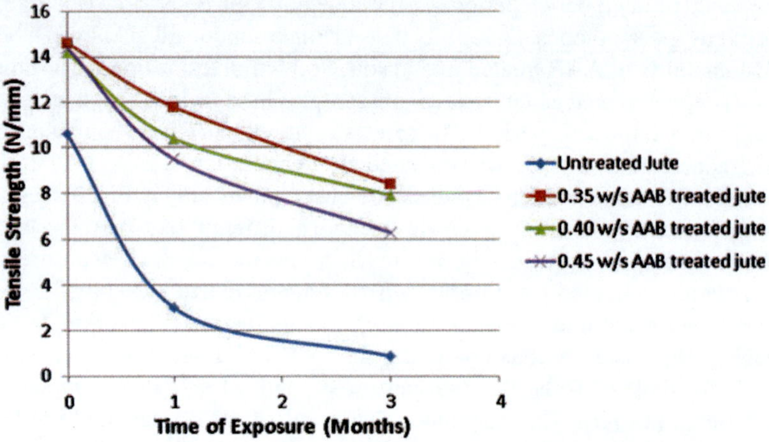

Fig. 6 Relationship between tensile strength and time of exposure of AAB treated JGT

sample are considered as 50 mm and 200 mm, respectively. The grab length is kept as 25 mm. The tensile loading is maintained at a deformation rate of 100 mm/ min as per the specifications of ISO 13934–1:1999 (1999) standards to determine narrow strip tensile strength. The results of tensile strengths with respect to the exposure period are given in Fig. 6.

The tensile strength of 0.35 w/s AAB treated jute when exposed for one month dropped from 14 to 12 N/mm whereas that of untreated jute is dropped from 10 to 3 N/mm, and for an exposure of three months, the tensile strength of treated jute dropped to 8 N/mm and that of untreated jute is 1 N/mm. The above tensile strength test results clearly indicate the sustainability of treated jute in soil for longer time. So, AAB is also used to increase the degradation time of jute geotextile.

7 Conclusions

The subgrade stabilization by reinforcing AAB treated JGT is examined in this study. The mechanical properties and life expectancy of AAB treated jute in soil are found to be increasing when compared to conventional untreated jute. After conducting many experiments to compare load-bearing capacity, we can conclude that the subgrade strength of a pavement can be greatly increased, and the thickness of pavement can also be reduced by the reinforcement of JGT. So, reinforcement of JGT is better in both economic and ecological aspects. The durability of JGT can also be increased by treating with AAB when compared to untreated, but the tensile strength of treated jute may slightly decrease as exposure period in soil increases. The further study of this topic is with respect to finding resilient modulus (M_R) using triaxial apparatus. Resilient modulus (M_R) is the elastic modulus based on recoverable strain under repeated loads, and it is helpful for characterizing pavement construction materials including surface, base and subbase materials under a variety of temperatures and stress states that simulate the conditions in a pavement subjected to moving wheel loads.

References

Andersson M, Tillman AM (1989) Acetylation of jute: effects on strength, Rot resistance, and hydrophobicity. J Appl Polym Sci 37(12):3437–3447

ASTM D1195/D1195M-09 (2015) Standard test method for repetitive static plate load tests of soils and flexible pavement components, for use in evaluation and design of airport and highway pavements. West Conshohocken, PA: ASTM International. www.astm.org. https://doi.org/10.1520/D1195_D1195M-09R15 (2015).

Basu G, Roy AN, Bhattacharyya SK, Ghosh SK (2009) Construction of unpaved rural road using jute-synthetic blended woven geotextile–a case study. Geotext Geomembr 27(6):506–512

Butterfield R (1999) Dimensional analysis for geotechnical engineers. Geotechnique 49(3):357–366

Chakraborty S, Kundu SP, Roy A, Basak RK, Basu Majumder S, Adhikari B (2010) Processing and fabrication of chemically modified jute fiber reinforced cement concrete composite. In: National seminar on recent advances in chemical engineering (RACE 2010) in GIET Gunupur, Odisha

Chakraborty S, Kundu SP, Roy A, Basak RK, Sen R, Basu Majumder S, Adhikari (2008) Development of jute fiber reinforced cement concrete composites, Project report for National Jute Board Ministry of Textile, pp 8–10

Datta U (2007) Application of jute geotextiles. J Nat Fibers 4(3):67–82

Fakher A, Jones CJFP (1996) Discussion of bearing capacity of rectangular footings on geogrid reinforced sand, by Yetimoglu T, Wu JTH, Saglamer AJ. Geotech Eng ASCE 122(4):326–327

Indian Standard: 1888–1982 (1982) Method of load test on soils. Bureau of Indian Standards, New Delhi

ISO E (1999) "13934-1: 1999." Textiles-tensile properties of fabrics-part 1: 13934-1

Gupta S, GuhaRay A, Kar A, Komaravolu VP (2017) Alkali activated binder treated jute reinforced soil: a preliminary study. In: Indian geotechnical conference

Gupta S, GuhaRay A, Kar A, Komaravolu VP (2018) Performance of alkali-activated binder-treated jute geotextile as reinforcement for subgrade stabilization. Int J Geotech Eng

Kar A, Halabe UB, Ray I, Unnikrishnan A (2013) Nondestructive Characterizations of Alkali Activated Fly Ash and/or Slag Concrete. Eur Sci J 9(24):52–74

Kar A, Ray I, Halabe UB, Unnikrishnan A, Dawson-Andoh B (2014) Characterizations and quantitative estimation of alkali activated binder paste from microstructures. Int J Concr Struct Mater 8(3):213–228

Khan AJ, Rahman MM (2009) Road subgrade reinforcement using jute geotextiles. In: *Proceedings of Bangladesh geotechnical conference*

Milligan GWE, Fannin RJ, Farrar DM (1986) Model and fullscale tests of granular layers reinforced with a geogrid. In: Proceedings of third international conference on geotextiles, Vienna, vol 1, pp 61–66

Provis JL, Van Deventer JSJ (eds) Geopolymers: structures, processing, properties and industrial applications. Elsevier (2009).

Ravi Shankar AU, Suresha SN (2006) Strength behaviour of geogrid reinforced shedi soil subgrade and aggregate system. Road Mater Pavement Design 7(3):313–330. https://doi.org/10.1080/14680629.2006.9690040

Rao AS (2003) Jute geotextile application in Kakinada port area. In: Proceedings of national seminar on jute geotextile and innovative jute products

Saha P, Roy D, Manna S, Adhikari B, Sen R, Roy S (2012) Durability of transesterified jute geotextiles. Geotext Geomembr 35:69–75

Sanyal T, Chakraborty K (1994) Application of a bitumen-coated jute geotextile in bank-protection works in the Hooghly estuary. Geotext Geomembr 13(2):127–132

Surendra PJ, Damgir RM (2011) Use Of jute geo textile for strengthening of sub grade of road work. Innovative systems design and engineering, vol 2, no 4. ISSN 2222-1727 (Paper), ISSN 2222-2871 (Online)

Uddin MK, Khan MA, Ali KI (1997) Degradable Jute plastic composites. Polym Degrad Stab 55(1):1–7

Wang WM, Cai ZS, Yu JY, Xia ZP (2009) Changes in composition, structure, and properties of jute fibers after chemical treatments. Fibers Polym 10(6):776–780

Variation of Swelling Characteristics of Bentonite Clay Mixed with Jarofix and Lime

G. Santhosh and K. S. Beena

Abstract Expansive soils undergo considerable amounts of volume changes due to moisture content fluctuations, and it causes problems to infrastructures on them. Also, due to industrialization, the production and the accumulation of various wastes have created serious problems of handling and disposal. One of the possible solutions of it is to utilize the waste materials for the improvement of soil. Jarofix is one of such waste material from zinc industries. So, in this study, the utilization of Jarofix to control the expansive nature of clay mixing along with lime has studied. For that, the variation of free swell index, swell potential and swell pressure of bentonite clay was studied by adding different percentages of Jarofix. The variation of the same has studied by adding different percentages of lime also. The results reveal that the Jarofix and lime can be used as an effective agent for reducing the expansive nature of soil.

Keywords Swelling characteristics · Bentonite · Jarofix · Lime · Moisture content fluctuations · Expansive soil

1 Introduction

Expansive soil underlined by the rock exists more than one-third of world's land surface. In India, almost 20% of the total area is covered by expansive soil (Jagtap et al. 2015; Verma and Maru 2013). Expansive soils are worldwide problems that pose several challenges for civil engineers. The soils which increase its volume when the water content increases are classified as expansive soils, and usually, their load carrying capacity is very low. The heaving and the settlement to this soil cause damage to the structures built on such soils. Expansive soils are also known as shrink–swell

G. Santhosh (✉) · K. S. Beena
School of Engineering, CUSAT, Kochi, Kerala, India
e-mail: santhoshg@scmsgroup.org

K. S. Beena
e-mail: beenavg@gmail.com

M. Latha Gali and R. R. P. (eds.), *Problematic Soils and Geoenvironmental Concerns*, Lecture Notes in Civil Engineering 88,
https://doi.org/10.1007/978-981-15-6237-2_30

soils because it tends to swell and shrink when the moister content varies. Evaporation of water from an exposed soil surface usually results in soil volumetric shrinkage. Drying of the soils and the cracking due to this are one of the crucial issues. Cracking is one of the causes for substantial damage to the foundation-supported structures. Shrinkage also can adversely influence the engineering properties and behavior of the soils. The adverse effects include decreased strength of the cracked soils and increased flow through the soils. Problem of expansive soils has appeared as cracking and breakup of pavements, railways, highway embankments, roadways, building foundations, channel and reservoir linings, irrigation systems, water lines and sewer lines (Bhuvaneshwari and Sowbi 2017). The problems associated with expansive clays are worldwide, having occurred in countries like USA, China, Australia, India, Canada and regions in Europe. Expansive soils are usually fine-grained soils. Not only such soils are characterized by large volume change, they are also usually having a high moisture-holding capacity, low bearing capacity, low strength and low permeability. Clay minerals, especially the smectite (montmorillonite) group which increase in volume during wetting and decrease in volume during drying are responsible for the expansive nature of this category of soils (Geeta Rani et al. 2017; Dang et al. 2017). The montmorillonite is hydrated aluminum silicates with three-layered lattice structure in which interparticle space when occupied water molecules leads to swelling and related phenomena. Construction engineers usually avoid the use of expansive soils because they are usually difficult to work on such soils. But due to rapid industrialization and huge population growth of our country, there is a scarcity of good land to meet the needs. The objective of stabilization of expansive soils is usually to increase their strength characteristics and to overcome the disruptive and harmful properties such as swelling and shrinkage. Various methods of foundation design and construction for buildings which may prove successful in expansive soils are as follows: (i) removing the expansive soil up to the firm stratum and backfilling with granular material; (ii) providing a reinforced concrete raft, inverted T-beam or a deep reinforced concrete beam; (iii) designing the structure flexible enough to tolerate differential movements; (iv) designing the structure rigid enough to move as a unit; (v) increasing the foundation pressure to counteract swelling pressure; (vi) providing a pile and beam construction with the pile installing up to the zone of inappreciable movement. Out of the above methods, the raft foundation and the pile and beam construction appear to be the most satisfactory methods.

Every year, damage to buildings, roads, pipelines and other structures constructed over expansive soils is much higher. The hazards due to expansive soils usually develop gradually, and it causes threat to the life rarely. Hence, they have received only limited attention, in spite of their severe effects on the economy. Most of the damages related to expansive soils are not due to a lack of appropriate engineering solutions, whereas it is due to the non-recognition of expansive soils at the site and expected magnitude of expansion early stages of construction. The most common and economical method for stabilizing these soils is using admixtures that prevent volume changes.

One of the methods to improve expansive soils is chemical stabilization. Chemical stabilization includes the mixing or injecting of chemical substances used for stabilization into the soil. Portland cement, lime, asphalt, calcium chloride, sodium chloride, etc., are the common chemical stabilization agents for improving the weak soils. The selection of a particular additive depends on costs, benefits, availability and practicality of its application. In recent years, many researchers have attempted to improve the soils by using different types of industrial wastes which are having cementitious value. Such utilization is effective for reducing disposal problems (Akinwumi et al. 2017).

Industrialization and urbanization have good and bad impacts. On the evil side, all over world, huge quantities of hazardous wastes are generated from different industries (Aswathy et al. 2016). Improper management of these wastes causes adverse effects on the society which may also lead to cause possible outbreak of epidemics. Hence, disposal of these different waste materials is a serious problem facing nowadays all over the world. Presently, dumping over the useful land has been the major option available for the disposal. This has created non-utilization of useful land and causes other environmental problems. Approximately, 960 million tonnes of solid waste are being generated annually in India from various sources. Of which, 4.5 million tones are hazardous in nature (Nitisha et al. 2004). The utilization of these industrial wastes in the construction field is one of the best solutions to dispose the waste. Presently, various wastes like fly ash, slag, mine tailing, pond ash, rice husk ash, cement kiln dust limestone dust, etc., have already been blended with lime, and cement is used to improve the geotechnical properties soil (Muntohar et al. 2013).

Jaroste is one of such solid wastes from zinc industries. Jarosite is a by-product of hydro-metallurgical process, used to produce zinc and lead. It is acidic in nature. Due to the presence of toxic substances like zinc, lead, cadmium, copper and other metallic and non-metallic oxides, it is subjected to leaching which may create adverse effects on environment and human health. So, Jarosite falls under hazardous waste category as per the prevailing Hazardous Waste (Management, Handling and Trans-Boundary Movement) Rules, 2008 (Nitisha et al. 2004). Hence, its disposal has become a major environmental concern in all zinc industries. When Jarosite is mixed with lime and cement, the resulting material is called Jarofix which is more stable than Jarosite (Sinha et al. 2015). At present, the total accumulated Jarofix in India is about 15 million tons and the annual production is about 5 million tons (Sinha et al. 2012). Like other waste materials, this is also occupying costly agricultural and useful lands and has become an environmental hazard in almost all zinc industries. So, the best way to solve the disposal problem of Jarofix is to decrease the quantity for disposal by utilizing in various civil engineering fields. Only a few literatures are available regarding the utility of Jarofix in various applications.

Presently, lime stabilization of soils is widely used in several structures in civil engineering field. When solid waste is added to the soil found to be ineffective for improvement of weak soil and if it can be effectively used along with lime for soil modification, it will also be beneficial to handle the problems created due to waste disposal (Dash and Hussain 2012). The strength of lime–soil mixture is influenced by several factors such as soil type, amount of lime added, moisture content, curing period, unit weight of soil and time elapsed between mixing and compaction. Various studies revealed that the addition of quicklime (CaO) or hydrated lime [$Ca(OH)_2$] to expansive soil reduces swelling and swelling pressure and improves the strength of the soils. The improvement of lime treated soils may be attributed due to cation exchange, flocculation; carbonation and agglomeration and pozzolanic reactions (Kate 2009).

The amount of strength increase in soil is highly dependent on pozzolonic characteristics of the soil. When the pozzolonic content in the soil is less, the improvement in strength obtained by adding lime is also less. The immediate increase in strength due to lime treatment results from flocculation–agglomeration reaction and leads to better workability, whereas long-term strength gain is due to pozzolanic reactions. Various studies have conducted the utilization of various wastes along with lime for weak soil improvement all over the world.

2 Experimental Work

2.1 Materials

In this paper, the utilization of Jarofix along with lime to reduce the swelling characteristics of expansive soils has studied. Bentonite is a highly expansive soil. So, different experiments have done using bentonite clay, bentonite mixed with Jarofix and the bentonite mixed with Jarofix and lime. The laboratory tests were done to evaluate physical and engineering properties of the materials used. The following are the description of materials used in this study.

Bentonite. Commercially available bentonite was used in this study. Bentonite is a type of clay which contains a very high proportion of clay mineral montmorillonite. Bentonite is highly plastic and water absorbent clay and has high shrinkage and swelling characteristics with high ion exchange capacity and very low water permeability (Varghese et al. 2016). When it is mixed with water, it rapidly swells. Bentonite slurry is often used to solve problems in the construction of borings and excavating trenches in water-saturated soils. Also, due to the binding properties of bentonite, it is used for retaining the soil inside the bore hole at the time of piling. Bentonite used in this study is of gray in color, very fine, odorless and smooth texture. As per IS classification based on liquid limit and plasticity index, the soil belongs to CH.

Jarofix. Jarosite is the major by-product at Binani Zinc Ltd., Binanipuram, Kochi, which is a major zinc manufacturing company in Kerala. Jarosite is a toxic material and is converted into Jarofix by adding cement and lime which is dumped over a large area inside the company. Jarofix which is used in the study is collected from there and is brought to laboratory. The physical appearance of the material is fine powder, light brown in appearance, and the lumps were broken by pulverizing between the fingers. After air drying and pulverizing, the Jarofix samples were transferred to plastic bags and stored in airtight containers at room temperature. For getting the nature of the material, the index properties were done. The grain size analysis showed that the material consists of mainly fine particles, and the silt and clayey fraction come to be 96.38%, and remaining 3.62% is sandy fraction. As per IS classification, the materials belong to MH. The chemical characteristics revealed that the main constituents of Jarofix are SO_3, CaO, SiO_2 and Fe_2O_3 and heavy metals like lead, cadmium and zinc of which zinc and cadmium are poisoners. The geotechnical properties of bentonite and Jarofix were determined as per IS specifications and are as shown in Table 1. The chemical properties of Jarofix are as shown in Table 2.

Lime. Lime stabilization is one of the oldest and common processes for improving the engineering properties of expansive soils (Dang et al. 2016). Lime is calcium oxide or calcium hydroxide. Lime used in this study was obtained in the form of burned shells from local shops. It was sprinkled with water and let to rest for a day, and after which, the powered form of lime was obtained. It was sieved through 425 microns before mixing it with other materials. To reduce the carbonation effect due to humidity, the lime was kept in an airtight plastic container.

Table 1 Geotechnical properties of bentonite and Jarofix

Properties	Bentonite	Jarofix
Specific gravity	2.42	2.31
IS classification	CH	MH
Liquid limit (%)	330	61
Plastic limit (%)	40.5	44
Plasticity index (%)	289.5	17
Shrinkage limit (%)	13	32
Optimum moisture content (%)	37	48.3
Maximum dry density (kN/m^3)	12.4	11.6

Table 2 Chemical composition of Jarofix[a]

Chemical components	Range (%)
SO_3	19–31
CaO	4–18
SiO_2	3–12
Al_2O_3	<12
Fe_2O_3	18–36
Na_2O	<4
Cadmium	<0.08
Lead	<5.5
Zinc	<3.5

[a]As supplied by Binani Zinc Ltd, Binanipuram., Kochi, Kerala

2.2 Experimental Studies

All the samples used in this study were air dried because oven dry may cause the variation of properties. Air-dried soil has been dry mixed with various percentages of additives. Sufficient quantity of distilled water has been added to bring the moisture content to the desired level. The index properties were determined to assess the properties and for the classification of materials. The aim of the experimental works is the utilization of Jarofix for reducing the compressive characteristics of expansive soil combined with lime, thereby reducing the problems created due to dumping of Jarofix in zinc manufacturing industries. So, in this study, bentonite was used as it is one of the most swelling clays. Initially, the experiments to measure the expansive nature of soil such as free swell index (FSI), swell potential and swelling pressure were found as per IS specifications using the bentonite clay and the bentonite mixed with 10, 20, 30, 40 and 50% of Jarofix. Later, to study the effect of lime on each bentonite–Jarofix mixture and to find the optimum percentage of lime in each mixture, all the mentioned experiments were done with varying percentage of lime. Different percentages of lime added were 2.5, 5, 7.5, 10 and 12.5, and all the results were analyzed, and the results are discussed in the following sessions. In this paper, the nomenclature B, J and L stands for bentonite, Jarofix and lime, respectively.

3 3. Results and Discussions

3.1 Free Swell Index (FSI)

The test for free swell index was done as per IS 2720 (Part 40)—1977. At room temperature, ten grams of the dried and pulverized sample fraction passing 425 μ sieve were poured into two graduated cylindrical jars of 100 mL capacity. One of the cylindrical jars is filled with the distilled water and the other with kerosene. After 24 h, the final volumes of the sample in the two cylinders were noted. FSI was obtained as the ratio of the difference in the final volumes of soil in water (V_w) and in kerosene to the final volume in kerosene (V_k) expressed as a percentage. It is written as shown in Eq. 1.

$$\text{FSI}\,(\%) = \frac{V_w - V_k}{V_k} * 100 \tag{1}$$

Figure 1 shows the variation of free swell index (FSI) of bentonite with different Jarofix contents. The free swell index was found to decrease from 671 to 250% when the percentage of Jarofix added to the bentonite clay increases to 50%. The percentage reduction in FSI is 62.7%. The FSI of Jarofix is 25%. The addition of more and more material having less free swell index to the bentonite may be the reason for decrease in the value.

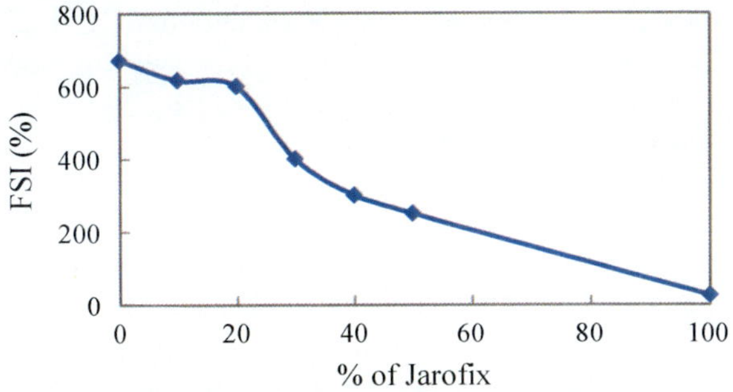

Fig. 1 Variation of free swell index of bentonite with different percentages of Jarofix

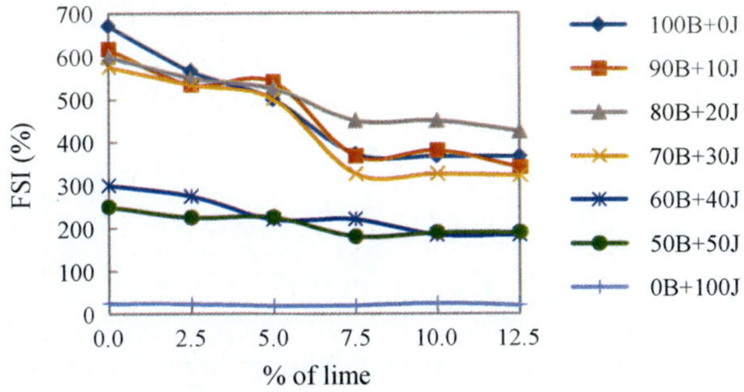

Fig. 2 Variation of FSI with different percentages of lime

To study the effect of lime on bentonite–Jarofix mixture, 2.5%, 5%, 7.5%, 10% and 12.5% of lime were added to all combinations. Results showed that the presence of lime decreases the free swell index. The FSI of bentonite is 671% without lime, and it decreases to 367% when the percentage of lime increases to 12.5% and the corresponding percentage decreases in FSI is 45.3%. For the mixture of 90% bentonite and 10% Jarofix (90B + 10 J) and the mixture of 80% bentonite and 20% Jarofix (80B + 20 J), the values decrease from 617 to 340% and from 600 to 425%, respectively. The corresponding percentage decrease is 44.9% and 29.2%. The FSI values decrease with the presence of lime. But it was not considerable after 7.5% of lime. The same trend has obtained for 70B + 30 J, 60B + 40 J and 50B + 50 J mixes also. The variations of FSI with lime for different mixtures are shown in Fig. 2.

3.2 *Swell Potential*

One of the typical characteristics of clay minerals is the presence of the negative charge on their surfaces. The quantity of the exchangeable cations required to balance that negative charge is called cation exchange capacity (CEC). If the CEC is higher for a soil, then its activity of the clay particles and the swelling potentials are high. (Nelson and Miller 1992).

Swell potential of a soil specimen is measured as the ratio of the increase in thickness to the original thickness expressed as a percentage when the material specimen compacted at OMC in a consolidation ring and soaked under a surcharge load. The dried soil (bentonite) passing through IS 4.75 mm sieve was used. Different soil–Jarofix mixtures were compacted in four layers each in the consolidation ring of 20 mm thickness and 60 mm diameter. Initial token surcharge pressure of 5 kPa was applied on the specimen after setting the dial gauge reading to zero (Phanikumar and Sharma 2007). The free swell method was adopted for the determination of swell potential. In this method, the sample is completely inundated in water and is allowed to swell freely under a token surcharge. Dial gauge readings were taken until equilibrium. The increase in the thickness (H) of the sample was noted after saturation. Swell potential (S %) is taken as the ratio of the increase in thickness to the original thickness (H) and is expressed as a percentage (Seed et al. 1962). It can be written as in Eq. 2.

$$S\,(\%) = \frac{\Delta H}{H} * 100 \tag{2}$$

The variation of swell potential of bentonite with different percentages of Jarofix is as shown in Fig. 3. As the percentage of Jarofix increases, the swell potential of soil decreases from 37 to 28.9%. The percentage decrease in swell potential is 21.9. The volume occupied by the clay minerals is replaced by the Jarofix particle which is having less swell potential compared to bentonite, may be the reason for decrease in value.

Figure 4 shows the variation of swell potential of different bentonite–Jarofix mixes when the lime increases from 0 to 12.5%. Results showed that the presence of lime decreases the swell potential. The swell potential of bentonite is 37% without lime, and it decreases to 23.3% when the percentage of lime increases to 12.5%. The

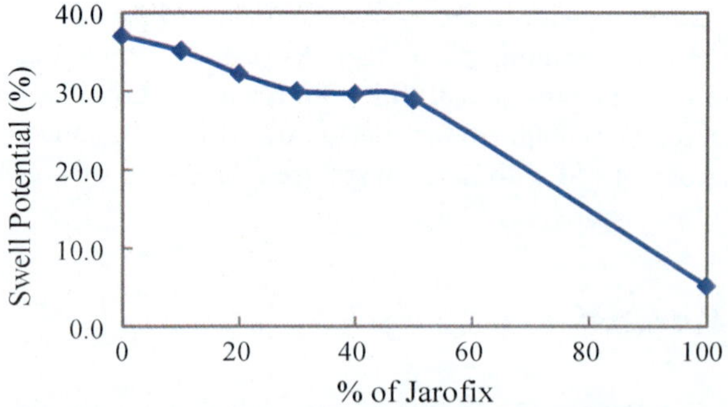

Fig. 3 Variation of swell potential of bentonite with different percentages of Jarofix

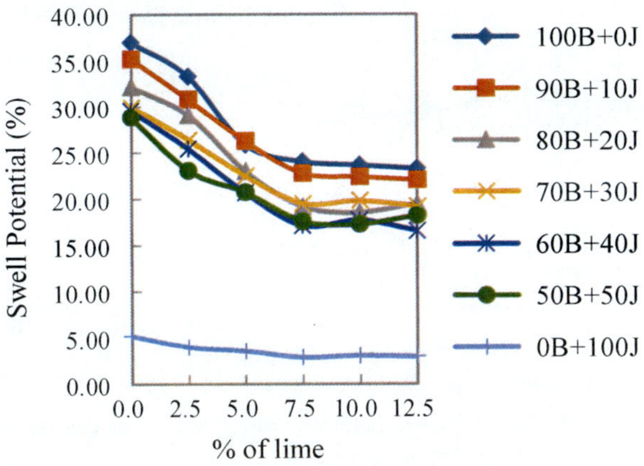

Fig. 4 Variation of swell potential with different percentages of lime

percentage decrease in swell potential is 37. For the mixture of 90% bentonite and 10% Jarofix (90B + 10 J), the swell potential decreases from 35.2 to 22.1%. The same for the mixture of 80% bentonite and 20% Jarofix (80B + 20 J) changes from 32.2 to 19.5%, respectively. The corresponding percentage decrease in swell potentials is 37.3 and 39.4, respectively. So, the swell potential values decrease with the presence of lime. But it was not considerable after 7.5%. The same trend has obtained for other mixes also.

The decrease in the percentage swell and free swell index due to the addition of lime is attributed to the fact that bentonite cations are substituted for by calcium, leading to the formation of calcium silicate and aluminate hydrates. The formation of cementitious compounds increases the strength and resists swelling of expansive soils and thus decreases the percentage swell and free swell index (Kumar et.al 2014).

3.3 Swell Pressure

When an expansive soil contacts with water, a pressure known as swelling or expansion pressure builds up in the soil and it is exerted on the overlying materials and structures. Swelling pressure can be defined as the maximum pressure that needs to be placed over a swelling soil to prevent volume increase. The swelling pressure is not a unique parameter for a swelling soil. It is significantly influenced by several factors, and the most important of them is the initial moisture content. For a given dry density, lower the initial moisture content, greater the water thirst and higher the swelling pressure upon saturation. In this study, the swell pressure was determined as per IS2720 (Part X41)-1977 using the consolidometer method. Samples of diameter of 60 mm and thickness 20 mm were used and were prepared at max dry density and optimum moisture content. During testing, sample was kept always submerged

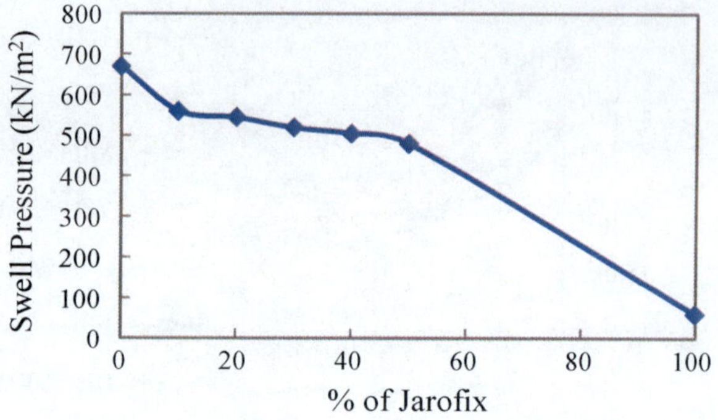

Fig. 5 Variation of swell pressure of bentonite with different percentages of Jarofix

in water. Free swelling of the sample was allowed under the seating load of 5 kPa. After the samples reached maximum swell, they were subjected to incremental loads and the total loads required to bring the sample to zero swell were used to determine the swell pressure of samples. The variation of swell pressure of bentonite with different percentages of Jarofix is as shown in Fig. 5. When the percentage of Jarofix increases, the swell pressure of the soil decreases from 670 to 480 kN/m^2 and the corresponding percentage in decrease is 28.4.

Figure 6 showed that similar to the free swell index and swell potential, the presence of lime decreases the swell pressure also. The swell pressure of bentonite is 670 kN/m^2 without lime, and it decreases to 181 kN/m^2 when the percentage of lime increases to 12.5%. The decrease in percentage is 73. The swell pressure of the mixture of 90% bentonite and 10% Jarofix (Jarofix 90B + 10 J) decreases from 560 to 174 kN/m^2. The same for the mixture of 80% bentonite and 20% decreases (80B + 20 J) from 545 to 136 kN/m^2. The corresponding percentage decreases are

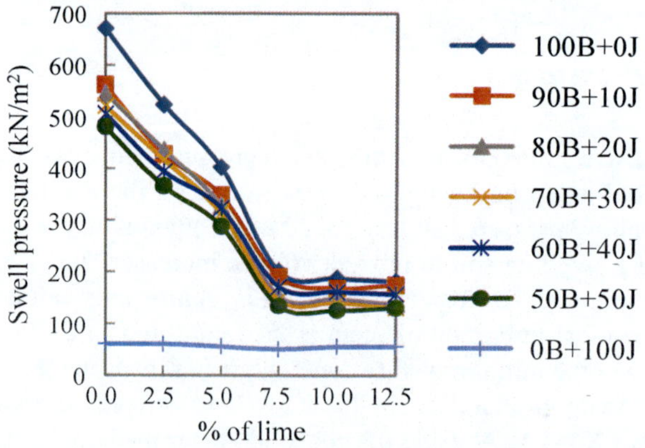

Fig. 6 Variation of swell pressure with different percentages of lime

70 and 75. The swell pressure values decrease with the presence of lime. But it was not considerable after 7.5%. The other mixes also showed the same trend similar to the free swell index and swell potential.

Hence, the study showed that the Jarofix is an effective material for reducing the swelling characteristics of expansive soil. The free swell index, swell pressure and swell potential of the bentonite were found to be decreased in the presence of Jarofix. Later, the addition of lime to all the mixes mentioned earlier leads to further reduction in swelling characteristics, and it was observed that the maximum reduction in swelling was corresponding to 7.5% lime. After that, the reduction in swelling with the addition of lime was not considerable.

Also, while analyzing the effect of lime on the free swell index, swell pressure and swell potential of various bentonite–Jarofix–lime mixtures with different percentages of Jarofix, it can be observed that the behavior of the mixes was same as that of bentonite–Jarofix mixtures without the presence of lime. Or in other words, the swelling characteristics of bentonite were found to be decreased with increase in percentage Jarofix even in the presence of lime.

4 Conclusions

Expansive soils are worldwide problems that pose several challenges for civil engineers. Some partially saturated clayey soils are very sensitive to variations in water content and show excessive volume changes. Such soils which increase in volume because of an increase in their water contents are classified as expansive soils. Also, the utilization of the waste material from various industries in engineering field plays an important role for protecting the environment from pollution. This study demonstrates the effect of the waste material Jarofix, which is the waste product from zinc industries combined with lime for reducing the swelling behavior of an expansive soil. Based on the results from the study, it can be concluded that Jarofix alone and Jarofix along with lime are suitable for reducing the swelling characteristics of expansive soils. The following are the conclusions.

- As the percentage of Jarofix increases the FSI, swell pressure and swell potentials of bentonite clay decrease.
- For each mix as the percentage of lime increases, the FSI, swell pressure and swell potentials of bentonite decrease.
- The optimum percentage of lime was obtained as 7.5%.

More studies are needed to understand the complete behavior of the soil–Jarofix mix along with lime, and the studies are continuing to explore more application of these materials.

References

Akinwumi II, Ojuri OO, Ogbiye AS, Booth CA (2017) Engineering properties of tropical clay and bentonite modified with sawdust. Acta Geotech Slov J, 47–56

Aswathy V, Salini U, Gayathri VG (2016) Utility of lime and red mud in clay soil stabilization. In: Indian geotechnical conference IGC2016 15–17 Dec 2016, IIT Madras, Chennai, India

Bhuvaneshwari S, Sowbi R (2017) Stabilization of expansive soils-evaluation of the behaviour with lime. Int J Civ Eng Technol (IJCIET) 8:1003–1013

Dang LC, Fatahi B, Khabbaz H (2015) Behaviour of expansive soils stabilized with hydrated lime and bagasse fibres. In: The 3rd international conference on transportation geotechnics (ICTG 2016), vol 143, pp 658–665

Dang LC, Khabbaz H, Fatahi B (2017) An experimental study on engineering behaviour of lime and bagasse fibre reinforced expansive soils. 19th international conference on soil mechanics and geotechnical engineering, pp 2497–2500

Dash SK, Hussain,M. :Lime Stabilization of Soils: Reappraisal. Journal of materials in civil engineering, ASCE :707–714(2012).

Geeta Rani T, Tulasi K, Sai Rama Krishna Y (2017) Ground granulated blast furnace slag as an expansive soil stabilizer. VFSTR J STEM 3:2455–2062

IS 2720 (Part 40)—1977

IS2720 (Part X41)—1977

Jagtap AS, Deshmukh AM, Waghmare KD, Jadhav AB, Humbe SD (2015) Improving engineering properties of soil by using sugarcane bagasse ash and lime. Int J Sci Res Dev 3:633–636

Kate JM (2009) Behavior of expansive clays treated with lime fly ash admixtures. In: IGC 2009, Guntur, India, pp 88–91

Kumar S, Dutta RK, Mohanty B (2014) Engineering properties of bentonite stabilized with lime and phosphogypsum. Slov J Civ Eng 22:35–44

Muntohar AS, Widianti A, Hartono E, Diana W (2013) Engineering properties of silty soil stabilized with lime and rice husk ash and reinforced with waste plastic fiber. J Mater Civ Eng ASCE, 1260–1270

Nelson JD, Miller DJ (1992) Expansive soils, problems and practice in foundation and pavement engineering. Wiley, USA

Nitisha R, Patil MP, Devendra D (2004) Utilization of jarosite generated from lead-zinc smelter for various applications: a review. Int J Civ Eng Technol IJCIET 5:192–200

Phanikumar BR, Sharma RS (2007) Volume change behavior of fly ash-stabilized clays. J Mater Civ Eng 19:67–74

Seed HB, Woodward RJ, Lundgren R (1962) Prediction of swelling potential for compacted clays. J Soil Mech Found Div 88(3):53–87

Sinha AK, Havanagi VG, Arora VK, Mathur S (2012) Design, construction & evaluation of Jarofix embankment and sub grade layers. Int J Environ Eng Res 1:97–103

Sinha AK, Havanagi VG, Arora VK (2015) Stress-strain behaviour of stabilised Jarofix waste material. In: 50th Indian geotechnical conference, Pune, Maharashtra, India

Varghese R, Chandrakaran S, Rangaswamy K (2016) Effect of organic content on geotechnical properties of bentonite clay. In: Indian geotechnical conference IGC2016, IIT Madras, Chennai, India

Verma SK, Maru S (2013) Behaviourial study of expansive soils and its effect on structures—a review. Int J Innov Eng Technol 2:228–238

Influence of Fly Ash Mixed with Bentonite and with Lime on Plasticity and Compaction Characteristics Including XRD and SEM Analysis

Nabanita Datta and Sujit Kumar Pal

Abstract This paper mainly deals with the fly ash modified with 20–50% of bentonite and 5–10% of quicklime (CaO = 84.94%) by dry weight basis. The specific gravity of fly ash, bentonite, fly ash–bentonite mix and fly ash–lime mix is 2.21, 2.80, 2.30–2.51 and 2.23–2.21, respectively, and liquid limit, plastic limit and plasticity index found for fly ash–bentonite mixin between 85.02–98.02, 17.55–25.23 and 67. 47–72.79%. OMC and MDD of fly ash and bentonite are 26.20% and13.00 kN/m^3 and 33.60% and 13.72 kN/m^3, respectively; fly ash–bentonite mix and fly ash–lime mix are in between 23.40–21.20% and 13.76–14.68 kN/m^3 and 22.30–23.40% and 13.33–13.09 kN/m^3, respectively, which shows increasing trend in MDD in case of fly ash–bentonite mix and decreasing trend for fly ash–lime mix. To confirm the mineralogical and microstructural changes of particles together with the spectrum of all elements and to validate the results, energy dispersive spectroscopy (EDS) is carried out where different peaks of different components observed with different shapes and sizes of structures are also identified from X-ray diffraction and scanning electron microscope (SEM). The results are reliable for field applications like landfilling purposes.

Keywords Fly ash · Bentonite · Lime · Compaction · XRD · SEM

1 Introduction

Safe disposal of municipal, industrial and hazardous wastes through landfilling is the most effective method. Liners were treated as low permeable barriers also known as an integral part of landfills, which minimize the migration of contaminants to surrounding geo-environment and groundwater (Sharma and Reddy 2004). To improve the engineering as well as environmental properties of Class F fly ash, 'attempts are made to stabilize fly ash with lime' widely used as barriers because of their cost-effectiveness and high contaminant reduction in the absence of natural

N. Datta (✉) · S. K. Pal
NIT Agartala, Jirania, Barjala, Tripura, India
e-mail: nabanitadatta@ymail.com

M. Latha Gali and R. R. P. (eds.), *Problematic Soils and Geoenvironmental Concerns*, Lecture Notes in Civil Engineering 88,
https://doi.org/10.1007/978-981-15-6237-2_31

clays, have been widely used as contaminant barriers. Fly ashes containing higher amounts of free lime which causes self-hardening characteristics find extensive bulk and potential applications in civil engineering and construction field largely because of their ability to develop considerable strength due to pozzolanic reactions with the available reactive silica only upon addition of certain cementing agents such as cement and lime (Gray and Lin 1972). According to Ingles (1987), a good rule of thumb in practice is to allow 1% by weight of lime for each 10% of clay in the soil.

Class C fly ash with high calcium content undergoes high reactivity with water even without addition of lime (Parsa et al. 1996). To improve the engineering properties of Class F fly ash which contains lower percentages of lime, attempt had been made to stabilize fly ash with lime or cement (Ghosh 1996; Ghosh and Subbarao 2007). As liners were used as barrier systems in landfill design to minimize the escape of contaminants from landfills and mitigate their impact on public health and the environment, expansive soil like high swelling bentonite with 2:1 expandable layers, high layer charge, high base exchange capacity, very thin flakes, high surface area, high absorption capacity, high viscosity and thixotropic had been used as barrier clays (Bell and Coultherd 1990); after modifying the properties of bentonite soils with some additives such as lime which in fact, when expandable clays tend to react readily with lime, as reported by Keith and Murray (1994) to make them suitable for landfilling material at site by waste materials in which the functional integrity of landfills depends heavily on the densification of the barrier systems, where bentonite provide the better density while packed together by rearranging of particles and fill up the voids after proper compaction treatment by stabilizers in the presence of optimum water content. The high swelling sodium bentonite was commonly used in earthen structures such as dams, to seal irrigation ditches, to prevent seepage of water from ponds and impounds and to prevent water from entering basements of homes by the mechanism of swelling for water impedance and to fill the pores and voids in the material into which it was incorporated preventing water or other liquids from moving through the barrier (Sherwood 1993).

When lime is added to a clay soil, it had an immediate effect on the properties of the soil as cation exchange begins to take place between the metallic ions associated with the surfaces of the clay particles, and the calcium ions of the lime alter the density of the electrical charge around the clay particles which leads to them being attracted closer to each other to form flocs; the process is being termed flocculation which is primarily responsible for the modification of the engineering properties of clay soils when they were treated with lime (Hilt and Davidson 1960) as well as a lime fixation point which represents the optimum addition of lime needed for maximum modification of the soil and corresponds with the point where further addition of lime did not bring about further changes in the plastic limit when lime was added to a clay soil (Kumar and Sharma 2004). Pal and Ghosh (2013) observed that fly ash addition to expansive soil reduced the optimum moisture content (OMC) and increased the maximum dry density (MDD) of the mixes. Conversely, Brooks et al. (2011) noted that ion-exchange reactions occurring on mixing fly ash with clay soils produced flocculated and agglomerated soil fabric that lowered the MDD and increased the OMC of the mixes. Sivapullaiah and Moghal (2011) obtained the improved values of MDD and OMC after compaction test using two different types of fly ashes with various lime percentages up to 10% which revealed the addition of lime imparts plasticity, otherwise non-plastic fly ashes resulting in marginal increase in dry density and moisture content values. Sivapullaiah and Jha (2014) reported after compaction test of soil–fly ash–lime mixes where the MDD was increased after the addition of fly ash to soil and OMC decreases which might have been attributed to improvement in gradation of soil with addition of fly ash which was predominated with silt-sized particles and MDD decreased with an increased in OMC after addition of varying lime content to soil–fly ash mixtures due to the enhanced flocculation and cementation of clay particles counteracting the effect of compactive effort. Similar observations had been found out by several researchers (e.g., Bell 1996; George et al. 1992; Kumar et al. 2007). McCarthy et al. (2012) reported similar kind of result for sulphate soil stabilized with lime and fly ash.

In the present study, attempts have been made to address some of the issues:

a. In this study, the stabilizing materials such as lime and bentonite (different percentages of up to 50%) are used as an effective liner material to prevent contamination of tracer material by providing the maximum dry density (MDD) and optimum moisture content (OMC) at standard compaction test in laboratory which can be applicable in field and also to protect the environment and human health.
b. The percentages of lime used in the present study are higher in quantity to get the best effect in practical field by compaction test to densify the filled material as well as liner which is the key requirement of providing liner at the landfill site.
c. Compaction of these materials was properly made to get the idea of OMC and MDD of mixes which is directly correlated with permeability of liner used in the landfill.
d. Proper compaction and different stabilized materials are the fundamental need of environmental protection and economical too by means of waste and industrial filled materials in the construction site, earth embankment as well as in dumpsite.

2 Materials, Sample Preparation and Programme

2.1 Materials

To recycle this waste, materials like fly ash, different additives, viz., bentonite and locally available lime are mixed in various mix proportions to modify waste containments considering the economic aspect. In this entire paper, all mixes are designated with common coding process as mentioned: FA stands for fly ash; B stands for bentonite; L stands for lime. The Arabic numerals before FA, B and L indicate their respective weight percentages in the mix. The test results on physical properties in Table 1 as per ASTM guidelines and chemical compositions in Table 2 of the fly ash

Table 1 Physical properties of materials

Properties	Fly ash	Bentonite
Specific gravity	2.21	2.80
Sand size (4.75–0.075 mm)	6.00	1.00
Silt size (0.075–0.002 mm)	84.00	26.80
Clay size (<0.002 mm)	10.00	72.20
Liquid limit (%)	NP	226.30
Plastic limit (%)	NP	32.00
Plasticity index (%)	–	193.70
Optimum moisture content (%)	26.20	33.60
Maximum dry density (kN/m^3)	13.00	13.72

Table 2 Chemical compositions of materials

Chemical composition	Fly ash (%)	Bentonite (%)	Lime (%)
Loss on ignition	1.14	9.52	7.85
SiO_2	60.36	51.43	2.77
Fe_2O_3	6.04	11.68	0.19
Al_2O_3	28.21	19.16	1.66
CaO	1.22	0.89	84.94
MgO	0.87	2.70	1.75
SO_3	0.65	Nil	0.11
Na_2O	Few traces	3.49	Nil
K_2O	Few traces	0.19	Nil
pH	6.99	7.75	Nil

Table 3 Properties of water

Descriptions	pH	Conductivity (mS/cm)	Total dissolved solids (g/L)	Dissolved oxygen (mg/L)
Water	6.67	0.061	0.040	5.67

sample, bentonite and lime are presented. Properties of tap water used in this study are shown in Table 3.

2.1.1 Fly Ash

The fly ash sample procured from Kolaghat Thermal Power Station West Bengal, India. According to (Book and of ASTM Standards 2012), this fly ash may be categorized as Class F type.

2.1.2 Bentonite

Commercially available bentonite powder is collected from the local market. According to IS: 6186 (1971), the soil is classified as highly plastic clay with a high degree of expansion.

2.1.3 Lime

According to (Annual Book of ASTM Standards 2011), analytical grade quicklime is used throughout the test programme for stabilizing the fly ash with bentonite and collected from the local market.

2.2 *Sample Preparation*

Six (6) different types of samples are prepared by various mix proportions where fly ash is first mixed with varying percentages of bentonite (i.e. 20, 30, 40 and 50%) on a dry weight basis; after that, fly ash is stabilized with 5 and 10% lime which increased the durability with the formation of ettringite. In this way, to get the paramount effects in the field, six mix designations are provided, and a total of eight numbers of compaction tests are conducted including fly ash and bentonite.

2.3 *Programme*

2.3.1 Experimental Programme

Atterberg's limits test of fly ash and bentonite including fly ash with lime of three different percentages are carried out according to respective ASTM guidelines. Particle size analysis of fly ash and bentonite as well as the specific gravity test of fly ash and bentonite including all combinations is carried out as per ASTM guidelines. The MDD and OMC values corresponding to key materials and each combination are determined by carrying out standard Proctor compaction test as per ASTM (D 698) guidelines.

2.3.2 Analytical Analysis

The analytical analysis (i.e. XRD, SEM) is carried out on selective samples of mix with raw materials collected from fractured pieces of oven-dried sample taken after knowing the MDD and OMC of raw materials like fly ash, bentonite, lime and with few mix proportions of materials like (80FA + 20B), (70FA + 30B), (60FA

+ 40B), (50FA + 50B); (80FA + 20B), (70FA + 30B), (60FA + 40B) and (50FA + 50B), (95FA + 5L), (90FA + 10L) for SEM analysis using standard Proctor compaction tests. SEM and XRD techniques are employed to study the change in surface morphology and the chemical composition of the samples. XRD is accomplished with Bruker D8 Advance model diffractometer to identify changes in mineralogy and crystalline phases. XRD is carried out with Cu-K_α radiation ($\lambda = 1.5406$ Å) at 1.2°/minute speed for Bragg's angle 2θ ranges 5° to 80°. The diffractometer is operated at an applied voltage of 40 kV and the applied current of 40 mA, where Joint Commity on Powder Diffraction Standards (JCPDS) software issued to analyse data obtained from diffractometer. The procedure mode of XRD is browsing incidence with a step size of 0.02° per second within $2\theta = 5°–80°$ scan range. Microstructure and images of samples are performed with SEM technique, respectively. The constitutive elemental analysis is carried out by attaching an energy dispersive spectrometer to the scanning electron microscope. This focuses an electron beam on a point on the sample. X-ray photographs generated by the sample are converted into electrical impulses, and the chemical elements presented are determined. The JEOL JSM-7600F FEG-SEM with resolution 1.0–1.5 nM at 15–1.0 kV and voltage of 0.1–30 kV is used for the SEM studies. Before the microstructural examination, the fractured surfaces of the samples for SEM study are gold coated, and the sample is coated with 10 mA thin layer gold–palladium for 300 s using a sputter coater, and the software used for image analysis is PC-SEM.

3 Results and Discussion

3.1 Experimental Results

The specific gravity values of fly ash, bentonite and their four mixes including three numbers of fly ash–lime mixes are within the ranges between 2.21–2.80, and details of all the values of specific gravity are shown in Table 4. The LL, PL and PI values of bentonite and all the mixes of three materials are within the ranges between 85.02–226.30%, 17.55–32.00% and 67.47–193.70% with details of all values of plasticity index shown in also Table 4.

OMC and MDD values at standard Proctor compaction tests are obtained from the mixes of fly ash stabilized with bentonite percentages of 20, 30, 40 and 50% on the dry weight of fly ash; the range of MDD values is 13.76–14.68 kN/m^3, and OMC shows in between 23.4–21.6%. OMC and MDD values of the mixes of fly ash stabilized with lime percentages of 5 and 10 are in the ranges between 22.3–23.4% and 13.33–13.09 kN/m^3, respectively. All the values are shown in the plots.

Table 4 Values of specific gravity and plasticity index

Mix proportions	Specific gravity	Liquid limit	Plastic limit	Plasticity index
Fly ash	2.21	NP	NP	NP
Bentonite	2.80	226.30	32.00	193.70
80FA + 20B	2.30	85.02	17.55	67.47
70FA + 30B	2.42	90.83	20.88	69.95
60FA + 40B	2.51	93.8	23.12	70.68
50FA + 50B	2.49	98.02	25.23	72.79
95FA + 5L	2.23	–	–	–
90FA + 10L	2.21	–	–	–

NP Non-plastic, *FA* Fly ash, *B* Bentonite, *L* Lime

3.2 Discussions on Experimental Results

3.2.1 Effect of Specific Gravity on Fly Ash Stabilized with Bentonite and Lime

The effects of specific gravity which is denoted by G on fly ash stabilized with bentonite from 20 to 50% are discussed based on the test results as shown in Table 4. From the table, it is noticed that obtained results of G for up to 50% bentonite with fly ash range from 2.30–2.51 which reveals that values of G increase with the increase of bentonite percentages. With the increment of bentonite percentages with fly ashes, the specific gravity gradually increases which may be due to when same mixes are compacted, the inter-connectivity of the pore channels decreases when water is added for specific gravity test, which causes the reduction in void spaces with the increase in content of particles.

From Table4, results of G observed are within the ranges of 2.23–2.21 on fly ash stabilized with lime of 5 to 10%. With the addition of lime content up to 10%, specific gravity gradually decreases with fly ash; due to the higher amount of lime percentages, calcium ions of lime reduce plasticity, and hence, different chemical reactions occur and cation exchange increases with flocculation occurred. Due to production of flocs between the particles in less time duration, agglomeration takes place which results the reduction of specific gravity, i.e. 2.23–2.21.

3.2.2 Effect of Plasticity Index on Fly Ash Stabilized with Bentonite and Lime

The values of liquid limit, plastic limit and plasticity index of fly ash-bentonite mixes including raw material fly ash are shown in Table 4, and the ranges between 85.02–226.30%, 17.55–32.00% and 67.47–193.70%, respectively, at which addition of bentonite to the mixes are more prominent where this implies that with the large amount of lime, i.e. 10% of lime addition, increased the pH value of the soil–water system where high pH creates strong alkaline condition around the silica present in the soil which undergoes accelerated pozzolanic reactions yielding cementitious gels of calcium silicate hydrate (C–S–H) is being highly porous and holds large quantity of water into it. The water attached to the gel (called gel water) can be as high as 28% of the volume of the gel, where this gel water along with the physicochemical adsorbed water on the clay surface had caused the increase in the LL of the samples. As the fly ash contains a relatively large amount of silica, an increase in its quantity in the soil mix has produced an increased quantity of silica gel, leading to increased LL. It is observed that the PL of all the mixes has increased due to bentonite of 20–50% for treatment. PL is the lowest water content at which the cohesion between the soil particles should be low enough that the particles can have relative movements, but same cohesion is high enough to maintain the soil mass in the remoulded position. The PI of mixes indicates the range of water content within the mixes which remains plastic. It determines the workability of soil (Sivapullaiah et al. 1996). Since it is the difference between LL and PL, the factors influencing LL and PL of the samples also influence PI. The PI of all the additive mixes with fly ash is found to increase almost for all except few cases. The PI increases with bentonite, where the effect of bentonite addition is more prominent and significant for the process. With the percentages of bentonite increase in the mixes, more quantity gel products form. Therefore, the increase in PI is more prominent for higher percentages of bentonite content.

3.2.3 Effect of Bentonite on MDD and OMC of Fly Ash–bentonite Mixes

In Fig. 1, 20–50% bentonite clay is mixed with fly ash on a dry weight basis to stabilize

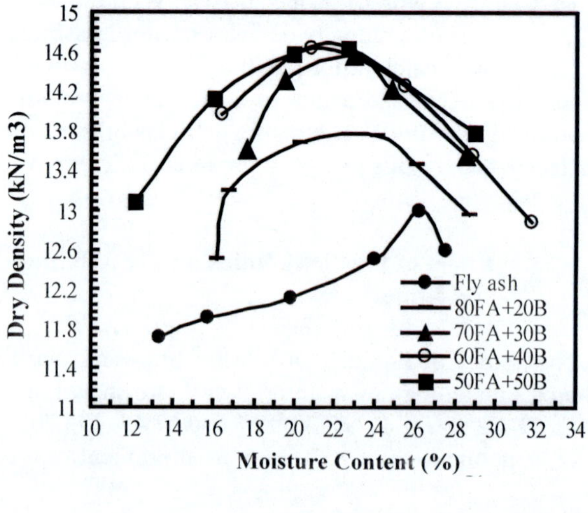

Fig. 1 Compaction results of fly ash stabilized with bentonite

fly ash, where bentonite clay plays an important role in controlling the compaction results, which shows the variation in OMC and MDD. Results show that OMC lies between 23.4–21.6% and MDD 13.76–14.68 kN/m^3 compacted at standard Proctor compaction energy. It implies that after addition of bentonite from 20 up to 50% with fly ash, the increasing value of MDD and the decreasing value of OMC are found with the increase of bentonite in the mixes. A similar trend was observed by Kumar and Sharma (2004). It reveals that ion-exchange reactions occurring on mixing fly ash with clay soils produced flocculated and agglomerated state of soil that lowered the MDD and improved the OMC of the mixes which were also found by Brooks et al. (2011). This may be attributed to improvement in a gradation of fly ash with the addition of bentonite which is predominated with silt-sized particles of fly ash. It also may be noticed that preliminary pH of bentonite soil with the addition of fly ash increases marginally. This may be due to the presence of little amount of free lime present in fly ash. Further the enhancement in MDD may be due to the filling of voids by the finer particles of bentonite and the reduce in OMC with the increase in moisture content, the friction between the particles of fly ash–bentonite decreases; in that way, particles come closer, and also, at the same time with increase in bentonite clay content, the air voids between the particles of fly ash of mixes are reduced, and as a result of which air content decreases, and the degree of saturation increases which was also reported by Pal and Ghosh (2013). OMC reduced, and the irregular shape of fly ash particles and the presence of holes and crevices in the clinker like particles resulted in air being entrapped which helps the particles to come closer.

3.2.4 Effect of Lime on MDD and OMC of Fly Ash–Lime Mixes

In Fig. 2, fly ash is stabilized with quicklime (CaO = 84.94%) of 5–10% to accelerate

Fig. 2 Compaction results of fly ash stabilized with lime

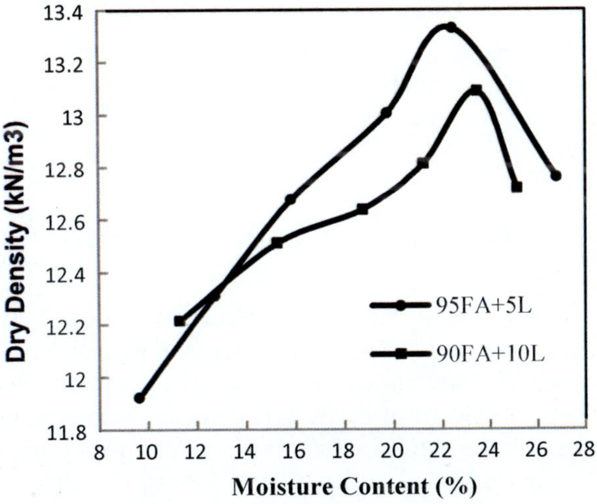

the fixation process of compaction matrix from the locally available market. Fly ash is a pozzolanic material, which is defined as siliceous, or siliceous and aluminous, and therefore, its engineering behaviour can be improved by the addition of cement or lime. Here, the plot depicted that after the compaction of fly ash stabilized with lime, the MDD of mixes range between 13.33–13.09 kN/m^3 and OMC 22.30–23.40% which attributed that MDD decreases with increase in OMC up to some extent, i.e. from 5 to 10% lime addition. It may be revealed that the fall in density is more significant at 10% lime addition than at other percentages of lime in between 5 to 10% of lime. In specific ranges, i.e. 10% lime reacts quickly with the fly ash and due to base exchange aggregation and flocculation, changes were found resulting in an increased void ratio of the mix, leading to a decrease in the density of the mix. Addition of lime beyond this value (lime fixation point, i.e. 4% lime) is mainly utilized for pozzolanic reactions. The increment in optimum moisture content is probably on account of additional water held within the flocs resulting from flocculation due to the lime reaction. Similar observations were noticed by Bell (1996), George et al. (1992) and Kumar et al. (2007) for different percentages of admixtures. The presence of higher lime content having 10% may be the cause for this increased dry density due to more reduction in void spaces occupied by lime contents after compaction, reacting with previously untreated lime content chemically, i.e. 5% lime content, but without any additional water absorbance. The increment in optimum moisture content can be attributed to the increasing amount of fines which require more water content because of their larger surface area, but the fines already inhabited the water into the pore spaces during the addition of 5% lime which does not require further any additional water; the same trend was obtained by Bell (1996), George et al. (1992) and Kumar et al. (2007).

3.3 Analytical Results

After getting the experimental values, analytical analysis is made to validate the results and to find the internal changes of particle and chemical compound produced after the experiment.

3.4 Discussion on Analytical Analysis

The fractured surfaces of compacted oven-dried samples are examined for the analytical investigation, and it is found from Figs. 3 and 4, 5, 6, 7, 8, 9, 10, 11, 12 and 13, X-ray diffraction and scanning electron microscope, the mineralogical and microstructural development of fly ash particles ensure mainly aluminium oxide (Al_2O_3), i.e. mullite, silicon dioxide (SiO_2), i.e. quartz, and aluminium silicate (Al_2SiO_5) compound from all the major peaks as fly ash is a by-product of powdered coal

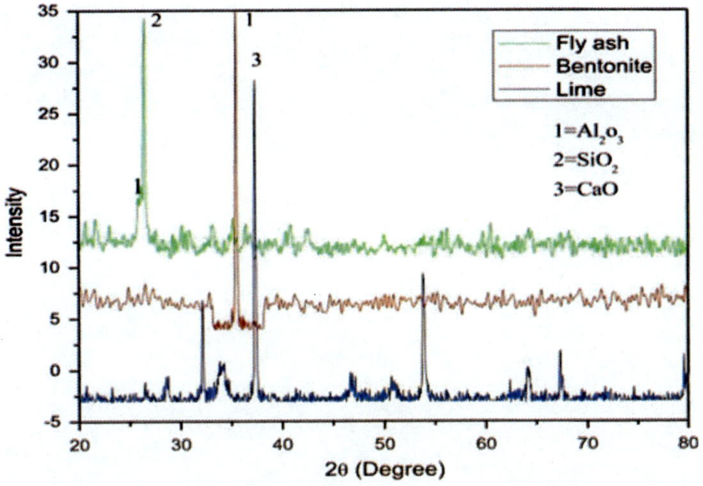

Fig. 3 XRD analysis of fly ash, bentonite and lime

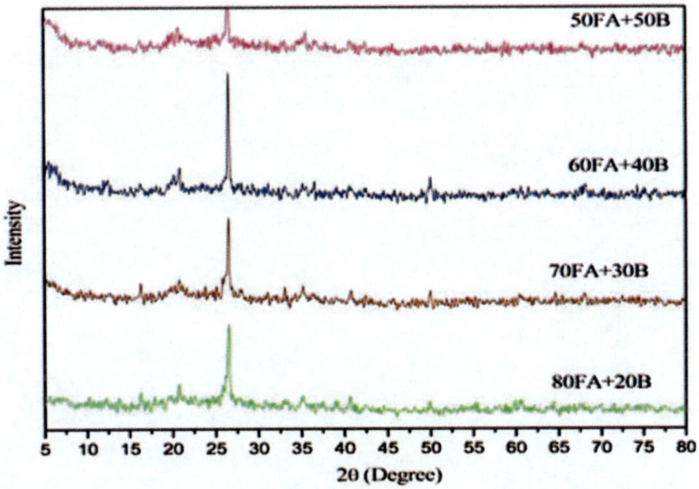

Fig.4 XRD analysis of fly ash mixed with bentonite

combustion and also exhibits pozzolanic activity from the large presence of alumina–silica glass; the particles are maximum round and oval shaped which occupy the coarser parts of fly ash which is found from SEM analysis. Bentonite confirms the only mullite with spherical-shaped structures. The X-ray diffraction (XRD) analysis of dry powder lime confirmed that all major peaks indicate the presence of hydrated calcium hydroxide [$Ca(OH)_2$] with the formation of calcium oxide (CaO) and calcium sulphate ($CaSO_4$) present in lime with a different flower-like structure in bentonite. From the mixes, i.e. fly ash stabilized with bentonite and fly ash stabilized

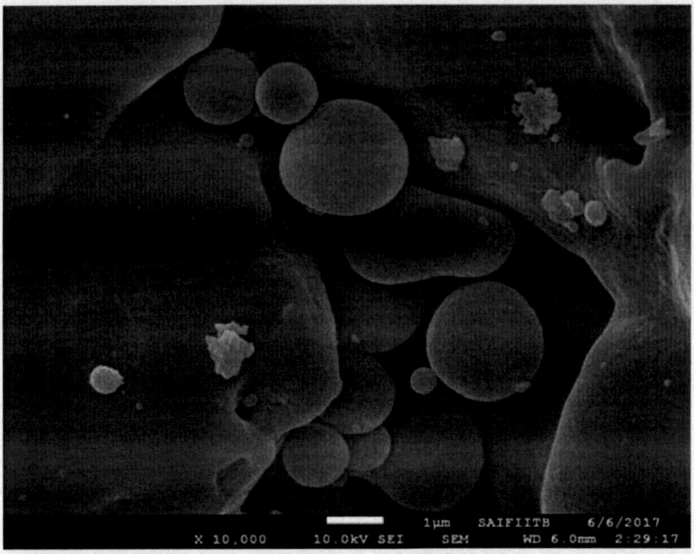

Fig. 5 SEM analysis of fly ash using spectrum

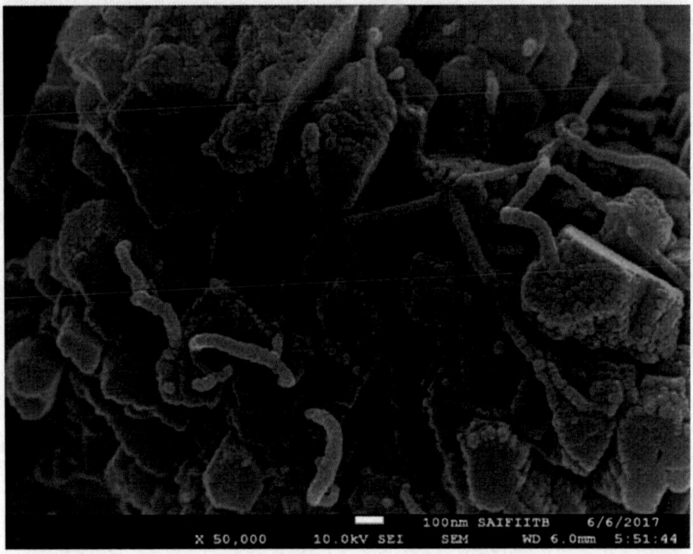

Fig. 6 SEM analysis of bentonite using spectrum

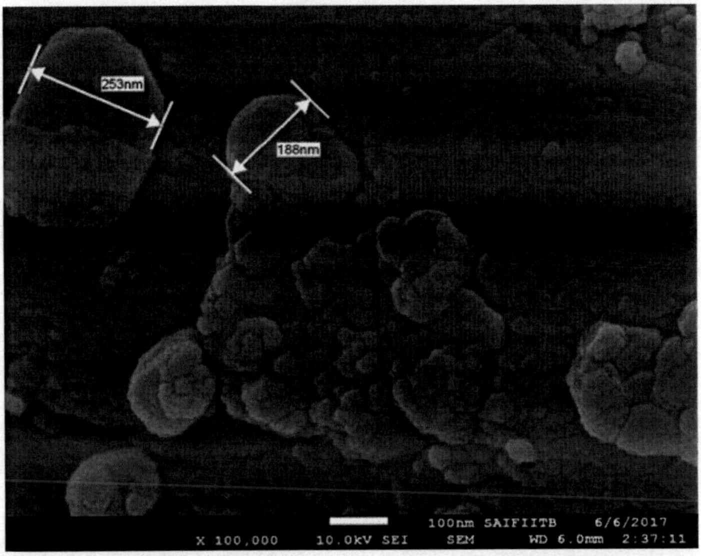

Fig. 7 SEM analysis of lime using spectrum

Fig. 8 SEM analysis of 80FA + 20B using spectrum

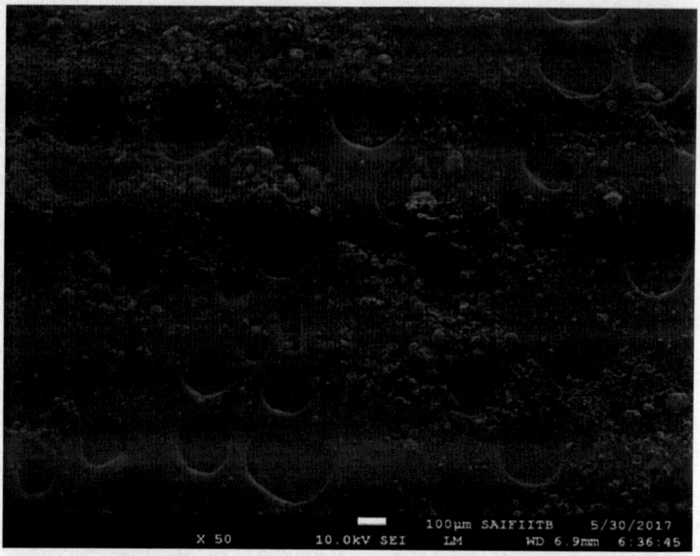

Fig. 9 SEM analysis of 70FA + 30B using spectrum

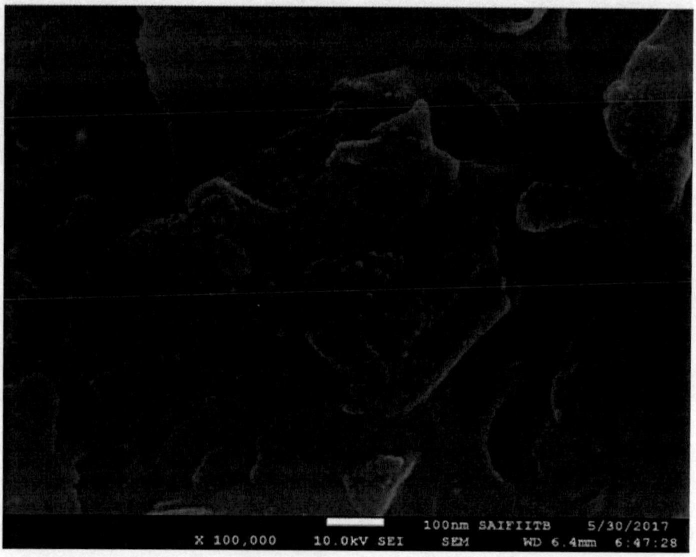

Fig. 10 SEM analysis of 60FA + 40B using spectrum

Fig. 11 SEM analysis of 50FA + 50B using spectrum

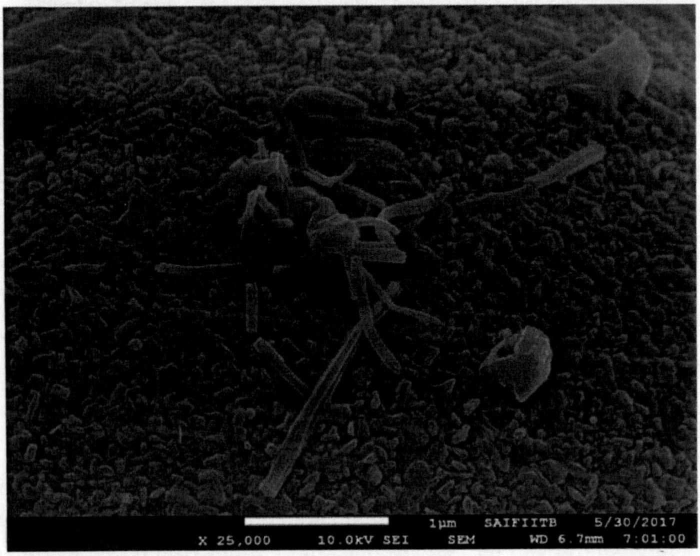

Fig. 12 SEM analysis of 95FA + 5L using spectrum

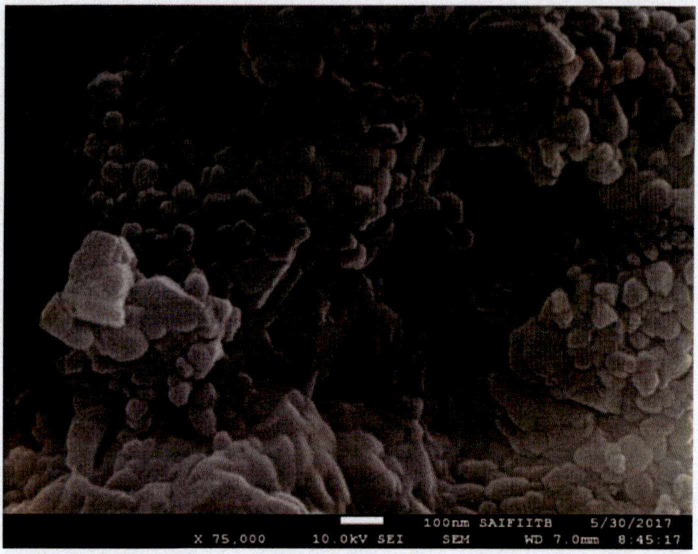

Fig. 13 SEM analysis of 90FA + 10L using spectrum

with lime describes that clay minerals are natural pozzolanas that react with lime to produce cementitious calcium silicate hydrate (C–S–H) and calcium aluminate hydrate (C–A–H), and similar observation is highlited by Greaves (1996). The reaction of fly ash with lime produces calcium silicate hydrate similar to that encountered in soil–lime reactions (Fraay et al. 1989). The compounds of XRD analysis of fly ash, bentonite and lime are mainly showing mullite, quartz, $Ca(OH)_2$ and $CaSO_4$.

4 Conclusion

An experimental study is carried out to investigate the potential of fly ash after stabilized with bentonite and modified with lime. The compaction and plasticity characteristics of fly ash stabilized with bentonite (20–50%) and lime (5–10%) including raw materials are studied through laboratory experiments. Compaction tests are conducted by standard Proctor compaction energy. The mineralogical and microstructural analysis of these materials after compaction is made as an analytical analysis by XRD and scanning electron microscope (SEM) for the detailed study of mineralogy and microstructure. Based on the experimental as well as analytical results, the following conclusion may be outlined:

- The specific gravity of raw materials like fly ash, bentonite and lime with their mixes such as fly ash–bentonite (20–50%) and fly ash–lime (5–10%) plays an important role correlated with MDD of mixes which may be considered for landfill.

- The obtained dry densities with optimum moisture contents are not abrupt; this nature is functional for field application and in a construction site.
- In general, MDD shows increasing trend of all the fly ash–bentonite mixes and decreasing trend for the fly ash–lime mixes which is correlated with flowing properties of soil media to provide a liner in the landfilling site to prevent the contamination of leachate materials from the ground surfaces; OMC does not follow any specific trend with respect to MDD as well as the mixes. The variation of dry density with moisture content of fly ash stabilized with bentonite of 20–50% and lime of 5–10% by dry weight shows more efficient results compared to that of unstabilized fly ash of this study.
- The mineralogical and microstructural analysis of raw materials with few mixes of fly ash–bentonite and fly ash–lime shows the structural position and changes of minerals and chemical components produced with different shapes and sizes at different peaks.
- Different percentages of stabilizers such as bentonite (20–50%) and lime (5–10%) selected for stabilization are on the basis of compaction effect, economy, time duration and environmental impact.
- The addition of lime only affects the whole process of forming alkalinity minimizing migration.

References

Annual Book of ASTM Standards (2011) ASTM C1529. Standard specification for quicklime, hydrated lime, and limestone for environmental uses

Annual Book of ASTM Standards (2012) ASTM C618 Standard specification for coal fly ash and raw or calcite natural pozzolan for use in concrete

Bell FG (1996) Lime stabilization of clay minerals and soils. Eng Geol 42(4):223–237

Bell F, Coultherd J (1990) Stabilization of clay soils with lime. Municipal Engineer 7:125–140

Brooks R, Udeyo FF, Takkalapelli KV (2011) Geotechnical properties of problem soils stabilized with fly ash and limestone dust in Philadelphia. J Mater Civ Eng 23(5):711–716

Fraay ALA, Bijen MJ, de Haan YM (1989) The reaction of fly ash in concrete a critical examination. Cem Concr Res 19:235–246

George SZ, Ponniah DA, Little JA (1992) Effect of temperature on lime-soil stabilization. Construct Build Mater 6(4):247–252

Ghosh A (1996) Environmental and engineering characteristics of stabilized low lime-fly ash. Ph.D. Dissertation, Indian Institute of Technology, Kharagpur, India

Ghosh A, Subbarao C (2007) Strength characteristics of class F fly ash modified with lime and gypsum. J Geotech Geoenviron Eng 133(7):757–766

Gray DH, Lin YK (1972) Engineering properties of compacted fly ash. Soil Mech Found Eng (Engl Transl) 984:361–380

Greaves H (1996) An introduction to lime stabilization. In: Proceedings of seminar on lime stabilization. Thomas Telford, London, pp 5–12

Hilt GH, Davidson DT (196) Lime fixation of clayey soils Highway Research Board Bulletin 262, Washington, D.C., pp 20–32

Ingles OH (1987) Soil stabilization, Chapter 38. In: Bell FG (ed) Ground engineer's reference book. Butterworths, London, pp 38/1–38/26

IS: 6186 (1971) Specification for bentonite
Keith KS, Murray HH (1994) Clay liners and barriers. In: Carr DD (ed) Industrial minerals and rocks, 6th edn. Society for Mining, Metallurgy and Exploration, Littleton, CO, pp 435–462
Kumar P, Sharma RS (2004) Effect of fly ash on engineering properties of expansive soils. J Geotech Geoenviron Eng 130(7):764–776
Kumar A, Walia BS, Bajaj A (2007) Influence of fly ash, lime, and polyester fibers on compaction and strength properties of expansive soil. J Mater Civ Eng 19(3):242–248
McCarthy MJ, Csetenyi LJ, Sachdeva A, Dhir RK (2012) Identifying the role of fly ash properties for minimizing sulfate-heave in lime-stabilized soils. Fuel 92(1):27–36
Pal SK, Ghosh A (2013) Volume change behavior of fly ash-montmorillonite clay mixtures. Int J Geomech 14(1):59–68
Parsa J, Munson-McGee SH, Steiner R (1996) Stabilization/solidification of hazardous wastes using fly ash. J Environ Eng ASCE 122(10):935–940
Sharma DH, Reddy RK (2004) Geoenvironmental engineering: site remediation, waste containment and emerging waste management technologies. Wiley, New York
Sherwood PT (1993) Soil stabilization with cement and lime. Transport research laboratory state of the art review. HMSO, London
Sivapullaiah PV, Jha AK (2014) Gypsum induced strength behaviour of fly ash-lime stabilized expansive soil. Geotech Geol Eng 32(5):1261–1273
Sivapullaiah PV, Moghal AAB (2011) Role of gypsum in the strength development of fly ashes with lime. J Mater CivEng 23(2):197–206
Sivapullaiah PV, Sridharan A, Stalin VK (1996) Swelling behaviour of soil-bentonite mixtures. Can Geotech J 33(5):808–814

Load–Settlement Behavior of Soft Marine Clay Treated with Metakaolin and Calcium Chloride

D. Venkateswarlu, M. Anjan Kumar, G. V. R. Prasada Raju, and R. Dayakar Babu

Abstract A developing country like India has to show its development by leaps and bounds in the fields of infrastructure development, transportation, and communication system. The peninsular India is having a lengthy coastal belt of around 7700 km; except north all three sides of India is surrounded by Bay of Bengal in the east, Indian Ocean in the south, and Arabian Sea on the west. The areas surrounded by the coast are being utilized since ages. So, inevitably several national and international authorities are doing research and development on coastal structures. The various potential problems associated with this marine clay are land slippage slope stability, poor foundation support, and poor drainage. Owing to such soils of poor engineering properties, a great diversity of ground improvement techniques such as soil stabilization and chemical treatment is employed to improve their mechanical behavior, thereby enhancing the reliability of construction. Metakaolin is applied in soil stabilization for foundations or road subgrade. However, metakaolin along with calcium chloride treatment for these poor soils, as an alternative to the traditional "remove and replace" strategies commonly utilized was found to be satisfactory in the laboratory evaluation. Hence, the authors arrived at an optimum content of metakaolin mixed to soft marine clay, and then further, it is chemically treated with calcium chloride. The optimum dosage of calcium chloride in combination with the optimum content of metakaolin was evaluated from the tests conducted in the laboratory. Further, the authors studied the performance of soft marine clay beds prepared with different

---

D. Venkateswarlu (✉)
Godavari Institute of Engineering and Technology (A), Rajahmundry, Andhra Pradesh 533296, India
e-mail: dumpa.venkateswarlu@gmail.com

M. Anjan Kumar
BVC College of Engineering, Palacharla, Rajahmundry 533104, India

G. V. R. Prasada Raju
Civil Engineering, JNTUK Kakinada, Kakinada, India

R. Dayakar Babu
Civil Engineering, KITS Divili, Divili 533433, India

M. Latha Gali and R. R. P. (eds.), *Problematic Soils and Geoenvironmental Concerns*, Lecture Notes in Civil Engineering 88,
https://doi.org/10.1007/978-981-15-6237-2_32

alternative treatments by conducting load tests in the laboratory. The test results indicated that the load-carrying capacity of the metakaolin-treated marine clay bed was improved and further it was increased by chemical treatment with calcium chloride.

Keywords Marine clay · Stabilization · Chemical treatment · Metakaolin · Calcium chloride

1 Introduction

Soft clays are highly compressible and undergo large deformations under little superimposed loads, under undrained and drained conditions. Presence of poor subgrade conditions is the major cause of pavement failures and soft clay subgrade is one such challenging situation. In many civil engineering constructions, soft and weak soils are often treated with ordinary Portland cement (OPC) and lime. The manufacture processes of conventional stabilizers are energy intensive and emit a large quantity of CO_2. It has been found that the beneficial properties of metakaolin (MK) may be used in improving the mechanical properties of soils and can be used as structural material (Kolovos et al. 2013). In this study, a weak marine clay was stabilized with metakaolin (MK) + calcium chloride ($CaCl_2$) with its high strength, low cost, low energy consumption, and less Co_2 emissions during production. The properties of lateritic soil can be improved by adding metakaolin for barrier system for the containment of municipal solid waste (Umar et al. 2015). The strength, stiffness, and ductility of the soil were improved after the stabilization with metakaolin-based geopolymer; also the shrinkage of the soil was reduced (Zhang et al. 2013). Metakaolin is an effective agent for cemented soils, and the addition of MK mainly changes the pore volume distribution and develops more CSH/Aft/CASH bonding and less fissures (Zhang et al. 2014). Metakaolin was beneficial in decreasing the leaching of heavy metals (Cyr et al. 2012). MK efficiently decreases the hydraulic conductivity of cemented soils, and addition of MK leads to less total porosity and throat pore diameter (Deng et al. 2015).

2 Materials Used

2.1 Soft Marine Clay

The soft clay used in this investigation was collected from a depth of 3.5 m below the ground level from Kakinada Port area in East Godavari District of Andhra Pradesh, India. I.S. code test procedures are adopted for the determination of properties of the materials used in the present study. The properties of the soil are specific gravity—2.68, grain size distribution (sand 1%, silt 27%, clay 72%), compaction parameters (maximum dry density—14 kN/m^3), OMC—27%, Atterberg limits (liquid limit—78%, plastic limit—28%, plasticity index—50%, shrinkage limit—12%), IS classification—CH, free swell index—42%, and CBR soaked—1.79%.

2.2 *Metakaolin (MK)*

Metakaolin (MK) is a thermally activated aluminosilicate material, white in color with a dull luster, obtained by calcining kaolin clay within the temperature range 6500–8000 °C. In the present investigation, metakaolin marketed by Jeetmull Jaichandlall Pvt. Ltd. Chennai, Tamilnadu, was used. The physical and chemical characteristics furnished by the manufacturer are moisture content of 0.18%, specific gravity of 2.65, bulk density of 710 kg/m^3, and pH of 7.0. Metakaolin consists majorly of SiO_2, Al_2O_3, and Fe_2O_3 contributing 53.7%, 39.2%, and 3.84% of the total. The next most abundant component is titanium oxide, TiO_2 (5.97%). According to ASTM standard specification (C618-2012), the sum of SiO_2, Al_2O_3, and Fe_2O_3 be $\geq$70% for any material to be used as a pozzolana.

2.3 *Calcium Chloride ($CaCl_2$)*

Commercial grade calcium chloride (consisting of 58% calcium chloride + 42% magnesium chloride) was used in this study. The quantity of calcium chloride was varied from 0 to 2% by dry weight of soil.

3 Laboratory Investigations

CBR tests were conducted on soft marine clay, marine clay stabilized with different percentages of metakaolin, and marine clay stabilized with different percentages in combination of metakaolin + calcium chloride to obtain optimum dosages of stabilizing agents. From the test results, it was observed that mix having 15% of metakaolin in marine clay is giving a soaked CBR value of more than 8% which is acceptable for subgrade material as per IRC:37-2012 specifications. Further, the CBR tests were conducted on marine clay stabilized with metakaolin + calcium chloride by adding different percentages of metakaolin + calcium chloride, and it was observed that 9% of metakaolin + 1% of calcium chloride in marine clay is giving a soaked CBR value of more than 8% which is acceptable for subgrade material as per IRC:37-2012 specifications. Optimum percentages of metakaolin and metakaolin + calcium chloride mixtures corresponding to 8% or more values of soaked CBR are furnished in Table 1.

The optimum dosages of stabilizing alternatives such as metakaolin (15%MK) and metakaolin + calcium chloride (9%MK + 1%$CaCl_2$) were added to the marine

Table 1 Different percentages of stabilizing alternatives used for marine clay

Material	Stabilizing agent	Different percentages of stabilizing alternatives (% by weight)	Soaked CBR	Optimum percentages of stabilizing alternatives (% by weight)
Soft marine clay	Metakaolin	0	1.79	15
		3	2.16	
		6	3.24	
		9	4.68	
		12	6.82	
		15	8.22	
Soft marine clay	Metakaolin + calcium chloride	3%MK + 0.5%$CaCl_2$	1.85	9%MK + 1.0% $CaCl_2$
		3%MK + 1.0%$CaCl_2$	1.92	
		3%MK + 1.5%$CaCl_2$	2.30	
		3%MK + 2.0%$CaCl_2$	2.60	
		6%MK + 0.5%$CaCl_2$	2.86	
		6%MK + 1.0%$CaCl_2$	3.54	
		6%MK + 1.5%$CaCl_2$	4.98	
		6%MK + 2.0%$CaCl_2$	6.45	
		9%MK + 0.5%$CaCl_2$	7.64	
		9%MK + 1.0%$CaCl_2$	8.86	
		9%MK + 1.5%$CaCl_2$	7.84	
		9%MK + 2.0%$CaCl_2$	7.24	

clay and are used as foundation beds material. The load tests were conducted on all the model foundation beds and the settlements are recorded. The details of foundation beds are shown in Table 2.

Table 2 Construction details of model test tanks

Foundation bed no.	1	2	3
Material	Untreated soft marine clay	Metakaolin-treated soft marine clay	Metakaolin + calcium chloride-treated soft marine clay

3.1 Construction of Foundation Beds

Soft marine clay was pulverized and the material passing through 4.75 mm IS sieve was used for this study.

Foundation Bed 1

For this, water was mixed directly with the soil corresponding to optimum moisture content of the natural soil, and it was compacted in layers of 3 cm thickness, at its OMC and MDD, to a total compacted thickness of 30 cm.

Foundation Bed 2

For the soil + metakaolin mix, the water content corresponding to OMC of the natural soil was taken and the required quantity of metakaolin (15%) was added to the soil and compacted in layers of 3 cm thick to a total compacted thickness of 30 cm.

Foundation Bed 3

For the soil + metakaolin + calcium chloride mix, the water content corresponding to the OMC of natural soil was taken, and the required quantity of the calcium chloride was dissolved in it. The solution was added to the metakaolin-treated marine clay to form a soil + metakaolin + calcium chloride mix. This mix was compacted in layers of 3 cm thickness to a total thickness of 30 cm.

3.2 Load Tests

Load tests were carried out on all the foundation beds prepared in a circular mild steel tank of diameter 60 cm. The loading was done through a circular metal plate of 100 mm diameter laid at the top of the soil bed and is shown in Plate 1.

The tank with the soil bed was mounted on the pedestal of a compression testing machine. Two dial gauges of least count 0.01 mm were arranged for obtaining the deformations. A hydraulic jack was placed on the loading plate, and it was connected to the proving ring as shown in Plate 2. Load tests were carried out at OMC and at saturated state. Loading was done in increments of 100 kPa till the ultimate load was reached. Settlements corresponding to each load were recorded, and the results are tabulated in Table 3.

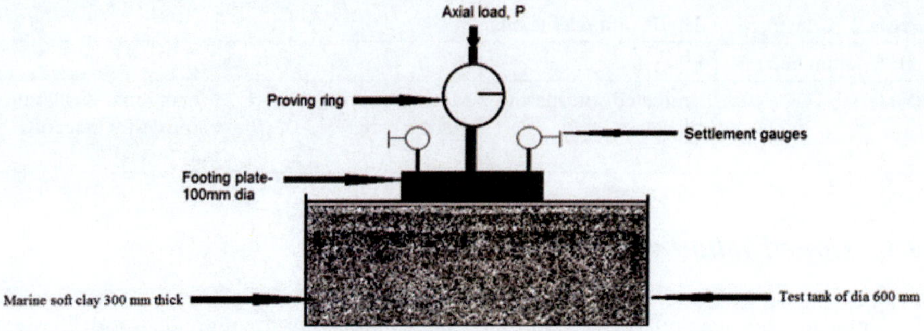

Plate 1 Schematic diagram of experimental setup

Plate 2 Laboratory experimental setup for conduct of load test (The Author and the laboratory technician (white shirt) are conducting the load test. The consent of the laboratory technician was taken and approved in the departmental committee for publishing in the paper along with the author as in the present case, the corresponding author)

4 Results and Discussion

Figures 1 and 2 clearly show that the soaked CBR was increased with the increase in percentage of stabilizing agents. This shows that the improvement in soft marine clay properties improved the performance of marine clay foundation beds.

From Tables 3 and 4, it is evident that the ultimate load-carrying capacity of the untreated soft marine clay foundation bed has increased from 900 to 1600 kPa at OMC and from 300 to 1100 kPa at FSC when soft marine clay was treated with metakaolin and 900–1900 kPa at OMC and 300–1300 kPa at FSC when the soft marine clay

Table 3 Load versus settlement values for different foundation beds at OMC and MDD

Pressure (kPa)	Soft marine clay test bed 1	Soft marine clay + metakaolin test bed 2	Soft marine clay + metakaolin + calcium chloride test bed 3
	Settlement (mm)	Settlement (mm)	Settlement (mm)
0	0	0	0
100	2.23	0.31	0.22
200	4.06	0.64	0.43
300	6.46	1.02	0.64
400	9.24	1.98	0.86
500	12.22	2.74	1.24
600	14.42	3.2	1.96
700	17.68	4.48	2.24
800	20.28	5.76	2.76
900	24.86	6.26	3.24
1000		7.14	3.88
1100		8.84	4.82
1200		10.46	5.44
1300		12.78	6.86
1400		14.24	8.26
1500		16.25	9.24
1600		18.24	10.46
1700			11.92
1800			13.02
1900			13.96

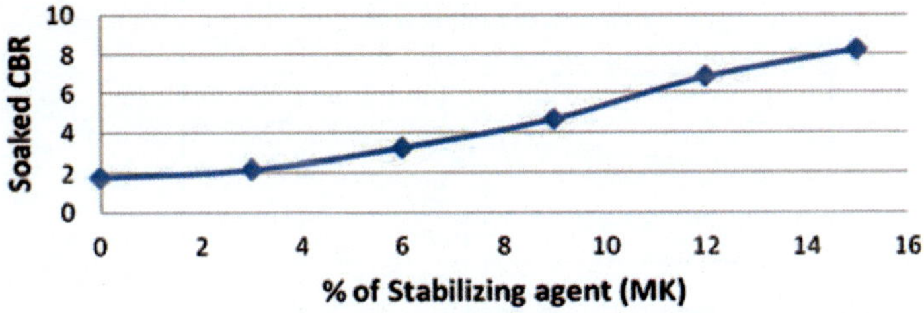

Fig. 1 Variation of soaked CBR values for soft marine clay stabilized with different percentages of metakaolin

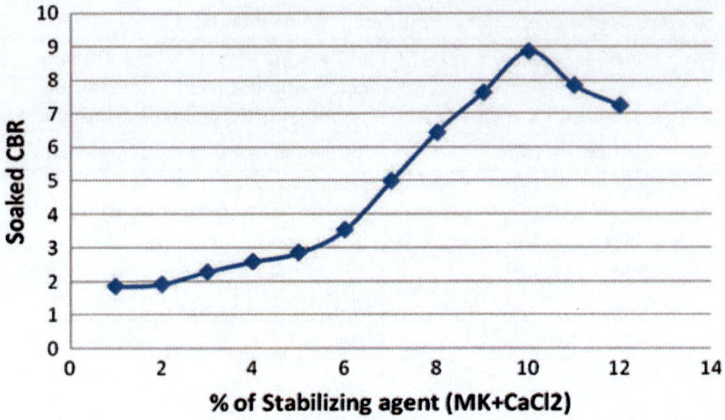

Fig. 2 Variation of soaked CBR values for soft marine clay stabilized with different percentages of metakaolin + calcium chloride

Table 4 Load versus settlement values for different foundation beds at full saturated condition (FSC)

Pressure (kPa)	Soft marine clay test bed 1	Soft marine clay + metakaolin test bed 2	Soft marine clay + metakaolin + calcium chloride test bed 3
	Settlement (mm)	Settlement (mm)	Settlement (mm)
0	0	0	0
100	6.47	0.34	0.26
200	15.84	0.86	0.64
300	29.22	1.44	0.86
400		1.92	1.24
500		2.42	1.82
600		3.68	2.14
700		4.86	2.86
800		6.46	3.64
900		8.22	4.46
1000		10.64	5.44
1100		13.26	6.62
1200			8.24
1300			10.28

was treated with metakaolin + calcium chloride. The addition of metakaolin has resulted in pozzolanic reactions between chemical additives and marine clay. The enhancement in load-carrying capacity is mainly because of production of CSH and CASH agents which are responsible for the better binding properties and sealing the

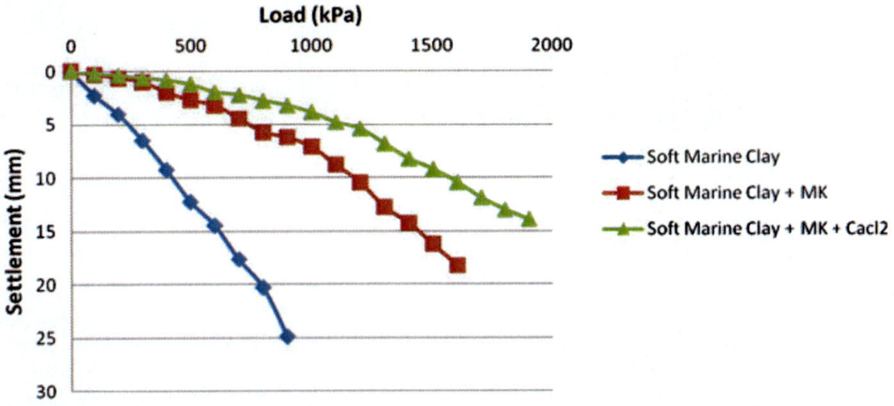

Fig. 3 Variation of load–settlement for soil treated with different additives tested at OMC

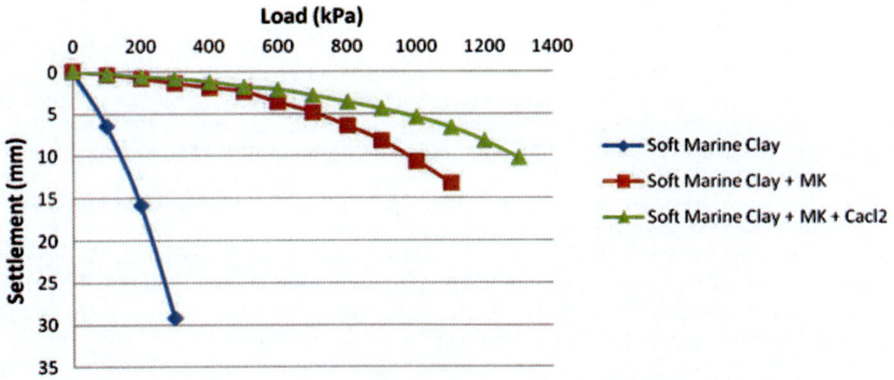

Fig. 4 Variation of load–settlement for soil treated with different additives tested at FSC

voids if any. The reduction in settlement values for the treated foundation beds may be attributed to the chemical reactions between $CaCl_2$ and metakaolin.

Hence, the authors effectively initiated the research in improving the weak soft marine clay to cater the foundation requirement and successfully arrived with a suitable solution (Figs. 3 and 4).

5 Conclusions

The following conclusions are drawn based on the laboratory test results.

1. The soil is pure soft clay and classified as CH.
2. When the soft marine clay is treated with 15% metakaolin, the percentage improvement in soaked CBR is 78% and treated with 9%MK + 1%$CaCl_2$

metakaolin + calcium chloride and the percentage improvement was further increased to 80%.

3. It was noticed from the laboratory tests conducted that the ultimate load-carrying capacity of the untreated soft marine clay foundation bed was improved by 44% at OMC and 73% at FSC when soft marine clay was treated with metakaolin.
4. It was observed from the laboratory investigations of the load test results that the ultimate load-carrying capacity of the untreated foundation bed was improved by about 53% at OMC and 77% at FSC when the soft marine clay was treated with metakaolin + calcium chloride.
5. Finally, the present study yielded an optimum use of metakaolin and calcium chloride in improving the weak marine clay suitable for good foundation material.

References

Cyr M, Idir R, Escadeillas G (2012) Use of metakaolin to stabilize sewage sludge ash and municipal solid waste incineration fly ash in cement-based materials. J Hazard Mater 243:193–203

Deng Y, Yue X, Liu S, Chen Y, Zhang D (2015) Hydraulic conductivity of cement-stabilized marine clay with metakaolin and its correlation with pore size distribution. Eng Geol 193:146–152

Kolovos KG, Asteris PG, Cotsovos DM, Badogiannis E, Tsivilis S (2013) Mechanical properties of soilcrete mixtures modified with metakaolin. Constr Build Mater 47:1026–1036

Umar SY, Elinwal AU, Matawal DS (2015) Hydraulic conductivity of compacted lateritic soil partially replaced with metakaolin. J Environ Earth Sci 5(4)

Zhang M, Guo H, El-Korchi T, Zhang G, Tao M (2013) Experimental feasibility study of geopolymer as the next-generation soil stabilizer. Constr Build Mater 47:1468–1478

Zhang T, Yue X, Deng Y, Zhang D, Liu S (2014) Mechanical behaviour and micro-structure of cement-stabilised marine clay with a metakaolin agent. Constr Build Mater 73:51–57

Stabilization of Clayey Soil Using Enzymatic Lime and Effect of pH on Unconfined Compressive Strength

Dani Jose and S. Chandrakaran

Abstract The performance of the pavement is dependent on the type and properties of the sub-grade soil. Various techniques are used for the stabilization of sub-grade soils having inadequate properties. Combination of Enzyme and lime found to be effective in improvement of strength of the soft soils. In the present work, the behaviour of enzymatic lime stabilized soil under varying environmental conditions is analysed. An extensive study was carried out on a soil sample having pH 3, 7 and 10 using optimum dosages of lime, enzyme, and enzymatic lime. The variation in the unconfined compressive strength was investigated after 3, 7, 14 and 28 days of curing. By using enzymatic lime accelerated enhancement of properties were obtained in comparison to lime and enzyme. Enzymatic stabilization found to be suitable for all conditions of soil while lime and enzymatic lime stabilization enhance properties much better in a neutral environment than the acidic and alkaline environment.

Keywords Soil stabilization · Enzymatic lime · Unconfined compressive strength · pH

1 Introduction

A pavement is a relatively stable crust constructed over the natural soil for supporting and distributing wheel loads and for providing an adequate wearing course. Change in the soil properties and the repeated application of wheel loads may result in excessive settlement leading to deterioration of pavements. Moisture variation, frost action, increase or decrease of water content in the clay soil causes problems to the pavement which necessitates frequent repair and modification. Soil stabilization will improve the soil properties and make them resistive to these deformations and deteriorations. Common methods in practice are stabilization using additives like lime, cement, fly ash, rice husk ash, crumb rubber, bio-enzyme, geosynthetics, biopolymer and different commercially available chemicals. Stabilization can improve stability,

D. Jose (✉) · S. Chandrakaran
National Institute of Technology Calicut, Calicut, Kerala 673601, India
e-mail: danijose92@gmail.com

M. Latha Gali and R. R. P. (eds.), *Problematic Soils and Geoenvironmental Concerns*, Lecture Notes in Civil Engineering 88,
https://doi.org/10.1007/978-981-15-6237-2_33

shearing resistance and bearing capacity, thus, control the undesirable effects associated with clay. Among these methods, lime stabilization is the common practice for ground improvement.

Utilization of various additives along with soil-lime mix found to be effective for treatment of problematic soil. The effect of climatic change, environmental concerns and increase in cost of materials necessitates an innovative ecofriendly method over these traditional methods. An economically feasible solution is the introduction of enzymatic soil stabilization.

Enzymes are biological catalyst obtained from plants and animals including microorganisms by extraction using suitable solvent. These are large protein molecules which are more efficient than inorganic catalyst. Enzyme can increase the reaction rate by a factor of 106–1012. They usually catalyse unique reactions; therefore, enzymes do not produce side reaction. Enzymes are temperature sensitive and work around temperature (35 °C) and loss their effectiveness at higher temperature. Also, they are pH sensitive too and work good at pH value 7 (Norris 2011).

Bio-enzyme is a naturally obtained, nontoxic, nonflammable and noncorrosive liquid enzyme formulation fermented from vegetable extracts. They catalyse the reactions between the clay and the organic cations by accelerating the cationic exchange. Enzyme provides higher soil compaction densities and enhances strength and stability of the soil (Shankar et al. 2009). Utilization of bio-enzyme in combination with lime will also give enhanced improvement in strength. The catalytic action of enzymes in the presence of lime brought a significant improvement of strength (Eujine et al. 2017a, b). The enzymatic lime is more effective in soil samples with higher clay fraction (Eujine et al. 2017a, b).

A very few researches have been conducted using enzymatic lime, and scope for detailed study on its influence on soil properties is there. The present work aims to study the behaviour of enzymatic lime-stabilized soil under various environmental conditions. The purpose of this research is to assess the strength enhancement due to curing and to study the variation in properties with respect to pH.

2 Materials and Methodology

Soil used for the work is collected from Vytilla, Ernakulam. The bio-enzyme using is an extract from sugar molasses and was obtained from Avijeet Agencies in Chennai, Tamil Nadu, under the chemical name TerraZyme. Lime used for the study is purchased from local agent at Kunnamangalam, Calicut. The properties of soil and enzyme are given in Tables 1 and 2.

The collected soil is highly plastic and of medium consistency in nature. From UCS, it is found that the soil has got medium strength, and so, the strength can be improved. According to IS Classification, the soil is highly compressible clay.

All specimens using in this study were prepared and tested using standard procedures described by the Bureau of Indian Standards. Soil sample used for the test was air dried, pulverized manually, sieved through 425 μ sieve and preserved in large

Table 1 Properties of soil

Properties	Obtained values
Specific gravity	2.62
Liquid limit (LL) (%)	76
Plastic limit (PL) (%)	30
Shrinkage limit (%)	13
Plasticity index (PI) (%)	46
Free swell (%)	14.3
pH	7.74
Bulk density (kN/m^3)	17.82
Max. dry density (kN/m^3)	13.86
Optimum moisture content (%)	27
Unconfined compressive strength (kN/m^2)	106.61
Cohesion (kN/m^2)	53.31
Clay fraction (%)	38.76
Silt (%)	34.52
Sand (%)	26.72
Soil classification	CH

Table 2 Properties of enzyme (Eujine et al. 2017a, b)

Properties	Obtained values
Boiling point	100 °C
pH	2.8–3.5
Vapour pressure: mmHg	As water
Melting point	Liquid
Vapour density (air = 1)	1
Solubility in water	Infinite
Evaporation rate	As water
Appearance and odour	Light gold liquid, characteristic odour
Production	Extracted from sugar molasses
Major constituents	Rhodasurf B1, calcium chloride

containers. TerraZyme was preserved in an airtight bottle in its original liquid form. Lime was sieved using a 425 μ sieve and preserved in an airtight container.

From the literature, it is found that the optimum lime content for soil stabilization varies from 2 to 6% by weight, whereas higher percentage may require for soil having higher clay fraction. The range of dosage of lime to be used was determined from liquid limit tests. Then, the soil is treated with 2–3.5% lime content, and unconfined compressive strength tests were conducted to verify and confirm the actual optimum lime content. Three samples of the same mix type were tested after 3, 7, 14, 21 and 28 days, and the average of the values was recorded.

A dilution ratio chart provided by manufacturer (that calculated the required dosage of TerraZyme for a soil based on particle size and plasticity index) is used to determine the range of optimum enzyme content. Soil samples were mixed with TerraZyme dosages between 110 and 130 ml/m^3, cured up to four weeks in airtight bags. The same tests were repeated on the samples, and the variation in properties was studied. From the results, the actual optimum enzyme content is determined.

The dosage of enzyme is fixed at optimum enzyme dosage, and the lime content is varied. The combinations involve 1.5, 2 and 2.5% of lime with 120 ml/m^3 of enzyme. The variation of properties on the soil lime enzyme system is studied for 3, 7, 14 and 28 days. From the obtained results, the optimum enzymatic lime content is fixed.

The pH of the untreated soil was varied, and all the above tests were conducted by changing soil pH. pH 3, 7 and 10 were considered for the study. The pH of the soil is changed by adding 1 N HCl to make it acidic and 1 N NaOH for making it alkaline. The soil with pH 3, 7 and 10 was treated with the optimum percentage of lime and enzyme, and the variation in properties at different pH was studied. The variation of liquid limit, plastic limit, plasticity index and unconfined compressive strength was observed. Based on the analysis of obtained results, the behaviour of enzymatic lime-stabilized soil under varying environmental condition is examined.

3 Results and Discussions

Based on the values of tested parameters, the optimum dosage of additives was fixed. Addition of 3% lime imparts maximum improvement in strength as well as reduction in plasticity. So, the optimum dosage of lime is fixed as 3% by weight of soil. The optimum dosage of enzyme was found as 120 ml/m^3 based on the analysis of results. It was observed that addition of 2% of lime with 120 ml/m^3 of enzyme will impart maximum improvement of properties. So, the optimum enzyme and lime content for enzymatic lime treatment are found as 120 ml/m^3 of enzyme and 2% of lime. From the test results, the optimum lime content is found as 3% by weight of soil, and the optimum enzyme content is fixed at 120 ml/m^3. The variation in the unconfined compressive strength of soil by the addition of enzymatic lime is studied in detail. The results are shown in Tables 3 and 4.

The enzymatic lime can improve the strength characteristics of soil effectively. The strength was increased to 367.9 kN/m^2 and 397.2 kN/m^2 after a curing period of 14 days and 28 days. From Fig. 1, it is clear that an improvement 3.45% was obtained after 14 days and 3.7% increase after 28 days. The variation of unconfined compressive strength is shown in Fig. 1.

The addition of 120 ml/m^3 enzyme and 2% lime gives maximum strength improvement after 28 days, which is considered as the optimum enzymatic lime concentration. The curing period increases the strength of soil, and the addition of enzyme

Table 3 Unconfined compressive strength of soil treated with enzymatic lime

	3 days (kN/m^2)	7 days (kN/m^2)	14 days (kN/m^2)	28 days (kN/m^2)
1.5% lime + 120 ml/m^3 enzyme	146.4	192.9	264.7	293.1
2% lime + 120 ml/m^3 enzyme	206.7	280.8	364.9	397.2
2.5% lime + 120 ml/m^3 enzyme	191.5	267.6	302.9	330.6

Table 4 Unconfined compressive strength of enzymatic lime-treated soil at different pH

	3 days (kN/m^2)	7 days (kN/m^2)	14 days (kN/m^2)	28 days (kN/m^2)
pH 3	170.2	229.9	283.6	334.6
pH 7	206.7	280.8	364.9	397.2
pH 10	182.9	253.6	317.5	367.2

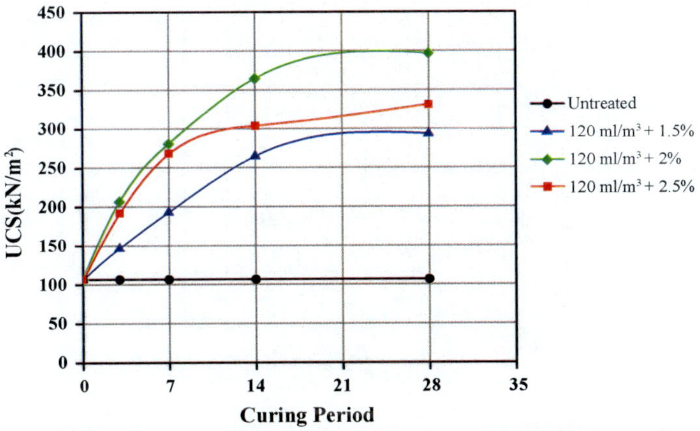

Fig. 1 Variation of UCS of soil treated with varying percentages of enzymatic lime

accelerates the strength improvement. The stress–strain curve for different curing periods are shown in Figs. 2, 3, 4 and 5.

The mechanism hypothesized was that enzyme modifies the clay, and it was the modified clay that interacted with lime, leading to an amplified improvement to the soil properties. Mechanism of improvement involves a primary mechanism of formation of a newly formed enzyme-modified clay that reacted with lime at an increased rate and the secondary mechanism. The secondary mechanism comprises further lime reactions with calcium aluminate and calcium alumina silicate by replacing the aluminium cations with calcium cations in the presence of the active agent in the enzyme Rhodasurf $R(OC_2H_4)_nOH$ to form stronger compounds (Eujine et al. 2017a, b).

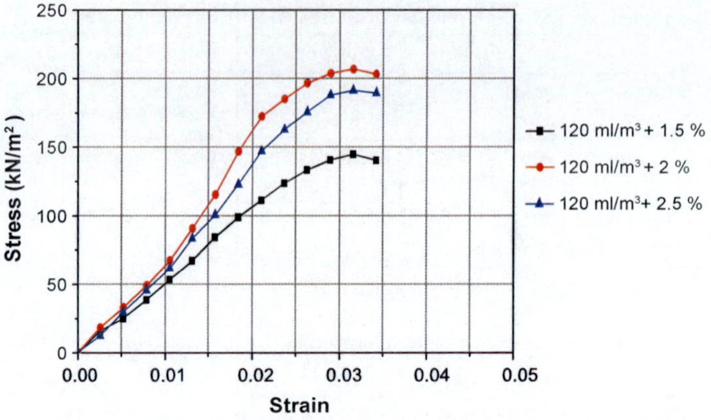

Fig. 2 Stress–strain curve of soil treated with varying dosages of enzymatic lime after 3 days of curing

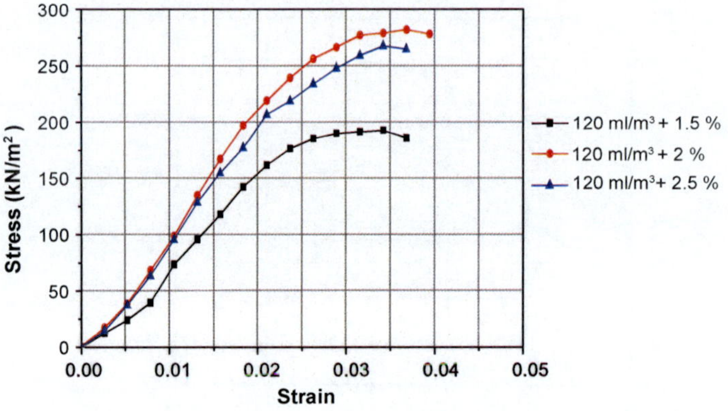

Fig. 3 Stress–strain curve of soil treated with varying dosages of enzymatic lime after 7 days of curing

Enzymatic lime stabilization of soil is found to be very effective in soil treatment. The results show that addition of 2% lime and 120 ml/m^3 of enzyme will improve the strength within one week itself by substantial amount. After 14 days, the soil attained the strength nearly the strength attained by lime and enzyme alone after 28 days. By addition of enzymatic lime to soil, both strength and plasticity characteristics have been improved. The lime content in optimum concentration of enzymatic lime is less than optimum lime content. This is significant result because the reduction in lime content can reduce the impact of lime to the environment. Thus, the obtained results show that the enzyme used with lime will improve the properties as well as we can reduce the lime content and curing period making method more economical as well as ecofriendly.

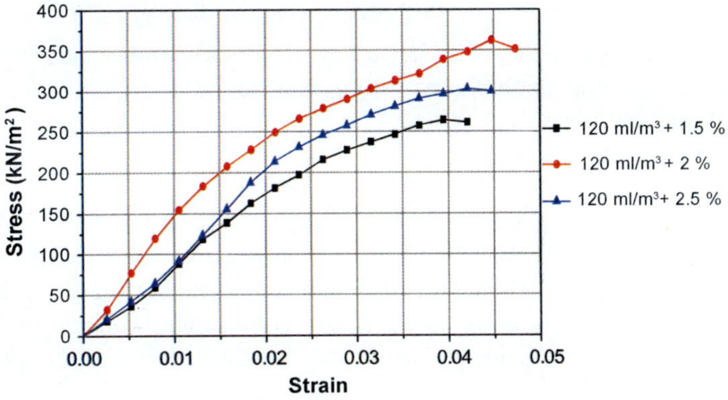

Fig. 4 Stress–strain curve of soil treated with varying dosages of enzymatic lime after 14 days of curing

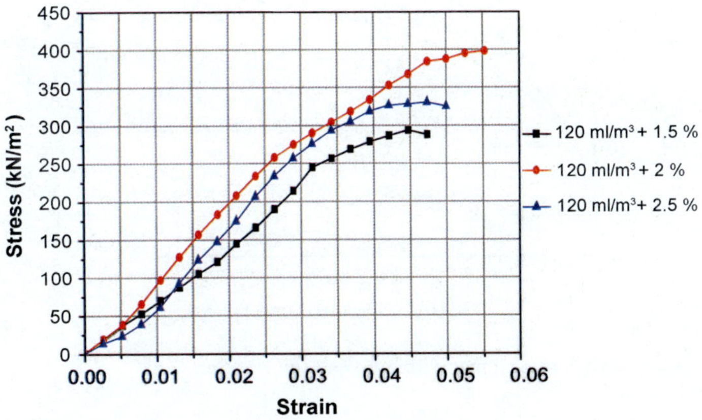

Fig. 5 Stress–strain curve of soil treated with varying dosages of enzymatic lime after 28 days of curing

3.1 Behaviour of Treated Soil Under Varying pH

In actual practice, one had to deal with contaminated sites like acidic soil as well as alkaline soils, etc. Contamination can lead to damages in buildings and other structures and can make the site unsuitable for the implementation developmental projects. A geotechnical engineer is concerned about the impact of soil as most of the effects of soil contamination are mainly due to changes in the geotechnical behaviour of foundation soil. So, the soil samples with pH 3 and pH 10 are treated using optimum dosage of enzymatic lime to examine the behaviour under different environmental conditions. The pH of soil is changed by adding 1 N HCL and 1 N NaOH to make soil acidic and alkaline. The variation of LL, PL, PI and UCS for a curing period of

3, 7, 14 and 28 days is observed. The obtained results are analysed and compared to derive the conclusion of the work carried out. It was found that the enzymatic lime will improve unconfined compressive strength of soil in three conditions. But pH of 7 is more favourable than pH 3 and 10. The result figures out that lime and enzymatic lime treatment is more suitable for soil in neutral environment than the soil in acidic and alkaline environment.

The strength improves with addition of enzymatic lime, and maximum improvement is obtained for pH 7. The strength was increased to 397.27 kN/m^2 at pH 7, while for soil having pH 10 the UCS is 367.28 kN/m^2 and for 3, the obtained value is 334.57 kN/m^2. The variation of UCS with curing time is shown in Fig. 6. The stress–strain curve for different curing periods is shown in Figs. 7, 8, 9 and 10.

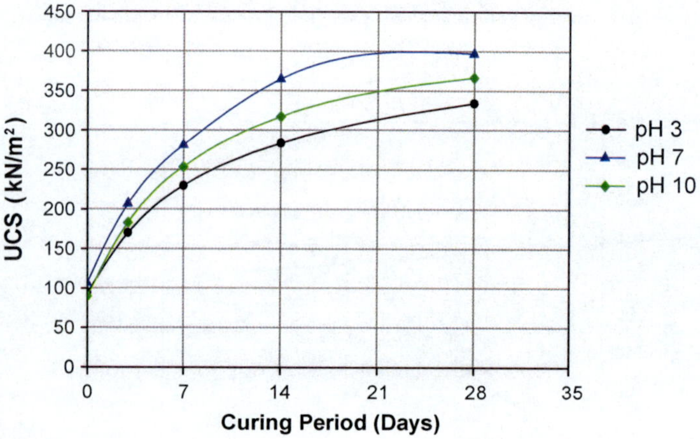

Fig. 6 Variation of UCS of enzymatic lime-treated soil at different pH

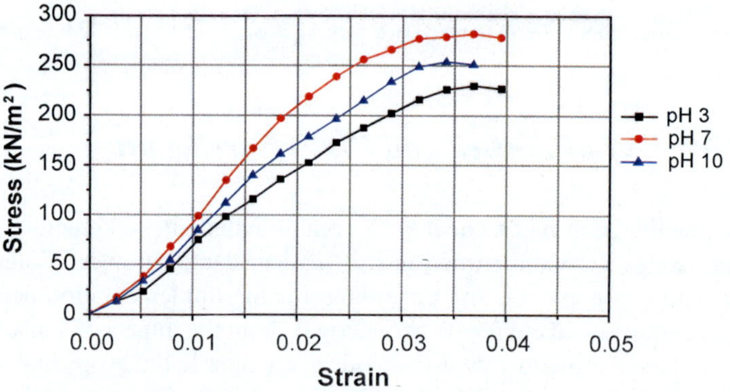

Fig. 7 Stress–strain curve of enzymatic lime treated soil at different pH after 3 days of curing

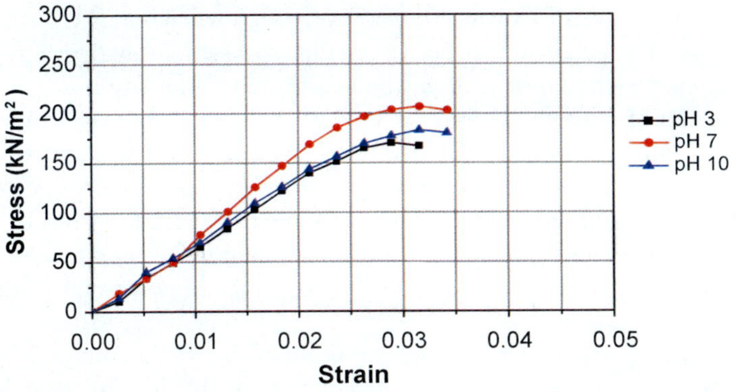

Fig. 8 Stress–strain curve of enzymatic lime treated soil at different pH after 7 days of curing

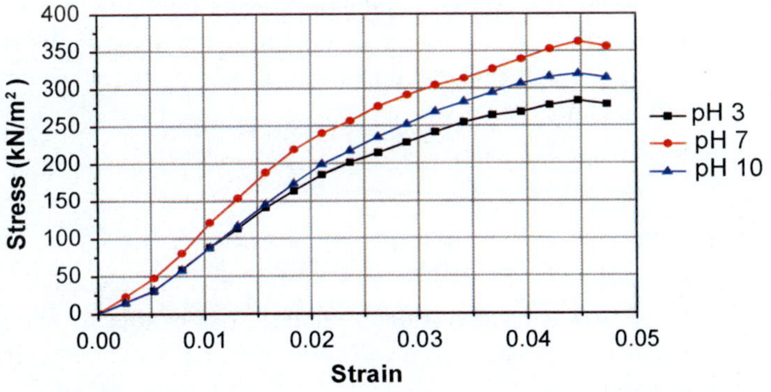

Fig. 9 Stress–strain curve of enzymatic lime treated soil at different pH after 14 days of curing

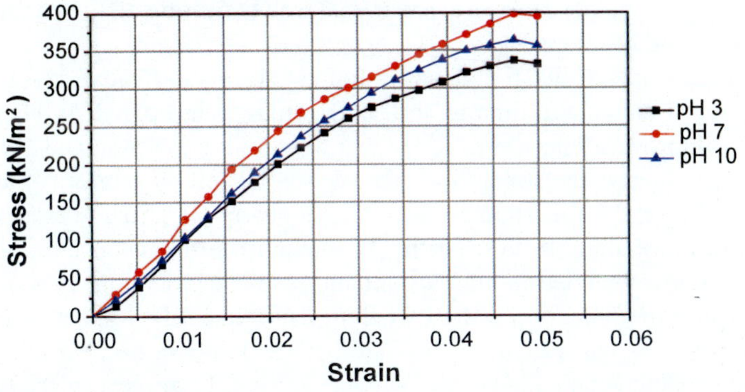

Fig. 10 Stress–strain curve of enzymatic lime treated soil at different pH after 28 days of curing

The obtained value of UCS is comparatively high when compared with lime and enzyme alone. The obtained value of UCS of the soil samples with different pH after addition of enzyme and lime is high when compared to the virgin soil.

On addition of optimum dosage, the pH of the system increases to 12, and this condition favours the formation of silica and alumina. The silica and alumina react with calcium to form calcium silicates hydrate gel if the high alkaline condition is maintained. The catalytic action of enzyme increases the rate of reaction and thus gives accelerated improvement of strength. The action of enzyme is not affected by the change in pH. The high strength obtained for alkaline and acidic strength substantiates the same.

In the acidic environment, silica does not dissolve from soil, and hence, lime silica reaction to produce calcium silicate gel was not able to take place. Insufficient CSH gel may be the reason behind the low strength in acidic environment compared to alkaline and neutral environment. In the alkaline environment, the high pH of soil due to lime content may impart a reduction in strength than that of pH 7. But the reduction in strength is only by a small amount which may not be taken under consideration. The results indicate that both neutral and alkaline environment favour the enzymatic lime stabilization.

4 Conclusions

An extended study was carried out on a natural soil sample to understand the behaviour of enzymatic lime-stabilized soil under varying environmental condition. In this work, the variation of unconfined compressive strength on addition of enzymatic lime is studied. The optimum dosage of additive was used for amending soil with pH 3, 7 and 10, and the variation of properties was observed. The effect of curing in improvement of properties was analysed by testing all the parameters after 3, 7, 14 and 28 days of curing.

Enzymatic lime stabilization of soil is found to be very effective in soil treatment. By addition of enzymatic lime to soil, both strength and plasticity characteristics have been improved. The results show that addition of 2% lime and 120 ml/m^3 of enzyme will improve the strength within one week itself by substantial amount.

After 14 days, the soil attained the strength nearly the strength attained by lime and enzyme alone after 28 days. The lime content in optimum concentration of enzymatic lime is less than optimum lime content for soil stabilization. This is significant result because the reduction in lime content can reduce the impact of lime on the environment to a small extent.

Addition of enzymatic lime to soil with pH 3, 7 and 10 gives accelerated improvement of properties compared to individual stabilizers. On comparing acidic, alkaline and neutral conditions, neutral condition favours the enzymatic lime stabilization. The obtained results indicate that there is significant improvement in properties at different pH compared to other stabilizing agents.

The catalytic action of enzyme makes enzymatic lime suitable to soils of different environmental conditions by giving comparatively good properties.

References

Eujine GN, Chandrakaran S, Sankar N (2017a) Accelerated subgrade stabilization using enzymatic lime technique. J Mater Civ Eng 29(9), Paper 04017085 (American Society of Civil Engineers)
Eujine GN, Chandrakaran S, Sankar N (2017b) The engineering behaviour of enzymatic lime stabilized soils. Proc Inst Civ Eng Ground Improv, Paper 1600014
Norris R, Ryan L, Acaster D (2011) Cambridge international AS and A level chemistry coursebook. Cambridge University Press, UK, pp 413–414
Shankar AU, Rai HK, Mithanthaya R (2009) Bio-enzyme stabilized lateritic soil as a highway material. J Indian Roads Congr, Paper No: 553
Wright KM, Mohai A, McQuaid G, Case study of the use of an enzyme soil stabilizer in Central and Eastern

Comparative Study on Stabilization of Marine Clay Using Nano-silica and Lime

M. R. Joju and S. Chandrakaran

Abstract This paper reports the experimental results of using two chemical additives to enhance the properties of marine clay collected from Vypin, Cochin area. Modifications in properties of marine clay by lime are investigated to compare the nano-silica particles' effect in mechanical properties of clay. The engineering properties studied included Atterberg limits, maximum dry unit weight, optimum moisture content and unconfined compressive strength. Lime to soil ratios of 3, 4 and 5% were used. Nano-silica used for the study varied from 0.5, 0.8 and 1%. Atterberg limits variation and unconfined compressive strength variation were evaluated for the curing period of 3, 7, 14 and 28 days. Strength is enhanced three times for amended clay with nano-silica. Optimum lime content for lime stabilized clay was obtained as 4% and nano-silica as 0.8%. Plasticity characteristics are remarkably modified in the lime-amended clay compared to nano-silica additive. Unconfined compressive strength improvement shows similar variations for both nano-silica and lime.

Keywords Marine clay · Lime · Nano-silica

1 Introduction

Vypin is one of the most densely populated islands in Asia. It has become the major industrial hub and fastest growing suburb in Kerala. Soft marine clay deposits have widely found in this region. Marine clays are highly compressible in nature. They are generally associated with high moisture content so obviously weak in shear strength. The infrastructure development in this region demands costlier approaches to tackle the geotechnical issues. So, modification of properties of soil is necessary for improving the engineering behavior. Soil stabilization can be done by chemical, mechanical and physical methods. Lime and cement are two promising chemical additives for soft clay. Rapid development in nanotechnology field added a new dimension in geotechnical engineering field. Many researchers have been studying

M. R. Joju (✉) · S. Chandrakaran
Civil Engineering, National Institute of Technology Calicut, Calicut, Kerala 673601, India
e-mail: jojumariyam@gmail.com

M. Latha Gali and R. R. P. (eds.), *Problematic Soils and Geoenvironmental Concerns*, Lecture Notes in Civil Engineering 88,
https://doi.org/10.1007/978-981-15-6237-2_34

the effect of nanoparticles in properties of soil and alteration of properties and microstructure of soil.

Lime stabilization reactions mainly improve the inter-particle bond by cementation, which finally improves the strength of clay and reduces the compressibility of clayey soil (Jose et al. 1988). Moreover, lime addition creates noticeable changes in consistency limits of clay. Liquid limit of soils initially decreases with an increase in lime content. The reason pointing out this decrease is due to the reduction in diffused double layer thickness, and this credited to increase in electrolyte concentration in pore fluid (Dash and Hussain 2012). Electrolyte concentration improvement imparts strong inter-particle movement resistance, and thereby, plastic limit will increase (Rajasekharan and Rao 1996).

In leading edge, nanotechnology-based studies are done in geotechnical engineering. Nanomaterials are substances with size in nanometer range with high specific surface area (Majeed and Taha 2011). These materials marked attention in wastewater treatments and water purification. Recently, studies show that it can be used for the modification of engineering properties of soil. Later, it will be substantially utilized as the chemical additives for the soil improvement (Pham and Nguyen 2014). Plasticity index, compressibility and unconfined compressive strength of soil have been beneficially improved for the clayey soil by the addition of nano-silica as concluded in a study (Changizi and Haddad 2017).

The rapid strength enhancement and compressibility reduction by nano-silica are useful for the expansive soils, where settlement failures of foundations are prominent (Ghasabhkolae et al. 2017). Lime addition improves the strength with increase in curing period. Lime stabilization has proven as an effective additive for marine clay by various studies. This paper presents a comparative study of stabilization using nano-silica and lime. As a recent material, nano-silica effect on natural clay properties can be compared with the already proven lime additive.

2 Materials Used

The main additives used for stabilization of marine clay from Cochin are nano-silica and lime. The procurement of materials was the first phase of the work.

2.1 Marine Clay

Marine clay for the work was collected from Vypin, Ernakulam. The grain size distribution, specific gravity, pH analysis and Atterberg limits were all done to study the properties of the materials. The determination of the percentage of clay, sand and silt sized particles within the sample was utilized for the work. Basic properties of natural clay soil are presented in Table 1, and grain size distribution of clay is shown

Table 1 Characteristics of marine clay

Characteristics	Obtained value
Specific gravity	2.65
Free swell	50%
Liquid limit	80%
Plastic limit	34%
Shrinkage limit	14%
Sand	11%
Silt	39%
Clay	50%
UCC[a]	97 kPa

[a]Unconfined compressive strength

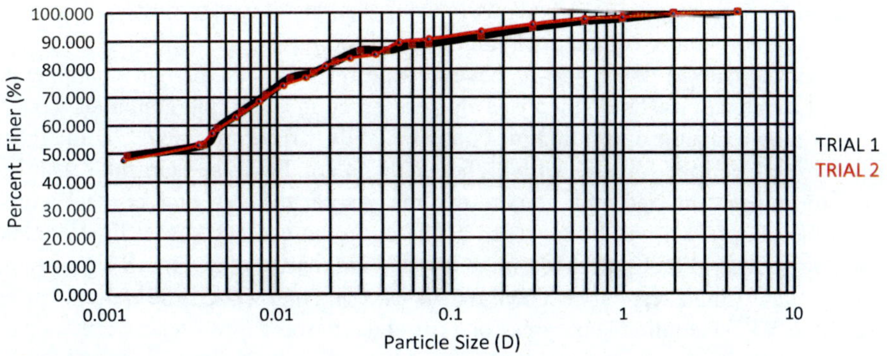

Fig. 1 Grain size distribution of clay

in Fig. 1. Marine clay is classified as clay with high compressibility (CH). Almost 50% of the soil constitutes clay particles.

2.2 Additives

Lime and nano-silica are the two additives used for the stabilization of natural marine clay. Lime was collected from local shop. Distilled water was used in preparing the specimens and curing. Nano-silica was collected from ASTRA chemicals, Chennai. The properties of nano-silica are shown in Table 2. Nano-silica was obtained in white-colored powder form. Size of the nano-silica is about 20 nm.

Table 2 Properties of nano-silica

Properties	Value
Specific surface area (m^2/g)	202
Specific gravity	2.2–2.4
pH	3.7–4.5
Particle size	17 nano
SiO_2 content (%)	99.88

3 Specimen Preparation and Test Methods

All the specimens tested in this study were prepared and tested using standard procedures described in the Bureau of Indian Standards—IS 2720 (Part 4):1985, IS 2720 (Part 5):1985, IS 2720 (Part 10):1991 and IS 2720 (Part 15):1965. Soil sample was air dried for a week, pulverized manually using weights, sieved through 425-μ sieve and preserved in large containers in an enclosed room.

Lime was sieved using 425-μ sieve and preserved in an airtight container to prevent carbonation. Nano-silica was also preserved in an airtight container.

It is known that the optimum lime content of soils varies from 2 to 6% with higher percentages required for soils with higher clay content. Eades and Grimm's test was used to find out the optimum content of lime. As per the test, lime content which gives the pH of 12.4 will be regarded as the optimum content. 4% of lime content was shown the pH of 12.4 based on the test done in marine clay. Figure 2 shows the Eades and Grimm's test curve. Then, for the experimental work, lime content varied from 3 to 5%. Unconfined compression tests and Atterberg limits tests were done on soils mixed with lime in the ranges 3–5% and cured up to 28 days in airtight bags.

Since tests were performed under constant moisture content and due to the absorption of water by nano-silica, mixing of soil with more than 1% nano-silica content led to a lack of water in the soil/nano-silica mixture; consequently, the soil became less compressible (Changizi and Haddad 2017). It was thus decided that 1% nano-silica content would be the maximum value in this study. Soil specimens were mixed with nano-silica in the ranges 0.5–1%, cured up to four weeks in airtight bags and tested for unconfined compressive strengths and Atterberg limits.

4 Results and Discussions

4.1 Effect of Lime on Properties of Clay

Effect of lime addition on marine clay can be found out through the determination of properties such as liquid limit, plastic limit and free swell and unconfined compressive strength. The Atterberg limits and unconfined compressive strength of lime-stabilized soil are found out for 3, 7, 14 and 28 days of curing time.

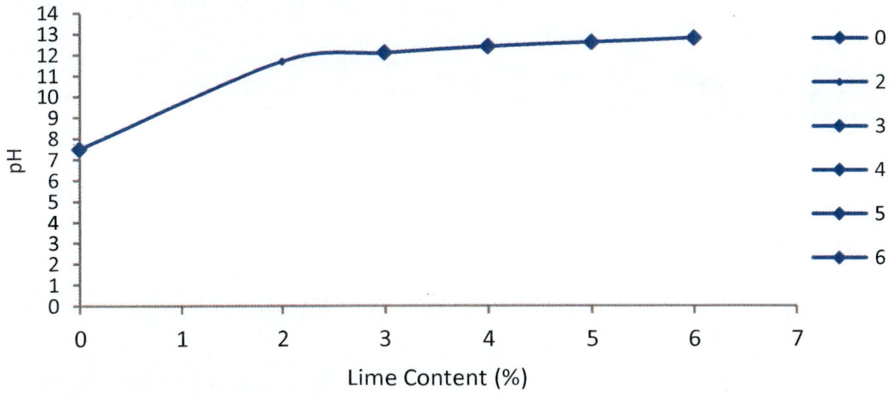

Fig. 2 Eades and Grimm's test curve

4.1.1 Effect on Liquid Limit

Liquid limit tests were done in Casagrande apparatus, after mixed with the required percentage of lime with clay and kept it in desiccators for the curing periods and tested in the curing periods of 3, 7, 14 and 28 days. Influence of lime and curing time on liquid limit is depicted in Fig. 3.

Even if there is a decreasing trend in liquid limit, 5% lime content reduction in liquid limit is not a drastic one. Reduction rate is lower than that of 4% lime. Reduction of liquid limit is due to the decrease in thickness of diffused double layer of clay.

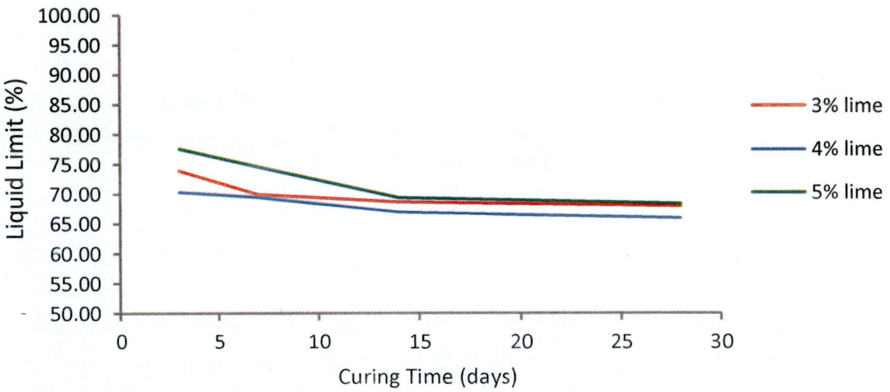

Fig. 3 Variation of liquid limit with curing time in lime-amended clay

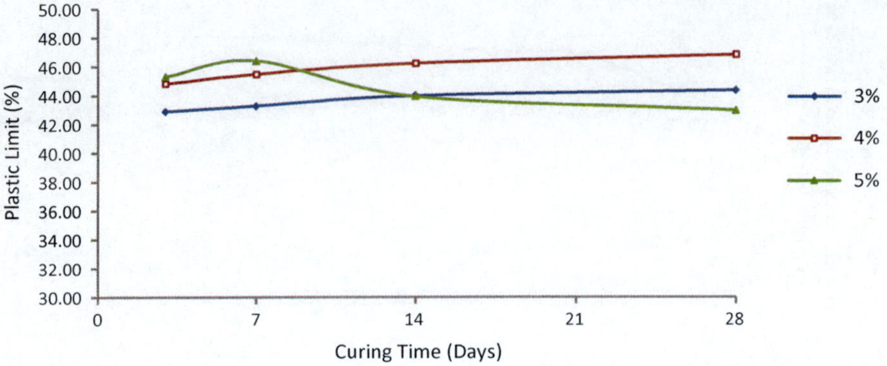

Fig. 4 Variation of plastic limit with curing time in lime-amended clay

4.1.2 Effect on Plastic Limit

Figure 4 depicts the effect of quantity of lime added and curing period on plastic limit of soil. The plastic limit of amended clay is increasing with curing time in 3%, 4% of lime. A noticeable increase in plastic limit can be observed. But, in 5% of lime-amended clay for the last days of curing time, it seems to be decreasing. The rate of increase in plastic limit is also moderate. The addition of lime tends to reduce the thickness of the diffuse double layer, which increases the charge concentration and thereby the viscosity of the pore fluid. Similar to liquid limit, plastic limit also does not change as much after the addition of 4% of lime. So, 4% lime can be fixed as lime fixation point.

4.1.3 Effect on Plasticity Index

For all percentages of lime, there is an immediate decrease of plasticity index which can be observed due to the reduction in liquid limit and increase in plastic limit. Figure 5 depicts the variation of plasticity index.

4.1.4 Effect on Shrinkage Limit

Shrinkage limit of lime amended soil is decreasing with increase in lime. Figure 6 represents the variation of shrinkage limit at 28 days.

4.1.5 Effect on Compaction Characteristics

Increase in lime content decreases dry density in lime-treated soil. Maximum dry density of lime-treated soil is decreased for higher concentration of lime. In higher

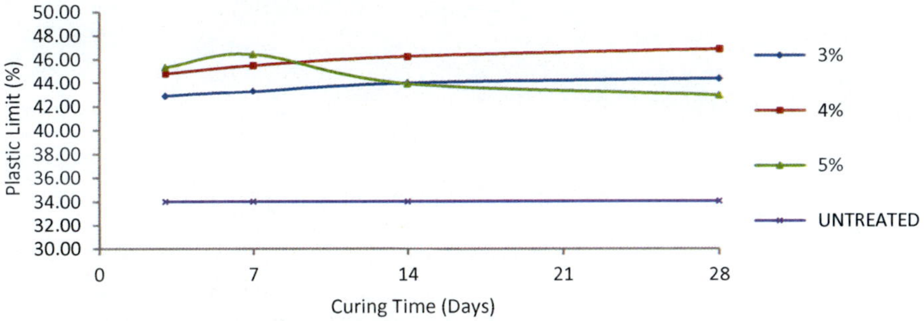

Fig. 5 Variation of plastic limit with curing time in lime-amended clay

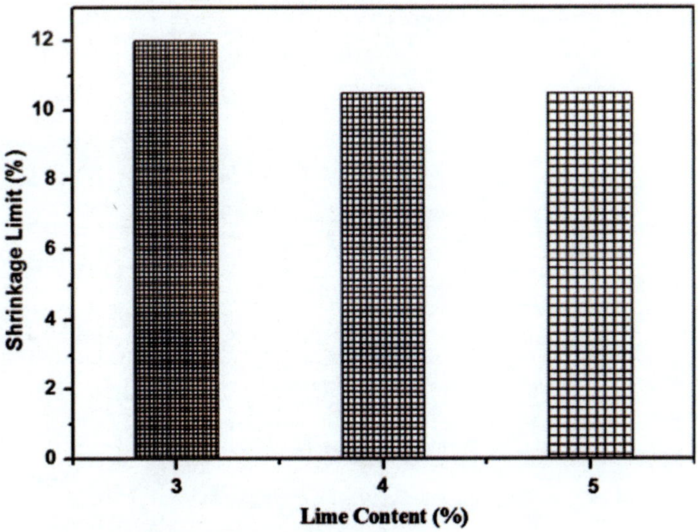

Fig. 6 Variation of shrinkage limit in lime amended soil

concentration of lime, electrolyte concentration increases and reduces thickness of the diffuse double layer, hence, producing a flocculated structure and resisting the compactive effort. Compaction curve is depicted in Fig. 7.

4.1.6 Effect on Unconfined Compressive Strength

Strength variation of lime amended soil is depicted in Fig. 8. With increased lime content, pozzolanic reactions enhanced and to form cementitious compounds result in peak in strength. From the graph, it can be observed that for 3 and 4% composition of lime, peak of strength is obtained. But, for 5%, a reduction in strength can be

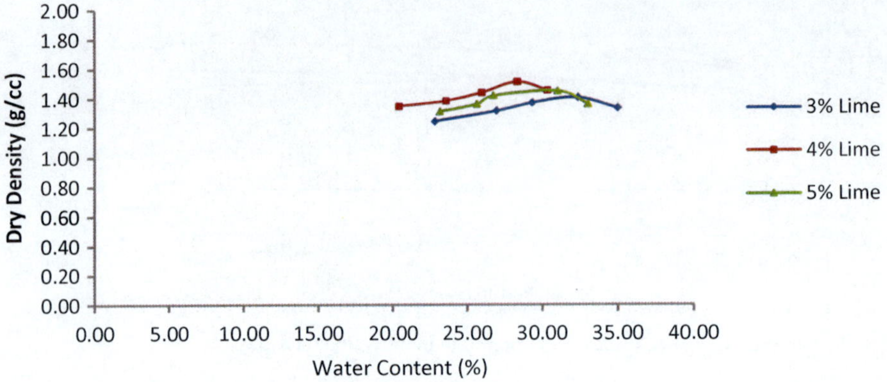

Fig. 7 Compaction curve of lime-amended clay

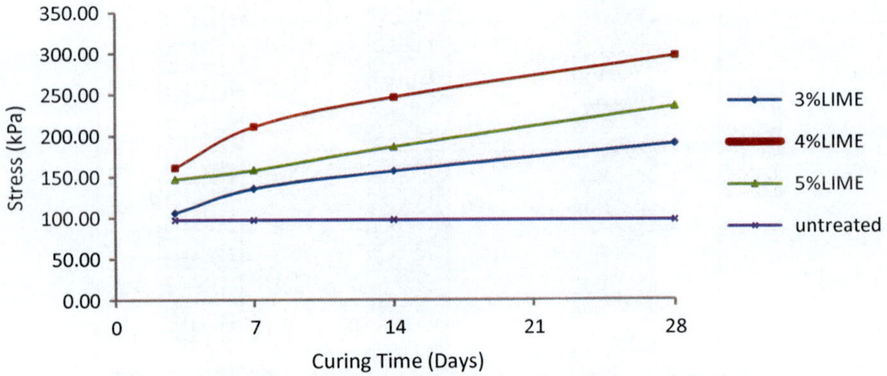

Fig. 8 Variation of UCC of lime-amended clay with curing time

analyzed. This is due to the factor that excessive lime will act as a lubricant for soil particles and results in reduction of strength.

4.2 *Effect of Nano-silica on Properties of Clay*

Influence of nano-silica stabilization on marine clay can be evaluated through the determination of geotechnical properties such as liquid limit, plastic limit and unconfined compressive strength. The variation of Atterberg limits and unconfined compressive strength of nano-silica stabilized soil is found out for 3, 7, 14 and 28 days of curing time. Marine clay is amended with 0.5%, 0.8% and 1% of nano-silica by weight of dry soil and cured for the above-mentioned days.

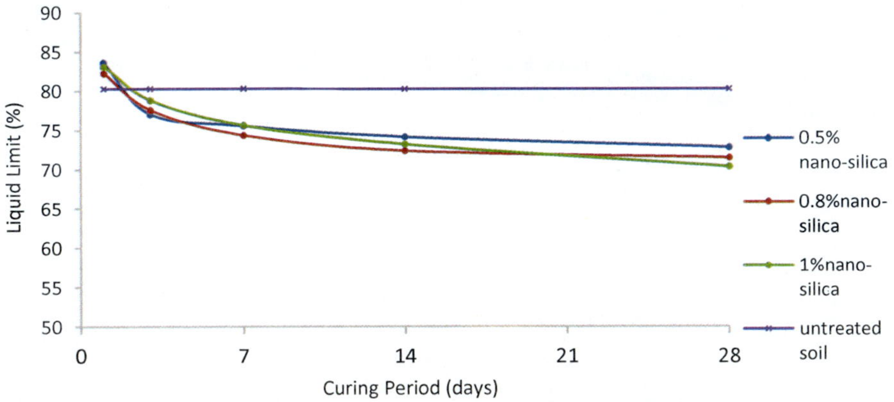

Fig. 9 Variation of liquid limit with curing time in nano-silica-amended clay

4.2.1 Effect on Liquid Limit

Influence of nano-silica and curing time on liquid limit is illustrated in Fig. 9. The figure shows that in all these three contents of nano-silica, liquid limit is decreasing. It can be concluded that at the first day of testing, liquid limit is increasing than untreated soil, due to the absorption of water by nano-silica.

But, thereafter, it can be seen the decreasing trend of liquid limit. Compared with natural clay, liquid limit reduces about 11% for 1% nano-silica after 28 days. Similarly, a same reduction can be observed in 0.8% nano-silica-amended clay. Hence, 0.8% can be taken as optimum as liquid limit variation comparing to 1% nano-silica which is same.

4.2.2 Effect on Plastic Limit

Figure 10 depicts the effect of quantity of nano-silica added and curing period on plastic limit of soil. The figure indicates that increase in nano-silica increases plastic limit. The rate of increase in plastic limit is also moderate with curing time. Similar to liquid limit, plastic limit also shows similar closeness of plastic limit for 0.8 and 1%. Hence, 0.8% can be taken as optimum point.

4.2.3 Effect on Plasticity Index

From Fig. 11, one can conclude that in all percentages of nano-silica, there is an immediate decrease of plasticity index which can be observed due to the reduction in liquid limit and increase in plastic limit. Compared with natural clay, plasticity index reduces 34%, 38% and 41%, respectively, for 0.5%, 0.8% and 1% of nano-silica.

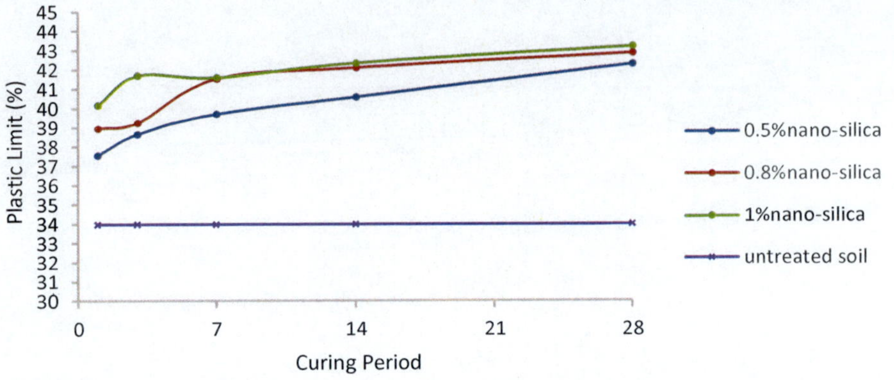

Fig. 10 Variation of plastic limit with curing time in nano-silica-amended clay

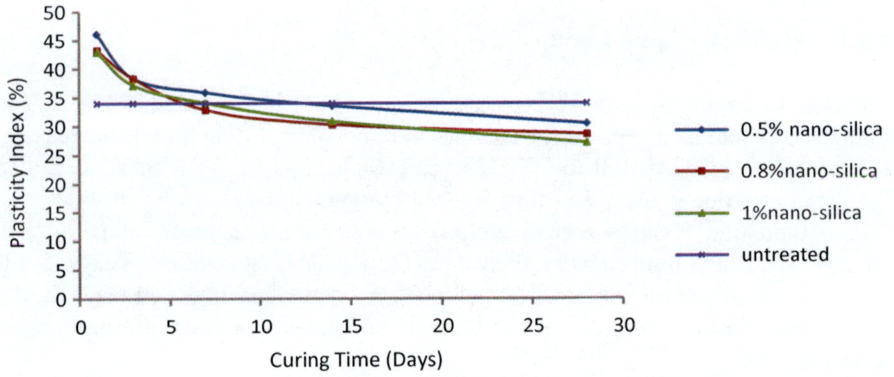

Fig. 11 Variation of plasticity index with curing time in nano-silica-amended clay

4.2.4 Effect on Shrinkage Limit

The shrinkage limits of soil stabilized with varying nano-silica contents obtained from shrinkage limit tests are presented in Fig. 12. The figure indicates that the shrinkage limit of stabilized soils decreases with an increase in nano-silica content. 28-day shrinkage limit is presented in Fig. 12. Clay surrounded with nano-silica particle improves the interfacial bond parameters which led to the reduction in shrinkage limit of soil.

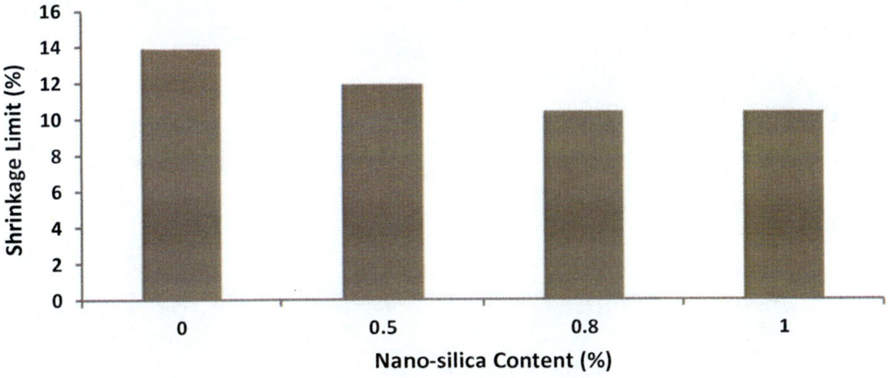

Fig. 12 Variation of shrinkage limit in nano-silica-amended soil

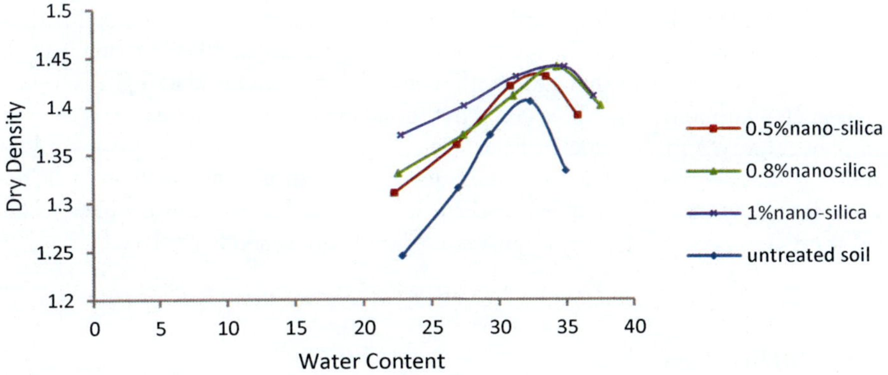

Fig. 13 Compaction curve of nano-silica-amended clay

4.2.5 Effect on Compaction Characteristics

Figure 13 depicts the dry density–water content relationship of varying percentages of nano-silica in clay. Increase in nano-silica content increases dry density in nano-silica-treated soil. The resulted changes are owing to the replaced air between the soil grains by nanomaterials and the effect of moisture absorbed by the nano-silica. Due to the effect of moisture absorbed by the nano-silica, the optimum moisture content of the stabilized soil increases.

4.2.6 Effect on Unconfined Compressive Strength

The result of unconfined compressive strength of nano-silica-amended soil is depicted in Fig. 14. The UCS of soil increases with increase in nano-silica content.

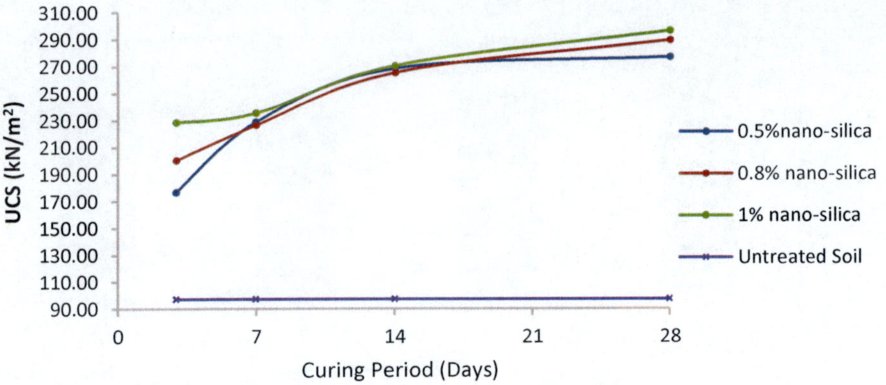

Fig. 14 Variation of UCC of nano-silica-amended clay with curing time

From the graph, it can be observed that for 0.5, 0.8 and 1% composition of nano-silica, strength is increasing. The UCS values are similar for 0.8% and 1%. The tendency of the UCS curves indicates that adding more nano-silica could lead to higher levels of resistance. Optimum nano-silica could not found out because increase in nano-silica amount improves the resistance value. The improvement of strength of clay with addition of nano-silica is due to the viscous gel formation by the nano-silica. When water is added to the clay with nano-silica, nano-silica absorbs the water and viscous gel formation occurs. This gel improves the bond between clay particles.

5 Conclusions

The study has been performed to assess how the nano-silica and lime influenced the properties of marine clay. Detailed comparison of study has been accorded based on the properties of soil.

- The optimal content of nano-silica is obtained as 0.8% and optimal content of lime is obtained as 4%.
- Liquid limit of clay reduced about 18% by the addition of lime and that of by nano-silica reduction was about only 11% at the optimal content after 28 days. Plastic limit of clay increased about 35% by the addition of lime and nano-silica addition improved only about 25%.
- Plasticity index of clay reduced from 46 to 19% at the optimal content of lime after 28 days, and plasticity index of nano-silica amended clay reduced to 27% only. Reduction of diffused double layer of clay by the lime addition is attributed to the good plasticity characteristics of clay. Meanwhile, the plasticity characteristics of clay amended with nano-silica were not remarkable one compared to lime addition.

- Both lime and nano-silica decreased the shrinkage limit of clay after 28 days of curing. Reduction rate shows similarity.
- Lime addition increased the OMC but decreased the dry density in higher percentages. But the nano-silica increased the dry density of soil with increase in nano-silica content and also the optimum moisture content of soil. The water absorption of nano-silica enhanced the OMC.
- Unconfined compressive strength of lime-amended clay at the 4% addition of lime to clay after 28 days was obtained as 300 kPa. The hydrated silica gel attributed the strength of lime addition in soil.
- Unconfined compressive strength of nano-silica-amended clay with 0.8% nano-silica after 28 days was shown 305 kPa. The 14-day strength of nano-silica-amended clay is more than that of lime-amended clay. So, the short-term strength was noticeable. Viscous gel improves the cohesion between particles of clay and thereby the strength of clay. Lime and nano-silica improved the strength of clay in different mechanisms.
- Plasticity characteristics changes were remarkable in the lime-amended clay compared to nano-silica-amended clay. But the strength of clay was enhanced by both. Minute quantity of nano-silica particles significantly improved the strength of clay.
- Lime makes the soil more workable by altering plasticity characteristics, and nano-silica improves the strength in short term. In the future era of geotechnical development in ground improvement, nano-silica and lime combination will be better chemical additives. The cost of nanomaterials is the main limitations facing nowadays. Nanopollution is a recent type of pollution generated by nanomaterials in electronic devices. Some future applications of nanotechnology have the potential to benefit the environment. Nano-silica has been proven as a better chemical additive for soil stabilization.

References

Changizi F, Haddad A (2017) Improving the geotechnical properties of soft clay with nano-silica particles. Proc Inst Civ Eng Ground Improv 170:62–71

Dash SK, Hussain M (2012) Lime stabilization of soils: reappraisal. J Mater Civ Eng ASCE 24(6):707–714

Ghasabhkolae N, Choobbasti AJ, Roshan N, Ghasemi SE (2017) Geotechnical properties of the soils modified with nano-materials: a comprehensive review. Arch Civ Mech Eng 17(3):639–650

Jose BT, Sridharan A, Abraham BM (1988) A study of geotechnical properties of Cochin marine clays. Mar Geotechnol 7(3):189–209

Kadivar M, Barkhordari K, Kadivar M (2011) Nanotechnology in geotechnical engineering. Adv Mater Res 261–263:524–528

Majeed ZH, Taha MR (2011) Effect of nanomaterial treatment on geotechnical properties of a Penang soft soil. J Asian Sci Res 2(11):587–592

Pham H, Nguyen QP (2014) Effect of silica nanoparticles on clay swelling and aqueous stability of nanoparticle dispersions. J Nanopart Res 16:569–574
Rajasekharan G, Rao SN (1996) The microstructure of lime stabilized marine clay. Ocean Eng 23:325–355

Mechanical Behavior of Boulder Crusher Dust (BCD)-Stabilized Dredged Soil

B. A. Mir and Kh Mohammad Najmu Saquib Wani

Abstract Dredged soil is a waste of sediments including organic matter, soils etc., excavated from a river and has low shear strength, low bearing capacity and high compressibility. Post-September 2014 floods in Srinagar–Kashmir, several lakhs of tones of dredged material was targeted for dredging out by 2017–18. However, dredged material posed serious disposal and environmental problem in and around the flood channels and Jhelum River in the capital city Srinagar. Therefore, in this study, boulder dust, a waste product of boulder crushing units which is left in huge quantities and poses enormous environmental, health as well as disposal problems was chosen as an additive for stabilization of dredged soils. Basic tests were performed on soil samples from three different sites on the flood channel running through HMT area in the outskirts of Srinagar–Kashmir. A comparison of the test results was run, and the weakest soil sample was selected for further research work of stabilization. The mechanical properties of the dredged soil were enhanced by using boulder dust in increments of 4%. Atterberg limits, compaction characteristics, CBR values and UC strength were studied. The experimental results revealed that the boulder dust is a promising additive for stabilization of the dredged soil keeping in view the environmental concerns and the economy of the material.

Keywords Dredged soil · Flood channel · Boulder dust · Stabilization · Mechanical properties

---

B. A. Mir (✉) · K. M. N. S. Wani
Department of Civil Engineering, National Institute of Technology Srinagar, Srinagar, Jammu and Kashmir 190006, India
e-mail: p7mir@nitsri.net

K. M. N. S. Wani
e-mail: sakibwani_02phd17@nitsri.net

M. Latha Gali and R. R. P. (eds.), *Problematic Soils and Geoenvironmental Concerns*, Lecture Notes in Civil Engineering 88,
https://doi.org/10.1007/978-981-15-6237-2_35

1 Introduction

Soft soil deposits are usually weak with lower values of shear strength and bearing capacity. Excessive settlement in such soils takes place over a significant stretch of time, thus making it vulnerable to catastrophes and calamities. Engineers face an improbable task of determining the mechanical behavior of such soil deposits. Using such materials as construction material, sub-grade material and foundation material or even as backfill poses a vindicating task for a geotechnical engineer (Mir and Mir 2004; Mir and Shah 2013; Mir 2015a; Mir et al. 2016; Wani and Mir 2019a).

Additives such as lime, cement, fly ash, lime–cement–fly ash admixture, cement kiln dust, emulsified asphalt, geo-fibers and polymer stabilizers have been used to enhance the engineering properties of weak soils and stabilize them, but the environmental concerns during the processing of such materials have always been in question. The options and efficiency of any material or stabilizer banks on the soil type and the field conditions encountered. The knowledge of mechanistic behavior of stabilized soil is equally an important factor as is the selection of the stabilizer (Wani and Mir 2019b). The main contributions of this study to practice are on quantifying improvement in mechanical behavior due to boulder crusher dust treatment. The UCS and CBR values have been seen to increase with addition of stone dust (Agarwal 2015; Bshara et al. 2014; Kumar and Biradar 2014; Sabat and Bose 2013; Ramadas et al. 2010; Suresh et al. 2009). It has been also reported that stone dust as an additive has improved swelling characteristics of black cotton soil (Patidar and Mahiyar 2014; Venkateswarlu et al. 2015).

This study deals with the soft soil deposits generated by dredging of Bemina–HMT flood channel of river Jhelum. The flood channel runs for a stretch of very nearly 80 km beginning from Padshahi Bagh to Wular Lake in north Kashmir. The water carrying capacity of this channel has been reduced considerably over a period of the last century mainly due to the infringements by local people and the sediment deposits over the years. Post-2014 September floods, the govt. made a cognizable move to restore the water carrying capacity of river Jhelum by dredging which would in turn increase the load on engineers to use the bulk sediments for construction purposes or fill materials. Also, the area falls under the boundaries of a bio-diverse area, "the Hokersar wetlands" which is an internationally renowned site. The reality of the fact is that the flood channel has not been dredged for almost a century now, and the water levels have severely reduced in turn affecting the ecosystem of the wetland. Hence, this research becomes a very important aspect in terms of both bulk utilization of the dredged materials/sediments from the flood channels and in restoring the ecological balance of the adjoining area that has been altered by humans and natural disasters like floods.

2 Materials and Testing Methodology

2.1 Materials

In this study, the dredged soil samples were collected from three distinctive test pits in HMT area of Srinagar, Kashmir, along the flood spill channel of river Jhelum. At each site, soil samplers were used to collect in situ samples and disturbed samples were collected by shovels in large bags. Samples were sealed and transported with extreme precaution for studying their various properties and behavior after treating with waste boulder crusher dust. The boulder dust was acquired from M.S Crushers, Lasjan Nowgam.

2.2 Testing Methodology

The disturbed samples were subjected to various soil tests like soil classification (IS 1498, 1970) specific gravity (IS 2720-Part 3, 1980), wet sieve analysis (IS 2720-Part 4, 1985), Atterberg limits (IS 2720-Part 5, 1985), light compaction tests (IS 2720-Part 7, 1980) etc. Unconfined compressive strength (IS 2720-Part 10, 1991), direct shear tests (IS 2720-Part 13, 1986), California bearing ratio (IS 2720-Part 16, 2002) and coefficient of permeability (IS 2720-Part 17, 1986) were conducted on in situ samples to determine shear strength parameters in the undisturbed state as per the Standard Codal procedures. The samples were prepared by varying the boulder crusher dust content by 4%, 8%, 12%, 16% and 20%, respectively, and a series of CBR tests, UCS tests, Atterberg limit tests and compaction tests were performed on the weakest sample at 0.95 $\gamma_{d(max)}$ and optimum moisture content. Table 1 gives different geotechnical properties of the weakest soil site, i.e., site 2 near Khushipora, HMT. The waste boulder crusher dust used in this research had a specific gravity of 2.77 and was used only after passing it through 0.427 mm sieve, so as to remove the bigger boulder pieces and other impurities.

3 Results and Discussions

3.1 Physical and Engineering Properties of Untreated Dredged Soil

The dredged soil used in this investigation was grayish brown in color with a lot of silt content, and the soil profile at the test pit showed a considerable variation at deeper depths (Mir et al. 2013; Mir 2015b; Mir et al. 2016). However, the soil used in this study was collected from a depth of 2.5 m below ground level. The specific gravity

Table 1 Physical properties of dredged soil

Properties	Site 2
Natural moisture content (%)	25
Bulk unit weight (kN/m^3)	19.1
In situ dry unit weight (kN/m^3)	15.4
Specific gravity (G) of dredged soil	2.66
Specific gravity (G) of boulder crusher dust	2.77
% Finer than 75 μm	95
Clay (%)	1
Silt (%)	94
Sand (%)	5
Gravel (%)	0
Coefficient of uniformity, C_u	4.38
Coefficient of curvature, C_c	0.80
Suitability number, S_n	281
Liquid limit (%)	28
Plastic limit (%)	24
Shrinkage limit (%)	17
Plasticity index (%)	4.52
P.I, A-line	6.17
P.I, U-line	18.40
Classification	ML
Clay mineral	Kaolinite
Flow index, I_f	4.94
Toughness index, I_t	0.91
Activity	4.52
Consistency index, I_c	0.807
Liquidity index, I_L	0.172
In situ unconfined compressive strength, q_u (kN/m^2)	15.5
Unconfined compressive strength at OMC, q_u (kN/m^2)	33.25
In situ cohesion by direct shear test, c_u (kN/m^2)	14.7
Cohesion by direct shear test at OMC, c_u (kN/m^2)	22.3
In situ angle of internal friction by direct shear, Φ_u (°)	25.6
Angle of internal friction by direct shear at OMC, Φ_u (°)	27
Optimum moisture content (%)	16
Maximum dry unit weight (kN/m^3)	17.5
CBR, un-soaked (%)	5.6

(continued)

Table 1 (continued)

Properties	Site 2
CBR, soaked @ 94 h (%)	1.3

of the sample varied between 2.6–2.8 at different locations. This is attributed to the time of deposition of sediments in the flood channel being well over 80 years. Since the selected dredged soil consists of 94% silt, liquid limit tests were also performed using cone penetration method for cross-validation. However, there was negligible variation between the test values, and the average values for each trial were finally taken.

From Table 1, it is clear that the said soil is not suitable as an engineered construction material and as such requires proper stabilization technique keeping in view the availability and economic cost of the additive.

3.2 Physical and Engineering Properties of Waste Boulder Crusher Dust Treated Dredged Soil

3.2.1 Effect of BCD on Compaction Characteristics of the Soil

Generally, a high compactive effort enhances the geotechnical parameters of the soil; in order to achieve a predetermined set of values, a certain degree of relative compaction is necessary (Nicholson et al. 1994). From Fig. 1, it can be seen that an increase in MDD is observed up to 16% BCD in treated soil and a reverse trend is observed afterward. The OMC decreases up to 16% BCD and follows the same trend as MDD (Indiramma and Sudharani 2016; Mishra and Mishra 2015). Figure 2

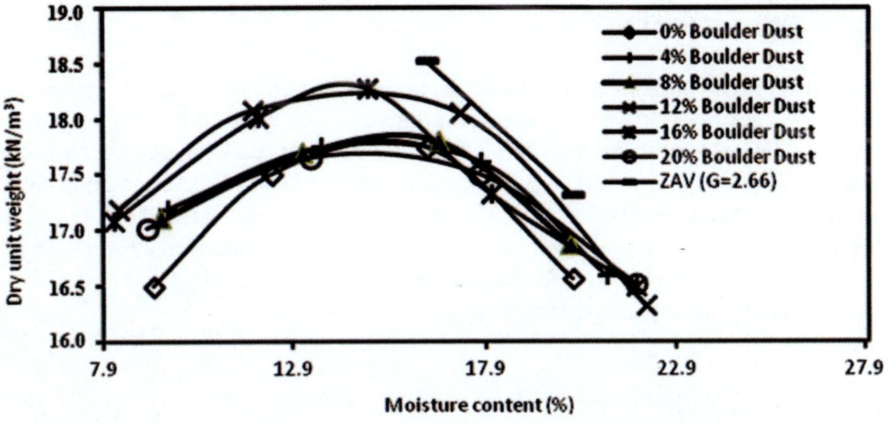

Fig. 1 Compaction curves for treated dredged soil with varying BCD content

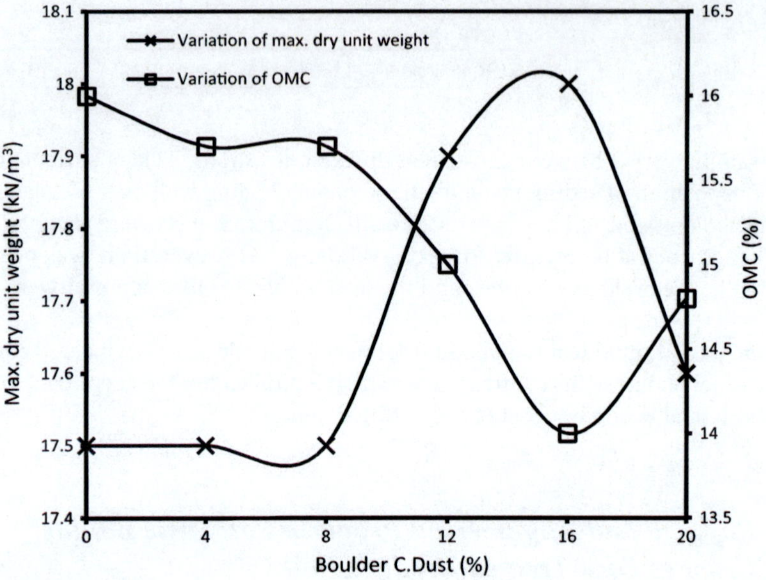

Fig. 2 Variation of compaction characteristics with varying BCD content

shows a contrast between MDD, OMC and BCD percentage for better understanding of the results. The increase in MDD is probably due to the reduction in percentage of voids and the improved binding between soil particles as a result of addition of BCD (Agarwal 2015). The reduction in OMC with addition of BCD was observed because of the reduction in surface area and probably the proper rearrangement of modified soil (Dixit and Patil 2016).

3.2.2 Effect of BCD on Consistency Limits

A series of liquid and plastic limit tests were performed on the untreated and BCD-treated soil samples. It was observed that the value of liquid limit and plastic limit kept on fluctuating with the increasing percentage of BCD content, and the plasticity index showed same results. The liquid limit of host soil was observed to be 28.56%, and it varied between 30.4 and 27.85% with 4% increments of BCD. The plastic limit of soil was 22.7% and it increased up to 12% and then showed a reverse trend. The plasticity index followed the same trend as LL and PL. Figures 3 and 4 show the fluctuating trend of consistency limits.

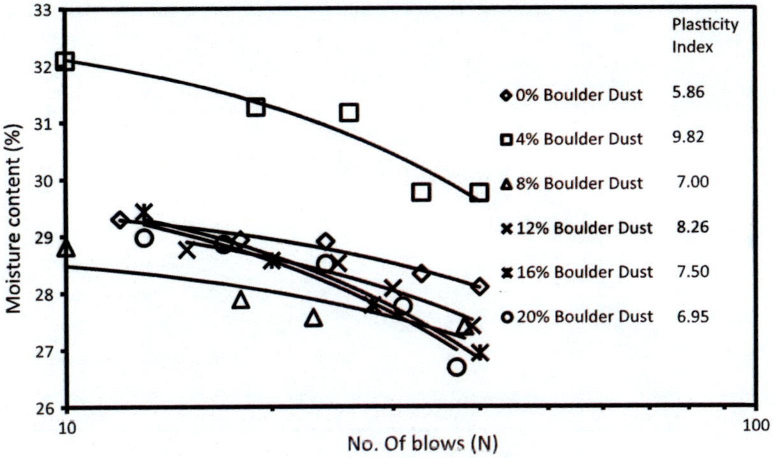

Fig. 3 Flow curves for BCD-treated dredged soil

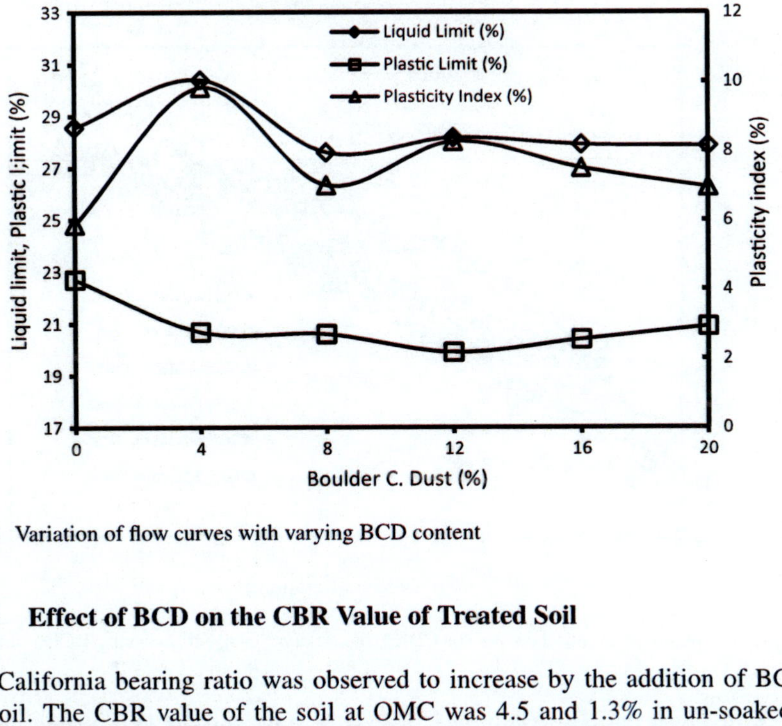

Fig. 4 Variation of flow curves with varying BCD content

3.2.3 Effect of BCD on the CBR Value of Treated Soil

The California bearing ratio was observed to increase by the addition of BCD to the soil. The CBR value of the soil at OMC was 4.5 and 1.3% in un-soaked and soaked condition simultaneously. The CBR value kept on increasing gradually as the percentage of BCD added to the soil was increased in increments of 4% as can be seen in Figs. 5 and 6; hence, the weak soil becomes suitable for use in pavements. Soil–BCD thicknesses are less than those required for granular bases carrying the

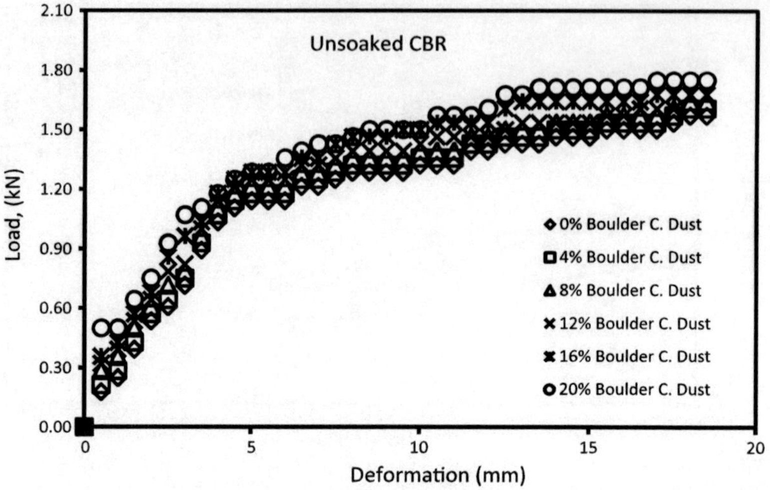

Fig. 5 Load versus penetration curves for CBR (un-soaked condition) with varying BCD content

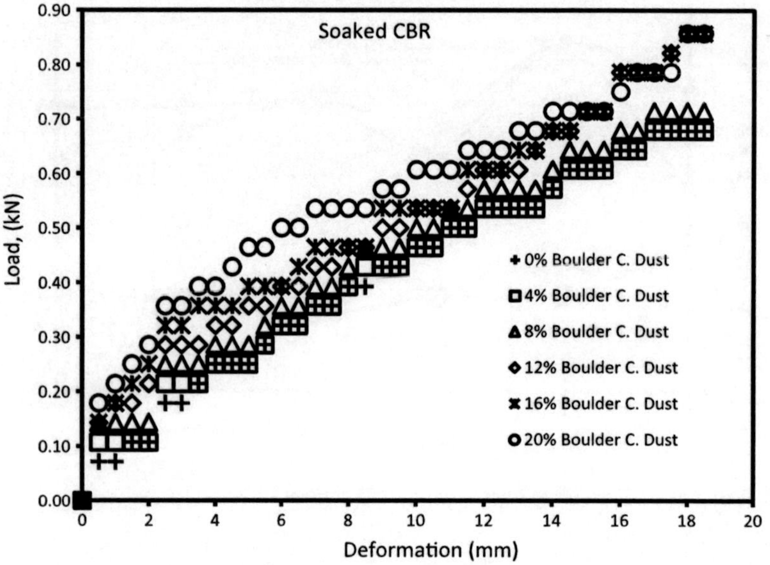

Fig. 6 Load versus penetration curves for CBR (soaked condition) with varying BCD content

same traffic over the same sub-grade. This is because soil–BCD is a rigid material that distributes loads over a broader area. A clear contrast between soaked and un-soaked CBR values can be seen in Fig. 7. The 160% rise in CBR value at 20% BCD can be attributed to the 0.425 mm size of dust used. This increase in the CBR value is not however significant as it has been observed by other researchers in different types

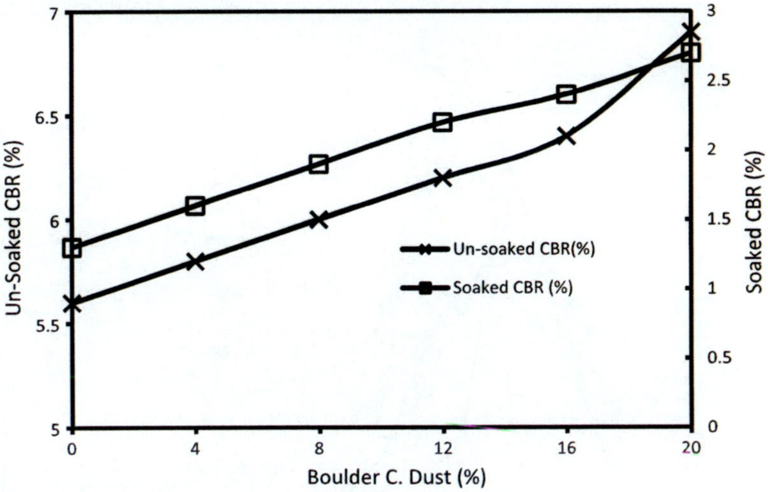

Fig. 7 Variation in un-soaked and soaked CBR values with varying BCD content

of materials like geo-synthetics, geo-grids, geo-textiles or even chemical stabilizers such as cement and other binders. It was concluded that when the size of dust was increased, the CBR value also increased up to a maximum point (Suresh et al. 2009). Hence, the improvement in CBR can be attributed to the size of the boulder crusher dust used.

3.2.4 Effect of BCD on the UCS Value of Treated Soil

Unconfined compression test is the simplest and quickest method to determine the shear strength parameters of soils. Test specimens were prepared, compacted under standard compaction at 0.95 γ_{dmax} and optimum moisture content. The samples were tested immediately and the test results revealed that addition of BCD has significant effect on the strength gain of the dredged soil. The optimum BCD content was found to be 16% as it showed the highest value of UC strength after which UCS value decreased. The decrease in the UCS value can be attributed to an increase in the angle of internal friction that might have reduced the overall shear strength.

The unconfined compressive strength was observed to increase by about two times with the addition of BCD to the soil. The UCS value of the soil increased from 46–83 kN/m^2. The percentage of BCD added to the soil was in increments of 4, 8, 12, 16 and 20% as can be seen in Fig. 8.

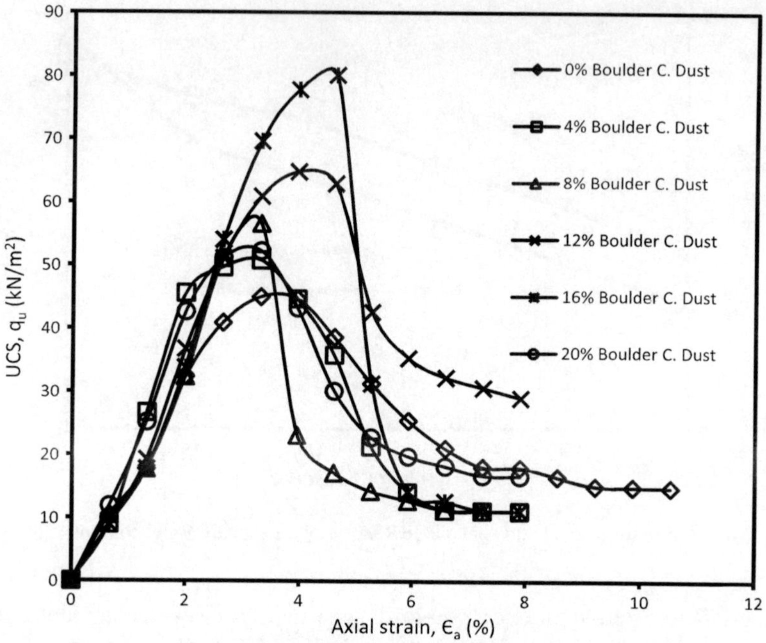

Fig. 8 Effect of BCD on stress–strain behavior of dredged soil with varying BCD content

4 Conclusions

The stabilization of the solid waste soil such as dredged soil with boulder crusher dust (BCD) is an effective means of stabilization of soils. It is seen that engineering properties of dredged soil have significantly improved by the addition of BCD.

- The CBR values increased making it an effective method in increasing the stability of the sub-grade and thus make the design more economical by reducing the overall thickness of the sub-grade layer. The use of coarser BCD wastes can further increase the CBR.
- The UC strength also increased, which clearly indicates that the material can be useful in practical application such as slope stability, as a fill and backfill material.
- The reduction in OMC with addition of BCD was observed because of the reduction in surface area and probably proper rearrangement of modified soil.
- The increase in MDD was due to the reduction in percentage of voids and the improved binding as a result of addition of BCD as it might contain some cementitious compounds.

5 Limitations and Future Scope of Work

Additional tests can be performed on the samples to further study the impact of BCD used on various other properties like shear strength, bearing capacity, permeability, etc. Tests like DST, plate load, etc., can be further carried out in laboratories to study the effect of addition of BCD to soil. The effect of the groundwater ecology should be given utmost priority so that the natural water composition does not get altered. Since very little work has been done on the field applications of using boulder dust, we could perform triaxial testing in laboratory in order to determine the field applications and limitations for the proper usage of BCD in the field. The size of the boulder dust can be increased and a number of trials conducted, so that the field application becomes broader, and we have a scope of using different sized waste dust for different engineering purposes.

Acknowledgements We wish to express our profound gratitude and thanks to the Dept. of Geotechnical Engineering, National Institute of Technology, Srinagar. We are extremely grateful to all the professors and the technical staff of geo-tech laboratory; without their guidance and support, this research could not have been accomplished.

References

Agarwal N (2015) Effect of stone dust on some geotechnical properties of soil. IOSR J Mech Civ Eng 12(1):61–64

Bshara AS, Bind EY, Sinha PK (2014) Effect of stone dust on geotechnical properties of poor soil. Int J Civ Eng Technol (IJCIET) 5(4):37–47

Dixit MS, Patil KA (2016) Utilization of stone dust to improve the properties of expansive soil. Int J Civ Eng Technol (IJCIET) 7(4):440–447

Indiramma P, Sudharani C (2016) Use of quarry dust for stabilising expansive soil. Int J Innov Res Sci Eng Technol (IJERSET) 5(1):1151–1157

IS 1498 (1970) Classification and identification of soils for general engineering purposes. Bureau of Indian Standards, New Delhi, India

IS 2720-Part 3 (1980) Method of soil tests: Determination of specific gravity. Bureau of Indian Standards, New Delhi, India

IS 2720-Part 4 (1985) Method of test for soils: Determination of grain size distribution for soil samples. Bureau of Indian Standards, New Delhi, India

IS 2720-Part 5 (1985) Method of test for soils: Determination of Atterberg limits (liquid limit and plastic limit) of soils. Bureau of Indian Standards, New Delhi, India

IS 2720-Part 7 (1980) Method of test for soils: Determination of water content dry density relation using light compaction. Bureau of Indian Standards, New Delhi, India

IS 2720-Part 10 (1991) Method of test for soils: Determination of unconfined compressive strength of soils. Bureau of Indian Standards, New Delhi, India

IS 2720-Part 13 (1986) Method of test for soils: Determination of shear strength parameter by Direct shear test. New Delhi, India

IS 2720-Part 16 (2002) Method of test for soils: Laboratory determination of CBR. Bureau of Indian Standards, New Delhi, India

IS 2720-Part 17 (1986) Method of test for soils: Laboratory determination of permeability of soils. Bureau of Indian Standards, New Delhi, India

Kumar UA, Biradar KB (2014) Soft subgrade stabilization with quarry dust-an industrial waste. Int J Res Eng Technol 3(08):409–412

Mir BA (2015a) Some studies on geotechnical characterization of dredged soil for sustainable development of Dal Lake and environmental restoration. Special issue of Int J Tech Res Appl 12:04–09

Mir BA (2015b) Some studies on the effect of fly ash and lime on physical and mechanical properties of expansive clay. Int J Civil Eng 13:3&4B, Transaction B: Geotech Eng 203–212

Mir BA, Mir FA (2004) Applications and geotechnical evaluation of dredged soil obtained from Dal Lake in Srinagar. In: Proceedings of national conference on soils and their applications in civil engineering, pp 26–37

Mir BA, Shah MY (2013) Need for dredging and dredged material characterization from Dal Lake Srinagar, an overview. In: Proceedings of IGC, vol 1, pp 1–9

Mir BA, Wani BA, Ahmad N, Ayoub R, Dar LA, Rashid SU, Aziz J (2013) Physical and compaction behaviour of dredged material from Dal Lake, Srinagar. Int J Civ Eng Appl 3(7):4–8

Mir BA, Amin F, Majid B (2016) Some studies on physical and mechanical behaviour of dredged soil from flood spill channel of Jhelum river, Srinagar. Acta IngenierÃa Civil 1(1):1–7. http://dx.doi.org/10.20936/AICV/160101

Mishra, RS, Mishra, B (2015) Improvement in characteristics of expansive soil by using quarry waste and its comparison with other materials like cement and lime being used for soil improvement. Int J Innov Res Sci Eng Technol 4(8):7416–7431

Nicholson PG, Kashyap V, Fujii CF (1994) Lime and fly ash admixture improvement of tropical Hawaiian soils. Trans Res Rec 1440:71–77. http://onlinepubs.trb.org/Onlinepubs/trr/1994/1440/1440-010.pdf

Patidar A, Mahiyar H (2014) An Experimental study on stabilization of black cotton soil using HDPE wastage fibres, stone dust & lime. Int J Adv Sci and Tech Res 6(4): 90–98

Ramadas TL, Kumar ND, Aparna G (2010) Swelling and strength characteristics of expansive soil treated with stone dust and fly Ash. In Indian geotechnical conference-2010, GEOtrendz, pp 557–560

Sabat AK, Bose B (2013) Improvement in geotechnical properties of an expansive soil using fly ash-quarry dust mixes. Elec J Geotech Eng 18, Bund Q 3487–3500

Suresh K, Padmavathi V, Sultana A (2009) Experimental study on stabilization of black cotton soil with stone dust and fibers. Indian geotechnical society, Guntur, A.P., pp 502–506

Venkateswarlu H, Prasad ACSV, Raju GVRR (2015) Study on behavior of expansive soil treated with quarry dust. Int J Eng Innov Tech 4(10):193–196

Wani KMN, Mir BA (2019a) Effect of biological cementation on the mechanical behaviour of dredged soils with emphasis on micro-structural analysis. Int J Geosyn Ground Eng 5(4):32. https://doi.org/10.1007/s40891-019-0183-9

Wani KMN, Mir BA (2019b) Influence of microbial geo-technology in the stabilization of dredged soils. Int J Geotech Eng 1–10. https://doi.org/10.1080/19386362.2019.1643099

Electro-osmosis: A Review from the Past

Amal Azad Sahib, I. Bushra, and G. Rejimon

Abstract Electro-osmosis is a powerful technique, as a means of dewatering soils of high compressibility and moisture content. Electro-osmosis is an established technique and has been investigated by many researchers as long as a century ago. The treatment factors that contribute to the effectiveness of electro-osmotic consolidation are type of electrode, voltage gradient, polarity reversal, current intermittence and duration of treatment. Copper, mild steel and stainless steel in different shapes and forms have been used as electrodes. Electrokinetic geosynthetics (EKG) used in electro-osmotic consolidation applications provide electrokinetic function in addition to the filtration and drainage functions. The EKG electrodes are less susceptible to corrosion due to the polymeric cover or treatment against corrosion. Even though most studies claim the effectiveness of this technique, the procedure is not widely accepted in the industry due to the risks and costs involved. This paper aims to review the suitability of this technique on soils around the world. Also, this paper looks closely into the properties of the soil that make it ideal for the success of this technique. With reference to Indian soils, the results of electro-osmosis studies on Kuttanad clay are presented.

Keywords Electro-osmosis · Electrode material · Kuttanad clay

A. A. Sahib (✉) · I. Bushra
Civil Engineering, TKM College of Engineering, Kollam, India
e-mail: azad52@yahoo.com

I. Bushra
e-mail: bushrasheik@gmail.com

G. Rejimon
Electrical and Electronics Engineering, TKM College of Engineering, Kollam, India
e-mail: reji_gopal@yahoo.co.in

M. Latha Gali and R. R. P. (eds.), *Problematic Soils and Geoenvironmental Concerns*, Lecture Notes in Civil Engineering 88,
https://doi.org/10.1007/978-981-15-6237-2_36

1 Introduction

From its inception since 1949, electro-osmosis is recognized as a technique that holds vast potential to improve waterlogged clayey soils (Casagrande 1949; Gray and Mitchell 1967; Golenko 1971; Chappell and Burton 1975; Wan and Mitchell 1976; Gray and Somogyi 1977; Lo et al. 1991; Shang and Masterson 2000; Hamir et al. 2001; Chew et al. 2004; Rittirong et al. 2008). Though the efficiency is highly questionable based on the parameters trialled for various soils, researchers around the world continue to bring out theories that influence the efficiency of this technique.

Electro-osmosis is unison of many mechanisms such as electrophoresis, dielectrophoresis, electrokinetic migration and electrochemical hardening. In an electrokinetic dewatering process, the two phases that dominate are electrophoretic sedimentation and electro-osmotic consolidation (Shang 1997). Electrophoresis is the phenomenon by which negatively charged ions (in this case clay) are attracted to positively charged electrode (anode). The movement of water, under a DC potential, from positive (anode) to negative electrode is termed as electro-osmosis.

Electro-osmosis (EO) is the most useful of the electrokinetic processes activated with EKG because it holds the potential to overcome the limitations of very slow and in some cases effectively zero hydraulic flow in fine-grained, low-permeability materials such as silts and clays.

2 Soil Parameters Governing EO

Electro-osmosis is fundamentally governed by the cation–anion distribution and water ion distribution. The better the co-ion (the ions in the external solution phase, a few of which invade the internal solution phase) exclusion, less affected is the electro-osmotic flow by salinity increases in pore water. Co-ion exclusion can be enhanced by a fixed charge density or cation exchange capacity and low water content. Active clays with high exchange capacity such as illitic and bentonitic soils can effectively exclude co-ions even at high salinity but low concentration of water itself reduces the amount of water moved per unit electrical charge. Inactive clays such as silty and kaolin clays with low exchange capacity at high water content and low salinity would ensure a high electro-osmotic transport but with increase in salinity, the co-ion exclusion falls off rapidly thus reducing the electro-osmotic efficiency (Gray and Mitchell 1967).

Electro-osmotic permeability differs from hydraulic permeability in that it is unrelated to grain size directly. The electro-osmotic flow rate can be 100–10,000 times greater than hydraulic flow rate in fine-grained soils. In the case where clay mineralogy details are unknown, particle size distribution may be used to determine the proportion of clay. Electro-osmotic process is not just limited to clayey soils. It has been reported in materials such as quartz powder, rock flour, ochre and alum sludge and in sewage streams including those comprising humic, anaerobic digested, surplus activated and primary sludges. The factors that influence the mechanism are the origin and magnitude of the negative surface charges and the salinity of the pore water which would act to compress the double layer and thus minimize the effectiveness of the surface charge (Jones et al. 2011).

The electro-osmotic flow, q_e, produced by an applied electric field is given by the expression,

$$q_e = k_e i_e A \tag{1}$$

where k_e is the electro-osmotic conductivity, i_e is voltage gradient and A is cross-sectional area.

The electro-osmotic flow velocity, v_e, is given by the expression,

$$v_e = \frac{D\xi \Delta E}{4\pi \eta \Delta L} \tag{2}$$

where v_e = electro-osmotic flow velocity, in centimetres per second, ΔE = applied voltage, in volts, ΔL = electrode spacing, in centimetres, η = viscosity, in centipoises, D = dielectric constant of the soil water and ξ = Zeta potential in volts.

Shang (1997) discusses the influence of zeta potential on the electro-osmotic permeability. By using the Stern–Gouy model, zeta potential can be measured and the effectiveness of a soil type to electro-osmosis can be evaluated.

2.1 Typical Electrode and Electrolyte Properties

Malekzadeh et al. (2017) conducted EK tests using polyaniline-coated galvanized steel electrodes. It was found that a 5 V application gave an improvement comparable to soil stabilized by uncoated electrodes with 30 V application, but a 30 V application did not give a proportionate improvement as the coating disintegrated at this voltage gradient.

Xue et al. (2015) tested on Dalian marine sludge using three electrode types and found iron electrodes to be most effective compared to copper and aluminium for this particular soil. This has been concluded based on the gradual decline of voltage and current, and the corrosion loss for iron was consistent with values calculated using Faraday's law. The strength improvement was attributed to a cementing action near the anode. The reason for the excess corrosion loss in the other two electrodes was assumed to be due to salinity and acidity. This could have been better supported with pH and conductivity measurements for this soil, which have not been mentioned in the paper. Also the same could be compared to treatments using painted or galvanized electrodes.

Citeau et al. (2011) conducted EK tests on a biscuit factory sludge, with varying conductivity (or salinity) and pH. They concluded that conductivity values around 0.3 mS/cm and slightly acidic pH of 5.3 gave the best results for EK treatment.

Soil cementation due to electrokinetic treatment which has a phenomenal role in the strength improvement is reported to be closely related to the electric field intensity, which in turn is controlled by the electrode layout and applied voltage (Mohamedelhassan et al. 2005; Rittirong et al. 2007). Rittirong et al. (2008) investigated the effects of the electrode configuration based on the electric field analysis of large-scale electrokinetic tests on model caissons embedded in calcareous sand. The pull-out resistance of the caissons increased by 90%, and it was concluded that increased electric field intensity on the exterior of the caisson, lowered electric field intensity on the interior and embedding the central electrode below the bottom of caisson can optimize the method.

Chen and Murdoch (1999) conducted EK field tests on illite–smectite clays of Cincinnati, Ohio. A mesh made from titanium, coated with metal oxides, 3.1 × 3.4 m, was used as the upper anode. The lower electrode was hydraulic fracture filled with granular graphite, 2.5 m radius. Using an average gradient of 20–31 V/m and current 42–57 A resulted in a total 2–4 L/h infiltration, 0.6–0.8 L/h of drainage due to electro-osmosis alone. The pH of soil reduced near the anode and slightly increased near the cathode, and the migration of pH was slower than similar fronts in laboratory experiments.

Chew et al. (2004) used electrokinetic geosynthetics (EKG) in lab and field trials on Singapore marine clay. 8 m deep soft clay underlying an 18 m sand fill was subjected to EO treatment using EK drains for two weeks showing ten-fold reduction in time to achieve the strengths compared to conventional PVDs. The total energy consumption was reported as 1.8 kWh/m^3. Further, an emphasis is laid on high conductance of the EVD, insulation of the shoes and provision of polarity reversal.

Rittirong et al. (2008) conducted a similar field trial on soils of Sarawak, Malaysia, to stabilize an access road, the total treatment area being 560 m × 4 m on each side of the existing road. It was divided into sections of 28 m × 4 m each for individual power supply. The electrification period was 20 h daily for five days in each section. Five power supplies were circulated over the treatment area. The total treatment period for the entire area was 14 days. Electro-osmotic flow was found to have ceased in 3 days of the power supply, though polarity reversal procedures were systematically followed. The gas generated at the soil/electrode interface is believed to have acted as an electrical insulator that increases the electrical resistivity which probably diminished the effectiveness of the electrokinetic treatment. The energy consumption was 0.7 kWh/m^3. Table 1 gives a comparison of the power consumed in various EO field trials.

Intermittent current, a two-step procedure—electrophoretic consolidation by top cathode–bottom anode condition and electro-osmotic consolidation by top anode–bottom cathode condition, reversing polarity immediately after electrophoresis and reducing time of electro-osmotic consolidation are some measures to reduce energy consumption (Shang 1997).

Table 1 Energy consumption in field tests

Project	Electrode material	*W* (kWh/m^3)	References
Stabilization of excavation	Steel	17	Bjerrum et al. (1967)
Stabilization of shipyard embankment	Steel	0.5	Chappell and Burton (1975)
Slope stability at Kootenay River	Steel	6.7	Wade (1976)
Field test of strengthening sensitive clay	Copper	6.4	Lo et al. (1991)
Improvement of steel pile groups	Steel	1	Milligan (1994)
Tuas land reclamation	EVD	1.8	Chew et al. (2004)
Sarawak road construction	EVD	0.7	Rittirong et al. (2008)

3 Results from EO Studies on Kuttanad Clay

Kuttanad clay has gained much significance as a very soft soil with low bearing capacity and large compressibility values. The waterlogging nature has rendered the area uninhabitable, especially during severe monsoons. The monsoons of 2018 have been particularly the most disastrous in the last 30 years. The authors attempted electro-osmotic studies in the laboratory using copper rods (Amal and Bushra 2014) and copper sheets (Amal et al. 2017) on Kuttanad soil. A 60 V/m application on a 30 cm × 30 cm test tank with 20 cm spacing of electrodes produced 20 kPa undrained shear strength when copper rods were used and 20–25 kPa when copper sheets covering the entire area was used (Fig. 1). This indicates that using intermittent

Fig. 1 Electro-osmotic set-up 30 × 30 cm

Table 2 Results of tests conducted for different voltage gradients on 30 cm × 30 cm mould (20 cm electrode spacing)

Voltage gradient (V/m)	Water collected (ml)		Shear strength @ anode before EO (kPa)	Shear strength @ anode after EO (kPa)	I^a (A)	P^a (kWh/m^3)
	After SS[a]	After EO				
20	105	78	2.2	18.3	0.08	0.56
40	64	34	1.1	21.8	0.1	1.4
60	295	165	2.1	27.3	0.8	17.1
80	300	190	2.9	28.3	1	28.4
100	282	152	1.1	31.5	1.2	42.6
120	172	85	2.4	41.3	2.1	89.6

[a]SS—self-stabilization, *I*—current, *P*—power consumed

electrodes (rods) would suit more than energy-intensive electrode sheets.

In the 30 cm × 30 cm mould, when the voltage gradients were varied in independent trials, keeping the spacing between electrodes fixed at 20 cm, there is a clear increase in the rate of improvement (Table 2). The shear strength for 120 V/m treated samples were about 1.3 times and 2.5 times more than the 60 V/m and 20 V/m treated samples. Therefore, it may be inferred that for smaller contact areas, a higher voltage gradient would yield better results, whereas for larger contact areas, the optimum voltage gradient need not be a higher value.

Even though the strength improvement is higher for higher voltage gradients in 30 × 30 mould, the water drained at the cathode reduces slightly with increase in voltage gradient. It may be inferred that the reduction in water content for higher voltages was accelerated due to heating, in addition to the drainage of water at the cathode. As can be seen from Fig. 2, the percentage of water drained by electro-osmosis (EO) to the total water drained by self-consolidation and EO is around 42% for 20 V/m EO, whereas it falls to around 32% for 120 V/m EO. The rate of heating

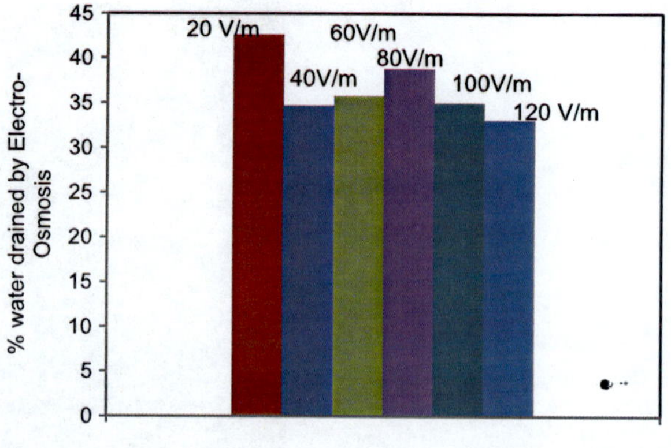

Fig. 2 Percentage water drained by electro-osmosis

Table 3 Results of tests conducted for different voltage gradients on 90 cm × 90 cm mould

Spacing of electrode (cm)	Voltage gradient (V/m)	Shear strength @ anode before EO (kPa)	Shear strength @ anode after EO (kPa)	*I* (A)	*P* (kWh/m^3)
20	60	2.9	158.4	2.6	6.2
	120	3.6	170.3	5.1	24.2
40	60	3.6	105.7	3.2	15.2
	120	3.9	93.4	6.5	61.6
60	60	2.6	92.5	3	21.3
	120	2.9	83.5	6.4	91
80	60	3.4	90.7	4.1	38.9
	120	3.6	130.9	8.7	165

of soils with increased potential gradients seems to be an interesting aspect to be investigated.

Table 3 shows the results from tests on a 90 cm × 90 cm mould using 60 and 120 V/m for different spacings of electrodes (sheets). The strength improvement for all the combinations is commendable but considering power loss, the voltage gradient of 60 V/m seems more reliable compared to 120 V/m.

The pH of Kuttanad soil in untreated form was 3.7. Table 4 shows the values of pH of the soil after electro-osmotic treatment. It is observed that near anode, the soil remains acidic and lower than 3.7 and near cathode, there is slight increase in the pH though it remains in the acidic range. However, this aspect corrodes the electrode material and hence deters electro-osmosis process. This is evident from the observations of immense power consumption. In field trials, this aspect has to be carefully handled.

While testing the pH of the Kuttanad soil, the values obtained clearly indicated that the soil is acidic in nature and after the electro-osmotic stabilization process, its acidity seems to be increasing. This led to the thought that the soil can be used as a battery electrolyte which can replace the present electrolytes that are not at all eco-friendly. Hence, an earth battery was prepared by using 1.89 kg of soil in liquid form having 66% water and by using copper sheet as anode and aluminium sheet as

Table 4 pH measurements after EO trials

Voltage gradient (V/m)	pH	
	Near anode	Near cathode
20	3.2	4.3
40	3.3	4
60	3.4	3.7
80	3.4	3.8
100	3.2	3.9
120	3.1	4.1

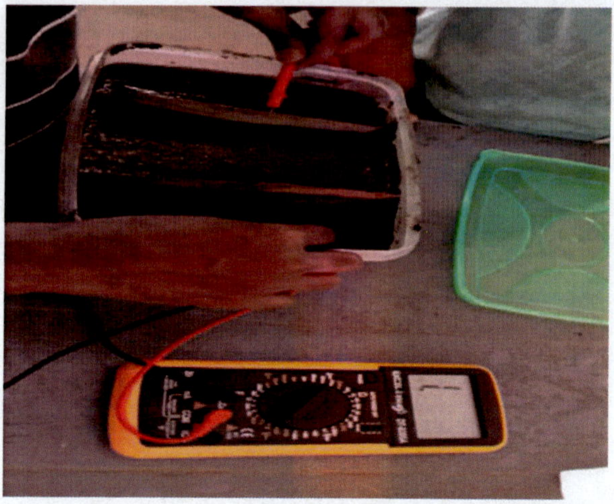

Fig. 3 Earth cell

cathode in a box of 20 × 15 × 7 cm, the total weight of cell was 3.5 kg (Fig. 3). The potential gradient in between the electrodes was measured with a multimeter. Initially from the cell, a potential difference of 0.75 V was obtained, i.e. without using concentrated sulphuric acid. Later, along with the soil, concentrated sulphuric acid was added by varying amounts starting from 15 to 30% by an increment of 5%. While adding 30% of sulphuric acid, 3.48 V was obtained which was comparatively higher than that of the commonly used batteries.

It is observed from Table 5 that as the concentration of sulphuric acid increases voltage obtained also increases. Usually, a small AA battery has 1.5 V and maximum current output of 1400 mA and has a weight of 23 g. Lead–acid battery has 12.0 V and maximum current output 8 A (with 38% of sulphuric acid) which has a weight of 4 kg and size of 15.1 × 9.8 × 9.4 cm. As compared to these batteries, the use of Kuttanad soil as a battery electrolyte is limited since the voltage and current obtained are not substantial compared to the size of the cell.

Table 5 Earth cell results

Concentration of sulphuric acid (%)	Voltage obtained (V)	Current obtained (mA)
0	0.75	40
15	2.1	110
20	2.7	155
25	3	160
30	3.5	186

4 Conclusions

In order to successfully implement the electro-osmotic technique on field, it is essential to conduct laboratory tests to determine a specific design standard considering various parameters such as pH, type of minerals, salinity, material and type of electrodes, electrode configuration and voltage gradient. Overall, the effectiveness of this technique can be determined from zeta potential values of the soil that give an indication of the electro-osmotic permeability.

With respect to Kuttanad soil, the effectiveness of the technique using two arrangements, (i) rods (ii) sheets, were evaluated. Both gave similar strength improvement, power consumption being less for the rods, hence making them more effective. Kuttanad soil that was tested here was already acidic and EO reduced the pH further (near the anode), thus corroding the electrodes and deterring the process.

Acknowledgements The authors would like to extend their sincere gratitude to TEQIP-II for funding this project under seed money and to the Department of Civil Engineering, Department of Electrical Engineering and Department of Mechanical Engineering, TKM College of Engineering, Kollam, for their assistance. We also appreciate the work of the eighth semester B.Tech. students of 2015–2016, who performed the laboratory tests as part of their final year project.

References

Amal AS, Bushra I (2014) Suitability of electro-osmosis in Kuttanad clay. In: Indian geotechnical conference, Kakinada, India

Amal AS, Bushra I, Rejimon G (2017) Electro-osmosis studies on Kuttanad clay. In: 6th Indian young geotechnical conference, pp 495–498

Bjerrum L, Moum J, Eide O (1967) Application of electro-osmosis to a foundation problem in a Norwegian Quick clay. Geotechnique 17(3):214–235

Casagrande I (1949) Electro-osmosis in soils. Géotechnique 1(3):159–177

Chappell BA, Burton PL (1975) Electro-osmosis applied to unstable embankment. J Geotech Eng Div ASCE 102(GT8):733–740

Chen J-L, Murdoch L (1999) Effects of electroosmosis on natural soil: field test. J Geotech Geoenviron Eng 125(12):1090–1098

Chew SH, Karunaratne GP, Kuma VM, Lim LH, Toh ML, Hee AM (2004) A field trial for soft clay consolidation using electrical vertical drains. Geotext Geomembr 22:17–35

Citeau M, Larue O, Vorobiev E (2011) Effect of electrolytes content on the electro-osmotic dewatering of agro-industrial sludge. In: Proceedings of the 11th international congress on engineering and food, May 22–26, Athens, Greece

Golenko NA (1971) Electro-osmotic and hydraulic flow in soils with a varying porosity. Fundamenty i MekhanikaGuntov 1:23–25

Gray DH, Mitchell JK (1967) Fundamentals aspects of electro-osmosis in soils. J Soil Mech Found Div ASCE 93(SM6):209–236

Gray DH, Somogyi F (1977) Electro-osmotic dewatering with polarity reversals. J Geotech Eng Div ASCE (GT1):51–54

Hamir RB, Jones CJPF, Clarke BG (2001) Electrically conductive geosynthetics for consolidation and reinforced soil. Geotext Geomembr 19:455–482

Jones CJFP, Lamont-Black J, Glendinning S (2011) Electrokinetic geosynthetics in hydraulic applications. Geotext Geomembr 29(4):381–390

Lo KY, Ho KS, Inculet II (1991) Electro-osmotic strengthening of soft, sensitive clays. Can Geotech J 28:62–63

Malekzadeh M, Sivakugan N, Kazum O, Mathan B (2017) Effect of polyaniline-coated galvanized steel electrodes on electrokinetic sedimentation of dredged mud slurries. Can Geotech J 54:1150–1157

Milligan V (1994) First application of electro-osmosis to improve friction capacity: three decades later. In: Proceedings of the 13th international conference on soil mechanics and foundation engineering, vol 5, New Delhi, India, pp 1–5

Mohamedelhassan E, Shang JQ, Ismail MA, Randolph MF (2005) Electrochemical cementation of calcareous sand for offshore foundations. Int J Offshore Polar Eng 15(1):71–79

Rittirong A, Shang JQ, Ismail MA, Randolph MF (2007) Effect of electric field intensity on electro-cementation of caissons in calcareous sand. Int J Offshore Polar Eng 17(1):74–79

Rittirong A, Douglas RS, Shang JQ, Lee EC (2008) Electrokinetic improvement of soft clay using electrical vertical drains. Geosynth Int 15(5):369–381

Shang JQ (1997) Electrokinetic dewatering of clay slurries as engineering covers. Can Geotech J 34(1):117–133

Shang JQ, Masterson KL (2000) An electro kinetic testing apparatus for undisturbed soils under in-situ stress conditions. Geotech Test J ASTM 23(2):215–224

Wade MH (1976) Slope stability by electro-osmosis. In: Proceedings of the 29th Canadian geotechnical conference, Vancouver, Canada, Section 10, pp 44–66

Wan T, Mitchell JK (1976) Electro-osmotic consolidation of soils. J Geotech Eng Div ASCE 102(GT5):473–491

ZhijiaXue XT, Yang Q, Wan Y, Yang G (2015) Comparison of electro-osmosis experiments on marine sludge with different electrode materials. Drying Technol 33(8):986–995

Probabilistic Performance Analysis of Prefabricated Vertical Drains on Soft Soils

T. G. Parameswaran, K. M. Nazeeh, and G. L. Sivakumar Babu

Abstract Properties of soft soils largely influence their behavior and the design scheme of the ground improvement that would be required. The time for 90% consolidation could be estimated theoretically with time factor, coefficient of consolidation, compressibility and permeability. As the reproducibility of these tests is very less, estimating the properties accurately is highly challenging. Repeating the test also would usually result in obtaining a wide range of values. Since the ground improvement scheme is largely dependent on these properties, probabilistic tools and reliability-based design must be used. In soft soils, prefabricated vertical drains (PVDs) along with preloading is a popular method of increasing the bearing capacity of soils wherein the improvement is achieved by accelerating the consolidation process. The spatial and temporal variability of the soil properties affects the performance of the PVDs installed in the ground, and the same is investigated using probability distribution of the various properties used in the design of the ground improvement scheme. Monte Carlo simulations using the probability distribution and statistics of the design properties are used to capture the effect of variability of the design properties on the time for consolidation.

Keywords Probabilistic analysis · PVD · Soft soils

1 Introduction

Huge amounts of consolidation settlements are commonly encountered in soft soils owing to their high compressibility. The consolidation settlements are usually large which pose damage to the structures, and hence, estimating their magnitudes or rates often plays a vital role in civil engineering projects. Consolidation settlements in clays are conventionally estimated with time factor, coefficient of consolidation, compressibility and permeability; the values of which are obtained from laboratory tests. Determining the values of these properties accurately from laboratory tests is

T. G. Parameswaran (✉) · K. M. Nazeeh · G. L. Sivakumar Babu
Department of Civil Engineering, Indian Institute of Science, Bangalore 560012, India
e-mail: parameswarantg@gmail.com

M. Latha Gali and R. R. P. (eds.), *Problematic Soils and Geoenvironmental Concerns*, Lecture Notes in Civil Engineering 88,
https://doi.org/10.1007/978-981-15-6237-2_37

highly challenging. Usually, a wide range of values are obtained when such tests are repeated. This would result in even more wider range of values for the settlements or the time rate of consolidation. Such range of values could even cause inappropriate designs of a ground improvement technique that might be required. In such cases, probabilistic tools and reliability-based designs serve as ideal methodologies.

In the present study, ground improvement through the application of prefabricated vertical drains (PVDs) and preloading is considered. PVDs with preloading improve the bearing capacity or the shear strength of the soil by accelerating the consolidation. As the properties of the soil surrounding the PVDs affect the consolidation, variation in the properties could lead to inappropriate design of the PVDs and affect their performance. Variation in the properties can occur from the inherent variabilities in the soil related to its formation, prevailed environmental or climatic factors. They could also arise from the sampling and testing process due to the inability of reproducing the exact sampling and testing conditions as well as other factors such as stress release and disturbances due to sampling tubes. Uncertainties in the values obtained would also be present due to the lesser reproducibility of the tests. Hence, variations in the properties estimated from lab/field testing are unavoidable. In soil consolidation, permeability (horizontal and vertical), compressibility and the bulk density of the soil play a very crucial role (Bari et al. 2013). Variability of these properties can alter the time for ultimate settlement and hence the design of PVDs. Inappropriate designs could lead to erroneous design times for ultimate settlement and would cause failures to the structures constructed.

Such variabilities and uncertainties arising from the above-mentioned sources or their effects cannot be addressed in a deterministic design perspective. At the same time, the use of probabilistic tools can help to quantify these variabilities and bring about the uncertainty in the design due to such variabilities. This will help to estimate the reliability of the ground improvement system to obtain the expected performance. Thus, an attempt is made in this paper to address such uncertainties and variabilities in soil consolidation and the PVD design scheme using reliability techniques.

2 Methodology

Reliability of a system can be defined as the probability of performing the required function under given conditions, without failure. Thus, under the specified conditions, the expected performance of the system can be quantified. The main advantage of reliability-based design (RBD) is the ability to incorporate the uncertainties or variabilities in the various factors or conditions considered in the design. In this paper, the effects of variabilities in the permeability (horizontal and vertical), compressibility and the bulk density of the soil on soil consolidation and the subsequent PVD design are being addressed.

Among the various methods available for modeling stochastic problems in geotechnical engineering, the use of Monte Carlo simulations has gained popularity in recent years (Elkateb et al. 2003). The method is versatile and mature enough to be

Table 1 Properties at the site

Variables	Parameters		
	Range	Mean	COV
Permeability in vertical direction, k_v (m/s)	10^{-9} to 10^{-10}	5.5×10^{-10}	0.5 and 1.0
Volume compressibility, m_v (m^2/MN)	0.7 to 1.0	0.85	0.1
Bulk density of the soil, γ (kN/m^3)	13 to 16	14.5	0.05

used for reliability analysis of large engineering structures (Haldar 2002; Schuëller 2002; Marek and Haldar 2003).

3 Site Details

The site taken for the current study is made up of a soft clay layer, 9 m thickness below which is 13 m thick sand. An embankment (7 m height, $\gamma_b = 18$ kN/m^3) is proposed to be constructed on the clay layer, and hence, the time for 90% consolidation at the center of this clay layer due to this embankment loading is being addressed. The design properties such as permeability in the vertical direction (k_v), volume compressibility (m_v) and the bulk density (γ) are considered as variables, and their statistics considered in the design are provided in Table 1.

The variations in the properties arise from the inherent variability in the soil properties at the site and from the uncertainties or errors that would be involved in the laboratory measurements. In such cases, probabilistic analysis serves as a better tool for addressing the variabilities.

4 Effect of Property Variation on Time for Settlement

In the current problem, if a deterministic approach is adopted, considering the average values for the properties, the value of coefficient of consolidation (c_v) would be 2 m^2/year and the time for 90% consolidation (t_{90}) would be coming close to 8 years. In cases where the variations in the properties are high, a deterministic approach may not represent the realistic behavior at the site. In such cases, probabilistic methods serve as a better tool for addressing such issues. Monte Carlo simulation, in which the analysis is carried out considering the properties as random variables and repeated to obtain converging results, is attempted here based on the properties and its statistics as shown in Table 1.

A lognormal distribution was selected for the properties (Bari et al. 2013) in Table 1. Also, a COV values of 0.5 and 1.0 were considered for the vertical permeability to address the uncertainty. The variation in the c_v and hence t_{90} for both the cases, after 10,000 simulations, is as shown in Figs. 1 and 2.

Fig. 1 Variation of c_v due to variation in COV of k_v

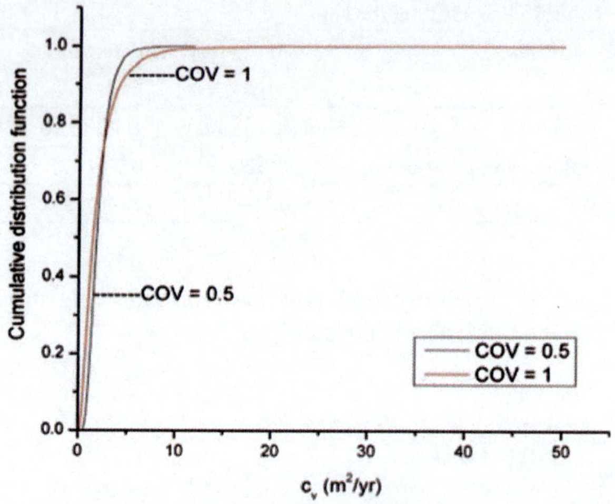

Fig. 2 Variation of t_{90} due to variation in COV of k_v

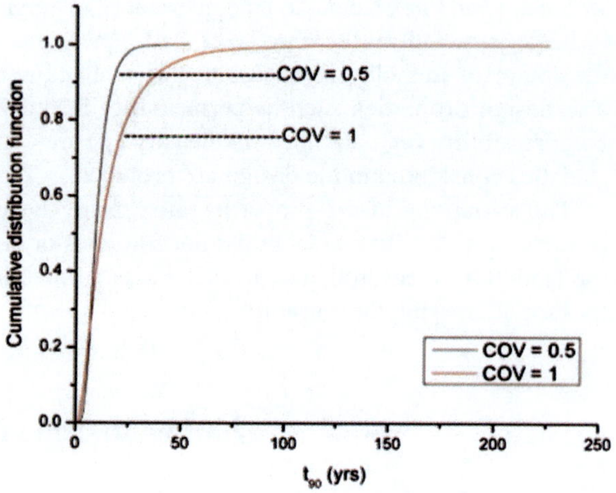

From the simulations, it was observed that the c_v values were ranging from 0.33 to 12.44 m^2/year for COV = 0.5 and 0.07 to 50.2 m^2/year for COV = 1.0. This would result in a t_{90} variation of 1.37–51 years for COV = 0.5 and 0.34–226 years for COV = 1.0. Table 2 shows the results of the c_v values and t_{90}.

The 95% confidence interval of t_{90} was observed to be [3.62, 24.06] years for COV = 0.5 and [2.23, 60.52] years for COV = 1.0. This means, there is 95% chance that the time for settlement will lie in between the ranges shown. Also, the t_{90} intercept from Fig. 2 for 0.95 probability is 20.49 years for COV = 0.5 and 45.84 years for COV = 1.0. This indicates, with 95% probability, it can be said that the time for settlement (t_{90}) will not go beyond the time specified for the considered cases.

Table 2 Variations in the c_v and t_{90} values due to variation in k_v COV

Outputs	For k_v COV = 0.5		For k_v COV = 1.0	
	Mean	COV	Mean	COV
Coefficient of consolidation c_v (m^2/year)	2.06	0.52	2.03	0.98
Time for 90% consolidation t_{90} (years)	10.47	0.5	16.62	0.96

Thus, when compared with the deterministic values, it can be seen that there is considerable variation in the c_v values and the resulting t_{90} when the variability in the properties is being addressed. As the performance of the PVDs is dependent on the c_v value of the soil, these variations should be addressed while designing their scheme.

5 Reliability-Based Design of PVDs

The design procedure suggested by Yeung (1997) is used in this study. The degree of consolidation (due to vertical and radial drainage) to be achieved is 90% in 2 months. The PVD size considered is 100 mm × 5 mm aligned in a triangular pattern.

Results of the Monte Carlo simulations suggest that the surcharge that needs to be applied (preloading) varied from 149 to 152 kPa. As there was little information available regarding the radial permeability (k_r), the design spacing was found out for $k_r = 1.0\ k_v$, $k_r = 1.5\ k_v$ and $k_r = 3.0\ k_v$. Table 3 shows the spacing values for the various cases considered.

The 95% confidence interval of obtained spacing values and the mean and COV values for different ratios of k_r/k_v are provided in Table 3. The analysis is repeated for different COV values of k_v to understand the effect of variation of k_v on spacing. The cumulative distribution function obtained for the different scenarios considered is shown in Figs. 3 and 4.

The observed variations suggest that it is important to consider the variability of the properties in the design and the design spacing to be provided in the site shall be decided based on the results obtained after the probabilistic analysis.

Table 3 Spacing of PVDs for various cases considered

Cases	For k_v COV = 0.5			For k_v COV = 1.0		
	Mean	COV	95% C.I[a]	Mean	COV	95% C.I[a]
$k_r = 1.0\ k_v$	0.77	0.18	[0.52, 1.06]	0.70	0.29	[0.37, 1.13]
$k_r = 1.5\ k_v$	0.88	0.17	[0.61, 1.12]	0.78	0.26	[0.44, 1.17]
$k_r = 3.0\ k_v$	1.03	0.12	[0.80, 1.18]	0.90	0.21	[0.56, 1.20]

[a]Confidence interval

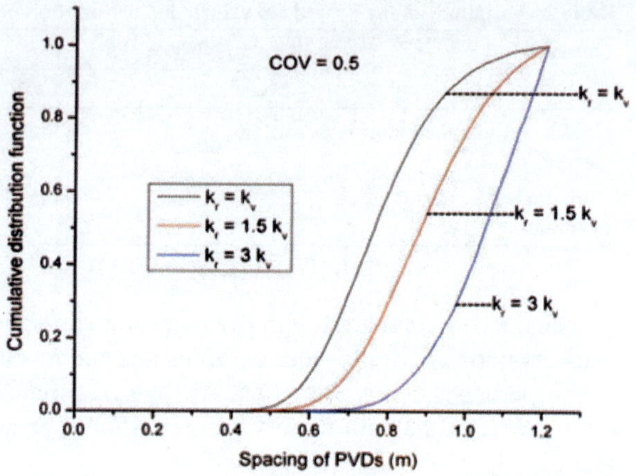

Fig. 3 Spacing of PVDs for different k_r/k_v ratios and k_v COV = 0.5

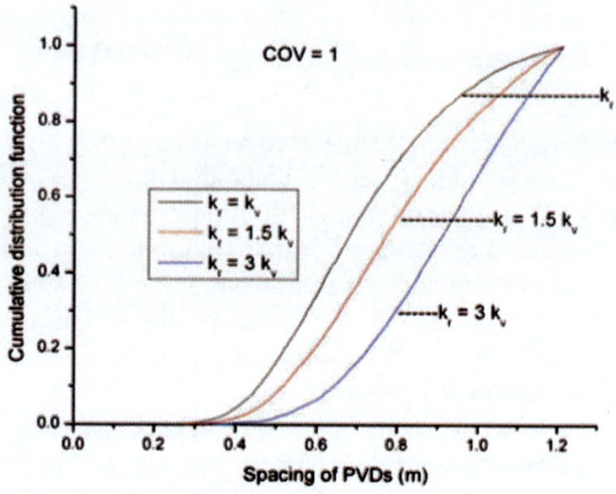

Fig. 4 Spacing of PVDs for different k_r/k_v ratios and k_v COV = 1.0

6 Conclusion

It can be evidently realized that variation in the properties propagates through the analysis process and in turn leads to the variation in the output. If a deterministic approach is selected for the current problem, the spacing of the PVDs would correspond to 0.93 m. But when the variations in the soil properties are being addressed, the spacing obtained is a distribution of values ranging from 0.30 to 1.23 m. Thus, the designer should be more careful in selecting the design spacing that would be required to achieve the target degree of consolidation and the time for ultimate settlement.

Reliability-based design emphasizes more on the variability or the uncertainty in the soil properties that would affect a design. If they are not addressed, the decisions could be erroneous and may lead to inappropriate designs causing failure in meeting the expected criteria or performance. Reliability methods often serve as a value addition to a deterministic design, and hence, it is suggested that the design is selected considering both the methodologies.

References

Bari MW, Shahin MA, Nikraz HR (2013) Probabilistic analysis of soil consolidation via prefabricated vertical drains. Int J Geomech 13(6):877–881

Elkateb T, Chalaturnyk R, Robertson PK (2003) An overview of soil heterogeneity: quantification and implications on geotechnical field problems. Can Geotech J 40(1):1–15

Haldar A (2002) Basic simulation concepts applicable in codified design. In: Euro-SiBRAM colloquium proceedings, ITAM Academy of Sciences of Czech Republic

Marek P, Haldar A (2003) Simulation as an alternative to codified risk-based engineering design. In: Proceedings of 4th international symposium on uncertainty modelling and analysis, p 222

Schuëller GI (2002) Past, present & future of simulation-based structural analysis. In: Euro-SiBRAM colloquium proceedings, ITAM Academy of Sciences of Czech Republic

Yeung AT (1997) Design curves for prefabricated vertical drains. J Geotech Geoenviron Eng 123(8):755–759

3-D Finite Element Study of Embankment Resting on Soft Soil Reinforced with Encased Stone Column

B. K. Pandey, S. Rajesh, and S. Chandra

Abstract This paper presents the outcomes of a 3-D finite element analysis performed to study the time-dependent behavior of embankment resting on the geosynthetic-encased stone column (GESC). The numerical analysis is carried out on both fixed and floating GESC. The results of the study show the effect of encasement stiffness, encasement length and length of the geosynthetic encased stone column on time-dependent behavior of the system. The use of GESC has provided significant improvement in reducing the generation and dissipation of excess pore water pressure, settlement and lateral displacement of the column along its length. The study indicates that the higher stress concentration in the case of GESC results in better time-dependent behavior. Additionally, the results also confirm that there is an optimum value of encasement stiffness, encasement length and length of geosynthetic-encased column beyond which no substantial improvement is attained.

Keywords Soil improvement · Encased stone column · Geogrid

1 Introduction

The massive infrastructure projects require suitable ground conditions to supports its foundation. There are different ground improvement techniques available to mitigate the problems of the compressible soils. In the improvement of soft soil for projects like embankment fill supports, LNG/Oil storage tanks, railroad, and other miscellaneous structures, the use of stone columns has proven efficacious both regarding economy and performance. Moreover, it is very successful because of its advantages. It helps in slope stability improvement, increasing bearing capacity and time rate of settlement, reduction in total and differential settlement, reducing liquefaction potential Barksdale and Bachus (1983). The application of stone column serves it purpose if the undrained shear strength (C_u) of soil lies between 15 and 50 kPa. However, if the undrained shear strength of the soil is <15 kPa, then the confinement provided

B. K. Pandey (✉) · S. Rajesh · S. Chandra
Department of Civil Engineering, IIT Kanpur, Kanpur, Uttar Pradesh 208016, India
e-mail: balbir@iitk.ac.in

M. Latha Gali and R. R. P. (eds.), *Problematic Soils and Geoenvironmental Concerns*, Lecture Notes in Civil Engineering 88,
https://doi.org/10.1007/978-981-15-6237-2_38

by the surrounding soil to the stone column is not adequate, and hence it fails in excessive settlement due to bulging. Therefore, reinforcing the stone column either by encasement or by placing the horizontal strip of geosynthetic within column body at regular interval provides the extra resistance against radial bulging of the column as reported by Alexiew et al. (2012) and Ali et al. (2012). In last two decades, various projects for the railroad and road embankment, bridge ramp, flood protection dike were accomplished using encased stone columns in Germany, Netherlands, Sweden and India (Raithel et al. 2005; Mahajan et al. 2016).

The idea of encasing the column in full or partially is studied experimentally by numerous researchers like (Raithel et al. 2004, 2005; Ayadat and Hanna 2005 ; Murugesan and Rajagopal 2006, 2007) under undrained condition. Furthermore, Najjar et al. (2010) studied the mode of failure and failure envelope of geosynthetic-encased stone columns (GESCs). Likewise, Dash and Bora (2013) reported the deformed shape of floating GESCs. Gu et al. (2016) quantified the effect of encasement length on stress concentration and radial strain. Miranda et al. (2017) explored the drained behavior of GESCs by considering the slice of the unit cell. Debnath and Dey (2017) reported increase in bearing capacity for geogrid-reinforced sand over GESC. In discovering the behavior of GESCs, the numerical modeling has also been extensively used. Yoo and Kim (2009) adopted unit cell (both 3D and axisymmetric) and full 3D model to compare the different modeling approach. Yoo (2010) by modeling quarter of column in 3D column model and full 3D model reported the effects of various parameters on GESCs behavior. Khabbazian et al. (2010) and Elsawy et al. (2010) observed that the encasement type and stiffness play an essential role in the response of the GESCs. Majority of the studies focus on unit cell approach (either 3D or axisymmetric or quarter column) by loading the column only or entire unit cell (Raithel et al. 2004; Elsawy et al. 2010; Castro and Sagaseta 2011; Fattah and Majeed 2012; Rajesh and Jain 2015; Rajesh 2017; Castro 2017 and others) or unit cell with embankment loading (Almeida et al. 2013; Elsawy 2013; Zhang and Zhao 2014; Hosseinpour et al. 2014, 2015). However, only limited study has been conducted in details considering the encasement parameter's (length and stiffness) and also GESC length on time-dependent responses under embankment loading. Therefore, there is a need for a comprehensive study to understand the modeling details as well as the behavior of GESC under the embankment loading.

This study focuses on understanding the effect of encasement length, encasement stiffness and length of GESC on the behavior of GESC under embankment loading through 3-D unit cell model. The impact of the parameters mentioned above is evaluated regarding settlement reduction, generation and dissipation of excess pore water pressure, lateral deformation of the column along its length and the stress concentration ratio.

2 Geometry and Model Description

Numerical analysis is performed using PLAXIS 3D, a finite element package (Brinkgreve et al. 2013). A hypothetical case study is formulated in the present study. The foundation soil consists of 9.5 m of soft clay underlying a 0.5 m of the sand platform. The sand platform serves as the drainage blanket and is laid before the embankment construction. The ground water table lies at the top of the clay surface, i.e., 0.5 m below the ground. The 3D unit cell is conceptualized from the central part of the embankment. The size of the unit cell model adopted is 2 m × 2 m × 10 m ($L \times B \times H$) with column installed at the center of the unit cell. The diameter (d) of GESC was worked out to be 0.8 m by considering spacing of GESC of 2 m arranged in a square pattern. Figure 1 shows the details of the GESC-reinforced soft ground considering 3D unit cell model. In the present study, the length of GESC (L_{GESC}) is varied between 5 and 10 m, and the axial stiffness of encasement is varied between 500 and 6000 kN/m (Murugesan and Rajagopal 2006; Khabbazian et al. 2010). The total height of the embankment is 6 m and is constructed in three stages; each stage is made of 2 m height, similar to the work carried out by Yoo (2010). The duration of each stage consists of two parts: one is construction period (t_{c1}, t_{c2}, and t_{c3}) and the other is rest period (t_{r1} and t_{r2}). The construction period is 20 days, and the rest period is 30 days. The rest period is provided for a partial dissipation of excess pore water pressure.

The displacement and hydraulic boundary conditions for the model are chosen based on the symmetry. Therefore, considering symmetry into account no displacement in the perpendicular direction of the symmetry plane and to the base is allowed (Yoo 2010). Congruently, at the symmetry plane, no flow at the boundary will take place. Therefore, the symmetry plane has been kept as a closed consolidation boundary. Moreover, the bottom boundary is also held closed consolidation boundary considering the impervious layer below the soft clay (Rajesh 2017). The soft soil is modeled using soft soil constitutive model material model. However, for the embankment, stone column and sand platform, a linear-elastic perfectly plastic model with Mohr–Coulomb failure criterion was used. Hardening soil can be used

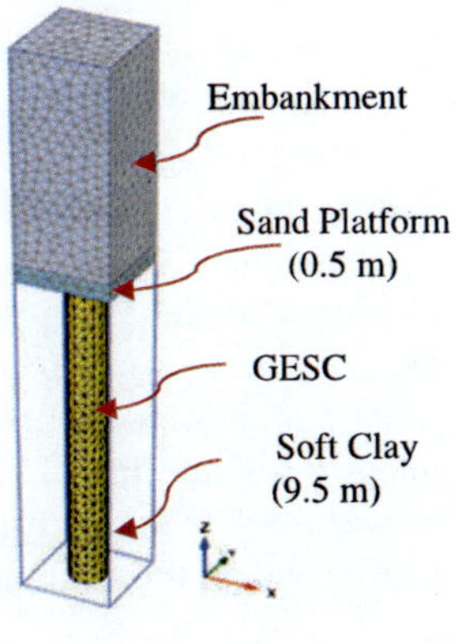

Fig. 1 Generated mesh connectivity plot and details for 3D unit cell

Table 1 Summary of the constitutive model and related parameter

Properties	Column (Mohr–Coulomb)	Soft clay (soft soil)	Platform (Mohr–Coulomb)	Embankment (Mohr–Coulomb)
	Chen et al. (2015)	Hosseinpour et al. (2014)	Huang et al. (2009)	Aljanabi et al. (2013)
γ (kN/m^3)	22	14.50	20	20
E (kPa)	40,000	750	20,000	20,000
ν	0.30	0.28	0.33	0.30
c' (kPa)	0.50	4	5	5
φ' (0°)	38	27.50	32	30
Ψ (0°)	10	–	–	–
C_c	–	1.26	–	–
C_s	–	0.097	–	–
k_h (m/day)	10.368	7.50×10^{-5}	1	1
k_v (m/day)	10.368	8.40×10^{-6}	1	1

Note E = tangential elastic modulus; ν = poisson's ratio; γ = soil unit weight; c' = cohesion of soil; φ' = soil friction angle; ψ = dilation angle; C_c = compressibility index; C_s = swelling index; k = permeability (subscript h and v corresponds to horizontal and vertical respectively); – = data not given

to represent the effect of strain rate on the behavior of granular materials. However, use of hardening soil model caused slow convergence problem while performing the parametric study. Hence, the modeling of stone column is limited to Mohr–Coulomb model. Table 1 summarizes the constitutive model and related parameters used in the analysis. The geogrid is used in the study and is modeled as an elastic material having orthotropic behavior. These are slender structures that can sustain only tensile force and no compression.

3 Mesh Sensitivity and Validation

The soft soil, sand platform and embankment were modeled using 10-node tetrahedral elements. The geogrid was modeled using 6-node triangular surface element. For modeling soil–geogrid interaction, joint elements (interfaces) were added to geogrid and surrounding soil. The interfaces comprised of 12-node interface elements with zero thickness. In the case of the material models like soft soil model and Mohr–Coulomb model, the primary interface parameter is the strength reduction factor (R_{inter}). The R_{inter} adopted in the present study for soft soil, sand platform, embankment and stone column are 1 as reported by Aljanabi et al. (2013), whereas for geogrid soil interface it is 0.7 as reported by Chen et al. (2015). The mesh generation process used in the study is fully automated. However, it also takes into account of soil profile and all the structural members as well as geometric and boundary conditions

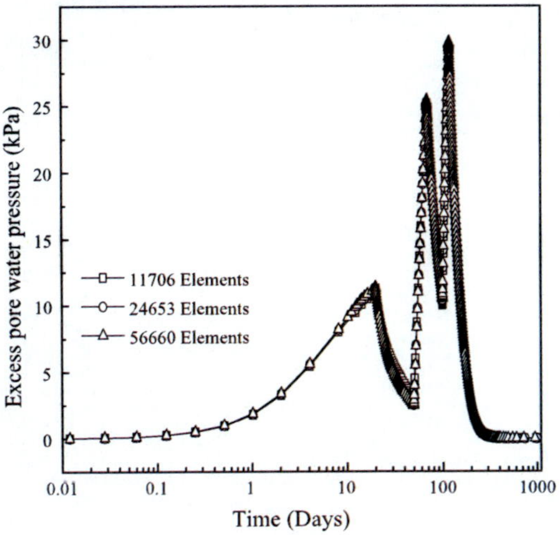

Fig. 2 Mesh sensitivity analysis with a variety of elements for GESC

(Brinkgreve et al. 2013). From the previous studies, it has been observed that mesh should be adequately fine to get the precise results (Castro 2017). Figure 2 shows the generation and dissipation of excess pore pressure with time for the embankment resting on GESC. From Fig. 2, it can noticed that the fine mesh having 24,653 elements predicts 0.5–1.01% lower excess pore pressure as compared to very fine mesh with 56,660 elements. Moreover, the time taken for the analysis in the case of 56,660 elements is thrice as compared to 24,653 elements. Hence, for a lesser computational time without compromising with the accuracy, the fine mesh with numbers of element lying between 20 and 25 thousand has been chosen for further studies.

The present study is validated with the study conducted by Rajesh (2017) in which the surface load is applied on the top of an axisymmetric problem. However, in the present study, 3D unit cell having a square cross-sectional area of dimension 2.25 m × 2.25 m is used as reported by Rajesh (2017). All the material properties and the dimensions were kept intact, except properties of the plate. A steel plate of thickness (t_p) 30 mm, unit weight of steel (γ_{steel}) = 78.5 kN/m^3 and $E = 4.44 \times 10^6$ kN/m^2 is used for the loading. The linear isotropic constitutive model has been adopted for steel plate. The details of the constitutive model and related parameters for other materials can be found in the referred paper. Figure 3 shows the result of 3D unit cell compared with axisymmetric model of Rajesh (2017) for soft soil reinforced with ordinary stone column (OSC). The results show that model is in good agreement with the referred study.

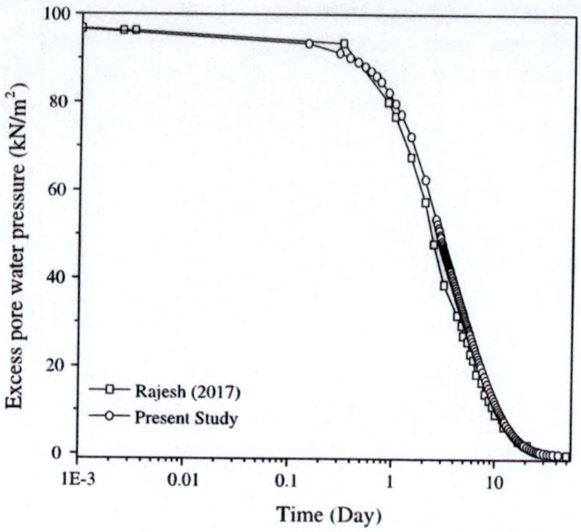

Fig. 3 Validation with Rajesh (2017) for OSC (very fine mesh)

4 Results and Discussions

The primary emphasis of the study is to understand the effect of parameters like the effect of the encasement, length of the encasement, axial stiffness of the encasement and length of the geosynthetic encased stone column on time-dependent behavior of GESC. The length of the encasement (L_{ESC}) is normalized by the length of the ordinary stone column (L_{OSC}) and is expressed as, $\zeta = L_{ESC}/L_{OSC}$. Furthermore, the length of geosynthetic-encased stone column (L_{GESC}) is similarly normalized by the depth of the treatment ($H_{Treatment}$) which is expressed as $\xi = L_{GESC}/H_{Treatment}$. Finally, the results of the study have been discussed in detail for the different parameters in the ensuing sections below.

4.1 *Effect of Encasement*

The first parametric study emphasizes on the influence of using different column type, i.e., no column (directly on soft soil), ordinary stone column (OSC) and GESC beneath an embankment. The influencing effect is appraised regarding generation and dissipation of pore water pressure and settlement with time. The fully encased column having encasement stiffness of 2000 kN/m has been used in the case of GESC below the embankment. The monitoring point in the case of excess pore water pressure is at mid-depth in soft soil region. In Fig. 4, excess pore water pressure is plotted against time. This result displays that the excess pore water pressure generated at the end of construction period (120 days) varies for different cases. It is on the higher side, i.e., 117.112 kPa for the embankment on the soft soil, whereas 49.201 and 29.479 kPa for embankment on OSC and GESC, respectively. Further, it can be seen that the time taken to dissipate the excess pore water pressure to a level of 0.1 kPa for

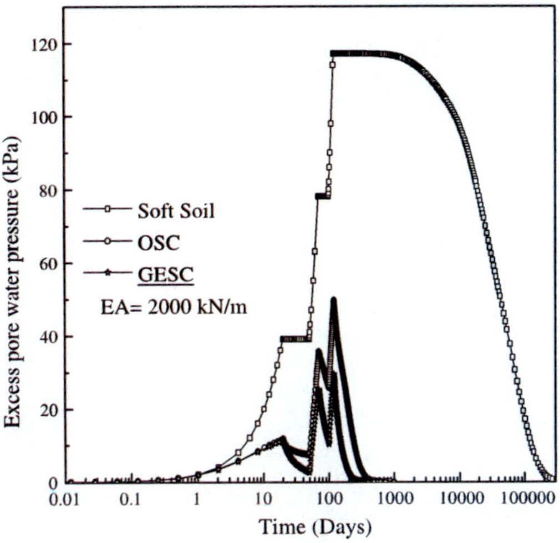

Fig. 4 Excess pore pressure variation with time for soft soil, OSC and GESC

embankment resting on soft clay, OSC and GESC was 337, 255, 534 and 317 days, respectively. The lower value of pore pressure generation and faster dissipation in the case of GESC and OSC are attributed to different factors as reported by Han and Ye (2001). The factors influencing excess pore water pressure include reduction due to a decrease in vertical stress, increment due to increase in the lateral stress from the column and reduction due to drainage of water from the soil. Considering this, it is evident for embankment on GESC to perform better as compared to that on OSC because of additional stiffness provided by the encasement. Nevertheless, it is also perceived that the excess pore water pressure generation for the initial 2–3 days is insignificant because at this point the reinforced and unreinforced soil has negligible stress concentration ratio as they behave as an undrained condition.

In Fig. 5, settlement at the bottom of the embankment is plotted against the time (including construction and service time) for embankment resting on soft clay, OSC and GESC. The variation of settlement for different types of embankment support can be clearly understood. From the figure, it is observed that the settlement at the end of the consolidation period is 1.63, 1.13 and 0.44 m for embankment resting on soft clay, OSC and GESC, respectively, which foresees the settlement reduction factor (β) of 0.69 and 0.27 for OSC and GESC, respectively. The settlement reduction factor is the ratio of settlement at the surface after treatment (δ_{at}) w.r.t that of before treatment (δ_{bt}) and is expressed as $\beta = \delta_{at}/\delta_{bt}$. Furthermore, the difference in the settlement for soft soil, OSC and GESC in the initial 10 days is almost insignificant with maximum settlement of roughly 11 mm. Thus, it envisages the improved performance of GESC in reducing the settlement. The reason can be attributed to the lesser excess pore water pressure generation (as from Fig. 4) and increased stiffness due to encasement.

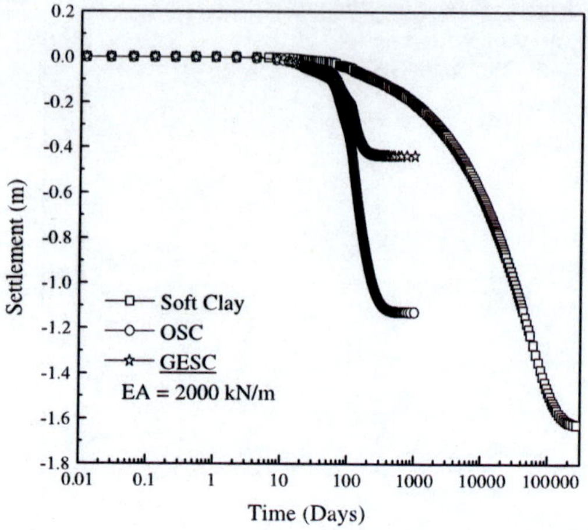

Fig. 5 Settlement variation with time at the surface for Soft clay, OSC and GESC

4.2 *Effect of Encasement Stiffness*

To study the influence of axial stiffness of different encasement stiffness varying between 500 and 6000 kN/m is adopted. In Fig. 6, excess pore water pressure generation and dissipation variation has been plotted with time for different stiffness of encasement. The figure displays that the excess pore water pressure generated at the end of the construction period varies between 20.26 and 41.55 kPa for different stiffness of encasement. The lower value of 20.26 kPa is observed for 6000 kN/m and 41.55 kPa for 500 kN/m stiffness. The variation of the generated excess pore water pressure is because lesser stresses are transferred to the soft clay because of the stiffer encasement. The difference is considerably lower (<5%) for stiffness of encasement lying between 4 and 5 thousand kN/m. Congruently, the time taken for

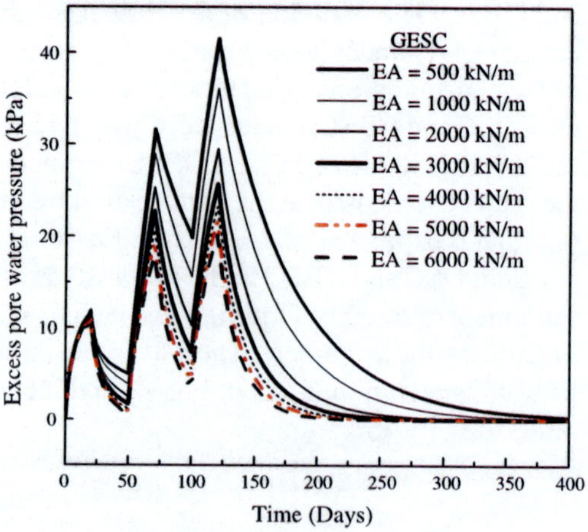

Fig. 6 Excess pore pressure variation with time for GESC having the different stiffness of encasement

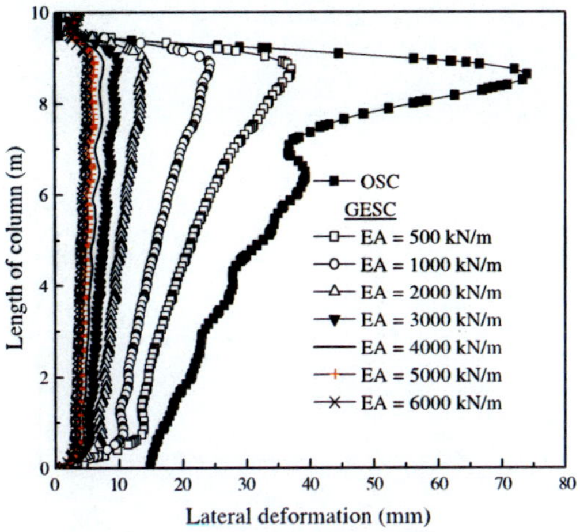

Fig. 7 Variation of lateral deformation of the column for OSC and GESC of different encasement stiffness

dissipating excess pore water pressure to 0.01 kPa is nearly 638 and 240 days, for stiffness 500 kN/m and 6000 kN/m, respectively. Moreover, the maximum difference in the time for stiffness lying between 4000 and 6000 kN/m is <16%. From this, we can say that the optimum stiffness of encasement can be 5000 kN/m for complete dissipation of excess pore water pressure. Using the encasement stiffness >5000 kN/m, there is no substantial enhancement in the performance is observed.

Figure 7 illustrates the lateral deformation of the OSC and GESC having the different stiffness of the encasement. As expected the lateral deformation in the case of OSC is 14.43 times higher as compared to GESC with stiffness of 6000 kN/m; however, it is 1.18 times for the encasement stiffness 5000 kN/m as compared to 6000 kN/m. From the previous studies (Barksdale and Bachus 1983; Christoulas et al. 2000; Murugesan and Rajagopal 2010; Malarvizhi and Ilamparuthi 2007; Yoo and Kim 2009), the maximum lateral deformation occurs at the depth varying between 2 and 3 * *D* from the top of the column. This maximum deformation depth is dependent on the loading area as well as the strength of the surrounding soil. In the current study as can be seen from Fig. 7, this depth is nearly 1.7 * *D*.

In Fig. 8, settlement reduction factor (β) is plotted against GESC improved ground having different encasement stiffness. However, from this figure we can clearly see that the difference in the settlement reduction by using GESC with encasement having stiffness 6000 kN/m is inconsequential (approx. 8%) as compared to encasement having 5000 kN/m stiffness. Therefore, we can evidently state that using encasement stiffness above 5000 kN/m has no significant effect on the settlement reduction.

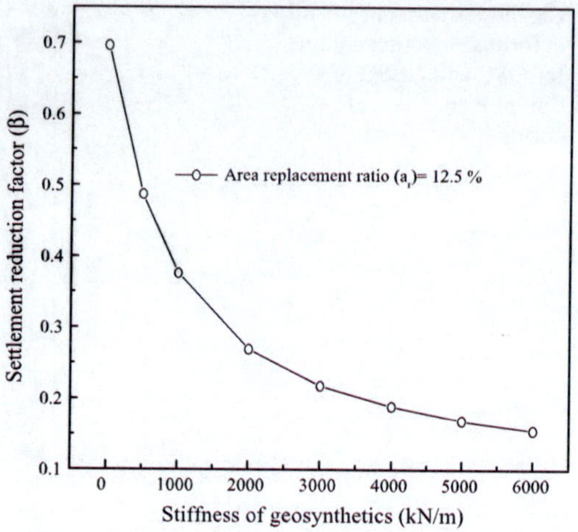

Fig. 8 Settlement reduction factor for GESC improved ground having varying encasement stiffness

4.3 Effect of Encasement Length

The influence of encasement length on behavior of GESC is analyzed by varying encasement length (ζ) from 0 to 1. The reference encasement stiffness used for this parametric study is 2000 kN/m. In Fig. 9, the variation of excess pore water pressure is plotted against time. The excess pore water pressure generation at the end of the construction period, i.e., 120 days is maximum for encasement length (ζ) of 0.50, whereas the difference in the generated excess pore water pressure is insignificant after $\zeta = 0.75$. Moreover, the time taken to dissipate the generated excess pore water pressure to a level of 0.1 kPa shows a slight variation of less than 2% for encasement length (ζ) beyond 0.75. Additionally, the lateral bulging of the column

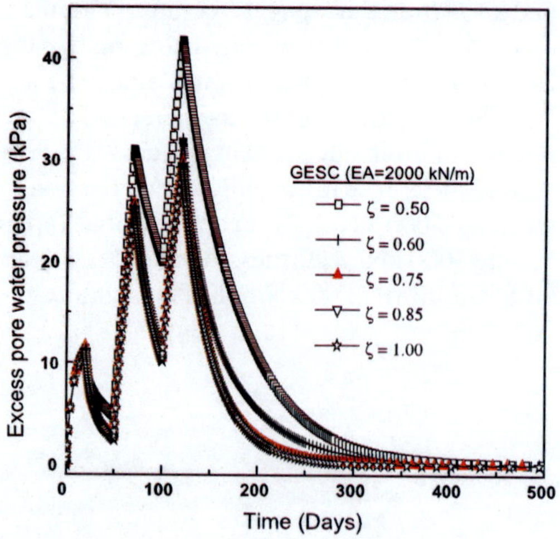

Fig. 9 Excess pore pressure variation with time for GESC having different length of the encasement

is dominant at the location where encasement terminates at the bottom as can be noticed from Fig. 10. In this figure, lateral deformation against depth from the top of the column for GESC having different encasement length is plotted. However, this lateral deformation in the case of the encased column is almost twice as compared to column without encasement (i.e., OSC) at the point of termination of the encasement (example let's say at 5 m from the top for $\zeta = 0.50$).

Similarly, in Fig. 11, settlement reduction factor (β) is plotted against embankment resting on GESC improved ground having varying encasement length. From this figure, it can be observed that full encasement of the column results in significant reduction in the settlement (>61%) as compared to the column without encasement. Moreover, the variance in the settlement reduction (β) for $\zeta = 0.75$ and $\zeta = 0.85$ w.r.t, $\zeta = 1$ is approximately 21% and 16%, respectively. Seeing this variation, one

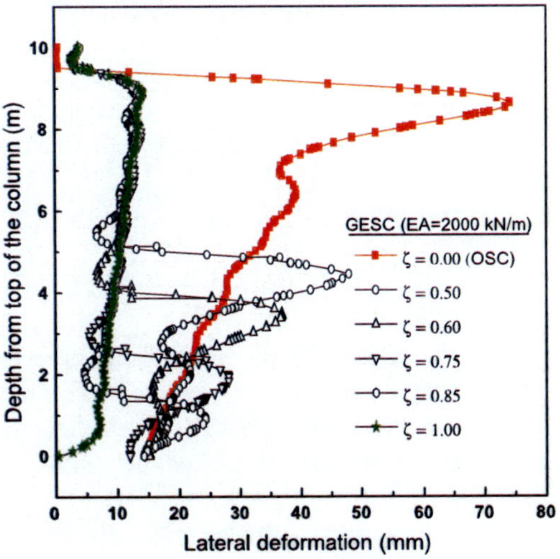

Fig. 10 Lateral deformation of GESC having different encasement length

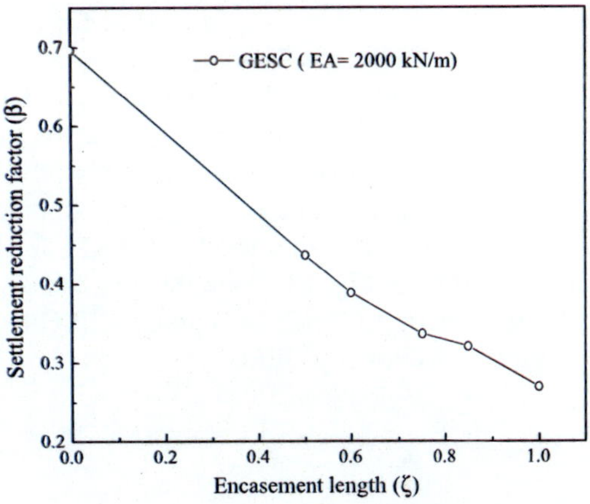

Fig. 11 Settlement reduction factor for GESC having varying encasement length

Fig. 12 Settlement reduction factor for embankment resting on GESC having varying length of GESC

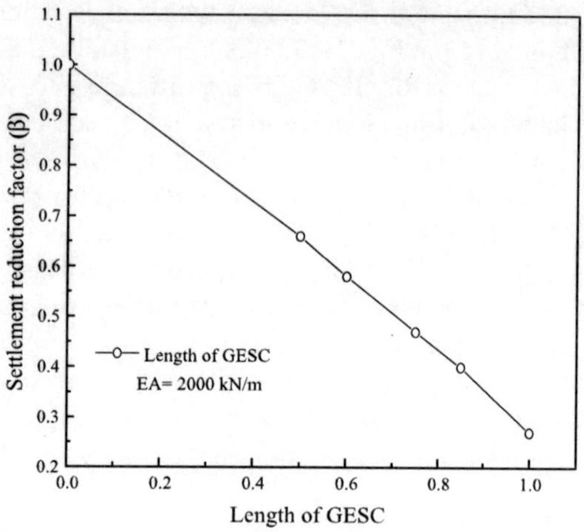

can state that the fully encased stone column contributes significant reduction in the settlement when compared to partially encased GESC.

4.4 *Effect of Length of GESC*

As already mentioned, the above discussions were based on the influence of properties of encasement considering fixed end column. However, the behavior of embankment resting on the floating GESC is reported in this segment. In Fig. 12, settlement reduction factor (β) is plotted against GESC improved ground having varying length of GESC.

The encasement stiffness was kept fixed at 2000 kN/m. From the figure, it is noticeable that the settlement reduction in the case of GESC length (ξ) of 1.00 is approximately 73% as compared to without improvement. Furthermore, the difference in the reduction is roughly 15% for GESC length (ξ) of 0.85 as compared to GESC length (ξ) of 0.75. This reduction in settlement values indicates that using GESC length (ξ) of 1.00, i.e., the end bearing GESC is more useful as compared to floating GESC. Similarly, the generated excess pore water pressure at the end of the construction period (120 days) is nearly 58% lower for GESC length (ξ) of 1.00 as compared to 0.5. Correspondingly, the time taken to dissipate the generated excess pore water pressure to a level of 0.1 kPa is also reduced by almost 94% as can be perceived from Fig. 13. From this discussion, we can infer that the performance of the end bearing GESC, i.e. having GESC length (ξ) of 1.00, is relatively high as compared to floating GESC.

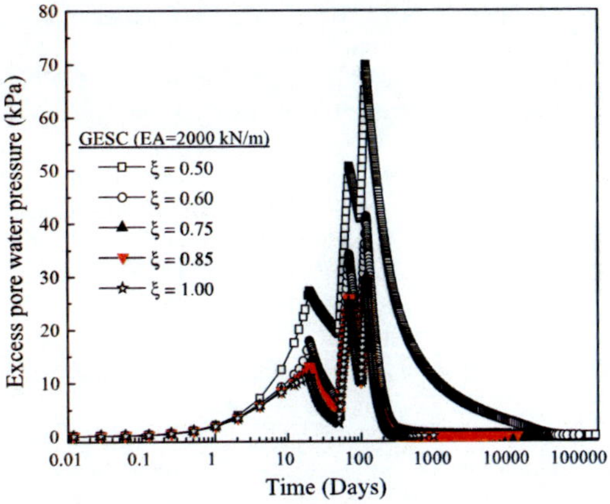

Fig. 13 Excess pore pressure variation with time for GESC having different length of GESC

5 Conclusion

The paper presents an organized numerical analysis of embankment resting on a geosynthetic encased stone column. The area replacement ratio in this study is kept constant at 12.5%. From the study, following points are concluded.

- Encasing the stone column significantly helps in (75%) reduction in generation of excess pore water pressure and at the same time (73%) reduction in the settlement as compared to embankment resting on soft soil, for the case analyzed in this study.
- The stiffness of the encasement plays a critical effect on the time-dependent behavior of the embankment resting on the GESC. However, the use of encasement stiffness greater than 5000 kN/m doesn't show significant improvement.
- The fully encased stone column shows substantial improvement regarding the time-dependent behavior of embankment resting on GESC as compared to partial encasement of GESC.
- The performance of fixed end GESC column outweighs the floating GESC supporting an embankment.

The presented study is primarily based on finite element simulations; further experimental investigations are required to confirm the outcomes.

References

Alexiew D, Raithel M, Kuster V, Detert O (2012) 15 Years of experience with geotextile encased granular columns as foundation system. In: ISSMGE - TC 211 international symposium on ground improvement IS-GI, Brussels

Ali K, Shahu JT, Sharma KG (2012) Model tests on geosynthetic-reinforced stone columns: a comparative study. Geosynth Int 19(4):292–305

Aljanabi QA, Chik Z, Kasa A (2013) Construction of a new highway embankment on the soft clay soil treatment by stone columns in Malaysia. J Eng Sci Technol 8:448–456

Almeida MSS, Riccio M, Hosseinpour I (2013) Performance of a geosynthetic-encased column (GEC) in soft ground: numerical and analytical studies. Geosynth Int 20:252–262

Ayadat T, Hanna AM (2005) Encapsulated stone columns as a soil improvement technique for collapsible soil. Proc Inst Civ Eng Ground Improv 9:137–147

Barksdale RD, Bachus RC (1983) Design and construction of stone column, vol 1. U.S Department of Transportation, Report No—FHA FHWA/RD-83 (1983)

Brinkgreve RBJ, Engin E, Swolf WM (2013) Plaxis 3D 2013 manual. Plaxis Bv, The Netherlands

Castro J (2017) Groups of encased stone columns: influence of column length and arrangement. Geotext Geomembr 45:68–80

Castro J, Sagaseta C (2011) Deformation and consolidation around encased stone columns. Geotext Geomembr 29:268–276

Chen J-F, Li L-Y, Xue J-F, Feng S-Z (2015) Failure mechanism of geosynthetic-encased stone columns in soft soils under embankment. Geotext Geomembr 43:424–431

Christoulas S, Bouckovalas G, Giannaros C (2000) An experimental study on model stone columns. Soil Found 40(6):11–22

Dash SK, Bora MC (2013) Improved performance of soft clay foundations using stone columns and geocell-sand mattress. Geotext Geomembr 41:26–35

Debnath P, Dey AK (2017) Bearing capacity of geogrid reinforced sand over encased stone column in soft clay. Geotext Geomembr 45(6):653–664

Elsawy MBD (2013) Behavior of soft ground improved by conventional and geogrid-encased stone columns, based on FEM study. Geosynth Int 20:276–285

Elsawy M, Lesny K, Richwien W (2010) Performance of geogrid-encased stone columns as a reinforcement of soft ground. In: Benz T, Nordal S (eds) Numerical methods in geotechnical engineering. Taylor and Francis Group, London UK, pp 875–880

Fattah MY, Majeed QG (2012) Finite element analysis of geogrid encased stone columns. Geotech Geol Eng J 30:713–726

Gu M, Zhao M, Zhang L, Han J (2016) Effects of geogrid encasement on lateral and vertical deformations of stone columns in model tests. Geosynth Int 23:100–112

Han J, Ye S-L (2001) Simplified method for consolidation rate of stone column reinforced foundations. J Geotech Geoenviron Eng 127(7):240–248

Hosseinpour I, Riccio M, Almeida MSS (2014) Numerical evaluation of a granular column reinforced by geosynthetics using encasement and laminated disks. Geotext Geomembr 42:363–373

Hosseinpour I, Almeida MSS, Riccio M (2015) Full-scale load test and finite-element analysis of soft ground improved by geotextile-encased granular columns. Geosynth Int 22:428–438

Huang J, Han J, Oztoprak S (2009) Coupled mechanical and hydraulic modeling of geosynthetic-reinforced column-supported embankments. J Geotech Geoenviron Eng 135(8):1011–1021

Khabbazian M, Kaliakin VN, Meehan CL (2010) Numerical study of the effect of geosynthetic encasement on the behaviour of granular columns. Geosynth Int 17:132–143

Mahajan R, Korulla M, Rimoldi P (2016) Geotextile encased columns design and installation. In: Proceedings of the 6th Asian regional conference on geosynthetics-geosynthetics for infrastructure development, New Delhi, India, pp 188–194

Malarvizhi SN, Ilamparuthi K (2007) Comparative study on the behavior of encased stone column and conventional stone column. Soils Found 47(5):873–885

Miranda M, Costa DA, Castro J, Sagaseta C (2017) Influence of geotextile encasement on the behaviour of stone columns: laboratory study. Geotext Geomembr 45:14–22

Murugesan S, Rajagopal K (2006) Geosynthetic-encased stone columns: numerical evaluation. Geotext Geomembr 24:349–358

Murugesan S, Rajagopal K (2007) Model tests on geosynthetic-encased stone columns. Geosynth Int 14(6):346–354

Murugesan S, Rajagopal K (2010) Studies on the behavior of single and group of geosynthetic encased stone columns. J Geotech Geoenviron Eng 136(1):129–139

Najjar SS, Sadek S, Maakaroun T (2010) Effect of sand columns on the undrained load response of soft clays. J Geotech Geoenviron Eng 136:1263–1277
Raithel M, Küster V, Lindmark A (2004) Geotextile-encased columns—a foundation system for earth structures, illustrated by a Dyke project for works extension in Hamburg. In: Proceedings of the Nordic geotechnical meeting NGM, Ystad, Sweden, pp 1–10
Raithel M, Kirchner A, Schade C, Leusink E (2005) Foundation of construction on very soft soil with geotextile encased column—state of the art. In: Proceedings of geo-frontiers 2005, Austin, Texas, United States, pp 1–11
Rajesh S (2017) Time-dependent behaviour of fully and partially penetrated geosynthetic encased stone columns. Geosynth Int 24:60–71
Rajesh S, Jain P (2015) Influence of permeability of soft clay on the efficiency of stone columns and geosynthetic encased stone columns—a numerical study. Int J Geotech Eng 9(5):483–493
Yoo C (2010) Performance of geosynthetic-encased stone columns in embankment construction: numerical investigation. J Geotech Geoenviron Eng 136:1148–1160
Yoo C, Kim S-B (2009) Numerical modeling of geosynthetic-encased stone column-reinforced ground. Geosynth Int 16:116–126
Zhang L, Zhao M (2014) Deformation analysis of geotextile-encased stone columns. Int J Geomech 15:1–10

Geotechnical and Physicochemical Properties of Untreated and Treated Hazardous Bauxite Residue Red Mud

Arvind Kumar Jha and Dhanraj Kumar

Abstract The storage of hazardous bauxite residue mud obtained from the Bayer's process in alumina industry leads to serious environmental impact on land, air and water. The detailed laboratory works and micro-analyses have been performed in the present paper to characterize the red mud and to investigate the effect of class F fly ash on the geotechnical properties of red mud for possible use as a construction material. The characterization of materials shows the higher value of specific gravity, liquid limit, optimum water content (OWC), maximum dry unit weight (γ_{max}), pH value and electrical conductivity of red mud as compared to fly ash. The micro-analyses show the presence of sodalite, hematite, goethite, calcite in red mud; and mullite and quartz in fly ash as predominant minerals. Further, treatment with various fly ash contents (0–80%) has observed a significant effect on the plasticity, compaction characteristics, alkalinity and electrical conductivity of red mud.

Keywords Compaction · Fly ash · Physicochemical · Red mud

1 Introduction

Red mud produces as a byproduct during the alumina extraction from bauxite in the Bayer's process which involves reaction with sodium hydroxide (NaOH) at high temperature and pressure. The red mud is alkaline in nature with high pH value ranging from 10 to 13. Due to its hazardous corrosive nature, it's posing a very serious and alarming environmental problem. The red mud produces approximately 90 million tons/year throughout the worldwide. In India, more than 4 million tons of red mud is being generated annually. The amount of the red mud generated per ton

A. K. Jha (✉)
Indian Institute of Technology (IIT) Patna, Bihta, Patna, Bihar 801103, India
e-mail: jhaarvind@iitp.ac.in

A. K. Jha · D. Kumar
Manipal University Jaipur, Jaipur, Rajasthan, India

M. Latha Gali and R. R. P. (eds.), *Problematic Soils and Geoenvironmental Concerns*, Lecture Notes in Civil Engineering 88,
https://doi.org/10.1007/978-981-15-6237-2_39

Table 1 Typical chemical composition of red mud (Sahu 2009)

Composition	Percentage (%)
Fe_2O_3	30–60
Al_2O_3	10–20
SiO_2	10–20
Na_2O	2–10
CaO	2–8

of the alumina processed varies greatly with the type of the bauxite ore used from 0.3 to 2.5 tons for high- and very low-grade bauxites, respectively (Cablik 2007).

It is reported that red mud contains silicon, aluminum, iron, calcium, titanium, sodium as well as an array of minor elements, namely K, Cr, V, Ba, Cu, Mn, Pb, Zn, P, F, S, As, etc. Due to significant amount of iron oxide in red mud (up to 60%), it derives its color and name (Kumar et al. 2006). The typical chemical composition of red mud is presented in Table 1.

Most of the aluminum extraction refineries is facing problem related to the safe disposal of red mud. Unscientific disposal of red mud possesses several environmental problems in terms of air pollution and ground surface and sub-surface contamination due to its highly alkaline behavior and the presence of toxic elements. However, due to its unique physical and chemical properties, red mud can be used in several aspects such as: (i) neutralizing acidic soils, (ii) production of crude and fine ceramics; (iii) bricks manufacturing; (iv) road construction; (v) cement production; (vi) use as an additive in ferrous metallurgy; (vii) treatment of water and sewage and (vii) various industrial purpose (rubber, paint, manufacture of adsorbents and catalysts) (Hannachi et al. 2010).

Further, fly ash is a solid waste generated by thermal power plants as a byproduct of coal burning. Fly ash is considered as an attractive construction material due to the presence of reactive silica and availability of free lime. Several attempts have been made by using fly ash to neutralize the heavy element from hazardous waste (Rios et al. 2008; Mohan and Gandhimathi 2009; Querol et al. 2006). Further, fly ash is used as a low-cost adsorbent material for the adsorption of heavy metal ions (Zn, Pb, Cd, Mn and Cu) present in the solid waste materials (Mohan and Gandhimathi 2009). The fly ash has also an ability to reduce the high alkalinity of red mud.

Previously, research works have been done to improve the properties of red mud using chemical or, solid industrial waste such as fly ash for its utilization in various construction activities (Table 2). It has been reported that red mud amended with fly ash is used to prepare a geopolymeric material (He et al. 2013). However, effective and bulk utilization of red mud as a construction material for different civil engineering projects are still a matter of concern and a recent research topic.

Hence, the present work is aimed to characterize the red mud and to investigate the potential of class F fly ash content (0–80%) to improve the geotechnical properties of red mud for its possible use as a construction material. To achieve the objectives, detailed characterization of red mud and fly ash has been done by performing various geotechnical tests, physicochemical tests and micro-analyses (XRD, SEM, EDAX

Table 2 Some findings on improvement of red mud with additives

Author	Description
Nikraz et al. (2007)	Compared the various properties of untreated and treated red mud
Deelwal et al. (2014)	Index and engineering properties were conducted at 1 and 7 days curing periods only with different percentage of lime up to 12% and by adding 1% gypsum. It has been concluded that red mud stabilized with lime can be used as sub-base, base course and also sub-grade material for road construction
Sabat and Mohanta (2015)	Studied the effects of fly ash and dolime fine stabilized red mud cushion on unconfined compressive strength (UCS), soaked California bearing ratio (CBR), percent swell and durability of an expansive soil at different curing periods. The maximum benefits of cushioning are observed at equal ratio of the height of the cushion that of soil at 28-day curing
Singh et al. (2017)	Worked on the geopolymerization of red mud using ground granulated blast furnace slag. It has been reported that red mud is highly alkaline and hence, it is possible to stabilize red mud by adding industrial waste materials rich in aluminosilicates by geopolymerization. It can be concluded that red mud can be stabilized effectively by the addition of little amount of alkali and ground granulated blast furnace slag

and FTIR). Further, effect of fly ash on the specific gravity, Atterberg's limits, swell index, compaction characteristics, alkalinity and electrical conductivity of red mud has been investigated in detail.

2 Materials Used and Methodologies Followed

2.1 Materials Used and Its Characterization

2.1.1 Red Mud

The geotechnical properties of red mud and fly ash are presented in Table 3. The red mud used in the present work was collected from Chhota Muri which is 65 km away from Ranchi (capital of Jharkhand state), India.

The particle size analysis of red mud shows the presence of predominant amount of sand-sized particle (4.75–0.075 mm) of 84.25%, silt-sized particle (0.075–0.002 mm) of 6.70% and clay-sized particle (<0.002 mm) of 9.04%. The specific gravity of red mud is obtained to be 2.98. The liquid limit and plastic limit of red mud are found to be 38.0 and 27.14%, and hence, the difference of liquid limit and plastic limit, i.e., plasticity index (PI) is 10.86%. The shrinkage limit of red mud is 26.52%. Based on the Indian Standard plasticity chart for soil classification, red mud is classified as inorganic silt of low plasticity. The differential free swell index (%) and modified

Table 3 Geotechnical properties of red mud (RM) and fly ash (FA)

Property	Red mud	Fly ash
Particle size analysis		
Sand (4.75–0.075 mm), %	84.25	63.62
Silt (0.075–0.002 mm), %	6.70	22.21
Clay (<0.002 mm), %	9.04	14.16
Specific gravity		
Specific gravity	2.98	1.95
Atterberg's limits		
Liquid limit, %	38.0	33.0
Plastic limit, %	27.14	–
Plasticity index, %	10.86	–
Shrinkage limit, %	26.52	–
Classification as per IS plasticity chart	Inorganic silt of low plasticity	–
Swell properties		
Differential free swell index, %	29.41	–
Modified free swell index	2.27	–
Compaction characteristics		
Optimum water content, %	29.1	22.51
Max. dry density, gm/cm^3	1.62	1.347
Physicochemical properties		
pH value	10.88	8.20
Electrical conductivity, mS/cm	4.01	0.45

free swell index of red mud are 29.41% and 2.27, respectively. As per classification system given by Sivapullaiah et al. (1987) based on modified free swell index, the red mud is classified as a soil having negligible swelling potential.

The pH and electrical conductivity are done to know the physicochemical properties of red mud. It is observed that red mud has high alkalinity having pH value of 10.88 and electrical conductivity value of 4.01 mS/cm (Table 1).

The X-ray diffraction (XRD) analysis of red mud (Fig. 1a) shows the presence of quartz, sodalite (aluminum sodium silicate chloride), goethite (iron oxide hydroxide), hematite (iron oxide), rutile (titanium oxide) and aluminum fluoride as predominant minerals. The microstructural examination of the red mud shows the small, apparently aggregated particles and a layer of material coating most of the particles surface (Fig. 2a). Further, the aggregated particles represent most likely the presence of hematite, confirming the XRD analysis, whereas the presence of flake/needle shaped particles shows the existence of some impurities, probably gypsum (He et al. 2013). The elemental analysis of red mud by energy-dispersive X-ray spectrometer (EDAX) is presented in Table 4. It is observed that red mud contains the presence of aluminum, sodium and silica with minor amount of calcium and chloride.

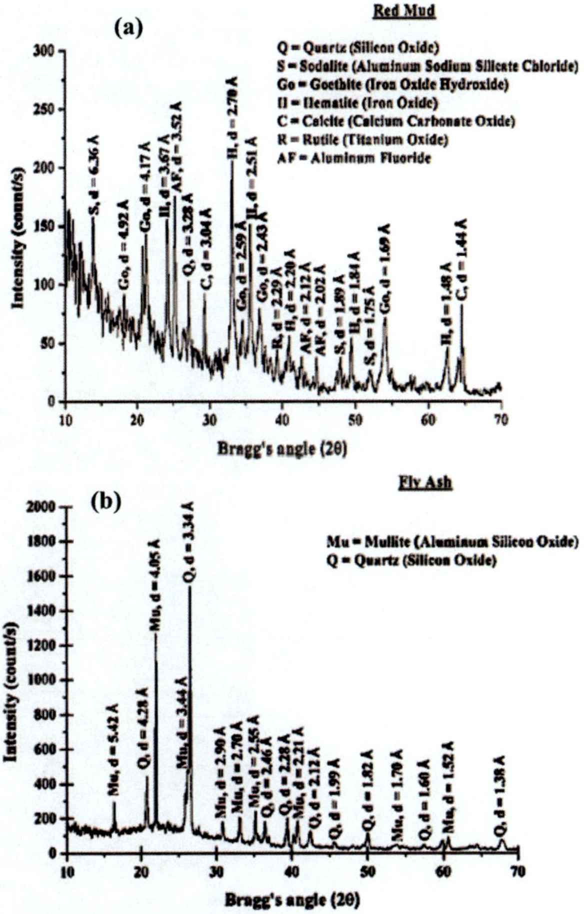

Fig. 1 Mineralogical analysis of red mud and fly ash

The Fourier transform infrared spectroscopy (FTIR) spectrum of RM is shown in Fig. 3a. It is observed from the figure that the broadbands occurred around 991 cm^{-1}. This shows the characteristics bands of Si–O and O–Si–O group, confirming the presence of silicate groups (Si–O and Si(Al)–O) (Smičiklas et al. 2014). The presence of band around 1440 cm^{-1} confirms the presence of carbonate groups. The minor broadband occurred around 2900 cm^{-1} illustrates the stretching vibrations of O–H and H–O–H bonding of interlayer-adsorbed H_2O molecule.

2.1.2 Fly Ash

The fly ash used in the present work was obtained from Kota thermal power plant and, is class F fly ash. The fly ash is pozzolanic in nature and contains less than 7% lime (CaO). The physical properties of fly ash are presented in Table 3.

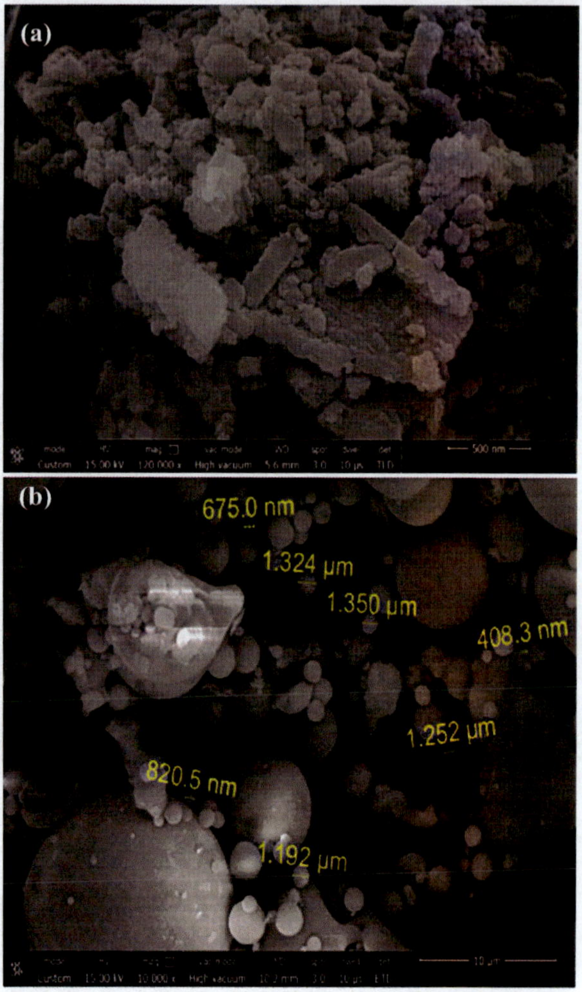

Fig. 2 Microstructural examination of **a** red mud and **b** fly ash

Table 4 Chemical composition of red mud and fly ash

Elements	Red mud (atomic %)	Fly ash (atomic %)
Silica (Si)	2.24	16.75
Aluminum (Al)	5.84	14.34
Calcium (Ca)	0.44	0.07
Ferrous (Fe)	6.60	0.84
Sodium (Na)	4.98	0.26
Titanium (Ti)	9.66	0.67

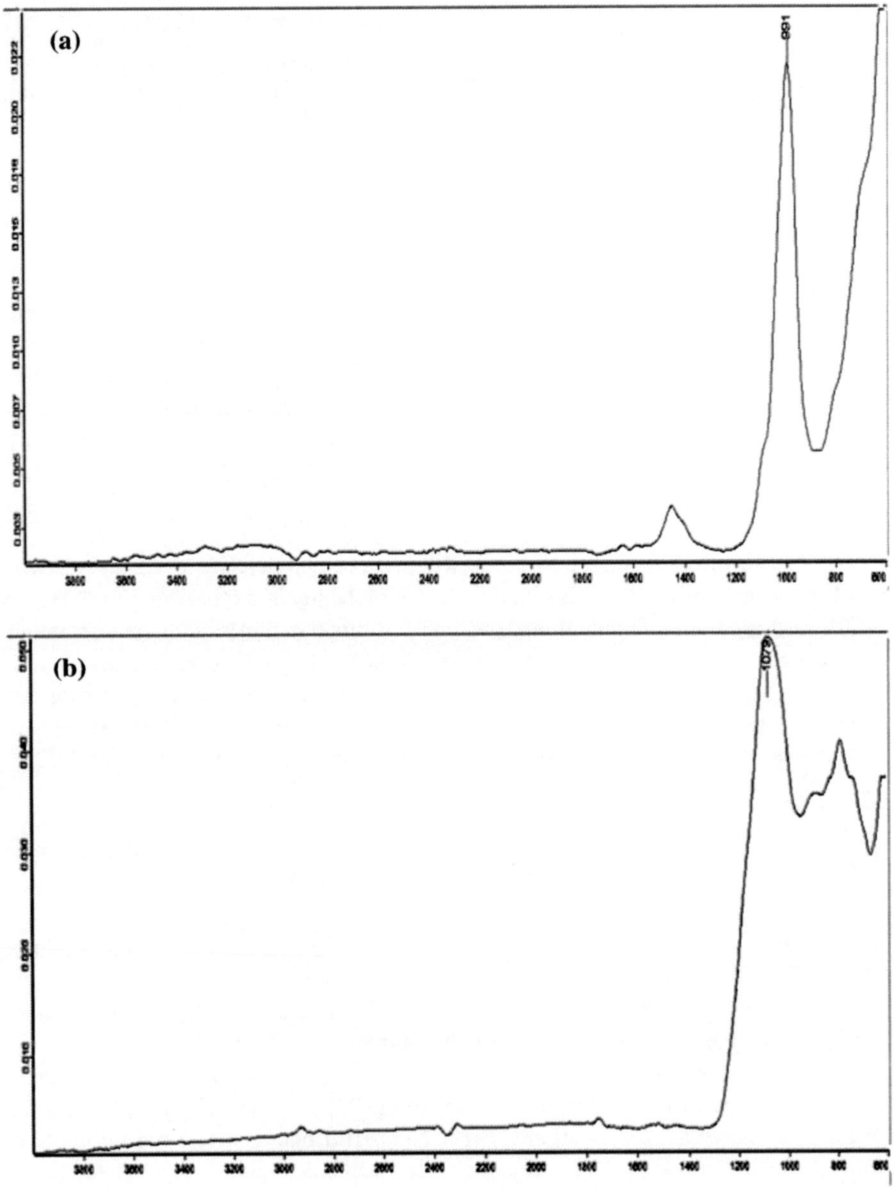

Fig. 3 FTIRR of red mud and fly ash

The particle size analysis of fly ash shows the presence of predominant amount of sand-sized particle (4.75–0.075 mm) of 63.62%, silt-sized particle (0.075–0.002 mm) of 22.21% and clay-sized particle (<0.002 mm) of 14.16%. The specific gravity of fly ash is obtained to be 1.98. The liquid limit of fly ash is found to be 33.0% (Table 3). Fly ash is found to be a non-plastic in nature. The optimum water content and maximum dry density of fly ash are found to be 22.51% and 1.347 gm/cm^3, respectively. The pH and electrical conductivity of fly ash are observed to be 8.20 and 0.45 mS/cm which is lower than red mud (Table 3).

The mineralogical analysis by X-ray diffraction (XRD) (Fig. 1b) shows the presence of quartz and mullite (aluminum silicon oxide) as predominant minerals in the fly ash. The microstructural analysis of the fly ash (Fig. 2b) shows the presence of various sizes of particles, i.e., 675.0, 408.3 and 820.5 nm. The shape of particle is observed to be rounded and spherical. The elemental analysis of fly ash by energy dispersive X-ray spectrometer (EDAX) is presented in Table 4. It is observed that fly ash contains significant amount of aluminum and silica with minor amount of calcium. The amount of calcium compared to silica and aluminum confirms that fly ash is of class F.

FTIR absorption spectroscopy is well known to characterize the materials with short-range structural order. The main adsorption bands of fly ash are 750, 1035 and 1072 cm^{-1} (Fig. 3b). The peak at 1036 cm^{-1} is attributed to asymmetric stretching vibrations of Al–O/Si–O bonds, while the Si–O–Si/Si–O–Al bonding band can be seen at 747 cm^{-1} (Guo et al. 2010). These bands are common in ring silicates and provide an indication of the degree of amorphization of the material, since its intensity does not depend on the degree of crystallization (Swanepoel and Strydom 2002).

2.2 *Methodologies Followed*

The experimental works have been performed by following the standard procedures. The particle size distribution and specific gravity of parent materials (red mud and fly ash) are done as per procedure given by Indian Standard (IS)—2720 (Part 4) (1985) and IS-2720 (Part 3) (1985), respectively. Further, Atterberg's limits (liquid limit, plastic limit and shrinkage limit) of samples are determined by standard procedure of IS-2720 (Part 5) (1985); IS-2720 (Part 6) (1972), respectively. Differential free swell, also known as free swell index (FSI), is performed as per as per Indian Standard (IS) 2720 Part (40) (1977). Modified free swell index has been calculated for the samples as per Sivapullaiah et al. (1987). The compaction characteristics (optimum water content and maximum dry density) of parent materials and mixes are determined by carrying out mini-compaction test procedure developed by Sridharan and Sivapullaiah (2005).

The pH test of parent materials and mixes is conducted according to IS-2720 (Part 26) (1987). The instrument was calibrated with standard buffer solution of pH 4.0, 7.0 and 9.0 prior to determine pH value of all samples. The same samples are used to determine the electrical conductivity.

XRD is accomplished with a PANanlytical model X'Pert powder diffractometer to identify the changes in the mineralogy and crystalline phases. XRD is carried out with CuKα radiation with $\lambda = 1.54$ Å at 1°/min speed for Bragg's angle range of 3–90°.

Changes in fabric and chemical composition of samples are performed with SEM and EDAX techniques, respectively. Quanta 200 FESEM coupled with energy-dispersive X-ray spectrometer is used for the SEM and EDAX studies. A piece of the dried samples was mounted on the specimen holder and coated with a 100 Å thin-layer gold palladium for 38 s using a sputter coater, poloran E5100 at 10^{-3} Torr vacuum prior to the microstructural examination. The gold coating is done in order to avoid charging problems during the imaging.

FTIR is most useful for identifying chemicals that are either organic or inorganic. FTIR relies on the fact that the most molecules absorb light in the infrared region of the electromagnetic spectrum. This absorption corresponds specifically to the bonds present in the molecule. The frequency range is measured as wave numbers typically over the range 4000–600 cm^{-1}.

3 Results and Discussion

3.1 Geotechnical Properties of Treated Red Mud with Fly Ash

3.1.1 Specific Gravity

The effect of fly ash on the specific gravity of red mud is shown in Fig. 4. It is observed that the specific gravity of the red mud reduces significantly with increase in fly ash content up to 80%. The continuous reduction in specific gravity of red mud is due to the replacement of material having higher specific gravity (red mud) with lower value of fly ash content. Further, it is well known that the fly ash is cenosphere particles

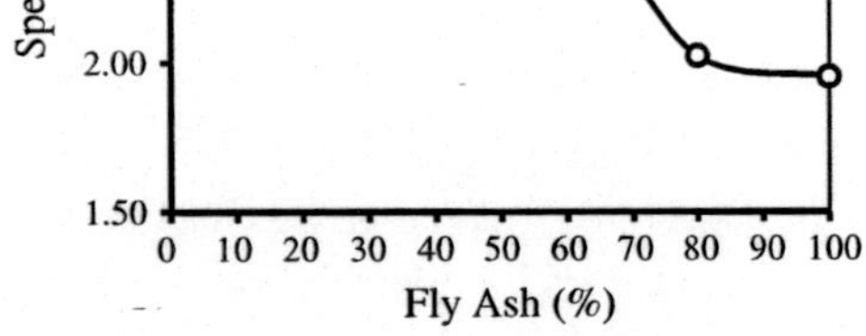

Fig. 4 Plot showing specific gravity of red mud with varying fly ash content

having cavity which causes reduction in its specific gravity. The replacement of red mud with fly ash also leads to decrease in the amount of ferrous (higher specific gravity) in red mud-fly ash mix.

3.1.2 Atterberg's Limits

Influence of fly ash on soil plasticity (liquid limit, plastic limit and plasticity index) of red mud is shown in Figs. 5 and 6.

Result shows that liquid limit of red mud decreases and plastic limit increases; hence, the derivative of liquid limit and plastic limit, i.e., plasticity index decreases with an increase in fly ash content (Fig. 7). Also, the shrinkage limit of red mud has marginal effect with an addition of fly ash (Fig. 8). The reductions in liquid limit and

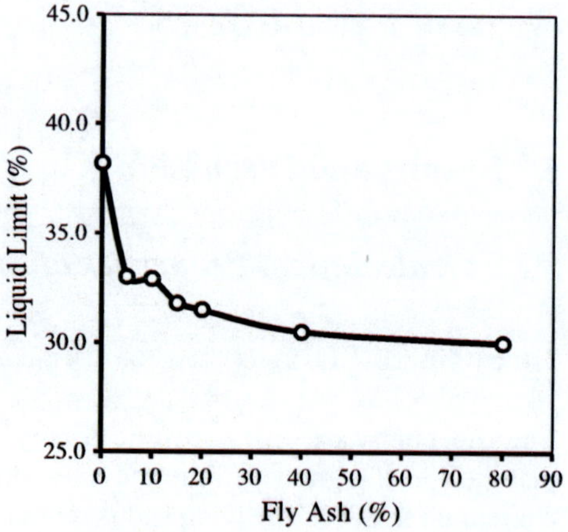

Fig. 5 Plot showing liquid limit of red mud with varying fly ash content

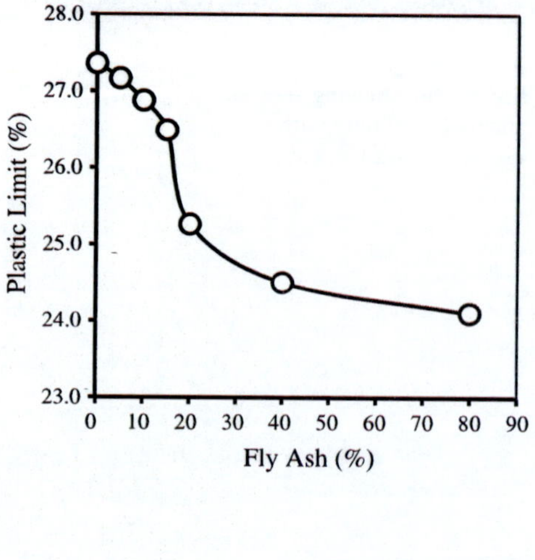

Fig. 6 Plot showing plastic limit of red mud with varying fly ash content

Fig. 7 Plot showing plasticity index of red mud with varying fly ash content

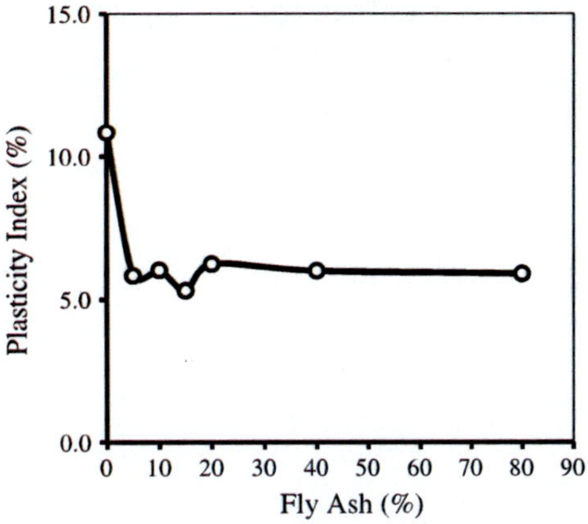

Fig. 8 Plot showing shrinkage limit of red mud with varying fly ash content

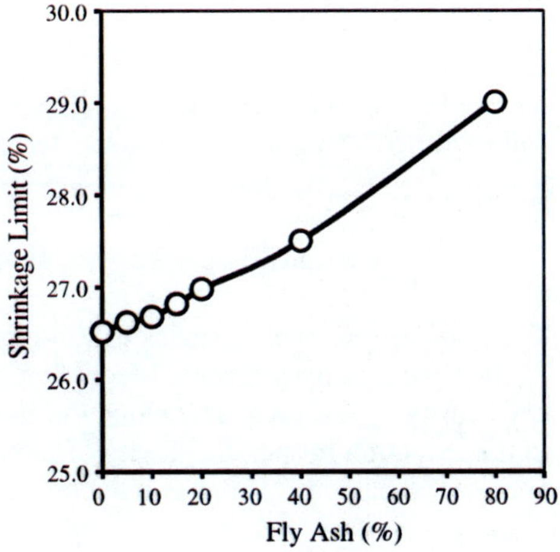

plastic limit are mainly due to the addition of non-plastic material (i.e., fly ash) to the red mud.

3.1.3 Free Swell Index

To capture the swell parameter of red mud with addition of fly ash, the modified free swell index (MFSI) has been performed as per procedure given by Sivapullaiah et al. (1987) to avoid the limitations of IS 2720 Part 6 (1972). The result shows that the increase in MFSI of red mud with addition of fly ash content up to 10% and reduces drastically after further addition of fly ash content up to 80% (Fig. 9). The increase in MFSI of red mud with lower fly ash content is due to the flocculation of red mud, whereas reduction in MFSI of red mud is due to the reduction in specific gravity of

Fig. 9 Modified free swell index (MFSI) of the red mud with fly ash content

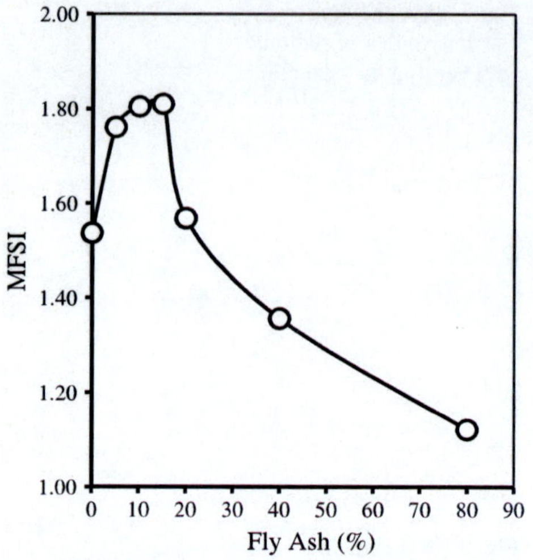

red mud-fly ash mix. The reduction in specific gravity results in the volume of soil solids ($W_s/(G_s * \gamma_w)$), leading to the decrease in the MFSI of red mud.

3.1.4 Compaction Characteristics of Red Mud with Fly Ash Content

The compaction characteristics of red mud-fly ash mixes are shown in Fig. 10. It is observed that fly ash causes reduction in optimum water content and maximum dry density of red mud. The reduction in maximum dry density of red mud is due to the reduction in specific gravity of mix, flocculation of particles due to the presence of free calcium available in the fly ash and formation of compacted matrix with

Fig. 10 Compaction characteristics of red mud with different fly ash content

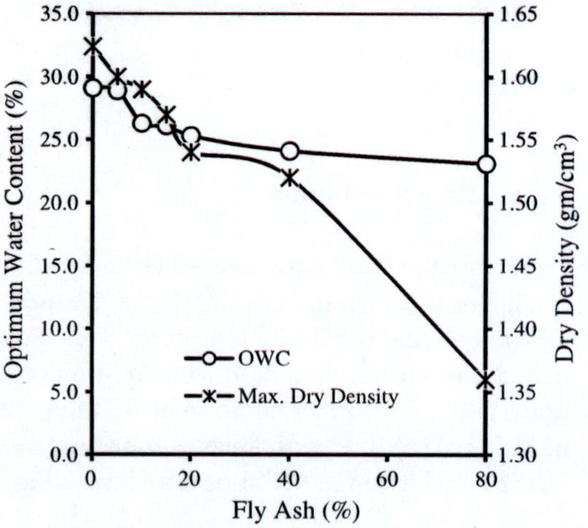

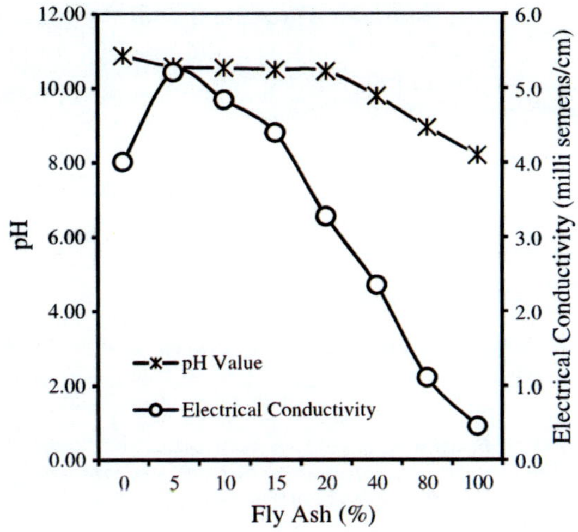

Fig. 11 pH and electrical conductivity of red mud-fly ash mixes

replacement of sand-sized particles with silt-sized particles of fly ash. This formation of compacted flocculated matrix counteracts the compactive effort and hence, reduction in dry density. However, reduction in optimum water content of red mud may be due to the non-plastic behavior of fly ash content.

3.2 Physicochemical Properties of Red Mud-Fly Ash Mixes

The physicochemical properties of red mud-fly ash mixes have been measured in terms of pH and electrical conductivity and are shown in Fig. 11. It is observed that the pH value and electrical conductivity of red mud reduce drastically with addition of fly ash content up to 80%. This confirms that fly ash has capacity to neutralize the higher alkalinity of red mud. Also, the reduction in electrical conductivity of red mud with addition of fly ash reflects the neutralization and the adsorption of the heavy element from red mud (Rios et al. 2008; Mohan and Gandhimathi 2009; Querol et al. 2006).

4 Conclusion

The details experimental works have been performed to find out the effect of fly ash (up to 80%) on the behavior of red mud. Based on the present study, following conclusions can be drawn:

1. The substitution of red mud with fly ash reduces the liquid limit, plastic limit, plasticity index and specific gravity. However, shrinkage limit increases with addition of fly ash content.
2. The swell index of red mud decreases drastically with an addition of higher fly ash content.
3. The optimum water content and maximum dry density of red mud decrease with an increase in fly ash content.
4. The alkalinity and electrical conductivity of red mud reduce significantly with addition of fly ash content.

Further work can be extended to find out the engineering behavior of red mud-fly ash mix and to synthesize the mix for preparing geopolymeric material.

Acknowledgements Authors appreciates the financial support provided by the Research and Development (R&D) division of Manipal University Jaipur (MUJ) under SEED grant [Project Number: MUJ/REGR1435/07].

Conflict of Interest Authors declare about no any kind of conflict of interest related to present work.

References

Bureau of Indian standards IS 2720 (Part 4) (1985) Methods of test for soils: grain size analysis, New Delhi

Bureau of Indian Standards IS 2720 (Part 26) (1987) Methods of test for soils: determination of pH value (second revision), New Delhi, India

Bureau of Indian Standards IS 2720 (Part 40) (1977) Methods of test for soils: determination of free swell index of soil, New Delhi, India

Bureau of Indian Standards (first revision) IS 2720 (Part 6) (1972) Methods of test for soils: determination of shrinkage factors, New Delhi, India

Bureau of Indian Standards (second revision) IS 2720 (Part 3) (1980) Methods of test for soils: determination of specific gravity, New Delhi, India

Bureau of Indian Standards (second revision) IS 2720 (Part 5) (1985) Methods of test for soils: determination of liquid limit and plastic limit, New Delhi, India

Cablik V (2007) Characterization and applications of red mud from bauxite processing. Gospodarka Surowcami Mineralnymi 23(4):27–38

Deelwal K, Dharavath K, Kulshreshtha M (2014) Evaluation of characteristic properties of red mud for possible use as a geotechnical material in civil construction. Int J Adv Eng Technol 7(3):1053

Guo X, Shi H, Dick WA (2010) Compressive strength and microstructural characteristics of class C fly ash geopolymer. Cement Concr Compos 32(2):142–147

Hannachi Y, Shapovalov NA, Hannachi A (2010) Adsorption of nickel from aqueous solution by the use of low-cost adsorbents. Korean J Chem Eng 27:152–158

He J, Jie Y, Zhang J, Yu Y, Zhang G (2013) Synthesis and characterization of red mud and rice husk ash-based geopolymer composites. Cement Concr Compos 37:108–118

Kumar S, Kumar R, Bandopadhyay A (2006) Innovative methodologies for the utilisation of wastes from metallurgical and allied industries. Resour Conserv Recycl 48(4):301–314

Mohan S, Gandhimathi R (2009) Removal of heavy metal ions from municipal solid waste leachate using coal fly ash as an adsorbent. J Hazard Mater 169(1–3):351–359

Nikraz HR, Bodley AJ, Cooling DJ, Kong PYL, Soomro M (2007) Comparison of physical properties between treated and untreated bauxite residue mud. J Mater Civ Eng 19(1):2–9

Querol X, Alastuey A, Moreno N, Alvarez-Ayuso E, García-Sánchez A, Cama J, Simón M (2006) Immobilization of heavy metals in polluted soils by the addition of zeolitic material synthesized from coal fly ash. Chemosphere 62(2):171–180

Rios CA, Williams CD, Roberts CL (2008) Removal of heavy metals from acid mine drainage (AMD) using coal fly ash, natural clinker and synthetic zeolites. J Hazard Mater 156(1–3):23–35

Sabat AK, Mohanta S (2015) Efficacy of dolime fine stabilized red mud-fly ash mixes as subgrade material. J Eng Appl Sci

Sahu SK (2009) Research paper on Indian primary aluminium market. Analysis of Indian primary aluminium industry from a microeconomic point of view. https://www.scribd.com/doc/19149792/Indian-Aluminium-Industry.

Singh SP, Samantasinghar S, Sindhuja D (2017) Stabilization of red mud using ground granulated blast furnace slag by geopolymerization technique. In: International congress and exhibition sustainable civil infrastructures: innovative infrastructure geotechnology. Springer, Cham, pp 338–350

Sivapullaiah PV, Sitharam TG, Rao KS (1987) Modified free swell index for clays. Geotech Test J 10(2):80–85

Smičiklas I, Smiljanić S, Perić-Grujić A, Šljivić-Ivanović M, Mitrić M, Antonović D (2014) Effect of acid treatment on red mud properties with implications on Ni (II) sorption and stability. Chem Eng J 242:27–35

Sridharan A, Sivapullaiah PV (2005) Mini compaction test apparatus for fine grained soils. Geotech Test J 28(3):240–246

Swanepoel JC, Strydom CA (2002) Utilisation of fly ash in a geopolymeric material. Appl Geochem 17(8):1143–1148

Durability of Cementitious Phases in Lime Stabilization: A Critical Review

Dhanalakshmi Padmaraj and Dali Naidu Arnepalli

Abstract Soil–lime interactions involve concomitant short-term and long-term alterations of the fine-grained soil resulting in the formation of a workable material bonded by various pozzolanic compounds. These pozzolanic compounds being cementitious in nature are expected to hold the soil particles together and bring long-term strength and stability to the soil–lime composites. However, the durability of cementitious phases formed due to pozzolanic reactions is highly subjective owing to the variations in the moisture and physiochemical factors like pH under diverse environmental conditions. The relative humidity and presence of atmospheric gases like carbon dioxide have a significant impact on the performance of the stabilized system. Carbonation of reaction products, as well as the effects of seasonal moisture fluctuations, can cause the decalcification of the cementitious phases and further degradation in the stabilized system. However, the type of reaction products and their chemical composition, which is a function of the mineralogy of the soil, will determine their durability in adverse conditions. The present study attempts to review the chemistry of reaction products formed in view of its inherent mineralogy. In addition, the degradation nature of the soil–lime composites under adverse conditions like moisture ingress and carbonation is evaluated for their long-term performance.

Keywords Durability · Lime stabilization · Carbonation · Leaching

1 Introduction

Chemical modification of fine-grained soils primarily involves the alteration of their inherent mineralogy to a refined form to attain the desired engineering performance. Among all different chemical stabilization techniques, lime stabilization is being successfully adopted in different parts of the world, owing to its simplicity and cost-effectiveness (Cherian 2018). The short-term soil–lime interactions include cation exchange reactions, agglomeration, and flocculation, thereby modifying the soil into

D. Padmaraj (✉) · D. N. Arnepalli
Indian Institute of Technology Madras, Chennai, India
e-mail: dhanalakshmip93@gmail.com

M. Latha Gali and R. R. P. (eds.), *Problematic Soils and Geoenvironmental Concerns*, Lecture Notes in Civil Engineering 88,
https://doi.org/10.1007/978-981-15-6237-2_40

a more workable engineered material. However, stabilization, in addition to the modification, is considered to commence by the formation of pozzolanic compounds from the aluminosilicates dissolved in the soil at high pH and calcium ions supplied by the lime (Eades 1962). Hence, the strength and stiffness of the new matrix formed increase with time owing to the binding action of the reaction products (Dash and Hussain 2011). However, the stabilization will be considered adequate if it can guarantee long-term strength and stability to the soil-lime composites.

The conventional method of estimation of optimum lime content for achieving the desired improvement relies on the amount of clay size fraction in the soil as well as variation in the plasticity index upon lime content. However, in the present scenario, a majority of the practicing companies and regulatory authorities employ pH or compressive strength testing as the main criteria for determining optimum lime content. The pH-based test devised by Eades (1962) for determining the optimum lime content (OLC) is widely accepted owing to its simplicity but has found to reflect the buffering capacity of soil matrix rather than its reactive nature (Cherian and Arnepalli 2015). Nonetheless, it is very well known that strength is not a reliable criterion for compacted chemically stabilized soils, but long-term durability is far more significant for ensuring lasting performance. Hence, additional research is needed to ensure that durability is also confirmed at the design soil–lime mix. Durability confers to the permanency of chemical stabilization, i.e., the ability of the soil particles and the stabilized products to remain intact for the design period (Chittoori et al. 2018). Hence, while ensuring the long-term performance, it is necessary to identify different kinds of environmental changes and their implications on the soil–lime composites (Hino et al. 2012). Different causes of degradation of chemically treated soils include seasonal wetting and drying cycles, leaching of calcium, as well as the carbonation of the free lime and cementitious compounds (McCallister and Petry 1992; Paige-Green 2008; Chakraborty and Nair 2018).

To assess the durability of the stabilization design, chemically treated soil samples are subjected to cycles of wetting and drying to replicate the moisture and temperature fluctuations occurring in the field. Also, a leachability analysis is carried out to determine the residual calcium removal due to rainfall infiltration (Chittoori et al. 2018; Deneele et al. 2016). In addition to the above, the carbonation of the treated soil–lime system can seriously deteriorate its long-term performance (Bagonza et al. 1987). Most of the laboratory and field-based pavement design studies involving lime additives demonstrate that when proper attention is paid to materials' design and durability, lime-stabilized subgrades outperform when compared to unstabilized areas (Mallela et al. 2004). Therefore, a fundamental understanding of the mechanism and possible degradations of soil–lime composites from a soil mineralogical perspective is essential. Given the above, the present study tries to discuss in brief, the stability of soil–lime composites by reviewing the chemistry of reaction products and possible degradations associated with them.

2 Reaction Products in Lime–Soil Composites

The rapid, short-term changes in terms of plasticity properties associated with the addition of lime modify the soil to be more workable (Diamond and Kinter 1965). This is achieved by the replacement of monovalent ions by calcium ions, which further compresses the diffused double layer, resulting in the reduction of water holding capacity. The clay particles come together and promote flocculation and agglomerations, to increase the workability. The long-term interactions, on the other hand, are classified into two categories: through solution mechanism and adsorption mechanism (Cherian 2018). The former mechanism involves the dissolution of aluminosilicates of clay mineral under highly alkaline conditions and their reaction with free calcium ions in the pore fluid to form cemented gels that crystallize with time (Eades 1962). The pH of pore fluid, dissolution of silica and alumina, clay mineralogy, and curing conditions affect the final composition of these compounds. The latter mechanism is concerned with the physical adsorption of lime onto the surface of clay particles, which is independent of the magnitude of the dissolution of aluminosilicates. It rather depends on the specific surface area as well as the electronegativity (Diamond and Kinter 1965). However, the nature and properties of cementitious phases formed in the long-term are responsible for the longevity of the soil–lime mixture. These long-term compounds are time- and temperature-dependent with temperature accelerating the reaction kinetics (Cherian et al. 2018).

The products of long-term pozzolanic reactions fall broadly into three types: hydrates of calcium silicates (CSH), calcium aluminates (CAH), and calcium aluminosilicates (CASH) (Croft 1967; Rao and Rajasekaran 1996). The letters C, S, A, H denote CaO, SiO_2, Al_2O_3, and H_2O, respectively, as per standard notations of cement chemistry. Calcium silicate hydrates are characterized by poor crystallinity and fibrous nature. They include three types: CSH gel (also called tobermorite gel), CSH (I) and CSH (II). Tobermorite gels are rich in calcium and are commonly seen in cement concrete. The remaining phases of CSH are characterized by the CaO/SiO_2 (C/S) ratio (CSH-I for C/S < 1.5 and CSH-II for C/S ratio > 1.5). An increase in the C/S ratio brings a stark difference in the morphology from the plate-like to fibrous nature (Diamond and Kinter 1965; Gard and Taylor 1976). These variable morphologies result in variable porosity and surface area and hence their moisture affinity and consequent durability of the soil–lime composites. The CSH (II) phase shows better resistance to moisture attack, and its formation is facilitated by higher lime content and increased curing period (Chakraborty and Nair 2018). The second type, calcium aluminate hydrates (CAH), belongs to a member of the tetra calcium aluminate hydrate group with hexagonal thin sheet morphology. They are characterized by better crystallinity, partially hydrated states as well as carbonate groups instead of hydroxyls. The well crystalline nature makes it readily identifiable under X-ray diffraction (Diamond and Kinter 1965).

Moreover, the basic mineralogy and consequent reactive nature also determine the mode of reaction (Eades 1962; Cherian and Arnepalli 2015). In the case of kaolinitic clays, the attack of lime is particularly at the edges owing to the relatively higher edge surfaces (Sloane 1965). However, long-term studies indicate lesser reactivity of kaolin in improving the strength. This has been attributed to the pH-dependent charges formed at the edges that facilitate adsorption of calcium and hydroxyls, blocking further pozzolanic reactions (Chemeda et al. 2018). Montmorillonite-rich soils, on the other hand, have relatively more silica content and react with lime to form hydrates of silicates than aluminates. This is justified by the dissolution studies on soils of different mineralogies by Cherian (2018) in which bentonite samples rich in montmorillonite showed preferential dissolution of silica, whereas kaolin samples showed a higher dissolution of alumina. However, the stability of siliceous phases of montmorillonite is highly sensitive to moisture attack resulting in its decalcification (Diamond and Kinter 1965). Besides, the higher cation exchange capacity and specific surface area of montmorillonite are associated with enhanced lime consumption reducing the pH of pore fluid, which can, in turn, commence the decalcification process (Yu et al. 2014).

Cementation of the precipitated gels and formation of inter-aggregate and intra-aggregate bonding due to crystalline intergrowth and drying give rise to an enormous increase in the strength upon curing. This further results in a reduction in the permeability and porosity of clay–lime mixtures, thereby transforming the less workable soil into an intact mass with improved mechanical properties. However, these changes are possible only when the system is subjected to an environment that continuously favors cementation and concomitant crystallization.

3 Calcium Leaching

Since the mechanism of clay–lime interactions and consequent transformations to the fabric are well established, researchers are currently focused on understanding their longevity under diverse environments. The use of lime-treated soils in areas where water-table fluctuations are significant emphasizes the importance of understanding their physiochemical and mechanical changes upon moisture movements. Though several studies have concluded that the cementitious matrix is impervious and stable to leaching, a fundamental understanding of the mechanism, causes, and effects of moisture movement in the lime-treated soils of varying mineralogy are necessary to predict their long-term performance.

A primary initiative to study the susceptibility of lime-treated clays to leaching was done by McCallister and Petry (1992) using a flexible-wall multiple leach cell apparatus. As per his study, soils treated with lime content corresponding to the pH method (OLC_{pH}) observed minimum calcium removal. This must be owing to the enhanced clay–lime interaction and simultaneous adsorption of lime on the electronegative clay platelets at the saturated calcium hydroxide solution. On the other hand, considering continuous consumption, the addition of lime contents higher than OLC_{pH} can promote continuous silica and alumina dissolution and further pozzolanic reactions. These massive reaction compounds can reduce the porosity, thereby blocking the moisture flow (Deneele et al. 2016; Khattab et al. 2007). Altogether, it must be understood that the pozzolanic reactions are time dependent and require an ambient condition without disturbing the clay–lime matrix for long-term performance. Furthermore, at a lower curing period, pozzolanic reactions are incomplete and create a favorable condition for the release of calcium via unreacted lime as well as unstable cementitious gels by a process called decalcification (Chakraborty and Nair 2018). Following, Chittoori et al. (2018) reported that soils rich in montmorillonite are highly susceptible to leaching and showed a reduction in the retained strength without considerable stabilizer loss. The destabilization of CSH formed from siliceous materials like montmorillonite into unstable forms can be a probable reason (Dash and Hussain 2011; Chakraborty and Nair 2018). However, the underlying mechanism that degrades the soil–lime system in view of its inherent mineralogy is not extensively studied. As a whole, it can be inferred that the optimum lime content determination criteria of the soil that favors maximum strength and durability should be considered with prime importance to its mineralogy and consequent pozzolanic reactivity (Cherian and Arnepalli 2015).

Together with lime content requirements, the compaction conditions achieved in the field also affects the clay–lime reactions. The placement moisture content and compaction effort play an important role in the longevity of soil–lime composites. At dry of optimum, the increased permeability of the flocculated mixture can enhance the rate of moisture flow, which promotes the easy removal of calcium (Le Runigo et al. 2011; Chittoori et al. 2011). Furthermore, the improved density achieved by better compaction reduces the porosity and further ingress of moisture and other contaminants. Hence, compaction preferably at wet of optimum with a higher compactive effort is found to be a stable state for overall improvement in the durability performance of treated soil.

In addition to all the reasons described above, the leaching rate is also influenced by the chemistry of circulating fluid. Infiltration test studies by Kamon et al. (1996) on lime- and cement-stabilized clay suggest that long-term exposure with acid rain can affect the engineering and physiochemical properties of the stabilized soil. At lower concentrations, calcium ion leaching proceeds that affect the alkalinity of stabilized soil impeding further pozzolanic reactions. At lower pH of the leachant, the release of adsorbed and even hydrated calcium degrades the strength of the stabilized soil. Nonetheless, the formation of secondary reaction products during leaching can fill the pores resulting in decreased porosity (Kamon et al. 1996).

As a whole, the performance of lime-treated soil under fluctuating moisture conditions occurring in the field is a function of its inherent soil mineralogy, lime content, compaction conditions, curing period as well as the pore fluid chemistry of leachant.

4 Carbonation

In areas where proper curing of the stabilized layer is difficult, the carbonation of lime and reaction products is commonly observed (Bagonza et al. 1987). The reaction occurs in the liquid phase where the dissolved carbon dioxide reacts with Ca^{2+} in the pore fluid to produce calcium carbonate. In addition, cementitious phases also undergo carbonation to a reasonable degree (Rezagholilou et al. 2017; Beardmore 2018). The phenomenon was initially considered to contribute toward strength owing to the partial cementation of soil particles by calcium carbonate. However, it was later disregarded as lime required for pozzolanic reactions are consumed for carbonation (Eades 1962; Diamond et al. 1964).

The failure of forty-four different cases of soil–lime base courses in South Africa due to carbonation has triggered the pavement and geotechnical engineers in understanding its mechanism, causes, and consequent effects. Consequently, the work conducted by Netterberg and Paige-Green (1984) has laid a foundation for the research area on the carbonation of lime-modified pavement layers. The improper sealing of lime-stabilized layer was found to be a significant cause of these deleterious effects. The effects of carbonation have far-reaching consequences that it largely nullifies the modification and cementation reactions increasing the plasticity properties and reducing the long-term strength. The strength gained after seven years of lime treatment has reduced by 45–70% when subjected to carbonation (Bagonza et al. 1987).

The main factors associated with carbonation as observed by concrete research fraternity are relative humidity, carbon dioxide concentration, compaction conditions, binder content, curing temperature as well as curing period (Goodbrake et al. 1979; Yeo and Nikraz 2012). Besides, the pozzolanic reactivity of the material, calcium dissolution rate, as well as the specific surface area of lime also dictates the ability of the material to carbonate (Cizer et al. 2010, 2012). In similar lines for soil–lime composites, the dominant mineralogy of the soil determines the lime reactivity and affinity for free lime carbonation (Bandipally et al. 2018). In addition, curing temperature affects the kinetics of the pozzolanic reaction decelerating the deleterious carbonation reaction (Cherian 2018). Hence, carbonation during the initial periods occurs in the decreasing order of the pozzolanic reactivity of the soil material. Also, the water holding capacity of clay minerals determines the transport properties of the gas. The higher water retention capacity of montmorillonite together with its highly reactive nature can prevent the entry and further interaction of CO_2 into the system (Ho et al. 2017). On the other hand, the residual lime available after short-term reactions is higher in kaolin and illite than montmorillonite clays, thereby expecting enhanced carbonation (Al-Mukhtar et al. 2014). These observations manifest the importance of involving the role of mineralogy in predicting the carbonation rate in soils.

The effects of carbonation depend on the nature of the reaction mechanism. Carbonation of unreacted lime observed in the early days of curing results in expansion owing to the higher molar volume of calcium carbonate in comparison with lime. Hence, the tensile stress induced in the soil–lime mix can result in microcracking decreasing their strength (Paige-Green 2008; Ashraf 2016). On the contrary, it is also observed that initial lime carbonation increases the strength of the soil due to carbonate precipitation (Cizer et al. 2012). This can happen if there are sufficient voids for calcium carbonate to precipitate and further form bonding. In the long-term, the stability of cementitious composites determines their affinity toward carbonation

(Chakraborty and Nair 2018; Cizer et al. 2012). CSH with low C/S is more vulnerable and the carbonation of the same results in the decalcification and formation of weak silica gels with Si–O–Si bonds. These siloxane bonds form a bridge between neighboring particles, thereby shrinking and pulling them together—a phenomenon called CSH shrinkage (Chen et al. 2006; Jones et al. 2010). The variation in the macropores and micropores associated with carbonation can explain the pore-filling mechanism and CSH shrinkage to a greater extent.

Hence, for achieving long-term durability from the carbonation, soil with higher pozzolanic reactivity and lime content to form stable reaction products should be selected.

5 Conclusion

The chemistry and morphology of various reaction products are very well discussed in the past literature in view of their inherent mineralogy. However, recent research is focused mainly on the stability of these compounds and their sustainability under adverse environmental conditions. More research should be carried out to understand the stability of reaction products under fluctuating moisture and carbonating environments for soils of varying mineralogy. It helps in evaluating the durability of any chemical stabilization given the soil chemistry toward the formation of a stable soil–lime matrix.

References

Al-Mukhtar M, Lasledj A, Alcover JF (2014) Lime consumption of different clayey soils. Appl Clay Sci 95:133–145

Ashraf W (2016) Carbonation of cement-based materials: challenges and opportunities. Constr Build Mater 120:558–570

Bagonza S, Peete JM, Freer-Hewish R, Newill D (1987) Carbonation of stabilized mixtures. HPTRC Transport and Planning Summer Annual Meeting, University of Bath, London, United Kingdom, pp 29–48

Bandipally S, Cherian C, Arnepalli DN (2018) Characterization of lime-treated bentonite using thermogravimetric analysis for assessing its short-term strength behaviour. Indian Geotech J 48(3):393–404

Beardmore RL (2018) Carbonation of cement stabilized materials in pavement layer. Doctoral dissertation, Stellenbosch University, South Africa

Chakraborty S, Nair S (2018) Impact of curing time on moisture-induced damage in lime-treated soils. Int J Pavement Eng 1–13

Chemeda YC, Deneele D, Ouvrard G (2018) Short-term lime solution-kaolinite interfacial chemistry and its effect on long-term pozzolanic activity. Appl Clay Sci 161:419–426

Chen JJ, Thomas JJ, Jennings HM (2006) Decalcification shrinkage of cement paste. Cem Concr Res 36(5):801–809

Cherian C (2018) Evaluation of lime stabilization mechanisms from chemico-mineralogical perspective. Doctoral dissertation, Indian Institute of Technology, Madras, India

Cherian C, Arnepalli DN (2015) A critical appraisal of the role of clay mineralogy in lime stabilization. Int J Geosynth Ground Eng 1(8):1–20

Cherian C, Kollannur NJ, Bandipally S, Arnepalli DN (2018) Calcium adsorption on clays: effects of mineralogy, pore fluid chemistry and temperature. Appl Clay Sci 160:282–289

Chittoori S, Pedarla A, Puppala AJ, Hoyos LR, Nazarian S, Saride S (2011) Leachate studies on lime and Portland cement-treated expansive clays. In: Geo-frontiers 2011: advances in geotechnical engineering. ASCE, Texas, United States, pp 4479–4488

Chittoori BC, Puppala AJ, Wejrungsikul T, Hoyos LR (2018) Experimental studies on stabilized clays at various leaching cycles. J Geotech Geoenviron Eng 139(10):1665–1675

Cizer Ö, Van Balen K, Van Gemert D (2010) Competition between hydration and carbonation in hydraulic lime and lime-pozzolana mortars. Adv Mater Res 133–134:241–246

Cizer Ö, Rodriguez-Navarro C, Ruiz-Agudo E, Elsen J, Van Gemert D, Van Balen K (2012) Phase and morphology evolution of calcium carbonate precipitated by carbonation of hydrated lime. J Mater Sci 47(16):6151–6165

Croft JB (1967) The structures of soils stabilized with cementitious agents. Eng Geol 2(2):63–79

Dash SK, Hussain M (2011) Lime stabilization of soils: a reappraisal. J Mater Civ Eng 24(6):707–714

Deneele D, Le Runigo B, Cui YJ, Cuisinier O, Ferber V (2016) Experimental assessment regarding leaching of lime-treated silt. Constr Build Mater 112:1032–1040

Diamond S, Kinter EB (1965) Mechanisms of soil-lime stabilization. Highway Res Rec 92:83–102

Diamond S, White JL, Dolch WL (1964) Transformation of clay minerals by calcium hydroxide attack. In: Proceedings of 12th national conference on clays and clay minerals, Atlanta, Georgia, pp 359–379

Eades JL (1962) Reaction of $CaOH_2$ with clay minerals in soil stabilization. Doctoral dissertation, University of Illinois at Urbana-Champaign

Gard JA, Taylor HFW (1976) Calcium silicate hydrate II (CSH II). Cem Concr Res 6(5):667–677

Goodbrake CJ, Young JF, Berger RL (1979) Reaction of hydraulic calcium silicates with carbon dioxide and water. J Am Ceram Soc 62(9–10):488–491

Hino T, Jia R, Sueyoshi S, Harianto T (2012) Effect of environment change on the strength of cement/lime treated clays. Front Struct Civ Eng 6(2):153–165

Ho LS, Nakarai K, Ogawa Y, Sasaki T, Morioka M (2017) Strength development of cement-treated soils: effects of water content, carbonation, and pozzolanic reaction under drying curing condition. Constr Build Mater 134:703–712

Jones D, Rahim A, Saadeh S, Harvey JT (2010) Guidelines for the stabilization of subgrade soils in California. University of California Pavement Research Center

Kamon M, Ying C, Katsumi T (1996) Effect of acid rain on lime and cement stabilized soils. Soils Found 36(4):91–99

Khattab SA, Al-Mukhtar M, Fleureau JM (2007) Long-term stability characteristics of a lime-treated plastic soil. J Mater Civ Eng 19(4):358–366

Le Runigo B, Ferber V, Cui YJ, Cuisinier O, Deneele D (2011) Performance of lime-treated silty soil under long-term hydraulic conditions. Eng Geol 118(1–2):20–28

Mallela J, Quintus HV, Smith KL (2004) Consideration of lime-stabilized layers in mechanistic-empirical pavement design. National Lime Association, pp 200–208

McCallister LD, Petry TM (1992) Leach tests on lime-treated clay. Geotech Test J 15(2):106–114

Netterberg F, Paige-Green P (1984) Carbonation of lime and cement stabilized layers in road construction. Soil Engineering Group and National Institute for Transport and Road Research, South Africa

Paige-Green P (2008) The durability of stabilized materials. In: 12th International conference of international association for computer methods and advances in geomechanics, pp 1–6

Rao SN, Rajasekaran G (1996) Reaction products formed in lime-stabilized marine clays. J Geotech Eng 122(5):329–336

Rezagholilou A, Papadakis VG, Nikraz H (2017) Rate of carbonation in cement modified base course material. Constr Build Mater 150:646–652

Sloane RL (1965) Early reaction determination in two hydroxide-kaolinite systems by electron microscopy and diffraction. In: Proceedings of 13th national conference on clays and clay minerals, pp 331–339

Yeo YS, Nikraz HR (2012) Evaluation of water ingress in cement-treated material for durability assessments. Aust J Civ Eng 10(2):117–128

Yu H, Huang X, Ning J, Zhu B, Cheng Y (2014) Effect of cation exchange capacity of soil on stabilized soil strength. Soils Found 54(6):1236–1240

Effectiveness of Cow Dung for Rammed Earth Application

H. C. Darshan, K. H. Mamatha, S. V. Dinesh, and B. M. Latha

Abstract Global warming is a very common problem worldwide. One of the reason is due to use of unsustainable natural resources for construction activities. This situation has forced engineers to think of alternative construction materials and methods of construction to minimize adverse impacts and protect environment from natural disasters. In this context, rammed earth is one of the sustainable and eco-friendly alternative construction techniques. In this study, two local soils were selected for rammed earth application which possesses varied colour. The soils were mixed in 1:1 proportion for aesthetic appearance. Initially, the basic properties of both the soils were determined, and then the properties of the mix were determined. To further improve the strength property of the soil, cow dung was selected as admixture. The cow dung dosage was varied from 2 to 8% with an increment of 2%. A series of compressive strength tests were carried out on unstabilized and cow dung stabilized soil blocks with varied percentages of cow dung and varied curing period. Based on the test results, 2% cow dung is found to be optimum. The compressive strength of the rammed earth is increased by 1.5–2 times that of the unstabilized soil block with a curing period of 28 days. Thus, cow dung can be an eco-friendly and economic alternative for rammed earth application.

Keywords Cow dung · Rammed earth · Compaction characteristic · Compressive strength

1 Introduction

Soil being eco-friendly and its use does not have adverse impact on the environment. Rammed earth (RE) is a monolithic construction formed by compacting processed

---

H. C. Darshan · B. M. Latha
Department of Civil Engineering, KVG College of Engineering, Sullia, Karnataka, India

K. H. Mamatha · S. V. Dinesh (✉)
Department of Civil Engineering, Siddaganga Institute of Technology, Tumakuru, Karnataka, India
e-mail: dineshsv204@gmail.com

M. Latha Gali and R. R. P. (eds.), *Problematic Soils and Geoenvironmental Concerns*, Lecture Notes in Civil Engineering 88,
https://doi.org/10.1007/978-981-15-6237-2_41

soil in progressive layers in a formwork (Venkatarama Reddy and Prasanna Kumar 2011a). RE was used in ancient years for shelter. RE can be used for load bearing as well as non-load bearing walls. Seamless wall surface, scope for adjusting the surface texture and colour, flexibility in wall thickness, etc. are some of the major advantages of rammed earth construction (Venkatarama Reddy and Prasanna Kumar 2011b). RE can be divided into two subcategories, i.e. stabilized rammed earth and unstabilized rammed earth. In general, unstabilized rammed earth lacks strength and durability. In the recent years, there has been an increase in the number of RE buildings stabilized with cement in some countries (Verma and Mehra 1950; Easton 1982, 2008; Hall 2002; Houben and Guillaud 2003; Walker et al. 2005). Cement is inorganic and produces greenhouse gasses during its production. Approximately, about 900 kg of carbon dioxide accounts for the production of 1 ton of cement. Also, cement-stabilized RE fails to give thermal comfort inside the building. Therefore, there is a need to find sustainable, organic and eco-friendly solution to address this problem. Monolithic earth walls perform better under earthquake conditions than walls made of separate bricks or blocks (Khadka and Shakya 2016). Some of the RE constructions around the world are—seven-storey load-bearing RE building in Germany, five-storey RE chalk houses in UK, RE Basgo Fort in Ladakh, India, etc. (Dulal Tripura and Darunkumar Singh 2016). The dry compressive strength of the rammed earth should be greater than 10 N/mm^2, and the plasticity index should lie between 2 and 30% (Walker et al. 2005).

Cow dung is available in large quantity in every corner of this world due to large number of locally available cows. The main constituents of cow dung are plant fibres (composed of cellulose, hemicellulose and lignin), amine organic compounds and fragments of intestinal tissues (Millogo et al. 2016). Cow dung actively reacts with kaolinite and fine quartz producing silicate amine which is insoluble. This insoluble silicate amine acts as a glue between soil particles and binds them together. The presence of significant amount of fibres in cow dung helps in preventing the propagation of cracks in the adobes and thereby reinforces the material. This phenomenon results in adobe microstructure homogeneous with an apparent reduction in porosity. Moreover, addition of cow dung results in increased water resistance of the soil, and therefore, cow dung is suitable for use in building construction (Millogo et al. 2016).

There are several studies available on the influence of straw debris, cement, lime and natural plant fibres on the physical and mechanical properties of the rammed earth (Mesbah et al. 2004; Minke 2006; El-Shekeil et al. 2012; Pacheco-Torgal and Jalali 2012; Morel et al. 2013; Millogo et al. 2014, 2015; Danso et al. 2015). But there are only few studies available on the effect of cow dung on the physical and mechanical properties of rammed earth (Ngowi 1997; Vilane 2010; Millogo et al. 2016). Therefore in this study, the effectiveness of cow dung stabilized rammed earth is investigated through laboratory experiments for the construction of walls.

2 Materials

2.1 Soils

In this study, two types of locally available soil samples were collected. Both the soil samples were having similar grain size and possess different colour. The soils were tested for their geotechnical properties, and all the tests were carried out as per relevant IS codes. The soils are designated as soil 1 and soil 2. The geotechnical properties of both soils are shown in Table 1. Based on the grain size distribution and plasticity characteristics of the soils, soil 1 and soil 2 are classified as silty sand with group symbol SM as per IS classification system. For soil 1, the maximum dry unit weight ranged from 19.2 to 20.2 kN/m^3 and optimum moisture content ranged from 9.2 to 9.9% under standard and modified proctor conditions. Similarly for soil 2, the maximum dry unit weight ranged from 17.6 to 18.5 kN/m^3 and optimum moisture content ranged from 12.9 to 13.6% under standard and modified proctor conditions. Figure 1 shows the particle size distribution curves of the soils. Figures 2 and 3 show the compaction curves of soil 1 and soil 2, respectively.

Table 1 Geotechnical properties of soils

Properties	Soil 1	Soil 2
Specific gravity	2.65	2.67
Grain size distribution (%)		
Gravel	3	7.5
Sand	62.5	60.8
Silt and clay	34.5	31.7
Atterberg's limit (%)		
Liquid limit	26	43
Plastic limit	NP	NP
Soil classification		
IS classification	SM	SM
Compaction characteristics		
Light compaction		
Maximum dry unit weight (kN/m^3)	19.2	17.6
Optimum moisture content (%)	9.9	13.6
Heavy compaction		
Maximum dry unit weight (kN/m^3)	20.2	18.5
Optimum moisture content (%)	9.2	12.9

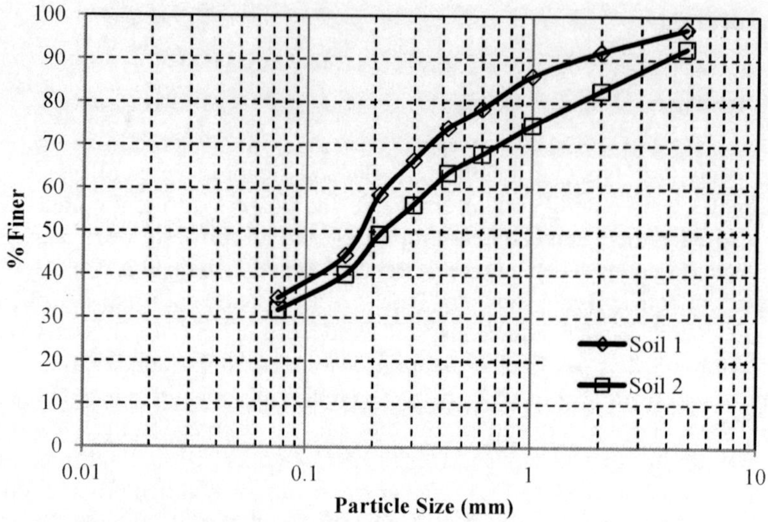

Fig. 1 Particle size distribution curves of the soils

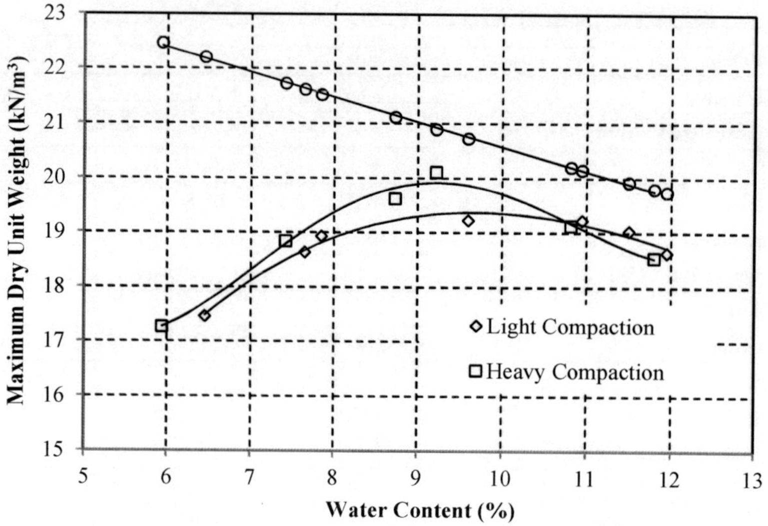

Fig. 2 Compaction curves of soil 1

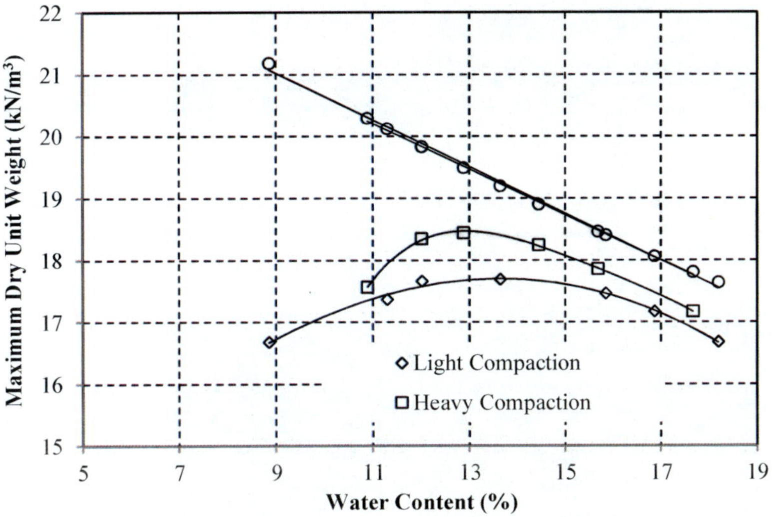

Fig. 3 Compaction curves of soil 2

2.2 *Cow Dung*

Cow dung was used as stabilizer. Locally available fresh cow dung was collected from a nearby village. The collected fresh cow dung was sun-dried for about 7–10 days. The sun-dried cow dung was then crushed into a powder in such a way that there are no lumps.

3 Sample Preparation

In order to attain good aesthetic appearance, collected soils were blended and used for stabilization. Since both the soils possess similar grain size distribution, they were mixed in 1:1 proportions. To this mix, the cow dung was added in varied percentages and the compaction characteristics of the stabilized mix under modified proctor condition was determined. For determining the compressive strength, cubes of size 150 mm × 150 mm × 150 mm were cast by maintaining the respective maximum dry unit weights and optimum moisture contents obtained from the compaction test under unstabilized and cow dung-stabilized conditions. While casting the cubes, the mass of dry soil required to attain the desired density was determined. The cow dung was added in varying percentages, viz. 2, 4, 6 and 8%, and mixed thoroughly to obtain a homogeneous mixture. Then to this dry mixture, desired amount of water which was determined based on optimum moisture content was added. An excess of 2% of water was added to account for evaporation losses during mixing. The mix was then placed in the mould and compacted by using a rammer of 16 kg in three

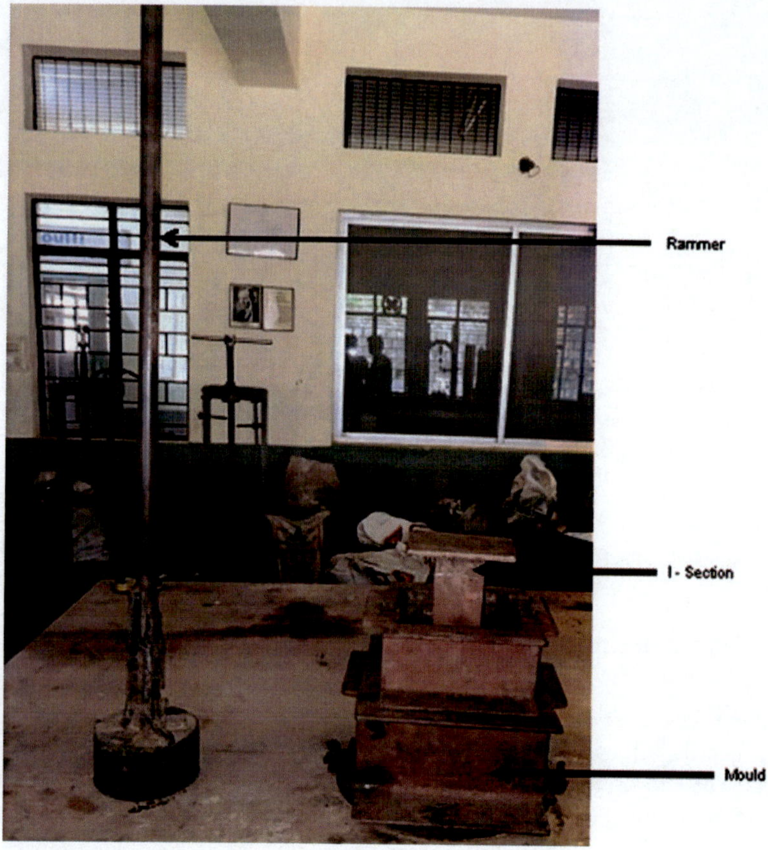

Fig. 4 Rammer, mould, collar and I-section used for compaction

layers. An I-section tool to transfer the ramming actions from rammer to the soil mix was fabricated in such a way that it is having square plates at top and bottom and a connecting C-channels were welded to both top and bottom plates. The rammer, mould, collar and I-section used for compaction is shown in Fig. 4. For each test condition, a minimum of three samples were prepared and tested. After casting, the cubes were air-dried in the laboratory. The cubes were cured for a period of 7 and 28 days and then tested.

4 Test Procedure

The compressive strength of the prepared samples was determined using compression testing machine (CTM). The test was carried out as per IS: 516—2004. The samples were tested after 7 and 28 days of curing in order to determine early and 28 days strength development. The specimen was kept in CTM in such a direction that the load is applied in the direction of ramming (vertical load). The load was continuously applied until the test specimen failed.

5 Results and Discussions

5.1 Effect of Cow Dung on Compaction Characteristics

The results of compaction test under unstabilized and cow dung-stabilized soil mix under modified proctor condition are shown in Table 2. Figure 5 shows the compaction curves of unstabilized and cow dung stabilized soil mix under modified proctor condition. It was observed that the addition of cow dung reduced the maximum dry unit weight and increased the optimum moisture content of the soil

Table 2 Compaction characteristics of the unstabilized and cow dung-stabilized soil mix under modified proctor condition

% of Cow dung	OMC (%)	Max. dry unit weight (kN/m^3)
0%	8.81	20.21
2%	10.30	18.90
4%	10.79	18.25
6%	12.27	17.56
8%	13.77	16.58

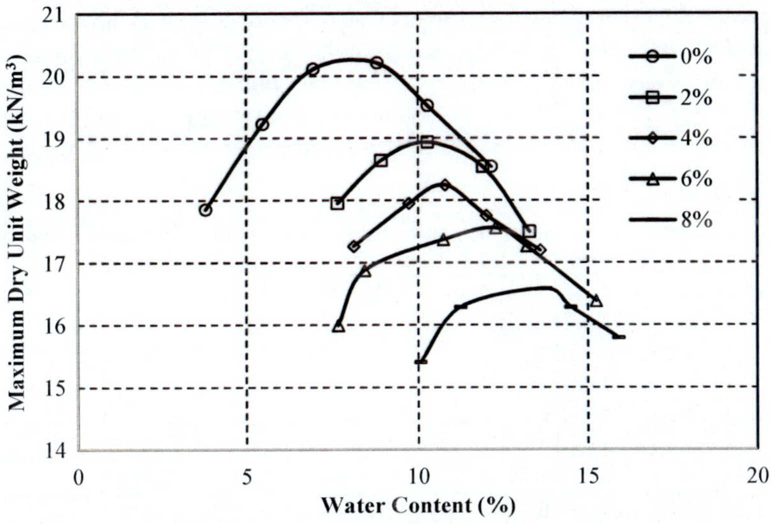

Fig. 5 Compaction curves of unstabilized and cow dung-stabilized soil mix under modified proctor condition

mix. With the increased percentage of cow dung, the density further reduced and optimum moisture content further increased. The reduced density and increased optimum moisture content are attributed to increased specific surface area of the mix with the addition of cow dung fines to the soil mix.

5.2 *Visual Observation*

After air dry curing in laboratory conditions for 28 days, there was fungal development on the external faces of the samples and the effect was little more pronounced in cubes prepared with higher percentage of cow dung. Both unstabilized and cow dung-stabilized cubes need to be handled with care; otherwise, the edges will be subjected to wear and tear. However, the cow dung-stabilized samples were sturdy in nature.

5.3 *Compressive Strength*

Table 3 shows the test result of compressive strength test carried out on both unstabilized and cow dung-stabilized soil mix for curing periods of 7 and 14 days, and the same is shown in graphical format in Fig. 6. Addition of cow dung showed an increase in the compressive strength. About 50% of the strength development was observed with 7 days of curing period. With 2% of cow dung, the compressive strength increased by 1.5 and 2 times when compared to that of the unstabilized rammed earth, respectively, for curing period of 7 and 28 days. From the test results, it is evident that 2% of cow dung is optimum for stabilizing the selected soil for rammed earth application. The reduced strength with increased cow dung dosage (beyond 2%) is attributed to the development of more fungus and bacterial activities. But at these dosages, compressive strength is more than unstabilized blocks.

Table 3 Test results of compressive strength test carried out on unstabilized and cow dung stabilized soil mix with a curing period of 7 and 28 days

% Cow dung	Compressive strength (MPa)	
	7 days	28 days
0%	0.685	1.23
2%	1	2.48
4%	0.825	1.93
6%	0.94	1.62
8%	0.76	1.351

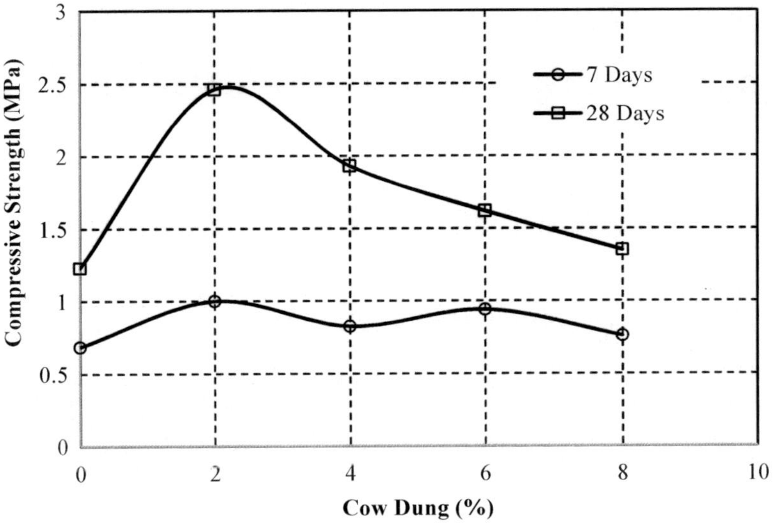

Fig. 6 Variation of compressive strength of rammed earth with percentage of cow dung and curing period

6 Conclusions

The effectiveness of cow dung for stabilizing soil for rammed earth application was investigated through laboratory study. The compaction characteristics and compressive strength under unstabilized and cow dung stabilized conditions were evaluated. Based on the test results, the following conclusions are drawn.

- Cow dung proves to be effective in stabilizing the soil for rammed earth application, and this technique proves to be eco-friendly and economic alternate solution to the construction industry.
- The addition of cow dung reduces the maximum dry unit weight and increases the optimum moisture content attributed to increased specific surface area due to cow dung fines.
- For 7 days curing period, about 50% of the 28 days compressive strength development was observed irrespective of the cow dung dosage.
- Among the various dosages of cow dung considered, 2% proves to be very effective and optimum for maximum strength development.
- With 2% of cow dung, the compressive strength of rammed earth increased by 1.5–2 times compared to that of unstabilized soil block.

References

Danso H, Martinson B, Ali M, Mant C (2015) Performance characteristics of enhanced soil blocks: a quantitative review. Build Res Inf 43(2):253–262

Dulal Tripura D, Darunkumar Singh K (2016) Behaviour of cement stabilized rammed earth circular column under axial load. Mater Struct 49:371–382

Easton D (1982) The rammed earth experience, 1st edn. Blue Mountain Press, Wilseyville

Easton D (2008) The industrialisation of monolithic earth walling for first world applications. In: Proceedings of 5th international conference on building with earth (LEHM 2008), Koblenz, Germany, pp 90–97

El-Shekeil YA, Sapuan SM, Abdan K, Zainudin ES (2012) Influence of fibre content on the mechanical and thermal properties of Kenaf fibre reinforced thermoplastic polyurethane composites. Mater Des 40:299–303

Hall M (2002) Rammed earth: traditional methods, modern techniques, sustainable future. Build Eng 77(11):22–24

Houben H, Guillaud H (2003) Earth construction—a comprehensive guide. Intermediate Technology Publications, London

IS: 516: Indian Standard (2004) Method of tests for strength of concrete. Bureau of Indian Standards, New Delhi

Khadka B, Shakya M (2016) Comparative compressive strength of stabilized and unstabilized rammed earth. Mater Struct 49:3945–3955

Mesbah A, Morel JC, Walker P, Ghavami K (2004) Development of a direct tensile test for compacted soil blocks reinforced with natural fibres. J Mater Civ Eng 16:95–98

Millogo Y, Morel JC, Aubert JE, Ghavami K (2014) Experimental analysis of pressed adobe blocks reinforced with *Hibiscus cannabinus* fibres. Constr Build Mater 52:71–78

Millogo Y, Aubert JE, Hamard E, Morel JC (2015) Properties of Kenaf fibres from Burkina Faso used as reinforcement of earth blocks. Materials 8:2332–2345

Millogo Y, Aubert J-E, Sere AD, Fabbri A, Morel J-C (2016) Earth blocks stabilized by cow dung. Mater Struct 49:4583–4594

Minke G (2006) Building with earth. Birkha User—Publishers for Architecture, Basel

Morel JC, Aubert JE, Millogo Y, Hamard E, Fabbri A (2013) Some observations about the paper "Earth construction: lessons from the past for future eco-efficient construction" by Pacheco-Torgal F, Jalali S, Constr Build Mater 44:419–421

Ngowi AB (1997) Improving the traditional earth construction: a case study of Botswana. Constr Build Mater 11:1–7

Pacheco-Torgal F, Jalali S (2012) Earth construction: lessons from the past for future eco-efficient construction. Constr Build Mater 29:512–519

Venkatarama Reddy BV, Prasanna Kumar P (2011a) Cement stabilized rammed earth. Part A: Compaction characteristics and physical properties of compacted cement stabilized soils. Mater Struct 44:681–693

Venkatarama Reddy BV, Prasanna Kumar P (2011b) Cement stabilized rammed earth. Part B: Compressive strength and stress-strain characteristics. Mater Struct 44:695–707

Verma PL, Mehra SR (1950) Use of soil–cement in house construction in the Punjab. Indian Concr J 24:91–96

Vilane BRT (2010) Assessment of stabilization of adobes by confined compression tests. Biosyst Eng 106(4):551–558

Walker P, Keable R, Martin J, Maniatidis V (2005) Rammed earth design and construction guidelines. BRE Bookshop, Watford

Geopolymerization of Expansive Black Cotton Soils with Alkali-Activated Binders

Mazhar Syed, Anasua GuhaRay, G. S. S. Avinash, and Arkamitra Kar

Abstract Black cotton soil (BCS) is highly expansive in nature when it is exposed to moisture. This property renders BCS unsuitable for use in geotechnical applications. Cement stabilization is one of the most popular methods for reducing the swelling properties of BCS. However, the production of cement leads to the emission of greenhouse gases, which is a threat to modern society. Hence, the present study aims to make use of two waste materials, fly ash and ground granulated blast furnace slag (GGBS) for stabilizing BCS. The study proposes a method of geopolymerizing BCS with alkali-activated binders (AAB). AAB is produced by the reaction of an aluminosilicate precursor (fly ash and/or GGBS) with an alkaline activator solution containing sodium silicate and sodium hydroxides. The water-to-solid (*w*/*s*) ratio is varied from 0.3 to 0.5 in this study. To identify the variations in chemical characteristics and surface morphology for both untreated BCS and BCS treated with AAB, mineralogical and chemical characterizations are performed through X-ray diffraction (XRD) and Fourier transform infrared (FTIR) spectroscopy. The index and shear strength properties before and after the treatment of BCS with AAB are compared. It is observed that the effect of geopolymerization in BCS significantly increases the UCS and CBR value and reduces the free swell and plasticity index. Recommendations on the practical implementation of this technique for stabilization of expansive soils are proposed based on the findings of this study.

Keywords Black cotton soil · Geopolymerization · Geotechnical · Microstructure

1 Introduction

Black cotton soil (BCS) is a boon for cultivation but becomes precarious to structure when it is exposed to moisture amendment (Katti 1978; Sivapullaiah et al. 2009). Usually, they exhibit highly hydrophilic plastic nature due to the presence of smectite group and high mineralogical concentration of montmorillonite that predominate a

---

M. Syed · A. GuhaRay (✉) · G. S. S. Avinash · A. Kar
BITS Pilani Hyderabad Campus, Secunderabad, Telangana 500078, India
e-mail: guharay@hyderabad.bits-pilani.ac.in

M. Latha Gali and R. R. P. (eds.), *Problematic Soils and Geoenvironmental Concerns*, Lecture Notes in Civil Engineering 88,
https://doi.org/10.1007/978-981-15-6237-2_42

high degree of expansiveness (Ola 1978). This swelling and shrinkage of soil can be resolved through the proper mechanism of chemical composition and mineralogical behavior (Ogundalu and Oyekan 2014). Poor quality of soil is prone to volumetric instability and owing to excessive deflections and differential movements arise which directly regulates the assertion of an octahedral sheet of montmorillonite minerals in the soil layers (Shukla and Parihar 2016). The oppressive nature and heave of BCS can be controlled through chemical soil stabilization by preserving one or more soil properties which indirectly relies on plasticity, specific surface area, chemistry, morphology, particle size and bonding (Stephenson et al. 1989; Sivapullaiah et al. 2009; Oza and Gundaliya 2013). The aptness of chemical additives into soil results in its permanent physical and chemical alterations through controlling setting and curing time (Das 2003; Zhao et al. 2014; Ural 2015). This proficiency aids to attain maximum shear strength, dust control on low volume roads, water erosion control, soil waterproofing, and leaching control of waste and recycled materials (Petry and Little 2002). Utilization of envirosafe cementitious binder is the new approach for stabilizing the weak geotechnical properties, and it shows the renewed interest to attain sustainable ecological and economic solution (Pourakbar et al. 2015; Gupta et al. 2018). Concretization of calcium-based traditional binders (i.e., lime, cement and fly ash) helps to upgrade the serviceable characteristics through pozzolanic action (Vitale et al. 2017a, b). Lime and cement are the most utilizable chemical binders for boosting the bearing capacity and durability of soil in several structures such as canal lining, railway base, subgrade layer of pavement and foundation (Al-Mukhtar et al. 2012; Jha and Sivapullaiah 2015). A mixture of lime in clayey soil tends to emend the soil workability by reducing the plasticity of montmorillonite; this cationic exchange influences due to the formation of cementitious gel through pozzolanic reaction and flocculation (Sherwood 1993; Bell 1996; Qubain et al. 2000; Khemissa and Mahamedi 2014). However, these traditional binders contribute to the ejection of greenhouse gases through hydration and generate around 7% of artificial CO_2 emissions due to carbonate decomposition (Gartner 2004; Al-Rawas et al. 2005; Ouhadi and Yong 2008).

Substitution of industrial solid wastes (fly ash, ground granulated blast furnace Slag), bagasse (waste from sugar industry), volcanic ash, bio-wastes (Zycosil, Fujibeton, Terrazyme, Permazyme), chemicals like sodium chloride, sodium silicate, chrome lignin, sodium hydroxide, calcium chloride, polymers, and their combination are the auxiliary key to obtain sustainable domain by switching the conventional additives through ignoble greenhouse gases' emission. It promotes to reuse the secondary by-products that degrade the carbon footprint of cementitious binder and aids to achieve desirable shear strength by stiffening the geomechanical traits. It also lessens the disposable issues and landfilling costs (Das and Sobhan 2013; Eberemu and Sada 2013; Thomas and Peethamparan 2015; Faisal and Muhammad 2016; Maneli et al. 2016; Pourakbar and Haut 2017; Gupta et al. 2018). Fly ash is a superfluous coal ingredient yield from power plants and effectually adopts as the replacement of conventional cementitious materials (according to Central Electricity Authority, New Delhi). Every year power plants produce 95 million tons of fly ash and majorly 80% are very fine (IRC-SP-20-2002). Due to the influence of pozzolanic action of fly ash in the clayey soil tends to control the excessive swelling behavior and imparts to raise the soil durability properties through flocculation reaction with a combination of cementitious chemical binder in the soil compound. This mechanism gained popularity owing to provincially available green compositional construction material through geopolymer function (Ural 2015; Lin et al. 2013). Quite recently, utilization of industrial by-product of precursor-based chemical binder for ground improvement divulges a renewed attention toward the concept of alkali mixture of polymerization (Gupta et al. 2018). Therefore, the application of fly ash-based geopolymer binder helps to improve the engineering behavior and attempts to reduce the disposal of waste solid material through oligomers technique (Davidovits 1994).

Geopolymerization is a novel technique for producing a new cementitious binder by mixing alkalis with a burnt mixture of kaolinite, limestone and dolomite (Davidovits 1983, 2013; Abdel-Gawwad and Abo-El-Enein 2016). Quite often the intrusions of alkali-activated binder (AAB) in the weak BCS show significant improvement on geoengineering properties. It aids to modify the soil as a cementing agent with a superior grade of serviceability by maintaining low atmosphere degeneration (Pourakbar et al. 2015; Miao et al. 2017). Dual characteristics of expansive soil can be restrained by fly ash bases of AAB, resulting in accretion in bearing capacity of the soil, early strength, preventing the formation of sulfate cracks and corrosion resistance (Gupta and Sharma 2016; Sofi et al. 2007). The present research proposes a method of fly ash-based geopolymerizing of the black cotton soil with alkali-activated binders (AAB) in order to control the heave and dual nature of the soil. The hydration reaction mechanism among the conventional cement and AAB is dissimilar, but their mechanical behaviors are the same and attain maximum load-carrying strength through polymeric reaction (Provis and van Deventer 2009; Pourakbar and Huat 2017). AAB is composed of sodium silicate and sodium hydroxides which are alkaline solutions and produce exothermic reactions when an aluminosilicate precursor (fly ash and or GGBS) is mixed with different water-to-solid ratio (0.3–0.5%). The microstructural and geotechnical tests are carried out for both treated and untreated soils at different curing periods of 3, 7, 21 and 28 days, and the relative effects are compared. The present study evaluates the optimum percentage of AAB with various proportions of fly ash in the soil, thereby proposing an ecological and economically sustainable stabilization procedure.

2 Materials

2.1 Soil

Black cotton soil (BCS) is majorly found in middle Africa, Southern Asia and other tropical areas including China (Das 2003; Patel and Shahu 2015). It is one of the penurious types of soil with micro-fine clayey particles of more than 60% under 75 microns sieve (Bell 1996). Typically, they are black in color due to fusion of iron and aluminum mixture. They are slender in nitrogen, phosphoric acid and organic matter but rich in calcium potash and magnesium. Usually, BCS confirms the elements (iron oxide and calcium carbonate and organic matter like humus) and clay minerals (Montmorillonite, Illite and Kaolinite) (Miao et al. 2017, 2018). BCS adopted in the present research is collected from the Nalgonda district of Telangana State in the southern part of India. This is a residual disturbed soil, and it is excavated at 30 cm depth below the natural ground surface in order to avoid grabbing of roots and vegetation or any organic matter. The soil is then placed in air-tight sacks and transported to the laboratory. It consists of dark grayish heavy clay loam particles, and it is oven-dried before lumps are processed by pulverizing with the help of a wooden mallet. The processed samples are subjected to various geotechnical investigations to establish their index, mechanical and engineering characteristics. The different physicochemical along with the index and engineering properties of BCS are provided in Table 1. The tests are conducted according to ASTM 4318, ASTM

Table 1 Properties of black cotton soil

Properties	Values
• Calcium	4.832%
• Sodium	0.780%
• Potassium	2.832%
• Magnesium	0.644%
• Silica	42.67%
• Aluminum	13.82%
Index and engineering properties	
Specific gravity	2.59
Liquid limit (LL)	62%
Plastic limit (PL)	20%
Plasticity index (PI)	42%
Shrinkage limit (SL)	21%
Grain size distribution	
• Gravel (>4.75 mm)	1%
• Sand (0.075–4.75 mm)	27%
• Silt and clay (<0.075 mm)	72%
Soil classification (according to USCS)	CH
Free swell index (FSI)	76.3%
Maximum dry density (MDD)	16.5 kN/m^3
Optimum moisture content (OMC)	24.6%
California bearing ratio (CBR)	
• Unsoaked	5.67
• Soaked	1.92
Unconfined compressive strength	185 kN/m^2

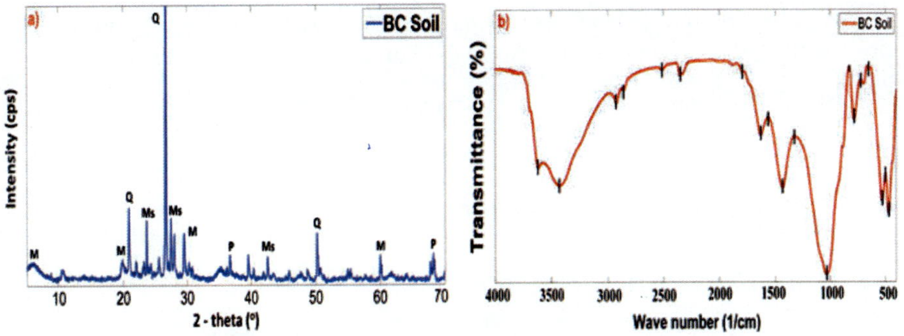

Fig. 1 **a** X-ray diffraction and **b** FTIR images of untreated BCS sample

D1557-12e1 and ASTM D2166/D2166M-16. The XRD and FTIR results of the raw BCS sample are shown in Fig. 1a, b. The untreated soil reveals the crystalline peaks of clay minerals such as quartz (Q), montmorillonite (M) and muscovite (Ms) The molecular vibrations are observed at 2920, 2350, 1600, 1015 and 779 cm^{-1}, respectively (Kar et al. 2013; Gupta and Sharma 2016; Vitale et al. 2017a; Gupta et al. 2018).

2.2 *Alkali-Activated Binder (AAB)*

Recent advancements in the field of soil chemical stabilization have led to a revived interest by applying fly ash-based geopolymerization techniques on expansive soil through alkali-activated binder. AAB is produced by mixing the activating solution of sodium silicate, sodium hydroxide and water until a clear blend is obtained. The exothermic reaction takes place due to the presence of an oligomer matrix between the sodium hydroxide pellets and sodium silicate solution. This alkaline solution (Na_2SO_3 and NaOH) is obtained from HYCHEM chemicals. The purity of sodium hydroxide pellets is 99%. The sodium silicate solution is composed of 55.9% water, 29.4% SiO_2 and 14.7% Na_2O. The amount of water-to-solid ratios (*w*/*s*) is varied from 0.3 to 0.5 in preparing the solution binders and optimum will be proposed. Literature suggests that workability reduces for *w*/*s* ratio below 0.3. For *w*/*s* ratio greater than 0.5, the lubrication increases to a great extent and produces a considerable amount of apparent cohesion (Miao et al. 2017; Gupta et al. 2018). The mass ratio of fly ash to sodium hydroxide to sodium silicate is maintained at 400:10.57:129.43 with varying water-to-solid ratio to prepare the solution (Kar et al. 2013). Additional water is combined with these blends to reduce the storage time from 24 to 6 h before using it to react with the fly ash in order to dissipate any residual heat that can hasten the alkali activation of fly ash and lead to a flash set. The percentages of fly ash are also varied, and an optimum solution is proposed. BCS is uniformly mixing with 5%, 10% and 15% of the proposed binder at different curing periods of 3, 7, 21 and 28 days, respectively.

3 Experimental Investigations

3.1 Chemical Characterization

X-ray Diffraction (XRD). XRD technique reveals to identify the crystalline minerals for both treated and untreated soil samples at different AAB proportions by maintaining 0.3, 0.4 and 0.5 percentages of water-to-solid (*w*/*s*) ratios at various curing periods. The incident X-rays strike the plane of atoms in the solid sample which characterized the sequence of peaks that are unique to particular minerals (Rees et al. 2004). Powdered soil samples in the form of XRD peak pattern are analyzed using RIGAKU Ultima IV diffractometer through CuKα rays by generating 40 mA and 40 kV. Peak operating 2θ range is maintained from 0° to 100° with a step 0.02° for 2θ values and integrated at the constant rate of 2 s per step in the diffractometer. Figure 2 shows the XRD peaks of treated and untreated dry BCS samples by uniformly adding 10% (by weight) of AAB mixture in the raw soil, and at definite curing periods (0, 3, 7, 21 and 28 days) the XRD crystalline mineral peaks are identified by maintaining 0.5% *w*/*s* ratio. The raw BCS confirms the presence of clay minerals such as quartz (Q), montmorillonite (M) and muscovite (Ms) in the sample. The addition of a fly ash-based geopolymer binder in the untreated soil shows the additional peaks of minerals due to polymeric reaction induced in the clay minerals. It shows the new crystalline patterns at different curing periods are mullite (Mu), augite (Au), hydroxy sodalite (S), portlandite which are the characteristics of the hardened AAB

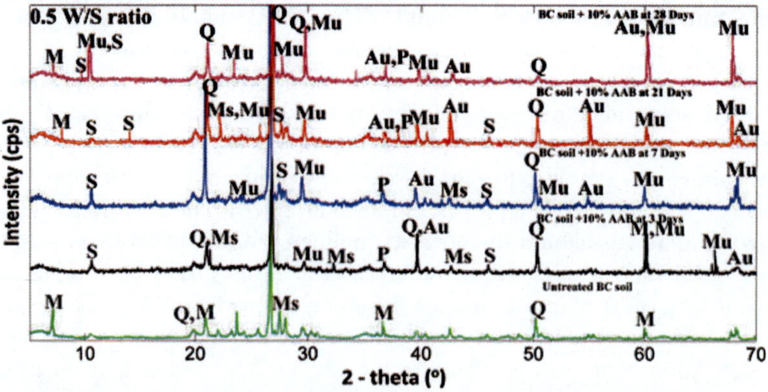

Fig. 2 XRD of BCS treated with 10% AAB at 0.5 *w*/*s* ratios at various curing periods

paste (Kar et al. 2013; Miao et al. 2017; Gupta et al. 2018). However, after 21 days of curing the relative peakedness of montmorillonite (M) is weakened, which indicates the reduction of troublesome of clay minerals in the modified soil due to AAB polymerization through stabilization technique. Since the individual peaks demonstrate the availability of crystalline minerals for both treated and untreated soil, it can be concluded that AAB remained in conjunction with BCS due to fly ash pozzolanic chemical reactions with varying percentages of AAB and *w/s* ratios in the samples.

Fourier Transform Infrared (FTIR) Spectroscopy. FTIR spectroscopy analyses are performed using a JASCO FTIR 4200 setup through K.Br. pellet arrangement. The wide spectral range is maintained between 4000 and 400 cm^{-1} based on infrared spectroscopy. Spectroscopy (absorption/transmission) of infrared beams identified the molecular vibrational characteristics between the molecules. Usually, the transmittance infrared spectroscopy is conducted for both treated and untreated soil samples in the dry condition in order to preserve fineness form at various curing periods. Soil samples are oven-dried at 105 °C for a minimum period of 1 h before mixing into K.Br. powder to remove the residue wetness and make the pellets through the hydraulic compressor. The molecular vibration characteristics of raw and treated BCS are obtained through infrared transmittance spectroscopy by observing O–H stretching (in portlandite, $[Ca(OH)_2]$ at 3640 cm^{-1} and asymmetric vibrations are found between 3435 and 3616 cm^{-1}. Figure 3 shows the FTIR spectroscopy of BCS treated with 10% AAB at 0.5 *w/s* ratio at various curing periods. Untreated BCS shows peaks at 2924 cm^{-1} for C–H (asymmetric stretching), and treated soil shows C–H stretching at 2855 similarly at 2350 for C=O (symmetric), 1633 for C=O (stretching), 1439 for C–H (bending), 1033 for Si–O–Si (antisymmetric stretch), for Al–O (stretching) at 785 and at 527 for Si–O–Al (bending) vibrations (Kar et al. 2013; Miao et al. 2017). After 3 days, polymeric reaction in the treated soil identifies a new bond of C=O carbonyl at 1740 cm^{-1} The presence of sodium carbonate in the clay minerals shows the CH_2/CH_3 of bending vibration at 1460 cm^{-1}. The

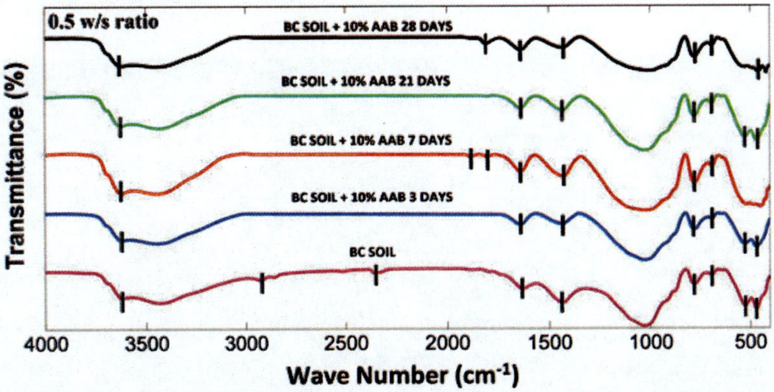

Fig. 3 FTIR spectroscopy of BCS treated with 10% AAB at 0.5 *w/s* ratios at various curing periods

presence of water-to-solid ratios in the modified soil increases the intensity of the band at 1700 cm^{-1} after 28 days of curing and loss of Si–O–Al band at 455 cm^{-1}. However, similar bonds are identified for the AAB-treated BCS samples at definite curing periods, but majorly shows chemical shifts, indicating the impermeable of BCS due to pozzolanic reaction in the sample.

Scanning Electron Microscopy (SEM). SEM is an imaging technique that produces the image of the sample through a focused beam of an electron that scrutinizes the surface of the specimen (Davidovits 1994). SEM–EDS study is operated using Thermo Scientific Apreo SEM provided by FEI. Apreo SEM is basically a wide range of compound lenses and specially designed by using integrating electrostatic and magnetic immersion technology. The inclusion of an electron beam enables the top-surface imaging for both treated and untreated BCS through excitation voltage with various spot sizes.

A series of electromagnetic lenses supervises the electron toward the target location and the electron beam screens the surface through a small aperture. Very high excitation voltage would blur the image due to the electron cloud. The target location is chosen randomly with various magnifications. Micrographs are captured at three different spots in each sample by maintaining 1000×, 1250× and 5000 magnifications. Typical surface image of untreated BCS shows in Fig. 4a which reveals a formation of flaky-type microstructure indicating the presence of montmorillonite clay minerals (Sivapullaiah and Sridaran 2017; Parhi et al. 2017). The unreacted fly ash particles are clearly visible in the form of sphere surface in Fig. 4b after 3 days of polymeric reaction. At 28 days of curing, the fly ash particles are reacted in the soil and form the cementitious surface shown in Fig. 4c.

Energy-Dispersive X-ray Spectroscopy (EDS). EDS is an elemental analysis used to measure the chemical composition in a broad range (Davidovits 2013). EDS also conducted in the same SEM machine using Thermo Scientific Apreo SEM provided by FEI. The EDS spectra are analyzed with the help of a software system that measures the relative abundance of emitted X-rays along with their energy. For

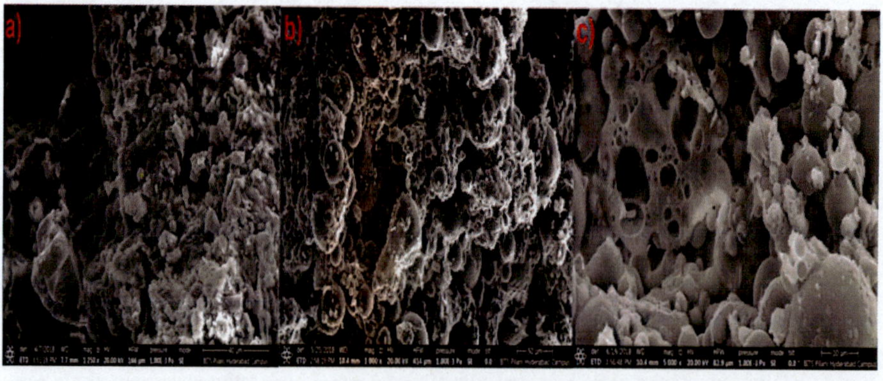

Fig. 4 SEM images of raw and modified BCS with AAB at various curing periods

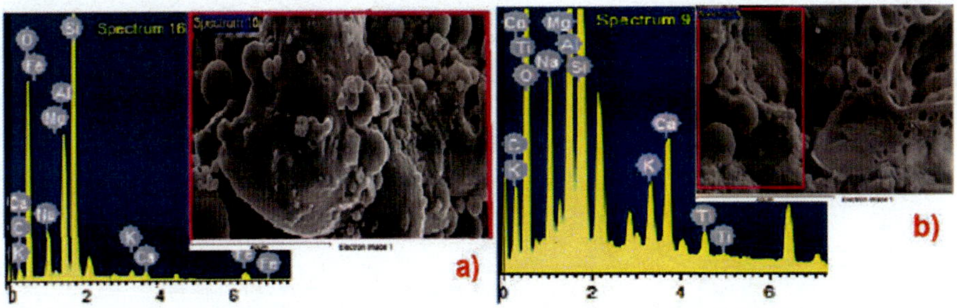

Fig. 5 EDXS images of AAB treated soil at **a** 3 days and **b** 28 days curing periods

the adequate working of the EDS analyzer, the probe current and working distance are maintained at 65.4–67.0 μA and 8–15 mm. Modified BCS sample confirms the presence of elements are calcium (Ca), aluminum (Al), sodium (Na), silicon (Si), potassium (K), magnesium (Mg), ferrous (Fe), carbon (C) and Titanium (Ti). Figure 5a, b show the EDXS images after 3 and 28 days of curing.

3.2 Geotechnical Characterization

A series of geotechnical tests are performed for both treated and untreated soil samples according to ASTM standards. After uniformly mixing the chemical binders in the raw BCS with various proportions of AAB and maintaining different water-to-solid ratios, the engineering properties of soil are determined at different curing. Atterberg's limits are measured according to ASTM D4318. The liquid limit (LL) and plasticity index (PI) of untreated BCS is found to be approximately 62% and 42%, respectively, indicating the high plasticity of the soil. Figure 6a shows the variation of plasticity index of treated soil with different percentages (5, 10, 15) of AAB at various curing periods of 3, 7, 21 and 28 days. It may be inferred that the fly ash-based alkali compound steadily induced the polymeric reaction in the clayey particles and stuffed the voids through natural pozzolana that indirectly leads to diminished the moisture-holding capacity around the clayey surface area. It is observed that the addition of up to 10% AAB tends to reduce the PI of treated soil and transformed it into a less plastic soil. As the percentage of AAB is increased in the soil layers by maintaining 0.5% of water-to-solid ratio, the percentage of the swelling index also reduced significantly, as shown in Fig. 6b. Standard Proctor's compaction is carried out on both untreated and treated BCS to find out the optimum moisture content (OMC) and maximum dry density (MDD) according to ASTM D698-12e2. A positive variation of dry density and moisture content is observed with respect to different curing periods. A maximum dry density of 19.7 kN/m^3 with 19.80% of OMC is obtained by mixing 10% (by weight) of AAB to raw BCS after 28 days of geopolymerization.

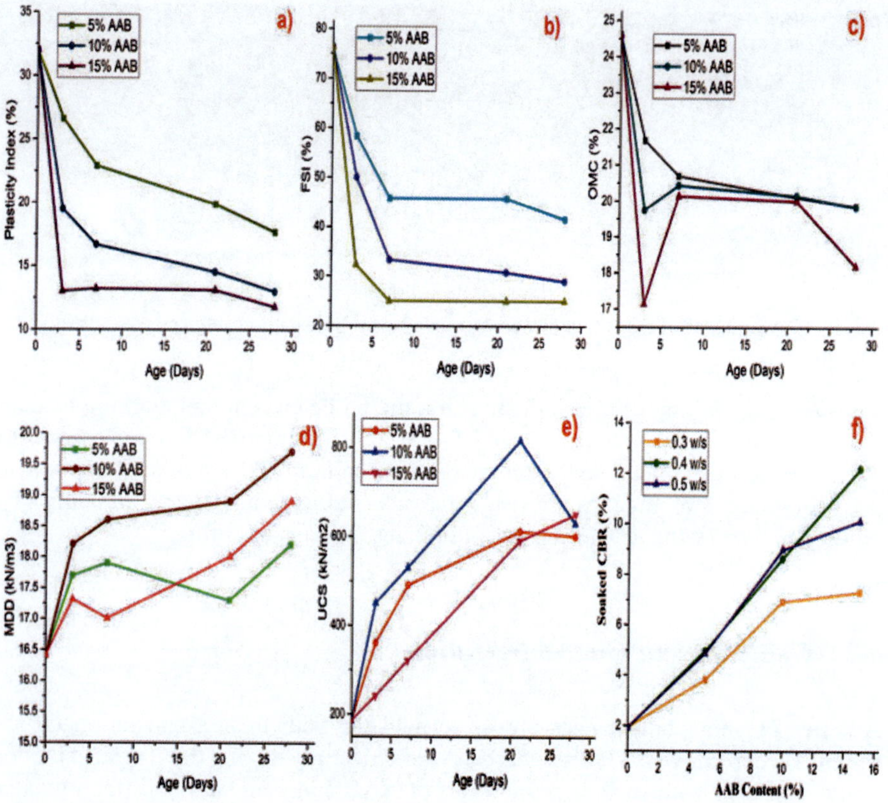

Fig. 6 Variation of **a** plasticity index, **b** FSI, **c** OMC, **d** MDD, **e** UCS, **f** CBR values of BCS mixed with various % of AAB at different curing periods

Figures 6c, d show the variations of optimum water content and maximum dry density for BCS treated with different percentages of AAB at different curing periods. The MDD curves of treated soil increase with respect to the percentage of fly ash-based binder and attained maximum at 10% of AAB; similarly, the OMC values depreciate from 24.6 to 19% as the dosage of AAB increases. The reduction of OMC attributes to the replacement of fly ash in the voids due to cationic exchange between the clayey soil and fly ash (Zha et al. 2008; Lin et al. 2013; Kolay and Ramesh 2016).

The undrained shear strength and CBR tests are conducted according to ASTM D2166/D2166M-16 and ASTM D1883-16, respectively, for both soaked and unsoaked soil samples in the laboratory. Unconfined compressive strength (UCS) tests are performed under a fixed strain rate of 1.2 mm/min (Das 2003). Treated and untreated soils are molded in a 3.8 cm diameter by maintaining the height of twice the diameter. The samples are kept in an airtight vacuum desiccator for 24 h for maturation and preserving moisture content (Pourakbar et al. 2015; Shukla and Parihar 2016). Figure 6e shows the variation of UCS values of soil modified with various proportions of AAB under different curing periods (0, 3, 7, 21 and 28 days). The addition of a chemical binder to BCS increases the compressive strength from 185 to 600 kN/m^2. The soaked and unsoaked CBR values also increase due to enhanced mechanical binding properties in the treated soil (Gupta and Sharma 2016). The enhancements of strength ratio attribute to the pozzolanic reactions and catalyze the property of fly ash during the period of soaking. Figure 6f represents the soaked CBR values after 3 days of geopolymer stabilization and attains the optimum value as 12% at 0.4 *w/s* ratio.

4 Summary and Conclusions

The present study proposes to utilize the envirosafe alkali-activated binder (AAB) as an alternate way of preserving one or more geotechnical properties through chemical soil stabilization techniques. Exclusion of carbon emission traditional binder aids to preserve the sustainable environment through the utilization of locally available waste and industrial by-products in an efficient way. The current research focuses on how the treated soil will alter the engineering behavior through microscopic physicochemical, mineralogical and microstructural surface morphology analysis. The important findings and main conclusions can be summarized as follows.

- AAB intensifies the mechanical improvement under potentially ionized conditions through cationic exchange over the clayey particles and fly ash-based aluminosilicate in the reinvent soil across the multiple functions of curing periods and chemical binder content.
- Microstructural analysis shows a significant reactivity by promoting new mineralogical phases and various molecular bonding between the particles, which are confirmed through XRD and FTIR analyses.
- EDS and SEM micrograph images of treated and untreated soil reveal the surface morphology formation of cementitious products of calcium silicate hydrates which comprise the microstructure alteration in the inter-grain porosity.
- The increase in the proportion of chemical binder content in the soil decreases the OMC from 24.6 to 19%. As the OMC of treated soil reduces, it indirectly leads to enhance the dry density to an optimum value of 21.3 kN/m^3 at 10% (by weight) of AAB.
- The consistency of treated soil shows positive trends by reducing the plasticity index from 42 to 15% after 28 days of geopolymerization, which highlights the treatment efficiency in the range of 0.3–0.5 *w/s* ratios.
- Due to chemical reaction induce between an aluminosilicate precursor (primarily Class F fly ash and/or slag), and an alkali activator solution of sodium hydroxide and sodium silicate will drastically boost the CBR value and undrained shear strength of the soil.
- Pozzolanic reaction surges the soil strength and bearing capacity by elevating physicochemical properties by altering the microscopic behavior that linked to

experimental evidence at micro-level chemical composition through a qualitative and quantitative study.

- Geopolymerization process decreases specific surface area (SSA) and water retention abilities through the coexistence of calcium silicate hydrate (C–S–H) and sodium aluminosilicate hydrate (N–A–S–H) gel formation commanding flocculation reaction contributes to diminishing the swell–shrink behavior by accretion of UCS values across the time-dependent curing.
- The contribution of the present research will help to identify an envirosafe sustainable binder that revamps the geotechnical properties through the manipulation of locally available industrial waste by-products. 10% mix of AAB at 0.4 *w/s* ratio is proposed as optimum to increase the shear strength characteristics of weak BCS in an economically viable way.

References

Abdel-Gawwad HA, Abo-El-Enein SA (2016) A novel method to produce dry geopolymer cement powder. HBRC J 12(1):13–24

Al-Mukhtar M, Khattab S, Alcover JF (2012) Microstructure and geotechnical properties of lime-treated expansive clayey soil. J Eng Geol 139:1727

Al-Rawas AA, Hago AW, Al-Sarmi H (2005) Effect of lime, cement, and Sarooj (artificial pozzolan) on the swelling potential of an expansive soil from Oman. J Build Environ 40:681–687

ASTM D2166/D2166M-16. Standard test method for unconfined compressive strength of cohesive soil

ASTM D1883-16. Standard test method for California bearing ratio (CBR) of laboratory-compacted soils

ASTM D698-12e2. Standard test methods for laboratory compaction characteristics of soil using standard effort (12 400 ft-lbf/ft3 (600 kN-m/m3))

ASTM D4318. Standard test methods for liquid limit, plastic limit, and plasticity index of soils

ASTM D1557-12e1. Standard test methods for laboratory compaction characteristics of soil using modified effort (56,000 ft-lbf/ft3 (2,700 kN-m/m3))

Bell FG (1996) Lime stabilization of clay minerals and soils. Eng Geol 42(4):223–237

Das BM (2003) Chemical and mechanical stabilization. Transportation Research Board

Das BM, Sobhan K (2013) Principles of geotechnical engineering. Cengage Learning, Boston

Davidovits J (1983) Geopolymer chemistry and properties. In: Proceedings of geopolymer '88, Compiegne, France, pp 1–3

Davidovits J (1994) Properties of geopolymer cements. In: First international conference on alkaline cements and concretes. Kiev State Technical University, Ukraine. Scientific Research Institute on Binders and Materials, vol 1, pp 131–149

Davidovits J (2013) Geopolymer cement. A review. Geopolymer Institute, Technical papers 21:1–11

Eberemu AO, Sada H (2013) Compressibility characteristics of compacted black cotton soil treated with rice husk ash. Niger J Technol 32(3)

Faisal M, Muhammad K (2016) Synthesis and characterization of geopolymer from bagasse bottom ash, waste of sugar industries and naturally available china clay. J Cleaner Prod 129:491–495

Gartner E (2004) Industrially interesting approaches to 'low-CO_2' cements. Cem Concr Res 34(9):1489–1498

Gupta S, GuhaRay A, Kar A, Komaravolu VP (2018) Performance of alkali-activated binder-treated jute geotextile as reinforcement for subgrade stabilization. Int J Geotech Eng 1–15

Gupta C, Sharma RK (2016) Black cotton soil modification by the application of waste materials. Periodica Polytechnica. Civil Engineering 60(4):479
Jha AK, Sivapullaiah PV (2015) Susceptibility of strength development by lime in gypsiferous soil—a micro mechanistic study. Appl Clay Sci 115:39–45
Kar A, Halabe UB, Ray I, Unnikrishnan A (2013) Nondestructive characterizations of alkali activated fly ash and/or slag concrete. Eur Sci J 9(24)
Katti RK (1978) Search for solutions to problems in black cotton soils. Indian Institute of Technology, Bombay
Khemissa M, Mahamedi A (2014) Cement and lime mixture stabilization of an expansive overconsolidated clay. Appl Clay Sci 95:104–110
Kolay PK, Ramesh KC (2016) Reduction of expansive index, swelling and compression behavior of kaolinite and bentonite clay with sand and class C fly ash. Geotech Geol Eng 34(1):87–101
Lin B, Cerato AB, Madden AS, Elwood Madden ME (2013) Effect of fly ash on the behavior of expansive soils: microscopic analysis. Environ Eng Geosci 19(1):85–94
Maneli A, Kupolati WK, Abiola OS, Ndambuki JM (2016) Influence of fly ash, ground-granulated blast furnace slag and lime on unconfined compressive strength of black cotton soil. Road Mater Pavement Des 17(1):252–260
Miao S, Shen Z, Wang X, Luo F, Huang X, Wei C (2017) Stabilization of highly expansive black cotton soils by means of geopolymerisation. J Mater Civ Eng 29(10):04017170
Miao S, Shi J, Sun Y, Zhang P, Shen Z, Nian H, Zhang P (2018) Mineral abundances quantification to reveal the swelling property of the black cotton soil in Kenya. Appl Clay Sci 161:524–532
Ogundalu AO, Oyekan GL (2014) Mineralogical and geotechnical characterization of maidaguri black cotton soil by X-ray diffraction (XRD), x-ray photoelectron (XPS) and scanning electron spectroscopy (SEM). Int J Eng Technol 4(6):345–353
Ola SA (1978) Geotechnical properties and behavior of some stabilized Nigerian lateritic soils. Q J Eng Geol Hydrogeol 11(2):145
Ouhadi VR, Yong NR (2008) Ettringite formation and behaviour in clayey soils. Appl Clay Sci 42:258–265
Oza JB, Gundaliya PJ (2013) Study of black cotton soil characteristics with cement waste dust and lime. Procedia Eng 51:110–118
Parhi PS, Garanayak L, Mahamaya M, Das SK (2017) Stabilization of an expansive soil using alkali activated fly ash based geopolymer. In: International congress and exhibition, sustainable civil infrastructures: innovative infrastructure
Patel S, Shahu JT (2015) Engineering properties of black cotton soil–dolime mix for its use as subbase material in pavements. Int J Geomate 8(15):1159–1166
Petry TM, Little DN (2002) Review of stabilization of clays and expansive soils in pavements and lightly loaded structures—history, practice, and future. J Mater Civ Eng 14(6):447–460
Pourakbar S, Asadi A., Huat BB, Fasihnikoutalab MH (2015) Stabilization of clayey soil using ultrafine palm oil fuel ash (POFA) and cement. Transportation Geotechnics 3:24–35
Pourakbar S, Huat BBK (2017) Laboratory-scale model of reinforced alkali-activated agro-waste for clayey soil stabilization. Adv Civ Eng Mater 6(1):83–105
Provis JL, van Deventer JSJ (2009) Geopolymers: structure, processing, properties and industrial applications. Woodhead, Oxford; CRC Press, Boca Raton, FL
Qubain B, Seksinsky E, Li J (2000) Incorporating subgrade lime stabilization into pavement design. Transp Res Rec J Transp Res Board 1721:3–8
Rees C, Lukey GC, Van Deventer JSJ (2004) The role of solid silicates on the formation of geopolymers derived from coal ash. In: International symposium of research students on material science and engineering, pp 20–22
Sherwood P (1993) Soil stabilization with cement and lime. State of the art review. Transport Research Laboratory, London
Shukla RP, Parihar NS (2016) Stabilization of black cotton soil using micro-fine slag. J Inst Eng (India) Ser A 97(3):299–306

Sivapullaiah PV, Prasad BG, Allam MM (2009) Effect of sulfuric acid on swelling behavior of an expansive soil. Soil Sediment Contam 18(2):121–122

Sivapullaiah PV, Sridaran AK (2017) Swelling behaviour of expansive soil treated with fly ash GGBS based binder. Geomech Geoeng 12(3):3

Sofi M, Van Deventer JSJ, Mendis PA, Lukey GC (2007) Engineering properties of inorganic polymer concretes (IPCs). Cem Concr Res 37

Stephenson RW, Dempsey BA, Heagler JB (1989) Chemically induced foundation heave. In: Foundation engineering: current principles and practices, pp 163–165

Thomas RJ, Peethamparan S (2015) Alkali-activated concrete: engineering properties and stress–strain behavior. Constr Build Mater 93:49–56

Ural N (2015) Effects of additives on the microstructure of clay. J Road Mater Pavement Des 10:1–16

Vitale E, Russo G, Dell'Agli G, Ferone C, Bartolomeo C (2017a) Mechanical behaviour of soil improved by alkali activated binders. Environments 4(4):80

Vitale E, Deneele D, Paris M, Russo G (2017b) Multi-scale analysis and time evolution of pozzolanic activity of lime treated clays. Appl Clay Sci 141:36–45

Zha F, Liu S, Du Y, Cui K (2008) Behavior of expansive soils stabilized with fly ash. Nat Hazards 47(3):509–523

Zhao H, Ge L, Petry TM, Sun YZ (2014) Effects of chemical stabilizers on an expansive clay. KSCE J Civil Eng 18(4):1009–1017

Stabilization of Soil Using Rice Husk Ash and Fly Ash

N. Srilatha and B. R. Praveen

Abstract A number of studies have been carried out to investigate the effect of addition of waste materials on modifying the properties of soil. The present study shows the modification of soil properties by adding locally available materials such as rice husk ash and fly ash. The cost of stabilization may be reduced by replacing by a good proportion of stabilizing agent using rice husk ash. The soil used in the present study is clay with high compressibility, which needs to be strengthened to minimize volume changes in the soil. The soil is stabilized by varying different percentages of rice husk ash and also fly ash. Observations are made for changes in compaction characteristics of the soil, unconfined compressive strength and California bearing ratio values of the soil. From the results, it is observed that the maximum dry density is increased with increase in percentages of rice husk ash, and the corresponding optimum moisture content is decreased and vice versa with the addition of fly ash to the soil. The optimum amount of rice husk ash and fly ash for improving the properties of the soil are discussed in detail.

Keywords Rice husk ash · Fly ash · Unconfined compressive strength

1 Introduction

Engineering properties of soils are improved by carrying out the process called stabilization. In professional practice, the stabilization of soils is required to carry out often as the soils available for construction may not be suitable for the purpose it is intended. In general, the highly plastic soils exhibit certain undesirable engineering properties under the application of the load. These soils tend to have low shear strength and upon wetting or other physical disturbances the shear strength of the soils is lost further (Mitchell 1986). Owing to these undesirable engineering properties, the plastic soils are more prone to shear failure due to the application of constant load over a period of time, and which is why these soils are considered to be not

N. Srilatha (✉) · B. R. Praveen
Ramaiah Institute of Technology, Bangalore, Karnataka 560054, India
e-mail: srilathanbr@gmail.com

M. Latha Gali and R. R. P. (eds.), *Problematic Soils and Geoenvironmental Concerns*, Lecture Notes in Civil Engineering 88,
https://doi.org/10.1007/978-981-15-6237-2_43

suitable for foundations (Liu and Evett 2008). Use of cement, lime and fly ash, etc., can be chemically transformed unstable soils into structurally strong foundations. Mixing stabilizers in a particular amount in clay induce textural change and give better improved strength. In the present study, two additives, namely rice husk ash (RHA) and fly ash are considered as stabilizers separately and with the mixing of the additives, the improvement in the soil properties such as compaction characteristics, unconfined compressive strength and California bearing resistance are investigated. An attempt has been made to investigate the improvement in the soil properties by adding the said additives. RHA with different proportions is mixed with soil samples, and the improvement in the stabilization is studied by examining the changes in the soil characteristics. Similar process is carried out by considering the fly ash (as additive) with different proportions.

2 Materials Used

The soil used in the present study was collected from a site in north Bangalore at 3 m depth below the ground level by making an open excavation. The properties of the soil used in the present study are shown in Table 1, and based on the gradation characteristics of the soil, it is classified as inorganic clays of high compressibility from the IS classification system (1985).

The Kengeri area which is located in Bangalore sub-urban is highly potential of rice husk. The husk collected from this area consists of 75% volcanic organic matter and rest 25% get converted into ash during the firing process known as rice husk ash (RHA). In general, RHA consists of considerable of amorphous silica which makes the RHA as a very good super pozzolana. Stabilizers with pozzolanic properties can bind the soil particles together and reduce the water absorption by clay particles resulting in increase the strength of the clayey soils. The RHA collected was ground and sieved through 0.075 mm aperture before mixing with the soil. Fly ash in powder

Table 1 Characteristics of test soil

Characteristics	Description
Specific gravity	2.67
Liquid limit (%)	51.74
Plastic limit (%)	27.30
Shrinkage limit (%)	15
Maximum dry unit weight (kN/m^3)	14.99
Optimum moisture content (%)	16.92
California bearing ratio (unsoaked) (%)	4.6
Unconfined compression strength (kN/m^2)	108.9

Table 2 Constituents of RHA

Characteristics (%)	RHA
SiO_2	72.3
Al_2O_3	5.1
Fe_2O_3	0.97
CaO	1.36
MgO	1.79
Loss of ignition (LOI)	16.78

Table 3 Constituents of fly ash

Characteristics (%)	Fly ash
SiO_2	55
Al_2O_3	20.3
Fe_2O_3	6.3
CaO	12
MgO	3.5
SO_3	1.5
Heavy metals	Trace

state is one of the India's major industrial solid waste which consists of pozzolanic properties in addition to certain self cementing properties. The constituents of RHA and fly ash are listed in Tables 2 and 3.

3 Methodology

The laboratory investigations were carried out on soil such as particle size distribution, consistency limits, compaction characteristics, California bearing resistance and unconfined compression tests. Soil specimens for the unconfined compressive strength and California bearing resistance (CBR) are prepared corresponding to the maximum dry density and optimum moisture content in all the tests. Firstly, the content of RHA is varied between 5 and 50% as additive to the soil, and the tests were carried out. Secondly, the content of fly ash as 1–20% as additive to the soil, and the similar tests were carried out. The various soil properties of the soil samples are studied separately with the mixing of two additives.

4 Results and Discussions

The results from the laboratory tests are examined thoroughly by plotting the graphs with additives as abscissa and various properties as ordinates and represented in the Figs. 1, 2, 3, 4, 5, and 6.

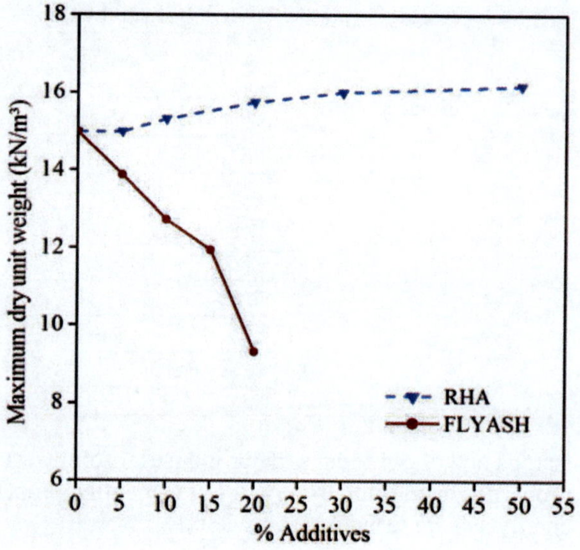

Fig. 1 Variation in maximum dry unit weight with different percentages of RHA and fly ash

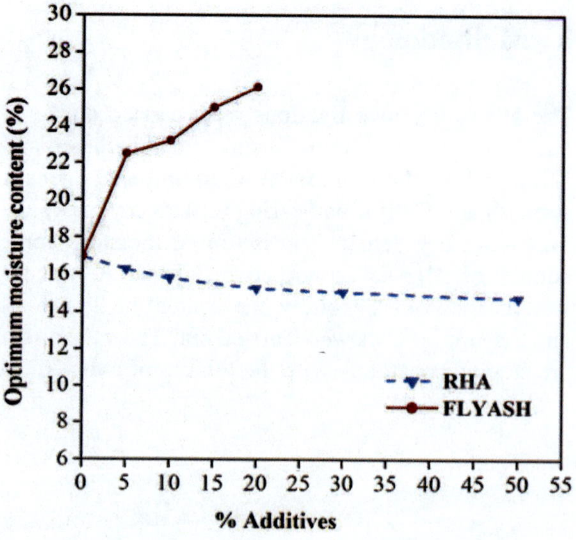

Fig. 2 Variation in optimum moisture content with different percentages of RHA and fly ash

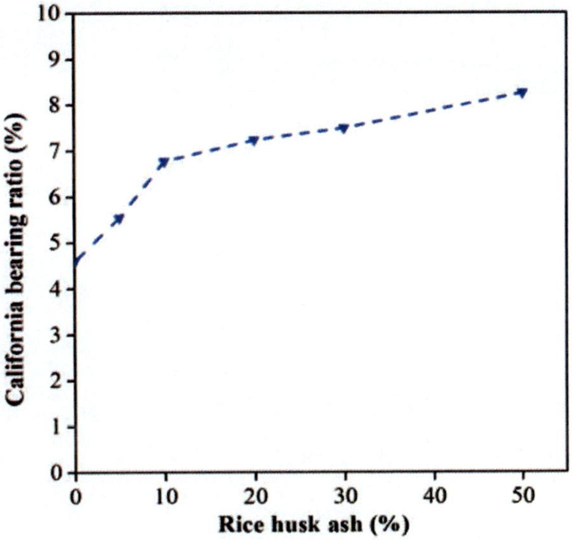

Fig. 3 Effect of RHA on CBR value of the soil

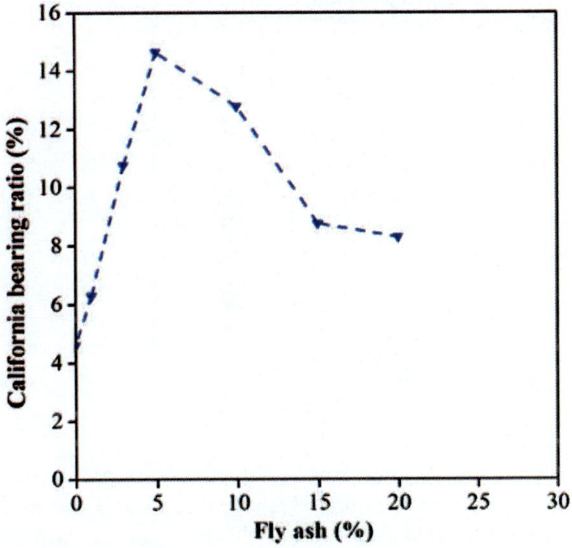

Fig. 4 Effect of fly ash on CBR value of the soil

4.1 *Effect of RHA and Fly Ash on Compaction Characteristics*

From Fig. 1, it is observed that with the addition of different percentages of RHA to the soil, the maximum dry density is increased. This may be due to the fact that clayey soil is having calcium oxide varying between 36 and 39%. The presence of

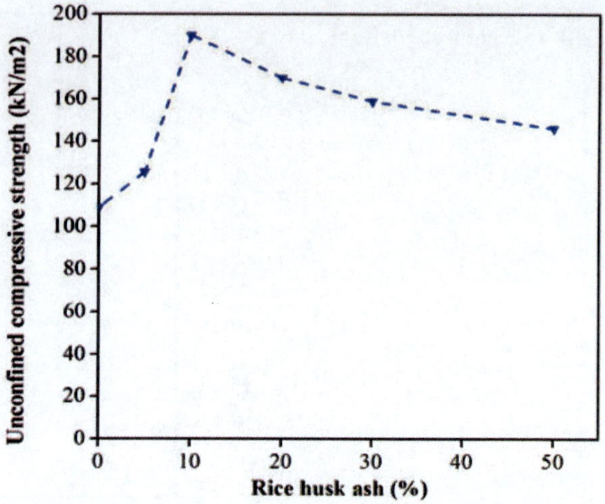

Fig. 5 Effect of addition of RHA on unconfined compressive strength of the soil

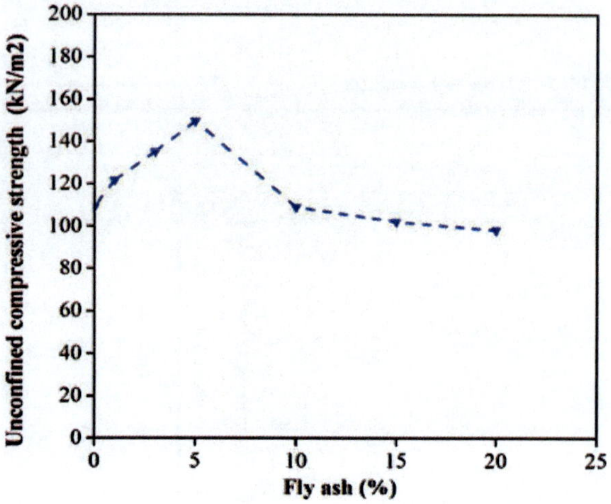

Fig. 6 Effect of addition of fly ash on unconfined compressive strength of the soil

divalent ca^{2+} might have induced the suppression of double layer, in turn resulted in flocculated structure. Hence, increase in dry density could be attributed due to the physicochemical interaction between the soil particles and RHA. Further, it is also seen that with the addition of different percentages of fly ash to the soil, the maximum dry density is observed to be decreased.

The variations in optimum moisture content with the addition of RHA and fly ash to the soil are shown in Fig. 2. From the plot, it is observed that with various percentages of RHA to the soil, the optimum moisture content is observed to be decreased. But with the addition of different percentages of fly ash to the soil the optimum moisture content is increased. As class 'C' fly ash has the self-cementation properties results in improving soil properties, including increasing stiffness, strength, freeze–thaw durability, reduced permeability and increased control of soil compressibility and moisture (Little et al. 2000).

4.2 *Effect of RHA and Fly Ash on CBR*

The CBR tests were conducted for the soil samples with varying percentage of RHA, and all the test samples were compacted to the maximum dry density and the corresponding optimum moisture content. From the compaction tests, it is clear that maximum dry density increased with increase in RHA content and optimum moisture content decreases with the same which implies that the strength of the soil also would also be relatively increasing. Figure 3 shows the effect of RHA on CBR value of the soil. From the figure, it is observed that CBR value keeps increasing with increase in RHA content to the soil.

Figure 4 shows the effect of fly ash on CBR value of the soil. From the figure, it is observed that CBR value keeps increasing with increase in RHA content to the soil. The CBR values of the soil–fly ash mixtures increased with increasing the fly ash content until 5%. Above 5% fly ash, the CBR value is observed to be decreased.

4.3 *Effect of RHA and Fly Ash on UCS*

The unconfined compression tests were conducted on soil–RHA mixtures, and all the samples were compacted to the corresponding maximum dry density and optimum moisture content. The RHA content is varied between 0 and 50%, respectively. The influence of RHA on unconfined compressive strength of the soil as shown in Fig. 5. From the figure, it is observed that unconfined compressive strength increased with increase in RHA content until 10% and the corresponding strength is about 190 kN/m^2. Above 10% RHA, the unconfined compressive strength value is observed to be decreased.

The influence of fly ash on unconfined compressive strength of the soil as shown in Fig. 6. The fly ash content was varied between 0 and 20%, respectively. From the figure, it is observed that unconfined compressive strength increased with increase in fly ash content until 5%, and the corresponding strength is about 149 kN/m^2. Above 5% fly ash, the unconfined compressive strength value is observed to be decreased.

5 Conclusions

The addition of RHA content to the soil increased the maximum dry density and decreased the optimum moisture content of the soil, whereas fly ash decreased the maximum dry density and increased the optimum moisture content of the soil. The optimum fly ash content was found at 5% for both CBR and unconfined compressive strength. The optimum RHA content was found at 10% for unconfined compressive strength. When the RHA content was increased from 0 to 50%, CBR improved by 56%.

References

IS (1985) Methods of test for soils. Bureau of Indian Standards, New Delhi (2720-1985)
Little DN, Males EH, Prusinski JR, Stewart B (2000) Cementitious stabilization. Transportation in the new millennium
Liu C, Evett J (2008) Soils and foundations, 7th edn. Prentice Hall International, Upper Saddle River, New Jersey
Mitchell JK (1986) Practical problems from surprising soil behaviour. Int J Geotech Eng 112(3):259–289

Influence of TerraZyme on Compaction and Consolidation Properties of Expansive Soil

Aswari Sultana Begum, G. V. R. Prasada Raju, D. S. V. Prasad, and M. Anjan Kumar

Abstract Augmentation as well as stabilization of soils is extensively used as a substitute due to the lacking of appropriate material on site. In this manuscript, universally available bioenzymes (TerraZyme) and their effect on engineering properties of soil are discussed. Differential Free Swell, Consistency Limit, Modified Compaction and Consolidation tests were conceded out in the laboratory for dissimilar mix proportions of TerraZyme with black cotton soil and from the results addition of the TerraZyme to the soil reduces the clay content and increases in the % of coarser particles, reduces Liquid limit values are decreasing and plastic limit increasing irrespective of the percentage of addition of TerraZyme. Maximum dry density increases and OMC goes decreasing with increase in % of TerraZyme. The consolidation parameters compressive index and coefficient of compressibility are decreased. From the above results, TerraZyme can be utilized for intensification of the expansive soil with a substantial save in cost of construction.

Keywords Consolidation characteristics · Expansive soil · TerraZyme · Free swell · Optimum moisture content · Maximum dry density

A. S. Begum · G. V. R. Prasada Raju
Department of Civil Engineering, JNTU College of Engineering, Kakinada, E.G.Dt., Andhra Pradesh, India
e-mail: aswari.sultana@gmail.com

G. V. R. Prasada Raju
e-mail: gvrpraju@gmail.com

D. S. V. Prasad (✉) · M. Anjan Kumar
Department of Civil Engineering, BVC Engineering College, Odalarevu, E.G.Dt., Andhra Pradesh, India
e-mail: drdsvp9@gmail.com

M. Anjan Kumar
e-mail: anjan_mantri@yahoo.com

M. Latha Gali and R. R. P. (eds.), *Problematic Soils and Geoenvironmental Concerns*, Lecture Notes in Civil Engineering 88,
https://doi.org/10.1007/978-981-15-6237-2_44

1 Introduction

In India enormous surface deposits are cover by Expansive soil, which have tendency to undergo volume change due to alter in moisture content with seasonal deviation. The setback by expansive soils has been recorded all over the planet. The method of improving the strength and durability of soil is known as soil stabilization. The main aim of stabilization is reduction in cost and to economically use the nearby existing material. Soil stabilization with bioenzyme is a revolutionary technology that is gaining recognition internationally. Numerous bioenzymes have been obtainable for soil stabilization, such as Renolith, Perma-Jaime, Terra-Jaime and Fujibeton. These enzymes have verified in the direction of active and very efficient and cost-effectively. The stabilization of soil through bioenzyme is an innovative technique which suitable accepted globally. Bioenzymes are organic, liquid, natural, non-toxic, non-flammable, non-corrosive and ecofriendly which are obtained from fermentation of organic matter. The thought of using enzyme stabilization used for roads was urbanized from the application of enzyme products use to take care of soil in order to improve horticultural applications. The enzyme allows soil materials to happen to additional with no trouble wet and further tightly compacted. The experiences of various researchers summarized below. Usha et al. (2018) premeditated expansive soil with stable limits, compressive properties and shear strength of the soil with four different doses of 200 ml/1.5 m^3, 200 ml/1.0 m^3, 200 ml/0.5 m^3, 200 ml/0.25 m^3. To study the effect of TerraZyme at 7, 14, 21 days curative period respectively. Application of 200 ml/0.25 m^3 TerraZyme reduced the liquid limit from 60.20 to 48% and the plasticity index from 33.13 to 24.50%. The compression test result shows that the liquid emulsion causes a slight increase in maximum compact density and a slight decrease in optimum humidity and increase in UCS value at 200 ml/0.25 m^3. This shows an improvement in shear strength percentage of 67.84% and 72.65% after 14 and 21 days, respectively. Unsoaked CBR for the dosage 200 ml/0.25 m^3 after 21 days of air dry curing is 480%. Anjali et al. (2017) observed that with the use of TerraZyme, the strength of the soil increases which is evident by the increase in UCS and CBR values. TerraZyme decreases the voids between the soil particles and thus increases the compaction and density of the soil. Optimum moisture content and consistency limits of the soil are decreased due to TerraZyme action as it increases the density of soil. It makes the soil water resistive by decreasing the permeability of the soil. TerraZyme reacts with the clay particles only before using it, and it is necessary to know the clay content of the soil. Effect of TerraZyme increases with increase in time, i.e., with time the strength of TerraZyme-treated soil increases. Sandee et al. (2017), an effective technique of land improvement using bioenzyme with different doses of TerraZyme was studied, in combination with 500, 700, 900 and 1000 ml/m^3 soil sample, the result being analyzed. The results show a significant increase in the CBR value of the soil sample as the dose of TerraZyme is increased. The composition of the TerraZyme local soil sample showed the optimal improvement in the stability limits, dry density and CBR values of the local soil sample. Bio-enzymatic soil treatment has been instrumental in improving the strength of the period. The best result of the CBR value was the two-week curing period with the third dose and the percentage increase of 131.49% compared to the native soil sample without TerraZyme. Baby et al. (2016) study, black cotton soil stabilized with TerraZyme of different dosages of 200 ml/3 m^3, 200 ml/2.5 m^3 and 200 ml/2 m^3 are tested after the curing period of 0 days, 14 days and 28 days. It is observed that liquid limit decreases from 60.2 to 56.53%, and the plastic limit also decreases from 32 to 30.09%. The UCS value increases from 3.57 to 8.72% when compared to actual soil with 4 weeks. CBR value increases from 1.2 to 5.67%. Swelling pressure reduces from 124 to 85.2 kN/m^2. Maximum dry density increases from 1.5 to 1.612 g/cm^3. Optimum moisture content decreases from 22.80 to 21.80%. Nandini et al. (2015), study the effect of TerraZyme dose on different dry concentrations to study its effect on the compressibility and strength properties of red soil. TerraZyme shows a significant increase in the strength of the treated red soil, and the amount of the dose depends on the dry density of the compacted soil and has a significant effect on red soil strength. The strength of TerraZyme treated red soil at all days of curing period, viz. 7, 15 and 30 days significantly larger than those obtained at 0 days curing as well as untreated red soil, and significant increase in UCS was observed at all dosage levels of TerraZyme. Rajoria and Kaur (2014) popular bioenzymes and their impact on the engineering properties of soil are discussed. Bioenzymes are non-toxic, organic and non-biodegradable. These chemicals do not harm humans, animals, fish or vegetation in general use and are made from organic matter and biodegradable. With the use of bioenzyme, total free pavement is possible because its use promotes the use of locally available materials. The use of bioenzymes increases the compressive strength and hardness of stabilized soil. Bioenzymes provide pavement flexibility and durability and also reduce crack formation. Bioenzymes reduce the swelling and contraction properties of a large amount of clay. The use of bioenzyme in pavement construction is proven to be very economical as compared to other traditional soil stabilization methods. The cost of construction project can be reduced considerably with the use of bioenzyme. Venkatasubramanian and Dhinakaran (2011) deliberate the strength distinctiveness on three soils for different dosages of bio-enzyme whose liquid limits of 46, 28 and 30% as well as plasticity index of 6, 6 and 5%. A strength characteristic in terms of unconfined compressive strength and CBR by 2 and 4 weeks of curing phase is premeditated. Surekha and Gangadhara (2010) conducted consolidation tests on expansive soil blend with different dosages of bioenzyme varied beginning 0.25–2% and conducted swelling potential and swelling pressure measured in the one dimensional consolidation using swell and load procedure, SEM studies and cation exchange capacity tests to examine the structural and CEC modification. Conclusion of examinations, bioenzyme treated expansive soil exhibit lesser percent of swell and swell pressures and curing period beyond 30 days did not give way any further noteworthy decline in swell properties. The arrangement of the soil changed from flocculated to dispersed. No significant changes were observed in cation exchange capacity values of bioenzyme treated soil specimen. Dhinakaran and Prasanna (2007) did experimental study on three different soils blend with different dosages of bioenzyme and found decrease in liquid limits and increase in plasticity index for three different soils, respectively, unconfined compressive strength is greater than before by

246–404% after 4 weeks of curing. The present work provides an effective technique of ground improvement using TerraZyme. In this examination, TerraZyme is used for improving the differential free swell, Atterberg limits, compaction and consolidation values of expansive soil. TerraZyme is a natural, non-toxic and liquid enzyme. Soil samples were prepared with four different dosages (0.0.0.1, 0.2 and 0.3 ml) of the TerraZyme adding together in expansive soil. Considerable augment is originated in the geotechnical properties of the soil sample as the dosage of TerraZyme has been increased.

2 Materials and Their Properties

Properties of various materials used during the laboratory experimentation are reported in the following section.

2.1 *Expansive Soil*

The soil used for this study was a typical black cotton soil which was brought from Tummalapalli, Amalapuram, Andhra Pradesh, shown in Fig. 1. The property of the expansive soil assessed based on relevant IS Code provisions. The physical properties of black cotton soil are specific gravity = 2.62, W_L = 86%, W_P = 32.01%, W_p = 35.44%, I.S. Classification = Clay of high compressibility (CH), OMC = 21.3%,

Fig. 1 Expansive soil

MDD = 16 kN/m^3, differential free swell = 140%, unsoaked CBR = 3.1%, coefficient of uniformity (C_U) = 0.66, coefficient of curvature (C_c) = 5.06, cohesion (C) = 16 kN/m^2, angle of internal friction (φ) = 20°, compression index (C_c) = 0.09, coefficient of permeability (K) = 3.2 × 10^{-3} and coefficient of compressibility (a_v) = 0.006.

2.2 *TerraZyme (TZ)*

It is a natural, non-toxic, corrosive and non-flammable liquid that is produced by the formation of vegetable and fruit extracts. The TerraZyme used in this study was obtained from Avijeet Agencies, Anna Nagar East, Chennai, India. TerraZyme is specifically designed to modify the engineering properties of the soil. Water dilution is required before application. The physical properties of the TerraZyme are presented in Table 1.

Table 1 Physical properties of TerraZyme

Identity	TerraZyme
Hazardous components	None
Boiling point	212 °F
Specific gravity	1.05
Melting point	Liquid
Evaporation rate	Same as water
Solubility in water	Complete
Appearance/Odor	Brown liquid, non-obnoxious
Special fire fighting procedures	None
Unusual fire/Explosion hazards	None
Unstable or stable	Stable
Conditions avoid	>45 °C; pH below 3.5, above 9.5
Incompatibility	Caustics, strong bases
Health hazards (acute and chronic)	None
Disposal method	Flush into any sewage system
Storing	Store at temperatures below 45 °C
Respiratory protection	Not required
Working	Normal good practices

3 Laboratory Tests

Various tests were carried out in the laboratory to find the indicator and engineering properties of the materials used during the study. Compression and consolidation tests were carried out using different percentages of TerraZyme mixed with abundant soil to find the optimum percentage of TerraZyme. Details of these tests are given in the following sections.

3.1 Index Properties

Related I.S. Learning codes [IS: 2720 (Part-5)-1985; IS: 2720 (Part-6)-1972], the index has been followed in finding the correct relations. In this research, liquid limit and plastic limit and samples were tried.

3.2 Compaction Properties

Optimum moisture content and maximum dry density of black cotton soil and TerraZyme mixes were determined according to I.S. heavy compaction test IS: 2720 (Part VIII).

3.3 Consolidation Properties

Consolidation Test: Different samples were prepared for consolidation test using expansive soil with different dosages of TerraZyme, and consolidation tests were conducted in the laboratory as per IS Code [IS: 2720 (Part-XV)-1986] as shown in Fig. 2.

4 Results and Discussions

A variety of tests have been carried out in the laboratory under the provisions of I S Code, and the test results are given below for the purpose of determining the optimum percentage. After mixing the TerraZyme with the expansive soil sample, combining different doses of TerraZyme with the expansive soil sample showed an optimal improvement in fixed limits, dry density and consolidation values.

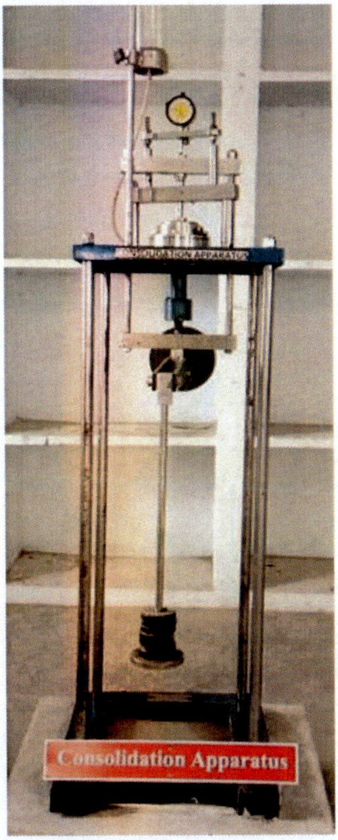

Fig. 2 I-D consolidation test apparatus

4.1 Differential Free Swell Test

The differential free swell tests result of expansive soil treated with different ml of TerraZyme was found to be decreased from 140%, 129%, 107% and 98%, respectively, with the addition of TerraZyme shown in Fig. 3.

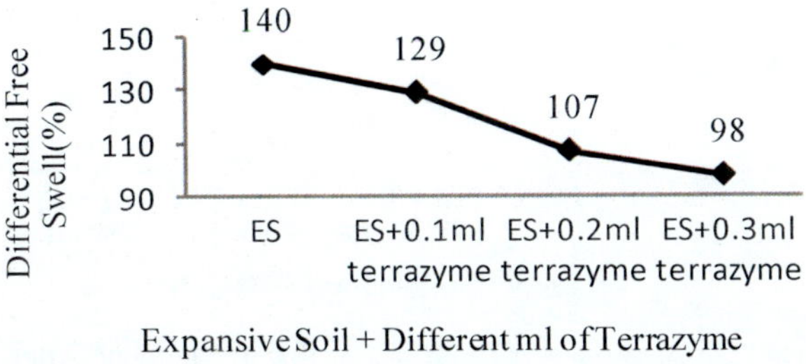

Fig. 3 Variation of differential free swell for expansive soil treated with different ml of TerraZyme

Table 2 Atterberg's limits

TerraZyme dosages (ml)	Liquid limit (%)	Plastic limit (%)	Plasticity index (%)
0	82.10	46.66	35.5
0.1	81.74	46.75	34.99
0.2	81.58	46.82	34.76
0.3	81.43	46.93	34.5

4.2 *Consistency Characteristics*

The Atterberg limit test was reduced on a soil sample treated with an increasing dose of TerraZyme. Liquid limit values are decreasing from 82.10, 81.74, 81.58 and 81.43, and plastic limit values are increasing from 46.66, 46.75, 46.82 and 46.93 by adding 0 ml, 0.1 ml, 0.2 ml and TerraZyme, respectively, when mixed with extensive soil as shown in Table 2.

4.3 *Compaction Properties*

As shown in Figs. 4 and 5, the optimum moisture content (OMC) of the treated soil samples decreased and the maximum dry density (MDD) increased as the dose of TerraZyme was increased. MDD and OMC values were 16 kN/m^3, 16.2 kN/m^3, 16.6 kN/m^3, 16.67 kN/m^3 and 21.3%, 20.9%, 20.4% and 20.01%, respectively. The increase in the density of the soil sample, and the increase in dose can be attributed to the accumulation and cementation of soil cells. Investigation can be concluded that the increase in MDD is caused by a decrease in the void ratio when the enzyme is added to the soil. It is observed that there is a decrease in OMC due to the efficient cation exchange process, which usually takes longer in the absence of such stabilizers.

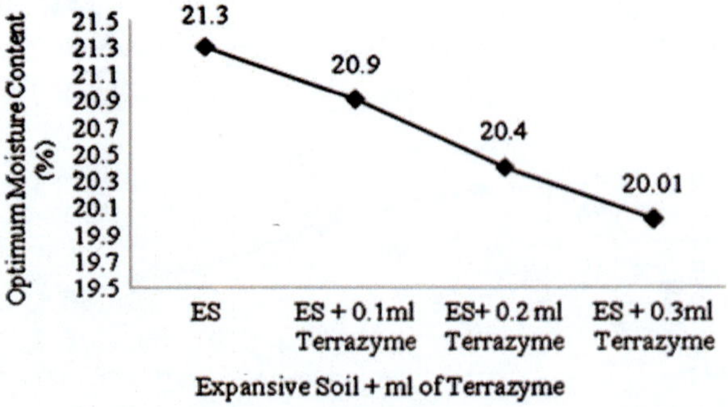

Fig. 4 Variation of OMC for expansive soil treated with different ml of TerraZyme

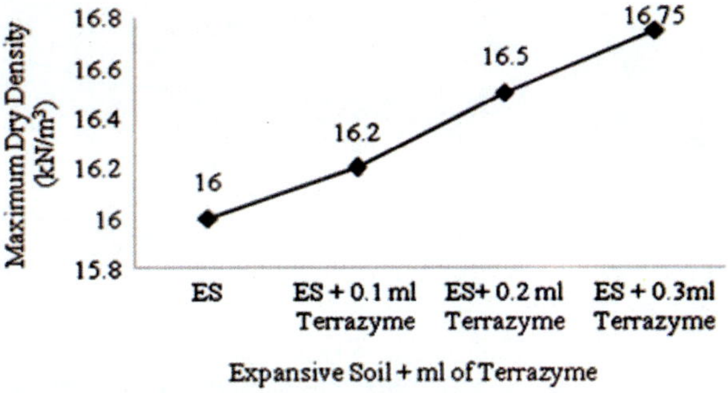

Fig. 5 Variation of MDD for expansive soil treated with different ml of TerraZyme

This is due to the reaction of the enzyme with the soil, leading to the effect of cementation.

4.4 Consolidation Test

The reduction in coefficient of compressibility for soil sample treated with TerraZyme, continuously decreased from 0.006, 0.004, 0.003 and 0.002 with 0%, 0.1%, 0.2% and 0.3% adding TerraZyme with an increase in applied pressure as shown in Fig. 6. The coefficient of permeability values are decreasing continuously from 3.2×10^{-3}, 2.10×10^{-3}, 1.44×10^{-3} and 7.7×10^{-4}, and the compressibility index (cc) also decreases from 0.09, 0.076, 0.063 and 0.046 by adding 0, 0.1 ml, 0.2 ml and 0.3 ml adding TerraZyme, respectively, shown in Figs. 7 and 8.

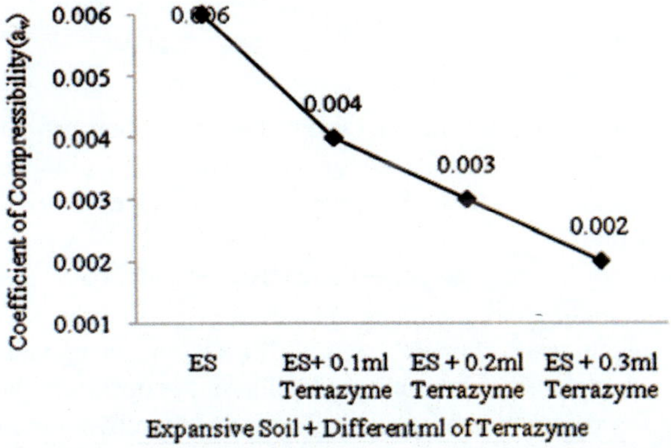

Fig. 6 Variation of coefficient of compressibility (a_v) for expansive soil treated with different ml of TerraZyme

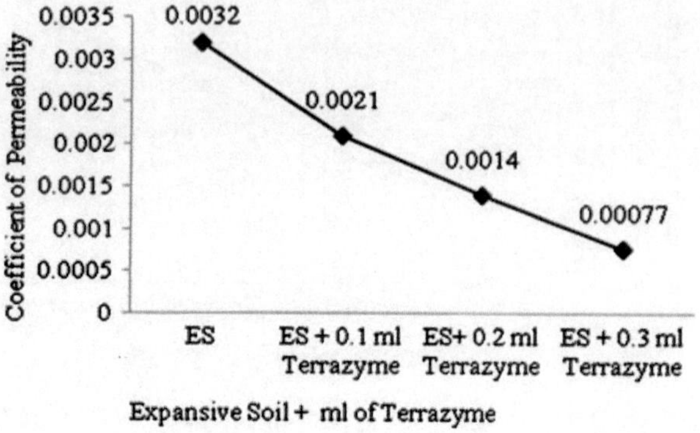

Fig. 7 Variation of coefficient of permeability for expansive soil treated with different ml of TerraZyme

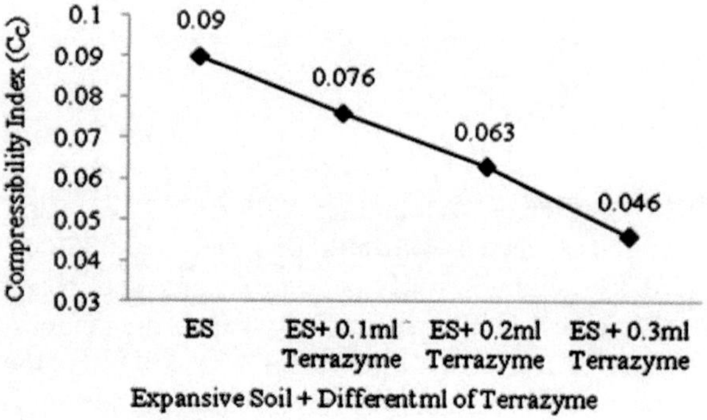

Fig. 8 Variation of compressibility index for expansive soil treated with different ml of TerraZyme

5 Conclusions

The aptness of TerraZyme for the alteration of geotechnical properties of gigantic soils is able to be fulfilled among the outcome of TerraZyme on index and engineering properties of plentiful soils. Based on the experimental outcome, the subsequent conclusions are tired.

Stabilization with TerraZyme proved excellent enhancement in the engineering character of black cotton soil.

Investigational end result showed that liquid emulsion causes a insignificant boost in the utmost packed density and a minor decline in the optimum dampness.

Lessening in the voids connecting the soil cells and thereby increases the compactness of the soil and optimum wetness at ease and consistency limits of the soil are decreased due TerraZyme which make the soil water resistive by means of diminishing the permeability.

The decrement in Coefficient of Compressibility for soil samples treated with TerraZyme continuously decreased with an increase in applied pressure and further it revealed that maximum decrease in Compression Index with degree of compressibility reduced slightly.

Enzyme is an organic fluid and is ecofriendly which has no harmful consequence on the environment.

References

Anjali G et al (2017) Review paper on soil stabilization by TerraZyme. Int J Eng Res Appl 7(4):54–57

Baby M, Gowshik A, Karthick RAV, Mohanasundram M (2016) Experimental study of expansive soil stabilized with terrazyme. Int J Eng Res Technol 5(01):897–899

Dhinakaran C, Prasanna KR (2007) Bioenzyme soil stabilization in road construction. Everyman's Sci XLI(6):397–400

IS: 2720 (Part 5)—1985 Indian Standard Code of practice for determination of liquid limit

IS: 2720 (Part 6)—1972 Indian Standard Code of practice for determination of plastic limit

IS: 2720 (Part VIII)—1983 Indian Standard Code of practice for determination of water content-dry density relation using heavy compaction

IS: 2720 (Part XV)—1986 Indian Standard Code of practice for determination of test for soils. Determination of consolidation properties

Nandini DN, Amate V, Prathap Kumar MT (2015) Compaction and strength characteristics of Terra-Zyme stabilized red soil. Int J Res Publ Eng Technol Manage 1(1):1–3

Rajoria V, Kaur S (2014) A review on stabilization of soil using bio-enzyme. Int J Res Eng Technol 03(01):75–78

Sandee P, Mohsin Khan M, Anurag S (2017) Stabilization of soil using bio-enzyme. Int J Civ Eng Technol 8(1):234–237

Surekha N, Gangadhara S (2010) Swelling properties of bio-enzyme treated expansive soil. Int J Eng Stud 2(2):155–159

Usha P, Shalini S, Shivani C (2018) Effect of bio enzyme—Terrazyme on compaction, consistency limits and strength characteristics of expansive soil. Int Res J Eng Technol 5(3):1602–1605

Venkatasubramanian C, Dhinakaran G (2011) Effect of bio-enzymatic soil stabilisation on unconfined compressive strength and California bearing ratio. J Eng Appl Sci 6(5):295–298

Predictive Models for Estimation of Swelling Characteristics of Expansive Soils Based on the Index Properties

S. Swapna Varma, Manish Gupta, and R. Chitra

Abstract The soils that exhibit volume changes with change in moisture content are called expansive or swelling soils. These soils are characterized generally by their blackish colour, high plasticity and the enriched presence of the clay mineral montmorillonite as the principal constituent. As the expansive soils have a tendency to change its volume with change in the moisture content, they can cause severe damage and distress to lightweight structures constructed over them due to the increased swell pressure evolved as a result of the swelling. Hence while designing the foundations on expansive soils, it is highly imperative to get an idea of the anticipated swell and the associated swell pressure that may damage the structural element. The various soil properties which clearly indicate the swelling characteristics of the expansive soils are free swell index, swelling index, shrinkage limit, swelling potential, swelling pressure, etc. The activity of clay derived from plasticity index and percentage of clay sizes present in the soil is also used as an indicator for identifying the expansive soil. If these swelling parameters can be derived from easily determinable index properties of soil, that would be really informative regarding the usefulness of the soil or the quantum of modification or improvement required for reclamation of such type of soil. A great deal of research has been done in correlating the swelling characteristics with the index properties and physical state of soil. The present work is an attempt to develop predictive models for the swelling characteristics based on their interrelationship with the index properties like liquid limit, plasticity index, shrinkage index, etc. and the physical properties like dry unit weight, grain size, especially the clay percentage, etc. The expansive soils investigated at Central Soil and Materials Research Station, New Delhi, are used to develop the predictive models. The applicability of the developed equations is finally checked by conducting validation study using three different data sets.

S. Swapna Varma (✉) · M. Gupta · R. Chitra
Central Soil and Materials Research Station, New Delhi, India
e-mail: sswapnavarma@gmail.com

M. Latha Gali and R. R. P. (eds.), *Problematic Soils and Geoenvironmental Concerns*, Lecture Notes in Civil Engineering 88,
https://doi.org/10.1007/978-981-15-6237-2_45

1 Introduction

Expansive soils will undergo volume changes with variation in their water content. Such type of soils swell during rainy seasons by absorbing water and at the same time shrink during summer season because of loss of water due to evaporation. The alternate swelling and drying can result in frequent volume changes leading to the instability of the soil as well as the structural instability. Hence, identification of expansive soils is very important to choose the correct remediation technique to improve the properties of the expansive soils to take in to account any future volume changes.

While designing the structural foundations, it is highly imperative to get an idea of the anticipated swell that can happen due to the expansive nature of the soil. Even though the oedometer test carried out on undisturbed samples is believed to be the most accepted and reliable technique for determination of swell pressure, it can also be derived from other interrelated soil parameters. The swelling or shrinkage behaviour of expansive soils is attributed mainly to its inherent mineralogical characteristics. The basic soil properties also can have an influence on soils swelling/shrinkage behaviour as the plasticity of soil is very important parameter in defining soils behaviour in presence of water.

To understand the swelling behaviour of expansive soils in deep, considerable research has been carried out worldwide, most of which has resulted in developing various empirical equations which relate the swelling parameters to certain physical soil properties like consistency limits, percentage clay content, initial water content, in situ density, etc. There are many studies carried out relating the soil's swelling characteristics to its index properties. The swelling characteristics include different parameters like swell potential, swelling pressure, free swell index, etc., out of which the free swell index parameter is selected for the present study, and its correlation with the other basic soil properties like percentage clay sizes, shrinkage index, shrinkage limit, liquid limit, plastic limit and plasticity index, etc., has been studied.

2 Objectives

The objectives of the present study can be summarized as:

- Laboratory testing of the selected soils for their index properties (liquid limit, plastic limit, plasticity index, shrinkage limit and shrinkage index) and physical properties (percentage of clay size). The free swell index test is done for studying the swelling characteristics of the selected soil samples.
- Development of predictive models by studying the statistical correlation between the swelling characteristics and selected index and physical properties, respectively.
- Validation study of the developed models using three different data sets collected from literature review.

3 Materials and Methods

The expansive soils are basically soils having high amount of clay minerals contained in it. Hence, the soil samples selected for the present study were six clayey-type soils from different national projects—four from Polavaram Irrigation Project, Andhra Pradesh and two from Wain Ganga Nal Ganga Project.

All the samples have been tested for their basic soil properties, i.e. mechanical analysis test to find out the physical property of percentage clay size (as per IS Code: IS 2720-4), Atterberg's limit test (as per IS Code: IS 2720-5) for determination of index properties like liquid limit, plastic limit and plasticity index, the wet pat-dry pat method for determination of shrinkage limit (as per IS Code: IS 2720-6). The swelling behaviour is determined by conducting the free swell index test (as per IS Code: IS 2720-40).

The differential free swell index test is carried out for this purpose. The oven-dried clay samples passing 425 μm sieve was collected, and 10 g of sample was placed, respectively, in two glass cylinders of 100 ml capacity. The cylinders were filled up to 100 ml with distilled water and kerosene oil, respectively. The samples were kept undisturbed for minimum 24 h, and the final volume of samples is noted down.

The free swell index (FSI) value can be calculated as given in Eq. 1 below:

$$\text{FSI} = \frac{V_\text{d} - V_\text{k}}{V_\text{k}} \times 100 \tag{1}$$

where V_d is the volume of soil in distilled water and V_k is the volume of soil in kerosene oil.

4 Results and Discussions

4.1 Laboratory Test Results

The main swelling parameter selected for the present study is the free swell index. Its correlation with each other selected soil properties, relevant to its expansive property, is studied. The consenses of the different laboratory test results are shown in Table 1.

4.2 Development of Predictive Models

The predictive models obtained for correlation of free swell index with liquid limit, plastic limit, plasticity index, shrinkage limit, shrinkage index and percentage clay size are shown in Figs. 1 and 2, respectively.

Table 1 Consenses of the laboratory test results

Characteristics (%)	S1	S2	S3	S4	S5	S6
Liquid limit	71.8	65.5	66.0	85.0	66.5	94.5
Plastic limit	32.0	34.6	31.3	41.4	32.9	48.4
Plasticity index	39.8	30.9	34.7	43.6	33.6	46.1
Shrinkage limit	12.6	9.6	10.0	14.5	13.3	16.6
Shrinkage index	19.4	25.0	21.3	26.9	19.6	31.8
Percentage clay size	40.0	49.9	34.8	71.4	56.2	64.8
Free swell index	54.5	36.4	36.4	108.3	53.9	116

All the models developed showed good values of R^2, ranging from 0.60 to 0.94. The developed equations are given below in Equations 2–7.

Equation obtained for prediction of Free Swell Index (FSI) with the Liquid Limit (LL) with R^2 value of 0.94:

$$\text{FSI} = 2.86\text{LL} - 146.67 \tag{2}$$

Equation obtained for prediction of Free Swell Index (FSI) with the Plastic Limit (PL) with R^2 value of 0.85:

$$\text{FSI} = 4.86\text{PL} - 111.04 \tag{3}$$

Equation obtained for prediction of Free Swell Index (FSI) with the Plasticity Index (PI) with R^2 value of 0.85:

$$\text{FSI} = 5.46\text{PI} - 140.49 \tag{4}$$

Equation obtained for prediction of Free Swell Index (FSI) with the Plasticity Index (PI) with R^2 value of 0.84:

$$\text{FSI} = 12.25\text{SL} - 89.15 \tag{5}$$

Equation obtained for prediction of Free Swell Index (FSI) with the Plasticity Index (PI) with R^2 value of 0.60:

$$\text{FSI} = 5.68\text{SI} - 68.44 \tag{6}$$

Equation obtained for prediction of Free Swell Index (FSI) with the Plasticity Index (PI) with R^2 value of 0.71:

$$\text{FSI} = 2.12\,(\%\text{Clay size}) - 44.59 \tag{7}$$

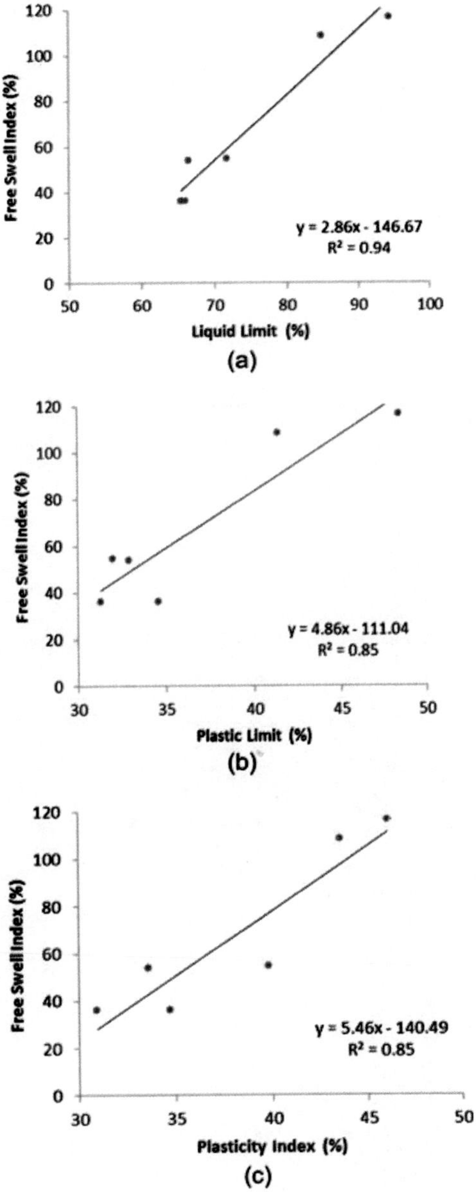

Fig. 1 **a–c** Predictive models obtained for free swell index based on liquid limit and plastic limit values, respectively

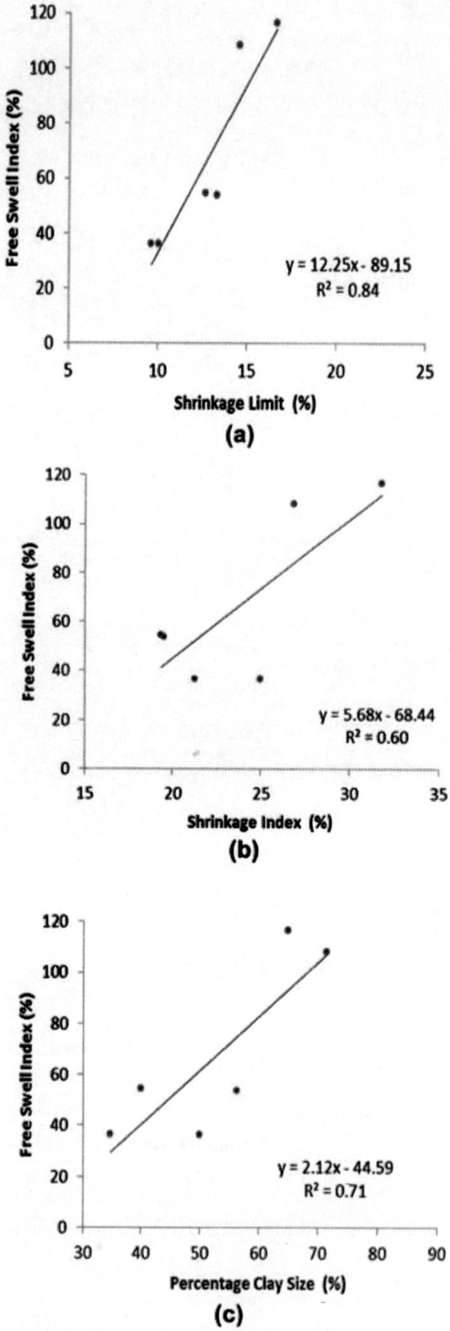

Fig. 2 **a–c** Predictive models for free swell index based on shrinkage limit, shrinkage index and percentage clay sizes values, respectively

4.3 Model Validation Study

For the validation of the developed predictive models, three different data sets of expansive soils were collected from literature review. The first set of data was selected from the study conducted by Rao et al. (2014), the second set of data from the study conducted by Jayasekera and Mohajerani (2003), and the last set of data was selected from the study of Patel et al. (2016).

The respective predictive model equations developed were validated using the selected three data sets separately. The correlation between the actual observed values and predicted values is given with the obtained R^2 values in Figs. 3, 4 and 5, respectively.

From the validation study results obtained for the first set of data as shown in Fig. 3a–f, it can be seen that the highest R^2 value of 0.82 is obtained for the shrinkage limit-based model for the prediction of free swell index value. The other correlation graphs showed a varying tendency in R^2 value.

Figure 4a–d shows the validation study results obtained for the second set of data, selected from the study of Jayasekera and Mohajerani (2003). They didn't select the shrinkage parameters for their study, and hence, only the plasticity characteristics and percentage clay size are considered for the study.

The results obtained in the validation study using the second data set indicate that for that particular data set, the correlation holds good between the predicted value and observed values for the models developed based on the liquid limit (with R^2 value 0.79), based on the plasticity index (with R^2 value 0.75) and the highest value for the percentage clay size-based model (with R^2 value 0.88).

The validation study results for the developed models obtained for the third set of data derived from the study of Patil et al. (2016) are shown in Fig. 5a–f, respectively.

The results obtained in the validation study using the third data set indicate that for that particular data set, all the correlation carried out between the predicted value and observed values for all the models developed are not holding good R^2 values (even the highest value of R^2 obtained being 0.40 is also not so good).

The consenses of the validation study is shown in Table 2, with the R^2 values obtained for the three different sets of data.

On comparison of the three different validation study results, it can be inferred that for the first set of data, the shrinkage limit-based model gave the maximum accuracy with R^2 value of 0.82, while the shrinkage index-based model gave R^2 value of 0.64. For the second set of data, though it didn't consider the shrinkage parameters, the maximum accuracy with R^2 value of 0.88 was obtained for the model based on clay size percentage. The liquid limit-based model and the plasticity index-based model also gave good results (R^2 value of 0.79 and 0.75, respectively). The third data set gave very poor results with the least accuracy compared to the other two sets.

However in correlation analysis of any developed empirical model, the accuracy of the model is greatly dependent on the range of the data selected for the study.

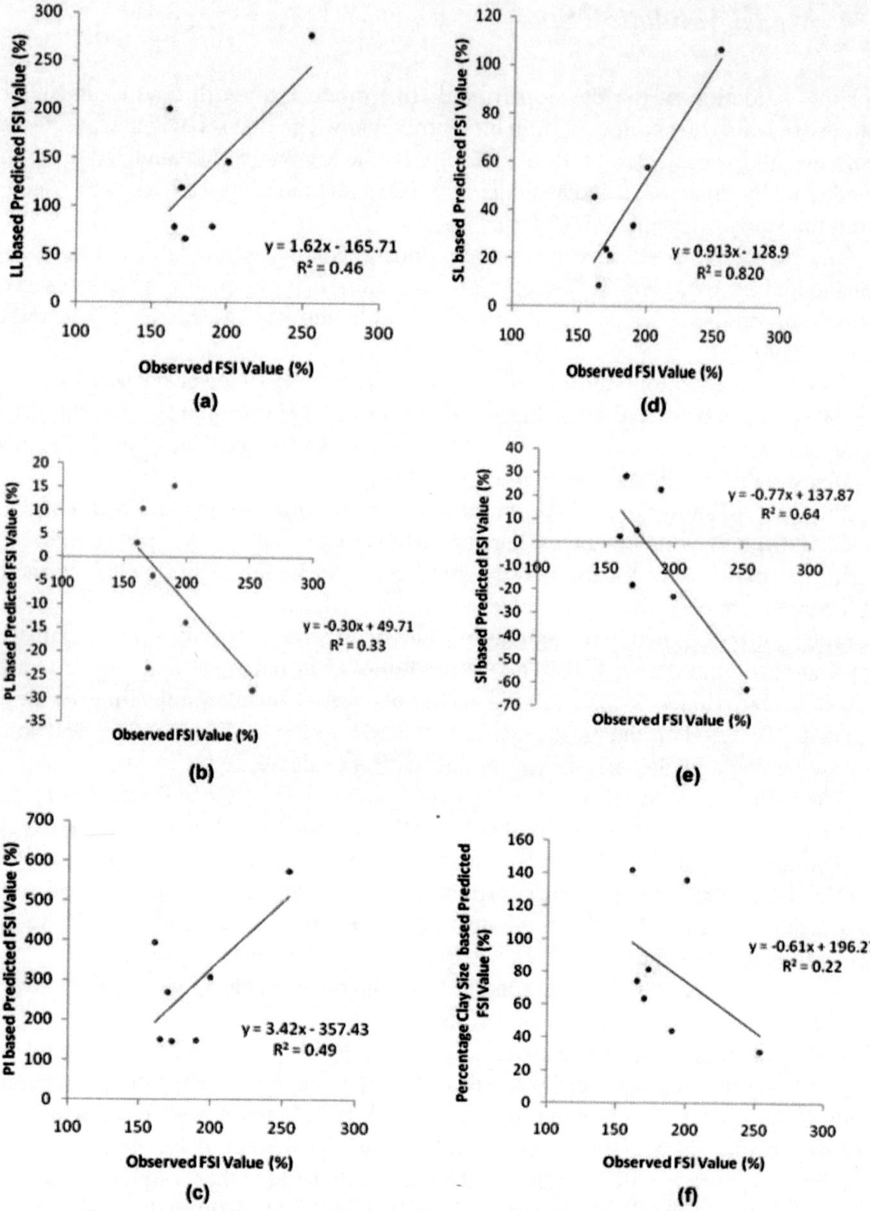

Fig. 3 **a–f** Validation study results obtained for the developed model using data set 1

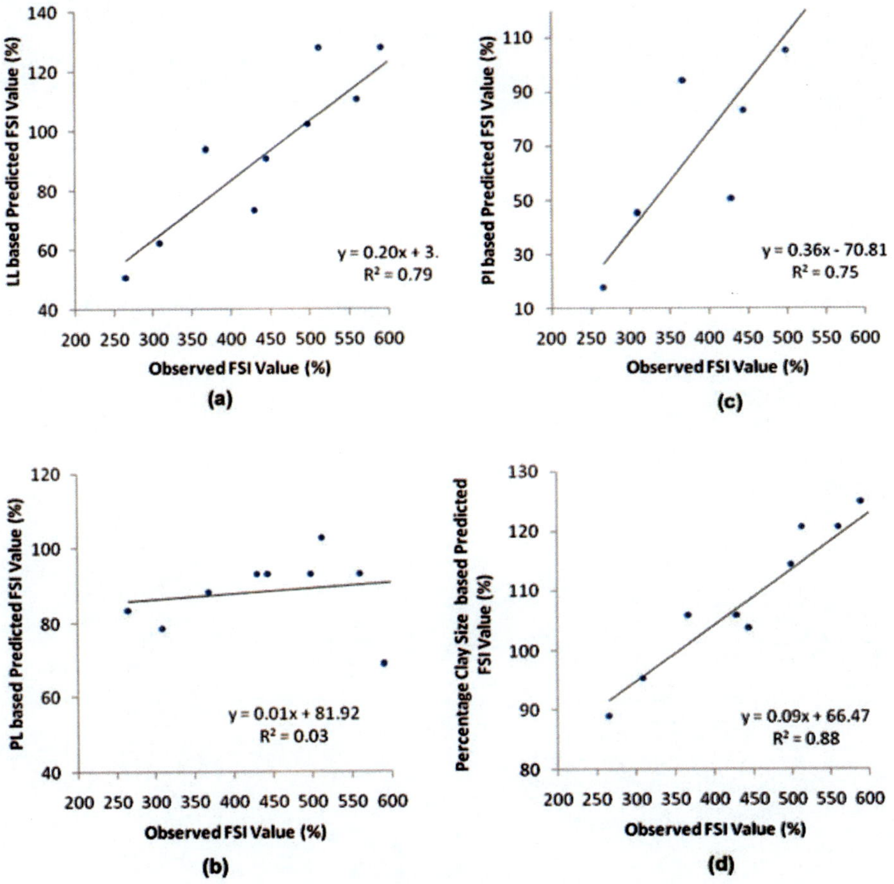

Fig. 4 a–d Validation study results obtained for the developed model using data set 2

Hence, the applicability of one set of equations is limited to that particular data set in its best possible manner, even though we can get some general inference about the swelling behaviour of the selected soil types and the parameters influencing the same.

5 Conclusions

As per the present study, the following major conclusions can be arrived at:

- Based on the laboratory tests conducted on six selected expansive soil samples, the predictive models were developed for estimation of free swell index parameter based on the six different dependent variables (i.e. liquid limit, plastic

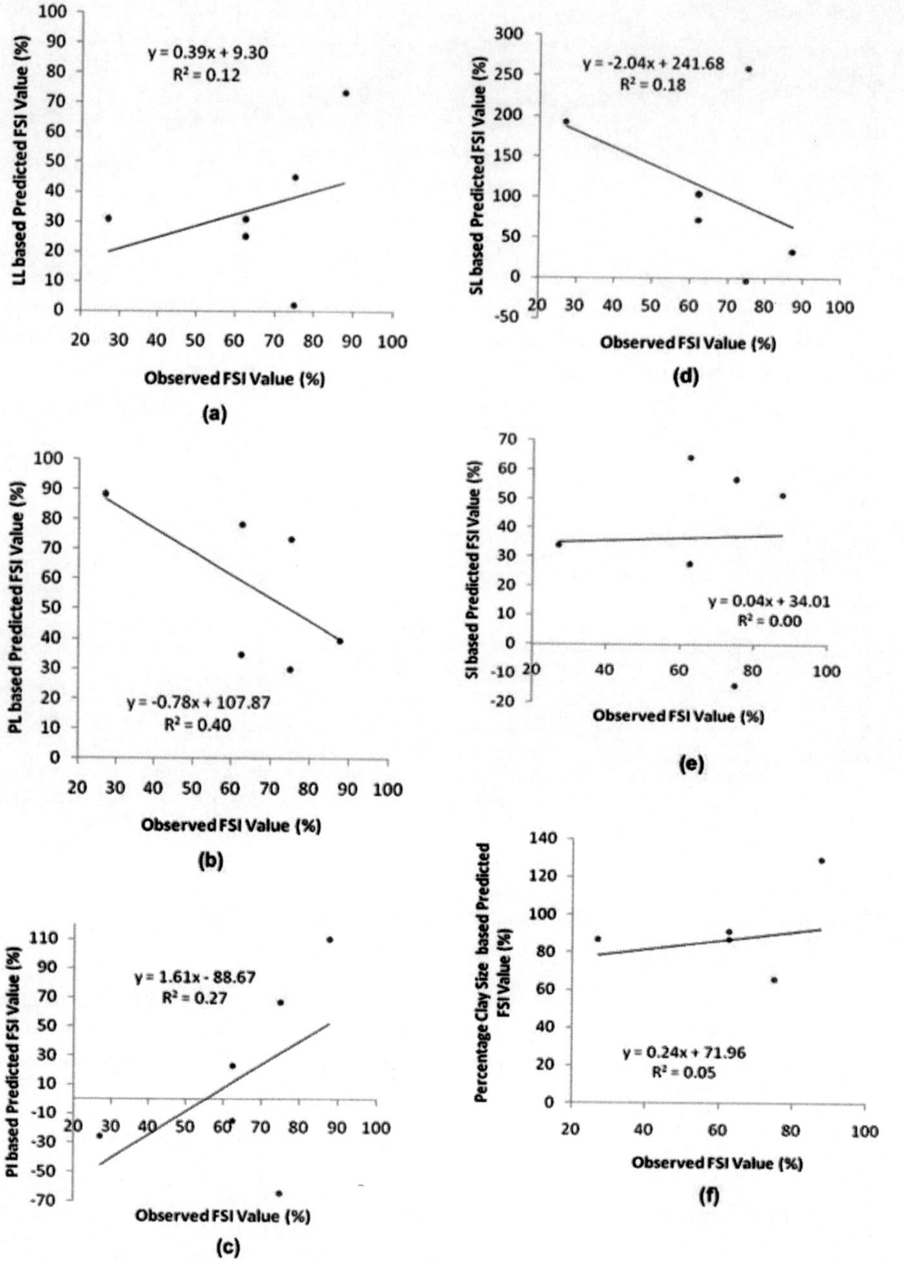

Fig. 5 a–f Validation study results obtained for the developed model using data set 3

Table 2 Consenses of the validation study

Details	Value of R^2 obtained		
	Data set 1 from study by Rao et al. (2014)	Data set 2 from study by Jayasekera and Mohajerani (2003)	Data set 3 from study by Patil et al. (2016)
For the model based on liquid limit	$R^2 = 0.46$	$R^2 = 0.79$	$R^2 = 0.12$
For the model based on plastic limit	$R^2 = 0.33$	$R^2 = 0.03$	$R^2 = 0.40$
For the model based on plasticity index	$R^2 = 0.49$	$R^2 = 0.75$	$R^2 = 0.27$
For the model based on shrinkage limit	$R^2 = 0.82$	–	$R^2 = 0.18$
For the model based on shrinkage index	$R^2 = 0.64$	–	$R^2 = 0.00$
For the model based on percentage of clay size	$R^2 = 0.22$	$R^2 = 0.88$	$R^2 = 0.05$

limit, plasticity index, shrinkage limit, shrinkage index and percentage clay size) successfully with good R^2 values. The developed equations are:

- FSI = 2.86 LL − 146.67
  (with $R^2 = 0.94$)
- FSI = 4.86 PL − 111.04
  (with $R^2 = 0.85$)
- FSI = 5.46 PI − 140.49
  (with $R^2 = 0.85$)
- FSI = 12.25 SL − 89.15
  (with $R^2 = 0.84$)
- FSI = 5.68 SI − 68.44
  (with $R^2 = 0.60$)
- FSI = 2.12 (%Clay Size) − 44.59
  (with $R^2 = 0.71$).

- The validation study for the developed model when carried out with three different data sets,
  - The first set of data gave the maximum accuracy for the model developed based on the shrinkage limit values for the prediction of FSI with R^2 value of 0.82.
  - The second set of data showed that maximum accuracy for the predicted values of FSI was obtained for the models developed based on the plasticity characteristics. For liquid limit-based model, R^2 value obtained was 0.79, for the plasticity index based model, the R^2 value obtained was 0.75, and for the clay size percentage-based model, the R^2 value obtained was the highest of 0.88.

- Unfortunately for the third set of data, none of the models gave good correlation for the predicted values of FSI with the actual observed values.
- The major limitation of any empirical model study is the model's dependency on the limited data set selected. Validation studies with good R^2 values give better confidence to the users to apply the equations effectively for another domain of data set.

Acknowledgements The authors are greatly acknowledged to Shri. Hassan Abdullah, Director, Central Soil and Materials Research Station, New Delhi, for his valuable guidance and motivational support during this work. The authors are also greatly indebted to those researchers, whose study data have been included in this study for the model validation analysis.

References

Jayasekera S, Mohajerani A (2003) Some relationships between shrink swell index, liquid limit, plasticity index, activity and free swell index. Aust Geomech 38(2):53–58

Patel AP et al (2016) Establishing relationship between swelling pressure and free swell index of soils—a case study. In: Proceedings of the conference on recent innovations in science, engineering and technology, organised at Pune, India, pp 4–7

Rao AS et al (2014) Prediction of swelling characteristics of remoulded and compacted expansive soils using free swell index. Q J Eng Geol Hydrol 37:217–226

Effect of Plastic Waste on Strength of Clayey Soil and Clay Mixed with Fly Ash

Mithun Mandal, Nagendra Roy, and Ramakrishna Bag

Abstract Due to rapid industrialization and development in urban areas, every year, huge quantity of plastic wastes are being generated throughout the world. Disposal of these wastes in landfills creates serious environmental problems. The study on the utilization of plastic waste with clayey soil and with clay–fly ash mixture is limited. This paper presents the effect of plastic waste on the strength of clayey soil and clay–fly ash mixture, and the results are presented in terms of compressive strength, shear strength, and California bearing ratio (CBR). The different percentage of plastic waste (0.5%, 1.0%, 1.5%, and 2.0% by dry weight) and 10% fly ash by dry weight were mixed into the clayey soil. From the experimental results, incorporation of the plastic waste into the clayey soil and clay fly ash mixture gives a decreasing trend of maximum dry density and increasing trend of optimum moisture content. The addition of 1% plastic waste in clayey soil and clay–fly ash mixture increases the unconfined compressive strength (UCS), internal friction angle and in the CBR value under soaked and unsoaked condition. The present study will help in consuming the considerable quantity of waste plastic, thereby reducing the environmental threat.

Keywords Clayey soil · Plastic waste · Fly ash · UCS · Shear strength · CBR

M. Mandal (✉) · N. Roy
Department of Civil Engineering, National Institute of Technology Rourkela, Rourkela, Odisha 769008, India
e-mail: mandalmithun659@gmail.com

N. Roy
e-mail: nroy@nitrkl.ac.in

R. Bag
Department of Civil and Environmental Engineering, Indian Institute of Technology Patna, Patna, Bihar 801103, India
e-mail: rkbag@iitp.ac.in

M. Latha Gali and R. R. P. (eds.), *Problematic Soils and Geoenvironmental Concerns*, Lecture Notes in Civil Engineering 88,
https://doi.org/10.1007/978-981-15-6237-2_46

1 Introduction

Due to rapid industrialization and development in urban areas, every year, large quantities of plastic wastes are being generated throughout the world. As the availability of land space in India as well as over the world is limited, the disposal of these plastic wastes in landfill is not a proper solution. Except this, disposal of these wastes in landfills creates serious environmental problems. This study reveals a simple way to use plastic waste as a strip form obtained from usable water bottle in the field of civil engineering as a reinforcing material. Nowadays, reinforced soil construction is an effective technique to improve the strength and stability of soils.

Nowadays, alarming growth of population is the major problem over the world. As there is limited space to live so need to improve the quality of soil in such a way that soft soil also can be effective to take higher load from the high rise building. Therefore, in this project, plastic wastes with fly ash were mixed to the soil to improve the strength criteria of soil. In this study, 10% fly ash and 0.5%, 1.0%, 1.5%, and 2.0% (by dry weight) plastic wastes were mixed.

Consoli et al. (2002) showed that polyethylene terephthalate (PET) fiber reinforcement increases the ultimate strength and peak strength of both cemented and un-cemented soil and also increases the friction angle of soil. According to Sobhan et al. (2002), incorporation of plastic strips obtained from postconsumer milk or water container in soil–cement–fly ash mixture results improvement of toughness index, compressive strength, split tensile strength and to delay the propagation of tensile cracks of the composite. As per Phani Kumar et al. (2004), with increase in the fly ash content to the soil plasticity characteristics, swelling properties, hydraulic conductivity of the soil was decreased and compressive strength of soil was increased. Dutta and Sarda (2008) evaluated a comparative study of compaction and CBR behavior of stone dust reinforced with LDPE (0.1–0.8%) and HDPE (0.25–2%) waste plastic strips. They stated that the addition of LDPE and HDPE waste plastic strips in stone dust resulted in marginal and appreciable, respectively, increase in the CBR and the secant modulus. Babu and Chouksey (2011, 2012) observed the use of plastic waste as a reinforced material in soil. From the test results concluded that addition of plastic waste into soil, helps to increase the compressive strength and reduce the compressibility of soil. They also proposed an analytical model for stress–strain response of plastic waste mixed soil. Choudhary et al. (2010, 2014) and Jha et al. (2014) observed that the addition of plastic waste to the fly ash and waste recycled product results in increase in CBR value and secant modulus. A few researchers (e.g., Akbulut et al. 2007; Senol et al. 2013) also have done such type of investigation to a limited extent in this direction.

From the study of all the literature reviews, it has been observed that, all the literature reviews based on plastic waste there have lack of use of plastic waste obtained from usable water bottle into clayey soil. So an experimental work was carried out to find out the effect of plastic waste on strength of clayey soil and clay mixed with fly ash. Therefore, the present study will help in consuming the considerable quantity of waste plastic and fly ash, thereby reducing the environmental threat. In this study, after collecting usable water bottle from market cut in the form of strips of length 12 mm and width 4 mm after that plastic waste (0.5%, 1.0%, 1.5%, and 2.0% by weight) and fly ash (10%) were mixed to the clayey soil. A series of compaction, direct shear, UCS, and CBR were performed on clayey soil and clay–fly ash mixture. The experimental test results obtained from all the tests are presented and discussed in this present work.

2 Materials and Methodology

This chapter discusses the materials and experimental methodology which were used in this study. In this present study, soil, fly ash and plastic waste were used to find out the effect of plastic waste on soil mixture. Different laboratory tests were carried out to characterize the material and find out the effect plastic waste on clayey soil and clay–fly ash mixture. Basic properties of the soil, fly ash and soil-fly ash mixture were carried out in the laboratory as per IS: 2720 (Part 6) 1972, IS: 2720 (Part 40) 1977, IS: 2720 (Part 3) 1980, IS: 2720 (Part 4) 1985, IS: 2720 (Part 5) 1985. A brief description about the material and methodology of test is discussed below.

2.1 Material Used

Soil. Soil used in the present study was taken from Rourkela, Odisha. Soil, which was used in this project, is slightly yellowish color and with high degree of expansiveness. A series of crushing, oven drying, and grinding process was conducted on soil. Basic properties of soil are discussed in Table 1. It was observed that soil basically consists of sand = 10%, silt = 56%, and clay = 34%. The grain size curve showed that D_{10} = 0.0011 mm, D_{30} = 0.0021 mm, and D_{60} = 0.020 mm. From this value of uniformity

Table 1 Basic properties of soil

Soil properties	Value
Specific gravity	2.65
Liquid limit, %	55.00
Plastic limit, %	22.00
Plasticity index, %	33.00
Shrinkage limit, %	12.00
Free swell index, %	60
Maximum dry density, kN/m^3	17.30
Optimum moisture content, %	17.40
ISCS	CH

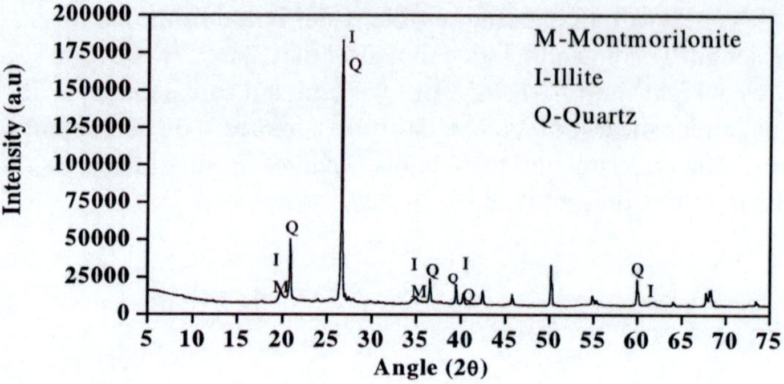

Fig. 1 XRD analysis of soil

coefficient (C_u) = 18.18 and coefficient of the curvature (C_c) = 0.20 < 1, it represents uniformly graded soil. From the X-ray diffractometer analysis (XRD), it was found that soil mainly consists of illite, montmorillonite, and quartz mineral are shown in Fig. 1. From scanning electron microscope (SEM), photographs observed that soil particles are circular and spherical in shape are shown in Fig. 3.

Fly Ash. In this study, fly ash was collected from the local vendor. It is slightly dark gray color. Basic properties of fly ash are described in Table 2. Nowadays, large amount of fly ash is generated from industry, therefore it is used in my project to reduce it by applying in geotechnical purpose. In this study, 10% fly ash was mixed to the clayey soil. The grain sieve analysis found that it mainly consists of sand = 9% and silt = 91%. From the curve, D_{10} = 0.004 mm, D_{30} = 0.008 mm, and D_{60} = 0.018 mm. From this value of uniformity coefficient (C_u) = 4.5 and coefficient of the curvature (C_c) = 0.888 < 1, it represents uniformly graded soil. From the XRD analysis, it has basically quartz, mullite, and magnetite minerals are shown in Fig. 2. From SEM, photographs observed that fly ash particles were rounded and spherical in shape are shown in Fig. 3.

Plastic Waste (PW). Plastic bottles obtained from usable water bottle were used in this study. These were collected from market and cut into pieces as a strip from of length 12 mm and width 4 mm. Plastic wastes at different percentage (0.5%, 1.0%, 1.5%, and 2.0% by dry weight) were mixed to the clayey soil and conducted the different test. To avoid the segregation, at first, plastic waste is randomly mixed to

Table 2 Basic properties of fly ash

Fly ash properties	Value
Specific gravity	2.32
Liquid limit, %	46.00
Free swell index, %	NA
Maximum dry density, kN/m^3	13.30
Optimum moisture content, %	24.70

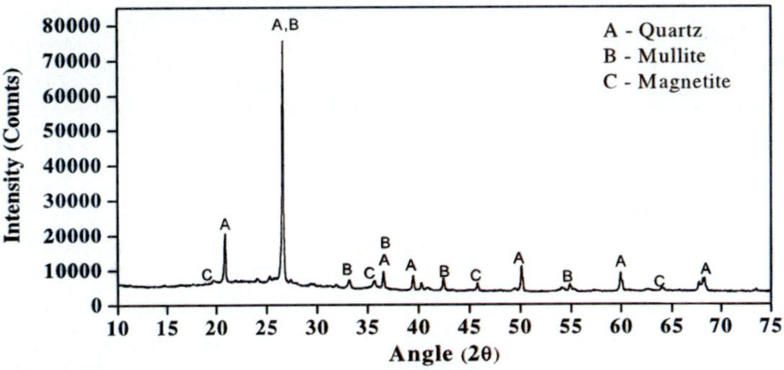

Fig. 2 XRD analysis of fly ash

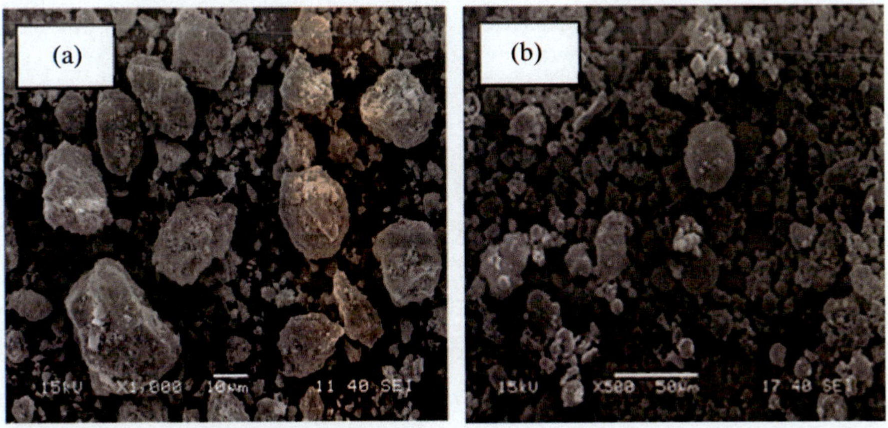

Fig. 3 SEM photographs. **a** Soil, **b** fly ash

the soil by hand mixing after that water was added. Basic properties of the plastic waste are shown in Table 3.

Soil–fly ash mixture. Fly ash is the non-swelling material. For that when fly ash was mixed to the soil, swelling properties of soil was reduced as well as Atterberg limit also be changed. In this present study, 10% fly ash by dry weight was mixed to

Table 3 Basic properties of plastic waste (Consoli et al. 2002; Khoury et al. 2008)

Plastic waste properties	Value
Specific gravity	1.06
Young modulus (E), GPa	2.0–3.0
Tensile strength, MPa	50–100
Linear strain, %	20–30
Density, kN/m^3	13.6

Table 4 Basic properties of soil–fly ash mixture

Soil properties	Value
Specific gravity	2.63
Liquid limit, %	51.00
Plastic limit, %	24.00
Plasticity index, %	27.00
Shrinkage limit, %	18.00
Free swell index, %	49
Maximum dry density, kN/m^3	16.60
Optimum moisture content, %	20.00

stabilize the soil. In this present study, Atterberg limit, FSI, and UCS test were also carried out by mixing 5 and 15% fly ash (by dry weight) to the soil. Test results show that soil with 10% fly ash provides good results as compared to 5 and 15% fly ash. For that, in this study 10% fly ash was taken. Effects of fly ash on soil are shown in Table 4.

2.2 *Experimental Methodology*

The objective of this present study is to find out the effect of plastic waste on the strength of clayey soil and clay–fly ash mixture. Basic material properties of soil and fly ash were determined as per IS 2720. To prepare the sample for every test, at first required amount of oven-dried soil and fly ash were taken and mixed thoroughly. Followed by to avoid the segregation of plastic waste, (0.5%, 1.0%, 1.5%, and 2.0%) as a strip form was randomly mixed with the soil and soil–fly ash mixtures by hand mixing after that water was added and mixed uniformly. The nomenclature of all samples which are used in the text below are shown in Table 5.

The prepared soil mixture is used for compaction, direct shear, unconfined compression strength (UCS) and California bearing ratio (CBR) test.

Determination of MDD and OMC value. In laboratory, the light compaction test was conducted as per IS: 2720 (Part 7)-1980, to find out the maximum dry density and optimum moisture content of a given soil sample. This test determines the optimum amount of water to be mixed with a soil in order to obtain maximum compaction for a given compaction effect (592 kJ/m^3). The soil with fly ash and plastic waste was compacted in a Proctor's mold by the standard procedure for light compaction by giving 25 blows to each layer. In this method, water amount was continuously increasing from 2 to 4% more than the previous water content and at different water content compaction test occurred to find out the dry density and moisture content of that soil sample. Repeat this process six to seven times. A curve between moisture content and dry density was drawn. From the curve, the water content at which dry

Table 5 Nomenclature of samples

Nomenclature	Soil (%)	Fly ash (%)	Plastic waste (%)
0.0% PW	100	0	0
0.5% PW	99.5	0	0.5
1.0% PW	99	0	1.0
1.5% PW	98.5	0	1.5
2.0% PW	98	0	2.0
FA + 0.0% PW	90	10	0
FA + 0.5% PW	89.5	10	0.5
FA + 1.0% PW	89	10	1.0
FA + 1.5% PW	88.5	10	1.5
FA + 2.0% PW	88	10	2.0

PW = Plastic waste

density will be maximum is called optimum moisture content and value of dry density corresponding to optimum moisture content is called maximum dry density.

Determination of shear parameter (c, φ) value. Direct shear test helps to determine the shear parameter like cohesion (*c*), friction angle (φ) of soil sample. In this test, a shear box of 6 cm × 6 cm × 2.5 cm was used, and the soil may be compacted at MDD and OMC obtained from light compaction test directly into the shear box after fixing the two halves of the shear box together by means of the fixing screws. After that box is placed into the shear apparatus, and test was conducted as per IS: 2720 (Part 13)-1986.

Determination of Unconfined Compressive Strength (UCS) value. The unconfined compressive strength (UCS) test was done on an unconfined cylindrical specimen (length = 10 cm and diameter = 5 cm; $L/D = 2$) of soil sample to find out the compressive strength of soil sample. Weight of the mixture was calculated by multiplying the MDD with the volume of sample, and soil sample was prepared at OMC. MDD and OMC values were calculated from light compaction test. Place calculated amount of mixture in UCS mold with the help of hydraulic jack, and then the specimen was extracted carefully to conduct the test. This test is strain controlled, and rate of strain is equal to 1.2 mm/min. Maximum load at which soil fails is called the compressive load, and dividing this load by area of specimen gives the compressive strength. In laboratory, unconfined compressive strength test was conducted by IS: 2720 (Part 10)-1991 at immediate and after 7, 14, and 28 days air curing periods.

Determination of California Bearing Ratio (CBR) value. California bearing ratio test is a penetration test, which was conducted to find out the bearing value of soil under unsoaked and soaked condition. In this test, CBR values are calculated at 2.5 and 5.0 mm penetration. Generally, CBR value of 2.5 mm penetration is more than the 5.0 mm penetration. If the CBR value for a penetration of 5.0 mm exceeds that for 2.5 mm, the test shall be repeated. If similar results follow, the bearing ratio shall be taken corresponding to 5 mm penetration for design. Weight of the soil mixture was calculated by multiplying the volume of CBR mold to the MDD value and sample was prepared at OMC. In laboratory, CBR test was conducted as per IS: 2720 (Part 16)-1987. In unsoaked condition, test was done immediately and in soaked condition put the CBR mold into water bath for 96 h for saturation after which the test was conducted.

3 Results and Discussions

Effects of plastic waste on strength criteria of clayey soil and clay–fly ash mixture are presented in terms of compaction, unconfined compression strength, shear strength, and CBR value. At first, plastic waste is mixed to the dry soil and soil–fly ash mixture by hand mixing to avoid the segregation of plastic waste, after that water was mixed to the soil–plastic mixture and conducted different test to find out the mechanical properties of soil mixture. Results obtained from different tests are discussed below.

3.1 Compaction Test

A series of compaction tests were performed of soil and soil–fly ash mixture with 0.5%, 1.0%, 1.5%, and 2.0% plastic waste and without plastic waste. The results of light compaction tests on soil and soil–fly ash mixture with and without plastic waste are presented in figures. Figures 4 and 5 show the light compaction curve of soil and soil–fly ash mixture with 0.5%, 1.0%, 1.5%, and 2.0% plastic waste and without plastic waste.

From the compaction test, Fig. 4 shows the continuous decrease of maximum dry density (MDD) along with an increase of optimum moisture content (OMC). Similar trend follows in case of soil–fly ash mixture. This is due to the replacement of soil with plastic waste, which has less specific gravity than soil and lubricating effect of adsorbed water by plastic waste also caused to decrease the maximum dry density of

Fig. 4 Light compaction curves for soil with plastic waste

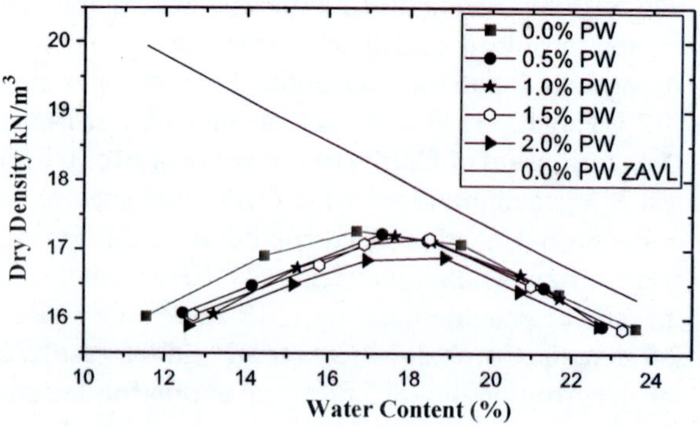

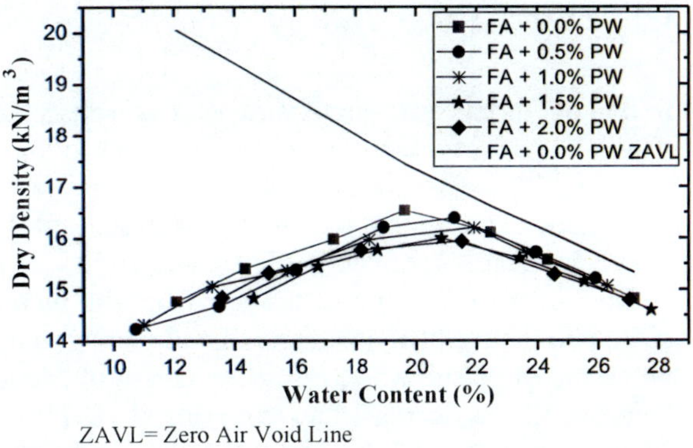

Fig. 5 Light compaction curves for soil–fly ash mixture with plastic waste

soil. The OMC and MDD value of different mix proportions are presented in Table 6.

Similar trend follows in case of soil–fly ash mixture. Figure 5 shows the continuous decrease of maximum dry density (MDD) along with an increase of optimum moisture content (OMC). This is because fly ash requires more water for inter-particle lubrication and at the same time replacement of soil with plastic waste.

It is also observed that mixing of 2% plastic in soil MDD value of soil decreased from 17.30 to 16.90 kN/m^3 and OMC value increased from 17.40 to 18.50%. Results also showed that mixing of 2% plastic waste and fly ash with soil MDD value of soil decreased from 17.30 to 15.90 kN/m^3 and OMC value increased from 17.40 to 21.40%.

Table 6 Variation of OMC and MDD of soil for different percentage of plastic waste and fly ash

S. No.	OMC, %	MDD, kN/m^3
0.0% PW	17.40	17.30
0.5% PW	17.50	17.20
1.0% PW	17.70	17.20
1.5% PW	18.20	17.10
2.0% PW	18.50	16.90
FA + 0.0% PW	20.00	16.60
FA + 0.5% PW	20.50	16.40
FA + 1.0% PW	20.80	16.30
FA + 1.5% PW	21.00	16.00
FA + 2.0% PW	21.40	15.90

3.2 Direct Shear Test

To find out the shear strength parameter as well as cohesion value and friction angle of soil, a series of direct shear test of soil and soil–fly ash mixture with 0.5%, 1.0%, 1.5%, and 2.0% plastic waste and without plastic waste were performed. The variation of cohesion value and friction angle of soil and soil–fly ash mixture with and without plastic waste are presented in Figs. 6 and 7.

The direct shear test results show that with increasing the percentage of plastic waste into the soil, friction angle of soil increased up to 1% plastic waste after that increasing of plastic waste results reduction of friction angle another side cohesion value of soil was decreased up to mixing of 1% plastic waste after that increased are shown in Figs. 6 and 7. In case of soil–fly ash mixture, friction angle of soil–fly ash

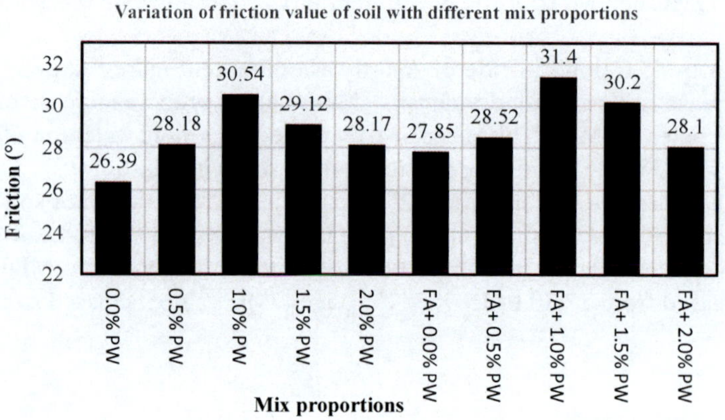

Fig. 6. Variation of friction angle of soil with different mix proportions

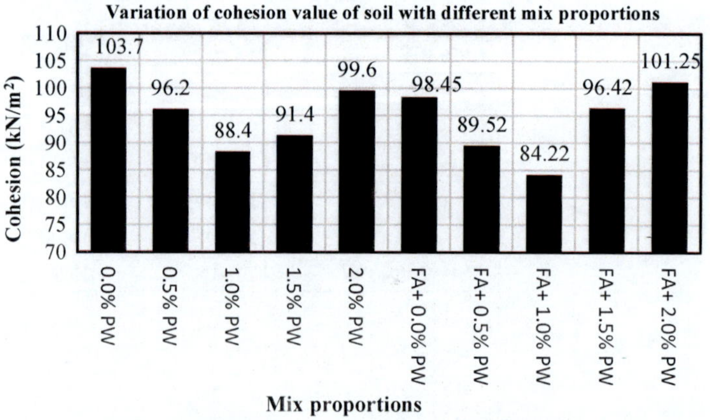

Fig. 7 Variation of cohesion value of soil with different mix proportions

mixture increased up to mixing 1% plastic waste to the soil–fly ash mixture after which friction angle decreased, and cohesion value of soil–fly ash mixture decreased up to mixing 1% plastic waste after which cohesion value increased.

Test results showed that mixing of 1% of plastic strip with soil friction angle increased from 26.39° to 30.54° and cohesion value decreased from 103.70 to 88.40 kN/m^2. This is due to good interlocking bond between soil and plastic waste and increased number of fibers intersecting the failure zone.

Other side due to mixing of 1% plastic strip and 10% fly ash with soil friction value of soil increased from 26.39° to 31.40° and cohesion value decreased from 103.70 to 84.22 kN/m^2. This is due to good interlocking bond between soil, fly ash, and plastic strips.

3.3 Unconfined Compressive Strength Test

A series of unconfined compression strength test were conducted to find out the unconfined compressive strength of soil and soil–fly ash mixture with 0.5%, 1.0%, 1.5%, and 2.0% plastic waste and without plastic waste at immediate, and after 7, 14 and 28 days curing period. The variation of compressive strength of different mixture is presented in the figures below. Fly ash is a pozzolanic material, which provides long-term reaction with the help of water, and this reaction is called pozzolanic reaction. This reaction helps to produce cementing compound on soil–fly ash mixture, for that strength of soil increases with combined effect of fly ash and plastic waste.

The UCS test results show that there is a good improvement in the strength of soil with the inclusion of plastic waste is shown in Fig. 8. Compressive strength of soil increases up to mixing of 1% plastic waste, and this is due to an increased number of fibers intersecting the failure zone, the tensile strength of fibers and better interlocking and intertwining of fibers with soil after which compressive strength decreased for replacement of soil with plastic waste.

Test results show a good improvement of compressive strength with curing periods when fly ash was mixed to the soil. In case of soil mixed with fly ash, there was a good improvement of compressive strength up to mixing of 1% plastic waste. This is due to formation of cementing agent with the help of fly ash and reinforcing effect of plastic waste. Beyond 1% of plastic waste caused reduction in compressive strength of soil–fly ash mixture. Table 7 shows the compressive strength of different mix proportions after immediate, 7, 14 and 28 days.

The test observed that due to addition of 1.0% plastic waste with soil, unconfined compressive strength increased by 394.5–560.7 kN/m^2 after 28 days air curing.

In another side addition of 1.0% plastic waste with soil and 10% fly ash, unconfined compressive strength increased by 394.5–880.2 kN/m^2 after 28 days air curing.

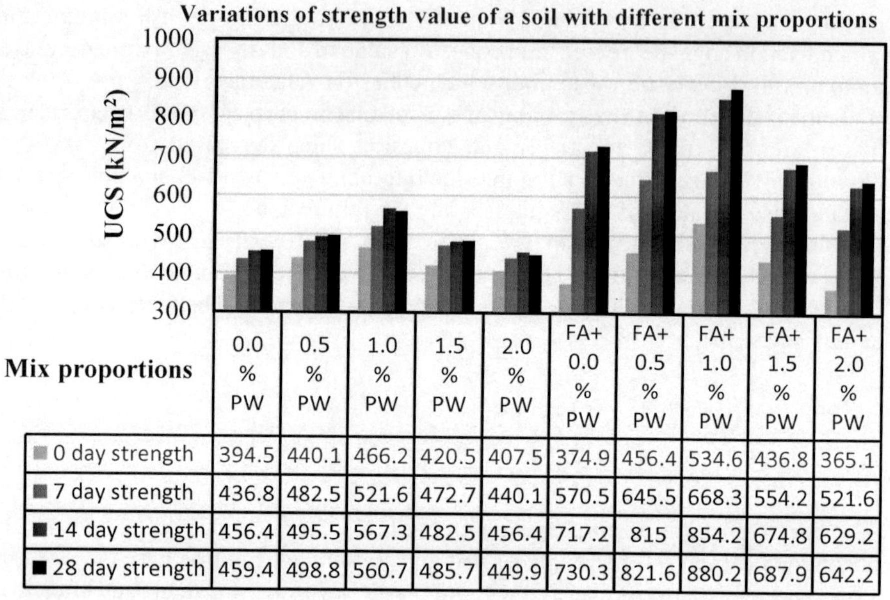

Fig. 8 Variation of compressive strength of different mix proportions at different curing period

Table 7 Compressive strength value of different mix proportions at different curing period

S. No.	Immediate strength (kN/m^2)	7 day strength (kN/m^2)	14 day strength (kN/m^2)	28 day strength (kN/m^2)
0.0% PW	394.5	436.8	456.4	456.4
0.5% PW	440.1	482.5	495.5	498.8
1.0% PW	466.2	521.6	567.3	560.7
1.5% PW	420.5	472.7	482.5	485.7
2.0% PW	407.5	440.1	456.4	449.9
FA + 0.0% PW	374.9	570.5	717.2	730.3
FA + 0.5% PW	456.4	645.5	815.0	821.6
FA + 1.0% PW	534.6	668.3	854.2	880.2
FA + 1.5% PW	436.8	554.2	674.8	687.9
FA + 2.0% PW	365.1	521.6	629.2	642.2

3.4 *California Bearing Ratio Test*

This present study involved to perform a series of laboratory CBR tests on soil and soil–fly ash mixture with 1.0 and 2.0% plastic waste and without plastic waste under unsoaked and soaked condition. This CBR value is most important to design the pavement or to design the thickness of pavement. CBR test results of different mix proportions are presented below.

Unsoaked Condition

Soaked Condition

Figures 9 and 10 show the load penetration curves of soil and soil–fly ash mixture with 1 and 2% plastic waste and without plastic waste under unsoaked and soaked condition, respectively. Test results observed that inclusion of plastic waste to soil helps to increase the CBR value of soil up to a certain limit after which the addition of plastic waste results in the reduction in CBR value is shown in Fig. 11.

CBR value of soil increased due to reinforcing effect of plastic waste, tensile strength of plastic waste and decreased for replacement of soil with plastic waste which have less specific gravity than soil. Inclusion of 1% of plastic waste with soil CBR value increased from 9.42 to 11.46% and 1.54 to 1.93% under unsoaked and soaked condition respectively.

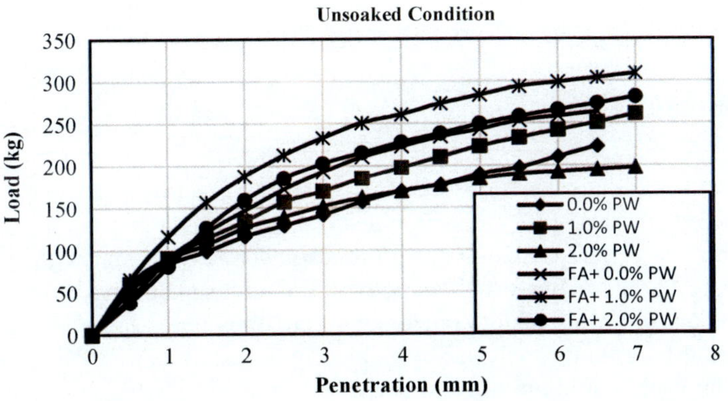

Fig. 9 Load penetration curve of soil with different mix proportions

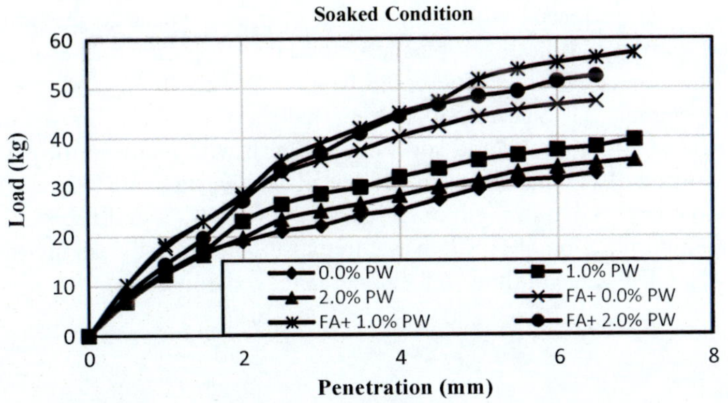

Fig. 10 Load penetration curve of soil with different mix proportions

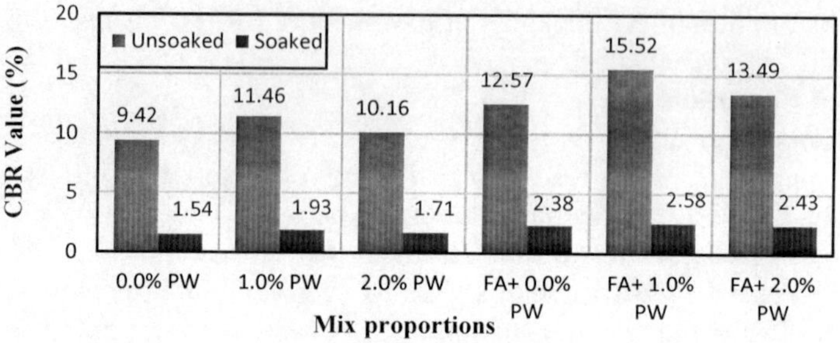

Fig. 11 Variation of CBR value of soil with different mix proportions under unsoaked and soaked condition

Another case due to mixing of 1% plastic strip and 10% fly ash with soil CBR value increased from 9.42 to 15.52% and 1.54 to 2.58% under unsoaked and soaked condition, respectively.

4 Conclusions

The effect of plastic waste on strength criteria of soil and soil–fly ash mixture has been studied in a series of compaction tests, unconfined compression strength tests, direct shear tests, and CBR tests. All the tests were carried out with 0.5%, 1.0%, 1.5%, and 2.0% plastic waste and without plastic waste. On the basis of experimental investigation following, major conclusions were drawn.

The Atterberg limits analysis shows that by the addition of 10% fly ash (by dry weight) into soil, liquid limit of soil decreased from 55 to 51%, plastic limit slightly increased, and shrinkage limit of soil increased from 12 to 18%. The free swell index of soil was reduced by 20% for inclusion of 10% fly ash into soil. Compaction characteristics show that optimum moisture content has increased and maximum dry density has decreased with the addition of plastic waste to the soil and soil–fly ash mixture. Addition of 1% plastic waste and 1% plastic waste + 10% fly ash with soil friction angle of soil increased from 26.39° to 30.54° and from 26.39° to 31.4°, respectively. UCS test results show that there is better improvement in the compressive strength of soil and soil mixed fly ash with inclusion of plastic waste. For the addition of 1% plastic waste and 1% plastic waste + 10% fly ash to the soil, unconfined compressive strength of soil has increased by 42 and 120% after 28 days air curing. Incorporation of plastic waste to the soil and soil–fly ash mixture helps to increase the CBR value both in soaked and unsoaked condition.

References

Akbulut S, Arasan S, Kalkan E (2007) Modification of clayey soils using scrap tire rubber and synthetic fibers. Appl Clay Sci 38(1–2):23–32. https://doi.org/10.1016/j.clay.2007.02.001

Babu GLS, Chouksey SK (2011) Stress-strain response of plastic waste mixed soil. Waste Manag 31(3):481–488. https://doi.org/10.1016/j.wasman.2010.09.018

Babu GLS, Chouksey SK (2012) Analytical stress-strain response of plastic waste mixed soil. J Hazard Toxis Radioact Waste 16(3):219–228. https://doi.org/10.1061/(ASCE)HZ.2153-5515.0000128

Choudhary AK, Jha JN, Gill KS (2010) A study on CBR behavior of waste plastic strip reinforced soil. Emirates J Eng Res 15(1):51–57

Choudhary AK, Jha JN, Gill KS, Shukla SK (2014) Utilization of fly ash and waste recycled product reinforced with plastic wastes as construction materials in flexible pavement. In: Geo-Congress 2014 Technical Papers, GSP 234 ASCE 2014, pp 3890–3902. https://doi.org/10.1061/9780784413272.377

Consoli NC, Montardo JIP, Prietto PDM, Pasa GS (2002) Engineering behavior of a sand reinforced with plastic waste. J Geotech Geoenviron Eng 128(6):462–472. https://doi.org/10.1061/(ASCE)10900241(2002)128:6(462)

Dutta R, Sarda V (2008) A comparative study of compaction and CBR behaviour of stone dust reinforced with waste plastic strips. Int J Geotech Eng 2(3):255–264. https://doi.org/10.3328/IJGE.2008.02.03.255-264

IS: 2720 (Part 6) (1972) Methods of test for soils, determination of shrinkage factors. Bureau of Indian Standards, New Delhi

IS: 2720 (Part 40) (1977) Methods of test for soils, determination of free swell index of soils. Bureau of Indian Standards, New Delhi

IS: 2720 (Part 7), (1980) Methods of test for soils, determination of water content and dry density using light compaction. Bureau of Indian Standards, New Delhi.

IS: 2720 (Part 3), (1980) Methods of test for soils, determination of specific gravity (Reaffirmed 1987). Bureau of Indian Standards, New Delhi

IS: 2720 (Part 4) (1985) Methods of test for soils, determination of grain size analysis. Bureau of Indian Standards, New Delhi.

IS: 2720 (Part 5) (1985) Methods of test for soils, determination of liquid and plastic limit. Bureau of Indian Standards, New Delhi

IS: 2720 (Part 13) (1986) Methods of test for soil, direct shear test. Bureau of Indian Standards, New Delhi

IS: 2720 (Part 16) (1987) Methods of test for soils, determination of california bearing ratio. Bureau of Indian Standards, New Delhi, India

IS: 2720 (Part 10) (1991) Methods of test for soils, determination of unconfined compressive strength. Bureau of Indian Standards, New Delhi

Jha JN, Choudhary AK, Gill KS, Shukla SK (2014) Behavior of plastic waste fiber-reinforced industrial wastes in pavement applications. Int J Geotech Eng 8(3):277–286. https://doi.org/10.1179/1939787914Y.00000000

Khoury NAM, Khoury CSM, Abousleiman Y (2008) Soil fused with recycled plastic bottles for various geo-engineering applications. In: GeoCongress 2008: Geotechnics of Waste Management and Remediation, pp 336–343. https://doi.org/10.1061/40970(309)42

Phani Kumar BR, Sharma RS (2004) Effect of fly ash on engineering properties of expansive soils. J Geotech Geoenviron Eng 130(7):764–767. https://doi.org/10.1061/(ASCE)1090-0241(2004)130:7(764)

Senol A, Etminan E, Yildirim H, Olgun CG (2013) Improvement of high and low plasticity clayey soils using polypropylene fibers and fly ash. In: ICSDEC 2012, ASCE 2013, pp 500–509. https://doi.org/10.1061/9780784412688.060

Sobhan K, Mashnad M (2002) Tensile strength and toughness of soil-cement-fly-ash composite reinforced with recycled high-density polyethylene strips. J Mater Civ Eng 14(2):177–184. https://doi.org/10.1061/(ASCE)0899-1561(2002)14:2(177)

Optimization of Buffer Layer Thickness Over Black Cotton Soil

M. Vinoth and P. S. Prasad

Abstract Black cotton (BC) soils are inorganic clays of medium to high compressibility and form a major soil group in India. They are characterized by high shrinkage and swelling properties. Due to its peculiar characteristics, it forms a very poor foundation material for road construction. R&D efforts have been made to improve the strength characteristics of BC soil by many researchers. Soil replacement with buffer layer remains is a very common technique but still there is no reliable formula or technique is available to estimate the required replacement depth. Therefore, this paper intends to optimize the replacement thickness of buffer layer required to maintain the heave within the acceptable range and to evaluate efficacy and impact of placing a very thin layer of stabilized BC soil (BC soil + 3% lime) or cohesive non-swelling material (Murram/Murram + 3%Lime) or impermeable geomembrane between the buffer layer and BC soil using laboratory studies.

Keywords Black cotton soil · Buffer layer · Stabilization

1 Introduction

Black cotton soils occur in the Central, Southern and Western parts of India. BC soils occurring above water table undergo volumetric changes with changes in moisture content. Increase in water content causes swelling and loss of strength, and decrease in moisture content brings about soil shrinkage. It has been observed that after drying, the BC soil develops cracks of varying depth. As a result of wetting and drying process, vertical movement takes place in the soil mass. Swelling and softening of BC soil cause excessive deflections, loss of subgrade support, pavement unevenness and longitudinal cracking of road pavements. Gradual intrusion of BC soil into subbase layer invariably leads to failure of drainage layer.

M. Vinoth (✉) · P. S. Prasad
CSIR-Central Road Research Institute, New Delhi, India
e-mail: vinothm.27@gmail.com

M. Latha Gali and R. R. P. (eds.), *Problematic Soils and Geoenvironmental Concerns*, Lecture Notes in Civil Engineering 88,
https://doi.org/10.1007/978-981-15-6237-2_47

BC soil predominantly consists of montmorillonite in structure (Terzaghi and Peck 1987). Mostly, BC soil particles have a size less than 0.001 mm. "According to Seehra (2008), at the liquid limit, the volume change is of the order of 200–300% and results in swelling pressure as high as 800–1000 kPa". Many researchers "Arafat and Ebid (2015), Eberemu et al. (2012), Wubshet and Tadesse (2014), Manikandan and Moganraj (2014), Singh and Vasaikar (2013), Tapas (2014), Kavish et al. (2014), Kawade (2014), Srinivas et al. (2016), Hammitt (1973), Kundan et al. (2015), Neelesh and Suneet (2016), Chandra and Venkatarathnam (2016), Rishi et al. (2016) and Stalin et al. (2010)" worked in this field to improve the strength characteristics of BC soil with technologies such as.

- Soil replacement/providing buffer layer (granular blanket)
- Alter the soil—with the addition of lime, cement, crushed glass cullet or other admixture
- Cohesive non-swelling soil (CNS) layer
- Encapsulation
- Heat treatment
- Load the soil to sufficient pressure intensity to balance swell pressure
- Granular piles, etc.

However, among the available options, soil replacement with buffer layer remains most common, economical and easily implementable solution if granular material is available locally. Granular layer reduces heave effect by reducing the depth of the swelling layer. Therefore, in this paper effect of varying the thickness of buffer layer with river sand, impact of very thin cohesive non-swelling layer (with and without stabilization) and geomembrane between the buffer layer and BC soil has been studied.

2 Materials

Black cotton soil and river sand were collected from Amaravati, Andhra Pradesh, India. Murram 125 micron thick geomembrane, 100% pure lime and OPC 45 grade cement were used in the present study. Physical properties and oxide composition of BC soil, river sand (RS) and murram (M) are given in Tables 1 and 2, respectively. Grain size distribution curves are shown in Fig. 1.

3 Methodology

The soil samples were compacted in layers, first BC soil was compacted to 95% corresponding to maximum dry density and optimum moisture content on top of it. Murram and river sand were compacted to 97% of MDD and OMC. Compaction of samples was carried out using static loading. Typical compaction procedure adopted

Table 1 Physical properties of BC soil, river sand and murram

Description	BC soil	RS	M
Specific gravity	2.70	2.64	2.66
Liquid limit (%)	78.00	NP	39
Plastic limit (%)	36.50	NP	18
Shrinkage limit (%)	12.97	–	–
I.S. classification	CH	SW	SW-SC
Maximum dry density (kN/m^3)	17.00	18.00	18.50
Optimum moisture content (%)	15.50	10.00	15.00
Free swell index (%)	74.36	–	–
Gravel	1.00	0.50	11.00
Sand (%)	5.00	98.50	83.00
Silt and clay (%)	94.00	1.00	6.00

Table 2 Oxides of BC soil, river sand and murram

Description	BC soil	RS	M
Silica (SiO_2)	68.86	89.91	46.19
Alumina (Al_2O_3)	17.44	7.19	16.71
Ferrous oxide (FeO)	8.19	–	35.47
Magnesium oxide (MgO)	3.46	1.71	–
Calcium oxide (CaO)	2.05	1.19	1.63

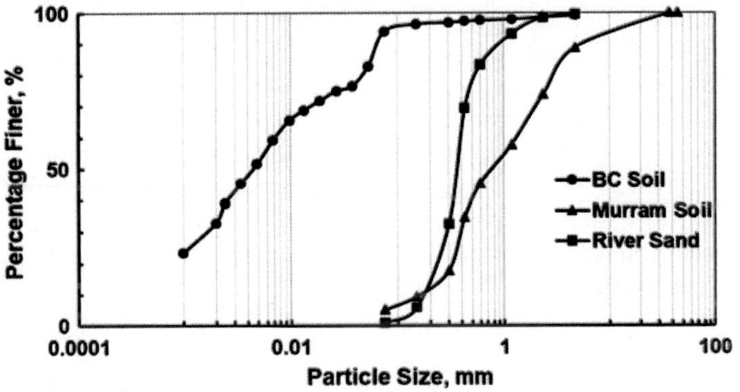

Fig. 1 Grain size distribution curves of soils used in this study

in the sample preparation is as follows: first BC soil was compacted to the required density with appropriate spacer arrangements, over which geomembrane was placed and sealed with wax. On top of geomembrane, stabilized murram layer was placed and finally on top of murram layer, river sand was placed and compacted to the required density. Geomembrane was cut into a size equal to the inner diameter of

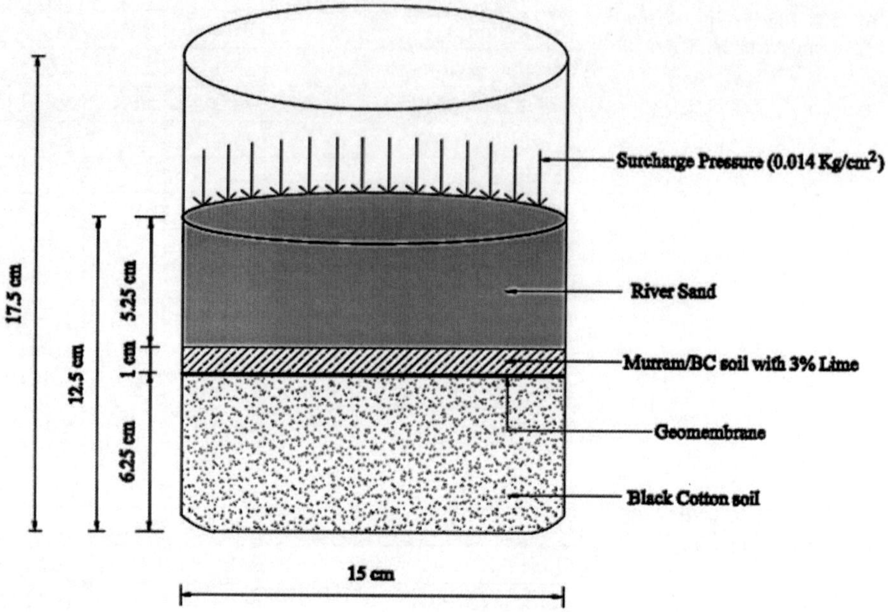

Fig. 2 Typical arrangement of samples in layer inside CBR mould

the mould. Typical experimental set-up is shown in Fig. 2. Expansion ratio (ER) was determined as per IS: 2720 (Part 16) 1987. The samples were submerged in water with a surcharge pressure of 0.014 kPa. At regular intervals, readings were noted down using the dial gauges and taken until equilibrium values reached.

Two samples were stabilized with 3% lime, in that one sample was cured in sand bath for 7 days and the second one was not cured for 7 days it was immediately saturated. Uncured sample was used to study the behaviour of stabilized sample in worst case condition. Subsequent to the completion of expansion ratio test, CBR test was conducted on these samples as per IS: 2720 (Part 16) 1987. After conducting CBR test, moisture content of BC soil was determined at three depths (top, middle and bottom) as per IS: 2720 (Part II) 1973.

3.1 Combinations

In order to optimize the buffer layer thickness, six different combinations have been prepared and tested for their performance. In the first case, BC soil was kept for the full height, and in the subsequent sample combinations, bottom half of the sample was filled with BC soil and variations were made in the top half of the sample height only. The main aim of the study is to control the swell by reducing the BC soil depth

Table 3 Sample combination details

S. No.	Sample ID	Composition	Height of each material (cm)
1	BC	100% BC soil	12.5
2	BC + RS	50% BC soil + 50% river sand	6.25 + 6.25
3	BC + M + RS	50% BC soil + 8% murram + 42% river sand	6.25 + 1 + 5.25
4	BC + BC3%L + RS (UC)	50% BC soil + 8% murram with 3% lime (uncured) + 42% river sand	6.25 + 1 + 5.25
5	BC + BC3%L + RS (C)	50% BC soil + 8% murram with 3% lime (cured) + 42% river sand	6.25 + 1 + 5.25
6	BC + G + M + RS	50% BC soil + geomembrane + 8% murram + 42% river sand	6.25 + 1 + 5.25

within the depth of influence and cutting off the water percolation from the ground surface. Details of these six combinations are given in Table 3.

4 Results and Discussion

Expansion ratio (ER) and CBR values of all the six combinations have been given in Table 4. From Table 4, it can be seen that without any buffer layer, BC soil swells more and has the least CBR value. This swell has to be controlled by using buffer layers with different combinations. In the first combination with river sand as buffer layer (BC + RS), the ER was brought down by 36% and CBR has been enhanced by five times. Still the swell has not been brought under control. In the second combination in addition to the river sand, a thin layer of murram has been introduced

Table 4 Expansion ratio and CBR for the different combination of samples

Sample ID	Expansion ratio (ER)	CBR after 25 days of soaking
BC	25.95	1.1
BC + RS	16.5	5.5
BC + M + RS	13.56	3
BC + BC3%L + RS (UC)	12.82	4.6
BC + BC3%L + RS (C)	9.09	8.2
BC + G + M + RS	4.85	4.9

which has a low permeability value. Using this BC + M + RS combination, ER value was reduced by 47.7% and CBR increased by 2.7 times. Marginal reduction in ER value can be noticed from earlier combination to this one but still not completely controlled.

In the next combination, murram was replaced with stabilized BC soil and over which river sand was filled. Two arrangements were tried within this combination: one sample was allowed for curing, and the other one was not allowed for curing. BC + BC3%L + RS (UC) and BC + BC3%L + RS (C) samples reduced ER by 50.5% and 64.4%, respectively. The increase in CBR value for the samples BC + BC3%L + RS (UC) and BC + BC3%L + RS (C) is 4.1 and 7.4 times. It can be seen that cured sample performed better than uncured sample. With all these combinations still, ER has not been brought down to negligible quantity. In the last combination, geomembrane was used between buffer layer and BC soil. ER value by using geomembrane in the combination (BC + G + M + RS) was down by 81.3%, and CBR was increased by 4.4 times. The plot between expansion ratio and time in days has been shown in Fig. 3.

From Fig. 2, it can be noted that trend of all the samples is similar. In all the combinations, 90% of the swell happens in the initial few days of saturation. There is a huge reduction in ER between the samples BC + M + RS and BC + G + M + RS, and this can be well explained using the moisture content data obtained after the CBR test. A plot between the depth of sample and moisture content has been shown in Fig. 4.

From Fig. 4, we can observe that moisture content at the top of samples BC + M + RS and BC + G + M + RS is 28.73% and 21.75% against the initial moisture content of 15%. So, from this, it can be seen that the geomembrane acted as a water barrier between BC soil and buffer layer and stopped water entering the BC soil and eventually controlled swell and reflected in low ER value. This can be validated by

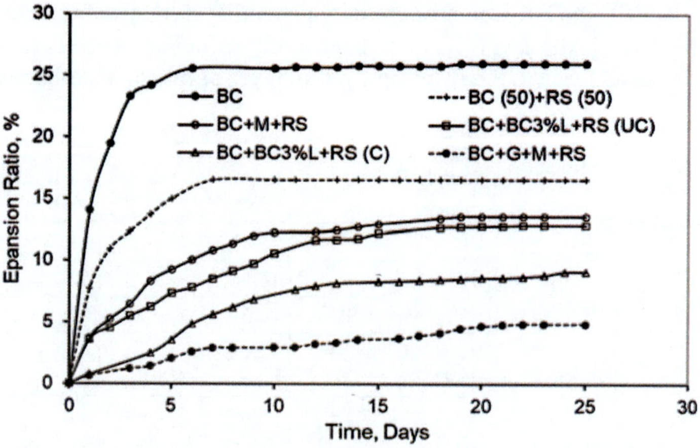

Fig. 3 Mould plot between ER and time for all samples

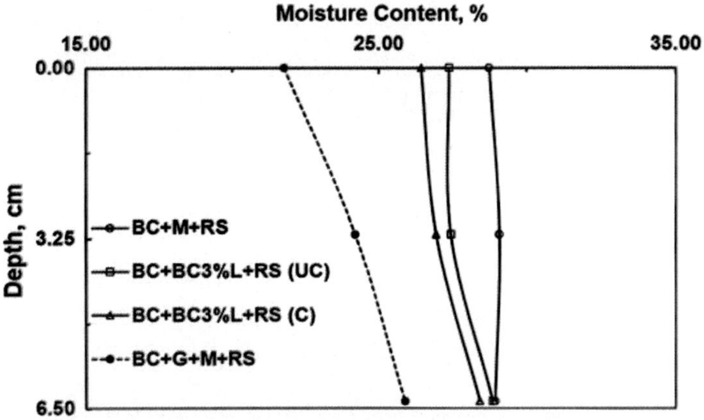

Fig. 4 Plot between moisture content and depth of sample

looking at the moisture content values of other two samples BC + BC3%L + RS (UC) and BC + BC3%L + RS (C) 27.38% and 26.43%, respectively, which are without geomembrane.

5 Conclusion

From the expansion ratio and CBR test conducted on BC soil with varied combinations of material and their thickness in buffer layer, the following conclusion are drawn.

- Sample with geomembrane (BC + G + M + RS) kept in between the BC soil and buffer layer performed better by controlling the swell. This combination not only controlled swell but all increased the CBR by 1.6 times than the sample without geomembrane (BC + M + RS).
- Upon curing the stabilized sample, better results were obtained. It is found that curing of stabilized sample must be carried out in order to get more effective outcomes.
- Only by adding buffer layers (BC + RS and BC + M + RS) did not reduce the ER by much. A water barrier (geomembrane) needs to be introduced in order to keep the swell in check along with the buffer layer.

Further studies are underway in optimizing the buffer layer thickness to completely nullifying the swell of BC soil.

Acknowledgements The author would like to thank the director of Central Road Research Institute for his permission to publishing this paper.

References

Arafat H, Ebid AM (2015) Optimum replacement depth to control heave of swelling clays. Int J Eng Innov Technol 4(9):73–81

Chandra PS, Venkatarathnam G (2016) Determining of thickness of cohesive non swelling soil layer for canal lining. Int J Mag Eng Technol Manage Res 3(5):629–634

Eberemu et al (2012) The geotechnical properties of black cotton soil treated with crushed glass cullet. Niger J Technol Res 23–30

Hammitt GM II, Richard GA (1973) Membrances and encapsulation of soil for control of swelling. In: Proceedings of workshop on expansive clays and shales in highway design and construction

IS: 2720 (Part 16) (1987) Laboratory determination of CBR. The Bureau of Indian Standards, New Delhi

IS: 2720 (Part II) (1973) Determination of water content. The Bureau of Indian Standards, New Delhi

Kavish SM et al (2014) Analysis of engineering properties of black cotton soil & stabilization using by lime. Int J Eng Res Appl 4(5):25–32

Kawade S et al (2014) Stabilization of black cotton soil with lime and geo-grid. Int J Innov Res Adv Eng 1(5):87–92

Kundan M et al (2015) Strength and settlement studies of black cotton soil reinforced with granular pile. Int J Eng Sci Res Technol 4(4):621–630

Manikandan AT, Moganraj M (2014) Consolidation and rebound characteristics of expansive soil by using lime and bagasse ash. Int J Res Eng Technol 3(4):403–411

Neelesh R, Suneet K (2016) A review on soil stabilization using RBI grade-81. Int Res J Eng Technol 3(7):213–217

Rishi S et al (2016) Effect of woven polyester geotextile on the strength of black cotton soil. Int J Innov Res Sci Eng Technol 5(7):12402–12408

Seehra SS (2008) Practical problems of highway construction in black cotton soil areas and in-place remedial measures: a case study. NBM Media Construction Portal. NBM Media Construction Portal

Singh S, Vasaikar HB (2013) Stabilization of black cotton soil using lime. Int J Sci Res 4(5):2090–2094

Srinivas K et al (2016) A Study on Improvement of expansive soil by using CNS (cohesive non swelling) layer. Int J Innov Res Technol 3(3):54–60

Stalin VK, Ravi E, Arun Murugan RB (2010) Control of swell: shrink behaviour of expansive clays using geosynthetics. In: Proceedings of Indian geotechnical conference, IIT Bombay, vol 1, pp 15–18

Tapas D (2014) Soil improvement by using jute geotextile and sand. Int J Sci Eng Technol 3(7):880–884

Terzaghi K, Peck RB (1987) Soil mechanics in engineering practice, 2nd edn. McGraw Hill, New York

Wubshet M, Tadesse S (2014) Stabilization of expansive soil using bagasse ash & lime. J EEA 32:21–25

Soil Stabilization Using Combined Waste Material

Uma Kant Gautam, Kumar Venkatesh, and Vijay Kumar

Abstract Weak soil is very problematic soil, and it has shrinking and swelling properties which can damage the structure constructed over it. Waste material on the daily basis is likely to become a problem for disposal. It creates environmental contamination and health risks. Hence, the utilization of waste material in the stabilization of weak soil effectively minimizes the negative effect on the environment. In this paper, the aim is to stabilize the weak soil using combinations of waste material. The waste materials used for the study are stone dust and solid waste from silica sand beneficiation plant. Stone dust is coming from polishing, cutting of stones, and cruising process during rock quarrying activities. Solid waste from silica sand beneficiation plant is a granular material, and it contains quartz and very less amount of clay, coal, and other minerals. Stone dust and solid waste from silica sand beneficiation plant are combined in different proportion with weak soil. Geotechnical properties of weak soil individually and in combination with varying proportion were investigated. The standard Proctor test and the California bearing ratio test have been performed. The results of these tests resemble that the combination of stone dust and solid waste from silica sand beneficiation plant is very effective for stabilizing the weak soil.

Keywords Stabilization · Solid waste of silica sand beneficiation plant · Stone dust · Combinations

U. K. Gautam (✉) · K. Venkatesh · V. Kumar
Department of Civil Engineering, MNNIT Allahabad, Allahabad, Uttar Pradesh, India
e-mail: umakantgautam24@gmail.com

K. Venkatesh
e-mail: venkatesh@mnnit.ac.in

V. Kumar
e-mail: vkr@mnnit.ac.in

M. Latha Gali and R. R. P. (eds.), *Problematic Soils and Geoenvironmental Concerns*, Lecture Notes in Civil Engineering 88,
https://doi.org/10.1007/978-981-15-6237-2_48

1 Introduction

Expansive soils are generally weak soils like black cotton soil, which are highly compressible clayey soil, grey to blackish in colour (Patidar 2014). They contain montmorillonite clay mineral which has highly expansive characteristics. It is highly sensitive to moisture changes, and if used as subgrade material, it tends to compress (Ali 2011). Different types of damages in the form of undulation, cracking, settlement, etc., are experienced by the buildings, sewer and water lines, irrigation canals, and roads, etc. Weak soil highly absorbs water to swell, becomes soft and loose strength, shrinks in volume, and develops cracks when it became dry during summer season (Ramadas 2010). These properties make them poor as construction material and for the foundation. Hence, the subgrade and its undesirable properties need to be modified by using a suitable stabilization technique. In subgrade construction of highway works, utilizing all types of natural materials is, mostly, unavoidable due to technical, economic, and environmental considerations (Ferber 2009). Subgrade properties control the structural design of highway pavements system (Shafabakhsh 2014). Soil stabilization is the process to provide strength to soil. When the admixture are added in the soil, the soil properties has been improved due to physical and chemical changes in the soil. Soil stabilization increases the geotechnical properties of soil, strength, bearing capacity, factor of safety against slope failure, earthen dams, and levees sliding. It reduces the swelling properties and plasticity index (Al-Rawas 2005). In soil stabilization, the use of industrial waste poses a disposal problem. Over last few years, environmental issues had affected humans to use waste materials as the substitute of some construction materials. Geotechnical engineers have examined waste materials to use in soil stabilization and improve the engineering properties of soil. Using waste material in soil stabilization may help alleviation of the environmental issues and also proves to be economical. The main objective of this study is the characterization of virgin soil and waste materials, combination of waste materials to examine the enhancement in the soil properties. The combination of waste materials was mixed with virgin soil, and the effect of stabilization on California bearing ratio value was observed. An enormous improvement of California bearing ratio value of soil is obtained which specifies the cost effectiveness of construction of pavement.

2 Materials and Methodology

2.1 Characterization of Virgin Soil

The soil sample had been collected from Chitrakoot, Uttar Pradesh, India, and various properties of virgin soil were determined in the geotechnical laboratory, MNNIT, Allahabad. Soil properties were shown in Table 1.

From Fig. 1 grain size analysis curve of virgin soil, Cu and Cc are found to be 17.14 and 0.58, respectively.

Table 1 Properties of virgin soil

Properties	Value
Specific gravity	2.742
Liquid limit (%)	40.29
Plastic limit (%)	14.09
Coefficient of curvature	0.5802
Coefficient of uniformity	17.14
Percentage fines	87.84
Classification (IS)	CI
OMC (%)	14.2
MDD (g/cc)	1.764
Free swell index (%)	55.56
CBR (%)	2.85

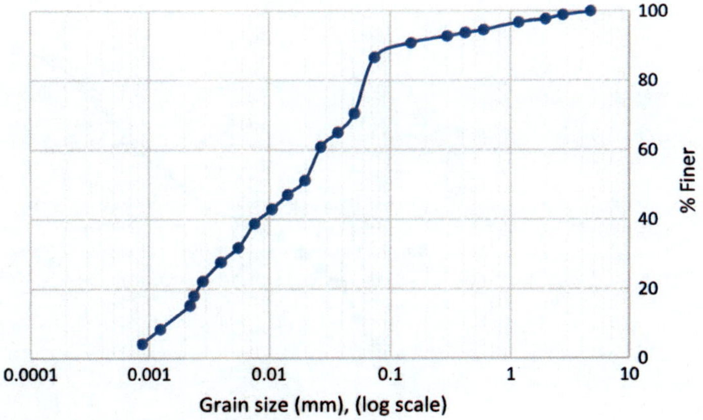

Fig. 1 Grain size distribution curve of virgin soil

The maximum dry density and optimum moisture content of virgin soil were evaluated by the compaction curve between dry density and water content. From Fig. 2, maximum dry density and optimum moisture content were found to be 1.764 g/cc and 14.2%, respectively.

From Fig. 3, the CBR value at 2.5 mm and 5 mm penetration was 2.85% and 2.61%, respectively. Hence, the CBR value of virgin soil was 2.85%.

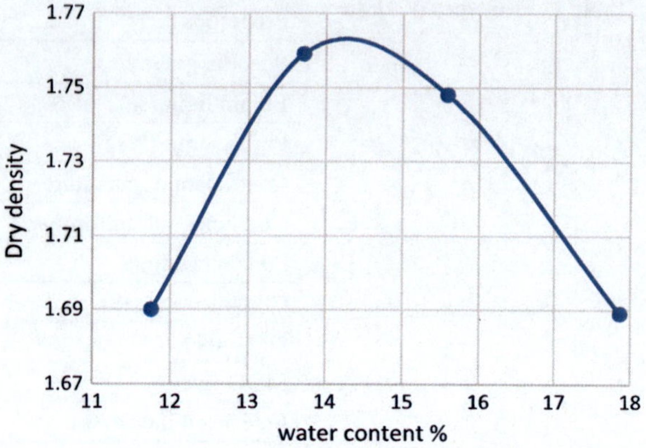

Fig. 2 Dry density versus water content curve for virgin soil

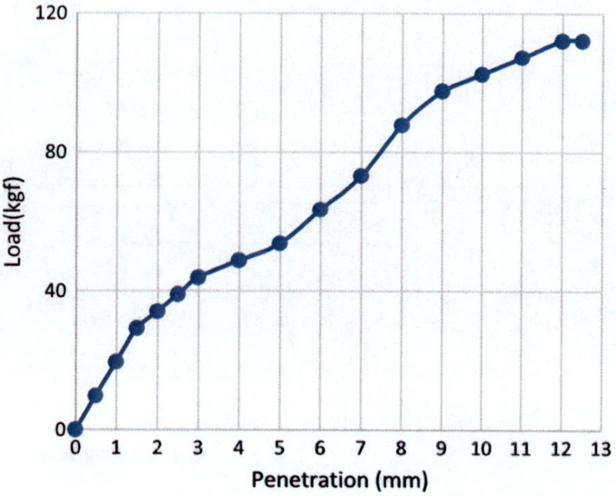

Fig. 3 Load versus penetration curve for virgin soil

2.2 *Stone Dust*

Stone dust is collected from SH-5 Markundi, Uttar Pradesh, India. It has latitude of 24.621081 °N and longitude 83.04703 °E. Various properties of stone dust determined in the geotechnical laboratory, MNNIT, Allahabad, were shown in Table 2.

From Fig. 4 grain size analysis of stone dust, Cu and Cc are found to be 3.556 and 1.255, respectively.

2.3 *Solid Waste from Silica Sand Beneficiation Plant*

The solid waste from silica sand beneficiation plant was taken from MMPL, Allahabad plant, Allahabad, Uttar Pradesh, India, and Various properties determined in geotechnical laboratory are shown in Table 3.

From Fig. 5 grain size analysis of solid waste from silica sand beneficiation plant, Cu and Cc are found to be 26.89 and 2.76, respectively.

2.4 *Methods*

To evaluate the effect of waste material combinations as a stabilizing additive in virgin soils, a series of tests were conducted on several combinations of waste materials and virgin soil in varying proportions by weight. During the performance of the experiments, the Indian Standard code had been followed (IS 1498 1970; IS 2720

Table 2 Properties of stone dust

Properties	Value
Specific gravity	2.655
D_{10} (mm)	0.183
D_{30} (mm)	0.388
D_{60} (mm)	0.655
Coefficient of curvature	1.255
Coefficient of uniformity	3.556
Percentage fines	3.3
Plastic limit (%)	Non
Classification (IS)	SP

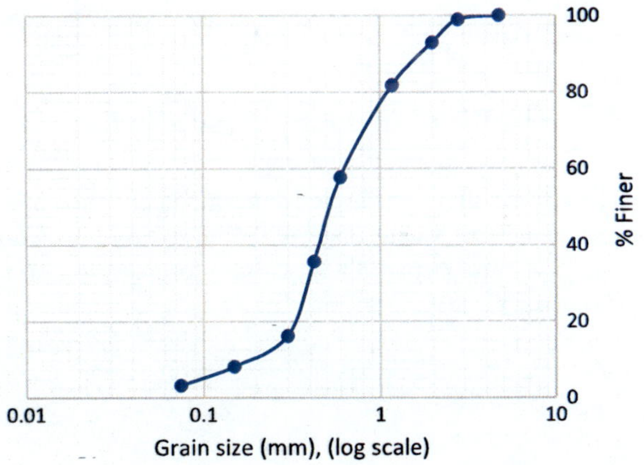

Fig. 4 Grain size distribution curve of stone dust

Table 3 Properties of solid waste from silica sand beneficiation plant

Properties	Value
Specific gravity	2.662
D_{10} (mm)	0.003
D_{30} (mm)	0.032
D_{60} (mm)	0.1027
Coefficient of curvature	2.75
Coefficient of uniformity	26.89
Percentage fines	48.59
Plastic limit (%)	Non

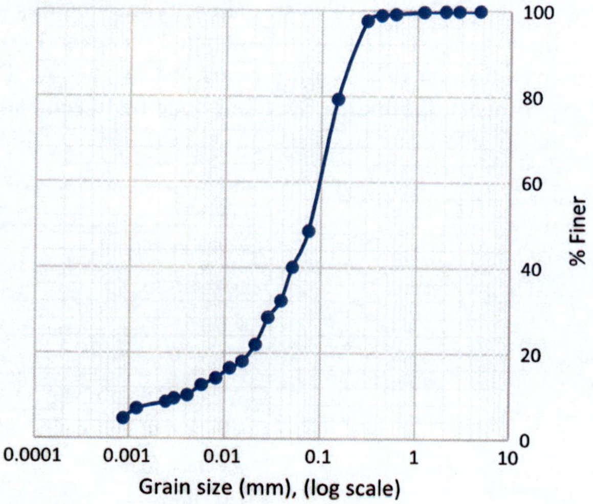

Fig. 5 Grain size distribution curve for solid waste from silica sand beneficiation plant

(Part 16) 1979; IS 2720 (Part 3, sec. 1) 1980; IS 2720 (Part 4) 1985; IS 2720 (Part 8) 1980).

2.5 Details of Sample Preparation

The virgin soil properties are determined and then mixed in given samples at different percentages of waste material combinations by weight of total solids. Totally ten samples were prepared with varying preparation on waste materials combination as shown in Table 4.

3 Results and Discussion

California bearing ratio test was performed on virgin soil with five varying proportion of waste material combination. Solid waste from silica sand beneficiation plant was added to soil keeping it fixed at 5%, but stone dust was varied as 5, 10, 15, 20 and 25%, respectively. Load versus penetration curve is shown in Fig. 6.

California bearing ratio test was performed on virgin soil with four varying proportion of waste material combinations. Stone dust was added to soil keeping it fixed at

Table 4 Details of combined proportions mixed with virgin soil specimen

Name of the sample	Virgin soil (%)	Solid waste (%)	Stone dust (%)
Virgin soil	100	0	0
S-5 D-5	90	5	5
S-5 D-10	85	5	10
S-5 D-15	80	5	15
S-5 D-20	75	5	20
S-5 D-25	70	5	25
D-5 S-10	85	10	5
D-5 S-15	80	15	5
D-5 S-20	75	20	5
D-5 S-25	70	25	5

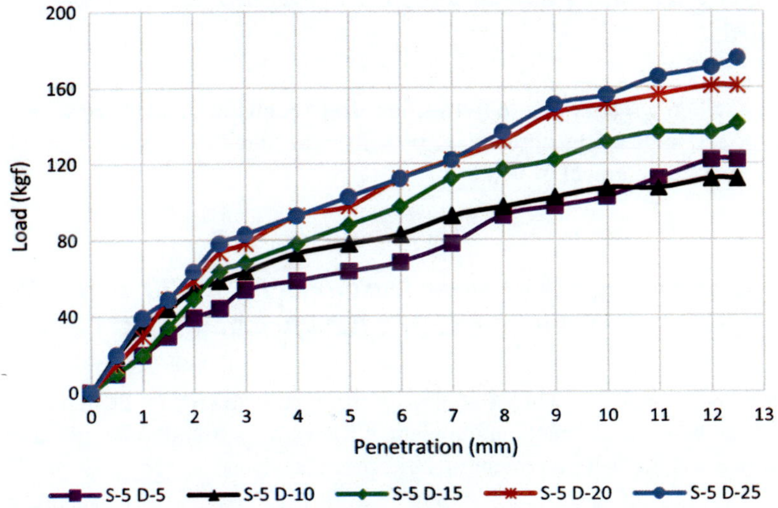

Fig. 6 Load versus penetration curve for various samples

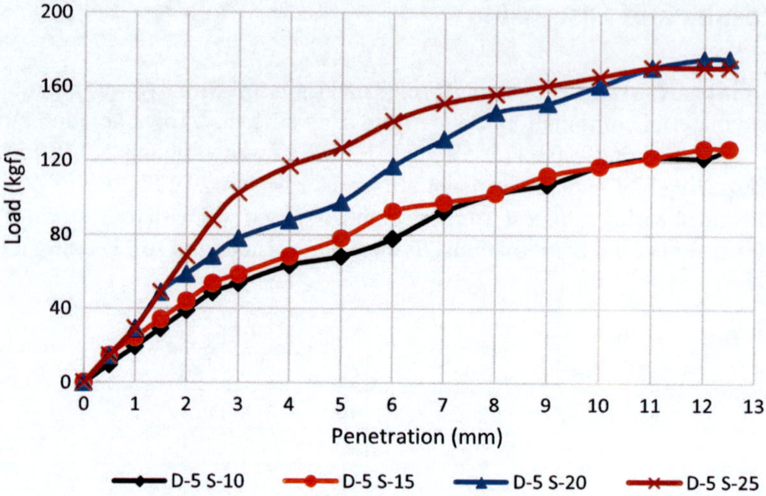

Fig. 7 Load versus penetration curve for various samples

5% while solid waste from silica sand beneficiation plant was added by varying as 10, 15, 20 and 25%, respectively. Load versus penetration curve is shown in Fig. 7.

3.1 After Mixing 5% of Solid Waste from Silica Sand Beneficiation Plant and (5, 10, 15, 20 and 25%) Stone Dust

In Fig. 8, when 5% solid waste from silica sand beneficiation plant was mixed with (5, 10, 15, 20 and 25%) stone dust, CBR value was found to increase from 2.85% to 5.69%, i.e. net increment of 99.65%.

3.2 After Mixing 5% of Stone Dust and (10, 15, 20, and 25%) Solid Waste from Silica Sand Beneficiation Plant

Figure 9 presented that when 5% stone dust is mixed with (10, 15, 20 and 25%) solid waste from silica sand beneficiation plant, CBR value is found to be increased from 2.85% to 6.4%, i.e. net increment of 124.56%.

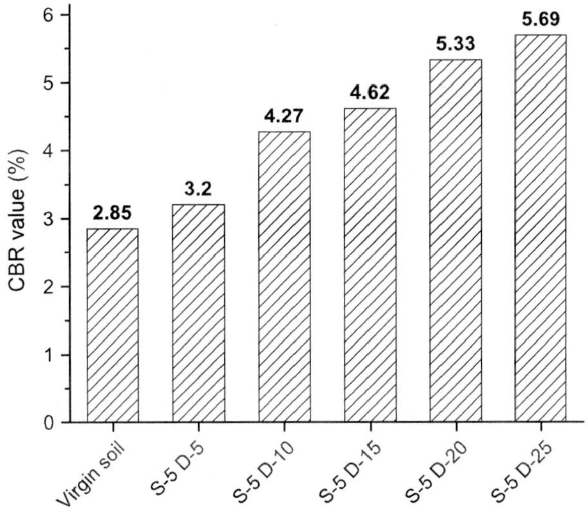

Fig. 8 CBR values for different combinations of soil–solid waste–stone dust

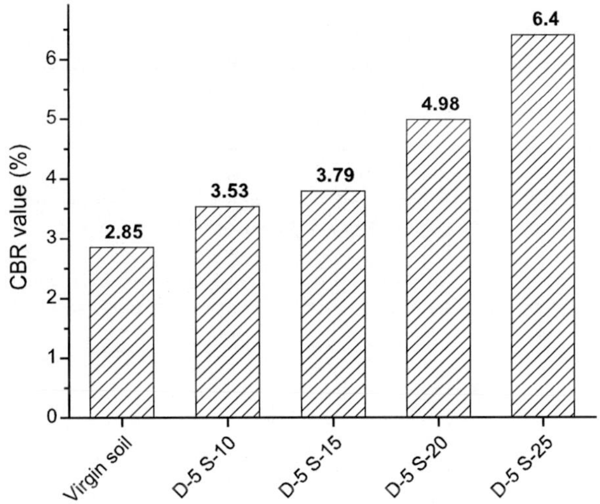

Fig. 9 CBR values for different combinations of soil–stone dust–solid waste

4 Conclusions

- As per IS soil classification system, soil was found to be CI, i.e. clay of intermediate plasticity.
- Waste material from silica sand beneficiation plant was found to be non-plastic in nature, and it was in the silty sand category.
- Characterization of stone dust was found to fall in the category of SP, i.e. poorly graded sand.

- The maximum CBR value was found to be at combination having 5% solid waste from silica sand beneficiation plant and 25% stone dust (S-5 D-25), and other the maximum CBR value was found be at the combination having 5% stone dust and 25% solid waste from silica sand beneficiation plant (D-5 S-25).
- The virgin soil mixed with the combination of 5% stone dust and 25% solid waste from silica sand beneficiation plant was very effective for soil stabilization.
- The disposal of the waste material problem is solved when waste material is used as a soil stabilizing material.

References

Ali MS (2011) Performance analysis of expansive soil treated with stone dust and fly ash. Electron J Geotech Eng 16:973–982

Al-Rawas AA (2005) Effect of lime, cement and Sarooj (artificial pozzolan) on the swelling potential of an expansive soil from Oman. Build Environ 40(5):681–687

Ferber V (2009) On the swelling potential of compacted high plasticity clays. Eng Geol 104(3–4):200–210

IS 1498 (1970) Classification of soil. Bureau of Indian Standards

IS 2720 (Part 16) (1979) Determination of California bearing ratio. Bureau of Indian Standards

IS 2720 (Part 3, sec. 1) (1980) Determination of specific gravity. Bureau of Indian Standards

IS 2720 (Part 4) (1985) Grain size distribution of soil. Bureau of Indian Standards

IS 2720 (Part 8) (1980) Determination of water content-dry density relation using standard compaction. Bureau of Indian Standards

Patidar A (2014) An experimental study on stabilization of black cotton soil using HDPE wastage fibres, stone dust and lime. Int J Adv Sci Tech Res 6(4)

Ramadas TL (2010) Swelling and strength characteristics of expansive soil treated with stone dust and fly Ash. In: Indian geotechnical conference 2010, GEOtrendz, pp 16–18

Shafabakhsh GH (2014) Case study of rutting performance of HMA modified with waste rubber powder. Case Stud Constr Mater 1:69–76

Characterization and Potential Usage of Stabilized Mine Tailings

Samir Kumar Sethi, Nagendra Roy, and G. Suneel Kumar

Abstract Stabilization is an established remediation for contaminated soils and for hazardous wastes. The main objective of this work is to study the behaviour of geotechnical characteristics of the tailing materials (TMs) which are stabilized using cement, lime and fly ash in different proportions and to propose an optimum percentage of blending agents. The geotechnical parameters like specific gravity, optimum moisture content, maximum dry density, California bearing ratio, unconfined compressive strength of mine tailing alone and mine tailing treated with pozzolanic reagents like cement, lime and fly ash in different proportions was studied. All the tests were performed as per the IS specifications and codal reference. The stabilized tailing material achieved considerable strength upto 1801.17 kN/m^2 at 28 days of curing period, thus making the stabilized TMs suitable for bulk fill construction. Fly ash can potentially stabilize the TMs treated along with cement and lime to reduce the construction cost.

Keywords Stabilization · Tailing materials · Pozzolanic reagents

1 Introduction

Tailings also called as mine dumps are the materials left over after the process of separating the valuable fraction from uneconomic fraction of an ore. Tailing consists of ground rocks and processed effluents, which are produced during the mine processing plants. Mechanical and chemical processes are used to extract the required material from the mine ore, and produced waste stream is called as tailings. This process of product extraction is never 100% efficient, nor is it possible to reclaim all recyclable and used processing reagents and chemicals. The unrecoverable and uneconomic

S. K. Sethi (✉) · N. Roy · G. Suneel Kumar
National Institute of Technology Rourkela, Rourkela, Odisha 769008, India
e-mail: samrinku86@gmail.com

N. Roy
e-mail: nroy@nitrkl.ac.in

M. Latha Gali and R. R. P. (eds.), *Problematic Soils and Geoenvironmental Concerns*, Lecture Notes in Civil Engineering 88,
https://doi.org/10.1007/978-981-15-6237-2_49

metals, minerals, chemicals, organics and process water are discharged, as slurry, to a dumping site generally known as a tailings management facility (TMF) or tailings storage facility (TSF). Not unexpectedly, the physical and chemical characteristics of tailings and their methods of handling and storage are of great and growing concern.

Tailings also called as mine dumps on which the materials left over after the process of separating the valuable fraction from uneconomic fraction of an ore. It consists of ground rock and effluents that are generated in a mine processing plant. Tailings are generally stored on the surface either within retaining structures or in the form of piles. Backfilling can provide ground and wall support, improve ventilation, provide an alternative to surface tailings storage and prevent subsidence.

1.1 Extraction of Tailing Material

Tailings are usually stored either within retaining structures or in the form of dry stacks on the surface. It can also be stored underground in mined-out voids by a process commonly referred to as backfill. Backfilling can provide ground and wall support, improve ventilation, provide an alternative to surface tailings storage and prevent subsidence.

The tailings result from the hydrometallurgical processing during the mineral extraction method. In many cases, tailings are presumed to be hazardous for the environment, health and economy. This is for the reason that, most of these tailings are rich in pyrite and pyrrhotite. The tailings have been traditionally stored in tailings disposal amenities (e.g. dams, ponds and other types of surface impoundments) situated at the mine surface in a variety of various methods.

However, the deprivation and/or failure of the aforementioned tailings disposal facilities can result in serious environmental and geotechnical problems with severe social and economic complications.

During the course of the mining of ores and minerals, a huge quantity of refuge is separated from the produced material during mineral processing, which is then usually broken and ground into micrometric size in order to recover useful components. In ore mines, it often happens that 95–98% of the rock quantity brought to the surface. In coal mines, this portion is 15–25% of the produced amount.

1.2 Stabilization/solidification

Stabilization/solidification is an established remediation technology for contaminated soils and treatment technology for hazardous wastes. Stabilization or modification of soil is referred to as a procedure in which cementing material or other chemical materials are these TMs can be used as a construction material in large quantities as a backfill, sub-grade in the construction of pavements and other construction materials (Aurora et al. 2010). The present study is on the stabilization of the TM using cement, lime, and fly ash (Basha et al. 2004; Desogus et al. 2013). So that the mobility of the contaminants is added to a natural soil material to improve its properties.

To overcome the above problems with contaminated mine tailings, stabilization is the best way, so that in the material should not move or to produce more chemically stable constituents; and solidification dim involves the addition of reagents to a contaminated material to impart physical stability to contain contaminants in a solid product and reduce access by external agents (e.g. air, rainfall) (Eren and Filiz 2009). As we are adding the cementing agents to the TM the sample is both stabilized and solidified (Glasser 1997).

1.3 Mine Tailing Characteristics

Tailing characteristics can vary greatly dependent on the ore mineralogy together with physical and chemical processes used to extract economic product. To established the design requirement of tailing storage facility following characteristics like physical properties, chemical composition, behaviour under pressure or consolidation rates, settling, drying time, and densification behaviour of mine tailings must be observed precisely.

2 Materials and Methodology

2.1 Mine Tailings

Tailings are the materials left over after the process of separating the valuable fraction from the uneconomic fraction of an ore. The composition of tailings is directly dependent on the composition of the ore and the process of mineral extraction used on the ore. Mine tailing used in this investigation was collected from a tailing dam situated in Kudremukh, Karnataka, India (13° 13′ 28″ N, 75° 13′ 36″ E). The samples were collected from a depth of 0.6 to 0.9 m below the natural ground surface. After collection, the samples were sealed in a plastic bag and transported to geotechnical laboratory. Laboratory tests were carried out on air-dried sample to study the basic properties of the tailing.

2.2 Cement

The manufactured Portland cement product is made from selected raw materials which are blended and reacted, usually in a rotary kiln, at temperatures exceeding

1400 °C. Cement we used in the tests is 43-grade ordinary Portland cement (OPC) which is locally available. It can be used either to modify and improve the strength of the soil or to transform the soil into a cemented mass with increased strength and durability (Singh and Pant 2006). The amount of cement used will depend upon whether the soil is to be modified or stabilized. The optimum usage of cement is 7–8%.

2.3 Lime

We used calcium oxide (Cao) as a stabilizing agent. We brought the calcium oxide which is locally available in the market. The stabilizing effects ultimately depend on chemical reaction by the lime on minerals in the soil to form cementitious compounds. In addition to lime to soil, two main types of chemical reaction occur.

- Alteration in the nature of the absorbed layer through base exchange phenomenon and
- Pozzolanic reaction.

The amount of lime required may be used on the unconfined compressive strength or the CBR test criteria. Normally, 2–8% of lime may be required for coarse-grained soils and 5–10% for plastic soils.

2.4 Fly Ash

Fly ash is a waste product generated from the combustion of fossil fuels in industrial operations such as power generation. It is present in flue gases collected by air pollution control devices, such as electrostatic precipitators. Fly ash particles mainly consist of silica and alumina. Other components may include carbon, phosphorus, sulphur, alkali compounds, and oxides of calcium, iron, magnesium, potassium, sodium, titanium and vanadium (Misra et al. 1996). Nowadays, fly ash is also using effectively in the stabilization of contaminated soils (Zhang et al. 2011).

To achieve the objectives set for the study, a proper planning as well as the knowledge of tests to be conducted becomes essential. Different geotechnical and chemical parameters like optimum moisture content, maximum dry density, specific gravity of mine tailings alone and mine tailings treated with pozzolanic reagents like cement, lime, fly ash in different proportions are found. All the tests were performed as per the IS specifications and the codal reference. Ordinary Portland cement (*C*), fly ash (FA), and lime (*L*) have been used for stabilization/solidification studies (Kolias et al. 2005).

3 Result and Discussion

3.1 Geotechnical Properties of the Tailing Material (TM)

The tailing samples were collected from a depth of about 0.5–0.6 m at locations (*L*1, *L*2, *L*3 and *L*4). After collecting the tailing material in core cutter, it is sealed in plastic bags and transported to the laboratory for testing. Index properties of mine tailings are determined (shown in Table 1). Location of sample collection points (L1, L2, L3 and L4) near tailing dam are shown in Fig. 1.

Table 1 Index properties of the tailing material (TM)

Property	Result
Specific gravity	3.24
Density index (I_d)	15.75
Field moisture content (%)	9.56
Liquid Limit	NP
Plastic Limit	NP
Plasticity index	0
Gravel (4.75–80 mm) (%)	0
Sand (75 μ–4.75 mm) (%)	95.00
Silt (2–75 μ) (%)	5.00
Clay (<2 μ) (%)	0
D_{10}	0.088
D_{30}	0.175
D_{60}	0.35
Coefficient of uniformity (C_u)	3.98
Coefficient of curvature (C_c)	0.99
Soil classification	SP

Fig. 1 Location of sample collection points (*L*1, *L*2, *L*3 and *L*4) near tailing dam

Based on the test results, the tailing material is classified as non-plastic poorly graded sand (SP) with plasticity index, $I_p = 0$. Particle size distribution curve of the mine tailing material is shown in Fig. 2. From the particle size distribution curve, it is observed that about 95% of the material is greater than 75 μ. The percentage of gravel, sand, silt and clay are presented in Table 1. The chemical composition of mine tailing material is shown in Table 2 (Sang-Hwan et al. 2011). pH of the TM determined using soil-to-water ratio 1:2.5 is about 7.45. Iron content of the TM is about 28.59%. Chemical composition of other additives used in this study is given in Table 2. The heavy metal content of tested mine tailing is presented in Table 3 (Jang and Kim 2000). Since tailing material contains heavy metals and to inhibit environmental leaching (e.g. due to infiltration of rainwater) remediation/stabilization of tailings is essential. Hence, this study is undertaken to stabilize the TM using commonly used additives like cement, lime and fly ash.

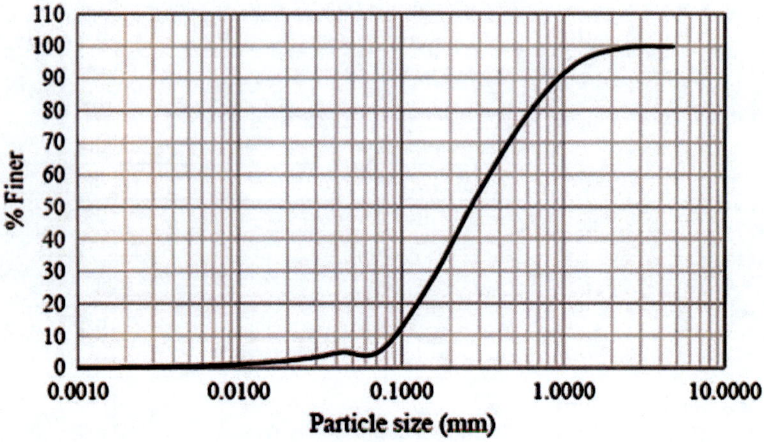

Fig. 2 Grain size distribution of TM

Table 2 Chemical characteristics of mine tailing (MT), cement (*C*), lime (*L*) and fly ash (FA)

Particulars	Mine tailing	Lime	Cement	Fly ash
pH	7.45	11.42	11.47	10.21
Electrical conductivity	145 μs	6.9 ms	6.95 ms	0.35 ms
Silica oxide SiO_2 (%)	62.06	11.36	19.48	72.08
R_2O_3 ($Al_2O_3 + Fe_2O_3$) (%)	31.67	5.27	10.25	5.72
Iron oxide Fe_2O_3 (%)	28.59	0.19	2.00	0.57
Aluminium oxide (Al_2O_3) (%)	3.08	5.08	8.25	5.15
Calcium oxide (CaO) (%)	0.073	72.90	56.08	12.34
Magnesium oxide (MgO) (%)	0.017	4.84	5.64	4.04
Loss on ignition (%)	2.32	4.24	4.77	0.76

Table 3 Heavy metal content of tailing material

Elements	Result (*w/w* %)
Lead (Pb)	0.00088
Zinc (Zn)	0.003
Chromium (Cr)	0.0001
Cadmium (Cd)	0.0001
Copper (Cu) 0.001	0.001
Cobalt (Co) 0.01	0.01
Magnesium (Mg) 0.08	0.08

NP—Non-plastic, D_{10} is the particle size in mm at 10% finer, D_{30} is the particle size in mm at 30% finer, D_{60} is the particle size in mm at 60% finer.

3.2 *Trial Mixes*

After sampling, tailing material was sealed in plastic bags and transported to laboratory for analysis. In laboratory, it was air-dried and compaction curves were developed for natural tailing material and its blended material. The compaction energy used is equivalent to standard Proctor test (IS 2720 Part-7) and modified Proctor test (IS 2720 part-8). The pozzolana dosing matrix is shown in Table 4. In Table 4, Mix ID MT/0/0/0 denotes 100% TM and MT/8/0/0 refers to 8% lime alone added to tailing material (dry) and MT/0/8/0 refers to 8% cement alone added to tailing material. Also, the complete details of other additives added to TM in different proportions are shown in Table 4. Relevant literature is also referred while choosing the dosing matrix for stabilization purpose where MT/0/0/0 denotes 100% TM.

Table 4 Admixture of dosing matrix

Mix ID	Lime (%)	Cement (%)	Fly ash (%)
MT/0/0/0	0	0	0
MT/8/0/0	8	0	0
MT/0/8/0	0	8	0
MT/8/8/0	8	8	0
MT/2.5/7.5/25	2.5	7.5	25
MT/2.5/7.5/50	2.5	7.5	50

3.3 Compaction Behaviour

Standard Proctor compaction tests were carried out on mine tailing material and its combinations after 24 h of curing with additives/stabilizing agents (i.e. *C*, *L*, and FA). The test results of proctor compaction are presented in Fig. 3 in the form of dry density versus moisture content. From Fig. 3, it is observed that the maximum dry density (i.e. $g_{d\max}$) and optimum moisture content (w_{opt}) values for natural MT are 20.65 kN/m^3 and 11.7%. When tailing material is mixed with 8% cement alone (dry basis), the maximum dry density and optimum moisture content obtained were 21.20 kN/m^3 and 10.1%.

Similarly for mix ID MT/8/0/0 (i.e. 8% lime), same results were obtained, i.e. $g_{d\max}$ = 20.4 kN/m^3 and w_{opt} = 10.5%. All the standard and modified compaction, test results of different mixes (MT/0/0/0, MT/8/0/0, MT/0/8/0, MT/8/8/0, MT/2.5/7.5/25, and MT/2.5/7.5/50) are shown in Tables 5 and 6, respectively. From Fig. 3, it is also observed that when tailing is mixed with 25% fly ash and 50% fly ash (i.e. mix

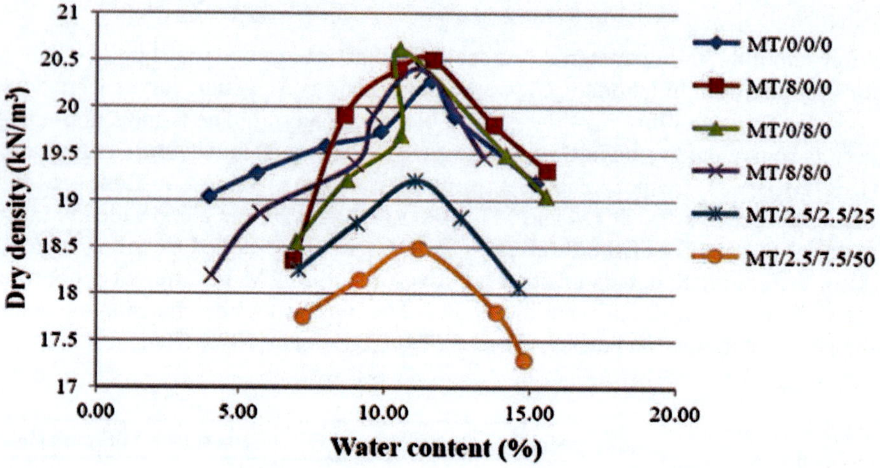

Fig. 3 Compaction curves for different trial mixes

Table 5 Standard proctor test results (tailing material and blends)

Mix ID	MDD (kN/m^3)	OMC (%)	*G*
MT/0/0/0	20.65	11.70	3.24
MT/8/0/0	20.4	10.5	3.10
MT/0/8/0	21.20	10.1	3.22
MT/8/8/0	21.10	11.0	3.27
MT/2.5/7.5/25	19.33	11.5	2.95
MT/2.5/7.5/50	18.54	12.2	2.76

Table 6 Modified proctor test results (tailing material and blends)

Mix ID	MDD (kN/m^3)	OMC (%)	*G*
MT/0/0/0	21.29	10.25	3.24
MT/8/0/0	21.78	8.25	3.10
MT/0/8/0	22.37	9.80	3.22
MT/8/8/0	21.58	9.65	3.27
MT/2.5/7.5/25	20.31	9.90	2.95
MT/2.5/7.5/50	19.62	9.80	2.76

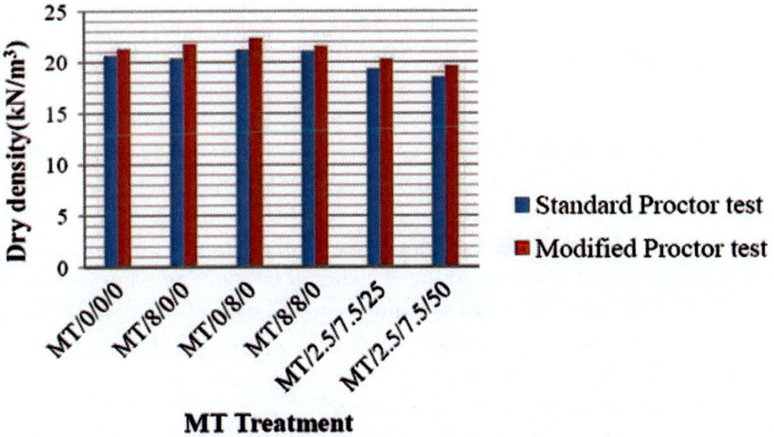

Fig. 4 Effect of additives on compaction characteristics

MT/2.5/7.5/25, MT/2.5/7.5/50), there is a reduction in maximum dry density value of about 10% (i.e. $g_{d\max} = 18.54$ kN/m^3).

However, this reduction is not significant. This reduction in dry density is attributed due to the addition of fines in the form of fly ash (Fig. 4).

3.4 *Unconfined Compressive Strength (UCS)*

UCS tests were conducted both on mine tailing and its blended soil sample. After mixing with lime, cement, and fly ash in different proportions, the UCS specimens were kept for curing period of 7 days, 28 days and 90 days. All the test specimens were compacted to maximum dry density and optimum moisture content ($g_{d\max}$ and w_{opt} were established initially). After the curing period, specimens were placed centrally on the lower plate of the compression testing machine and the load is applied to the ends of the specimen. During loading stage, the rate of deformation is kept uniform, approximately 1.25 mm/min. The maximum load (P) exerted by the machine is then

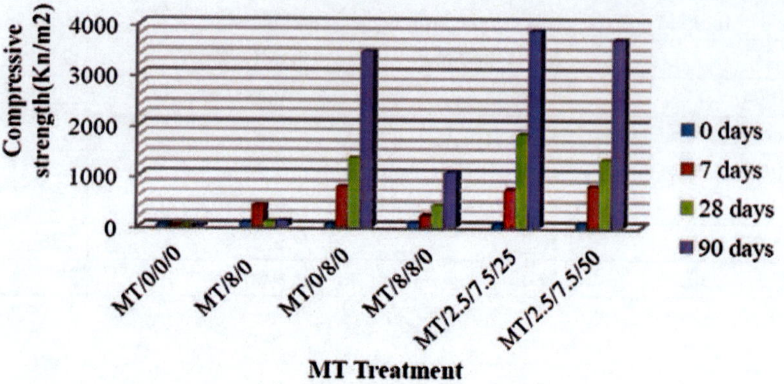

Fig. 5 Average UCS of different trial mixes

recorded. The unconfined compressive strength (p) is calculated as: $p = P/A_n$ where $P =$ maximum recorded load (kN) and A_n is the cross-sectional area (m^2). The results of UCS test specimens are plotted in Fig. 5. It is observed that there is a significant variation of compressive strength of blended test specimens when compared with test specimens prepared with tailings alone. The initial unconfined compressive strength of tailing is 17.22 kN/m^2, when mine tailing is mixed with 8% cement UCS increases with curing period (i.e. from an initial value of 31.76 to 1332.9 kN/m^2 after 28 days and 1721.35 kN/m^2 after 90 days).

Similarly when mine tailing is mixed with 8% lime, there is an increase in strength but not as significant as observed with cement mixed tailings (see Table 7, mix IDMT/8/0/0). However when mine tailing is mixed with cement, lime, and fly ash, maximum UCS value was about 1801.17 kN/m^2 after 28 days curing period and 3857.06 kN/m^2 after 90 days. This value of UCS on cured specimen for mix ID MT/2.5/7.5/25 is significant when compared with the initial UCS value of TM alone. All the details are shown in graphical form in Fig. 5.

Table 7 UCS of cured test specimens (TM blends)

Mix ID	Unconfined compressive strength (kN/m^2)			
	0 days	7 days	28 days	90 days
MT/0/0/0	17.22	–	–	–
MT/8/0/0	58.99	40.46	88.80	94.49
MT/0/8/0	31.75	764.5	1332.9	1721.23
MT/8/8/0	67.54	211.25	407.86	523.9
MT/2.5/7.5/25	39.82	713.46	1801.2	1928.53
MT/2.5/7.5/50	66.82	782.18	1292.5	1844

Table 8 Specific gravity of TM and blends

Mix ID	Specific gravity, *G*
MT/0/0/0	3.24
MT/8/0/0	3.10
MT/0/8/0	3.22
MT/8/8/0	3.27
MT/2.5/7.5/25	2.95
MT/2.5/7.5/50	2.76
CEMENT (alone)	3.11
LIME (alone)	3.04
FLY ASH (alone)	2.07

3.5 Specific Gravity

Specific gravity tests were directed on mine tailing, cement, lime, fly ash and blended mixtures. The test outcomes of the specific gravity are given in Table 8. From Table 8, it is observed that the specific gravity of MT is 3.24. After mixing with additives in different proportions to the MT, the specific gravity gradually reduced. The specific gravity of mix ID MT/2.5/7.5/50 is significantly reduced (i.e. from 3.24 to 2.76) because the MT is blended with the low specific gravity additives.

3.6 Permeability

Hydraulic conductivity results of mine tailing and blended mine tailing are shown in Table 9. It is observed that hydraulic conductivity of blended mine tailing decreased from an initial value of 2.4×10^{-3} cm/s (untreated MT) to $7.73 \text{ X } 10^{-5}$ cm/s (treated MT). The reduction is attributed to reduction in void ratio.

Table 9 Hydraulic conductivity of the TM and its blended sample

Mix ID	Permeability (cm/s)
MT/0/0/0	2.4×10^{-3}
MT/8/0/0	7.73×10^{-5}
MT/0/8/0	1.40×10^{-4}
MT/8/8/0	5.06×10^{-5}
MT/2.5/7.5/25	1.45×10^{-5}
MT/2.5/7.5/50	2.28×10^{-5}

Table 10 CBR results of TM and blends

Mix ID	Unsoaked CBR	Soaked CBR
MT/0/0/0	55.05	50.13
MT/8/0/0	58.21	63.93
MT/0/8/0	63.05	68.12
MT/8/8/0	65.01	69.13
MT/2.5/7.5/25	60.12	62.33
MT/2.5/7.5/50	56.02	58.20

3.7 *California Bearing Ratio Test*

The soaked and unsoaked CBR values of MT and treated MT are given in Table 10. It is observed from Table 10 when MT is mixed with cement, cement and lime, cement lime and fly ash, the soaked CBR value is greater than unsoaked CBR value (i.e. for MT/8/0/0, MT/0/8/0, MT/8/8/0, MT/2.5/7.5/25 and MT/2.5/7.5/50). This may be attributed due to pozzolanic reaction between the admixtures added and tailing.

3.8 *Direct Shear*

Direct shear tests were conducted on both MT as well as blended MTs. Results of direct shear test are presented in Table 11. From Table 11, we can observe that, in wet condition the cohesion of the MT is increasing from 7.8 to 14 kN/m^2 and the friction angle is decreasing from 42° to 36°. The cohesion in wet condition is more than the cohesion in dry condition this is because the water content is inducing some additional cohesion to the blends .The variation of shear strength parameters of mine tailing treated with different percentages of lime, cement and fly ash are depicted in Fig. 6.

Table 11 CBR results of TM and blends

Mix ID	c (kN/m^2)	φ
MT/0/0/0	2	40
MT/8/0/0	8	36
MT/0/8/0	8	37
MT/8/8/0	10	34
MT/2.5/7.5/25	12	30
MT/2.5/7.5/50	14	29

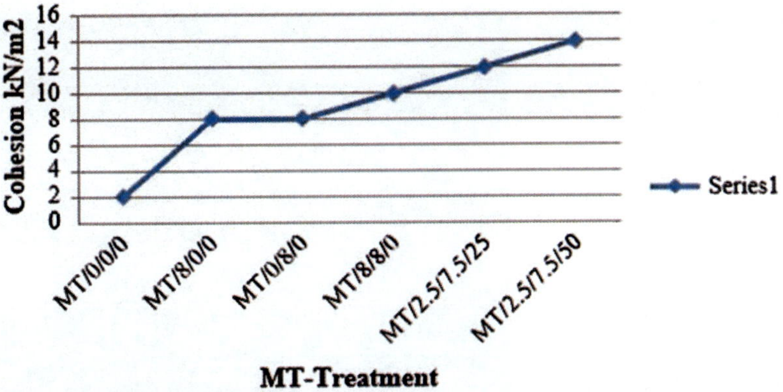

Fig. 6 The variation of shear strength parameters of mine tailing treated with different percentages of lime, cement and fly ash

3.9 SEM Results

The scanning electron microscopy was done for the specimens which are cured for 28 days; the images of the SEM analysis are shown in Figs. 7, 8, 9, 10, 11 and 12. From the Figures, it is noticed that, the formation of lumps in the treated TM is formed due to the pozzolanic action of the binding materials. The pozzolanic material in the treated TM collects and binds the heavy metal compounds to the walls of the soil particles. This formation of lumps will help in the inhibition of the leaching of heavy metals from the stabilized specimen.

Fig. 7 Microstructure of the mix MT/0/0/0

Fig. 8 Microstructure of the mix MT/8/0/0

Fig. 9 Microstructure of the mix MT/0/8/0

Fig. 10 Microstructure of the mix MT/8/8/0

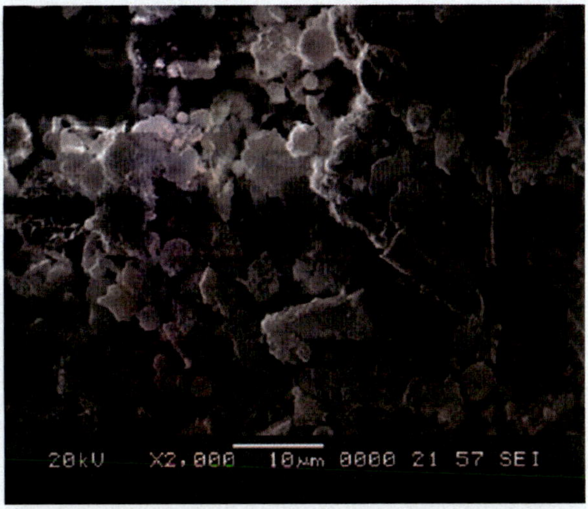

Fig. 11 Microstructure of the mix MT/2.5/7.5/25

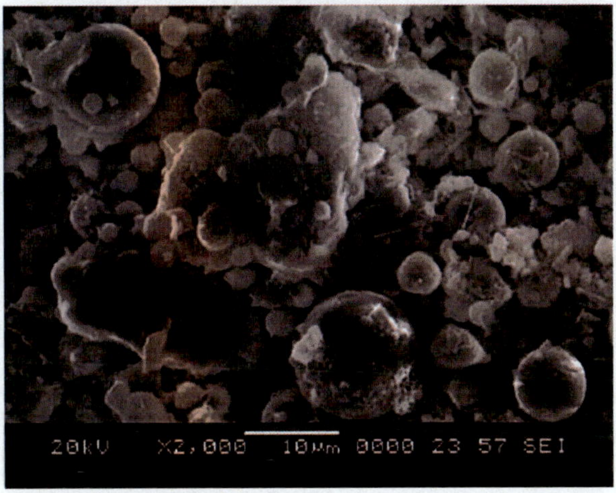

Fig. 12 Microstructure of the mix MT/2.5/7.5/50

4 Conclusion

The stabilized tailing material can have considerable strength (up to 1801.17 kN/m^2) at 28, thus making the stabilized tailing material suitable for bulk fill construction (construction of embankments, fill construction in urban areas, etc.). The use of fly ash in conjunction with cement and lime increased the UC strength beyond the values obtained by the use of lime and cement alone. The use of Portland cement alone in the stabilization process gave a very good result in immobilization of the heavy metals ions from the stabilized tailings. Addition of lime to TM does not give any significant results in compressive strength but it reduced the concentration of heavy metals. The tailing material stabilized with cement reach their ultimate strength within 28 days and a major portion of the ultimate strength (about 70%) is gained within only 7 days. The use of lime, cement and fly ash together resulted a better increase in compressive strength. Fly ash can potentially stabilize the TM treated along with cement and lime to reduce construction cost. The use of fly ash in conjunction with cement and lime also resulted in immobilization/encapsulation of the heavy metal ions from the stabilized tailings.

References

Aurora A, Colin DH, Paula JC, Kevin HG, Edward RB, Alison KC (2010) Long-term performance of aged wasteforms treated by stabilization/solidification. J Hazard Mater 181:65–73

Basha EA, Hashim R, Mahmud HB, Muntohar AS (2004) Stabilization of residual soil with rice husk ash and cement. Constr Build Mater 19:448–453

Desogus P, Manca PP, Orrù G, Zucca A (2013) Stabilization–solidification treatment of mine tailings using Portland cement, potassium dihydrogenphosphate and ferric chloride hexahydrate. Miner Eng 45:47–54

Eren S, Filiz M (2009) Comparing the conventional soil stabilization methods to the consolid system used as an alternative admixture matter in Isparta Darıderematerial. Constr Build Mater 23:2473–2480

Glasser FP (1997) Fundamental aspects of cement solidification and stabilization. J Hazard Mater 52:151–170

Jang A, Kim S (2000) Technical note solidification and stabilization of pb, zn, cd and cu in tailing wastes using cement and fly ash. Miner Eng 13:14–15

Kolias S, Kasselouri-Rigopoulou V, Karahalios A (2005) Stabilization of clayey soils with high calcium fly ash and cement. Cement Concr Compos 27:301–313

Zhang L, Ahmari S, Zhang J (2011) Synthesis and characterization of fly ash modified mine tailings-based geopolymers. Constr Build Mater 25:3773–3781

Misra M, Yang K, Mehta RK (1996) Application of fly ash in the agglomeration of reactive mine tailings. J Hazard Mater 51:181–192

Sang-Hwan L, Hyun P, Namin K, Seunghun H, Anna H (2011) Evaluation of the effectiveness of various amendments on trace metals stabilization by chemical and biological methods. J Hazard Mater 188:44–45

Singh TS, Pant KK (2006) Solidification/stabilization of arsenic containing solid wastes using Portland cement, fly ash and polymeric materials. J Hazard Mater B131:29–36

Influence of Processing Temperature on Strength and Structural Characteristics of Alkali-Activated Slag Lateritic Soil

T. Vamsi Nagaraju, D. Neeraj Varma, and M. Venkata Rao

Abstract Ground granulated blast furnace slag (GGBS)-based geopolymer is an effective binder that attains high strength by curing at different temperatures. This paper presents the experimental results obtained from tests conducted on alkali-activated GGBS lateritic soil blends. Unconfined compressive strengths were determined for alkali-activated blends to which GGBS was added in varying quantities (5, 10, 15, 20, and 25% by dry weight of soil). UCS samples were tested at different curing temperatures. The processing temperature is found to influence the development of compressive strength. Further, samples were conducted for SEM analysis to know the structural development. Test results indicated that unconfined compressive strength increased with increasing GGBS content and curing period. This study illustrated that GGBS-based geopolymer can be an effective soil stabilizer for lateritic soils.

Keywords Lateritic soil · Geopolymer · GGBS · Temperature · Strength

1 Introduction

Laterites are natural earth materials with unique properties. They are tropical weathered soils having rich minerals of iron oxides, quartz, and kaolinite (Lohnes and Demirel 1973). The process of natural weathering in laterites was leaching of silica and formation of iron oxides, aluminum oxides, and colloidal sesquioxides (Lohnes and Demirel 1973). Sherman (1949) found that there is a significant influence of the mineralogical composition of laterites by the intensity of moisture variation. Lateritic soils generally exhibit higher values of shear parameters (Fish 1971). This can be

T. Vamsi Nagaraju (✉) · M. Venkata Rao
Department of Civil Engineering, S. R. K. R. Engineering College, Bhimavaram 534204, India
e-mail: Varshith.varma@gmail.com

D. Neeraj Varma
Department of Civil Engineering, RVR & JCE, Guntur, India
e-mail: tvnraju@srkrec.edu.in

M. Latha Gali and R. R. P. (eds.), *Problematic Soils and Geoenvironmental Concerns*, Lecture Notes in Civil Engineering 88,
https://doi.org/10.1007/978-981-15-6237-2_50

attributed due to the presence of sesquioxides, which have a tendency of cementation (binding) among the individual soil particles (Baldovin 1969; Fish 1971; Lohnes and Demirel 1973; Paulson 1975).

Nowadays, in India, a huge network of rural roads is in execution for each and every corner under different schemes launched by the government including the prime minister's rural connectivity programme Pradhan Mantri Gram Sadak Yojna (PMGSY). There is a need of economical and promising material to fulfill requirements. Lateritic soil was a desirable material and most abundantly using in pavements and railway embankments. However, there is still a need to enhance strength characteristics. In the last few decades, chemical alteration of problematic soils like expansive clays and lateritic soils using with lime and cement was in vogue for quite long (Ola 1974, 1977; Portelinha et al. 2012; Amadi and Okeiyi 2017; Phanikumar and Nagaraju 2018). However, these are causing greenhouse effects. Therefore, utilization of geopolymers or alkali-activated aluminosilicate types of cement has been recognized as one of the most successful promising green materials (Vamsi Nagaraju 2018). In this regard, some research has been conducted, and a brief account of the results is presented below:

Patrick et al. (2014) studied the influence of processing temperature of Na-activated lateritic soil on compressive strength of lateritic soils. Test results indicated that specimens processed at 60–105 °C for 7 days and 250 °C for 3 h gave similar strengths. Alkali activation of lateritic soils can give rise to eco-friendly building materials. Phummiphan et al. (2017) presented details of strength properties of alkali-activated lateritic soils treated with fly ash. Unconfined compressive strength tests for 7 days soaked strength revealed that strength obtained for lateritic soil FA geopolymers meets the strength requirement for both light and heavy traffic pavement specified by Thailand National Authorities.

Phummiphan et al. (2018) conducted an experimental study on the fly ash geopolymer lateritic soil and ground granulated slag blends with the variation of $NaOH/Na_2SiO_3$, GGBS content, and curing time. It was reported that GGBS to be used as a replacement material with fly ash geopolymer to the development of a low-carbon stabilized pavement base.

This paper presents the experimental results obtained from tests conducted on alkali-activated GGBS lateritic soil blends. Unconfined compressive strengths were determined for alkali-activated blends to which GGBS was added in varying quantities. SEM analysis was carried on the lateritic soil and blended alkali-activated lateritic soil samples. This paper also explores the importance of geopolymer soils for sustainable development in ground engineering.

Table 1 Geotechnical properties of lateritic soil

Property	Lateritic soil
Specific gravity	2.72
Gravel (%)	17
Sand (%)	42
Fines (%)	41
Liquid limit (%)	51
Plastic limit (%)	29
Plasticity index (IP)	22
USCS classification	CH
Optimum moisture content (%)	14.4
Maximum dry density (kN/m^3)	22.1
Unconfined compressive strength (kg/cm^2)	1.25
California bearing ratio, CBR (%) (un soaked)	1.46

2 Materials and Methods

2.1 Materials

Ground granulated blast furnace slag (GGBS) used in this was obtained from Vespa ash Pvt Ltd, a supplier of GGBS in Vijayawada, India. Lateritic soil was collected from Tadepalligudem, India. The alkaline activator solution is composed of Na_2SiO_3, 7 molarities of NaOH, and a Na2SiO3/NaOH ratio of 2.5. Geotechnical characteristics of lateritic soil are shown in Table 1.

Based on the index properties and the grain size distribution, the soil is classified as CH (USCS).

2.2 Tests Performed

Compaction characteristics were performed for soil, GGBS soil blends in the standard Proctor test. Undrained shear strength (ASTM 2000, D2166) of the soil blended with alkali-activated GGBS was determined by preparing cylindrical samples (38 mm in diameter and 76 mm in length) at 14% liquid alkali activator. Unconfined compression tests are also conducted with varying temperature. The effect of curing period was also studied by curing the samples for 14 days and 40 days. Alkali activation is done by using sodium hydroxide solution (NaOH) with 7 molarity concentration and sodium silicate solution (Na_2SiO_3), keeping Na_2SiO_3/NaOH ratio as 2.5 and varying GGBS content in the blends.

Table 2 Effect of GGBS on compaction characteristics

GGBS (%)	OMC (%)	MDD (kN/m^3)
0	14.5	22.1
5	14.1	22.4
10	13.7	22.7
15	13.2	23.3
20	12.5	23.8
25	12.0	24

2.3 *Test Variables*

GGBS was added in various percentages such as 5, 10, 15, 20, and 25% to the dry soil in alkali-activated GGBS soil blends for performing the tests.

3 Results and Discussion

3.1 *Effect of GGBS on Compaction Characteristics of GGBS Lateritic Soil Blends*

Compaction characteristics, OMC, and MDD were determined for unblended clay and clay blended with various amounts of GGBS were tested. Table 2 shows the compaction characteristic test results.

Table 2 shows the Proctor compaction data for unblended lateritic soil and treated lateritic soil with GGBS. The test data indicated that both MDD and OMC were improved with increasing GGBS content. For instance, the MDD of untreated lateritic soil increased from 22.1 to 24.0 kN/m^3 for the treated blend having 24% GGBS content. Meanwhile, the OMC of untreated lateritic soil decreased from 14.5 to 12% for the treated blend having 30% GGBS content.

3.2 *Effect of Alkali Activator and GGBS Content on Strength Characteristics*

The effect of clay-GGBS geopolymers on strength characteristics were studied with varying amounts of GGBS such as 5%, 10%, 15%, 20% and 25% to the dry weight of soil. Table 3 shows the strength characteristics.

Figures 1 and 2 show the stress–strain characteristics of alkali-activated GGBS lateritic blends having varying GGBS content (0, 5, 10, 15, 20, and 25%) for the liquid alkali activator contents of 14%. The stress–strain data pertaining to 0 days

Table 3 Effect of GGBS on strength characteristics

Additive content %	Unconfined compressive strength (kg/cm^2) for curing periods of			
	14 days Ambient curing	40 days Ambient curing	3 h @ 250 °C	7-day @ 100 °C
0%	5.5	7	19.4	15.9
5% GGBS	10.4	11.3	35.1	25.7
10% GGBS	15.3	23.4	56.9	77.4
15% GGBS	25.2	43.6	85.3	55.5
20% GGBS	38.4	60.7	111.6	59.2
25% GGBS	50	80.5	128.9	115.8

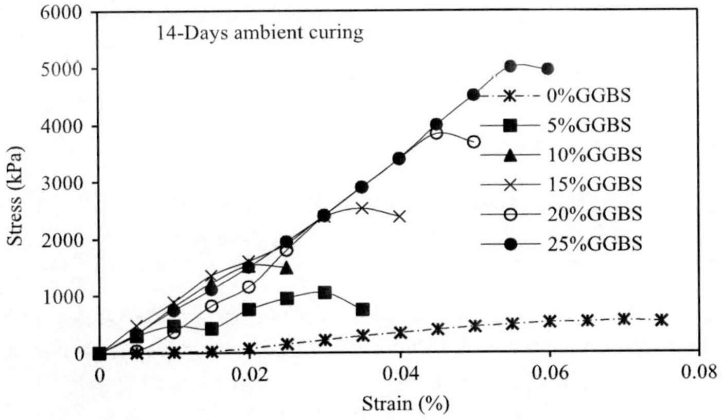

Fig. 1 Stress-strain curves at 14-days ambient curing

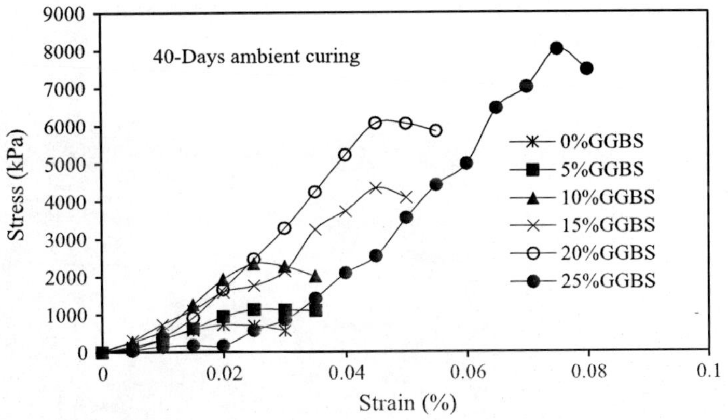

Fig. 2 Stress-strain curves at 40-days ambient curing

curing are not possible to being tested, because initial setting takes more duration. The strength development in the lateritic–GGBS geopolymer specimen is due to the sodium hydroxide solution (NaOH) leaches the silicon and aluminum in the amorphous phase of GGBS and the sodium silicate solution (Na_2SiO_3) acts as a binder. The variation of UCS with GGBS content in the GGBS–alkali-activated soil for the curing period of 14 days and 40 days was observed. UCS increased with increase in GGBS content. The insignificant strength development is because the amount of alkaline activator is not enough to leach the silica and the alumina in the amorphous phase of GGBS for the geopolymerization process.

Table 3 shows the effect of UCS due to temperature and GGBS content. UCS data of samples cured for 3 h at 250 °C indicate two times higher than that of samples cured for 40 days at an ambient condition. This indicates that increasing the curing temperature to gain on mechanical properties is expected to increase the energy and cost of the process. However, the gain time is observed when applying higher temperatures would more likely render the process more efficient.

3.3 Effect of Alkali Activator on Microstructural Characteristics of Alkali-Activated Blends

In order to better understand the effect of the binder on the original soil as well as the effect of geopolymers on the soil, selected specimens were analyzed (after submitted to the respective UCS test) using scanning electron microscopy (SEM).

Figures 3 and 4 show the variation of structural development, micrographs of original (untreated) soil exhibit pores on their surfaces which generally does not give enough resistance toward induced stresses. The structure of alkali-activated soil blends matrix is dense due to re-condensation of developed particles and the presence of the glassy phase.

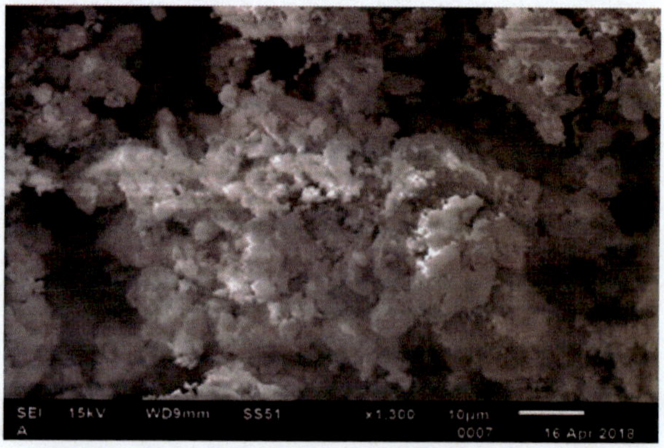

Fig. 3 SEM micrograph of lateritic soil

Fig. 4 SEM micrograph of alkali-activated soil with 20% GGBS content

4 Conclusions

In the present scenario, new technologies are needed to overcome the challenges possess by problematic soils. Geopolymer technologies in concrete and mortar are quite well known, and lot of work has been done, yet much work remains to be done in ground improvement for competence. Bulk utilization of industrial by-products or wastes could be possible in ground improvement by alkali activation of slags. The conclusions arising from this research are drawn as follows:

1. Addition of GGBS content enhances the Proctor compaction characteristics of the GGBS–soil blends.
2. Unconfined compressive strength increases by increasing the GGBS content in the GGBS-based alkali-activated soil blends. Moreover, strength characteristics are also influenced by processing temperature. The results clearly confirm that strength of the alkali-activated blends attains higher values with the expose of processing temperature.
3. In this study, liquid alkaline activator, NaOH with a concentration of 7 molars, and Na_2SiO_3/NaOH ratio of 2.5 can be considered as constant for the manufacturing of GGBS-based lateritic geopolymer.
4. The NaOH leaches the silicon and aluminum in the amorphous phase of GGBS and the Na_2SiO_3 acts as a binder.
5. Alkali activation of soils enhances soil engineering properties and its application in order to reduce the carbon footprint.

References

Amadi AA, Okeiyi A (2017) Use of quick and hydrated lime in stabilization of lateritic soil: a comparative analysis of laboratory study. Int J Geoeng 2017(8):3

Baldovin G (1969) The shear strength of lateritic soils. In: Proceedings of specialty session of engineering properties of lateritic soils, 7th international conference on soil mechanics and foundation engineering, Mexico City, vol 1, pp 129–142

Fish RO (1971) Shear strength and related engineering properties of selected Puerto Rican oxisols and ultisols. Unpublished M.S. thesis, Iowa State University, Ames, Iowa

Lohnes RA, Demirel T (1973) Strength and structure of laterities and lateritic soils. Eng Geol 7:13–33

Ola SA (1974) Need for estimated cement requirements for stabilizing lateritic soils. Transp Eng J ASCE 100(2):379–388

Ola SA (1977) The potentials of lime stabilization of lateritic soils. Eng Geol 11(4):305–317

Patrick NL, Achille BM, Elie K, Uphie CM, Marie PD, Hubert R (2014) Influence of the processing temperature on the compressive strength of Na activated lateritic soil for building applications. Constr Build Mater 65:60–66

Paulson SK (1975) The strength, structure, and mineralogy of selected Hawaiian lateritic soils. Unpublished M.S. thesis, Iowa State University, Ames, Iowa

Phanikumar BR, Nagaraju TV (2018) Engineering behaviour of expansive clays blended with cement and GGBS. Ground Improv. https://doi.org/10.1680/jgrim.17.00054

Phummiphan I, Suksun H, Tanakorn P, Arul A, Shui Long S (2017) Marginal lateritic soil stabilized with calcium carbide residue and fly ash geopolymers as a sustainable pavement base material. J Mater Civ Eng 29(2):04016195, 1–10

Phummiphan I, Suksun H, Runglawan R, Arul A, Shui Long S, Prinya C (2018) High calcium fly ash geopolymer stabilized lateritic soil and granulated blast furnace slag blends as a pavement base material. J Hazard Mater 341:257–267

Portelinha FMH, Lima DC, Fontes MPF, Carvalho CAB (2012) Modification of a lateritic soil with lime and cement: an economical alternative for flexible pavement layers. Soils Rocks Sao Paulo 35(1):51–63

Sherman GD (1949) Factors influencing the development of lateritic and laterite soils in the Hawaiian Islands. Pac Sci 3(4):307–314, 120

Vamsi Nagaraju T (2018) Potential of geopolymer technology towards ground improvement. In: 2nd international conference on advances in concrete, structural and geotechnical engineering. BITS Pilani, Rajasthan

Stabilization of Expansive Soil Using Lime Sand Piles—A Case Study

K. Premalatha and K. Sabarishri

Abstract This paper presents a case study of the stabilization of expansive soil using lime sand piles. The technique was used in the construction of flexible pavement on expansive soil of Dr. MGR Bus terminus of the Salem Municipal Corporation in Tamil Nadu, India. The soil profile at the site was a layered system with the first layer filled with earth from Salem quarry. The second layer is silty clay followed by silty sand in only certain locations. The fourth layer was gravelly sand followed by weathered rock. The filled up earth in this site is non-plastic silty sand. The silty clay layer possesses high plasticity which will undergo volume change due to moisture content. This clay was black in color and indicated the presence of organic content between 20 and 30%. This kind of soil is always referred to as a problematic soil and is locally known as black cotton soil. Generally, any new structure is to be laid on the surface of original ground level but as in this site the original ground is a problematic soil and would always cause the failure of lightly loaded structure it required stabilization. The stabilization of second silty layer which required improvement in strength and improvement of reduction in volume change characteristics was achieved using lime stabilization. Out of many different lime stabilization techniques, lime sand pile was preferred for the site because of its proven performance, durability, constructability, and low cost. The stabilization using lime sand piles was achieved methodically. After curing the site for maximum of 30 days, flexible pavement was laid on the prepared ground. The design field CBR value of 15% was ensured after ground improvement before laying the pavement. The selection of material and construction of pavement were as per MORTH revision 5 "Specification for road and bridge works," 2013, and design of pavement was as per IRC 37-2012.

Keywords Expansive soil · Stabilization · Lime sand piles · Case study

K. Premalatha · K. Sabarishri (✉)
Department of Civil Engineering, College of Engineering, Anna University, Guindy, Chennai, Tamil Nadu, India
e-mail: sabarikalyan@gmail.com

K. Premalatha
e-mail: kvpremalatha@yahoo.com

M. Latha Gali and R. R. P. (eds.), *Problematic Soils and Geoenvironmental Concerns*, Lecture Notes in Civil Engineering 88,
https://doi.org/10.1007/978-981-15-6237-2_51

1 Introduction

Construction of pavement on expansive soils posses risk due to the fact that it is susceptible to differential settlements, very low shear strength and high compressibility. The worst enemy of road pavements being water, particularly in expansive soils, when pavements are laid on these soil bases, they develop undulations at the surface due to the loss of strength of subgrade through softening during monsoons. Expansive soils have low bearing capacity and high shrinkage characteristics. Due to its peculiar characteristics, the expansive soil demonstrates an unsuitable behavior as a foundation material for pavement construction. CBR values of expansive soils are found to be generally in the range from 2 to 4%, which demands excessive pavement thickness in the design of flexible pavements. These shortcomings in the construction of flexible pavements on expansive soils can be overcome through the improvement of strength characteristics of expansive soils by soil stabilization techniques. Out of many stabilization techniques available, lime stabilization is a common and effective type of chemical stabilization which is in practice. Lime stabilization is unique to the fact that the lime actually reacts with expansive soil itself, rather than the soil and the stabilizer forming a two-material system. Traditionally, this has been exploited well in the construction of pavements where the in situ material stiffness and strength can be improved to a great extent. The use of lime as a stabilizer can be divided into three main groups, viz. lime columns, lime slurry pressure injection, and lime piles. Hydrated lime in the range from 3 to 5% appears to bring remarkable improvement in the engineering characteristics of expansive soils. In addition to being cost-effective, lime soil stabilization is also suited to manual methods of construction. The lime slurry injection technique involves the introduction of lime slurry into the pores and fissures of the expansive soil under pressure, forming matrix enclosing areas of untreated soil. This matrix is formed without disturbing the inherent soil structure in a significant manner. Lime slurry injection technique has been successfully used in India to treat expansive soils under railway tracts (Bhattacharya and Bhattacharya 1989). The authors have reported reduction in shrinkage cracks under the treated section of tracks. However, this method essential forms a two-material system, and hence, researchers have developed some other methods where lime will react with the soil. The similarities between lime column and lime pile technique have rendered confusion in the literature as they have been used interchangeably throughout. One must make sure to discover the method of construction referred to in each case. The lime column refers to the creation of deep vertical columns of lime stabilized material. This soil lime reaction results in columns with greater material strength and improved permeability. The lime pile technique basically involves forming of holes in the ground filled with lime. The method of forming the hole, the type of lime used and filling techniques are different in different countries. Conventionally, geotechnical properties of lime pile-treated soils are dependent on their ionic content, type of lime used, curing time, curing temperature, and water content. The effect of lime on the behavior of clay soils has been studied by many researchers (Locat et al. 1990; Nelson and Miller 1992; Little 1995; Graves 1996; Rogers and Glendinning

1997; Nalbantoglu and Tuncer 2001). Installation of lime pile is quick and easy with relatively less amount of vibration. The use of lime and the lime pile technique is common in Japan, USA, Turkey, and Scandinavia (Rogers and Glendinning 1997). There has been a good amount of research on the use of lime stabilized pavement in India. This paper present a case study of the stabilization of expansive soil using lime sand piles for the construction of flexible pavement in Dr. M.G.R. bus terminus in Salem municipal corporation.

2 Materials and Methods

This section discusses the soil conditions of the site and the necessity of soil stabilization for construction of pavement.

2.1 *Soil Profile*

The soil profile in the proposed site for the construction of pavement in the bus terminus was essentially a layered system. The first layer was filled with earth from Salem quarry. The second layer was silty clay. This silty clay layer, followed by silty sand, was found in some borehole locations, i.e., for borehole locations 2, 4, 6, 7, 8, and 9. In other borehole locations, this layer is not found. The fourth layer is gravelly sand, and the last layer is completely weathered rock. The variations in the thickness of the above layers are listed Table 1.

The properties of the soil are detailed in Table 2 and Table 3.

The filled up earth was silty sand and non-plastic. The silty clay layer possesses high plasticity and will undergo volume change due to variation in moisture content. This clay was black in color and indicated the presence of organic content. The percentage of organic content varied between 20 and 30%. This is always referred as problematic soil and locally known as black cotton soil. The silty sand and gravely sand layer were non-plastic and also in the compacted state. Generally, any new structure is to be laid on the surface of original ground level. But in this site, the original ground layer was problematic soil. The presence of this layer always causes failure of lightly loaded structure. Hence, this layer required stabilization.

Table 1 Thickness of layered soil of the site

Thickness (mm)	Soil layer
0.4–1.5	Filled up earth
0.5–2.5	Silty clay
0.4–1.5	Silty sand/gravelly sand

Table 2 Atterberg limits and free swell index

Bore hole no.	Depth (m)	Liquid limit (%)	Plastic limit (%)	Free swell (%)
1	0.3	–	N.P	18
	0.5	53.84	36.72	56
	1	53.9	36.78	67
	2	51.93	35.81	65
	3	–	N.P	0
	3.4	–	N.P	–
	0.5	–	N.P	18
	1	47.64	33.52	60
	2	–	N.P	4
2	2.5	–	N.P	5
	3	–	N.P	0
	3.5	–	N.P	–
	0.5	22.3	21.14	22
3	1	54.76	35.63	60
	2	54.93	35.81	55
	3	–	N.P	0
	3.5	–	N.P	–
	3.7	–	N.P	–
	0.5	–	N.P	10
	1	54.56	35.24	62
4	2.5	–	N.P	4
	3	–	N.P	0
	3.5	–	N.P	–
	0.5	18.54	N.P	16
	1	51.26	35.12	56
	1.4	–	N.P	3
5	2	–	N.P	3
	3	–	N.P	0
	3.5	–	N.P	–
	0.5	–	N.P	17
	1	54.93	35.81	67
	2	53.86	35.72	55
6	2.5	24.35	21.18	24
	3	–	N.P	0
	3.4	–	N.P	–
	0.5	–	N.P	10

(continued)

Table 2 (continued)

Bore hole no.	Depth (m)	Liquid limit (%)	Plastic limit (%)	Free swell (%)
	1	–	N.P	14
7	2	52.88	35.74	54
	3	51.9	36.81	50
	4.5	23.18	20.46	22
	5	–	N.P	–
	0.3	–	N.P	11
	1	51.92	35.79	56
8	1.5	37.58	25.42	52
	2	19.24	N.P	8
	3	–	N.P	3
	3.4	–	N.P	–
	0.3	–	N.P	4
	0.5–1.0	50.84	33.8	55
9	1	22.45	68.76	20
	2	18.3	N.P	8
	3	–	N.P	3
	3.3	–	N.P	–

3 Lime Sand Pile Installation and Recommendations

The second, silty clay layer was the layer which requires improvement in strength and improvement in the reduction in volume change characteristics, i.e., swell and shrinkage characteristics. This was achieved by lime stabilization. For the present site, condition lime sand piles were adopted for stabilization. The lime sand pile construction was carried out in the following manner.

The top made up soil was removed up to the silty clayey layer. A hole of 300 mm was drilled using an auger. The depth of the hole is up to the top of the underlying silty sand/gravelly sand layer. The spacing between the auger holes was 1500 mm. Pulverized lime powder ($Ca(OH)_2$) was mixed with sand in 1:1 ratio. Powered limestone was not used because it is not reactive. Only lime powder ($Ca(OH)_2$) was used for stabilization. The above mix was poured in layers inside an auger up to a height of 250 mm and compacted well to achieve the maximum density. The augured bore holes were watered after filling it with lime. The site was left for curing for about 30 days to allow the lime react with the soil. After the completion of previous steps, two layers of murram or sand–gravel mix (1:1), each of 150 mm thickness was laid. Each murram layer was compacted for 12 cycles with vibro max. Proper drainage arrangement was made in the site to ensure no stagnation of water before and during construction.

Table 3 Grain size analysis

	Depth (mm)	Gravel (%)	Sand (%)			Silt and clay (%)
			Coarse	Medium	Fine	
1	0.3	37	7	12	12	32
	0.5	0	0	3	3	94
	1	0	1	2	3	94
	2	0	0	3	9	88
	3	27	9	21	22	21
	3.4	43	11	17	14	15
	0.5	43	10	14	10	23
	1	2	2	11	10	75
2	2	20	11	27	13	29
	2.5	8	9	29	19	35
	3	3	10	37	26	24
	3.5	24	20	30	16	10
	0.5	25	7	14	15	39
	1	0	0	2	1	97
	2	0	0	1	1	98
3	3	47	12	19	12	10
	3.5	50	10	16	10	14
	3.7	54	13	15	8	10
	0.5	46	7	16	12	19
	1	0	0	2	2	96
4	2.5	14	10	24	24	28
	3	34	6	25	18	17
	3.5	56	14	13	9	8
	0.5	25	7	16	14	38
	1	0	2	6	6	86
	1.4	30	18	9	14	29
5	2	33	13	7	17	30
	3	26	11	17	27	19
	3.5	46	12	20	12	10
	0.5	27	10	20	16	27
	1	0	0	2	1	97
	2	0	0	2	3	95
6	2.5	4	11	25	24	36
	3	12	11	40	21	16
	3.4	66	6	12	8	8

(continued)

Table 3 (continued)

	Depth (mm)	Gravel (%)	Sand (%)			Silt and clay (%)
			Coarse	Medium	Fine	
	1	27	9	19	16	29
	2	0	0	2	5	93
7	3	0	0	1	10	89
	4.5	4	6	22	36	32
	5	6	4	28	28	34
	0.3	38	8	19	16	19
	1	3	2	5	4	86
8	1.5	3	8	14	13	62
	2	16	5	25	19	35
	3	7	5	33	29	26
	3.4	20	16	34	20	10
	0.3	40	6	13	15	26
	0.5 -1.0	7	1	4	8	80
	1	10	11	24	24	31
9	2	10	10	29	22	29
	3	32	13	19	16	20
	3.3	56	13	15	11	5

3.1 CBR Tests

For the design of flexible pavement, the field density of the compacted sand–gravel mix was conducted at least ten locations so that design CBR can be determined in the laboratory. The field execution is shown in the photographs as shown in Fig. 1. The design field CBR was ensured after the ground improvement. The CBR after ground improvement was 15%.

3.2 Design of Flexible Pavement

The flexile pavement was designed for a CBR of 15%. The initial traffic volume in terms of commercial vehicles per day is 750/day as given by the client. To compute the design traffic the data assumed as per IRC 37 2012 guidelines are $r = 0.05$, $n =$ 10 years, $P = 750$ vehicles/day $D = 2$ $F = 3.5$. The cumulative number of standard axles that is considered for design is 30 msa (traffic in msa). The thickness of different layers of pavement for the CBR of 6% and traffic of 30 msa as per IRC: 37–2012 are given in Table 4. The selection of materials and construction of pavement were as per MORTH revision 5 "Specifications for road and bridge works" 2013.

Fig. 1 Stabilization of expansive soil in Salem new bus stand

Table 4 Thickness of layers as per IRC 37-2012

Layer	Thickness in mm
Granular subbase	250
Granular base	150
Dense bituminous macadam	40
Bituminous/Semi-dense bituminous concrete	30

4 Conclusions

The stabilization using lime sand piles was achieved methodically for the Salem bus stand site. After curing the site for maximum of 30 days, flexible pavement was laid on the prepared ground. The design field CBR value of 15% was ensured after ground improvement before laying the pavement. The lime sand pile stabilization technique proved to give promising results in terms of improvement in the strength of soil.

Acknowledgements This work was done for Salem Corporation for the construction of bus terminus pavement. The authors acknowledge the officials of Salem Municipal Corporation for funding this work.

References

Bhattacharya P, Bhattacharya A (1989) Stabilisation of bad banks of railway track by lime slurry pressure injection. In: Indian geotechnical conference, Visakhapatnam, vol 1, pp 315–319

Graves HM (1996) An introduction to soil stabilization. In: Proceedings, seminar on lime stabilization, Loughborough University. Thomas Telford, pp 5–12

IRC 37-2012 Guidelines for the design of flexible pavements, Indian Roads congress, New Delhi

Little DN (1995) Handbook for stabilization of pavement subgrades and base courses with lime. National Lime Association, Kendall/Hunt Publishing Company, Dubuque

Locat J, Berube M-A, Choquette M (1990) Laboratory investigations on the lime stabilization of sensitive clays: shear strength development. Can Geotech J 27(3):294–304

MORTH revision 5 (2013) Specifications for road and bridge works

Nalbantoglu Z, Tuncer ER (2001) Compressibility and hydraulic conductivity of chemically treated expansive clay. Can Geotech J 38(1):154–160

Nelson JD, Miller JD (1992) Expansive soils: problems and practice in foundation and pavement engineering. Wiley, New York

Rogers CDF, Glendinning S (1997) Slope stabilization using lime piles. In: Ground improvement geosystems: proceedings of the third international conference on ground improvement geosystems, densification and reinforcement. Thomas Telford, London, pp 174–180

Application of Enzyme-Induced Carbonate Precipitation (EICP) to Improve the Shear Strength of Different Type of Soils

Alok Chandra and K. Ravi

Abstract Urease enzyme derived from the agricultural source precipitates calcium carbonate ($CaCO_3$) from an aqueous solution of urea and calcium chloride via urea hydrolysis, which strengthens the soil by cementing and bridging soil particles. Different combinations of urea, calcium chloride and urease enzyme will have varying influence on the mechanical properties of the treated soil. This work aims to analyse the efficiency of enzyme-induced carbonate precipitation (EICP) on three different soil types (silty sands, clayey sand and silt). The optimum combination of urea, calcium chloride and urease enzyme is studied by mixing different concentrations, and the combination of these reagents in beakers and optimum precipitation is evaluated by gravimetrically measuring the amount of $CaCO_3$ precipitated. The observed optimum combination is applied to three different soil types, and the improvement in the compressive strength of the soils specimens due to carbonate precipitation is observed by conducting unconfined compressive strength (UCS) tests. The precipitation experiments in the beakers reveal that the increase of the urea-$CaCl_2$ concentration may inhibit the activity of urease, thereby precipitating lower amount of $CaCO_3$. The results of the UCS tests show that the technique of enzymatic calcium carbonate precipitation improves the shear strength of the all the three types of soils, however, more strength gain was obtained in the case of clayey sand. The microstructural observations with the help of scanning electron microscopy (SEM) and X-ray powder diffraction (XRD) tests verify the existence of calcite in the pores of compacted soil specimens.

Keywords EICP · Precipitation · Enzyme · Calcite · Shear strength

A. Chandra (✉) · K. Ravi
Department of Civil Engineering, Indian Institute of Technology Guwahati, Guwahati, Assam 781039, India
e-mail: alok.ce@iitg.ac.in

K. Ravi
e-mail: ravi.civil@iitg.ac.in

M. Latha Gali and R. R. P. (eds.), *Problematic Soils and Geoenvironmental Concerns*, Lecture Notes in Civil Engineering 88,
https://doi.org/10.1007/978-981-15-6237-2_52

1 Introduction

One of the major set of challenges in the infrastructure development is expansion of the population all around the geographical boundaries of the major cities (Dejong et al. 2010). People have started migrating to the periphery of these major cities which were previously ignored because of poor ground condition for the infrastructure. Now construction activity has been started on the weak and problematic soil which needs a substantial stabilization. Traditional method for ground improvement employs mechanical stabilization and injection of grouts particularly (cement or polymers), out of which most of these methods are expensive, require heavy machinery, disturbing other urban infrastructure and involve toxic chemicals which have significant environmental impact (Karol 2003). Therefore, there is a demand of a new, eco-friendly and sustainable ground improvement technology that can meet the infrastructure demand of the society.

Recent innovative research on precipitation of calcium carbonate ($CaCO_3$) in granular soils by biological process (named as biocalcification or biogrouting or biomineralization) is an emerging and promising soil stabilization technique and has been extensively studied over two decades (Whiffin et al. 2007; Van Paassen et al. 2010; Chou et al. 2011). In this process, calcium carbonate is precipitated either by microbial means or non-microbial means such that carbonate minerals fill within the soil pores, cementing and bridging the soil grains, thus providing overall stability to the soil structure. Most of the published work on this research includes microbial-induced calcite precipitation (MICP) which includes incorporating bacteria (such as Sporosarcina pasteurii) to produce urease enzyme, which helps in hydrolysis of urea and leads to $CaCO_3$ precipitation in the presence of calcium ions at favourable condition (such as pH and temperature). The $CaCO_3$ crystals nucleate on the cell walls of the ureolytic bacteria as negatively charged cell wall binds with the Ca^{2+} ions (Stocks-Fischer et al. 1999).

Thus, in general, microbially treated soils show improvement in strength, stiffness (DeJong et al. 2006, 2010; Van Paassen et al. 2010), reduction in porosity, compressibility and hydraulic conductivity (Whiffin 2004; Whiffin et al. 2007; Chou et al. 2011). Furthermore, many of the investigations that are described in the available technical literature report show that calcium carbonate precipitation can solve various geotechnical problems such as bearing capacity of shallow foundation, slope stability, wind erosion and seismic settlement (Meyer et al. 2011; Burbank et al. 2011; Harkes et al. 2010; Ivanov and Chu 2008; De Muynck et al. 2010; Cheng et al. 2013).

Even though potential application of MICP has been successfully demonstrated in various laboratories, pilot scale and few large-scale experiments, its real application in the field is facing numerous challenges. Bacteria's cultivation and storage at a large scale are a complex process, which require a gigantic bioreactor and special environmental conditions for the production of large biomass (Zhu and Dittrich 2016). Micro-organism requires sufficient oxygen and nutrients to remain active for deep soil improvement, delay or insufficient supply of which can show detrimental effect in microbial activity at desired treatment depth. Importantly, bacteria are not expected to enter through the pore throats smaller than approximately 0.4 μm, limiting the applicability of microbial treatment mostly to gravel and sand with lesser fine contents (Mitchell and Santamarina 2005). In addition to this, the injection of bacteria and their nutrients into the ground can change indigenous bacterial population and can disturb the ecological balance, leaving behind its long-term impacts to the ecosystem (Gat et al. 2016; Akiyama and Kawasaki 2011).

Alternatively, $CaCO_3$ can be precipitated by non-microbial means by incorporating urease enzyme derived from the agricultural sources (such as jack beans, watermelon seeds) also known as enzyme-induced carbonate precipitation (EICP). In EICP, microbial-free urease enzymes, which are commercially available in the powder forms are used, which use the similar precipitation path, i.e. urea hydrolysis and precipitate calcium carbonate in the presence of calcium ions as per Eq. 1 mentioned below. Here, the soil grain surface itself acts as the nucleation site for the $CaCO_3$ crystals growth.

$$CO(NH_2)_2 + 2H_2O + Ca^{2+} \xrightarrow{\text{Urease}} 2NH_4^+ + CaCO_3 \quad (1)$$

Unlike microbial process, EICP is relatively simple and straightforward process that eliminates the concern of complex process of bacterial cultivation, nutrient supply for their survival, oxygen availability for deep soil treatment and microbial ecology misbalance at injection points. Besides, the free enzyme has some other advantages of its own. Free urease has a size in the order of 12 nm per subunit (Blakeley and Zerner 1983) which is easily soluble in water, facilitating its large-scale production at treatment site. Solubility of urease enzyme makes the EICP solution to achieve similar viscosity as that of water, which can penetrate into very small pore spaces making its applicable to much finer soils. Furthermore, it has a limited lifespan, and its activity decreases with time (Marzadori et al. 1998). This limited lifespan is potentially advantageous in some engineering applications as the enzyme can naturally degrade and eliminate long-term impacts to the ecosystem.

In recent years, enzyme-induced carbonate precipitation (EICP) has attracted various researchers who have successfully demonstrated its application through laboratory scale experiments. These include injecting or percolating the EICP solution into the sand column providing an improvement in the strength and reduction in permeability (Yasuhara et al. 2012; Neupane et al. 2013, 2015; Almajed et al. 2018; Dilrukshi et al. 2018; Putra et al. 2016). Thus, EICP columns can be installed in the form of piles to provide slope stability, to support foundation and embankment, to restrict lateral spreading in liquefiable soils and to remediate the settlement of the foundation under existing structures (Hamdan et al. 2013). Also, EICP solution can provide more durable and substantial wind erosion stabilization in very short period of time (Knorr 2014; Almajed et al. 2020; Hamdan and Kavazanjian 2016). EICP can be used for selective plugging of the fractures present in the damaged structure of oil extraction wells, thereby preventing transportation of sand during oil recovery (Nemati and Voordouw 2003) and also enhancing oil recovery from the chalk reservoirs (Larsen et al. 2008).

Earlier research on EICP were mostly focused to stabilize the cohesion less soils (Ottawa sand, silica sand), where treatment was provided by injecting or percolating the treatment solution (Almajed et al. 2018; Putra et al. 2017; Neupane et al. 2013). Alternatively, few researchers have adopted mix and compact method to treat poorly graded sand (Carmona et al. 2017), Silts, Silty sands, Organic soil (Oliveira et al. 2017) and Clayey sand (Chandra and Ravi 2020) as per their optimum moisture content (w_{opt}). As far the authors knowledge, data on Unconfined compressive strength (UCS) behaviour of silty sand (SM), clayey sand (SC) and low compressible silt (ML) treated with similar concentration of EICP solutions (ie. Urea-$CaCl_2$ and urease enzyme) has not been studied. Thus, considering the limitation in previous studies, the work reported in this paper explores the behaviour of EICP process in the following ways.

- Determination of an optimum and standard concentration of EICP treatment solution by conducting beaker experiment.
- Determination of unconfined compressive strength (UCS) of three different types of cohesive soil, mixed with standard EICP solution and compacted as per their respective optimum moisture content (w_{opt}) and maximum dry unit weight ($\gamma_{d\,\max}$).
- Presence of calcite crystal in the voids of the soil is confirmed on the basis of results obtained from scanning electron microscopy (SEM) and X-ray powder diffraction (XRD) test.

2 Materials and Methodology

2.1 Characteristics of the Soils

Three different types of soils are used in this work and are classified as silty sand (SM), clayey sand (SC), low compressible silt (ML) according to Unified Soil Classification System (ASTM D 2487 2017). Silty sand (SM) and clayey sand (SC) were obtained from the campus premises of Indian Institute of Technology, Guwahati. Silt (ML) soil was obtained from the side slope of the Brahmaputra river embankment. The grain size distribution curves and the main characteristics of the soils studied in this work are given in Fig. 1 and Table 1.

The compaction of the soils is characterized by a standard Proctor test (ASTM D698 2012), which provides a maximum dry unit weight ($\gamma_{d\,\max}$) and an optimum water content (w_{opt}), which are used to prepare the specimens of the soils used in this work. Specific gravity (G) of the soil was determined by means of water pycnometer (ASTM D854 2014). Liquid limit ($w_l\%$), plastic limit ($w_p\%$) and plastic index (PI %) of the soil were determined according to ASTM D4318 2017. pH of the soil suspension was determined using a pH-sensitive electrode system according to ASTM D 4972 2013.

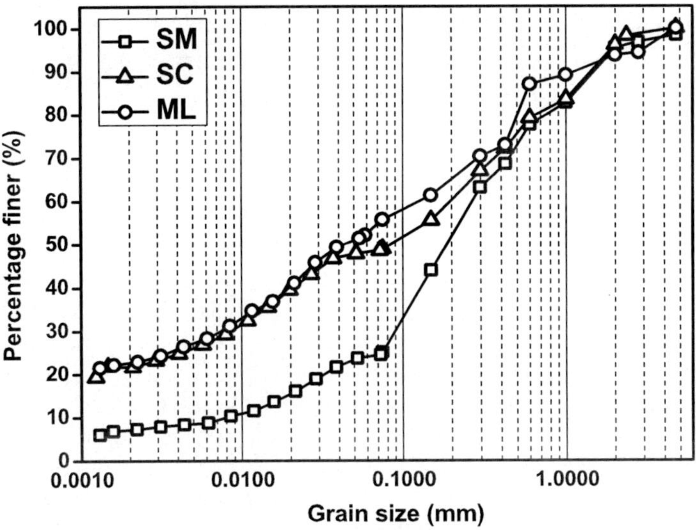

Fig. 1 Grain size distribution curves for the soils

Table 1 Main characteristics of the soils

Properties	Soil		
Soil classification	SM	SC	ML
Sand (%)	74.8	51	44.2
Silt (%)	17.92	27.28	32.84
Clay (%)	7.28	21.72	22.96
Specific gravity, *G*	2.62	2.73	2.64
Liquid limit, w_l (%)	35.35	44	49.38
Plastic limit, w_p (%)	28.5	23.70	22.51
Plastic index, PI (%)	36.1	20.3	26.87
Maximum dry density, $\gamma_{d\,\max}$ (g/cc)	1.81	1.74	1.74
Optimum water content, w_{opt} (%)	15.73	18	16.22
pH	6.5	7.1	5

2.2 Cementation Reagents and Enzyme

The basic treatment solution for the EICP process comprises the mixture of urease enzyme, urea ($CO(NH_2)_2$) and calcium chloride di-hydrate ($CaCl_2.2H_2O$) were obtained from Sigma-Aldrich company Ltd. The urease enzyme which helps in the hydrolysis of urea is derived from agricultural source, i.e. Canavalia ensiformis (jack beans), and is available in powder form with specific activity of 40,150 U/g (One micromolar unit will liberate 1.0 μmole of NH_3 from urea per min at pH 7.0 at

25 °C). The enzyme was tightly packed and stored in a refrigerator at 4 °C throughout the experiment plan.

2.3 Beaker Experiments

Under this test, the precipitation of carbonate minerals was directly evaluated in 100 ml transparent beakers. To prepare the cementation solution, equimolar concentration of (0.25, 0.5, 0.75, 1.0, 1.25 mol/l) urea-$CaCl_2$ was thoroughly mixed with distil water in various beakers. In a separate beaker, urease enzyme was also mixed with distil water to achieve a concentration of 8 kU/l (0.2 g/l as per the enzyme activity of 40,150 U/g). Then, 10 ml of urea-$CaCl_2$ solution and 10 ml of urease solution were mixed in the final beakers, resulting in a total solution volume of 20 ml and was allowed to cure for 14 days inside a controlled humidity chamber (95 ± 5%) and temperature (27 ± 3 °C). The beakers were sealed with thin plastic sheet to prevent any evaporation loss or spill of the solution. Curing period of 14 days was adopted to allow complete hydrolysis of urea (Carmona et al. 2016, 2017). Later on, the amount of material precipitated is then evaluated by filtering the solution in the final beaker through the filter paper and drying the beaker along with filter paper for 24 h at 60 °C (Neupane et al. 2013). Finally, the amount of calcium carbonate precipitation is gravimetrically measured as the sum of the mass of the calcium carbonate deposited on the filter paper and on the bottom of the beaker. Hence, more mass obtained from a particular combination of urea-$CaCl_2$ and urease concentration, more is its hydrolysing efficiency.

2.4 Specimen Preparation and UCS Testing

The soil specimens for unconfined compressive (UCS) test were prepared as per its maximum dry unit weight ($\gamma_{d\,\max}$) and an optimum water content (w_{opt}) obtained from standard Proctor test. The optimum solution mixed with oven-dried soil consisted of urea-$CaCl_2$ and urease solution as shown in Table 2. For comparison a control soil sample was also prepared with only distil water. The sample preparation consisted of the following steps: (1) to form the standard EICP solution (obtained from beaker experiment), the required amount of urea-$CaCl_2$ and urease solution was mixed together and was stirred for 5 min at 400 rpm; (2) the oven-dried soil was mixed with the standard EICP treament solution as per their respective optimum water content (obtained from the standard Proctor test) to obtain a lump-free homogeneous paste; (3) the paste was then kept in a humidity chamber (95 ± 5%) for 24 h to ensure even distribution of solution through the soil (ASTM D698 2012); (4) the paste was then compacted to form the cylindrical test specimen of 38 mm diameter and 76 mm height; (5) after preparation, the specimens were carefully put inside a plastic bag and cured for 14 days inside a controlled humidity chamber (95 ± 5%) and temperature

Table 2 Testing program

Treatment solution name and concentration				
	Control		0.5 M urea-$CaCl_2$	
	Urea-$CaCl_2$ (mol/l)	Urease (kU/l)	Urea-$CaCl_2$ (mol/l)	Urease (kU/l)
	0	0	0.5	8
Soil type	UCS	SEM and XRD	UCS	SEM and XRD
Silty sand (SM)	3	–	3	1
Clayey sand (SC)	3	–	3	1
Silt (ML)	3	–	3	1

Note Number of tests (UCS, SEM and XRD) carried out

(27 ± 3 °C); (6) after the curing time, each sample was removed from the plastic bag and tested for their unconfined compressive strength (UCS); (7) UCS tests were performed as per ASTM D2166 (2005) under a constant strain rate of 1 mm/min, and the data were recorded using an automatic data acquisition system. All the UCS tests were repeated twice with three replicates of each soil specimen type in each test to guarantee the reliability of the results.

2.5 Scanning Electron Microscopy (SEM) and X-ray Powder Diffraction (XRD) Test

Finally, small amount of soil sample was collected from the specimens of three types of soil and was used to examine the effect on morphology and mineralogical composition of the soil due to precipitated minerals using scanning electron microscopy (SEM) and X-ray powder diffraction (XRD) test, respectively. Dried soil samples were mounted on aluminium stubs, sputtered with gold coating (twice) and were imaged under (ZEISS Sigma 300) scanning electron microscope. XRD patterns were recorded using (Bruker D8 Advance) X-ray diffractometer over the range of 10°–60° 2θ at a step size = 0.05°, scan speed of 0.2°/min. Phase identification was made by searching the Inorganic Crystal Structure Database (ICSD) powder diffraction file database and published literature.

3 Results and Discussions

3.1 Beaker Experiments

While conduction of beaker expriments, it was observed that as soon as urease solution was mixed with the cementation solution of urea-$CaCl_2$, the final solution turned milky white indicating a quick start of urea hydrolyis. After completion of the 14 days curing period, on removing the plastic seal from the beaker, a pungent smell of ammonia (NH_3) was sensed. Based on the results of beaker experiments, Fig. 2 illustrates the effect of equimolar concentration of the urea-$CaCl_2$ solution on the mass of $CaCO_3$. The result shows that the urease concentration of 8 kU/l was able to hydrolyse almost all urea-$CaCl_2$ solution of concentration 0.5 mol/l and consequently a greater mass of $CaCO_3$ was precipitated as it touches the 100% precipitation line. 100% precipitation line presented in Fig. 2 is the line corresponding to the theoretical mass of $CaCO_3$, obtained when maximum possible $CaCO_3$ is precipitated, and can be calculated from Eq. 2.

$$\text{Mass of } CaCO_3 = C \times V \times M \tag{2}$$

where

C is the final concentration of solution (in mol/l),

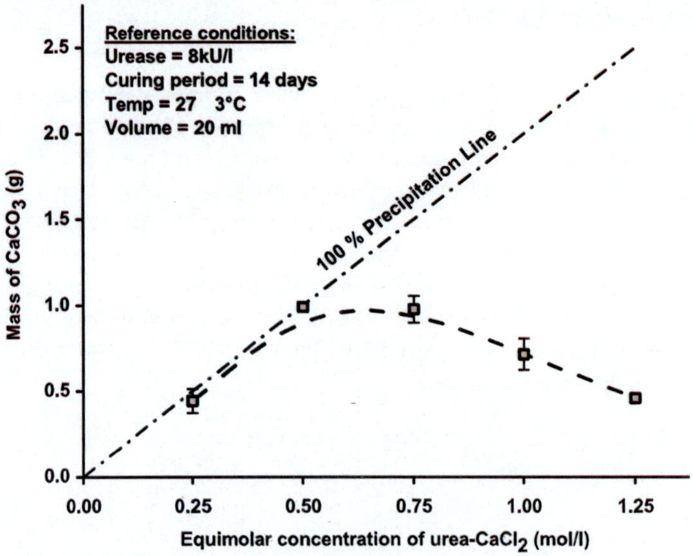

Fig. 2 Mass of calcium carbonate precipitated in beaker experiments

V is the final solution volume in beaker (in litres), and

M is the molar mass of $CaCO_3$ (100·087 g/mol).

The results also show that for 8 kU/l concentration of urease enzyme, with the increase in the urea-$CaCl_2$ concentration beyond 0.5 mol/l, lower quantity of urea was hydrolysed and lesser mass of $CaCO_3$ was precipitated. This behaviour may be due to high urea-$CaCl_2$ concentration which may inhibit the catalysation capacity of urease enzyme (Yasuhara et al. 2012; Carmona et al. 2016). Figure 2 also shows that urease concentration of 8 kU/l showed slightly detrimental effect on urea-$CaCl_2$ concentration of 0.25 mol/l, and lesser mass $CaCO_3$ was precipitation. This behaviour may be due to that urease concentration of 8 kU/l was excessive for hydrolysing urea-$CaCl_2$ concentration of 0.25 mol/l and catalysing efficiency might have suppressed (Neupane et al. 2013; Carmona et al. 2017). The results clearly show that for a particular urease concentration, there is an optimum urea-$CaCl_2$ concentration, which allows its maximum hydrolysis and precipitates maximum mass of $CaCO_3$. Hence, urea-$CaCl_2$ concentration of 0.5 mol/l and urease concentration of 8 kU/l were adopted as a standard concentration of the treatment solution in the current study.

3.2 *Unconfined Compression Strength (UCS) Test*

UCS test results show the effect on three different types of soil specimens treated with the equal standard concentration of treatment solution (urea-$CaCl_2$ of 0.5 mol/l and urease enzyme of 8 kU/l). SEM and XRD results confirm the presence of calcite by its rhombohedral shape and distinct peak, respectively (Figs. 7 and 8), which is the most stable polymorph of $CaCO_3$. Figures 3, 4 and 5 compares the stress–strain behaviour of untreated soil specimens with that of the treated specimens which show a clear improvement in compressive strength of the treated soil. The stress–strain curves of the all the three soils show a linear elastic response followed by a decrease in the strength after achieving peak strength. Figure 6 shows the effect of soil type on the unconfined compressive strength (q_u). A significant improvement of compressive strength was observed in the silty sand (SM) and clayey sand (SC) soil with an improvement ratio of 1.54 and 1.88, respectively. In contrast, silt (ML) shows lesser improvement in the compressive strength with an improvement ratio of just 1.24.

Most effective increase in strength and stiffness of the soil is observed when $CaCO_3$ is precipitated at interparticle contacts of sand grains to provide effective load transfer (Harkes et al. 2010; Mortensen et al. 2011). The presence of an optimum amount of fine content in the pores throat of sands facilitates a dense arrangement and provides further improvement in strength of the soil. In addition to this, presence of clayey particles offer more specific surface area for the calcite bond formation (Sharma and Ramkrishnan 2016). Similar effect can be seen in case of clayey sand (SC), where pores between the coarse grains were filled with the finer sized content

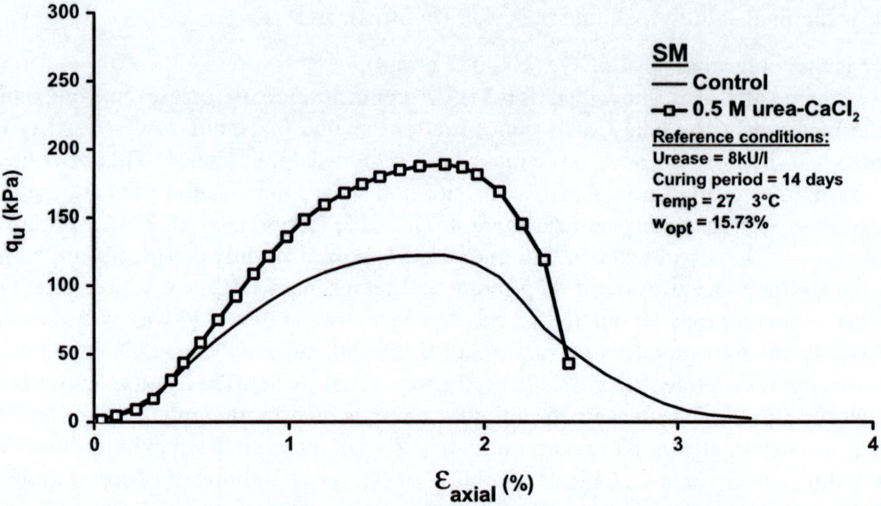

Fig. 3 Stress–strain curve of silty sand (SM) for control and 0.5 M urea-$CaCl_2$

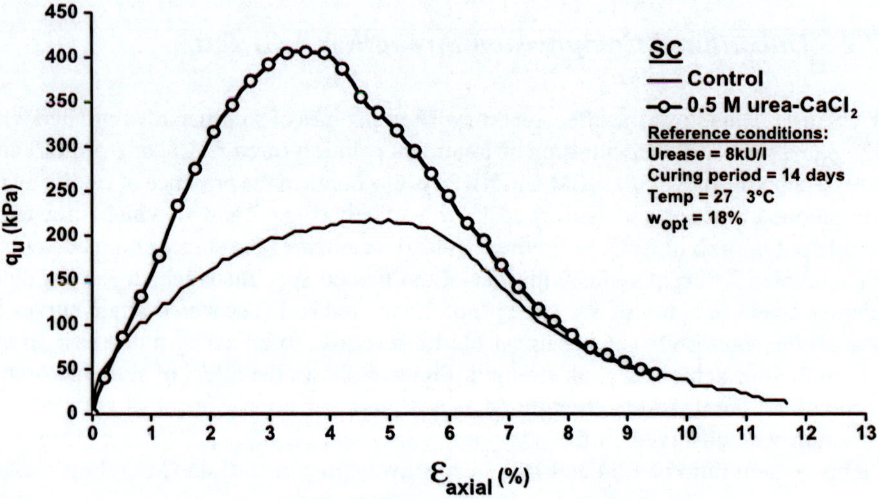

Fig. 4 Stress–strain curve of clayey sand (SC) for control and 0.5 M urea-$CaCl_2$

(49%), and thus resulted in greater particle contacts area for the formation of calcite bonds to improve the UCS of the soil (Soon et al. 2014).

On the other hand, silty sand (SM) contained more percentage of coarse grain particles, in comparison with fine content (25.2%). After applying the same concentration of EICP solution, silty sand (SM) was observed to show less improvement in UCS value due to the presence of more void space, which may have provided lesser

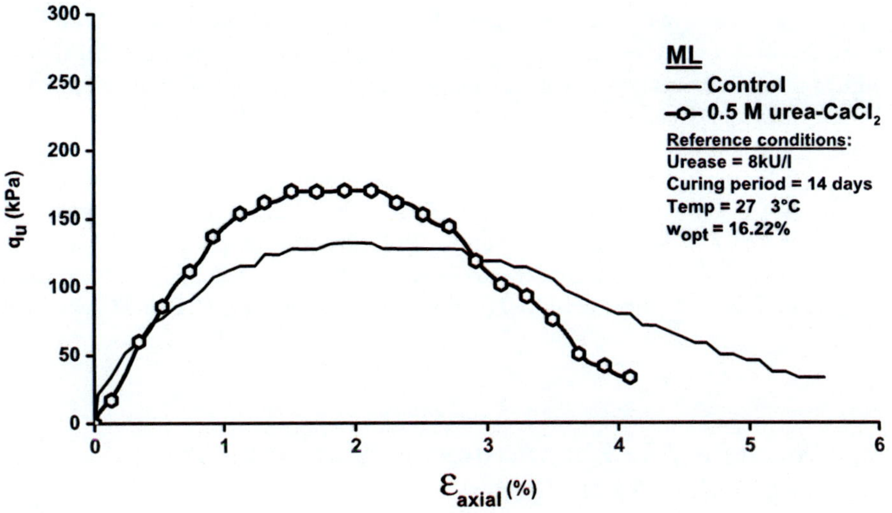

Fig. 5 Stress–strain curve of silt (ML) for control and 0.5 M urea-$CaCl_2$

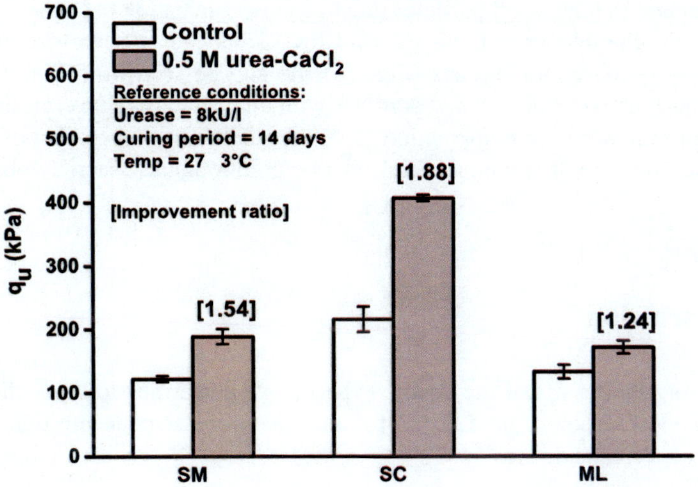

Fig. 6 UCS test: Effect of treatment on the unconfined compressive strength (q_u) of three different types of soil

possibilities for calcite crystal nucleation and point to point load transfer (Soon et al. 2014). The lowest improvement was observed in case of low compressible silt (ML) which contained more amount of fine content (58%). In addition to this, silt (ML) soil used in this study had a pH value of 5 which would have created a slight acidic environment, hindering the hydrolysis of urea and thereby precipitating less amount of calcite as compared to the other two soils, i.e. silty sand (SM) and clayey sand

(SC) which have a pH value of 6.5 and 7.1, respectively (Carmona et al. 2017; Chou et al. 2011; Burbank et al. 2011). The amount of calcite precipitated in the soil and its pH after treatment was not evaluated in this study. Therefore no relationship can be drawn between calcite content and improvement ratio of the three different soils. Nevertheless, it cannot be directly related, as these soil have different values of Atterberg limits, optimum moisture content, maximum dry density, pH and porosity. Carmona et al. (2017) and Soon et al. (2014) also observed a significant improvement ratio in the case of Silts and Silty Sands. While Kannan et al. (2020), Sharma and Ramkrishnan (2016), and Chandra and Ravi (2020) reported a considerable improvment in high compressible clays (CH), low compressible clays (CI) and Clayey sand (SC).

3.3 Scanning Electron Microscopy (SEM) and X-ray Diffraction (XRD) Tests

Figure 7 shows morphology of the treated soil sample was analysed using SEM. It can be observed from Fig. 7a that for silty sand (SM) soil, the silt particles are observed to be attached to large sand particles due to calcite nucleating between the adjacent surfaces. On the other hand, clayey sand (SC) soils can be seen forming a dense bonds/coating of calcite crystals (Fig. 7b). As can be seen from Fig. 7c, silt soil (ML) shows more of fine-grained particles with unattached calcite crystals. Figure 8 shows the mineralogical composition of the three different types of soils. In the all cases, the presence of calcite minerals can be predominantly seen by observing its peaks.

4 Conclusion

Considering the results of the beaker experiments performed to study the effect of various concentration of urea-$CaCl_2$ for a constant urease concentration, following observations can be drawn.

- An increase in concentration of urea-$CaCl_2$ for a constant concentration of urease enzyme and the efficiency of calcium carbonate precipitation is decreased as high urea-$CaCl_2$ concentration inhibit the catalysing capacity of urease enzyme.
- For a low concentration of urea-$CaCl_2$, urease enzyme can become excessive such that it may have suppressed the hydrolysis of urea, thereby resulting in lesser mass of calcium carbonate precipitation.

Based on the results of the UCS, SEM and XRD tests done to study the effect of soil type on the efficiency of the EICP process, the following conclusions can be drawn:

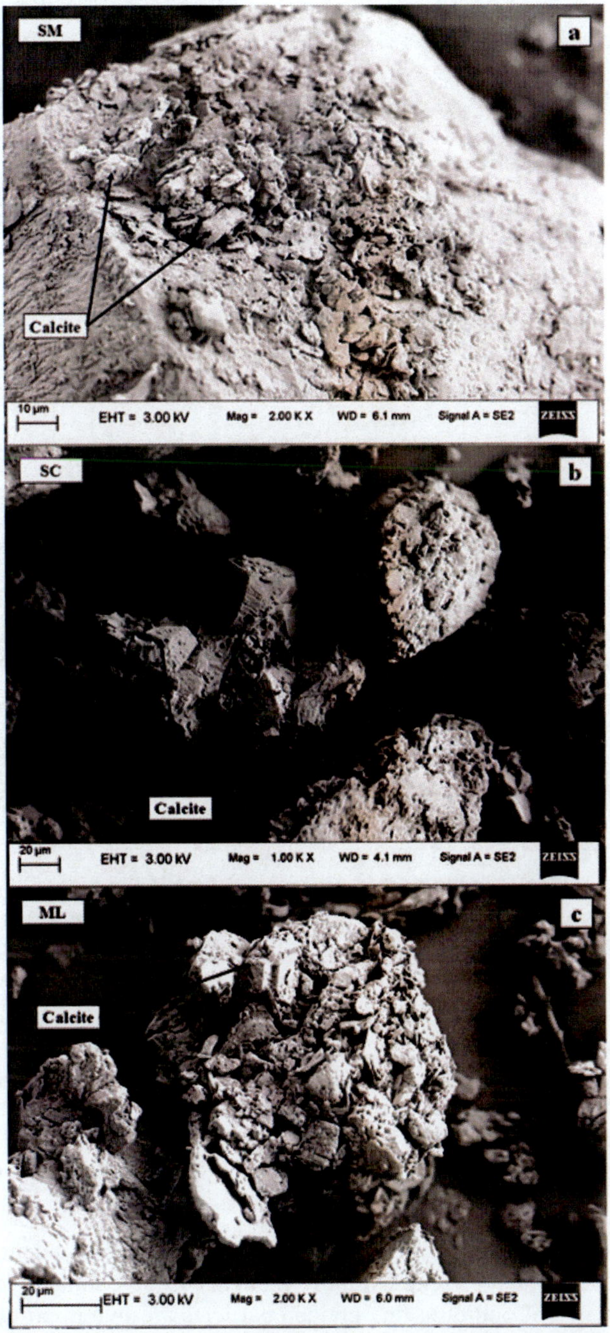

Fig. 7 SEM image of treated soil, **a** silty sand (SM); **b** clayey sand (SC); **c** silt (ML)

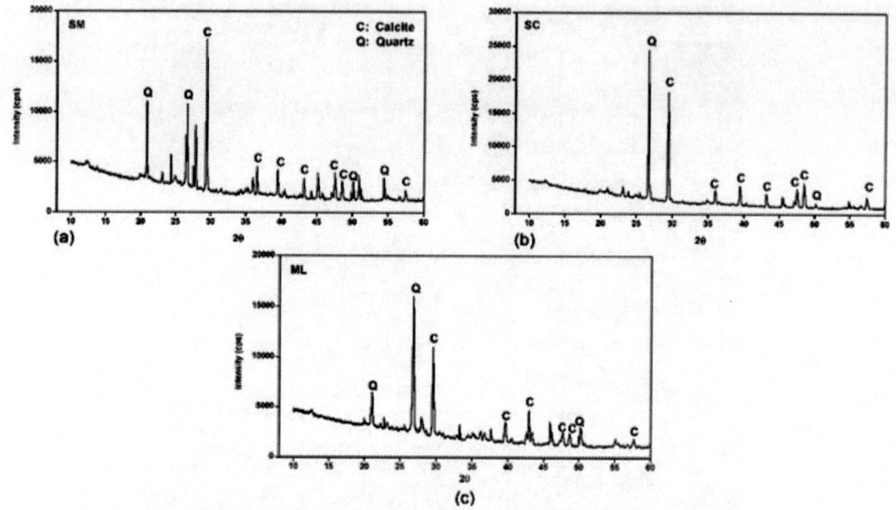

Fig. 8 XRD peak of three different types of soil, **a** silty sand (SM); **b** clayey sand (SC); **c** silt (ML)

- A significant improvement was observed in the unconfined compressive strength of silty sand (SM) and clayey sand (SC) soils, whereas lesser improvement was observed in the unconfined compressive strength of silt (ML) soil.
- The presence of optimum fine content in clayey sand (SC) promoted more contacts points for the calcite crystals nucleation and better establishment of effective bonds between the soil particles as compared to silty sand (SC). Further increase in fine content for silt (ML) increased the void ratio, thereby reduces calcite nucleation points decreasing the load transfer effects.
- Neutral to basic pH value of the soil favours the EICP process, thereby creating a better environment for urea hydrolysis and consequently more mass of calcium carbonate precipitation (as in case of SC and SM), and on the other hand low pH value of soil (as in case of ML) creates an acidic environment and hinders the hydrolysis of urea, thereby precipitating less amount of $CaCO_3$.
- SEM image shows the presence of $CaCO_3$ in all the three types of soil. It is clear from the image that $CaCO_3$ crystals facilitate load transfer between sand grains of both silty sand (SM) and clayey sand (SC) soils. Silt (ML) soil contains $CaCO_3$ crystals in unattached form.
- Results of XRD test confirm the presence of calcite minerals in all the three types of soils.

The above conclusion suggests that the enzyme-induced calcite precipitation (EICP) has a potential for strength improvement in cohesive soils. Alkaline pH of the soil and presence of optimum fine-grained particles are two important factors that promote the urea hydrolysis and effective bonding of $CaCO_3$ crystals to the soils, respectively.

References

Almajed A, Khodadadi Tirkolaei H, Kavazanjian Jr E (2018) Baseline investigation on enzyme-induced calcium carbonate precipitation. J Geotech Geoenviron Eng 144(11):04018081

Almajed A, Lemboye K, Arab MG, Alnuaim A (2020) Mitigating wind erosion of sand using biopolymer-assisted EICP technique. Soils Found

Akiyama M, Kawasaki S (2011) Fundamental study on new grouting material using calcium phosphate compounds—crystal precipitation test and unconfined compression test of sand test pieces—. Jpn Geotech J 6:341–350

ASTM D2166/D2166M-16 (2016) Standard test method for unconfined compressive strength of cohesive soil. West Conshohocken, PA. https://doi.org/10.1520/D2166_D2166M-16

ASTM D2487–17 (2017) Standard Practice for Classification of Soils for Engineering Purposes (Unified Soil Classification System), West Conshohocken, PA. https://doi.org/10.1520/D2487-17

ASTM D4318–17e1 (2017) Standard test methods for liquid limit, plastic limit, and plasticity index of soils. West Conshohocken, PA. https://doi.org/10.1520/D4318-17E01

ASTM D698-12e2 (2012) Standard test methods for laboratory compaction characteristics of soil using standard effort (12400 ft-600 kN-m). West Conshohocken, PA. https://doi.org/10.1520/D0698-12E02

ASTM D854-14 (2014) Standard test methods for specific gravity of soil solids by water pycnometer. West Conshohocken, PA. https://doi.org/10.1520/D0854-14

Blakeley RL, Zerner B (1983) Jack bean urease: the first nickel enzyme. Inorg Chim Acta 79:11

Burbank MB, Weaver TJ, Green TL, Williams BC, Crawford RL (2011) Precipitation of calcite by indigenous microorganisms to strengthen liquefiable soils. Geomicrobiol J 28(4):301–312

Carmona JP, Oliveira PJV, Lemos LJ (2016) Biostabilization of a sandy soil using enzymatic calcium carbonate precipitation. Procedia Eng 143:1301–1308

Carmona JP, Venda Oliveira PJ, Lemos LJ, Pedro AM (2017) Improvement of a sandy soil by enzymatic calcium carbonate precipitation. Proc Inst Civ Eng-Geotech Eng 171(1):3–15

Chandra A, Ravi K (2020) Effect of magnesium incorporation in Enzyme-Induced Carbonate Precipitation (EICP) to improve shear strength of soil. In: Advances in Computer Methods and Geomechanics. Springer, Singapore, pp 333–346

Cheng L, Cord-Ruwisch R, Shahin MA (2013) Cementation of sand soil by microbially induced calcite precipitation at various degrees of saturation. Can Geotech J 50(1):81–90

Chou CW, Seagren EA, Aydilek AH, Lai M (2011) Biocalcification of sand through ureolysis. J Geotech Geoenviron Eng 137(12):1179–1189

De Muynck W, De Belie N, Verstraete W (2010) Microbial carbonate precipitation in construction materials: a review. Ecol Eng 36(2):118–136

DeJong JT, Fritzges MB, Nüsslein K (2006) Microbially induced cementation to control sand response to undrained shear. J Geotech Geoenviron Eng 132(11):1381–1392

DeJong JT, Mortensen BM, Martinez BC, Nelson DC (2010) Bio-mediated soil improvement. Ecol Eng 36(2):197–210

Dilrukshi RAN, Nakashima K, Kawasaki S (2018) Soil improvement using plant-derived urease-induced calcium carbonate precipitation. Soils Found 58(4):894–910

Gat D, Ronen Z, Tsesarsky M (2016) Soil bacteria population dynamics following stimulation for ureolytic microbial-induced $CaCO_3$ precipitation. Environ Sci Technol 50(2):616–624

Hamdan N, Kavazanjian E Jr (2016) Enzyme-induced carbonate mineral precipitation for fugitive dust control. Géotechnique 66(7):546–555

Hamdan N, Kavazanjian E Jr, O'Donnell S (2013) Carbonate cementation via plant derived urease. In: Proceedings of the 18th international conference on soil mechanics and geotechnical engineering, Paris, Sept 2013

Harkes MP, Van Paassen LA, Booster JL, Whiffin VS, van Loosdrecht MC (2010) Fixation and distribution of bacterial activity in sand to induce carbonate precipitation for ground reinforcement. Ecol Eng 36(2):112–117

Ivanov V, Chu J (2008) Applications of microorganisms to geotechnical engineering for bioclogging and biocementation of soil in situ. Rev Environ Sci Bio/Technol 7(2):139–153

Kannan K, Bindu J, Vinod P (2020) Engineering behaviour of MICP treated marine clays. Mar Georesour Geotechnol 1–9

Karol RH (2003) Chemical grouting and soil stabilization, 3rd edn. M. Dekker, New York

Knorr B (2014) Enzyme-induced carbonate precipitation for the mitigation of fugitive dust. Arizona State University

Larsen J, Poulsen M, Lundgaard T, Agerbaek M (2008) Plugging of fractures in chalk reservoirs by enzyme-induced calcium carbonate precipitation. SPE Prod Oper 23(04):478–483

Marzadori C, Miletti S, Gessa C, Ciurli S (1998) Immobilization of jack bean urease on hydroxyapatite: urease immobilization in alkaline soils. Soil Biol Biochem 30(12):1485–1490

Meyer FD, Bang S, Min S, Stetler LD, Bang SS (2011) Microbiologically-induced soil stabilization: application of Sporosarcina pasteurii for fugitive dust control. In: Geo-frontiers 2011: advances in geotechnical engineering, pp 4002–4011

Mitchell JK, Santamarina JC (2005) Biological considerations in geotechnical engineering. J Geotech Geoenviron Eng 131(10):1222–1233

Mortensen BM, Haber MJ, DeJong JT, Caslake LF, Nelson DC (2011) Effects of environmental factors on microbial induced calcium carbonate precipitation. J Appl Microbiol 111(2):338–349

Nemati M, Voordouw G (2003) Modification of porous media permeability, using calcium carbonate produced enzymatically in situ. Enzyme Microbial Technol 33(5):635–642

Neupane D, Yasuhara H, Kinoshita N, Unno T (2013) Applicability of enzymatic calcium carbonate precipitation as a soil-strengthening technique. J Geotech Geoenviron Eng 139(12):2201–2211

Neupane D, Yasuhara H, Kinoshita N, Ando Y (2015) Distribution of mineralized carbonate and its quantification method in enzyme mediated calcite precipitation technique. Soils Found 55(2):447–457

Oliveira PJV, Freitas LD, Carmona JP (2017) Effect of soil type on the enzymatic calcium carbonate precipitation process used for soil improvement. J Mater Civil Eng 29(4):04016263

Putra H, Yasuhara H, Kinoshita N, Neupane D, Lu CW (2016) Effect of magnesium as substitute material in enzyme-mediated calcite precipitation for soil-improvement technique. Front Bioeng Biotechnol 4:37

Putra H, Yasuhara H, Kinoshita N, Hirata A (2017) Application of magnesium to improve uniform distribution of precipitated minerals in 1-m column specimens. Geomech Eng 12(5):803–813

Sharma A, Ramkrishnan R (2016) Study on effect of microbial induced calcite precipitates on strength of fine grained soils. Perspect Sci 8:198–202

Soon NW, Lee LM, Khun TC, Ling HS (2014) Factors affecting improvement in engineering properties of residual soil through microbial-induced calcite precipitation. J Geotech Geoenviron Eng 140(5):04014006

Stocks-Fischer S, Galinat JK, Bang SS (1999) Microbiological precipitation of $CaCO_3$. Soil Biol Biochem 31:1563–1571

van Paassen LA, Ghose R, van der Linden TJ, van der Star WR, van Loosdrecht MC (2010) Quantifying biomediated ground improvement by ureolysis: large-scale biogrout experiment. J Geotech Geoenviron Eng 136(12):1721–1728

Whiffin VS (2004) Microbial $CaCO_3$ precipitation for the production of biocement. Doctoral dissertation, Murdoch University

Whiffin VS, van Paassen LA, Harkes MP (2007) Microbial carbonate precipitation as a soil improvement technique. Geomicrobiol J 24(5):417–423

Yasuhara H, Neupane D, Hayashi K, Okamura M (2012) Experiments and predictions of physical properties of sand cemented by enzymatically-induced carbonate precipitation. Soils Found 52(3):539–549

Zhu T, Dittrich M (2016) Carbonate precipitation through microbial activities in natural environment, and their potential in biotechnology: a review. Front Bioeng Biotechnol 4:4

Improving the Strength of Weak Marine Clays by Treating with POFA and DRWP Inclusions

K. Ramu, R. Dayakar Babu, and K. Roja Latha

Abstract India, being a peninsular country, has a long coastline and also been the habitat for a good amount of the population. On the other hand, the accumulation of various waste materials is now becoming a major concern to the environmentalists. New methods and new materials of construction have been continuously explored, and hence, in order to achieve both the needs of improving the soft marine clays and also to make use of the industrial wastes in the best possible way. The present study was done by partially replacing the marine clay with palm oil fuel ash (POFA), a relative new agro-waste and further adding it with lime and discrete reinforcing waste plastic inclusions (DRWPI). The results obtained revealed that the penetration and the strength characteristics of POFA improved for modified soft marine clay and are further improved with an optimum percentage of lime and optimum dosage of discrete waste fiber.

Keywords Marine clay · Palm oil fuel ash · Discrete reinforced waste plastic inclusions

1 Introduction

The marine clays are widely spread over the east and west coast of India. These soils are having its natural moisture content very nearer to its liquid limit, and hence, these soils are very soft and sensitive. Flocculated structure is formed, which is a loose, open structure of clay particles, when the clay is deposited in oceans. Hence, these soils generally have low shear strength and high compressibility. Researchers gave several techniques to improve the engineering behavior of these soils. Chemical stabilization may be one of the well-suited methods to improve the engineering characteristics of

K. Ramu (✉) · K. Roja Latha
University College of Engineering, JNTUK University Kakinada, Kakinada 533003, India
e-mail: kramujntu@hotmail.com

R. Dayakar Babu
Kakinada Institute of Science and Technology, Divili 533433, India

M. Latha Gali and R. R. P. (eds.), *Problematic Soils and Geoenvironmental Concerns*, Lecture Notes in Civil Engineering 88,
https://doi.org/10.1007/978-981-15-6237-2_53

fine-grained soil. In the chemical stabilization, different types of additives are mixed with the soil to develop chemical reaction with the clay minerals and to form a stable compound to improve the engineering properties of the soil.

1.1 Usage of Waste Materials

Accumulation of various waste materials is now becoming a major concern to the environmentalists. Earlier, all the government policies had been planned to increase development of industries, but the environmental issues were not concerned. But now, the situation of environmental pollution has become alarming, and it is the foremost task to think regarding the disposal of waste materials which are generated along with the main product before finalizing any project. The generation, handling and disposal of solid wastes are now a grave concern in the country. In these circumstances, there is a need to explore the possibility of utilization of these waste materials in bulk.

1.2 Need for the Study

To fulfill the above objective, new methods and new materials of construction have been continuously explored, and hence, in order to achieve both the needs of improving the marine clays and also to make use of the industrial wastes in the best possible way. The present experimental study has been taken up by partially replacing the marine clay with palm oil fuel ash (POFA), a relative new agro-waste and further adding it with lime and discrete reinforcing waste plastic inclusions (DRWPI). The prime objective was aimed at evaluating the performance of stabilized marine clay with an optimum of POFA, as a replacement and further adding it with lime and discrete reinforcing waste plastic inclusions (DRWPI).

2 Review of Literature

Generally, clayey soils are stiff when they are dry and lose their stiffness as they become saturated. Clay beds are formed in a flocculated structure, when the clay particles are deposited in saltwater. A very loose card-like structure is formed with the clay, silt and fine sand particles are settled almost at the same time (Bjerrum and Heiberg 1971). Hence, these marine sediments usually formed with high void ratios pose problems associated with the fine-grained soils deposited at a soft consistency. The marine clay deposits are existing at high natural moisture content, which may be nearer to its liquid limit. Clays have their natural moisture content nearer to its liquid limit had low shear strength and high compressibility. The behavior of these soft marine clay deposits varies with the different environmental conditions.

These soft clay behaviors can be a challenge to the geotechnical engineers to build infrastructure on these deposits. Several researchers are recommended for different ground improvement techniques. Replacing the problematic soil by a suitable soil is one of the earliest methods in the ground improvement. The availability and the cost involved in this method made the researchers to identify alternative techniques.

Various remedial measures like chemical stabilization, preloading with and without vertical drains, soil reinforcement and densification have been practiced to varying degrees of success. The limitations of these techniques questioned their adaptability in all conditions. Hence, the researchers are being done all over, to evolve the most effective and practical treatment methods, to alleviate the problems caused to any structures laid on marine clay strata. The investigations on the physical, chemical and mineralogical properties of marine clay are reported in Shridharan et al. (1989) and Chew et al. (2004). Significant research on strength and stiffness of the marine clays is conducted by various researchers Koutsoftas et al. (1987) and Zhou et al. (2005). The load carrying capacity of the subgrade has been improved with the addition of GBFS stabilized with lime and cement (Kumar and Mehta 1998). Fly ash is observed to be a potential material to be utilized for clayey soil improvement in addition (Muntohar and Hantoro 2000). The lime fly ash admixtures reduced the water absorption capacity, and compressibility of soils was observed (Zhang and Cao 2002). The engineering properties of marine clay of the Eastern Coast of India are studied and reported in Basack and Purkayastha (2009). The effect of sodium lignosulfonate using as a stabilizing agent in the marine clay was reported in Mathew et al. (2017). The physical and engineering characteristics of soft clay improved with the compacted lime fly ash column are observed (Kalyana Krishna et al. 2015).

From the literature, stabilization of marine clays is an effective alternative for geotechnical engineers considering the economics of construction. Mechanical stabilization was used to alter the physicochemical properties of marine clay in order to permanently stabilize them. Based on this, the present work was carried out to study the performance of an agro-waste, palm oil fuel ash (POFA) as partial replacement in marine clay, lime as binder and discrete reinforcing waste plastic inclusions (DRWPI) as randomly distributed reinforcing elements in combination with the behavior of weak marine clay.

3 Methodology

The study is carried out on the weak marine clay by partially replacing it with palm oil fuel ash (POFA) and further adding it with lime and discrete reinforcing waste plastic inclusions (DRWPI). The following percentages presented in Table 1 were considered.

Table 1 Variable and percentages considered

Material	% Content
POFA as replacement	0, 10, 20, 30
Lime	0, 2, 4, 6
Discrete reinforcing waste plastic inclusions (DRWPI)	0, 1, 2, 4

3.1 Materials Used

Marine Clay. The soil used for the study has been collected at a depth of around 0.5 m to 1 m below the ground level at the port area in Kakinada. This soil is tested in the laboratory to access the properties of the soil based on the relevant IS code provisions and is presented in Table 2.

Palm Oil Fuel Ash (POFA). Palm oil fuel ash (POFA) used for this work was brought from the Ruchi Palm Oil Factory, Peddapuram, East Godavari District, Andhra Pradesh. As per IS code of practice, various tests were conducted to determine the properties of palm oil fuel ash (POFA) and are presented in Table 3.

Lime. Commercially available lime (Birla Company Lime) was used in this investigation.

Table 2 Physical properties of marine clay

S. No.	Property	Value
1	Specific gravity	2.62
2	Sand (%)	7
3	Silt (%)	38
4	Clay (%)	55
5	Liquid limit (%)	76.8
6	Plastic limit (%)	32.5
7	Plastic index (%)	44.3
8	Soil classification	CH
9	Differential free swell (%)	36
10	Maximum dry density (g/cc)	1.33
11	Optimum moisture content (%)	31.2
12	Cohesion (kPa)	35
13	Angle of internal friction (°)	0
14	CBR—Unsoaked (%)	2.8
15	CBR—soaked (%)	1.5

Table 3 Physical properties of POFA

S. No.	Property	Value
1	Specific gravity	1.91
2	Sand size particles (%)	73
3	Silt (%) and clay size particles (%)	27
4	Maximum dry density (g/cc)	1.28
5	Optimum moisture content (%)	14.3

Discrete Reinforcing Waste Plastic Inclusions (DRWPI). More than a 100 million tonnes of plastic is produced worldwide each year. Disposal of plastic through recycling, burning or land filling is a myth because it does not undergo bacterial decomposition. Once plastic is produced, and the harm is done once and for all. Plastic wastes clog the drain and thus hit especially urban sewage systems. So recycling of waste plastics was directly involved to protect the environment. The waste plastic water bottles were used by cutting them into discrete strips.

3.2 Sample Preparation and Laboratory Testing

The soil was initially air dried prior to the testing. The air-dried soil was pulverized and passed through a 4.75 mm sieve. The tests were conducted in the laboratory on the pulverized marine clay to study the behavior of marine clay, before and after treatment with the considered materials. The tests were conducted as per IS codes of practice. In the laboratory, various experiments were conducted by part replacing marine clay with different percentages of POFA and stabilizing it with lime as a binder and also further by discrete reinforcing waste plastic inclusions (DRWPI) as discrete reinforcement.

4 Discussion of Results

Plasticity, compaction, strength and CBR tests were conducted with a view to determine the optimum combination of palm oil fuel ash (POFA) as a replacement in marine clay.

4.1 Effect of % Palm Oil Fuel Ash (POFA) as Replacement in Marine Clay

From Fig. 1, we can ascertain the reduction in plasticity index by about 31.45%. This clearly gives an initial idea of the improvement of the weak marine clay when part is replaced with POFA. This is due to the replacement of plastic soil with a non-plastic waste material. Figures 2 and 3 show the variation of compaction parameters with the % replacement of POFA. From these graphs, it can be seen that with the inclusion of POFA, there is a decrease in maximum dry density corresponding to marginal decrease in optimum moisture content.

Figures 4, 5 and 6 also show clear improvement in the penetration and strength characteristics with a clear indication of imparting friction to the pure clayey soil.

The results show that there is an improvement of about 46.5% and 46% in CBR unsoaked and soaked values, respectively, up to 20% POFA as replacement, and on further replacement, there is a decrease in the values. A similar trend is seen in the strength values with an improvement of 26% for 20% POFA. This is due to the possible initial binding of particles with the presence of little amounts of calcium oxide in POFA and also as it is a pozzalonic material. Hence, from the above findings,

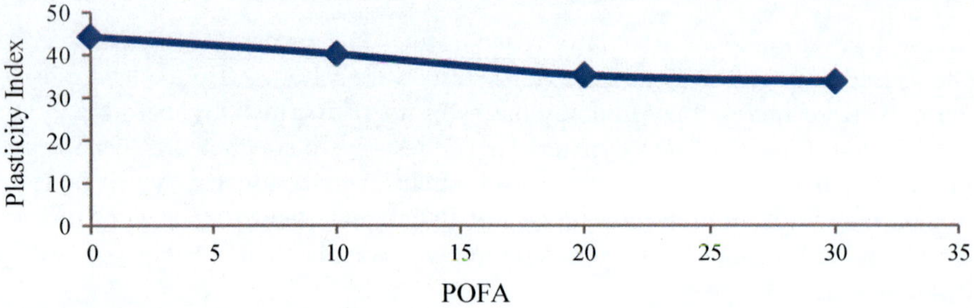

Fig. 1 Variation of plasticity index with the percentage replacement of marine clay with POFA

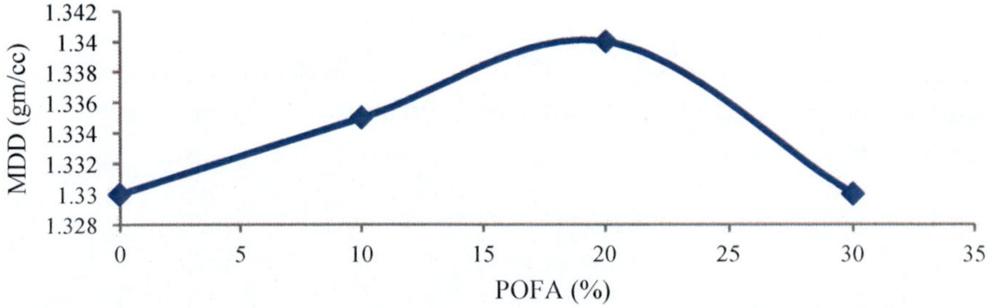

Fig. 2 Variation of maximum dry density with the percentage replacement of marine clay with POFA

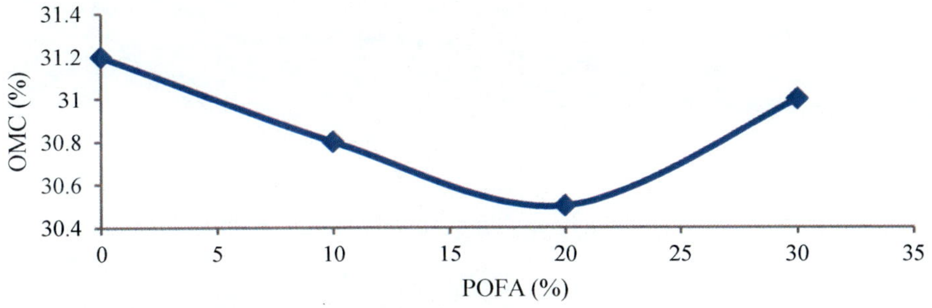

Fig. 3 Variation of optimum moisture content (OMC) with the percentage replacement of marine clay with POFA

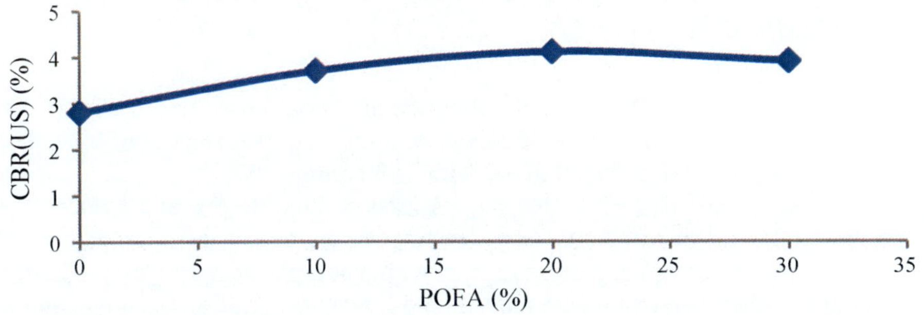

Fig. 4 Variation of unsoaked CBR with the percentage replacement of marine clay with POFA

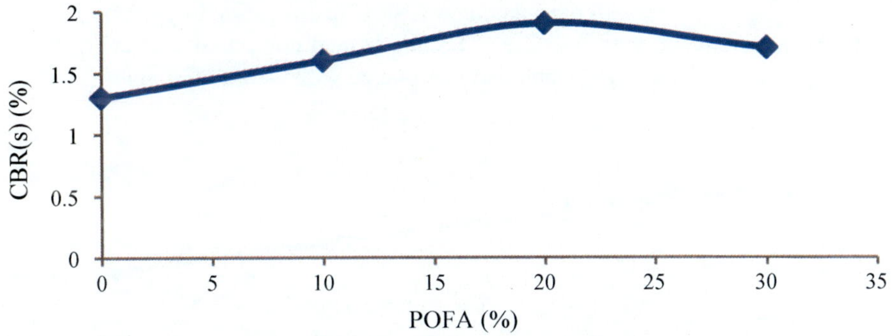

Fig. 5 Variation of soaked CBR with the percentage replacement of marine clay with POFA

the optimum percentage of palm oil fuel ash (POFA) as a replacement in marine clay is 20%.

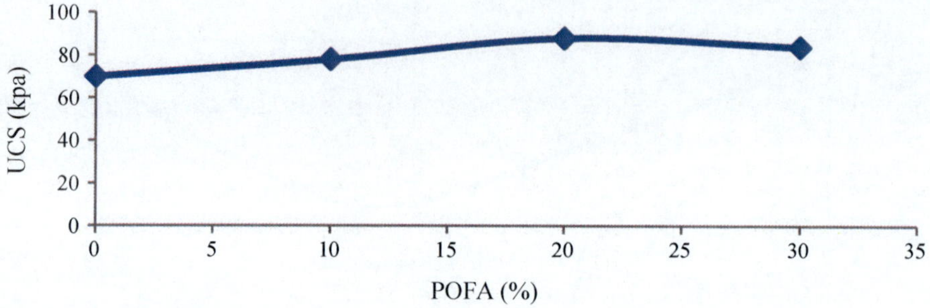

Fig. 6 Variation of UCS with the percentage replacement of marine clay with POFA

4.2 *Effect of Percentage of Lime on the Properties of POFA Modified Marine Clay*

Figures 7, 8, 9, 10, 11 and 12 show the variation of various properties of 20% POFA as a replacement in marine clay and mixed with different percentages of lime. The percentage of lime varied from 0 to 6% with an increment 2%.

From the graphs, it is clear that the addition of lime by 6% had reduced the plasticity index by about 72.5%. This clearly gives an idea of the improvement of the weak marine clay when part is replaced with a non-plastic agro-waste material, POFA and further blending it with binder, lime. The same trend was observed in the case of maximum dry density with an improvement of 6% and slight reduction in optimum moisture content by 4.5%. The results show a clear improvement in the penetration characteristics and the strength characteristics with the increment in % lime. The CBR values both unsoaked and soaked had increased by about 56% and 110.5%, respectively, with 6% lime. The unconfined compressive strength (UCS) had improved by about 51% when part is replaced with 20% POFA and 6% lime.

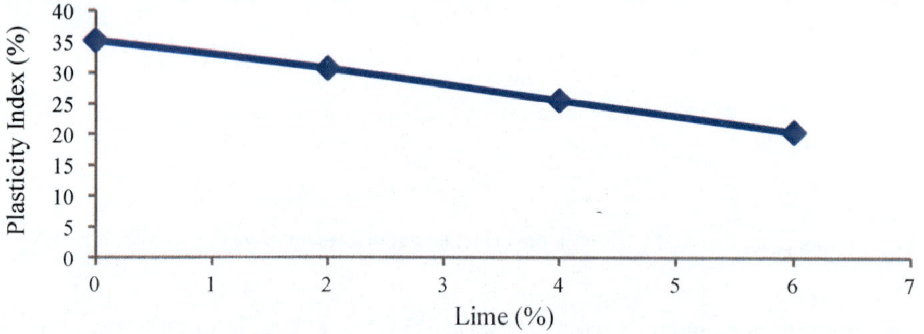

Fig. 7 Variation of plasticity index with the percentage of lime mixed to 20% replacement of marine clay with POFA

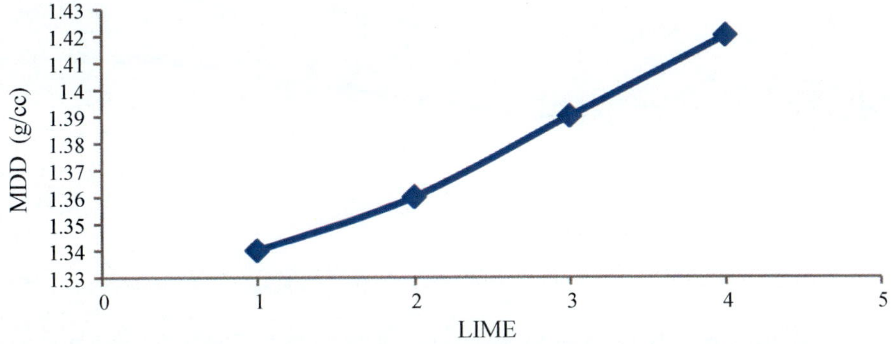

Marine Clay with POFA.

Fig. 8 Variation of maximum dry density with the percentage of lime mixed to 20% replacement of marine clay with POFA

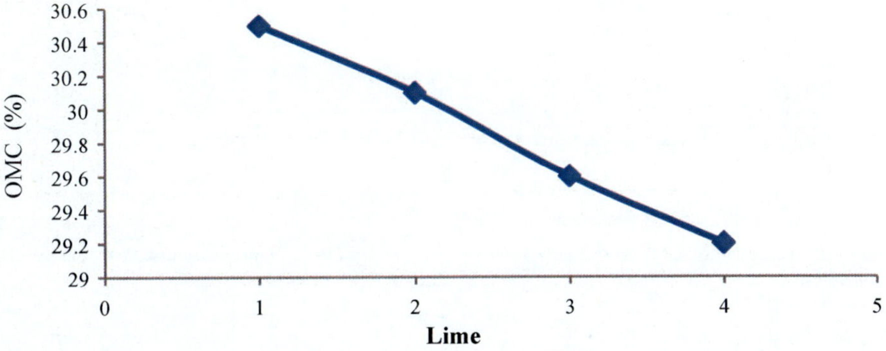

Fig. 9 Variation of optimum moisture content (OMC) with the percentage of lime mixed to 20% replacement of marine clay with POFA

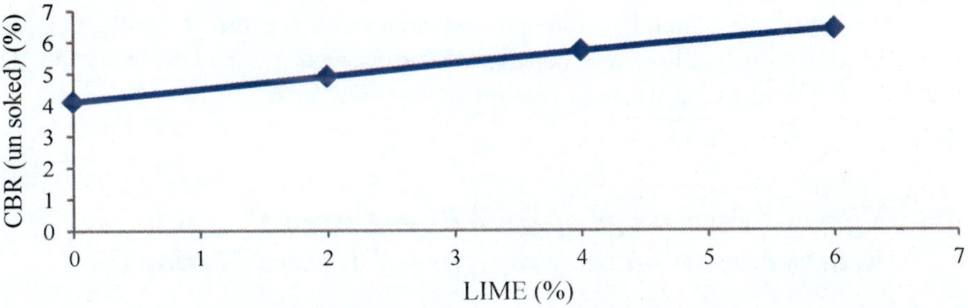

Fig. 10 Variation of unsoaked CBR with the percentage of lime mixed to 20% replacement of marine clay with POFA

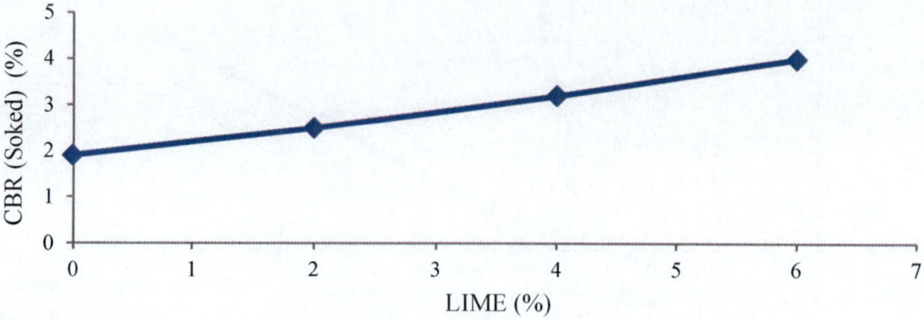

Fig. 11 Variation of soaked CBR with the percentage of lime mixed to 20% replacement of marine clay with POFA

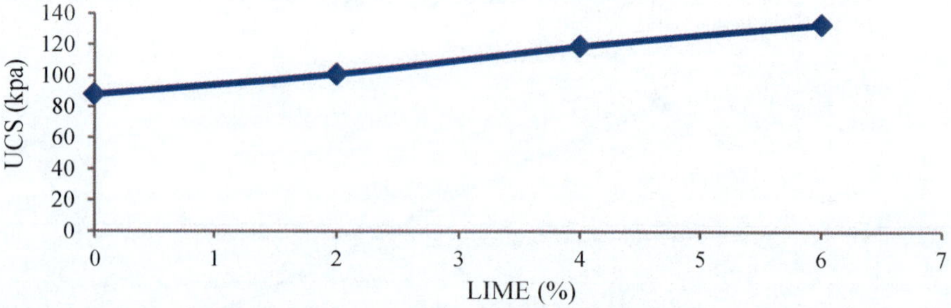

Fig. 12 Variation of UCS with the percentage of lime mixed to 20% replacement of marine clay with POFA

Hence, from the above discussions, it is evident that there is improvement in the characteristics of marine clay when replaced with 20% POFA and further mixed with 6% lime. From the trends exhibited in the graphs, further addition of lime might still improve the values, but based on the economic aspects and a motive to utilize the discrete waste plastic inclusions as a sustainable initiative, the percentage of lime was restricted to 6%, and it was considered for the next mode of stabilization step with discrete reinforcing waste plastic inclusions (DRWPI).

4.3 Effect of Percentage of DRWPI as Discrete Reinforcement on the Properties of Treated Marine Clay

The influence of the discrete reinforcing waste plastic inclusions (DRWPI), as discrete reinforcement on the behavior of palm oil fuel ash (POFA) and lime-treated marine clay is clearly presented in Figs. 13, 14, 15 and 16. The percentage of DRWPI was varied from 0 to 4%, i.e., 0, 1, 2 and 4%. From the graphs, it is clear that the addition of discrete reinforcing waste plastic inclusions (DRWPI) up to 2% had

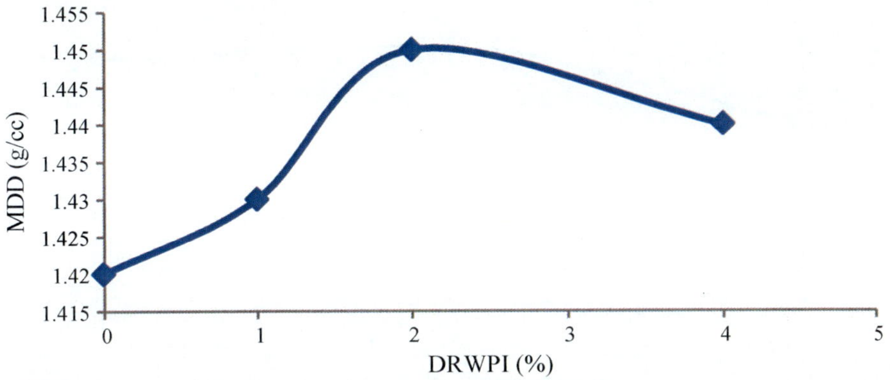

Fig. 13 Variation of maximum dry density with the percentage of DRWPI mixed to lime and POFA-treated marine clay

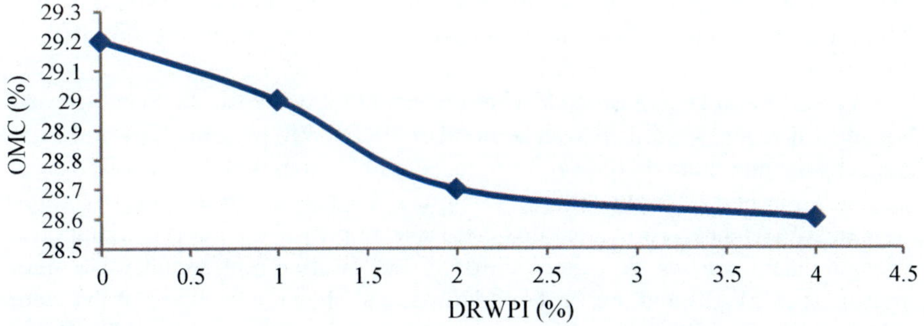

Fig. 14 Variation of maximum dry density with the percentage of DRWPI mixed to lime and POFA-treated marine clay

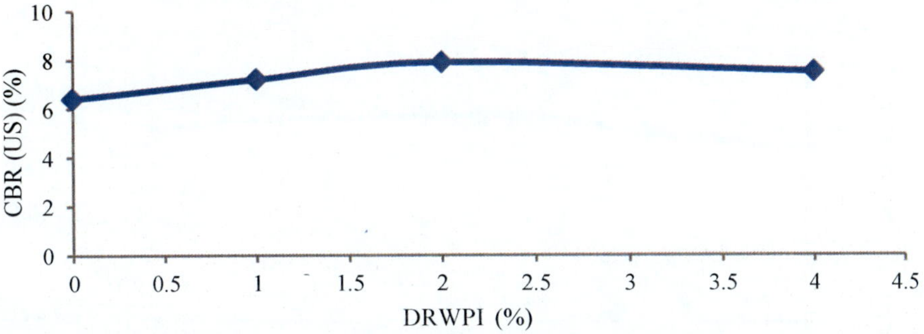

Fig. 15 Variation of unsoaked CBR with the percentage of DRWPI mixed to lime and POFA-treated marine clay

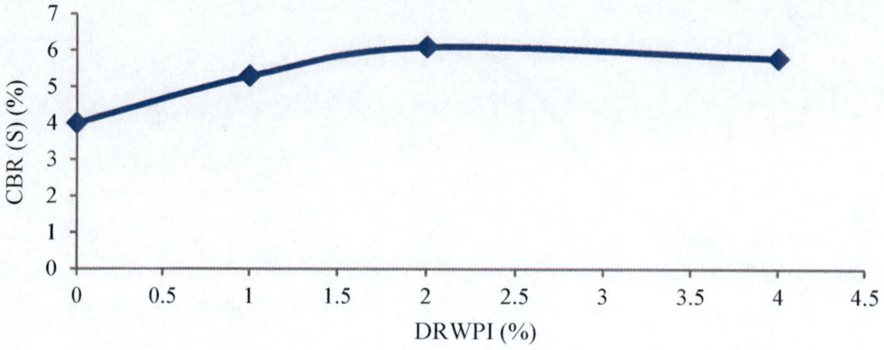

Fig. 16 Variation of soaked CBR with the percentage of DRWPI mixed to lime and POFA-treated marine clay

improved the maximum dry density by about 2.1% and the further addition of 4% DRWPI decreased the maximum dry density, whereas the corresponding optimum moisture content had shown a little decrease by above 2%.

Figures 15 and 16 present a clear improvement in the penetration characteristics both unsoaked and soaked up to an addition of 2% DRWPI by about 23% and 52.5%, respectively, and there on the values reduced with further addition. The same trend was shown in Fig. 17 for unconfined compressive strength (UCS) with an improvement of 20%. Hence, it is evident that there is an improvement in the penetration and strength characteristics with the addition of 2% discrete reinforcing waste plastic inclusions (DRWPI) and, on further addition, had shown a decrease in the values. This is due to the reinforcing action of the discrete elements in the treated soil matrix with a better orientation up to 2%, and on further addition, there was a distraction in the orientation due to more amounts of discrete elements intervening in the frictional mobilization.

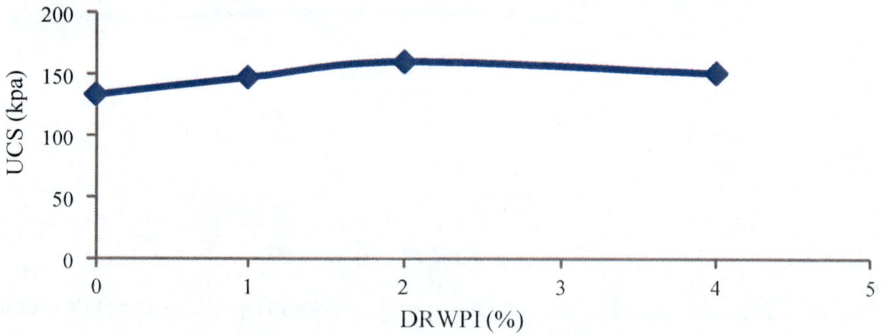

Fig. 17 Variation of UCS with the percentage of DRWPI mixed to lime and POFA-treated marine clay

4.4 *Curing Studies on POFA and Lime Modified Marine Clay Mixed with Optimum DRWPI Percentage*

Curing studies on samples prepared with 2.0% DRWPI + 6% Lime + 10% POFA as a replacement in marine clay were presented in Figs. 18, 19 and 20. From these results, it was observed that the treatment improved the marine clay with the time. Figures 18 and 19 present a clear improvement in the penetration characteristics for both unsoaked and soaked, upon 28 days curing, by about 24% and 34%, respectively. The same trend was shown in the unconfined compressive strength (UCS) with an improvement of 36%. The penetration and the strength characteristics of POFA modified marine clay further mixed with an optimum percentage of lime and optimum dosage of discrete waste fiber inclusions had shown pronounced improvement as the curing period increased. This is due to the development of strength due to the enhanced binding capacity of the C-S-H gel.

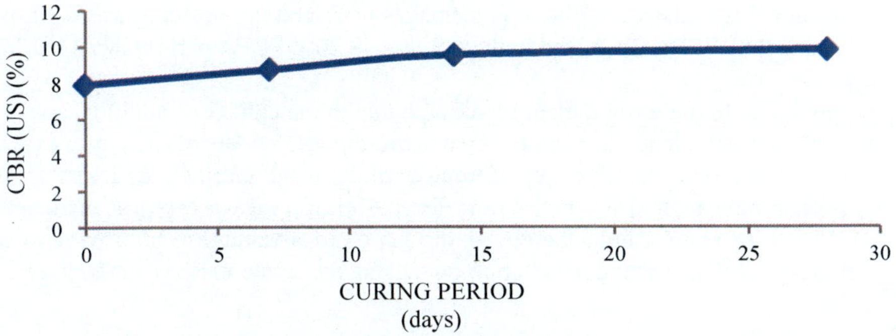

Fig. 18 Variation of unsoaked CBR with the curing period of the samples prepared with 2.0% DRWPI + 6% lime + 10% POFA as replacement in marine clay

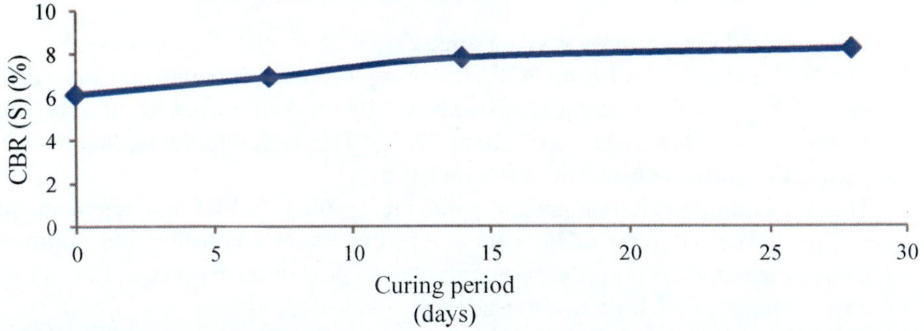

Fig. 19 Variation of unsoaked CBR with the curing period of the samples prepared with 2.0% DRWPI + 6% lime + 10% POFA as replacement in marine clay

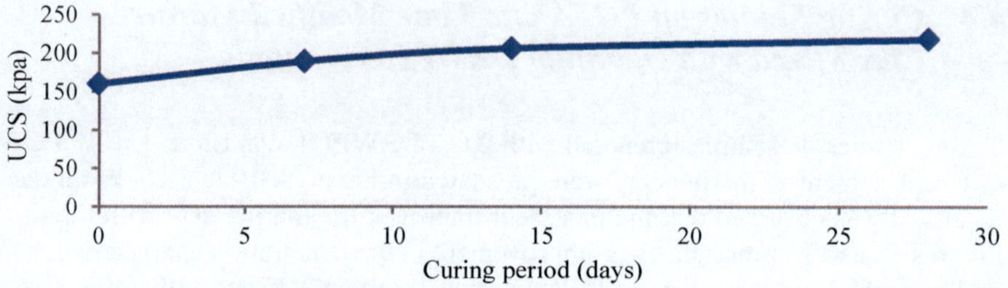

Fig. 20 Variation of unsoaked CBR with the curing period of the samples prepared with 2.0% DRWPI + 6% lime + 10% POFA as replacement in marine clay

Finally, the above discussions reveal that there is an improvement in all the properties of weak marine clay with the increase in % POFA as replacement, and for further mixing with 6% lime, there was considerable improvement in the values. Based on the economic aspects and a motive to utilize the discrete waste plastic inclusions as a sustainable initiative, different percentages of discrete reinforcing waste plastic inclusions (DRWPI) were mixed to the 6% lime + 20% POFA part replaced marine clay.

This is due to the development of strength due to the enhanced binding capacity of the C-S-H gel. Hence, it can be summarized that the results of the entire experimentation ascertain the objective of improving the weak marine clay by using an agro-waste, palm oil fuel ash (POFA) and discrete reinforcing waste plastic inclusions (DRWPI) with a binder lime, thereby giving a twofold advantage of improving weak marine clay with a sustainable solution by reusing the waste materials effectively.

5 Conclusions

The following conclusions were arrived with the laboratory study.

1. The soil used for the study was a marine clay with IS classification as CH.
2. The 20% palm oil fuel ash (POFA) as a replacement had reduced the virgin plasticity index of the marine clay by about 31%, and on further addition of lime by 6% it had further reduced by about 72.5%. This is due to the replacement of plastic soil with a non-plastic waste material.
3. There is a reduction in compaction parameters with 20% POFA as replacement, and on further addition of lime up to 6% there was a considerable improvement by about 6%. Hence, keeping in view the economic aspects and the above observation, the 6% lime was considered.
4. The CBR values for both unsoaked and soaked had increased by about 46.5% and 46%, respectively, with 20% POFA as replacement, and with further addition of lime by 6%, the values had further increased by about 56% and 110.5%, respectively.

5. The penetration and strength characteristics had increased up to an addition of 2% DRWPI with respect to POFA + lime-treated marine clay and there on the values reduced with further addition.
6. Finally, it can be summarized that the present study yielded a twofold advantage of improving problematic marine clay with a sustainable solution by reusing waste materials.

References

Basack S, Purkayastha RD (2009) Engineering properties of marine clays from the eastern coast of India. IETR 1(6):109–114

Bjerrum L, Heiberg S (1971) A review of the data on shear strength of the upper dried and weathered crust of Norwegian normally consolidated marine clays. Norwegian Geotechnical Institute, Oslo, Norway. Internal report 50901–02

Chew SH, Kamrazzuman HM, Lee FH (2004) Physicochemical and engineering behavior of cement treated clays. J Geotech Geoenviron Eng ASCE 130(7):696–706

Kalyana Krishna M, Joseph A, Chandrakaran S (2015) Ground improvement using compacted lime flyash column techniques for cochin marine clays. In: Proceedings of the IGC 2015, Pune

Koutsoftas DC, Foott R, Handfelt LD (1987) Geotechnical investigations offshore Hong Kong. J Geotech Eng ASCE 113(1):87–105

Kumar A, Mehta HS (1998) Laboratory investigations on use of stabilised granulated blast furnace slag in roads. Indian Highways, Dec 1998

Mathew AS, Nair AS, Krishnankutty SV (2017) Study on marine clay replaced with sodium lignosulfonate and cement. IJERT 6(3):496–498

Muntohar A, Hantoro G (2000) Influence of rice husk ash and lime on engineering properties clayey subgrade. EJGE 5:1–13

Shridharan A, Rao SM, Chandrasekaran S (1989) Engineering properties of cochin and mangalore marine clays. Indian Geotech J 1:265–278

Zhang J, Cao X (2002) Stabilisation of expansive soil by lime and flyash. J Wuhan Univ Technol-Mater Sci Ed 17(4)

Zhou C, Yin JH, Cheng CM (2005) Elastic anisotropic viscoplastic modeling of the strain-rate-dependent stress-strain behavior of $K0$-consolidated natural marine clays in triaxial shear tests. Int J Geomech ASCE 5(3):218–232

Plasticity and Strength Characteristics of Expansive Soil Treated with Xanthan Gum Biopolymer

Suresh Prasad Singh, Ritesh Das, and Debatanu Seth

Abstract The xanthan gum biopolymer is mixed with expansive soil in different proportions such as 0, 0.2, 0.5, 0.8 and 1.0% by weight of the dry soil mass. The plasticity, compaction and strength characteristics are studied by performing the liquid limit, plastic limit, shrinkage limit, linear shrinkage and unconfined compressive strength tests. The optimum moisture content (OMC) and maximum dry density (MDD) of different mixes are determined by performing light and heavy compaction tests. The UCS value of specimens compacted to MDD at OMC corresponding to heavy compaction effort is evaluated at curing periods of 0, 3, 7 and 28 days. With an addition of xanthan gum, the liquid limit and plastic limit values increase substantially, however, the plasticity index is found to decrease. An improvement in strength of the magnitude of 93% is observed at 1% xanthan gum. Finally, scanning electron microscope images depict the reason behind the UCS strength increment.

Keywords Expansive soil · Xanthan gum · Plasticity characteristics · Strength properties · Microstructure

1 Introduction

Expansive soils are those, which experience great changes in volume when their water content varies. These types of soil are widely distributed throughout the world, although they are especially abundant in arid zones, where conditions are suitable for the formation of clayey minerals of smectite group such as montmorillonite or some types of illites. These clay particles are characterizing by small particles size, large specific surface area and high cation exchange capacity. The swelling of this type of clay is related to three major factors such as geology, the engineering factors of the soil and local environmental conditions. Geology primarily determines the presence in the soil of these types of expansive clay minerals. Among the engineering factors included are the soil moisture content, plasticity and dry density. The most

S. P. Singh (✉) · R. Das · D. Seth
National Institute of Technology Rourkela, Rourkela, Odisha 769008, India
e-mail: spsingh@nitrkl.ac.in

M. Latha Gali and R. R. P. (eds.), *Problematic Soils and Geoenvironmental Concerns*, Lecture Notes in Civil Engineering 88,
https://doi.org/10.1007/978-981-15-6237-2_54

important local environmental conditions to consider are the amount of clay fraction in the soil, its initial moisture conditions and confining pressure (Sabtan 2005). Volume changes of these types of soils are a major cause of natural disasters, since they cause extensive damage to the structures and infrastructure on top of them. This has led some authors to emphasise there work on stabilization of expansive soil by lime, cement, fly ash and different type of chemicals are well documented in various papers. However, common chemical materials used for ground improvement have several environmental concerns such as high emission of greenhouse gases during their production and soil ecosystem disturbance after its application into the ground. In detail, ordinary cement production emits CO_2, a significant greenhouse gas, during chemical calcination and fuel burning. It has been reported that 5% of the global CO_2 emission is induced by cement industries (Worrell et al. 2001). Rapidly increasing demand for environmentally friendly and sustainable development has motivated the geotechnical engineers to find and develop non-conventional ground improvement methods (Dejong et al. 2006). Biopolymers are the environmentally friendly additives. These are produced by living organisms. Among various types of biopolymers, xanthan gum has been commonly used due to its appropriate strengthening efficiency and economic feasibility based on massive commercialization (Chang et al. 2015b). Xanthan gum has been introducing in to geotechnical engineering to decrease the hydraulic conductivity of silty sand via pore filling (Gates et al. 2009) as well as to increase the undrain shear strength of soil by increasing liquid limit (Nugent et al. 2009). Another recent study shows the possibilities of using xanthan gum as a soil strengthener, and showed xanthan gum preferentially forms firm xanthan gum clayey soil matrices via hydrogen bonding (Chang et al. 2015a). Qureshi et al. (2017) investigated the effects of wetting and drying cycles on xanthan gum treated sand and concluded that xanthan gum has good durability against slaking. A noticeable increase in shear resistance and decrease in permeability of sand and silt was reported by Ayeldeen et al. (2016). Moreover, it has been suggested that biopolymers could be applied for anti-desertification practices such as surface erosion reduction and afforestation programmes to resist desertification (Chang et al. 2015c). Labille et al. (2005) have presented a paper on the aggregation rate of montmorillonite aqueous suspension by xanthan gum and other six biopolymers. Results show that among seven bacterial polysaccharides succinoglycan and YAS34 show the best aggregation properties. On the other hand, Chen et al. (2013, 2014) have used xanthan gum and guar gum in mine tailing to improve undrain shear strength, dust control and erosion resistance.

This paper presents the plasticity, compaction and strength behaviour of the expansive soil treated with xanthan gum biopolymer. The plasticity characteristics of treated soil are examined by conducting liquid limit, plastic limit and volumetric shrinkage test with immediate (0 days) and 28 days curing period. Light and heavy compaction test has also performed to understand the compaction behaviour. The strength characteristics of treated soil are investigated by conducting unconfined compression test (UCS). The UCS test has conducted at different curing period such as 0, 3, 7, 28 days' interval. Furthermore, the microscopic change between treated and un-treated soil is examined by conducting scanning electron microscopic test (SEM) to interpret the plasticity and strength variation of treated expansive soil.

2 Materials and Methodology

2.1 Expansive Soil

The expansive soil used in this study is collected from Sector 14 of the Rourkela Steel Plant, Odisha. The soil is sundried and mixed thoroughly to bring homogeneity. Further, it is pulverized and dried in a hot air oven at 1100C for 24 h and kept in air tight containers for subsequent use. This soil possesses liquid limit of 86% and a differential free swelling value of 217%. The X-ray diffraction (XRD) result shown in Fig. 1 indicates that the soil has montmorillonite mineral, which demonstrate the high expansiveness of soil. The grain size analysis is shown in Fig. 2. From the plot, it is found that the soil contains 37% of clay, 55% of silt and 8% sand. So, from the grain size analysis, it is found that the soil is mainly silty clay and classified as MH as per Indian Standard Soil Classification System (ISSCS).

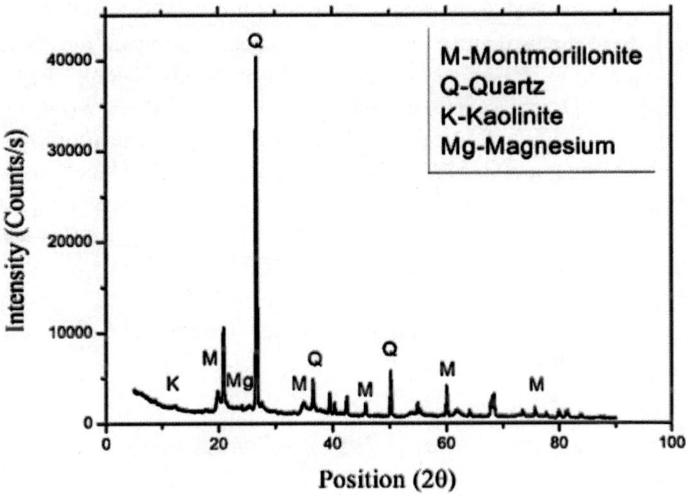

Fig. 1 XRD micrograph of the expansive soil

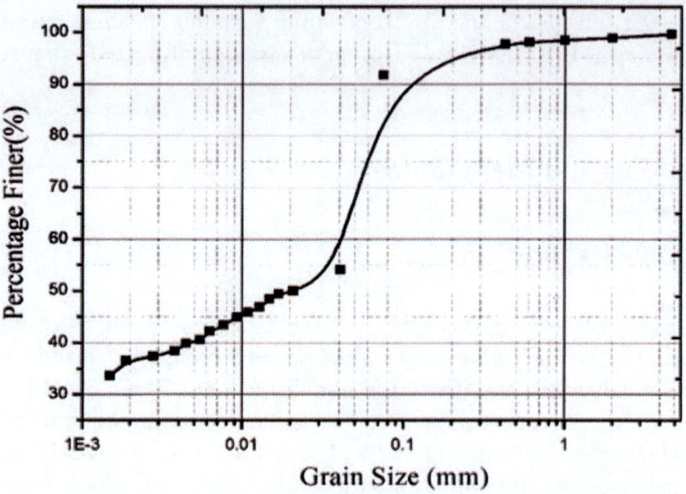

Fig. 2 Grain size distribution curve of the expansive soil xanthan gum

2.2 *Xanthan Gum*

Xanthan gum a polysaccharide mainly used as a food additive and rheology modifier is formed via the fermentation of glucose or sucrose using the Xanthamonous campestris bacterium (Chang et al. 2015). The resulting anionic polysaccharides are composed of D-uronic acid, D-mannose, pyruvylated mannose, 6-O—acetyl D-mannose and 1, 4-linked glucan (Cadmus et al. 1982). The simplified chemical structure of xanthan gum (C35H49O29) is a linear linked b-D glucose backbone with a tri-saccharide side chain and the backbone is in alignment, providing stability and conformity throughout the structure via the formation of hydrogen bonds. From the Loba Cheme xanthan gum container, it is found that the xanthan gum has a pH value in the range of 6–8 with a molecular weight of 1.0×10^6 g/mole (Bueno et al. 2013). The original powdered xanthan gum and chemical structure are given in Fig. 3a, b.

2.3 *Sample Preparation and Testing Programme*

2.3.1 Consistency Limits Test

Liquid limit and plastic limit, shrinkage limit tests and linear shrinkage are performed by following IS 2720 Part 5 (1985), IS 2720 Part 6 (1985) and IS 2720 Part 20 (1992), respectively. For each test, soil sample is first oven-dried and sieved through 425 µm. After that the soil is mixed with 0, 0.2, 0.5, 0.8, 1.0% xanthan gum by the weight of the dry soil mass. Required quantity of water is added to the xanthan gum soil mix to prepare a soil-water paste. The consistency values of different mixes are

(a)

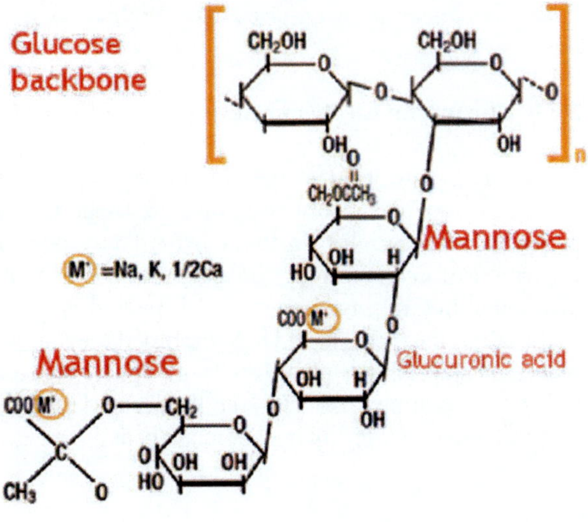

(b)

Fig. 3 Chemical structure of xanthan gum (skinchakra.eu/blog/archives/361-Xanthan)

found out immediately and after a curing period of 28 days. For determination of the consistency values at 28 days, the wet samples are sealed in plastic bags and kept it in a desiccator for 28 days. Thereafter, the tests are conducted as usual.

2.3.2 Compaction Test

The light and heavy compaction test is performed by following IS 2720 Part 7 (1980) and IS 2720 Part (1983) guidelines. Initially, air dried sample is mixed with xanthan gum at different concentration such as 0, 0.2, 0.5, 0.8, 1.0%. Hand mix is preferred for homogeneous mixing. Then mixed samples are kept in a polythene bag for 2–3 h for through distribution of water after that the compaction tests are performed.

2.3.3 Unconfined Compression Strength Test

Unconfined compression strength (UCS) testing of treated and stabilized specimens are performed in accordance with IS 2720 Part 10 (1991). Oven-dried soil samples are mixed and compacted to MDD at OMC of heavy compaction in the UCS mould (50 mm diameter × 100 mm height). The required compaction is performed by pressing the UCS mould in hydraulic jack. Hand mixing is preferred for better homogeneous mixing of UCS specimen. UCS samples are prepared for 3, 7, 28 days curing period. For each curing period and each mix, three numbers of UCS samples are prepared. Moulded UCS samples are then covered in wax for better moisture conservation. While testing a strain rate of 2% per minute is used. After testing, the failed specimens are oven-dried and weighted to determine the moisture content.

2.3.4 Scanning Electron Microscope Test (SEM)

SEM images are used to determine the microstructural properties of soil fabric because it provides information on size, shape and study of orientation and aggregation of soil particles. In this current study, this method is used to observe the formation of cementitious materials that are perceived to detect with other types of microstructural instruments. Each treated and un-treated samples are prepared by grinding the oven-dried UCS specimens. Each dried grounded specimen is then placed on aluminium stub covered with double-sided carbon tape and coated by sputtering with platinum for 120 s at 30 mA under high vacuum until it is completely covered and ready for microscopic analysis.

3 Results and Discussion

3.1 Consistency Limits

3.1.1 Liquid Limit and Plastic Limit

Figure 4 demonstrates the variation of liquid limit and plastic limit of the test soil sample with different curing period. Result shows that with increasing xanthan gum content the liquid limit increases. Because, due to the rheology modification nature of xanthan gum, it increases the density of diffuse double layer of the clay soil particles. As a result, soil particles required more amount of water to fulfil the diffuse double layer. Thus, liquid limit and plastic limit of the treated soil increases. Whereas, beyond 0.8% of xanthan gum the liquid limit of the treated soil is found to be decreased. The reduction is attributed to the reduction in the thickness of double layer due to the particle agglomeration and reduction in the specific surface area of the agglomerated soil. On the other hand, the increase in plastic limit leads to increase in shearing

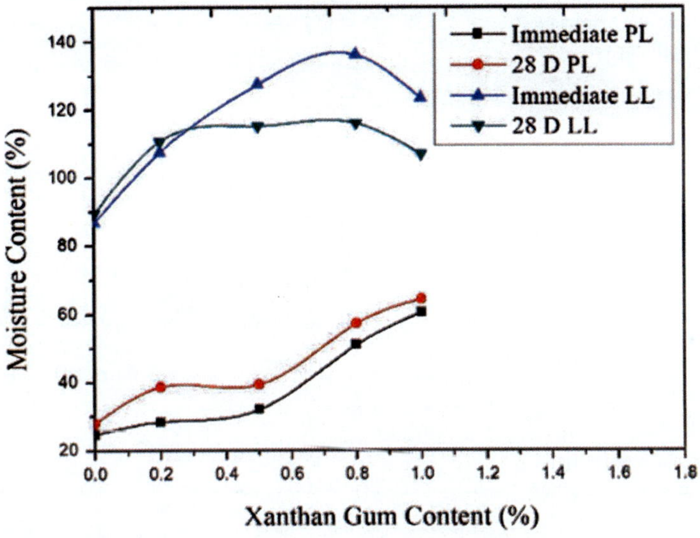

Fig. 4 Variation in liquid limit and plastic limit with saturation period

resistance. Depending upon the particle arrangement, size and shape of the pores vary. Thus, the flocculated structure will have higher plastic limit. The variation of plasticity index with xanthan gum content is presented in Fig. 5. It is noticed that the plasticity index value initially increases with the xanthan gum content up to 0.05% and thereafter there is a sharp fall in the plasticity index value. This is evident as the liquid limit value decreases beyond 0.6% xanthan gum, whereas the plastic limit value is found to increase continuously up to 1.0% xanthan gum.

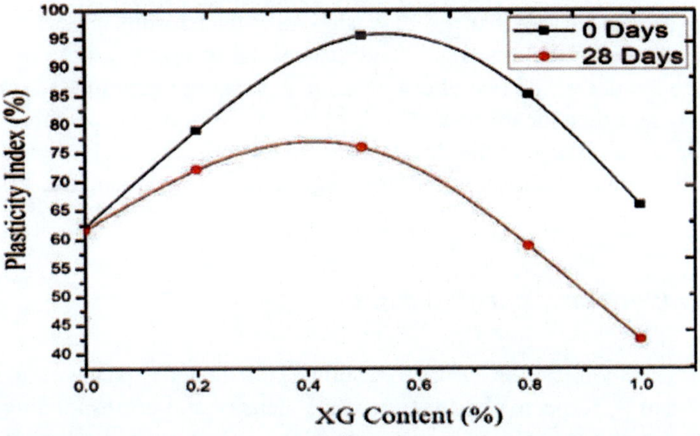

Fig. 5 Variation in plasticity index with saturation period

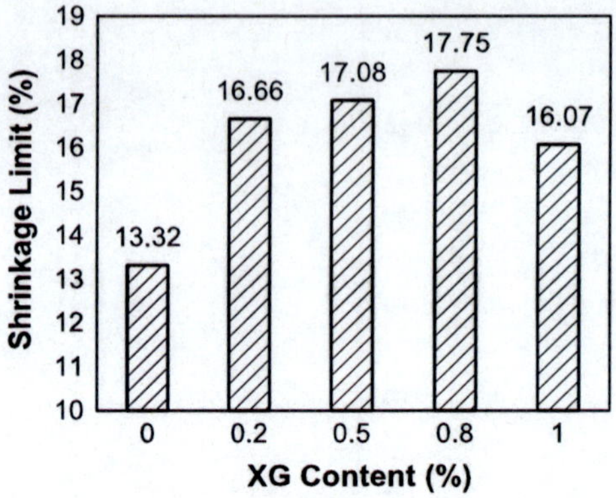

Fig. 6 Change in shrinkage limit with xanthan gum concentration

Table 1 Variation of linear shrinkage value with xanthan gum concentration

Soil mix	Linear shrinkage (%)
Virgin soil	24.4
Soil + 0.2% xanthan gum	24.2
Soil + 0.5% xanthan gum	23.9
Soil + 0.8% xanthan gum	23.9
Soil + 1.0% xanthan gum	23.7

3.1.2 Shrinkage Limit and Linear Shrinkage

Figure 6 depicts the variation in shrinkage limit with xanthan gum concentration. It shows that with increasing xanthan gum content the shrinkage limit of the treated soil increases. Xanthan gum addition increases the density of the pore fluid of the treated soil mass. Thus, at the time of dewatering, the high density pore fluid of treated soil evaporates at a slower rate than the un-treated soil, which results in decreasing rate of change in volume of the treated soil and reduces the shrinkage cracks in the treated soil. Due to the decreased rate of evaporation, the xanthan gum treated soil samples change very less than the un-treated soil specimen. Thus, experimental result shows that the linear shrinkage value decreases from 24.4 to 23.7% at 1% xanthan gum concentration. Table 1 shows linear shrinkage value with xanthan gum content.

3.2 Compaction Characteristics

Light and heavy compaction curves for xanthan gum treated expansive soil are shown in Figs. 7 and 8, respectively. Maximum dry density and corresponding optimum moisture content are evaluated with increasing xanthan gum concentration in the expansive soil. From the experimental result, it is found that dry density decreases from 16.09 to 15.91 kN/m^3 with increasing xanthan gum concentration in case of light

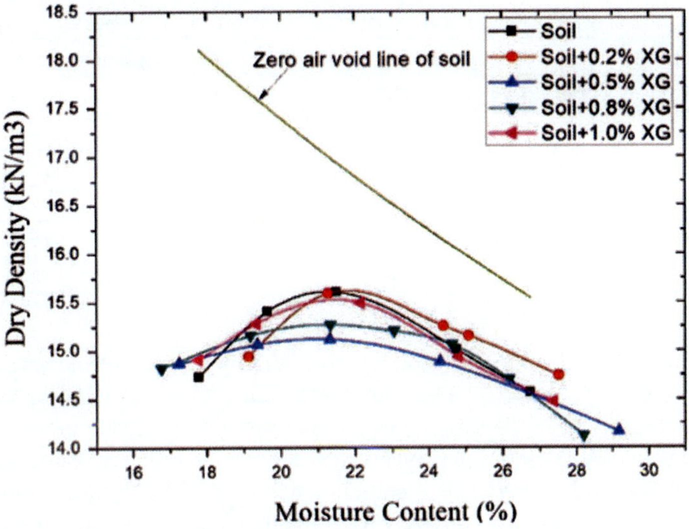

Fig. 7 Light compaction curves for xanthan gum added soil

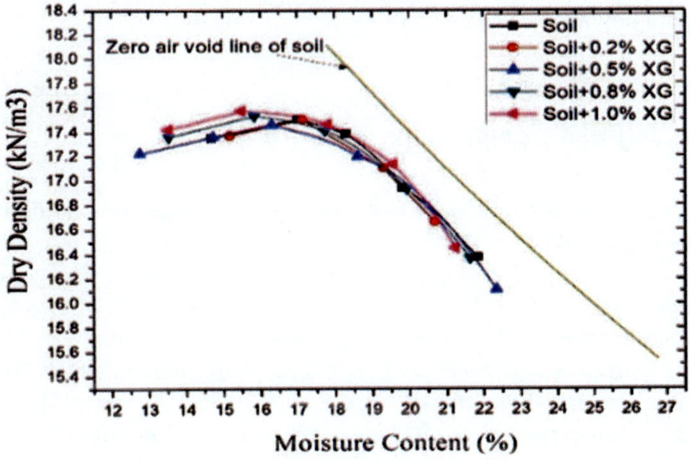

Fig. 8 Heavy compaction curves for xanthan gum added soil

compaction. While in case of heavy compaction, the dry density increases slightly from 17.43 to 17.61 kN/m^3 with increasing xanthan gum concentration from zero to 1%. In case of optimum moisture content, it decreases from 17.5 to 15.07% in heavy compaction and decreases from 21 to 22.1% in light compaction with similar increment in biopolymer concentration. Xanthan gum addition increases the density of the double layer of the soil particle. This high viscous double layer of the soil particle attracts other soil particles and form flocculated structure. This flocculated

Table 2 OMC and MDD values of soil amended with xanthan gum polymer

Soil description	Heavy compaction		Light compaction	
	OMC (%)	MDD (kN/m^3)	OMC (%)	MDD (kN/m^3)
Soil	17.5	17.43	21	16.09
Soil + 0.2% XG	16.9	17.47	21.5	15.85
Soil + 0.5% XG	16.31	17.46	21.34	15.18
Soil + 0.8% XG	16.10	17.53	21.7	15.47
Soil + 1.0% XG	15.70	17.61	22.1	15.91

structure restrains the effect of compaction thus more amount of water required to lubricate the soil particles. Thus, the OMC value increases with the increase in xanthan gum content in the mixture as observed in light compaction test. However, the higher viscosity of the double layer water prevents the soil particles to move closer under the imposed loading. This results in a decrease of MDD value with increase of xanthan gum content in the mixture. On the other hand, in heavy compaction, due to increased energy level the xanthan monomer acts as a lubricating material, and which helps the soil particle to become more oriented in nature and compact into a denser matrix. The OMC and MDD values of virgin soil and soil amended with xanthan gum polymer are given in Table 2.

3.3 *Unconfined Compressive Strength*

UCS tests are performed on treated and un-treated expansive soil sample compacted to their respective MDD at OMC. Typical stress–strain plots of test soil samples with different curing period is depicted in Fig. 9. A remarkable change in strength has observed. The average UCS strength of un-treated soil is 984.78 kPa, while xanthan gum treatment induced significant increase in UCS value. The UCS values of 0, 3, 7, 28 days cured specimens at 1% xanthan gum content is found to be 1170.99 kPa, 1573.17 kPa, 1669.37 kPa and 1900.85 kPa, respectively. There is almost 93% strength increment in 28 days cured sample. The strength increment of xanthan gum treated soil entirely depends on the strength and density of the xanthan gum soil linkages. Xanthan gum polymers have strong sorption and microstructural interaction with clay (Latifi et al. 2016) When xanthan gum mixed with clay and comes in contact with the water, the carboxyl group (–COOH) of xanthan monomer starts to release hydrogen ions (H^+) or share hydrogen with clay particle which create hydrogen bonding between the clay particle and xanthan gum monomer. On the other hand, the –COO group starts to attracts and share the cations such as Na^+, Ca^+, Al^+ with clay particles. These direct interactions of xanthan monomers create cation bridging or ionic bonding between clay particles and xanthan gum. With increase in xanthan gum concentration and curing period these, bonding increases and upturns the compressive strength accordingly. These increased xanthan gum linkages convert

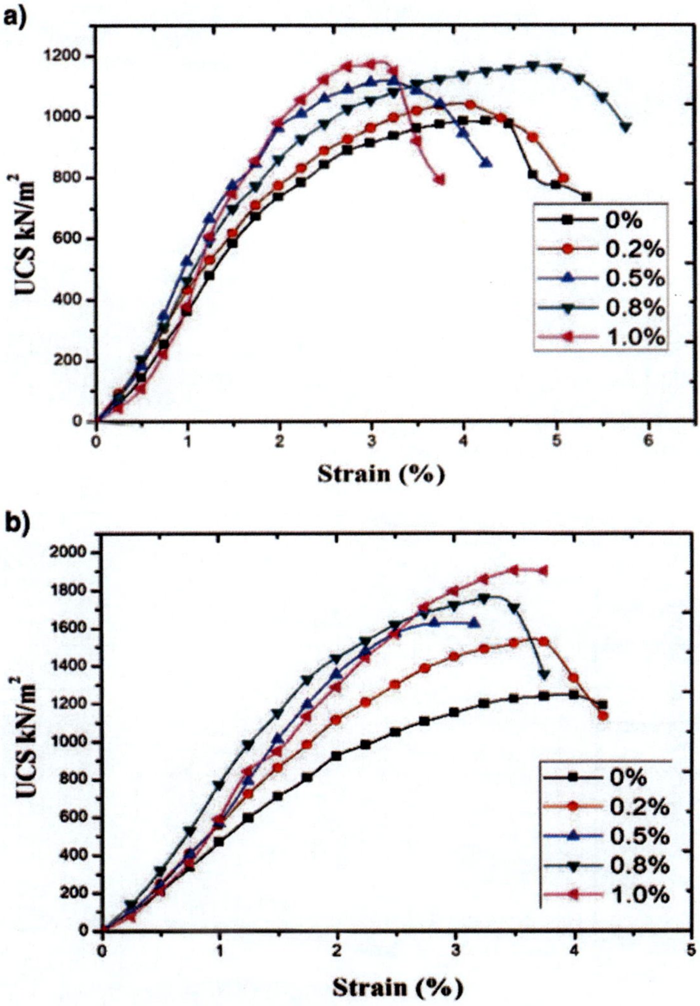

Fig. 9 **a** Stress–strain plots of immediately tested xanthan gum treated soils, **b** stress–strain plots for xanthan gum treated soils after 28 days of curing

the soil more brittle in nature. In Fig. 10, change in failure pattern is given. In Figs. 11 and 12, the variation in UCS strength and strain of different soil xanthan gum mix and curing period are given.

Fig. 10 Failure pattern of immediate and 28 days cured samples

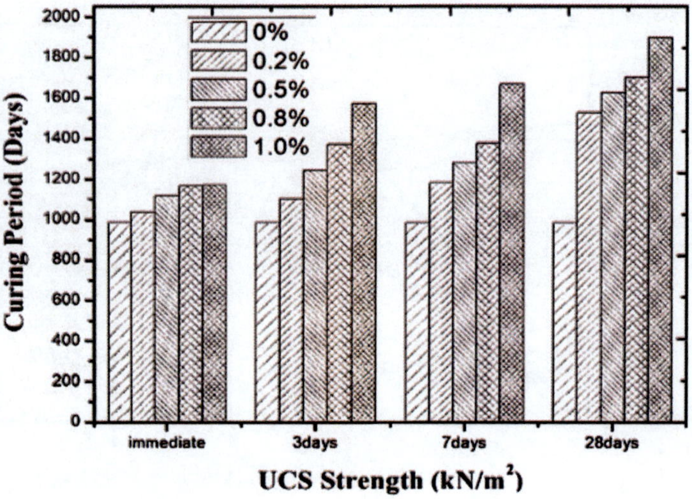

Fig. 11 UCS of xanthan gum treated soil

3.4 Microstructural Studies

The microstructural change in xanthan gum treated soil determined by using SEM test. This test is conducted on virgin soil, 1.0% xanthan gum treated 3 days and 28 days UCS samples. Due to electrically charged clay particles, xanthan gum monomer directly interacts with the flaky clay sheets with hydrogen bonding and cation bridging. Along with this direct connection with soil particle, the xanthan monomer also bridges between distant particles. In Fig. 13(a), the typically flaky particles of clay have shown at 3000× and 5000× magnification. SEM image of

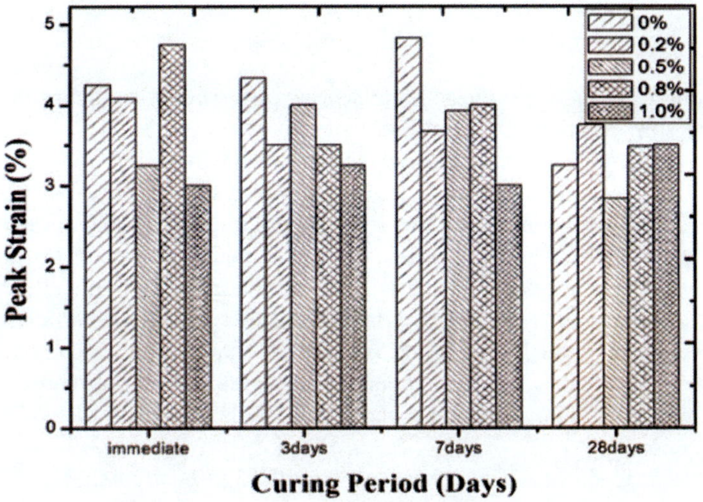

Fig. 12 Peak strain values with xanthan gum content and curing period

(a) (b) (c)

Fig. 13 SEM image of **a** 0 day, **b** 3 days, **c** 28 days cured specimens treated with 1% xanthan gum

the un-treated soil shows undulated, flaky soil particles with large amount of voids in it. Whereas, Fig. 13(b), (c) shows 3 days and 28 days cured samples at 1.0% xanthan gum mix, which depicts formation of large size soil particles with less voids in it. Which also shows that the small clay particles are coated by gum like cementitious material and forms large size of agglomerated clay particles. Which can be explained by direct interaction of clay particle and xanthan monomers by H-bonding and electrostatic bonding. These two bindings are believed to be the main factors for strength increment (Latifi et al. 2016). These results confirm that the strengthening mechanism and improvement in swelling characteristics of xanthan gum treated soil.

4 Conclusions

The present study examines the geo-engineering properties of xanthan gum stabilized expansive soil. Xanthan gum treatment in expansive soil shows some noticeable results. From the experimental results, the following conclusions have been made to satisfy the mechanism of the xanthan gum stabilized expansive soil:

The results show that with increase in biopolymer the liquid limit of the soil increase due to the increase in pore fluid density. Whereas, between 0.8% and 1.0% of xanthan gum concentrations, the particle accumulation is higher than the increase in viscosity of pore fluid. Thus, the specific surface area of the clay particles decreases which decreases liquid limit slightly. Similar mechanism has occurred in case of plastic limit also. Due to this the xanthan gum mix soil shows constantly increase in plastic limit for both immediate test and after 28th day test with increasing xanthan gum concentration.

In volumetric shrinkage, the increases pore fluid viscosity helps in reducing the evaporation rate of the treated soil. This decreases the rate of volume change and decrease volumetric shrinkage of the xanthan gum treated soil.

With increasing xanthan gum content in the soil the flocculation of the particle occurs by high viscous diffuse double layer, which resist the impact energy of the compacting rammer in standard compaction. Due to this OMC increase and MDD of the treated soil decreases, but in case of heavy compaction, the increased compaction energy renders the soil particles more oriented, which results in slight increase in MDD value and a reduction of the OMC value.

Xanthan gum is shown to have impressive strengthening effect on treated soil. It shows about 93% strength increment of 28 days' sample at 1% xanthan gum concentration. This strength increment is due to direct interaction of negatively charged soil particles and carboxylic group of xanthan gum (i.e., hydrogen bonding, cation bridging). With increasing curing period, these interaction increases which effect the strengthening property and stiffness also. These xanthan matrices increase the stiffness of soil and make the soil more brittle in nature.

SEM image confirms the formations of gum like cementitious products that result from chemical reactions between the xanthan gum and soil particles, which improves the soil strength behaviour.

From the experimental programme, it is found that xanthan gum addition to the expansive soil increases the strength abruptly with curing period. Although, it is also found that the plasticity index of the soil also decreases. So from the above conclusions, it can be understood that xanthan gum biopolymer is highly environmentally friendly and it can be used as an expansive soil modifier additive. There is another prominent advantage is that it will not cause any harmful effect in soil mineralogical properties.

References

Ayeldeen MK, Negm AM, El Sawwaf MA (2016) Evaluating the physical characteristics of biopolymer/soil mixtures. Arab J Geosci 9(5):371

Bueno VB, Bentini R, Catalani LH, Petri DFS (2013) Synthesis and swelling behavior of xanthan-based hydrogels. Carbohydr Polym 92(2):1091–1099

Cadmus MC, Jackson LK, Burton KA, Plattner RD, Slodki ME (1982) Biodegradation of xanthan gum by Bacillus sp. Appl Environ Microbiol 44(1):5–11

Chang I, Im J, Prasidhi AK, Cho GC (2015) Effects of Xanthan gum biopolymer on soil strengthening. Constr Build Mater 74:65–72

Chen R, Zhang L, Budhu M (2013) Biopolymer stabilization of mine tailings. J Geotech Geoenviron Eng 139(10):1802–1807

Chen R, Lee I, Zhang L (2014) Biopolymer stabilization of mine tailings for dust control. J Geotech Geoenviron Eng 141(2):04014100

Chang I, Im J, Prasidhi AK, Cho GC (2015a) Effects of Xanthan gum biopolymer on soil strengthening. Constr Build Mater 74:65–72

Chang I, Prasidhi AK, Im J, Cho GC (2015b) Soil strengthening using thermo-gelation biopolymers. Constr Build Mater 77:430–438

Chang I, Prasidhi AK, Im J, Cho GC (2015c) Soil strengthening using thermo-gelation biopolymers. Constr Build Mater 77:430–438

DeJong JT, Fritzges MB, Nüsslein K (2006) Microbially induced cementation to control sand response to undrained shear. J Geotech Geoenviron Eng 132(11):1381–1392

Gates WP, Bouazza A, Churchman GJ (2009) Bentonite clay keeps pollutants at bay. Elements 5(2):105–110

IS: 1498 (1970) Classification of soil. Bureau of Indian Standards, New Delhi

IS: 2720 (part 7) (1980) Determination of water content-dry density relation using Heavy compaction. Bureau of Indian Standards

IS: 2720 (part 3/sec. 1) (1980) Determination of specific gravity. Bureau of Indian Standards

IS: 2720 (part 4) (1985) Grain size distribution of soil. Bureau of Indian Standards.

IS: 2720 (part 5) (1985) Determination of liquid limit, plastic limit and shrinkage limit. Bureau of Indian Standards

IS: 2720 (part 15) (1986) Methods of test for soils part 15 determination of consolidation properties. Bureau of Indian Standards

IS 2720 part 10 (1991) Methods of test for soils, Part 10: determination of unconfined compressive strength. Bureau of Indian Standards, New Delhi

Labille J, Thomas F, Milas M, Vanhaverbeke C (2005) Flocculation of colloidal clay by bacterial polysaccharides: effect of macromolecule charge and structure. J Colloid Interface Sci 284(1):149–156

Latifi N, Horpibulsuk S, Meehan CL, Abd Majid MZ, Tahir MM, Mohamad ET (2016) Improvement of problematic soils with biopolymer—an environmentally friendly soil stabilizer. J Mater Civ Eng 29(2):04016204

Nugent R, Zhang G, Gambrell R (2009) Effect of exopolymers on the liquid limit of clays and its engineering implications. Transp Res Rec: J Transp Res Board 2101:34–43

Qureshi MU, Chang I, Al-Sadarani K (2017) Strength and durability characteristics of biopolymer-treated desert sand. Geomech Eng 12(5):785–801

Sabtan A (2005) Geotechnical properties of expansive clay shale in Tabuk, Saudi Arabia. J Asian Earth Sci 25(5):747–757

Worrell E, Price L, Martin N, Hendriks C, Meida LO (2001) Carbon dioxide emissions from the global cement industry. Annu Rev Energy Env 26(1):303–329

Strength Properties of Expansive Soil Treated with Sodium Lignosulfonate

Suresh Prasad Singh, Prasad S. Palsule, and Gaurav Anand

Abstract In India, nearly 23% of land surface is covered with expansive soil. The problems associated with such soils caused the tremendous cost to the project if proper geo-technical investigation and stabilization are not done. To prevent the failures of structures like foundations, retaining structures, slopes, lightweight structures, pavements, the improvement of swelling soil should be done which is economical and environmentally friendly. The use of lignin-based organic polymer which is a by-product of pulp industry is a sustainable and viable technique. An assessment of lignosulfonate as a stabilizer is necessary, and thus in the current study, the compaction, plasticity, swelling, and strength characteristics of expansive soil are checked with addition of sodium lignosulfonate in percentages varying from 0 to 12. The plasticity characteristics of expansive soil are seen to be improved significantly. However, a marginal increase in strength occurs with the addition of lignosulfonate which further increases with curing period.

Keywords Black cotton soil · Lignosulfonate · Stabilization

1 Introduction

Expansive soil causes structural damages mainly in lightweight structures such as residential buildings, pavements, road subgrades due to its volume change behavior upon moisture changes (Pasupuleti et al. 2012; Phani Kumar and Sharma 2004). The cost of repairing and maintenance is much higher and undesirable. The preventive measures are thus essential to take while constructing structure in expansive soils. The mainly expansive soils are residual soils and having natural deposit thickness around 3–10 m which represents the surficial deposits. The expansive soil is also known as the swelling soil, black cotton soil, 'regur.' The swelling soils are stabilized mainly to improve its volume change behavior. Generally, the chemical stabilization methods are used to modify the engineering properties of expansive soils.

S. P. Singh (✉) · P. S. Palsule · G. Anand
National Institute of Technology Rourkela, Rourkela, Odisha 769008, India
e-mail: spsingh@nitrkl.ac.in

M. Latha Gali and R. R. P. (eds.), *Problematic Soils and Geoenvironmental Concerns*, Lecture Notes in Civil Engineering 88,
https://doi.org/10.1007/978-981-15-6237-2_55

Different traditional stabilizers such as cement, fly ash, lime, chlorides are used frequently to improve expansive soil properties; however, these stabilizers cause the cost to environment either during their production or during the use in the form of CO_2 emission, increased pH of soil making it alkaline so there is a need to use non-traditional stabilizers like rice husk ash, lignin, xanthan, and guar gum. These non-traditional stabilizers stabilize the soil and also environmentally friendly. The lignosulfonate is water-soluble anionic polyelectrolyte polymers. They are by-product from the production of wood pulp using sulfite pulping. The lignosulfonate is used mostly in the construction industry as a plasticizer for concrete and dust suppressant for unpaved roads.

Different additives are used such as salt, polymers, cement, lime, fly ash to stabilize the expansive soil; these additives generally alter the mineralogical structure and thereby improve properties of soil such as strength and stiffness (Indraratna et al. 2012; Canakci et al. 2015). Nowadays, waste materials which are by-products of different industries are also studied with the purpose to improve the engineering properties of expansive soil. It is beneficial for environmental concerns to reduce the use of traditional stabilizers such as cement and lime (Canakci et al. 2015). Lignin is a class of complex organic polymers that form important structural materials in the support tissues of vascular plants and some algae (Martone et al. 2009). Lignin is particularly important in the formation of cell walls, especially in wood and bark because they lend rigidity and do not rot easily (Lebo et al. 2001). The natural lignin gets modified in various industrial processes, and the lignosulfonate is a modified lignin obtained from wood pulping operations (Ali 2015). The effect of lignin addition to soil is either direct or it may contribute to the formation of humic acid, which increases soil stability (Ceylan et al. 2015). Lignosulfonate is environmentally friendly, non-corrosive, non-toxic chemical that does not alter the soil pH upon treatment. Lignosulfonate belongs to the family of lignin-based organic polymers (Indraratna et al. 2012).

Various researchers showed that the addition of lignosulfonate to the expansive soil is effective in reducing the swelling potential of soil also increased the strength and stiffness of soil (Indraratna et al. 2012; Lekha et al. 2015; Ta'negonbadi and Noorzad 2017). The thickness of the diffused double layer reduced due to the ion exchange process happened between lignosulfonate and expansive soil minerals. The reaction between clay and LS depends on the charge of clay minerals and chargeability of admixture (Vinod et al. 2010). Alazigha et al. (2018a, b) stated the improvement in the expansive soil characteristics like strength, durability, swell potential, shrinkage limit. The coefficient of consolidation value determined prompted that the soil structure altered due to the chemical treatment done with lignosulfonate and changed from clayey to silty nature. Alazigha et al. (2018a, b) conducted different advanced tests like CT scan, XRD, SEM/EDS, FTIR, nuclear magnetic resonance, a specific surface area (SSA) which signify the reduction in water affinity of expansive soil particles due to the hydrophobic nature of aromatic carbon chain present in the lignosulfonate. The smearing effect observed in the soil structure restrained the water entry inside soil lattice.

2 Materials

2.1 Soil

The expansive soil used in study is obtained from Sector 14 area of Rourkela Steel Plant, Odisha. The same is dried in hot air oven at a temperature of 110 °C for 24 h. Further, it is pulverized in a ball mill and mixed thoroughly to bring homogeneity in the mass. The properties of expansive soil used in the study are presented in Table 1. The specific gravity of expansive soil is 2.74, and the liquid limit of the soil is 136% with plasticity index of 109%. It is classified as high plasticity clay 'CH' as per Indian Standard Soil Classification System. The differential free swell value for the soil is 1933%. The major minerals present in it are quartz and montmorillonite (Fig. 1).

Table 1 Properties of expansive soil

Index and engineering properties	
Specific gravity	2.74
Liquid limit (LL)	136%
Plastic limit (PL)	27%
Plasticity index (PI)	109%
Shrinkage limit (SL)	21%
Soil classification (according to USCS)	CH
DFS value	1933%
Maximum dry density (MDD)	15.2 kN/m^3
Optimum moisture content (OMC)	25.6%

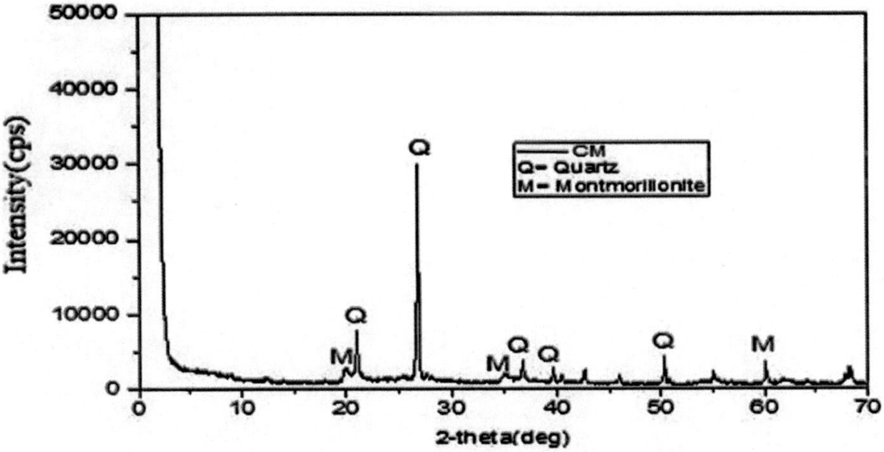

Fig. 1 X-ray diffraction image of untreated expansive soil sample

2.2 *Sodium Lignosulfonate*

The sodium lignosulfonate used as a stabilizer in the current study is obtained from the 'Triveni Chemicals,' Vapi, Gujarat. It has dark brown color and in powdered form. The measured pH value is 6.6, i.e., it has slightly acidic nature. Figure 2a shows the molecular formula and the brown-powdered sodium lignosulfonate used in study. The molecular formula suggested that the lignosulfonate is having poly-anionic chain with aromatic carbon chain and alcohol group, sodium cation and sulfonic group present.

3 Methodology

The experimental program consisted of the treatment of expansive soil with different percentages of sodium lignosulfonate. The percentages of lignosulfonate added decided on the basis of literature available. It is varying from 1 to 12%. The study is done to obtain the effect on compaction characteristics, plasticity characteristics, and strength characteristics of soil. The pH test is carried out to check the soil is being acidic or alkaline. All the tests are conducted as per Indian Standards.

Figure 3 shows the particle size distribution curve for expansive soil. The wet sieve analysis is done for particles greater than 75 μm, and for particles of silt and clay size, the hydrometer analysis is carried out to obtain the curve. The curve suggested that the soil consists of 9% sand-sized particles and 25% silt-sized particles while the clay-sized particles are the maximum, i.e., around 66%.

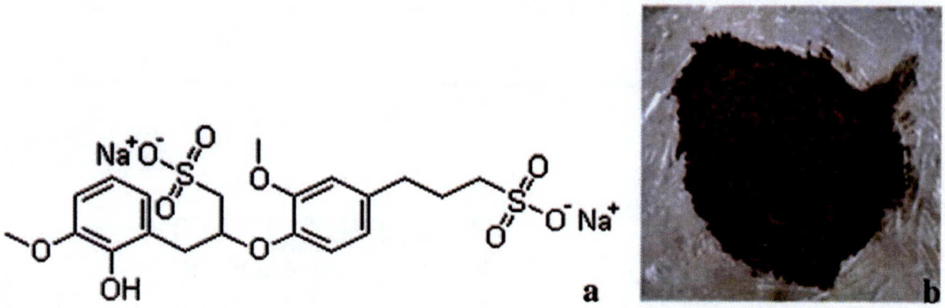

Fig. 2 **a** Structure of sodium lignosulfonate, **b** sodium lignosulfonate powder

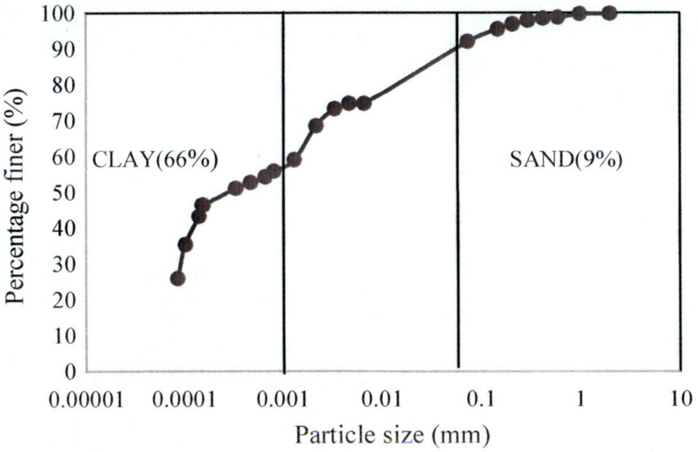

Fig. 3 Gradation curve of the expansive soil

4 Results and Discussion

4.1 Compaction Characteristics

The light compaction and heavy compaction tests are carried out to obtain the relationship between dry density and moisture content for varying percentages of sodium lignosulfonate. Figures 4 and 5 show the plots of light compaction and heavy compaction curves. Table 2 gives the values for maximum dry density (MDD) and optimum moisture content (OMC) of different soil mix proportions.

It can be seen that for light and heavy compactions, the MDD values are slightly improved and OMC values are slightly decreased but the variation in values is not significant. The poly-anionic nature of lignosulfonate makes the soil structure more

Fig. 4 Light compaction curves

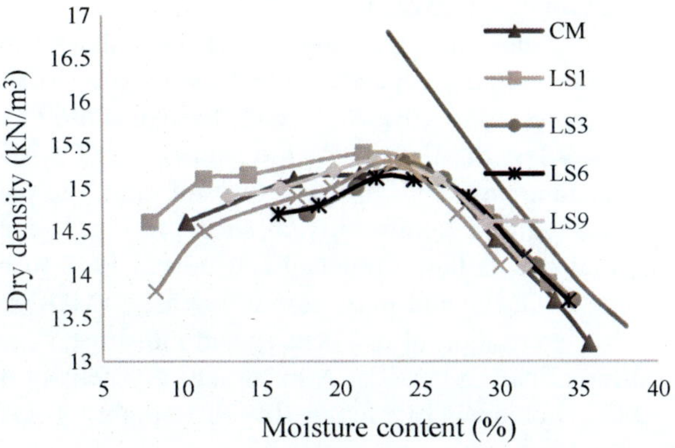

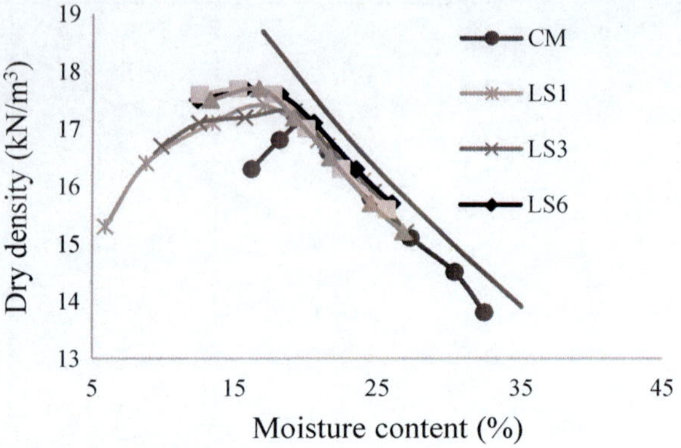

Fig. 5 Heavy compaction curves

Table 2 Compaction characteristics of lignosulfonate-treated soil

MIX	Light compaction		Heavy compaction	
	OMC (%)	MDD (kN/m^3)	OMC (%)	MDD (kN/m^3)
CM	25.6	15.2	19.8	17.1
LS1	21.5	15.4	17.1	17.5
LS3	23.9	15.3	19.3	17.3
LS6	22.3	15.1	15.7	17.7
LS9	22.3	15.3	15.2	17.7
LS12	23.4	15.3	16.7	17.7

dispersive when it gets adsorbed on the particle surface which helps in increasing the soil dry density, whereas simultaneously the electrolytic reaction between cations present in sodium lignosulfonate and negative soil charge caused the flocculation which resists the soil structure to be compacted. The adsorption of sodium lignosulfonate on the soil surface decreased the soil affinity toward water due to the hydrophobic nature of carbon chain present which reflected in slight decrease in optimum moisture content.

The voids through natural pozzolana that indirectly lead to diminished the moisture-holding capacity around the clayey surface area. It is observed that the addition of up to 10% AAB tends to reduce the PI of treated soil and transformed it into a less plastic soil. As the percentage of AAB is increased in the soil layers by maintaining 0.5% of water to solid ratio, the percentage of the swelling index is also reduced significantly, as shown in Fig. 6. Standard Proctor's compaction is carried out on both untreated and treated BCS to find out the optimum moisture content (OMC) and maximum dry density (MDD) according to ASTM D 698. A positive variation of dry density and moisture content is observed with respect to different curing periods. A maximum dry density of 19.7 kN/m^3 with 19.80% of OMC is obtained by mixing 10% (by weight) of AAB to raw BCS after 28 days of geopolymerization.

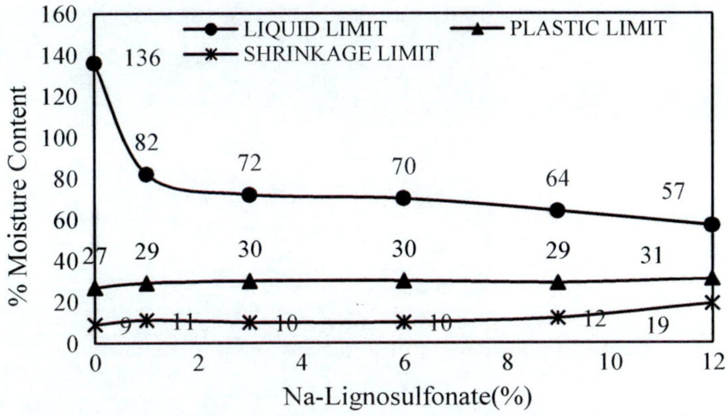

Fig. 6 Variation of liquid, plastic, and shrinkage limits with lignosulfonate content

4.2 *Plasticity Characteristics*

The plasticity characteristics are checked through Atterberg limits which are liquid limit and plastic limit. The shrinkage limit is also determined to understand the volume change behavior. Figure 6 shows the plasticity limits values and shrinkage limit. The high liquid limit of expansive soil is due to the mineral montmorillonite present in the soil in large quantity.

The liquid limit of expansive soil is decreased from 136 to 82 after 1% addition of chemical and decreased to 57% when the chemical added is 12%, which showed the decrease in liquid limit for small addition of sodium lignosulfonate is significant. This can be due to the adsorption of lignosulfonate particles on the soil particles' due to the adsorption of chemical the bonding of soil particles occurred caused the aggregation as well as the hydrophobic carbon chain present in lignosulfonate resisted the water entry inside the soil lattice structure. The increase in aggregation reduces the specific surface area which reduces the amount of moisture to satisfy the double layer and caused the reduction in liquid limit. The plastic limit of soil increased due to the increase in aggregation which increases the need of water to reach the plastic state. The plasticity index thus reduces continuously as the additive percentage increases. The trend of plasticity index is clearly shown in Fig. 6. Figure 7 also shows the variation of shrinkage limit with sodium lignosulfonate percentage. The shrinkage limit for the expansive soil is around 9% which showed the very high shrinking of soil after drying and creates the shrinkage cracks. For untreated soil, the shrinkage index is 127%, and for soil mix treated with 1% of sodium lignosulfonate is 71%, and for 12% of additive content it is 38%. Thus, it reflects the decrease in volume change behavior of expansive soil which is one of the indicators of stability induced in soil. The shrinkage limit is increased from 9 to 19%.

Fig. 7 Shrinkage limit test specimens

4.3 Strength Characteristics

The strength characteristics of untreated and treated soil mix proportions are assessed with the unconfined compression test and direct shear test. The UCS tests are carried out on a sample preparation of specimen with L/D ratio 2 having dimensions of length 10 cm and diameter 5 cm. The static compaction is done to achieve the desired light compaction density at optimum moisture content. The prepared samples are then coated with wax to prevent the moisture loss or gain from the compacted sample. The cylindrical samples are cured for 3, 7, 28 days curing period, and the unconfined compression test is carried out. The samples are also tested for the immediate effect of sodium lignosulfonate on mixing with expansive soil.

The direct shear test on soil samples done as per the IS 2720 code. The soil samples are prepared of size $6 \times 6 \times 2.5$ cm size cube. The direct shear test is carried out on 4 samples each with normal load increased from 50 to 200 kN/m^2. The graph between normal load and shearing load is plotted to get the required shear strength parameters in undrained conditions. The apparent cohesion and angle of internal friction are calculated from the obtained graph. The strain rate for both the tests is 1.25 mm/min. The strength values for untreated and treated soil samples are compared by varying both the percentage of additive and also the curing period.

4.3.1 Unconfined Compressive Strength

The unconfined compression strength test is a special kind of triaxial test. The unconfined strength of clay soil is mainly coming from the apparent cohesion developed in the soil due to the surface tension. The results of UCS tests gave the undrained shear strength of soil at MDD in optimum moisture content conditions. The curing of samples is done to check what are the changes occurring in the soil structure due to the leaching of sodium lignosulfonate into the soil. The results of UCS tests are given in Fig. 10. It is clearly seen from Fig. 8 that with increase in additive content up to 9% the unconfined strength of soil increased from 305 to 437 kPa for 28 days curing period. For all the curing periods, the trend is increasing of unconfined strength with additive content up to 9% and for 12% addition the strength is decreased. The trend of increase of strength with increasing sodium lignosulfonate is due to the aggregation of soil particles and decreased the soil particle's affinity to water.

It also showed in Fig. 9 that with curing period the strength of treated soil is also increased for all soil mix proportions. The curing period avails time to sodium lignosulfonate penetrate in soil structure, and new bonds are created between the cations of sodium lignosulfonate and the negative charge of expansive soil such as the accumulated negative charged surface and the charge deficiency in lattice structure. This proved the increase of strength with curing time; however, this increased strength with curing time is not much for treated samples compared to untreated soil samples.

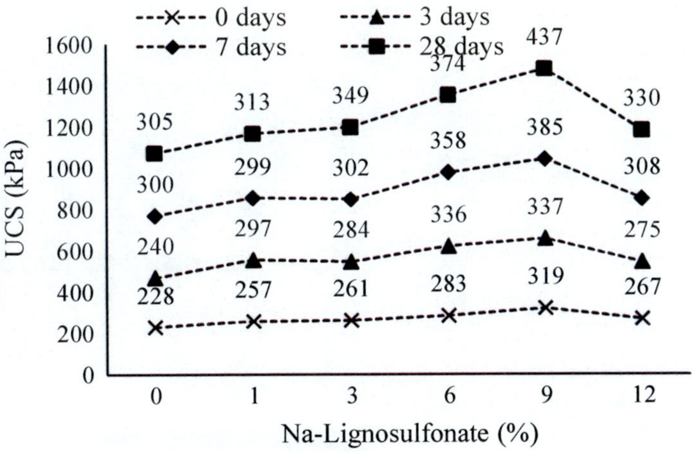

Fig. 8 Variation of UCS with additive percentage

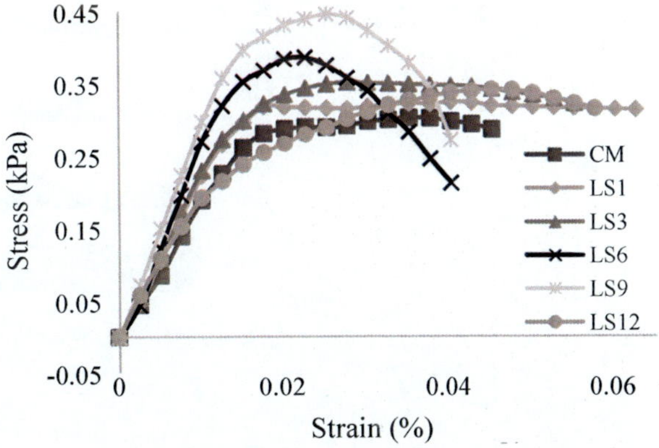

Fig. 9 Stress–strain behavior of soil–lignosulfonate mixes

This is because the permeability of soil is very low, and due to low permeability, the leaching process is very slow, and thus, it is needful to check the strength at larger curing periods. The stress strain relationship for 28 days of cured soil samples is shown in Fig. 9. The stress–strain graphs showed in Fig. 11 that for increasing additive content the behavior of soil mix is being more toward over-consolidated soil which is due to the increased stiffness of treated soil. For 12% sodium lignosulfonate, the stress–strain curve is again being like normally consolidated soil. This showed reduced stiffness due to the high concentration of lignosulfonate particles. 6% and 9% sodium lignosulfonate added soil mix gives the peak strength at lower strain values. Young's modulus of elasticity at failure and axial stiffness values is given in Table 3 (Fig. 10).

The modulus of elasticity at failure values increase from around 15.8 to 22 MPa and then again decreased for 12% additive content. The maximum strength obtained for 9% sodium lignosulfonate.

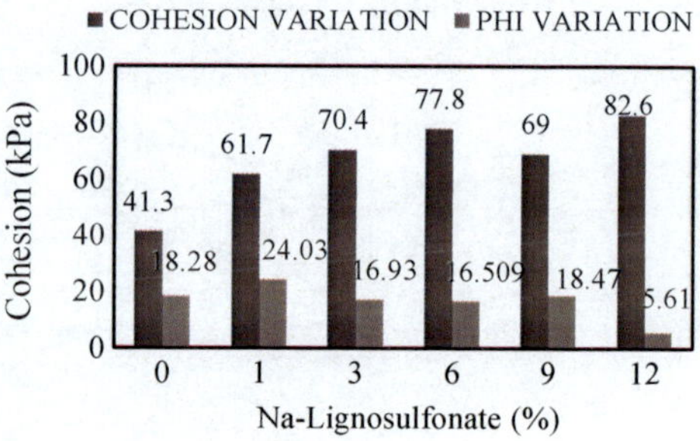

Fig. 10 Variation of shear strength parameters with percentage of lignosulfonate

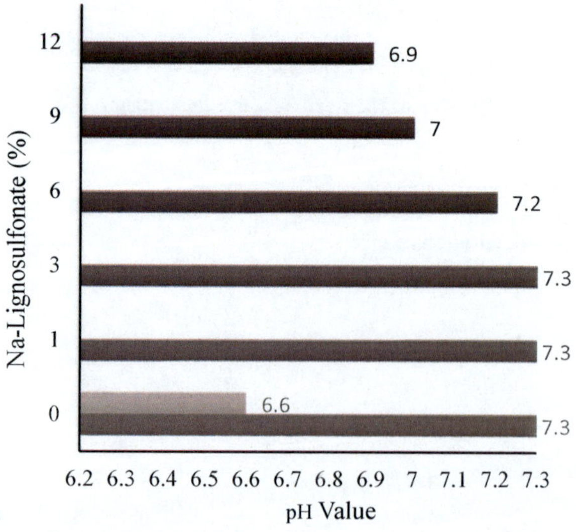

Fig. 11 pH variation with additive content

Table 3 Modulus of elasticity at failure and axial stiffness

Mix Id	CM	LS1	LS3	LS6	LS9	LS12
Modulus of elasticity, E (MPa)	15.83	17.77	18.40	21.47	22.05	13.58
Axial stiffness, k (N/mm)	311	349	361	422	433	267

4.3.2 Direct Shear Test

The direct shear test is conducted on different soil mix proportions in a direct shear box of size 6 × 6 cm with 2.5 cm depth. The soil samples are prepared with compacted dry density obtained from light compaction test and optimum moisture content is used during mixing. The results of direct shear tests are obtained for different normal loads. The shear strength parameters are calculated by plotting graphs between shear stress and the normal stress. The cohesion value of soil mix of untreated and treated soil is an indicator of stiffness of soil. The increased cohesion value for 12% additive content is not in accordance with the axial stiffness values obtained from stress–strain relationships of UCS results. This is due to the undrained condition followed during the shearing stage. The undrained unconsolidated results give the erratic results sometimes due to no volume change during shearing. The pore pressure generated during test changes with the increasing normal load. The angle of internal friction is reduced with increasing percentages of sodium lignosulfonate. This can be justified as the lignosulfonate particles coat the soil particles, and the aggregation caused in the soil mix reduced the contact points; thus, the frictional component reduced which reflects in decreased angle of internal friction.

4.4 pH Test

The pH is defined as the figure expressing the acidity or alkalinity of a solution on a logarithmic scale on which 7 is neutral and values lower than 7 are acidic and the higher values represent the alkaline solutions. The soil pH should be near to the neutral. The more acidic soils are corrosive in nature while the more basic soils that are not useful for vegetation can contaminate groundwater. After the treatment of expansive soil with any chemical, it is essential to check the change occurred in pH value of soil as chemical treatment can alter the soil chemistry. Thus after treating the expansive soil with sodium lignosulfonate, it is needed to obtain the variation in pH value of stabilized soil. The variation of pH with different percentages of sodium lignosulfonate is given Fig. 11. The sodium lignosulfonate has pH value of 6.6, which is slightly acidic, and the expansive soil has pH 7.3. When the additive percentage is increased in the soil mix from zero to 12%, the pH value decreases from 7.3 to 6.9. The soil pH is not altered much when the soil is treated with lignin-based organic polymer. The results of tests are assured and in accordance with the literature

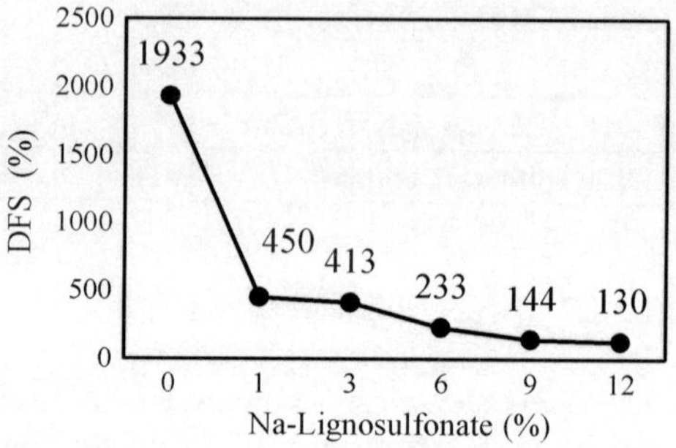

Fig. 12 Differential free swell value

available. The very little or insignificant change in pH observed proved the sodium lignosulfonate an environmentally friendly stabilizer.

4.5 Differential Free Swell

The expansive soil contains the montmorillonite mineral which expands its volume due to the affinity of negatively charged surface of soil particles toward bipolar water. The montmorillonite mineral has a very large specific surface area of around 800 m^2/g. Such a large specific surface adsorbs a very large amount of water into the soil lattice structure. This adsorption of moisture caused the expansion of volume of soil. This great volume changes from maximum expansion to shrinkage limit caused the distress in the structural elements. The seasonal temperature changes from drying to wetting caused change in swelling pressure exerted on structure. Such a stress occurred leads to the failure of structures in contact with soil. This swelling nature of expansive soil can be identified and assessed in terms of differential free swell. The differential free swell is the percentage increase in soil volume when in contact with water in comparison with the volume of soil mix in the kerosene. The results obtained in the DFS test are shown in Fig. 12. The graph plotted between the differential free swell value and the additive percentage is showed that the DFS value decreased from about 1930% for the expansive soil to 450% and to 130% for 1 and 12% addition of sodium lignosulfonate powder.

4.6 Scanning Electron Microscopy

The microstructural changes that occurred in soil after treating with sodium lignosulfonate are analyzed with the help of scanning electron microscopy. The SEM images obtained for 5000× magnification of dry soil powder. The powdered soil sample is obtained from the 28 days cured and tested UCS specimens. The UCS samples are broken down and powdered with the help of stone to form powder, which then

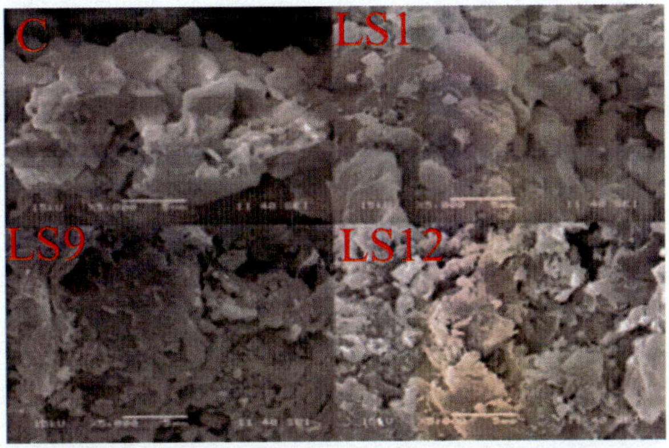

Fig. 13 SEM images for untreated and treated soil

stored airtight containers after being soaked with ethanol for SEM test. The images obtained in test are as shown in Fig. 13. The CM, LS1, LS9, and LS12 suggested the 0, 1, 9, 12% of additive content, respectively. It is clearly seen that in untreated soil mix the flaky particles of clay soil are present. The treated soil samples with the aggregation of soil particles are seen. As the lignosulfonate percentage increased, the aggregation of soil particles is more up to 9% addition; while for maximum additive content, the small-sized aggregates are formed more. This explained the reduced UCS value and axial stiffness of soil. For the LS1 and LS9, the continuous mass of soil can be observed which is an indicator for large aggregates.

5 Conclusions

The paper examined the plasticity and strength characteristics of expansive soil treated with sodium lignosulfonate. The lignosulfonate is added to the soil in five different proportions, and all the mixtures were tested for Atterberg limits, compaction characteristics, swelling index, pH, UCS, and direct shear parameters.

- The plasticity characteristics were improved significantly by adding Na-Lignosulfonate due to the adsorption of lignosulfonate on the particle surface and in between interlayers of soil particles which added water-proofing effect due to hydrophobic nature of carbon chain and limits water entry. A marginal increase in shrinkage limit value is also observed.
- No definite trend in the values of MDD and OMC is observed with addition of the lignosulfonate. The slight increase in MDD value is attributed to the anionic dispersion due to the poly-anionic nature of lignosulfonate.
- Reduction in swelling index and swelling potential is observed with increment in lignosulfonate percentage.
- The unconfined compression strength is increased after the treatment. The variation of UCS values with curing period and with increasing concentration of

sodium lignosulfonate suggested that the strength is contributed by the binding of soil particles due to aggregation, flocculation caused by ion exchange between soil surface and cations from lignosulfonate.

- The increased cohesion value in direct shear test indicates increased stiffness of the soil. The direct shear test is conducted in unconsolidated undrained condition. In this situation, the soil remained in partial saturation until the normal stress is enough to cause full saturation, due to which the shear parameters obtained were total shear parameters and gave nonzero angle of internal friction.
- The SEM image shows that the aggregation happens in soil structure from 1 to 9% additive percentage but for 12% the small-sized aggregates are observed which results in reduced strength.
- No significant change in pH value is observed upon treatment. From study, it is possible to conclude that sodium lignosulfonate can be used as a stabilizer for expansive soil effectively without any harmful impact on the environment.

References

Alazigha DP, Indraratna B, Vinod JS, Heitor A (2018a) Mechanisms of stabilization of expansive soil with lignosulfonate admixture. Transp Geotech 14:81–92

Alazigha DP, Vinod JS, Indraratna B, Heitor A (2018b) Potential use of lignosulfonate for expansive soil stabilization. https://doi.org/10.1680/jenge.17.00051

Ali UU (2015) Use of biofuel coproduct for pavement geo-materials stabilization. Procedia Eng 125:685–691

Arora KR (2008) Soil mechanics and foundation engineering. Standard Publishers Distributors, Delhi

Canakci H, Aziz A, Celik F (2015) Soil stabilization of clay with lignin, rice husk powder and ash. Geomech Eng

Ceylan H, Kim S, Ali UU, Yang B (2015) Strength performance of Iowa soils stabilized with bio-fuel industry co-product. Procedia Eng 125:317–323

Chen B (2004) Polymer-clay nanocomposites: an overview with emphasis on interaction mechanisms. Br Ceram Trans 103(6):241–249

Das CP, Patnaik LN (2000) Removal of lignin by industrial solid wastes. Pract Period Hazardous Toxic Radioact Waste Manage 4(4):156–161

Franco C, Gianluca C, Andrea N, Mattia G, Valentina C, Alessandro P, Lucia B (2014) Lignin as a co-product of second-generation bioethanol production from lingo-cellulosic biomass. Energy Procedia 45:52–60

Indraratna B, Mahamud MAA, Vinod JS (2012) Chemical and mineralogical behaviour of lignosulfonate treated soils. In: GeoCongress, ASCE, pp 1146–1155

IS: 2720-Part-5 (1970) Determination of liquid and plastic limits

IS: 2720-Part-2 (1973) Determination of water content

IS: 2720-Part-7 (1983) Determination of water content-dry density relation using light compaction

IS: 2720-Part-4 (1985) Grain size analysis

IS: 2720-Part-6 (1972) Determination of shrinkage factors

IS: 2720-Part-8 (1983) Determination of water content-dry density relation using heavy compaction

IS: 2720-Part-10 (1973) Determination of unconfined compressive strength

IS: 2720-Part-13 (1973) Direct shear test

IS: 2720-Part-40 (1977) Determination of free swell index of soil

IS: 2720-Part-26 (1987) Determination of pH value
IS: 2720-Part-3 Sect. 1 (1980) Determination of specific gravity-fine grained soils
Kamsuwan T, Srikhirin T (2014) Effect of lignosulfonate on mechanical and setting time properties of geopolymer paste, Researchgate/publication/228912027
Kim S, Gopalakrishnan K, Ceylan H (2012) Moisture susceptibility of subgrade soils stabilized by lignin-based renewable energy coproduct. J Transp Eng ASCE 138(11):1283–1290
Lebo SE Jr, Gargulak JD, McNally TJ (2001) Lignin. Kirk Othmer Encyclopedia of Chemical Technology. Wiley, New York
Lekha BM, Sarang G, RaviShankar AU (2015) Effect of electrolyte lignin and fly ash in stabilizing black cotton soil. Transp Infrastruct Geotech 2:87–101
Martone PT, Estevez JM, Lu F, Ruel K, Denny MW, Somerville C, Ralph J (2009) Discovery of lignin in seaweed reveals convergent evolution of cell-wall architecture. Curr Biol 19(2):169–175
Pasupuleti VKR, Kolluru SK, Blessingstone T (2012) Effect of fiber on fly-ash stabilized sub grade layer thickness. Int J Eng Technol 4:140–147
Phani Kumar BR, Sharma RS (2004) Effect of fly ash on engineering properties of expansive soils. J Geotech Geoenviron Eng ASCE 130(7):764–767
Santoni RL, Tingle JS, Nieves M (2005) Accelerated strength improvement of silty sand with non-traditional additives. J Transp Res Board 1936:34–42
Smith B (1999) Infrared spectral interpretation: a systematic approach. CRC Press, Florida
Ta'negonbadi B, Noorzad R (2017) Stabilization of clayey soil using lignosulfonate. Transp Geotech 12:45–55
Vinod JS, Indraratna B, Mahamud MAA (2010) Internal erosional behaviour of lignosulfonate treated dispersive clay. In: Ground improvement technologies and case histories, ISGI09, pp 549–554

Behavior of Industrial Waste Bagasse Ash and Blast Furnace Slag-Treated Expansive Clay for Pavement Subgrade

Akhilesh Singh, K. S. Gandhi, and S. J. Shukla

Abstract The stability of lightweight structure such as pavement is majorly influenced by subgrade soil. Expansive soil has high strength but it becomes problematic in the presence of water; it expands and shrinks during wet and dry conditions, respectively, because of its mineralogical composition. To minimize the cost of stabilization and to improve the load bearing capacity of such soil, these are the major concerns for problematic high plastic expansive subgrades. In this study, the experiment has been carried out to examine the feasibility of waste materials bagasse ash and ground granulated blast furnace slag as a soil stabilizer to improve subgrade. Consistency limits, California bearing ratio, unconfined compressive strength, and swelling pressure were studied to check the effect on high plastic clay when treated with waste material. The experimental results showed the improvement in soaked California bearing ratio and reduction in the swell–shrink behavior of soil when combined with bagasse ash and ground granulated blast furnace slag. Hence, these wastes can be used as a pozzolanic material to stabilize the high plastic expansive clay of the pavement subgrade.

Keywords Bagasse ash · Blast furnace slag · High plastic clay · Pavement subgrade · Stabilization

A. Singh (✉) · K. S. Gandhi · S. J. Shukla
Applied Mechanics Department, SVNIT Surat, Surat, India
e-mail: akhi.22oct@gmail.com

K. S. Gandhi
e-mail: khushbu.berawala@scet.ac.in

S. J. Shukla
e-mail: sdv@amd.svnit.ac.in

M. Latha Gali and R. R. P. (eds.), *Problematic Soils and Geoenvironmental Concerns*, Lecture Notes in Civil Engineering 88,
https://doi.org/10.1007/978-981-15-6237-2_56

1 Introduction

The arid and semiarid regions across the world are rich in expansive clay minerals where climate is characterized by rainfall and drought, subjected to wetting and drying cycle and cause tremendous instability and distress. Cracks and deformation occur due to swelling nature of expansive clays in road pavements, building foundations, irrigation structures, etc. In all such cases, soil stabilization using various waste materials such as fly ash, rice husk ash, cement kiln, bottom ash, and blast furnace slag is having been carried out as they are pozzolanic in nature and minimize the water absorption capacity by binding the clay particles (Hossain 2011). Sugarcane is a major crop, and it covers around 110 countries leading to total production of over 1500 million tons. Sugarcane production in India is above 300 million tons per year. After the optimize extraction of all economical sugar from sugarcane, about 40–45% left as fibrous residue, which is then reused in the same industry for heat generation as fuel in boilers and at last leave 8–10% ash as waste, known as sugarcane bagasse ash (SCBA). Bagasse is generally used as a preliminary fuel source in sugar mills when it is burned in large quantities; it produces enough heat energy to supply the needs of mill. Then dumping of these wastes in open land like any other waste causes a grave threat to society. James et al. (2017) highlighted that in the year 2011–12, sugarcane production reached at 361.04 million tonnes in India and became the highest cane producer in the world. Presently, India is the second-largest sugarcane producer in the list of producers that produces 341.2 million metric tonnes of sugarcane. Bagasse ash is a pozzolanic material which is rich in silica and aluminum oxides. Kiran et al. (2013), Chittaranjan and Keerthi (2011), and Kharade et al. (2014), investigated the effect of bagasse ash as a very potential stabilizer for expansive clay and revealed that it can be used as a stabilizer and improved swelling behavior of expansive soil can be observed. Bahurudeen and Santhanam (2015) discussed the mineralogical composition character of bagasse ash, which stated that bagasse ash is composed of majorly silica mineral such as quartz. Other minerals' presence was also reported in varying percent depending on the source of ash. ASTM C618 suggests that for any natural pozzolana; silica, alumina, and iron oxide should be at least 70%, and also, SO_3 content should be less than 4%. It has been reported in almost all the studies related to bagasse ash.

Blast furnace slag is a by-product of the furnaces producing iron. Chemically, slag is a mixture of silica, lime, and alumina, that is, the same oxides that makeup Portland cement. Cementitious properties are initialized on the reaction of ground granulated slag with water. Different studies have been carried out to improve the problematic soil with BFS reported by various researchers (Sharma and Sivapullaiah 2016; Pathak and Pandey 2014; Oormila and Preethi 2014; Yadu and Tripathi 2013; Phani Kumar and Sharma 2004).

Studies relating to the bagasse ash or blast furnace slag individually with some admixture have been carried out by few researchers. However, no studies on the joint action of bagasse ash and BFS as stabilizing agents for expansive soils have been published to date. The purpose of this study is to thoroughly investigate the combined effect of sugarcane industrial waste bagasse ash and granulated blast furnace slag on swell–shrink and strength properties of high plastic expansive clay for lightweight structure.

2 Materials and Methodology

2.1 Materials

Soil

Expansive clays under investigation were collected from the campus of the Sardar Vallabhbhai National Institute of Technology, Surat. Properties of collected clay samples are listed in Table 1.

Bagasse Ash

Bagasse ash was collected from the sugar factory, Bardoli, Gujarat. The chemical composition of BA is represented in Table 2 (Figs. 1 and 2).

Blast Furnace Slag

Blast furnace slag was collected from a cement manufacturing plant in Kim, Gujarat (Astro Cement Pvt. Ltd.). BFS is primarily made up of silica, alumina, calcium oxide, and magnesia. The chemical composition of BFS is given in Table 3.

Table 1 Properties of sample soil

Soil property	Values
Specific gravity	2.64
Grain size analysis	
Gravel content (%)	0
Sand content (%)	15
Silt content (%)	52
Clay content (%)	33
Consistency limits	
Liquid limit (%)	64.05
Plastic limit (%)	27.12
Shrinkage limit (%)	22.42
IS Classification	CH
Differential Free Swell Index (%)	57.89
Compaction	
Optimum moisture content (%)	22.7
Maximum dry density (kN/m^2)	1.58
Unconfined compressive strength q_u (kN/m^2)	244
CBR (%)	3.3

Table 2 Chemical constituents of BA

Compounds	% by weight
SiO_2	62.43
Al_2O_3	4.28
Fe_2O_3	6.98
CaO	11.8
MgO	2.51
SO_3	1.43
K_2O	3.51
LOI	4.73

Fig. 1 Bagasse ash

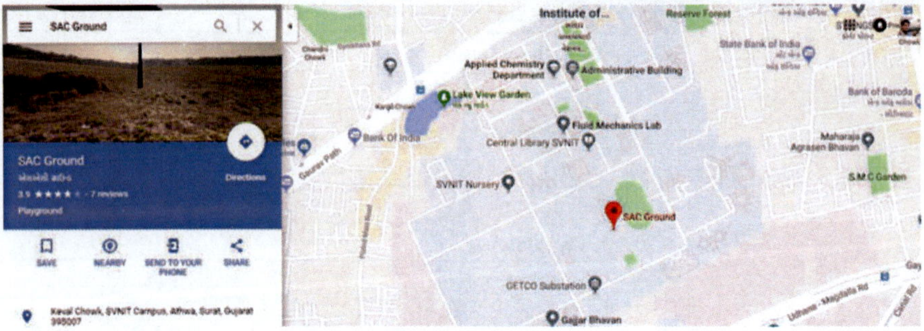

Fig. 2 SAC Ground, SVNIT Campus. *Source* Google Maps

2.2 Methodology

Mix selected for the BA: BFS ratios are 2:1 to 8:1, 1:1 to 4:1, and 2:3 to 10:3. A three-stage study of BFS content varying from 2.5 to 7.5%, and bagasse content varying from 5 to 25% in one case or another was carried out to obtain high strength for a combination.

Table 3 Chemical constituents of BFS

Compounds	% by weight
CaO	32.6
SiO_2	39.86
Fe_2O_3	1.46
Al_2O_3	23.13
Sulfide	1.96
MnO	0.09

Source Rabbani et al. (2012)

Table 4 Mix selected for study

Mix No.	Composition		
	Soil (%)	BA (%)	BFS (%)
1	100	0	0
2	90	10	0
3	92.5	5	2.5
4	87.5	10	2.5
5	82.5	15	2.5
6	77.5	20	2.5
7	90	5	5
8	85	10	5
9	80	15	5
10	75	20	5
11	87.5	5	7.5
12	82.5	10	7.5
13	77.5	15	7.5
14	72.5	20	7.5
15	67.5	25	7.5
16	92.5	0	7.5

Sample passing through 4.75 mm sieve was taken for compaction and then CBR testing (Table 4 and Fig. 3).

3 Test Result and Analysis

Liquid limit and plastic limit of untreated and treated soil were conducted following IS 2720 Part V. Special attention was given considering the fact that as consistency will decrease while the progress of experiment, i.e., water must not be added after experiment starts.

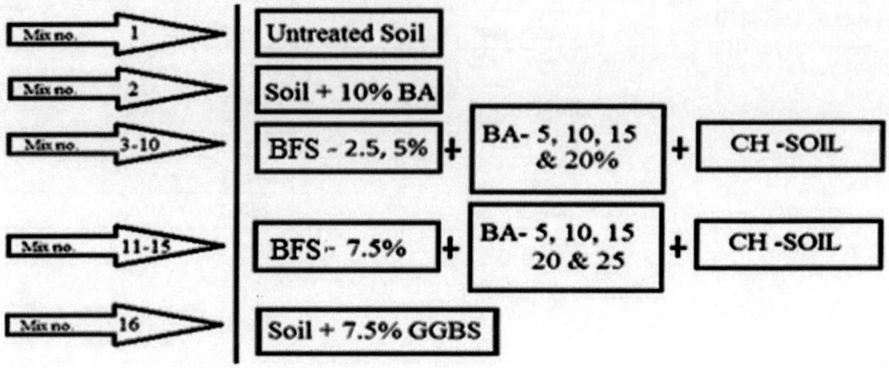

Fig. 3 Pictorial diagram of mix selected

Table 5 Atterberg limit results

Mix No.	S.L	P.L	L.L
1	22.42	27.12	64.05
2	25.76	30.84	58.64
3	23.45	28.31	62.35
4	23.62	28.78	58.07
5	23.76	29.13	57.34
6	31.16	36.51	56.78
7	20.56	27.13	62.48
8	29.33	35.24	60.54
9	37.48	42.51	58.89
10	36.11	41.32	56.32
11	27.13	32.07	59.64
12	23.01	29.61	57.05
13	19.65	25.14	56.49
14	23.22	29.84	55.98
15	24.52	30.41	55.12
16	26.87	32.51	54.68

Shrinkage factors were found out using Part VI of the IS 2720. All samples were added with water equal to or greater than liquid limit of the sample was added and kept for 24 h in humidity chamber to mitigate water evaporation; after filing of shrinkage dish with gently tapping, samples were weighted and kept for open drying first then for oven drying (Table 5).

Free swell index for all mixes was obtained using IS code provision. Free swell index was reduced primarily because non-expansive material BA replaced the fine particles of expansive clay.

Table 6 Swelling pressure and FSI for all samples

Mix No.	Swelling pressure (kg/cm^2)	FSI (%)
1	0.560	57.89
2	0.300	66
3	0.370	62
4	0.240	60
5	0.238	58
6	0.235	55
7	0.230	52
8	0.320	50
9	0.285	46
10	0.243	42
11	0.201	40
12	0.201	37
13	0.256	27
14	0.276	21
15	0.254	23
16	0.290	62.1

The swelling pressure test and FSI value for all samples were obtained following IS code methodology. Swelling pressure for all mixes was obtained using IS 2720 PART XXXXI. FSR got reduced because BA can provide sufficient divalent and trivalent cations such as Ca^{2+}, Al^{3+}, Fe^{3+} that increased flocculation and exchange of cations (Table 6).

Optimum moisture content and maximum dry density of all samples found out using light compaction following IS code. 5 kg sample was taken per water content with a difference of 2–4% depending upon sample. Light compaction results were obtained for oven-dried samples passing through sieve 4.75 mm; results show that OMC increases with bagasse content and MDD decreases with bagasse content; it is because of BA being finer and less dense material then soil occupies pores in soils and provides bondage to soil particle by adsorbing more water than soil. Pozzolanic reaction in bagasse ash with clay fraction of soil caused OMC to increase. The lower specific gravity of BA caused reduction in dry density, in comparison with the compacted natural soil. Proctor results show that MDD increases with BFS content and decreases with BA content, and OMC increases with BA and BFS content. BFS in denser powdered form acts as pozzolanic material and increases the MDD initially with the addition of bagasse and decreases thereafter, while OMC increases initially and then has a decreasing trend. Majorly the range of MDD and OMC is 1.4–1.5 g/cc and 25–30%, respectively. The flocculated and agglomerated fine particles that are caused by cation exchange occupied larger spaces influenced the dry density to drop (Table 7).

Table 7 OMC and MDD of samples

Mix No.	OMC (%)	MDD (g/cc)
1	23.747	1.504
2	24.77	1.405
3	24.094	1.477
4	25.323	1.465
5	25.6	1.423
6	27.664	1.387
7	24.166	1.455
8	25.269	1.462
9	26.597	1.454
10	27.082	1.462
11	24.95	1.499
12	25.749	1.431
13	26.295	1.425
14	26.396	1.451
15	29.227	1.385
16	24.502	1.543

California bearing ratio of all mixes was obtained for 0 days curing and 4 days soaking, and for mix 12 and 14 for 0, 7, 14, and 28 curing and 4 days soaking were obtained (as 0 days values were higher) following guidelines of IS code. Oven-dried samples passing through 4.75 mm sieve cast at OMC and MDD and soaked for 4 days were tested. Mix of soil, BA and BFS and their respective pozzolanic, cementation, and hydration reaction increased the CBR value. The mix with 10–20% BA with 7.5% BFS was found to be suitable as subbase material of pavement. CBR 2.5 values were greater than CBR 5.0 values. CBR values were increased from 3 to 16% for 0 days sample for mix 14 (72.5% soil, 20% BA, 7.5% BFS). The two high UCS value samples were also tested for 28 days and 4 days curing period. CBR 2.5 value was increased to 41% for sample 12 (Figs. 4 and 5; Table 8).

UCS results for samples of all mixes at 0, 7, 14, and 28 days were carried out. Initially, addition of bagasse has reduced strength for 0 days but had gained strength 350% of untreated soil at 28 days curing. Or 800% compared to 0 days strength of a mix. The pozzolanic and hydration reaction between soil, BA, and BFS influenced increase in UCS value, forming CSH and CAH, and they fill the voids and bind the particles of soils together and improving the strength of expansive soils (Table 9 and Fig. 6).

0 days' UCS value shows a decreasing trend as admixture added did not provide strength as the pozzolanic reaction may not have started or started giving compounds like CSH and CAH while 7 days' results show higher strength for BFS 7.5% and BA 20% (Figs. 7, 8 and 9).

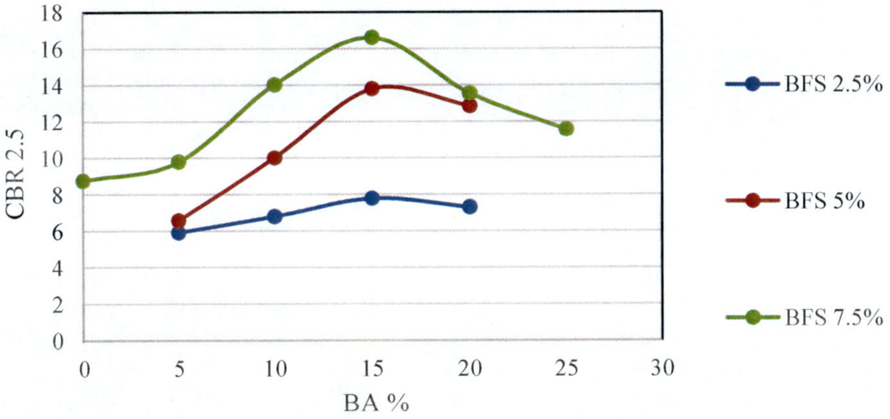

Fig. 4 CBR 2.5 values

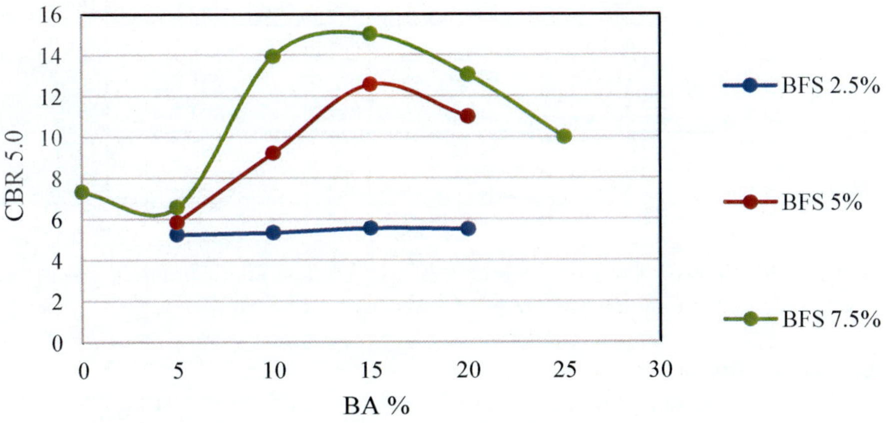

Fig. 5 CBR 5.0 values

Table 8 CBR test results for 7, 14, and 28 days

Mix No.	28 days		14 days		7 days	
	CBR 2.5	CBR 5.0	CBR 2.5	CBR 5.0	CBR 2.5	CBR 5.0
12	41.39	39.47	33.32	32.71	22.64	21.32
14	39.87	38.72	32.76	31.14	23.42	22.73

BA content between 10–20% is higher strength yielding range as further addition of BA, i.e., 25% results into a very less strength mix (Fig. 10).

Table 9 UCS test results for 0, 7, 14, and 28 days

Mix No.	0 days avg. UCS kN/m^2	7 days avg. UCS kN/m^2	14 days avg. UCS kN/m^2	28 days avg. UCS kN/m^2	% increase w.r.t 0 days	% increase w.r.t untreated soil
1	138.9	214.8	234.8	244.6	176.1	100.0
2	222.8	244.8	270.5	315.5	141.5	128.9
3	230.5	300.8	300.5	330.2	143.2	134.9
4	250.0	320.0	355.5	386.3	154.5	157.8
5	234.5	305.2	330.8	350.2	149.3	143.1
6	220.5	290.2	310.2	325.2	147.4	132.9
7	150.5	280.2	310.2	330.5	219.5	135.0
8	153.8	304.5	358.0	391.9	254.7	160.1
9	187.8	400.2	560.2	590.5	314.3	241.3
10	192.0	444.5	573.0	646.6	336.7	264.2
11	130.0	280.2	400.2	450.2	346.1	184.0
12	102.3	501.4	608.3	827.0	808.3	338.0
13	160.9	517.7	637.4	871.6	541.6	356.2
14	120.6	505.2	610.2	727.2	602.7	297.2
15	180.3	300.9	319.6	330.4	183.2	135.0
16	297.7	567.3	898.7	1052	353.3	333.4

14 and 28 days UCS value shows a higher UCS strength trend as pozzolanic reaction provided compounds like CSH and CAH.

Durability Studies

Alternate wetting and drying cycle (W/D cycle) of UCS sample cured for 0, 7, 14, and 28 days were done to check the durability of stabilized mix comparing to virgin soil by giving 5 h wetting cycle and 42 h drying cycle at 70-°C. Only samples of mix 12 came out to be best as its 7, 14, and 28 days samples remained for 2, 6, and 9 cycles (Figs. 11 and 12).

4 Conclusion

In this study, expansive clay has been stabilized with industrial waste bagasse ash and blast furnace slag. The following results were drawn from the experimental work.

1. Flocculated and agglomerated fine particles because of cation exchange occupied larger space influenced the MDD to fall. Reduction in dry density is also accompanied by lower specific gravity of BA as analyzed with compacted natural soil.

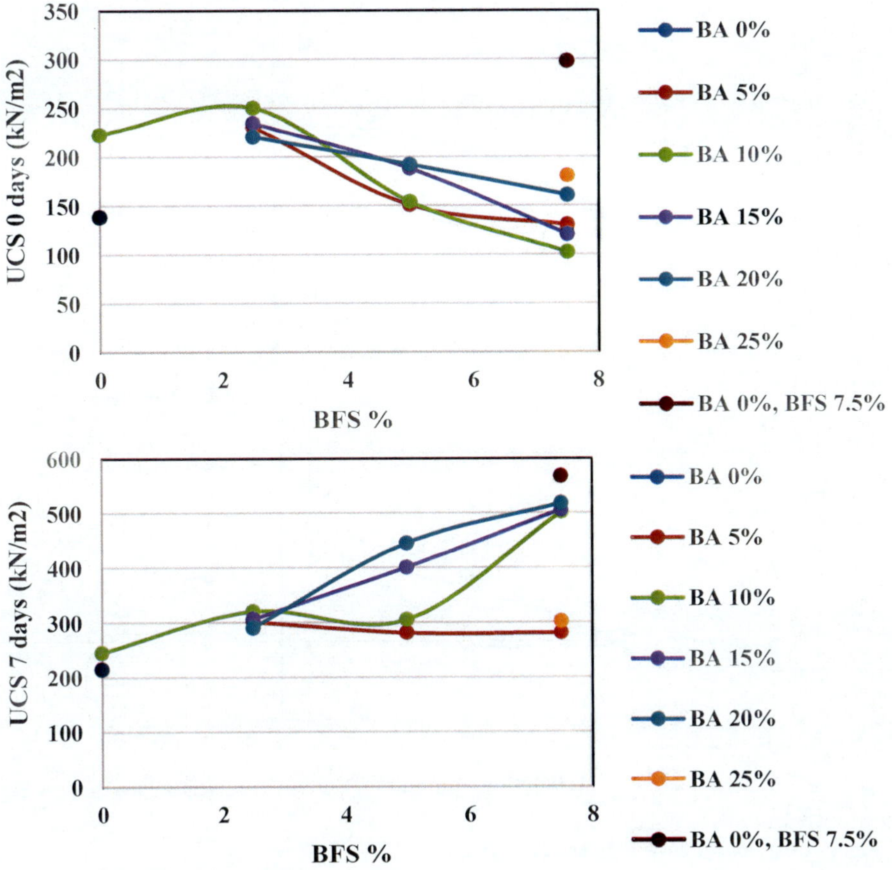

Fig. 6 Variation of UCS with percent of BFS and BA (0 and 7 days)

2. Swelling properties got reduced because BA can provide sufficient divalent and trivalent cations that increase flocculation and cation exchange. In addition to this, BA as non-expansive material replaced the fine particles of expansive clay contributing to fall in swelling.
3. Unconfined compressive strength increases with increase of BFS content from 10 to 20% bagasse ash content.
4. Progressive cementation of mix of soil, BA, and BFS as hydration and pozzolanic reaction influenced CBR value to increase.
5. Sample mix containing 7.5% BFS and 10–15% BA has shown a reduction in swelling pressure and high strength with good durability; hence, it can be used to stabilize high plastic expansive clay.

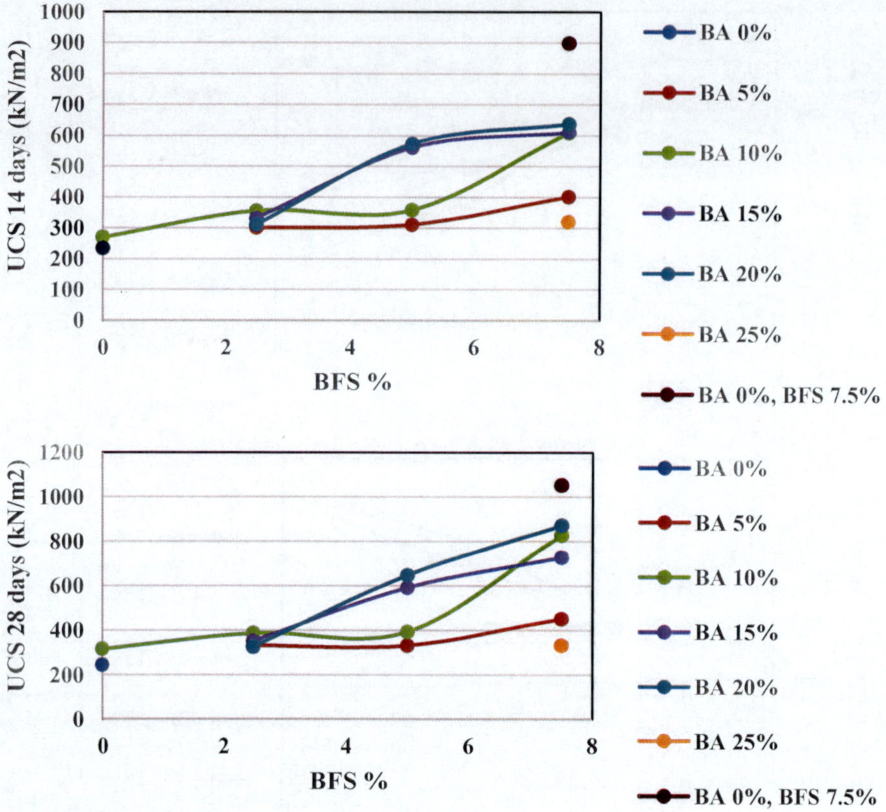

Fig. 7 Constant BA curves (14 and 28 days)

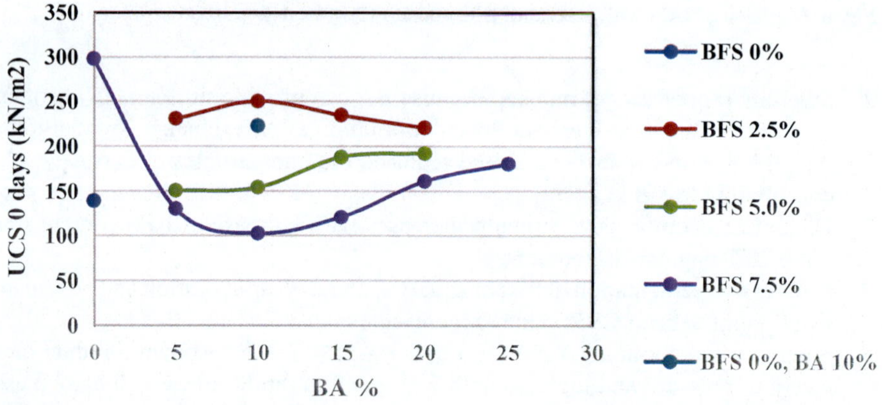

Fig. 8 Constant BFS curve 0 day

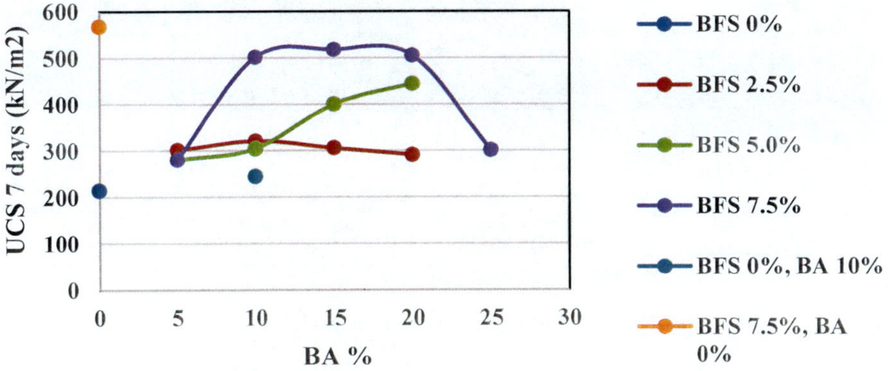

Fig. 9 Constant BFS curve 7 days

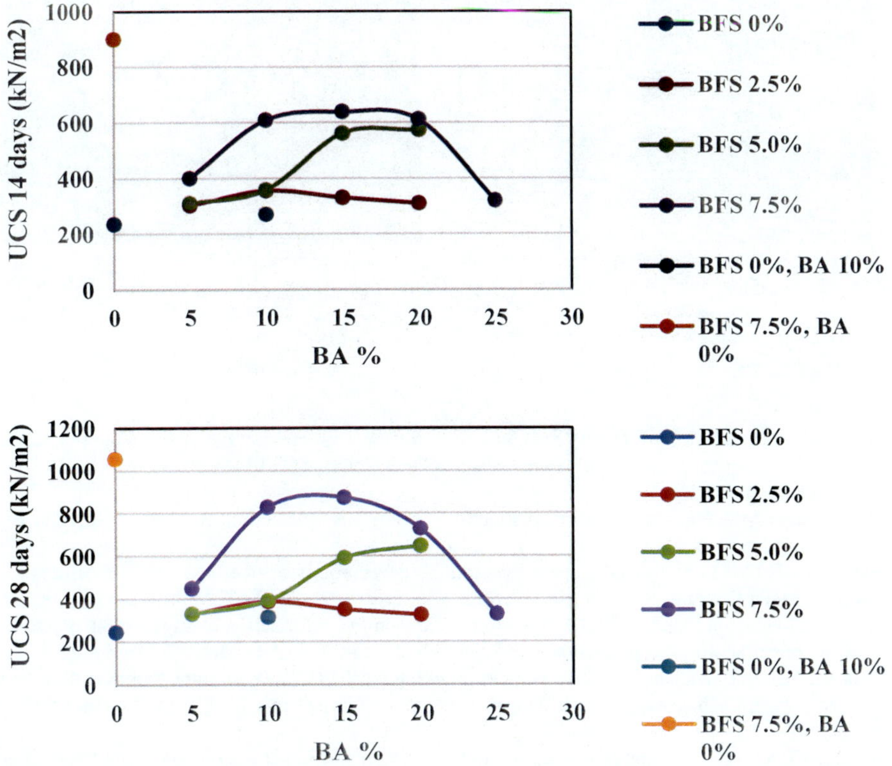

Fig. 10 Constant BFS curve (14 and 28 days)

Fig. 11 Wetting cycle of treated (mix 12) versus untreated Soil

Fig. 12 Samples after 1st and 6th cycle, respectively

References

Bahurudeen A, Santhanam M (2015) Influence of different processing methods on the pozzolanic performance of sugarcane bagasse ash. Cement Concr Compos 56:32–45. https://doi.org/10.1016/j.cemconcomp.2014.11.002

Chittaranjan MV, Keerthi D (2011) Agricultural wastes as soil stabilizers. Int J Earth Sci Eng 4(6):50–51

Hossain AK (2011) Stabilized soils incorporating combinations of rice husk ash and cement kiln dust. J Mater Civ Eng 23(9):1320–1327. https://doi.org/10.1061/(ASCE)MT.1943-5533.0000310

James J, Kasinatha Pandian P (2017) A short review on the valorisation of sugarcane bagasse ash in the manufacture of stabilized/sintered earth blocks and tiles. Adv Mater Sci Eng 1–15

Kharade AS, Suryavanshi VV, Gujar BS, Deshmukh RR (2014) Waste product 'Bagasse Ash' from sugar industry can be used as stabilizing material for expansive soils. IJRET Int J Res Eng Technol 1(1):506–512

Kiran RG, KL (2013). Analysis of strength characteristics of black cotton soil using bagasse ash and additives as stabilizer. Int J Eng Res Technol (IJERT) 2(7). ISSN: 2278-0181

Oormila TR, Preethi TV (2014) Effect of stabilization using flyash and GGBS in soil characteristics. Int J Eng Trends Technol 11(6):284–289

Pathak AK, Pandey V (2014) Soil stabilization using ground granulated blast furnace slag. Int J Eng Res Appl 4(4):164–171

Phani Kumar B, Sharma RS (2004) Effect of fly ash on engineering properties of expansive soils. J Geotech Geoenviron Eng 130(7):764–767

Rabbani P, Daghigh Y, Atrechian MR, Karimi M, Tolooiyan A (2012) The potential of lime and grand granulated blast furnace slag (GGBFS) mixture for stabilisation of desert silty sands. J Civ Eng Res 2(6):108–119
Sharma AK, Sivapullaiah PV (2016) Ground granulated blast furnace slag amended fly ash as an expansive soil stabilizer. Soils Found 56(2):205–212
Yadu L, Tripathi RK (2013) Effects of granulated blast furnace slag in the engineering behavior of stabilized soft soil. Procedia Eng 51:125–131 (International conference in chemical, civil and mechanical engineering)

Improvement of Soft Clay Bed Using Fibre-Reinforced Soil-Cement Columns

Lambture Mahesh, Rakesh J. Pillai, G. Sumanth Kumar, and V. Raman Murthy

Abstract Being a challenge to deal with structural foundations in soft clays, several techniques were promulgated across the world. This paper presents the load–settlement behaviour of soft clay provided with soil-cement columns in test tanks. Based on the experimental work performed on 10% cement mixed single clay column resting on hard stratum simulating end bearing condition, it is revealed that load-carrying capacity of clay bed increased by 3–4 times that of virgin clay. Also, the load-carrying capacity of clay bed increased 5–7 times that of virgin clay with the group of three columns arranged in triangular pattern with centre to centre spacing of two times the diameter of column. Group capacity of soil-cement column is 1.5–2 times of single column which is not multiplicative of single column. UCC tests were carried to find out the optimum fibre content by varying the fibre content from 0 to 3%. From UCC test, it is observed that 2% fibre content is suitable for the present study considering the strength improvement and ease of mixing. It is observed from the tests that with the addition of 2% fibre in soil-cement column, load-carrying capacity of clay bed increased by 5.5–7.5 times when tested with group of columns. There is 10–20% increase in ultimate load-carrying capacity of soil-cement column after addition of fibre and the mode of failure of columns changed from brittle to bulging failure.

Keywords Soft clay · Ground improvement · Cement columns · UCC · Fibre reinforcement

L. Mahesh (✉) · R. J. Pillai · G. S. Kumar · V. R. Murthy
Department of Civil Engineering, National Institute of Technology Warangal, Warangal, Telangana, India
e-mail: lambutremd@gmail.com

R. J. Pillai
e-mail: rakeshpilla@gmail.com

G. S. Kumar
e-mail: gundetisumanth@gmail.com

V. R. Murthy
e-mail: vrm@nitw.ac.in

M. Latha Gali and R. R. P. (eds.), *Problematic Soils and Geoenvironmental Concerns*, Lecture Notes in Civil Engineering 88,
https://doi.org/10.1007/978-981-15-6237-2_57

1 Introduction

The soft clay deposits spread along the shore lines of many world nations are characterized by high compressibility and low shear strength. This coupled weakness of these deposits pose a challenge to the geotechnical engineers while designing suitable foundation system to support the infrastructure. Several foundation systems were developed to support the structural loads. Most of these systems come partly underground improvement which subsequently supports the loads. Despite several advancements made on various aspects of improvement or supporting systems in these soft clay beds, there still exist several unknowns with respect to load transfer mechanisms and settlements. In the recent times, deep mixing using lime or cement has been widely used to improve these beds within short time periods (Faro et al. 2015). In this process, the cemented soil columns are surrounded by soft clay and understanding the overall improvement requires a great deal of experimental work (Yao et al. 2016). In the present work, an attempt is made to study the addition of fibre reinforcement along with cement on the load-carrying capacity of the improved system. This system is beneficial when the deep mixed columns are extended up to the stiffer stratum, whereby higher load-carrying capacity of deep mixed columns with fibre reinforcement can be utilized (Consoli et al. 2003; Xiao et al. 2013, 2015; Farouk and Shahien 2013). For this, the experimental studies are carried out in the test tanks with end bearing columns.

2 Experimental Study

2.1 Materials Used

Soil: the locally available soil is used to make soft clay. The properties of soil are: $G = 2.71$, gravel = 1%; sand = 30%; silt = 30%; clay = 39%; liquid limit = 63%; plastic limit = 20%; plasticity index = 43% and IS classification = CH.

Cement: OPC-53 grade cement is used in this study.

Fibre: polypropylene fibre is used. The properties of fibre are: length = 6 mm, diameter = 26 μm, aspect ratio = 231, $E = 7$ GPa, tensile strength = 540 MPa, ρ = 910 kg/m^3.

2.2 Test Procedure

The load tests were carried out in model test tanks of 30 cm diameter and 30 cm height and also in test tank of 50 cm × 50 cm × 60 cm. The test tank is placed on the pedestal of loading frame centrally. The soft clay is prepared corresponding to the required consistency indices of 0.1, 0.25 and 0.5. 50 mm diameter (d) and PVC pipes are used to form 25 cm long soil-cement columns within soft clay.

The soft clay prepared corresponding to a given consistency is placed in the test tank is layers of 5 cm by marking on the sides of test tank. PVC pipes are placed at the desired spacing (2d) and soil-cement mix or soil-cement-fibre mix as per the test condition is placed and pressed while gradually withdrawing the PVC pipe by rotation by maintaining adequate overlapping with respect to the placement of clay around the pipes. The process is repeated till the test specimen with soil-cement columns is formed. At the top of soil-cement columns, a levelling course of 2 cm thick sand is placed before placing the 12 cm diameter test plate over them. The system is allowed for 7 days curing by covering it with a polythene bag and at the end of curing period, load test was carried out. The load was applied in increments of and the corresponding settlements were recorded with the help of dial gauges. The load-settlement plots were drawn and the ultimate loads were obtained by drawing tangents to the initial and final straight-line portions (Fig. 1).

Fig. 1 Soil-cement column in small test tank under loading platform

3 Results and Discussion

The load tests were carried out on end bearing soil-cement columns of 50 mm diameter and 25 cm long at different consistencies of clay. The pressure-settlement curves are presented below.

3.1 *Influence of Different Fibre Content on the Soil-Cement Column*

For the calculation of optimum fibre content, UCC test is performed by varying fibre content from 0 to 3% by weight of soil. For this test, samples of size 50 mm × 100 mm are used. Different samples are prepared for different consistencies, i.e. $I_c = 0.1$, $I_c = 0.25$, $I_c = 0.5$. These samples are tested after 14 days of curing.

From Figs. 2 and 3, it can be seen that both strength and deformation is increasing after addition of fibre. Deformation is increased by about 1.2–2 times after addition of fibre. Strength is increased by about 20% after addition of fibre.

From Figs. 4 and 5, it can be seen that without fibre failure of sample is brittle

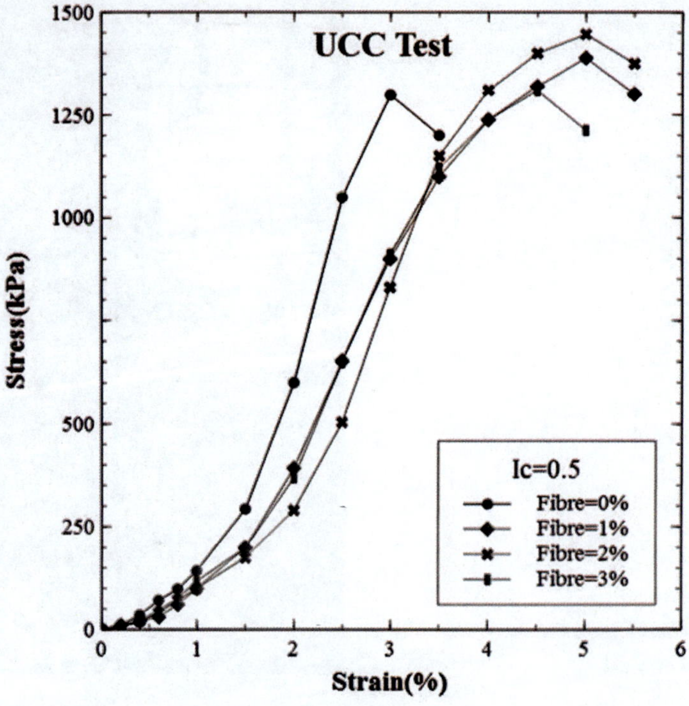

Fig. 2 UCC test result for $I_c = 0.5$ for different fibre content

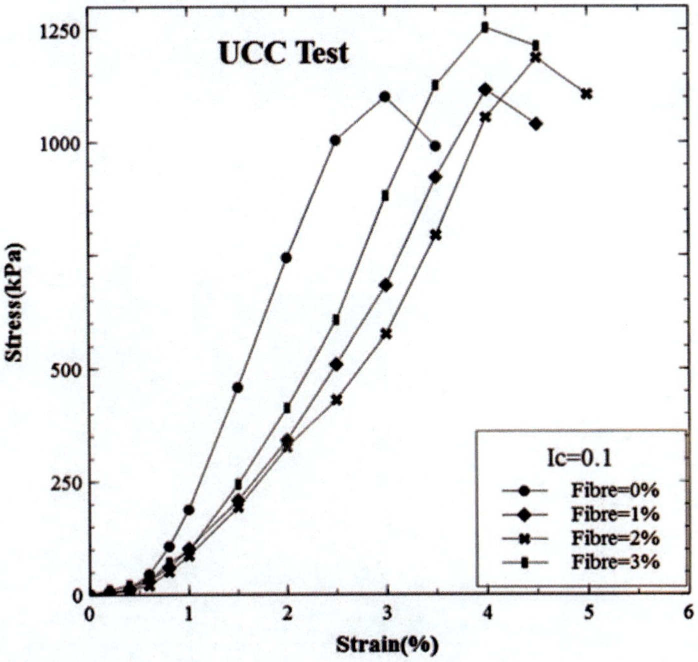

Fig. 3 UCC test result for $I_c = 0.1$ for different fibre content

Fig. 4 UCC test sample without fibre after test

Fig. 5 UCC test sample with fibre after test

and with addition of fibre, and sample fails due to bulging which is an indication of ductile behaviour. From all the above observations and considering the ease of mixing, 2% fibre content by weight was used in the present study.

3.2 *Influence of Consistency on Soil-Cement Column Capacity*

The load tests are carried out on soil-cement columns of 5 cm × 25 cm at different consistencies of clay.

It can be seen from this Fig. 6 that the settlements are considerably higher with decrease in consistency for any load increment. The ultimate load-carrying capacity values are obtained from the intersection points of tangents drawn along the initial and final straight-line portions of load-settlement plots (Table 1).

The pressure-settlement plots for a 3 column group at different soil consistency indices (I_c) are presented in Fig. 7. It can be observed from this figure that the initial consistency of clay has significant influence on the load-carrying capacity of column group. The ultimate load-carrying capacity of the group is about 180 kPa at $I_c = 0.5$ and its 120 kPa, 50 kPa at 0.25 and 0.1, respectively. These values indicate that the

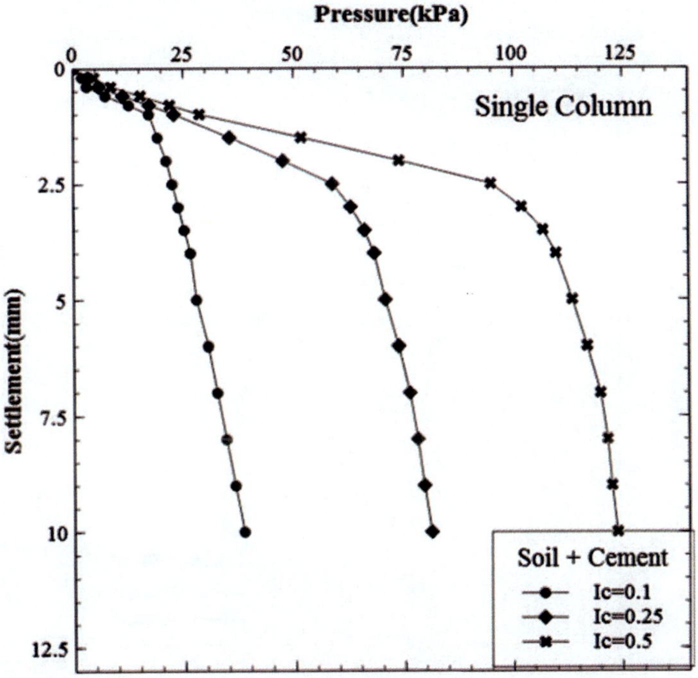

Fig. 6 Single soil-cement columns at different consistencies

Table 1 Ultimate load-carrying capacity of soil-cement columns for different soil consistencies

Columns	$I_c = 0.1$	$I_c = 0.25$	$I_c = 0.5$
Single column (kPa)	25	70	100
Group of columns (kPa)	50	120	180

ultimate capacity is increased around 3 times when the consistency index is increased from 0.1 to 0.5.

3.3 *Effect of Soil Consistency on Behaviour Soil-Cement Columns*

Pressure-settlement plot for the initial consistency of 0.1, 0.25 and 0.5 is shown in Figs. 8, 9 and 10 (Table 2).

From above observations, we can conclude that capacity of single column is around 3 times that of original soil and capacity of group of column of three is 5–7 times that of original soil. Also, the capacity of group of column of three is 1.7–2.1 times that of single column.

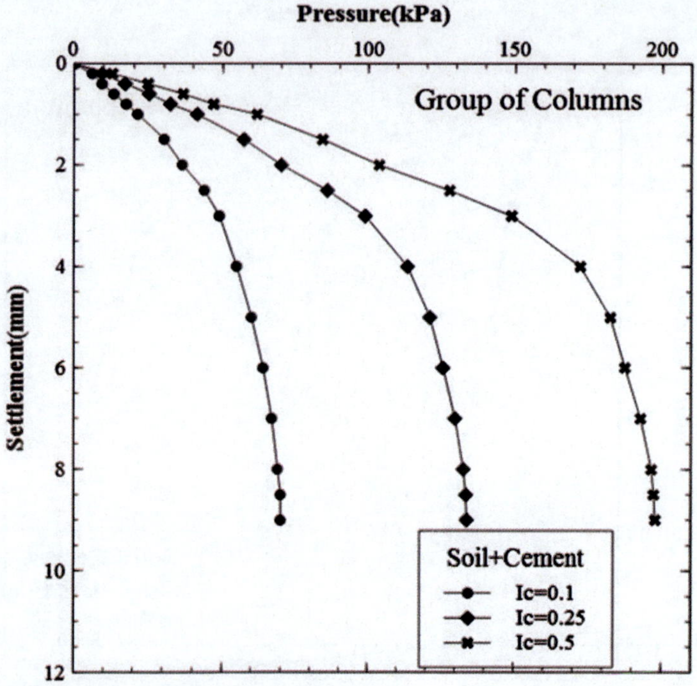

Fig. 7 Group of columns at different consistencies

From this, it can be concluded that group capacity is not multiplicative of single column.

3.4 *Effect of Fibre Addition on Soil-Cement Column Behaviour*

Pressure-settlement curves for the different initial consistencies with fibre-reinforced single soil-cement column and group of columns are plotted in order to compare its capacity with original soil is shown in Figs. 11, 12 and 13 (Table 3).

From this, it can be concluded that capacity of fibre-reinforced single column is 3.5–4 times of original soil and capacity of group of column is 5.5–7.5 times of original soil. Whereas capacity of fibre-reinforced group of column is around 1.5–2 times that of single column.

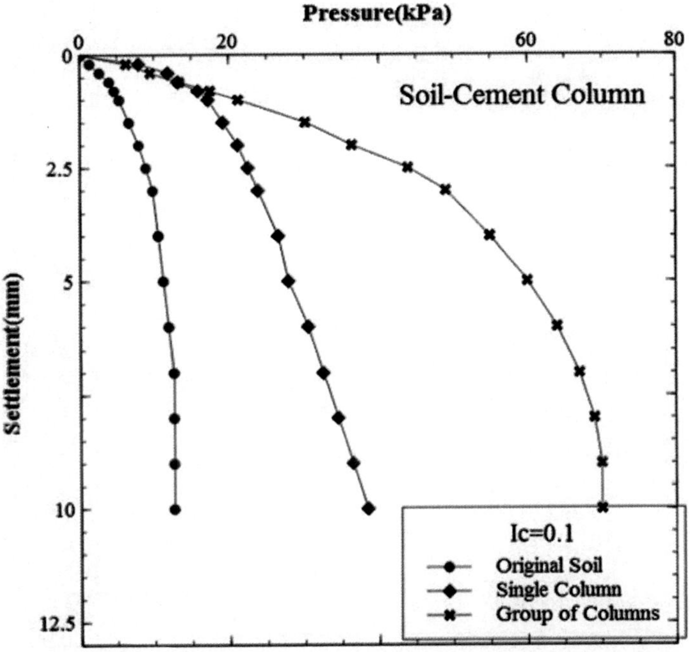

Fig. 8 Pressure-settlement plot for the capacity of single and group of column for $I_c = 0.1$

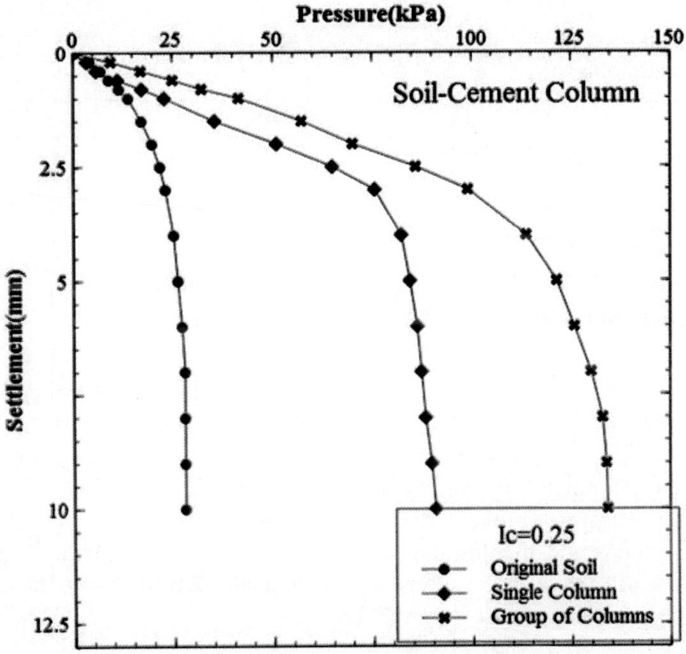

Fig. 9 Pressure-settlement plot for the capacity of single and group of column for $I_c = 0.25$

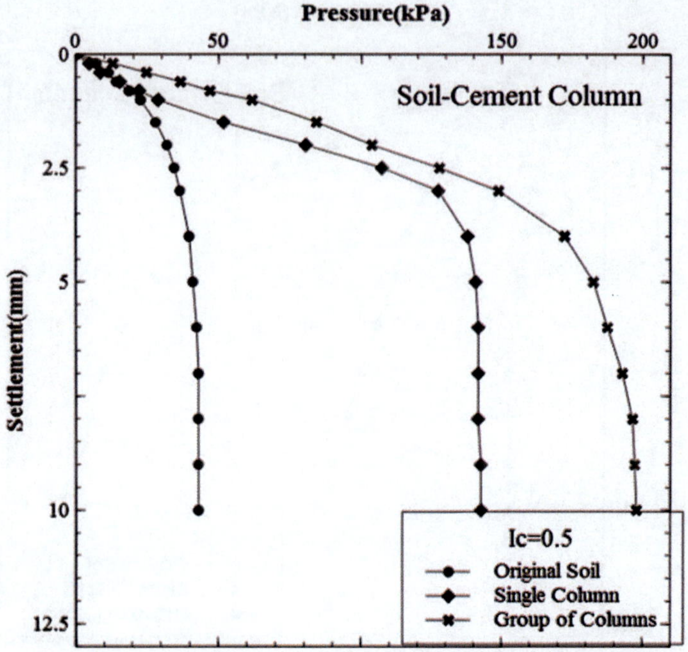

Fig. 10 Pressure-settlement plot for the capacity of single and group of column for $I_c = 0.5$

Table 2 Comparison of ultimate load-carrying capacity of soil

I_c	Original soil (kPa)	Single column (kPa)	Group of columns (kPa)
0.1	8	25	50
0.25	22	70	120
0.5	30	100	180

3.5 Comparison Between Soil-Cement Columns and Fibre-Reinforced Soil-Cement Columns

It can be observed from Table 1 that increase in ultimate load capacity of soil after addition of fibre is about 15–20%. Figure 14 indicates the failure pattern of soil-cement columns and Fig. 15 indicates the failure pattern of fibre-reinforced soil-cement column. From Fig. 14, it can be observed that failure of soil-cement column brittle as it was broken into pieces, whereas from Fig. 15, it can be observed that when fibre is added column is showing bulging failure which indicates ductile behaviour. From above observations, we can conclude that addition of fibre improves the ductility of soil-cement column (Table 4).

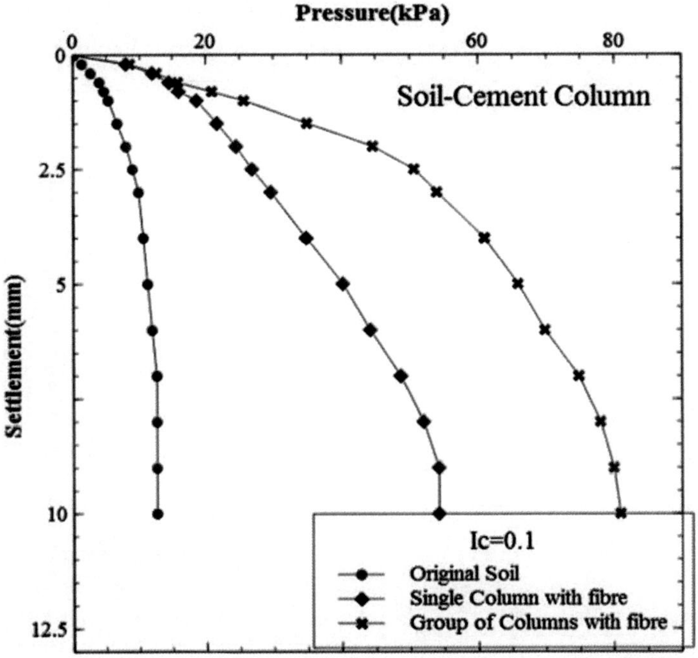

Fig. 11 Pressure-settlement plot for the capacity of single and group of column for $I_c = 0.1$

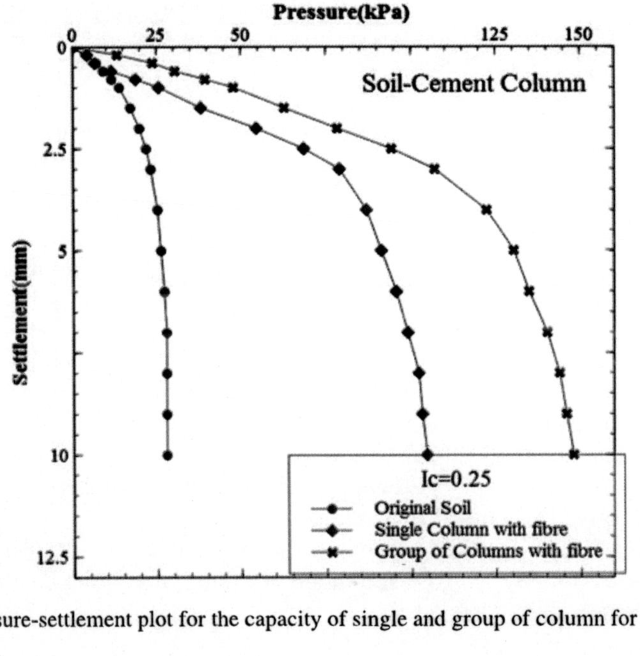

Fig. 12 Pressure-settlement plot for the capacity of single and group of column for $I_c = 0.25$

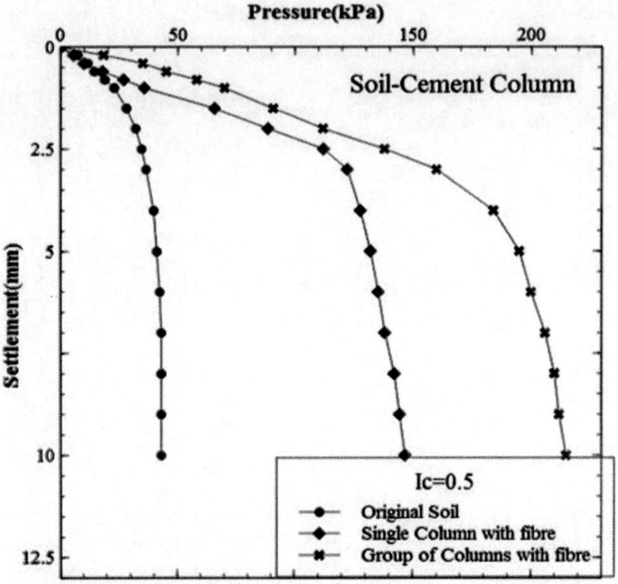

Fig. 13 Pressure-settlement plot for the capacity of single and group of column for $I_c = 0.5$

Table 3 Comparison of ultimate load-carrying capacity of soil with fibre-reinforced SCC

I_c	Original soil (kPa)	Single column (kPa)	Group of columns (kPa)
0.1	8	30	65
0.25	22	80	140
0.5	30	120	190

4 Conclusions

The following conclusions are drawn from the experimental work carried out in this study.

1. The load-carrying capacity of soil-cement columns is significantly increased with increasing initial soil consistency.
2. In case of soil-cement columns, the increasing cement content has shown marginal increase in load-carrying capacity.
3. The group capacity of end bearing soil-cement columns is found to be not multiplicative value of a single column capacity.
4. There is increase in ultimate load-carrying capacity of soil-cement column after addition of fibre by 15–20% and mode of failure of column changes to ductile from brittle.

Fig. 14 SCC after test

Fig. 15 SCC with fibre after test

Table 4 Comparison of ultimate load-carrying capacities of soil at different consistencies

I_c	Single SCC (kPa)	Single SCC with fibre (kPa)	Group of SCC (kPa)	Group of SCC with fibre (kPa)
0.1	25	30	50	65
0.25	70	80	120	140
0.5	100	120	180	190

References

Consoli NC, Vendruscolo MA, Prietto PDM (2003) Behavior of plate load tests on soil layers improved with cement and fiber. J Geotech Geoenviron Eng ASCE 129(1):96–101

Faro VP, Consoli NC, Schnaid F, Thomé A, Lopes LDS (2015) Field tests on laterally loaded rigid piles in cement treated soils. J Geotech Geoenviron Eng ASCE 141(6):601–605

Farouk A, Shahien MM (2013) Ground improvement using soil–cement columns: experimental investigation. Alexandria Eng J 52(4):733–740

Xiao HW, Lee FH, Zhang MH, Yeoh SY (2013) Fiber reinforced cement treated clay. In: Pierre D, Jacques D, Roger F, Alain P, Francois S (eds) 18th Soil mechanics and geotechnical engineering. Proceedings of the 18th international conference on soil mechanics and geotechnical engineering, Paris, vol 2–6, pp 2633–2636

Xiao HW, Lee FH, Goh SH (2015) Fibre distribution effect on behavior of fibre-reinforced cement-treated clay. In: Proceedings of 15th Asian regional conference of soil mechanics and geotechnical engineering, Nov, Fukuoka, Japan, vol 2(60), pp 2063–2068

Yao K, Yao Z, Song X, Zhang X, Huc J, Pan X (2016) Settlement evaluation of soft ground reinforced by deep mixed columns. Int J Pavement Res Technol 9:460–465

Influence of Soil–Cement Columns on Load-Deformation Behavior of Soft Clay

G. Sumanth Kumar, V. Ramana Murty, Lambutre Mahesh, and J. Rakesh Pillai

Abstract Being a challenge to deal with structural foundations in soft clays, several techniques were promulgated across the world. This paper presents the load–settlement behavior of soft clay provided with soil–cement columns in test tanks. This study revealed that the load-carrying capacity of clay bed with soil–cement columns increases with increasing initial clay consistency and it is increased by about 3.5 times when the consistency is increased from 0.1 to 0.5. The threshold cement content is found to be 15% for the floating soil–cement columns and a 25% replacement of cement with fly ash has shown a slight increase of about 1.15 times that of the capacity attained with cement alone. SEM and EDAX results support that there is an improvement of surrounding clay bed due to diffusion of calcium ions from the soil–cement columns.

Keywords Soil cement columns · Load deformation behaviour · Soft clay · Load carrying capacity · Diffusion of calcium ions · Consistency

1 Introduction

Large tracts of soft clays are present along the coastlines and estuaries of several world nations. In India also, considerable soft clay beds are present along the 7500 km long coastline. The soft clay deposits are characterized by high natural water contents

G. Sumanth Kumar (✉) · V. Ramana Murty · L. Mahesh · J. Rakesh Pillai
Department of Civil Engineering, National Institute of Technology Warangal, Warangal, Telangana, India
e-mail: gundetisumanth@gmail.com

V. Ramana Murty
e-mail: vrm@nitw.ac.in

L. Mahesh
e-mail: lambutremd@gmail.com

J. Rakesh Pillai
e-mail: rakeshpilla@gmail.com

M. Latha Gali and R. R. P. (eds.), *Problematic Soils and Geoenvironmental Concerns*, Lecture Notes in Civil Engineering 88,
https://doi.org/10.1007/978-981-15-6237-2_58

almost close to or sometimes more than liquid limit, making them unsuitable for any construction purpose (Porbaha 1998). However, keeping in view a great deal of economic and cultural activity along the coasts, these deposits need to be improved to support the infrastructure.

Several ground improvement techniques such as stone columns, preloading with vertical drains (Chu et al. 2006), electro-osmosis, and deep mixing using lime or cement have been promulgated to improve these soil beds (Broms 1979; Lorenzo and Bergado 2006; Liu et al. 2011; Puppala and Porbaha 2004; Farouk and Shahien 2013; Kitazume and Terashi 2013). Despite these developments, there are several unknowns to be further understood to make the techniques more efficient and feasible (Moseley 1994; Shen et al. 2003). In the present work, an attempt is made to study the influence of cement mixed soil columns on the overall improvement of soft clay bed. For this purpose, the experimental work was carried out in test tanks by varying the initial consistency of clay. The cement content of the formed columns is also varied to find the threshold value of cement content for maximum strength gain for a given initial consistency when the columns are floating. Further, a partial cement replacement with fly ash (Horpibulsuk et al. 2011) is also investigated.

2 Experimental Study

2.1 *Materials Used*

Soil: The locally available black cotton soil is used to make soft clay. The properties of soil are: $G = 2.71$, gravel = 1%; sand = 30%; silt = 30%; clay = 39%; liquid limit = 63%; plastic limit = 20%; plasticity index = 43%; and IS classification = CH.

Cement: OPC-53 grade cement is used in this study.

Fly ash: Fly ash collected from Ramagundam Thermal Power Plant is used for partial replacement of cement.

2.2 *Test Procedure*

The load tests were carried out in model test tanks of 30 cm diameter and 30 cm height and also in test tank of 50 cm × 50 cm × 60 cm. The test tank is placed on the pedestal of loading frame centrally. The soft clay is prepared corresponding to the required consistency indices of 0.1, 0.25, and 0.5. 50 mm diameter (d) PVC pipes are used to form 25 cm long floating soil–cement columns within soft clay.

The soft clay prepared corresponding to a given consistency is placed in the test tank in layers of 5 cm by marking on the sides of test tank. PVC pipes are placed at the desired spacing (2d) and soil–cement mix or soil–cement–fly ash mix as per the test condition is placed and pressed while gradually withdrawing the PVC pipes by rotation by maintaining adequate overlapping with respect to the placement of clay around the pipes. The process is repeated till the test specimen with soil–cement columns is formed. At the top of soil–cement columns, a leveling course of 2 cm thick sand is placed before placing the 12 cm diameter test plate over them. The system is allowed for 7 days curing by covering it with a polythene bag and at the end of curing period, load test was carried out. The load was applied in increments of 4 kPa and the corresponding settlements were recorded with the help of dial gauges. The load–settlement plots were drawn and the ultimate loads were obtained by drawing tangents to the initial and final straight line portions (Fig. 1).

Fig. 1 Soil–cement column in small test tank under loading platform

3 Results and Discussion

The load tests were carried out on floating soil–cement columns of 50 mm diameter and 25 cm long at different consistencies of clay. The pressure–settlement curves are presented below.

3.1 *Influence of Consistency on Soil–Cement Column Capacity*

Figure 2 shows the load–settlement plot for single-column system at different initial soil consistencies. It can be seen from this figure that the settlements are considerably higher with a decrease in consistency for any load increment. The ultimate load-carrying capacity values are obtained from the intersection points of tangents drawn along the initial and final straight line portions of load–settlement plots. The ultimate load-carrying capacities are 30 kPa, 25 kPa, and 8 kPa for 0.5, 0.25, and 0.1 consistency indices, respectively.

The load–settlement plots for the system with 3 column groups at different soil consistency indices (I_c) are presented in Fig. 3. The ultimate load-carrying capacity is obtained as mentioned previously. It can be observed from these figures that the initial consistency of clay has a significant influence on the load-carrying capacity of single or column group. The ultimate load-carrying capacities for a group of columns are about 90 kPa, 60 kPa, and 33 kPa at 0.5, 0.25, and 0.1 consistency indices, respectively. These values indicate that the ultimate capacity is increased

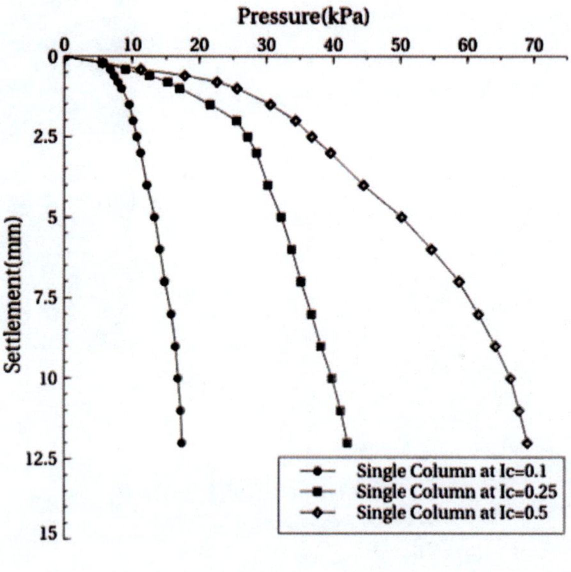

Fig. 2 Soil–cement columns at different consistencies with 10% cement content (single column)

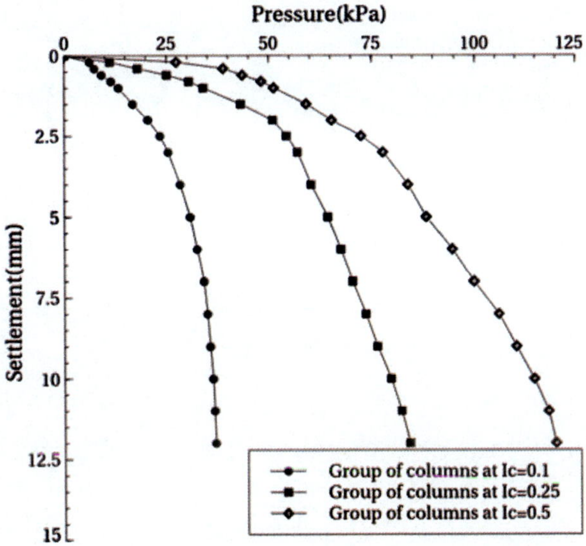

Fig. 3 Soil–cement columns at different consistencies with 10% cement content (group of columns)

around 3 folds when the consistency index is increased from 0.1 to 0.5. Further, it is noted that the group capacity is not the multiplicative value for a single pile system.

3.2 Influence of Cement Content on Capacity of Group of Soil–Cement Columns at Different Consistencies

The influence of cement content used in the group of soil–cement columns at different initial soil consistency is shown in Figs. 4, 5 and 6 and Table 1. From these figures, it can be inferred that the load-carrying capacity of the system is increasing with increasing cement content in the columns at all the soil consistencies. However, the increase in load-carrying capacity is only nominal beyond 15% cement content for the soil consistency up to 0.25. At initial consistency of 0.5, marginal increase in load capacity is observed at 20% cement content. From these observations, it is understood that the gain in strength by using higher cement contents is limited for low initial soil consistency.

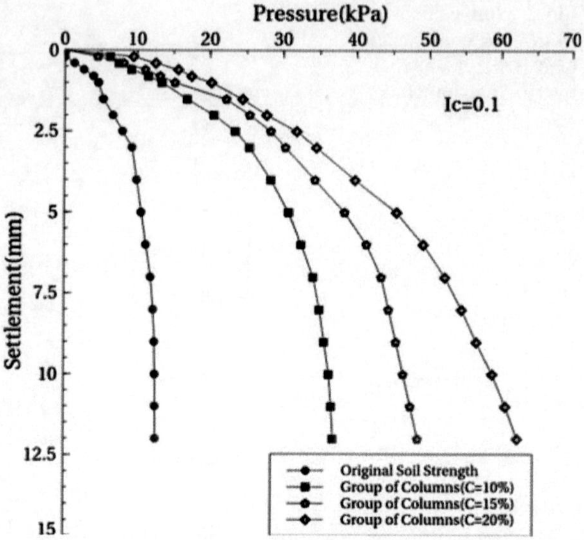

Fig. 4 Group of soil–cement columns at different cement contents for $I_c = 0.10$

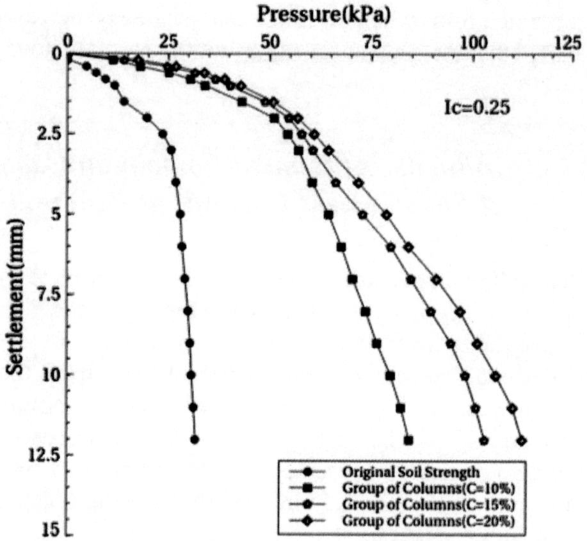

Fig. 5 Group of soil–cement columns at different cement contents for $I_c = 0.25$

3.3 Pressure–Settlement Plots of Soft Clay with Soil–Cement–Fly ash Columns at Different Consistencies

25% of cement content in soil–cement columns is replaced with fly ash and tests are repeated at different initial soil consistencies after 28 days of curing period.

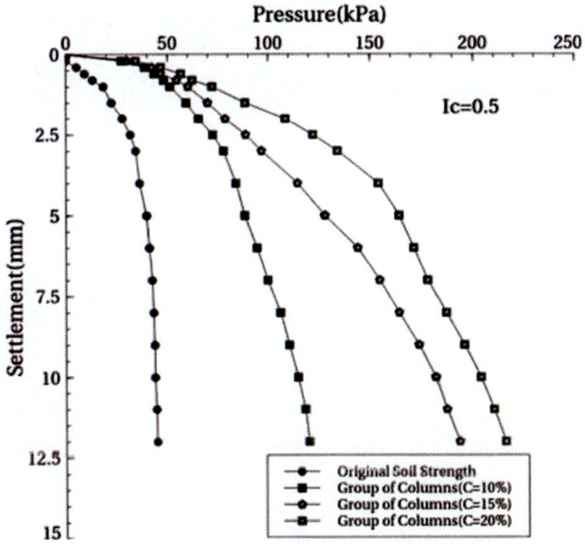

Fig. 6 Group of soil–cement columns at different cement contents for $I_c = 0.50$

Table 1 Ultimate load-carrying capacities of soil–cement columns at different cement contents

I_c	10% cement (kPa)	15% cement (kPa)	20% cement (kPa)
0.1	33	40	45
0.25	60	75	85
0.5	90	140	160

The pressure–settlement plots for the fly ash blended soil–cement column group at different consistencies for 28 days curing period are shown in Figs. 7 and 8 and in Table 2.

From these figures and table, it can be observed that the additional strength gain is nominal due to addition of fly ash. However, by such replacement of cement, considerable economy can be envisaged in practice.

3.4 Improvement of Surrounding Soil

Several researchers have observed the improvement of surrounding soil around the soil–cement column system (Shen et al. 2003). They have reported the change in the engineering properties of improved clay bed. Here, an attempt is made to understand the improvement of soil surrounding the soil–cement columns by SEM and EDAX analysis.

Fig. 7 Pressure–settlement plots for group of soil–cement columns with 10% cement content

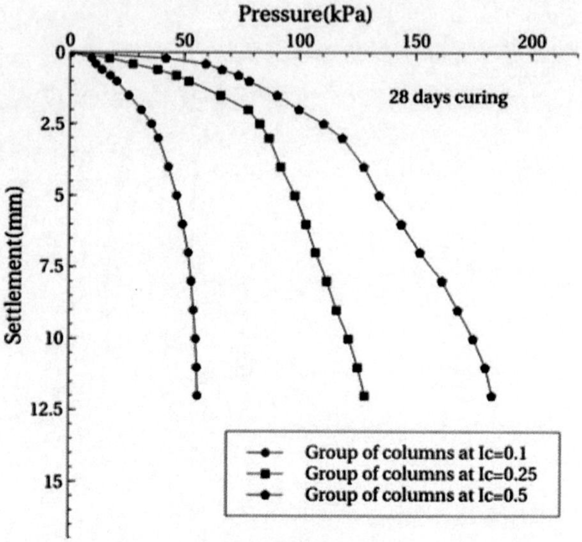

Fig. 8 Pressure–settlement plots for group of soil–cement–fly ash columns with 10% cement content

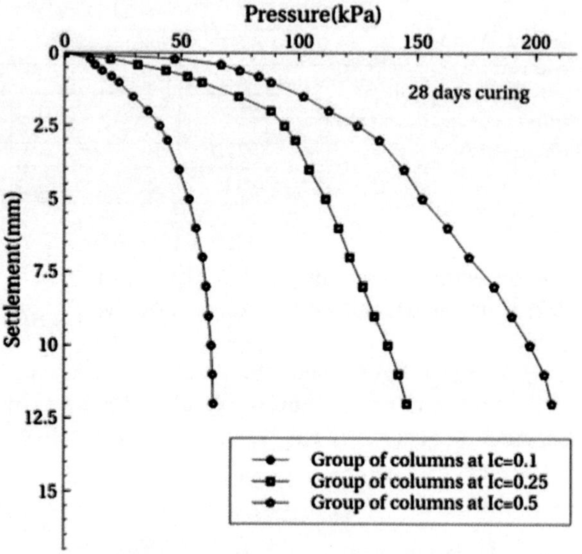

Table 2 Ultimate load-carrying capacities of soil–cement–fly ash columns at different consistencies

I_c	Soil + cement (kPa)	Soil + cement + fly ash (kPa)
0.1	50	58
0.25	91	99
0.5	137	158

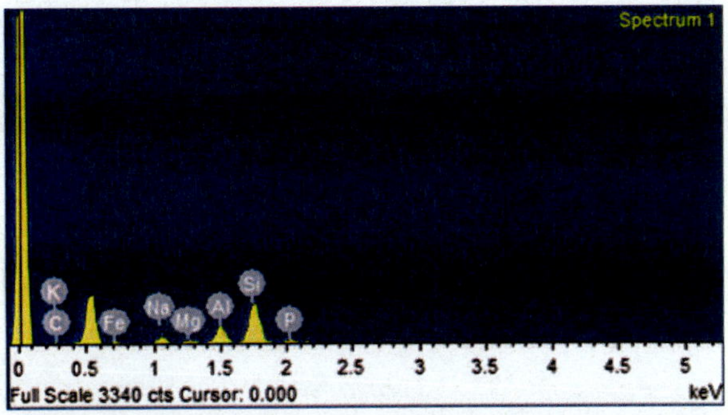

Fig. 9 Typical EDAX analyses of virgin soil

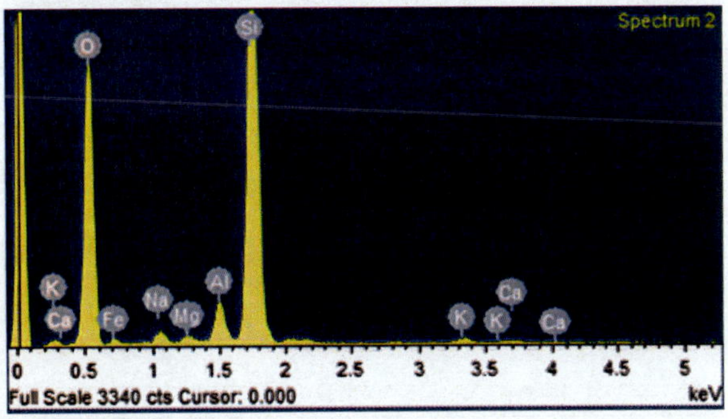

Fig. 10 Typical EDAX analyses of treated surrounding soil

3.4.1 Mineralogical Study

Mineralogical studies by scanning electron microscopy (SEM) and electron dispersive X-ray analysis (EDS or EDAX) were conducted on the samples of soil surrounding the group of soil–cement columns. EDAX helps in determining the elements/compounds formed at particle level and thereby the results can be used to summarize the possible formation of cementitious and pozzalonic compounds. SEM analysis provides qualitative understanding of the degree of improvement at microstructural level. The samples free from organic content were subjected to carbon coating and they were transferred to the SEM equipment for imaging. The carbon-coated samples were exposed to X-rays to obtain high-resolution and magnified images. EDAX analyses were attempted on the same samples.

EDAX Results

Figures 9 and 10 show typical EDAX test results with chemical elements identified. The chemical elements including calcium, silica, and aluminum and their presence strongly support the formation of cementitious compounds such as calcium silicate hydrate and calcium aluminate hydrates in the treated material.

Fig. 11 Typical SEM results of **a** virgin and **b** soil around the soil–cement columns

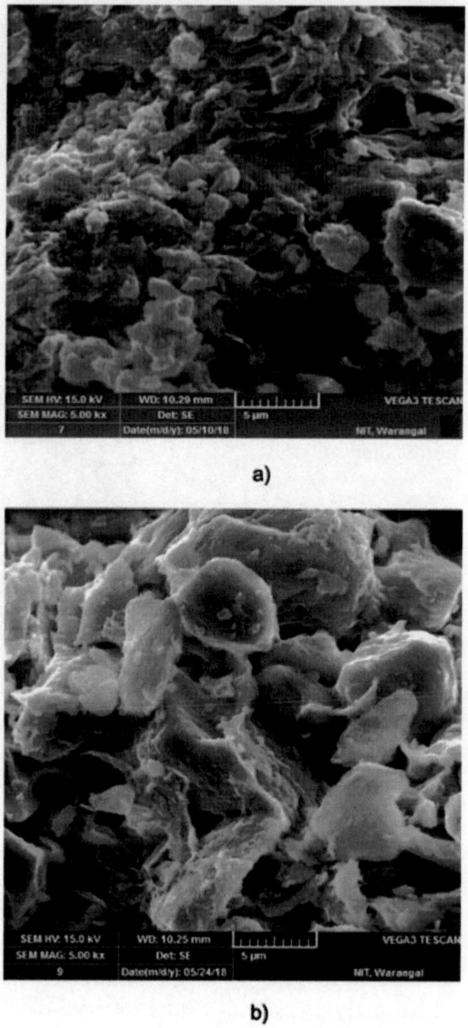

SEM Results

The scanning electron micrographs (SEMs) of virgin clay and the surrounding soil around the soil–cement columns are shown in Fig. 11a, b. It can be observed from these figures that the formation of the cementitious compounds can be supported by the flocculated structure of soil particles as can be seen in Fig. 11b. The presence of cementation compounds along with fine and cement-treated clay structure is known to enhance the soil properties. Overall, both SEM and EDAX studies confirm that there is some effect of cementation due to diffusion of cations which lead to improvement of clay bed.

4 Conclusions

The following conclusions are drawn from the experimental work carried out in this study.

1. The load-carrying capacity of soil–cement columns is significantly increased with increasing initial soil consistency.
2. In case of floating soil–cement columns, the increasing cement content has shown marginal increase in load-carrying capacity. At lower initial consistency, the gain in load-carrying capacity is only nominal even when the cement content is increased from 10 to 20%.
3. The group capacity of floating soil–cement columns is found to be not multiplicative value of a single column capacity.
4. The replacement of 25% of cement content with fly ash in the soil–cement columns can be attempted without any strength loss.
5. The clay around the soil–cement columns is modified by possible diffusion of cations from soil–cement columns as indicated by SEM and EDAX analysis.

References

Broms BB (1979) Lime columns a new foundation method. J Geotech Eng Div 105(4):539–556

Chu J, Bo MW, Choa V (2006) Improvement of ultra-soft soil using prefabricated vertical drains. Geotext Geomembr 24:339–348

Farouk A, Shahien MM (2013) Ground improvement using soil–cement columns: experimental investigation. Alexandria Eng J 52(4):733–740

Horpibulsuk S, Rachan R, Suddeepong A (2011) Assessment of strength development in blended cement admixed Bangkok clay. Constr Build Mater 25(4):1521–1531

Kitazume M, Terashi M (2013) Deep mixing method. CRC Press/Balkema Publishers, Netherlands

Liu S-Y, Du Y-J, Yi Y-L, Puppala AJ (2011) Field investigations on performance of t-shaped deep mixed soil cement column–supported embankments over soft ground. J Geotech Geoenviron Eng 138(6):718–727

Lorenzo GA, Bergado DT (2006) Fundamental characteristics of cement-admixed clay in deep mixing. J Mater Civ Eng 18(2):161–174

Moseley MP (1994) Ground improvement. CRC Press/Blackie Academic & Professional Publishers, USA

Porbaha A (1998) State-of-the-art in deep mixing technology. Part I: Basic concepts and overview of technology. Ground Improv 2(2):81–92

Puppala AJ, Porbaha A (2004) International perspectives on quality assessment of deep lime mixing. Proc. Geosupport 2004, Drilled Shafts, Micropiling, Deep Mixing, Remedial Methods, and Speciality Foundation Systems, ASCE, Orando, Florida

Shen SL, Miura N, Han J, Koga H (2003) Evaluation of property changes in surrounding clays due to installation of deep mixing columns. In: Grouting and ground treatment, pp 634–645

Analysis of the Influence of Polymeric Fabric Waste on Soil Subgrade

Deepak Chaudhary and R. P. Singh

Abstract This study focuses on the productive utilization of discarded and used polymeric fabric bags in the construction of rural roads. The paper describes the analysis of soil–plastic fiber waste composite. The materials used are locally available silty sand from NIT Jamshedpur campus and waste discarded cement bags. Experimental tests have been conducted to determine engineering properties of soil as envisaged by Indian standard codes. Apart from this, a series of California bearing ratio tests have been done on reinforced soil and un-reinforced soil under soaked and unsoaked conditions. Reinforcement of soil has been done in two different ways: small pieces mixed with soil sample and inserting layers of fabric in the specimen. Comparative study has been done to determine the optimum value of CBR for ascertaining the most effective soil–fiber mix. To evaluate the behavior of un-reinforced and reinforced soil subgrade, PLAXIS-2D software is used.

Keywords Geotextiles · Subgrade · Waste polymeric bags · Reinforcement · California bearing ratio (CBR) · FEM

1 Introduction

With the growth of cities and industrial areas, most of the land with appropriate strength has been consumed for construction. The geotechnical engineers have been forced to develop and strengthen the poor soil sites for various constructions needed for new developments. Among the various alternatives, for strengthening the existing weak soil, reinforcing the soils with some additive elements is one such successful alternative. Major geotechnical problems in construction involving silty–clayey soils

D. Chaudhary (✉) · R. P. Singh
Department of Civil Engineering, NIT Jamshedpur, Jamshedpur, Jharkhand, India
e-mail: deepakchdhr109@gmail.com

R. P. Singh
e-mail: rpsingh.ce@nitjsr.ac.in

M. Latha Gali and R. R. P. (eds.), *Problematic Soils and Geoenvironmental Concerns*, Lecture Notes in Civil Engineering 88,
https://doi.org/10.1007/978-981-15-6237-2_59

are due to their low strength, durability and high compressibility of soft soils, and the swell–shrink nature of the overconsolidated clay-dominated soils.

Road is the primary requirement of the rural population to get connected with the urban centers. So it is essential to adopt cost-effective and durable roads in our country. The prohibitive cost of commercially available geotextiles in India can overcome by the extensive use of discarded and used polymeric bags. These polymeric bags can be joined together and can be used as an alternative reinforcing component in place of commercially available geotextiles for stabilization of low volume roads. This study focuses on the productive utilization of discarded and used polymeric fabric bags in the construction of rural roads. Thus, the presented works mitigate the challenge of solid waste management as well. The paper describes the analysis of soil–plastic fiber waste composite.

Soil reinforcement is a technique which is mainly used to improve the stiffness and strength of soil using geo-engineering methods. This technique is necessary in the areas where chances of erosion are high. Soil from such areas is also highly susceptible to various environmental and natural factors such as high compressibility, poor shear strength, and temperature changes. So it is necessary to do soil reinforcement by using different engineering techniques to enhance the soil strength.

Many other publications on reinforcing soil with plastic waste are available like Athanasopoulos (1993), Benson and Khire (1994), Babu and Vasudevan (2008), Choudhary (2010), Consoli et al. (2010), Haneef (2014), Gray and Ohashi (1983), Harper (2006), Khabiri (2011), Kalumba and Chebet (2013), Ranjan and Rao (2007), Ranjan et al. (1994), Resl and Werner (1986), Sathiya Priya and Arunmairaj (2016), Yetimoglu and Salbas (2003) etc. In these findings, the following points may be noted. The use of plastic waste as reinforcing material in soils:

(i) Increases the strength of the soil,
(ii) Improves California bearing ratio of the soil,
(iii) Reduces the compressibility of the soil,
(iv) Decreases the coefficient of permeability, and
(v) Changes brittle cemented soil to a ductile state.

2 Materials and Methods

The materials used in this research include soil, water, and polypropylene fabric waste. The basic properties of these materials are described below.

2.1 Soil

The soil material used in this research was locally available soil from NIT Jamshedpur campus. The soil was well-graded fine silty sand. It was tested for its liquid limit,

Fig. 1 Fine silty sand from NIT Jamshedpur campus

plastic limit, and shrinkage limit, particle size distribution by sieve analysis, Proctor compaction test, shear strength, and California bearing ratio (CBR) test.

2.2 Polymeric Fabric Waste

Plastic fabric wastes of the type of polypropylene were used as reinforcing material in the present investigation. These were obtained from waste synthetic bags of cement, which are made up of strong, flexible and cast polypropylene. These empty cement bags were washed with clean running water and dried well. Reinforcement of soil has been done by two methods. In the first method, i.e., method-1 the soil was mixed with waste synthetic bag pieces of different sizes, i.e., 1 cm × 1 cm, 2 cm × 2 cm, 3 cm × 3 cm, and 4 cm × 4 cm in different proportions, i.e., 0.05, 0.1, and 0.15% by weight. In the second method (method-2), synthetic bag was used in the form of sheet layer for reinforcing the soil by varying the depth and number of layers, i.e., no layer, single layer, two layers, and three layers. Figures 1, 2 and 3 show the soil sample, the cut pieces of fabric used as layer in the specimen and the cut pieces as mixed with the soil specimen, respectively.

2.3 Method

Main objective of this study was to analyze the influence of fiber waste on soil subgrade. For this analysis, a series of CBR test have been done on reinforced soil and un-reinforced soil under soaked and unsoaked conditions. Reinforcement has been done by two methods as discussed earlier. A series of CBR test have been

Fig. 2 Small pieces of the waste cement bag

Fig. 3 Layers of waste discarded cement bag

conducted under soaked and unsoaked conditions on the various reinforced specimens and un-reinforced specimens referring IRC: 37-2001, IS: 1498:1970, IS: 2720 Part XVI: 1996. After this, comparative study has been done to determine the optimum value of CBR for ascertaining the most effective soil–fiber mix among the specimens prepared by the two methods with so many variations. The various mixes have been given definite identification codes as described next.

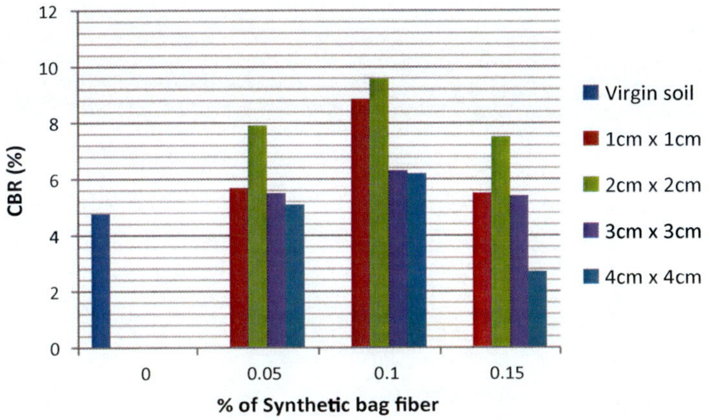

Fig. 4 Variation of CBR values against % of waste synthetic bag pieces for unsoaked condition

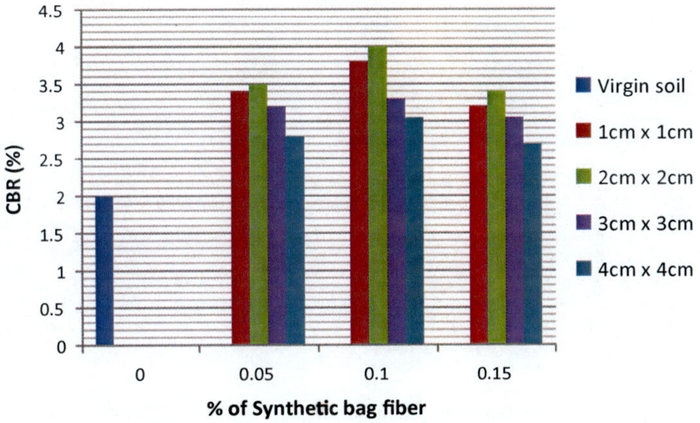

Fig. 5 Variation of CBR values against % of waste synthetic bag pieces for soaked condition

soaked condition, CBR value is 4% which is 100% more than the CBR value of virgin soil.

Figures 6 and 7 show the CBR values of specimens in unsoaked and soaked conditions, respectively, prepared by method-2, i.e., by placing layers of fabric in the specimens at different positions with varying the number of layers. The unsoaked specimens' CBR tests reveal the results of a pattern. For single-layer reinforcement, the highest CBR value is observed when the layer is placed in the middle of the specimen. For two layers reinforcement, the maximum CBR value is obtained when the layers are placed at top and bottom of the specimen. In case of three layers, the highest CBR value is obtained for the layers placed at top, middle, and bottom positions. For unsoaked condition, CBR value is 110.25% more for TMB mix than the CBR of virgin soil.

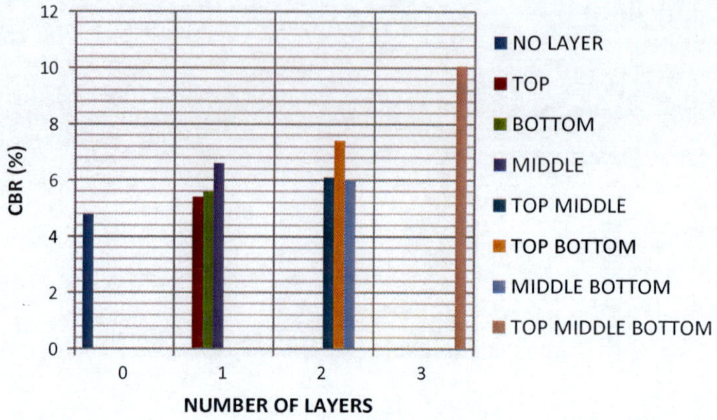

Fig. 6 Variation of CBR value against number of layers of waste synthetic bag for unsoaked condition

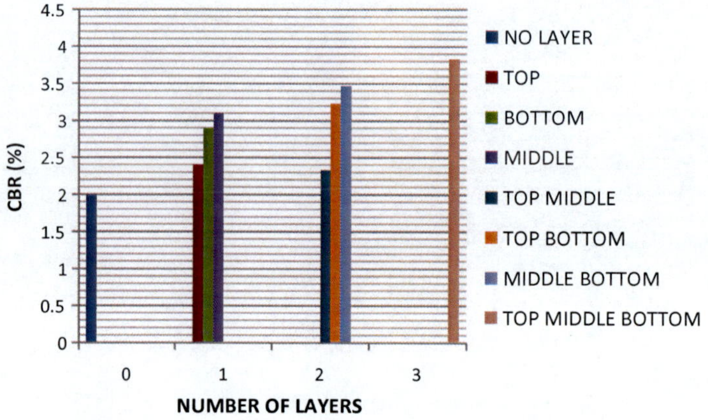

Fig. 7 Variation of CBR value against number of layers of waste synthetic bag for soaked condition

The CBR values of soaked specimens show the same pattern of increase due to the placement of single, double, and triple layers. For soaked condition, maximum CBR value with three layers of reinforcement is 90% more than CBR of virgin soil.

3.1 *Discussion*

The results obtained by a series of CBR tests as presented above clearly indicate that soil reinforcement with the cut pieces of used and discarded polymeric bags improves the strength of soil subgrade considerably. Since the reinforcement has been done by two different methods, it is aimed to obtain the optimum value of CBR for establishing the most effective way of soil–fiber mixing. For method-1, optimum value of CBR is obtained for mix M2-b which is 9.6 and 100% more than that of virgin soil. The mix M2-b contains 0.10% fiber waste by weight in the soil sample. While for specimens by method-2, the optimum value of CBR obtained was for mix TMB. In this case, CBR value obtained is 10.05% which is 110.25% more than the CBR value of virgin soil. So, overall optimum condition for maximum value of CBR is for mix TMB, i.e., the reinforced soil sample with three layers of fiber waste placed at top, middle, and bottom of the specimen in the form of sheets of fiber waste.

In addition to this, numerical analysis was conducted for the determination of total deformation of soil subgrade for the optimum conditions obtained during experimental work by using FEM-based software PLAXIS-2D.

4 Numerical Analysis

Finite element method based on PLAXIS-2D software was used to evaluate the total settlement of subgrade for the optimum conditions obtained during experimental work, and comparative studies have been done between un-reinforced and reinforced soil.

4.1 Materials and Modeling

The embankment was defined in the structure mode. Modeling of an embankment has been designed to evaluate the total deformation of subgrade. In this analysis, three calculation stages are used, i.e., mesh analysis, consolidation analysis, and calculation of total deformation (Fig. 8).

Table 5 shows the various geotechnical properties of material used during the modeling of an embankment.

The correlations between E and CBR obtained using FEM taken into account the rigid boundary condition of the CBR mold (Putri et al. 2012). Following Eq. 1 showing the correlation between Young's modulus of soil (E) and California bearing ratio (CBR) of soil for Poisson's ratio, $\mu = 0.3$

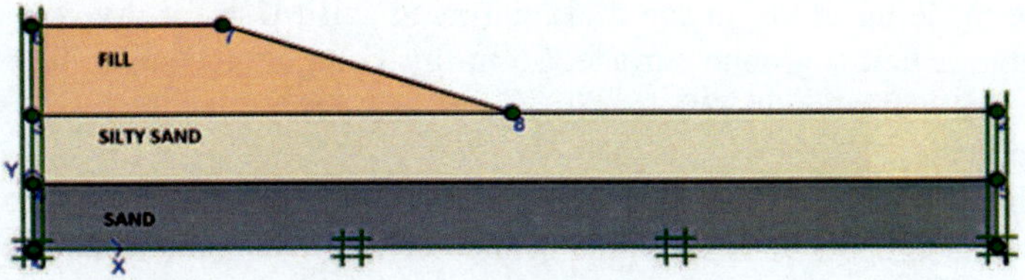

Fig. 8 Soil layer distribution in embankment model

Table 5 Material properties of the road embankment and sub-soil

Mohr-Coulomb	1 Sand	2 Fill	3 Silty sand
Type	Drained	Drained	Drained
γ_{unsat} (kN/m^3)	17.00	16.00	21.12
γ_{sat} (kN/m^3)	20.00	20.00	22.74
k_x (m/day)	1.000	1.000	0.001
k_y (m/day)	1.000	1.000	0.000
e_{init} (–)	1.000	1.000	0.500
c_k (–)	1E15	1E15	1E15
E_{ref} (kN/m^2)	13,000	8000	4033.7
μ (–)	0.300	0.300	0.300
G_{ref} (kN/m^2)	5000.000	3076.923	1551.348
E_{oed} (kN/m^2)	17,500.00	10,769.23	5430.943
c_{ref} (kN/m^2)	1.00	1.00	0.20
φ (°)	31.00	30.00	23.00
ψ (°)	0.00	0.00	0.00
E_{inc} (kN/m^2/m)	0.00	0.00	0.00
y_{ref} (m)	0.000	0.000	0.000
$c_{increment}$ (kN/m^2/m)	0.00	0.00	0.00
T_{str} (kN/m^2)	0.00	0.00	0.00
R_{inter} (–)	1.00	0.65	1.00
Interface permeability	Neutral	Neutral	Neutral

$$E = 840.53\,\mathrm{CBR}(\mathrm{kPa}) \tag{1}$$

4.2 Result and Discussion

Figures 9, 10 and 11 show the deformed mesh after undrained construction of an embankment for M1, M2-b, and TMB mix, respectively.

Figure 12 shows that the comparative study of maximum total deflection of subgrade made up of virgin soil and reinforced soil M2-b for dry condition when water table is below ground surface. So in this comparative study, it is concluded that the total deflection is 32% less in case of M2-b mix than the total deflection of virgin soil.

Figure 13 shows that the comparative study of maximum total deflection of subgrade made up of virgin soil and reinforced soil M2-b for wet condition when water table is at ground surface. It is concluded that the total deflection is 33% less in case of M2-b mix in comparison with virgin soil. The numerical analysis shows

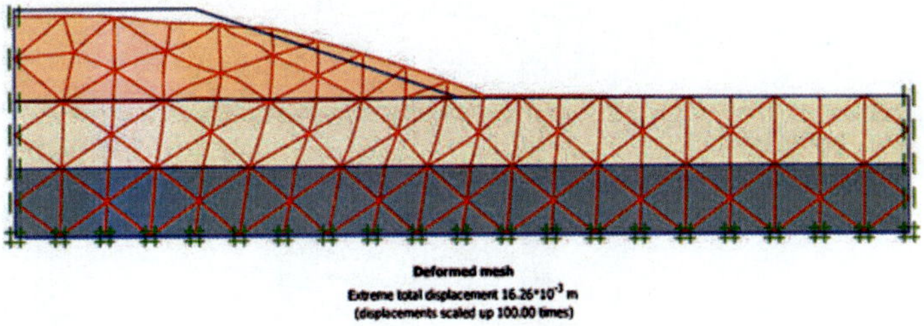

Fig. 9 Deformed mesh after undrained construction of embankment of virgin soil

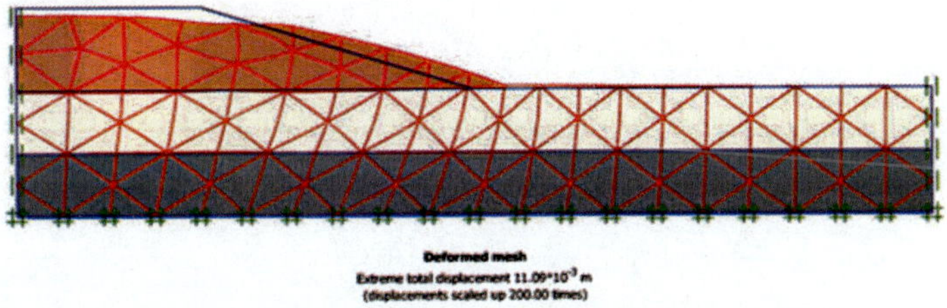

Fig. 10 Deformed mesh after undrained construction of embankment of soil reinforced with M2 mix at 0.10%

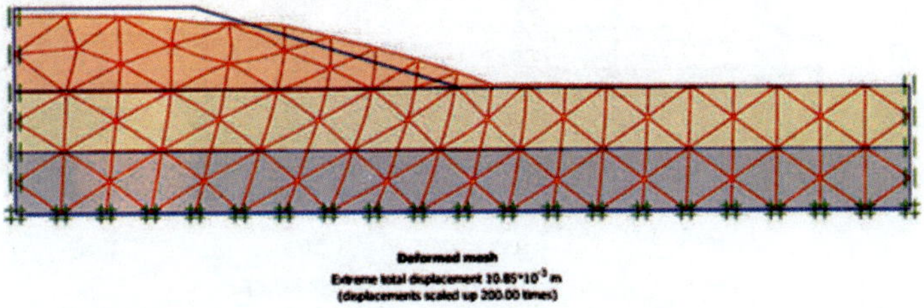

Fig. 11 Deformed mesh after undrained construction of embankment of soil reinforced with three layers of fiber waste at TMB

that the settlement which is more than 25 mm in the wet condition for virgin soil reduces to 18 mm if the soil is reinforced with 0.10% fiber pieces (1 cm × 1 cm) by weight.

Figure 14 shows that the comparative study of maximum total deflection of subgrade of virgin soil and soil reinforced with layers of waste fiber (M, MB, and TMB) for dry condition when water table is below ground surface. So it is concluded that the total deflection is 33.3% less in case of TMB mix in comparison with total deflection of virgin soil.

Figure 15 shows that the comparative study of maximum total deflection of subgrade of virgin soil and soil reinforced as M, MB, and TMB for wet condition

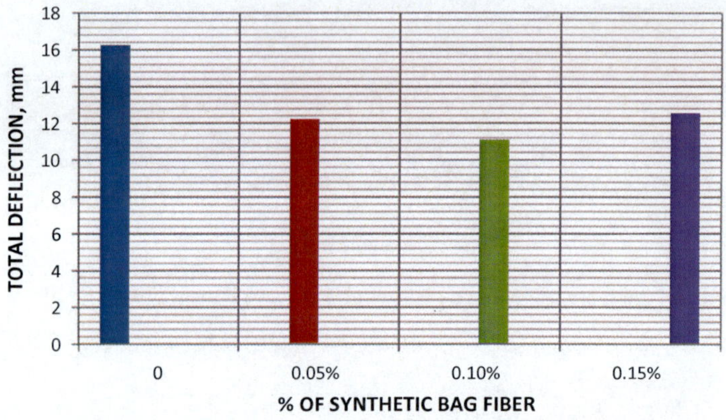

Fig. 12 Variation of total deflection in dry condition for different percentage of fabric waste of size 2 cm × 2 cm

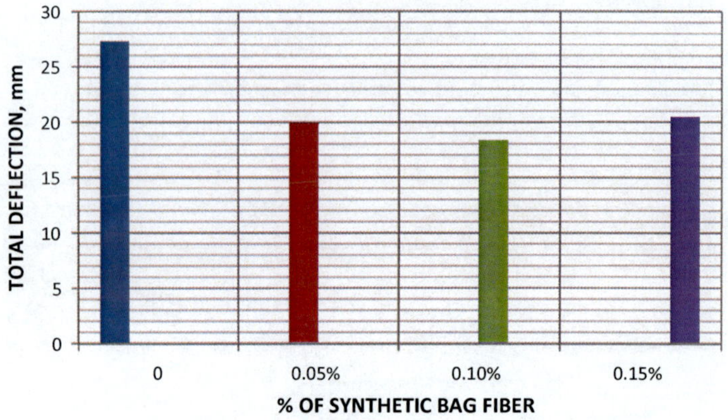

Fig. 13 Variation of total deflection in wet condition for different percentage of fabric waste of size 2 cm × 2 cm

when water table is at ground surface, and it is concluded that the total deflection is 33.3% less in case of TMB mix in comparison with total deflection of virgin soil.

5 Conclusion

Based on the experimental studies and analytical studies, the following conclusions have been drawn.

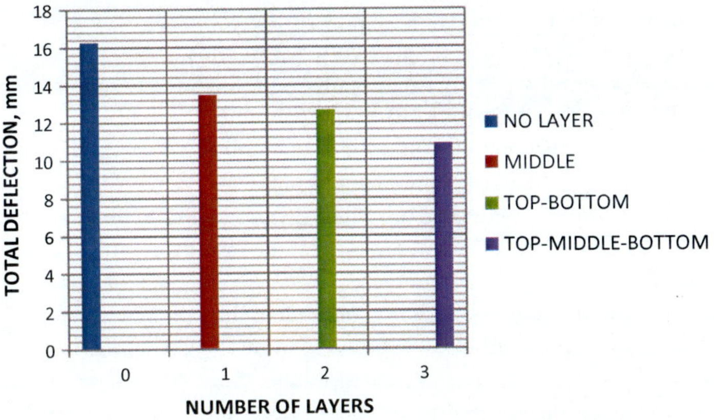

Fig. 14 Variation of total deflection in dry condition for different number of layers and positions of fabric waste

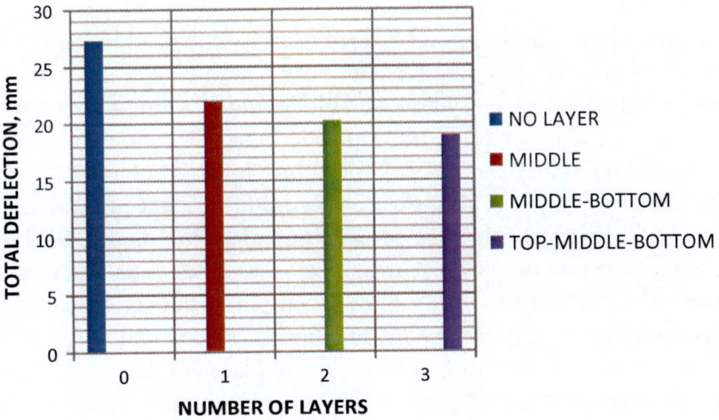

Fig. 15 Variation of total deflection for different number of layers and positions of fabric waste

1. The selected soil sample was found to have low strength with soaked and unsoaked CBR value of 1.2% and 4.8% respectively. It indicates that subgrade has to be improved.
2. Polymeric fiber-reinforced subgrade improves the strength properties of soil which makes the subgrade suitable for pavement, without capping layer.
3. The optimum polypropylene fiber content was found as 0.10% for mix M2 in case of reinforcing method-1, and optimum number of layers and position of fiber sheet was found as TMB in case of reinforcing method-2 on the basis of CBR test results.
4. The soaked and unsoaked CBR values of the subgrade soil were increased to 4.01% and 9.6% at M2 (0.10%), respectively, for method-1.

5. The soaked and unsoaked CBR values of the subgrade soil were increased to 3.9% and 10.05% at TMB, respectively, for method-2.
6. From experimental work, it was concluded that the subgrade strength can be increased up to 110.25% by reinforcing soil with waste fiber.
7. The PLAXIS analysis shows that there was reduction in deflection and strain values in artificial fiber-reinforced subgrade.
8. Based on the output obtained from PLAXIS, it was found that maximum deflection and vertical strain values are less for fiber-reinforced soil subgrade than the values of un-reinforced soil.
9. From analytical analysis, it was concluded that the total deflection in soil subgrade can be reduced to 33% by reinforcing it with waste bag fiber.
10. As the material was waste, and biodegradable therefore can be suggested to use for the construction of roads like village roads and temporary roads.

6 Scope of Future Study

1. Soil having some cohesion in it or clayey soil can be used instead of using pure cohesionless soil and comparative study can be done.
2. As reinforcing material was waste, and biodegradable limitations were there. So, instead of using this material, other plastic materials which are much stronger and non-biodegradable can be used for improving soil subgrade.
3. For further study, some admixtures can be used to enhance the strength of study material and some natural fibers can be used instead of using man-made fibers.
4. Laboratory research study is mainly limited to CBR tests in order to achieve the main objective of this study at constant OMC but for better analysis, it is recommended that varying OMC values should be calculated for different soil–plastic mix and CBR test should be conducted at that OMC.
5. Main objective of this study was to analyze the effect of polymer bag on soil subgrade only but further the effect of this soil–fiber mix can be analyzed for slope stability also.

Acknowledgements I would like to show my humble gratitude to Dr. R. P. Singh, Associate Professor, and Head of Department of Civil Engineering, National Institute of Technology, Jamshedpur, for nurturing me with proper knowledge and guidance without which the possibility of carrying out this research work was not possible. His constant supervision, encouragement, and care have made this project a reality. I shall always be indebted to him.

I would like to convey my regards to all the staff of Laboratories of Civil Engineering Department for their constant support. I am also very grateful to my friends for helping me while doing tests in laboratory. I am significantly obliged to thank my parents for their constant support and encouragement at each and every step of my life. Their blessings have empowered me to finish my work effectively.

References

Athanasopoulos GA (1993) Effect of particle size on the mechanical behaviour of sand-geotextile composites. Geotext Geomembr 12(3):255–273

Benson CH, Khire MV (1994) Reinforcing Sand with strips of reclaimed high-density polyethylene. J Geotech Eng ASCE 121(1994):390–412

Babu GLS, Vasudevan AK (2008) Strength and stiffness response of coir fiber-reinforced tropical soil. J Mater Civ Eng 20(9):571–577

Choudhary AK, Jha JN, Grill KS (2010) A study on CBR behaviour of waste plastic strip reinforced soil. Emirates J Eng Res 15(1):51–57

Consoli NC, Cassagrande MDT, Coop MR (2010) Performance of fibre-reinforced sand at large shear strains. Geotechnique 57(9):751–756

Gray DH, Ohashi H (1983) Mechanics of fiber reinforcement in sand. J Geotech Eng ASCE 1109(3):335–353

Haneef FM (2014) Effect of plastic granules on the properties of soil 4(4):160–164 (Version 1). www.ijera.com. ISSN: 2248-9622

Harper C (2006) Handbook of plastics technologies: the complete guide to properties and performance. In: Harper CA (ed). McGraw-Hill, New York

IRC: 37-2001, Guidelines for the design of flexible pavements (second revision). The Indian Roads Congress, New Delhi, India

IS: 1498:1970, Classification and identification of soils for general engineering purpose. Indian standard code of practice, Laboratory classification criteria for coarse grained soils

IS: 2720 Part XVI:1996, Methods of test for soil. Indian standard code of practice, Laboratory Determination of CBR

Kalumba D, Chebet FC (2013) Utilization of polyethylene (plastic) shopping bags waste for soil improvement in sandy soils. In: 18th international conference on soil mechanics and geotechnical engineering 2013

Khabiri MM (2011) Geosynthetic material suitable depth staying to control failure of pavement rutting. Adv Mater Res 255–260:3454–3458

Khanna SK, Justo CEG (2017) Highway engineering, 10th edn. Paperback

PLAXIS-2D. https://www.plaxis.com.

Putri EE, Rao Kameswara NSV, Mannan MA (2012) Evaluation of modulus of elasticity and modulus of subgrade reaction of soils using CBR test. J Civ Eng Res 2(1):34–40

Ranjan G, Rao ASR (2007) Basic and applied soil mechanics, 2nd edn. New Age International

Ranjan G, Vasan RM, Charan HD (1994) Behaviour of plastic-fibre-reinforced sand. Geotext Geomembr 13(8):555–565

Resl S, Werner G (1986) The influence of nonwoven needle-punched geotextiles on the ultimate bearing capacity of the subgrade. In: Proceedings of the third international conference on geotextiles, vol 4, Vienna, pp 1009–1013

Sathiya Priya V, Arunmairaj PD (2016) Experimental and analytical investigation of artificial fiber reinforced subgrade soil for flexible pavement. Int J Eng Trends Technol (IJETT) 35(4):180–184

Yetimoglu T, Salbas O (2003) A study on shear strength of sands reinforced with randomly discrete fibers. Geotext Geomembr 21(2):103–110

Strength and Durability Characteristic of Lime Stabilized Black Cotton Soil

Noolu Venkatesh, Danish Ali, Rakesh J. Pillai, and M. Heera Lal

Abstract Expansive soil which is found in several parts of India possesses high swelling and shrinkage properties. The volume change and uplift pressure generated in these soil deposits cause severe damage to the lightweight structures and pavements. In order to mitigate the problems associated with expansive soils, it is necessary to stabilize this soil. Among all the stabilization techniques, lime treatment is one of the best suitable methods for expansive soils. In the present study, lime stabilization technique is used to improve the engineering properties of black cotton soil including the resilient modulus value, which is important for mechanistic flexible pavement design. Optimum amount of lime required for stabilization was determined using Atterberg's limits. Considerable increment was observed in unconfined compressive strength values and California bearing ratio values of black cotton soil stabilized with 6% lime. Repeated load triaxial tests under different confining pressures and deviatoric stress levels were conducted on the treated samples in order to determine the resilient modulus. The effect of curing period and moisture content on the resilient modulus was investigated. In order to study the durability of lime stabilized clayey subgrade soil, the effect of wetting and drying cycles on the engineering properties of the treated material was examined. The results show that the strength and stiffness characteristics of lime treated clay have considerably reduced after five wetting and drying cycles.

Keywords Expansive soil · Lime stabilization · Resilient modulus · Wetting and drying

N. Venkatesh (✉) · D. Ali · R. J. Pillai · M. Heera Lal
Department of Civil Engineering, National Institute of Technology Warangal, Warangal, Telangana, India
e-mail: nooluvenkatesh29@gmail.com

R. J. Pillai
e-mail: rakeshpilla@gmail.com

M. Latha Gali and R. R. P. (eds.), *Problematic Soils and Geoenvironmental Concerns*, Lecture Notes in Civil Engineering 88,
https://doi.org/10.1007/978-981-15-6237-2_60

1 Introduction

Expansive soils, well known for their alternate swelling and shrinkage upon moisture fluctuations have occupied large tracks of land in many countries including India (Kulkarni and Patil 2014; Hamza and Halil 2016). Several remedial techniques were developed based on the extensive research carried out all over the world to safeguard the lightly loaded structures, which are highly susceptible for damage under cyclic volumetric changes of these deposits (Nelson and Miller 1992; Phanikumar and Sharma 2004; Nalbantoglu 2004; Cocka et al. 2009; Sharma et al. 2012). Chemical stabilization is the extensively used procedure to improve the engineering properties of black cotton soil. Several researchers used lime as an additive to improve the engineering properties of weak subgrade soils (i.e. plasticity and strength characteristics) (Kumar et al. 2007; Seco et al. 2011).

In recent years, usage of resilient modulus as a stiffness parameter of subgrade material is populated (NCHRP 2004; AASTHO 2008). Resilient modulus is defined as the ratio between cyclic deviatory stress to recoverable strain and it is used to characterize the elasto-plastic and nonlinear response of granular material. Resilient modulus can be determined by performing cyclic triaxle test on the soil specimen in the laboratory. Resilient modulus is divided into three categories based on its level of accuracy. Level 1 consists of determination of resilient modulus values from cyclic triaxle apparatus. Level 2 consists of resilient modulus values obtained from correlation of unconfined compressive strength and California bearing ratio. Level 3 resilient modulus possesses less accuracy and it is obtained from empirical relations of water content, clay content and index properties of soil. Several factors influence resilient modulus of subgrade soils. Among those, moisture content and stress levels are major influencing factors in determination of resilient modulus (Li et al. 1994; Zaman et al. 1994).

Croft (1967) studied the influence of lime on the engineering properties of expansive soil and found that at 4% of lime, there is drastic decrement of liquid limit, plasticity index and maximum dry density, and increment in plastic limit and optimum moisture content observed. Bell (1996) observed that 3% addition of lime results in increase in strength due to the pozzolanic reaction between ca^{2+} in lime and silica and aluminium of expansive soil. Edil et al. (2006) used fly ash as an additive to soft clay and observed nearly 1.8 times increment in resilient modulus. Rout et al. (2012) studied the influence of lime-cement on resilient modulus of clayey soil and observed nearly 2.5 times increment in resilient modulus.

Studies regarding resilient modulus of subgrade soils less mentioned in literature. In the present study, a series of tests conducted on determine the Atterberg's limits, California bearing ratio, unconfined compressive strength and resilient modulus of expansive soil and lime stabilized expansive soil. Mineralogical and morphological studies conducted in order to identify the modification structure and formation of new minerals after stabilization.

Table 1 Properties of black cotton soil

S. No.	Property	BC soil
1	Specific gravity	2.65
2	Grain size analysis	
	Gravel (%)	3
	Sand (%)	25
	Silt (%)	27
	Clay (%)	45
3	Liquid limit	64
4	Plastic limit	23
5	Plasticity index	41
6	IS soil classification	CH
7	Optimum moisture content	17
8	Maximum dry density (g/cc)	1.65

2 Experimental Study

2.1 Materials Used

The following materials were used in this investigation.

2.1.1 Soil

Locally available black cotton soil collected from a site near the north boundary of NIT Warangal campus is used in this study and the properties of the soil are listed in Table 1.

2.1.2 Lime

Lime is collected from Bayyaram, which is 60 km far away from Warangal, and it consists of 61% calcium. So, it can be used as additive for improving the properties of soil.

2.2 Methodology

An optimum percentage of lime obtained from Atterberg's limits. Compaction characteristics, unconfined compressive strength, California bearing ratio and resilient modulus of BC soil and lime stabilized BC soil are determined.

The procured BC soil is air-dried and stored before starting laboratory experiments. Modified Proctor tests are conducted as per IS 2720 Part 8 (1980) to determine the optimum moisture content and dry density of both BC and lime stabilized BC soil. UCS tests are performed on 7, 14 and 28 days cured samples as per IS 2720 Part 10 (1973). Soil sample having diameter 50 mm and height of 100 mm is used in this study of density and moisture content obtained from modified Proctor test. CBR test is conducted for both soaked and un-soaked condition of the samples cured at 28 days as per IS 2720 Part 16 (1973).

In the present study, an automated cyclic triaxial test apparatus is used to determine the resilient modulus of soil sample. Resilient modulus test is conducted under three confining pressures and five deviatoric stress levels as mentioned in AASTHO T-307. Samples are prepared with three different moisture contents (OMC, OMC + 2% and OMC + 4%) in order to verify the effect of moisture content on the resilient modulus.

Durability tests were conducted on lime stabilized BC soil as per IS: 2720 (Part 16) 1987. At optimum dosages of lime, UCS test samples were prepared at a specified diameter of 50 and 100 mm height so that an aspect ratio of 2 is maintained which were subjected to a curing period of 7 days. The 7 days cured samples were then subjected to alternate wetting and drying cycles. For wetting condition, the cured samples were to be immersed in potable water for 5 h at room temperature and then the samples are to be removed, whereas for drying condition, the soaked samples were to be placed in an oven for about 42 h at 71 °C. Thus, one wetting–drying cycle constitutes 5 h of wetting and 42 h of drying.

Scanning electron microscopy (SEM) and X-ray diffraction analysis are conducted on lime modified black cotton soil to observe the mineralogical and morphological variations occurred in the soil due to addition of stabilizer.

3 Results and Discussion

From Fig. 1, it can be observed that addition of lime to black cotton soil causes decrement in liquid limit and plasticity and increment in plastic limit due to the flocculation of clay particles and increase in the coarser particle content (Horpibulsuk et al. 2011). With addition of more than 6% lime to BC soil, there is no noticeable change observed in Atterberg's limits. It signifies that the BC soil can absorb maximum amount of calcium ions at 6% lime addition.

Modified Proctor test was conducted on BC soil, lime modified BC soil and the results are represented in Fig. 2. From the figure, it can be noted that optimum moisture content increase and maximum dry density decreases for both virgin and stabilized BC soil. Lesser specific gravity of lime causes reduction in the dry density, whereas flocculation of clay particles is responsible for increase in optimum moisture content (Horpibulsuk et al. 2013). Optimum moisture content and dry density, which were obtained from modified Proctor test, were used to prepare samples for UCS.

Figure 3 represents UCS values of virgin soil and lime stabilized soil for curing

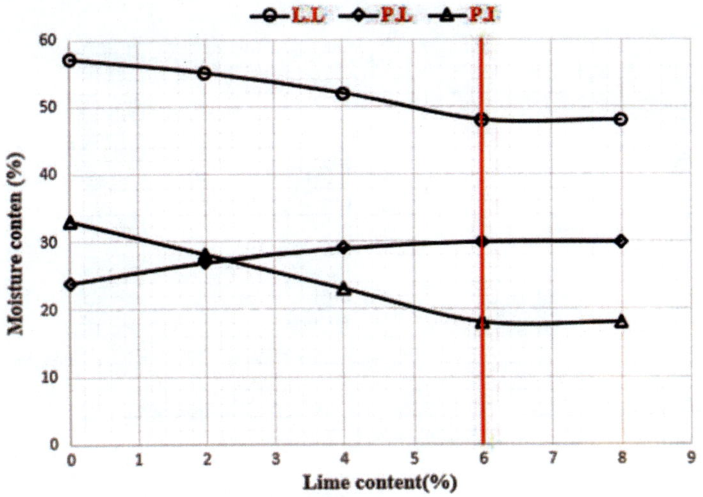

Fig. 1 Variation of Atterberg's limits with lime

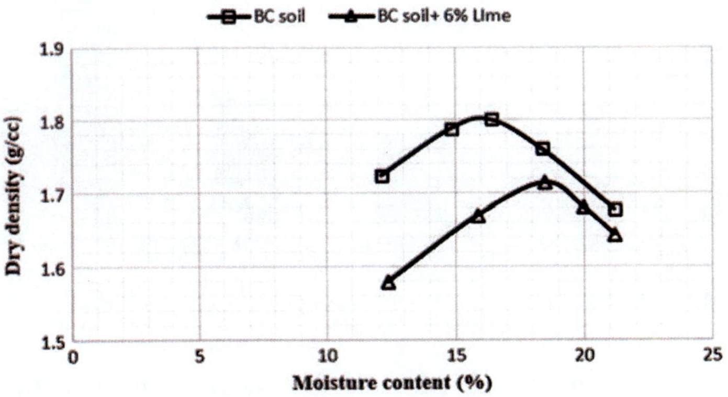

Fig. 2 Compaction curves of BC soil with and without addition of lime

period of 7, 14 and 28 days. It can be noted that lime modified BC soil sustain higher stresses than the BC soil and increase in curing period also enhances UCS value due to the pozzolanic reaction between silica and aluminium presented in BC soil and calcium presented in lime. CBR test was conducted on BC soil and lime modified BC soil. CBR test sample was prepared with density and moisture content obtained from modified Proctor test and tests were conducted under both soaked and un-soaked condition.

From Fig. 4, it can be observed that the addition of 6% of lime causes drastic increment in CBR value due to the pozzolanic reaction and 45–60% reduction observed for soaked samples compared to un-soaked samples.

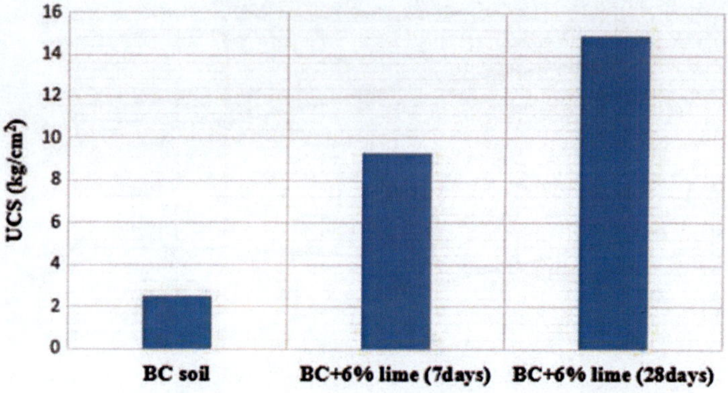

Fig. 3 Variation of unconfined compressive strength with and without lime

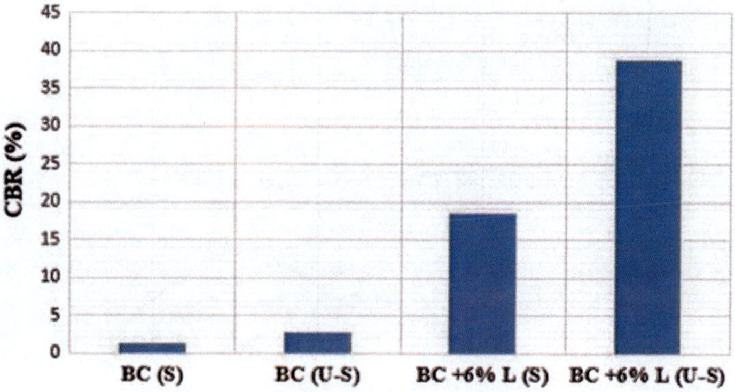

Fig. 4 Variation of CBR of BC soil with and without lime

Resilient modulus is determined as a ratio of cyclic deviatoric stress to recoverable strain and it calculated for average last five load cycles. Figure 5 shows the resilient modulus of BC soil and lime stabilized BC soil for samples prepared with OMC. From Fig. 5, it can be observed that addition of lime leads to an increase in resilient modulus by 2.5 times due to the flocculation of clay particles.

Resilient modulus decreases with increasing deviatoric stress for both virgin BC and lime stabilized BC soil due to strain softening. Figure 6 represents the resilient modulus of BC and lime stabilized BC soil for samples prepared with OMC + 3%. Increasing moisture content by 3% results in 5–80% decrement in resilient modulus due to lubricant effect. Resilient modulus increases with increasing confining pressure for both BC sand and lime stabilized BC soil.

Figure 7 shows the UCS values of each wetting and drying cycles. From Fig. 7, it is observed that after first W-D cycle, the UCS strength increased from 9.31 to 9.42 kg/cm^2, whereas from second W-D cycle the decrement in UCS value is observed. For the first W-D cycle, the increment in UCS value is observed on account

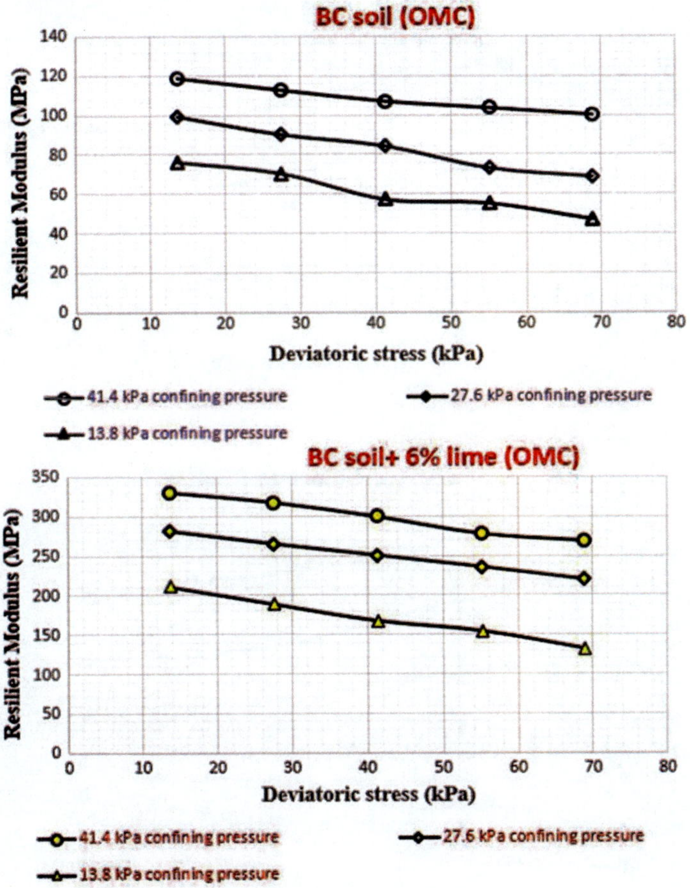

Fig. 5 Variation of resilient modulus with stress levels of BC soil with and without lime (OMC)

of the decrement in moisture content of the sample below 1%, thereby removing the pore water effect in the sample. For the subsequent W-D cycles, failure of samples was observed because of the penetration of water into the specimen on account of the presence of open cracks formed from the drying condition. From the generated cracks, water enters and exits developing substantial pressure and degrading the surrounding material (Pradhan and Bhattacharjee 2009).

3.1 Mineralogical and Morphological Studies

SEM and XRD studies were also performed in this study for detecting the changes in the chemical composition of the soil after a curing period of 28 days. The tests

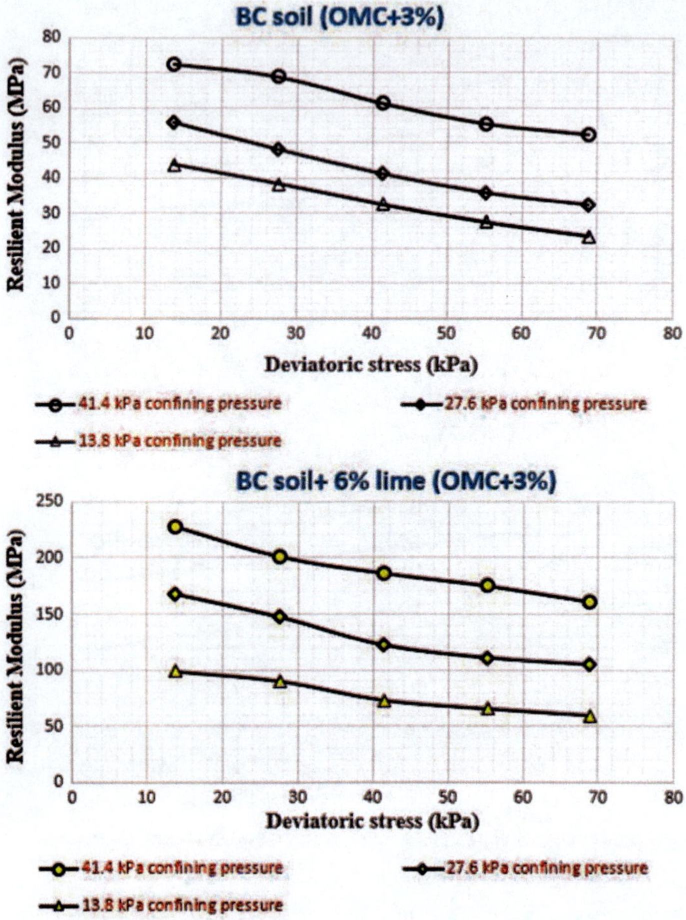

Fig. 6 Variation of resilient modulus with stress levels of BC soil with and without lime (OMC + 3%)

were mainly conducted in order to identify the formation of various cementitious compounds on stabilizing soil specimens with optimum % of lime; that is, for the sample that gave maximum UCS strength.

From Fig. 8, the XRD analysis of reaction products obtained on stabilization of soil samples with optimum percentage of GGBS after 28 day curing period is displayed. From XRD analysis, quartz, which is crystalline in structure, is detected with the help of sharp peaks. The hydration products anorthite ($CaAl_2Si_2O_8$) and gismondine ($CaAl_2Si_2O_84H_2O$) are mainly formed on account of the occurrence of the pozzolanic reaction. As curing period increases, cementitious products are formed on account of the reaction between silica and alumina present in the soil

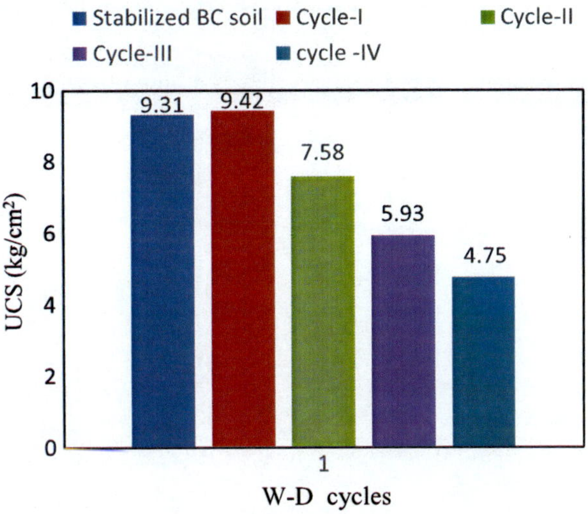

Fig. 7 UCS values after each W-D cycles lime stabilized BC soil

with the calcium hydroxide. The XRD analyses also demonstrate the formation of additional peaks of calcite ($CaCO_3$) on account of cementitious reactions (Solanki and Zaman 2012).

Figure 9 shows the SEM micrographs of BC soil and BC soil stabilized with optimum % of lime and cured for 28 days. With the addition of lime, large aggregations or flocs which are mainly formed due to the reduction of swell in the soil sample can be clearly visible from the SEM as shown in Fig. 9 (Al-Rawas 2002). From the micrograph, the hydration products such as C–S–H gel mixed with calcium hydroxide can be observed from the micrograph. The hydrated grains formation also reveals the occurrence of pozzolonic reaction in the soil–lime mixtures.

4 Conclusions

This paper deals with physical, mechanical, morphological and mineralogical studies on BC and lime stabilized BC soils and the following conclusions are made.

- The results indicated that 6% lime was found to be optimal for BC soil. The reduction in plasticity of soil is evident from these tests.
- Lime stabilization causes significant improvement in strength properties of BC soil. Stabilization of lime caused increase in unconfined compressive strength by 10–14.5 times compared to natural soil where as 14 times increment observed in CBR.
- Addition of lime caused nearly 2.5 times increment in resilient modulus. Resilient modulus increases with confining pressure, whereas decreases with deviatoric stress for both virgin BC soil and lime stabilized BC soil.

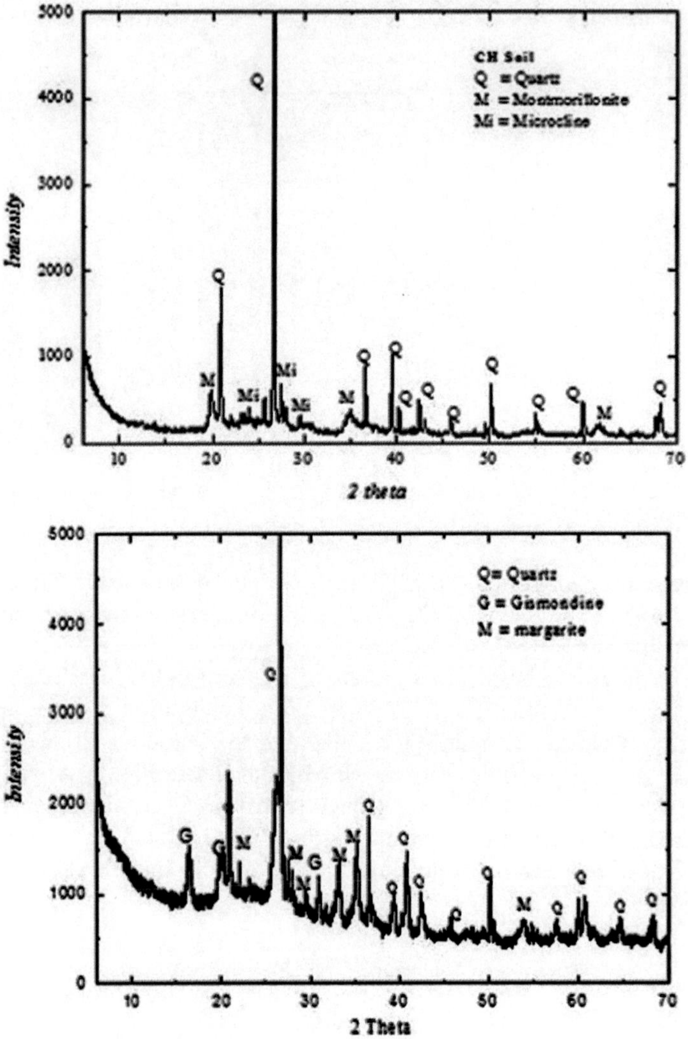

Fig. 8 XRD analysis of BC soil with and without addition of lime

- Durability studies performed on the BC soil stabilized with optimum dosages resulted in the increment in strength for first W-D cycle and continuous reduction in strength for second to fourth cycles which collapsed completely after fifth cycle.
- The formation of cementing agents such as calcium silicate hydrates and calcium aluminium silicates liable for improved strength was evidenced from mineralogical studies using XRD. The formation of pozzolonic compounds and flocculation is evident from the SEM images.

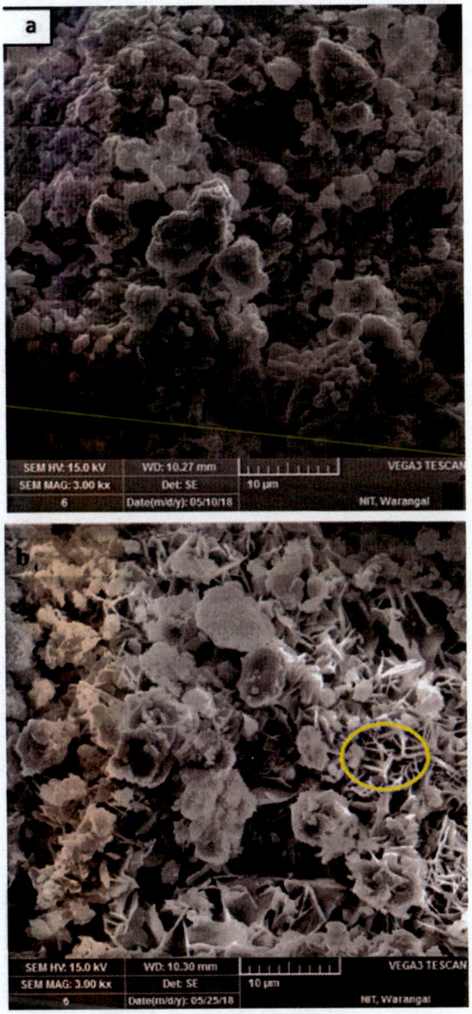

Fig. 9 SEM analysis of BC soil with and without addition of lime

References

AASHTO (2008) Mechanistic-empirical pavement design guide, Interim edn. A Manual of Practice, American Association of State Highway and Transportation Officials, Washington, DC

Al-Rawas AA, Taha R, Nelson JD, Al-Shab BT, Al-Siyabi H (2002) A comparative evaluation of various additives used in the stabilization of expansive soils. Geotech Test J 25(2):199–209

Bell FG (1996) Lime stabilization of clay minerals and soils. Eng Geol 42(4):223–237

Cokca E, Yazici V, Ozaydin V (2009) Stabilization of expansive clays using granulated blast furnace slag (GBFS) and GBFS-Cement. Geotech Geol Eng 27:489–499

Croft JB (1967) The influence of soil mineralogical composition on cement stabilization. Geotechnique 17(2):119–135

Edil TB, Acosta HA, Benson CH (2006) Stabilizing soft fine-grained soils with fly ash. J Mater Civ Eng 18(2):283–294

Hamza G, HaliIbrahim F (2016). Response surface methodology for optimization of stabilizer dosage rates of marginal sand stabilized with sludge ash and fiber based on ucs performances. KSCE J Civ Eng (Springer) 20(6)

Horpibulsuk S, Phetchuay C, Chinkulkijniwat A (2011) Soil stabilization by calcium carbide residue and fly ash. J Mater Civ Eng 24(2):184–193

Horpibulsuk S, Phetchuay C, Chinkulkijniwat A, Cholaphatsorn A (2013) Strength development in silty clay stabilized with calcium carbide residue and fly ash. Soils Found 53(4):477–486

IS 2720 (1980) Part 8: determination of water content dry density relation using heavy compaction. Bureau of Indian Standards, New Delhi, India

IS 2720 (1973) Part 10: determination of unconfined compressive strength of soil. Bureau of Indian Standards, New Delhi, India

IS 2720 (1973) Part 16: laboratory determination of C.B.R of soil (second revision). Bureau of Indian Standards, New Delhi, India

Kulkarni VR, Patil GK (2014) Experimental Study of stabilization of B.C. Soil by using slag and glass fibers. J Civ Eng Environ Technol 1(2):107–112

Kumar A, Walia BS, Bajaj A (2007) Influence of fly ash, lime, and polyester fibers on compaction and strength properties of expansive soil. J Mater Civ Eng 19(3):242–248

Li D, Selig ET (1994) Resilient modulus for fine-grained subgrade soils. J Geotech Eng 120(6):939–957

Nalbantoglu Z (2004) Effectiveness of class C fly ash as an expansive soil stabilizer. Constr Build Mater 18(6):377–381

NCHRP (2004) Guide for mechanistic-empirical design of new and rehabilitated pavement structures. Final Rep. No. NCHRP 1–37A, National Research Council, Transportation Research Board, Washington, DC

Nelson JD, Miller DJ (1992) The pozzolanic effect of fly ash on the CBR behaviour of black cotton soil. J Test Eval 31(6):1

Phanikumar BR, Sharma RS (2004) Volume change behavior of fly ash-stabilized clays. J Mater Civ Eng 19:67–74

Pradhan B, Bhattacharjee B (2009) Performance evaluation of rebar in chloride contaminated concrete by corrosion rate. Constr Build Mater 23(6):2346–2356

Rout RK, Ruttanapormakul P, Valluru S, Puppala AJ (2012) Resilient moduli behavior of lime-cement treated subgrade soils. In: Geo congress 2012: state of the art and practice in geotechnical engineering, pp 1428–1437

Seco A, Ramírez F, Miqueleiz L, García B (2011) Stabilization of expansive soils for use in construction. Appl Clay Sci 51(3):348–352

Sharma NK, Swain SK, Umesh C, Sahoo. (2012) Stabilization of a clayey soil with fly ash and lime. Geotech Geol Eng 30:1197–1205

Solanki P, Zaman M (2012) Microstructural and mineralogical characterization of clay stabilized using calcium-based stabilizers. Scanning electron microscopy, InTech

Zaman M, Chen DH, Laguros J (1994) Resilient moduli of granular materials. J Transp Eng 120(6):967–988

Experimental Studies on Lateritic Soil Stabilized with Cement, Coir and Aggregate

A. U. Ravi Shankar, B. A. Priyanka, and Avinash

Abstract The characteristics of subgrade soil play a vital role in designing the pavement structure so that the pavement has required support from the bottom layer. During adverse weather conditions and higher traffic loads moving on any pavement, it should be able to withstand the impact and perform well for longer duration. Load is transmitted from pavement to the subgrade layer and distributed evenly through the soil particles. All types of soil are not capable of handling such impacts by their own and needs additional stabilization processes. Several stabilization processes are available in which the best one has to be taken into consideration. Fibers such as coconut coir are important in giving extra stability to the soil particles. Cement is well-known material in construction sector along with aggregates. Lateritic soil is available abundantly in coastal areas of southern parts of India which has porous structure and demands stabilization when the intended purpose is specific and requires higher strength and durability. In this study, coconut coir along with cement and aggregate are taken as stabilization materials to stabilize lateritic soil. Initially, basic properties of soil like plastic limit, liquid limit and plasticity index are determined. Grain size analysis is done and modified Proctor test is conducted to determine the optimum moisture content (OMC) and maximum dry density (MDD) of the soil. Unconfined compression test (UCS), California bearing ratio test (CBR), flexural fatigue analysis, durability properties with respect to wet-dry cycles and freeze–thaw cycles are evaluated for untreated and treated soil specimen. As per UCS values, cement can be taken at an optimum dosage of 6%. The coir fibers from natural husk of coconut and aggregates of 10 mm below size were added to the soil–cement mixes and an optimum dosage of coir and aggregate is determined. The study showed positive results in terms of CBR values of cement-aggregate treated soil.

A. U. Ravi Shankar (✉) · B. A. Priyanka · Avinash
National Institute of Technology Karnataka, Surathkal, Mangalore, Karnataka 575025, India
e-mail: aurshankar@gmail.com

B. A. Priyanka
e-mail: priyankabiluve@gmail.com

Avinash
e-mail: avinash@gmail.com

M. Latha Gali and R. R. P. (eds.), *Problematic Soils and Geoenvironmental Concerns*, Lecture Notes in Civil Engineering 88,
https://doi.org/10.1007/978-981-15-6237-2_61

Keywords Lateritic soil · Stabilization · UCS · CBR · Durability · Fatigue

1 Introduction

Pavement structure is a combination of surface course upon which vehicle moves, supporting base and sub-base layers and the subgrade to which all the loads will be transferred. Therefore, the subgrade needs to be strong enough to withstand all the loads. At the same time, the chances of adverse traffic conditions may exist in most cases which demands for strong, durable and long-life pavement structures. In cases of higher level of groundwater, the chances of capillary rise and percolation of groundwater into pavement surface are more which results in early deterioration of pavement surface. Therefore, it is required to provide proper drainage and strong pavement surface which gets support from a stronger subgrade. Whenever the existing subgrade is weak in terms of clay content, swelling behavior and lack of gradation required for pavement construction, it becomes the need of the hour to stabilize the subgrade to modify its properties to suit the requirements.

The stability of underlying soil subgrade significantly affects the performance of pavement. The in situ subgrade does not possess adequate subgrade qualities and can give rise to various problems while construction, especially when they are wet. These soils can be soft, plastic and difficult to compact. If these soils are stabilized with cement, its engineering properties/subgrade qualities are improved to larger extent. When the groundwater aquifer is at higher level, the moisture infiltrates into the subgrade and base layers. The infiltration of water can also take place due to capillary action. This infiltration results in lower strength of subgrade, softening and reduces modulus of soil. If the soil is stabilized with cement, it reduces permeability. It helps to hold moisture and maintains high level of strength and stiffness even when saturated (McConnell 2009).

Lateritic soil is abundantly available in most parts of coastal regions of southern India. It is having better strength, but with porous structure. In this study, the materials such as cement, aggregate and coir fibers are taken to modify the properties of lateritic soil to suit the requirements. Main aim of the stabilization is to provide additional strength to the lateritic soil which is needed in deciding the design parameters (Khanna and Justo 2011).

2 Objectives

The main objectives of the present study are as follows:

- To evaluate the index properties of the lateritic soil.
- To determine the optimum dosage of cement that can be added to the lateritic soil to modify its strength properties.

Table 1 Physical properties of cement

Physical properties	Results
Compressive strength (28 days), N/mm^2	43
Specific gravity	3.12
Fineness (minimum specific surface), m^2/kg	301
Initial setting time, minutes	50
Final setting time, minutes	240
Soundness (Le Chatelier's test), mm	10
Normal consistency, %	28

- To evaluate the characteristics of lateritic soil with optimum dosage of cement and 10 mm down size aggregates (20, 25 and 30% of the soil weight).
- To determine the characteristics of lateritic soil with optimum dosage of cement along with different proportions of coir addition (0.5 and 1.0% of soil weight).

3 Experimental Investigations

3.1 Materials Used

The materials used in the present study are lateritic soil, cement, aggregates and coconut coir. The lateritic soil was collected from a borrow pit in the form of disturbed samples at NITK campus, Karnataka, India. The collected sample was dried in open air and then kept in oven for eliminating any moisture content present in it. The cement used was ordinary Portland cement (OPC) 43 grade. The basic properties of cement are tabulated in Table 1. Aggregate (10 mm down) was collected from nearby quarry. The husk of coconut naturally gives coir that can be extracted and used for various applications such as brush making, floor mats, mattresses and so on. For the present investigation, the aspect ratio (length/diameter) of coir is taken as 20. The length of coconut coir is 20 mm and the diameter is 1 mm.

3.2 Specimen Preparation

Initially, the untreated lateritic soil was tested to determine the basic properties including specific gravity (IS: 2720 (Part III) 1980), grain size distribution (IS: 2720 (Part IV) 1980), Atterberg's limits (IS: 2720 (Part V) 1985), standard Proctor test (IS: 2720 (Part VII) 1980) and modified Proctor compaction test (IS: 2720 (Part VIII) 1983). For UCS, CBR, durability and fatigue test, soil specimens were prepared by varying the cement contents as 3, 6, 9 and 12% of dry weight of soil (IS 2720 (Part I) 1983). Then, the specimens were prepared by taking optimum amount of cement

(6%) and varying coir fiber at 0.5 and 1.0% and with 20, 25 and 30% aggregates. Minimum of three specimens were prepared for each test and the average value was reported. Oven-dried soil was first dry mixed with various percentages of cement, coir and cement-aggregate mix and then sufficient quantity of water was added to evaluate the moisture content of the mixture needed to achieve the desired optimum moisture content (OMC) with respect to modified compaction. The mixtures were remolded into cylindrical specimens of 38 mm diameter and 76 mm height by static compression in compaction frame to the required maximum dry density (MDD). The compaction test is done immediately after treating soil with the stabilizer. Triaxial compression test was performed as per (IS: 2720 (Part XI) 1993) and the fatigue test was conducted on cylindrical specimens of length to diameter ratio 2. Effects of curing period on the UCS were examined for specimens treated with the different cement content after subjecting to 0, 7, 14, 28 and 60 days of curing period. During this period, each specimen was kept in desiccators in an attempt to minimize losses of moisture from it and the soil moisture constant was maintained during the curing days.

4 Results and Discussion

4.1 *Phase 1: Tests on Basic Properties of Lateritic Soil*

The results of tests on basic soil properties of lateritic soil are tabulated in Table 2.

4.2 *Phase 2: Tests on Laterite Soil + Cement*

4.2.1 Compaction Tests

The standard and modified compaction tests were conducted on both treated and untreated soil. For both standard and modified Proctor compaction which was conducted immediately after mixing, MDD increased till 6% of cement and later started decreasing. Vice versa OMC decreased till 6% of cement and later started increasing. The standard and modified Proctor test results for untreated soil and soil with different percentages of cement, coir and aggregates are presented in Tables 3 and 4, respectively.

4.2.2 UCS Test

The UCS test was conducted on treated soil specimens for 0, 7, 14, 28 and 60 days of moist curing in desiccators and the results are tabulated in Table 5. From the results,

Table 2 Basic properties of lateritic soil

Test	Results	Method
Specific gravity	2.661	IS:2720 (Part III)-1990
Grain size distribution (%)		IS:2720 (Part IV)-1985
Gravel	22	
Sand	45	
Silt	31	
Clay	2	
Soil classification	Silty sand	
Atterberg limits (%)		IS:2720 (Part V)-1986
Liquid limit	36	
Plastic limit	27	
Plasticity index	9	
Standard Proctor test		IS:2720 (Part VII)-1980
MDD (g/cc)	1.965	
OMC (%)	12.05	
Modified proctor test		IS:2720 (Part VIII)-1983
MDD (g/cc)	2.095	
OMC (%)	11.12	
UCS (kPa)		IS:2720 (Part X)-1991
OMC-2%	227	
OMC	589	
OMC + 2%	306	
CBR (%)		IS:2720 (Part XVI)-1987
Soaked	18	
Un-soaked	49	
Triaxial E-modulus (kPa)	27,619	IS:2720 (Part XI)-1971

it is observed that, UCS of soil samples increased with increase in cement contents, whereas it is decreasing with increasing in curing period. This may be due to the formation of dicalcium silicates in steps which is the result of hydration process of cement when it comes in contact with water. This leads to the strength gain with days. Untreated lateritic soil shows decrease in strength at longer curing periods which is due to the absence of cement in it. When the quantity of binder (cement) is very less (3%). Similar trend of lower strength is observed compared to the soil specimens with higher cement dosage.

Table 3 MDD and OMC of standard Proctor test immediately after mixing

Sample	OMC (%)	MDD (g/cc)
Untreated soil	12.05	1.965
Soil + 3% cement	11.9	1.98
Soil + 6% cement	11.7	1.995
Soil + 9% cement	13.6	1.975
Soil + 12% cement	14.2	1.97
Soil + 6% cement + 0.5% coir	12	1.96
Soil + 6% cement + 1% coir	11.80	1.982
Soil + 6% cement + 20% aggregates	12.1	1.97
Soil + 6% cement + 25% aggregates	11.5	1.982
Soil + 6% cement + 30% aggregates	11.1	1.991

Table 4 MDD and OMC of modified Proctor test immediately after mixing

Sample	OMC (%)	MDD (g/cc)
Untreated soil	11.12	2.0965
Soil + 3% cement	11	2.13
Soil + 6% cement	10.9	2.133
Soil + 9% cement	12.5	2.124
Soil + 12% cement	12.8	2.12
Soil + 6% cement + 0.5% coir	11.12	2.13
Soil + 6% cement + 1% coir	10.83	2.125
Soil + 6% cement + 20% aggregates	11.2	2.21
Soil + 6% cement + 25% aggregates	10.9	2.26
Soil + 6% cement + 30% aggregates	10.9	2.33

Table 5 Variation of UCS with different curing periods

Curing period (days)	0	7	14	28	60
	UCS (kPa)				
Untreated soil	589	556	366	275	242
Soil + 3% cement	684	934	1053	1124	795
Soil + 6% cement	716	1858	2038	2230	2376
Soil + 9% cement	749	2077	2354	2564	2668
Soil + 12% cement	857	2164	2439	2675	2931

Table 6 Variation of soaked CBR with different curing periods

Soaked CBR values (%)				
Curing period (days)	0	7	14	28
Soil	18	18	17	15
Soil + 3% cement	28	33	38	41
Soil + 6% cement	30	73	98	107
Soil + 9% cement	34	89	101	117
Soil + 12% cement	37	96	106	125
Soil + 6% cement + 0.5% fiber	34	75	88	109
Soil + 6% cement + 1% fiber	40	78	101	117
Soil + 6% cement + 20% aggregates	51	81	104	145
Soil + 6% cement + 25% aggregates	58	99	125	168
Soil + 6% cement + 30% aggregates	60	112	137	184

4.2.3 CBR Test

CBR test was conducted using standard test equipment on treated soil samples cured for 0, 7, 14 and 28 days of moist curing. The untreated soil was tested for both soaked and un-soaked condition for modified Proctor compaction densities. Whereas, treated soil was tested only for soaked condition. The curing was done using wet gunny bags by covering the CBR molds with soil in polyethylene covers. The gunny bags were made wet twice a day and the results of CBR test are tabulated in Table 6.

From Table 6, it can be seen that, CBR value of untreated lateritic soil was constant at 0 and 7 days of curing. However, after 14 days of curing, its value started decreasing. At all curing periods, the CBR values were increased with increase in dosage of cement. The addition of water to cement leads to hydration of cement giving rise to the strength gain. Increase in strength may be due to the formation of dicalcium and tricalcium silicates in cement matrix. Strength development is not observed in untreated specimens which does not had cement in its matrix. For different curing periods, it experienced alternate wetting and drying, and hence it weathered a bit, which in turn reduced the CBR value.

4.2.4 Durability Tests

According to Dempsey and Thompson (1968), durability is the capability of any material to retain its strength and to remain integral part of the matrix over a long period of time up to the life required to the particular material. Over a period of time, the material is subjected to the weathering changes and gets deteriorated upon losing its integrity and stability. Therefore, it is important to assess the durability properties of any material before saying it to be a perfect material for the intended use. Stabilizing additive needs to be stronger enough and also capable enough to withstand the cycles of weathering changes in the atmosphere. These changes in the atmosphere can be represented in terms of wet-dry cycles and freeze–thaw cycles. In the present study, the treated specimens were subjected to both the cycles 12 times as per ASTM D 559–2015 (standard test method for wetting and drying of compacted soil–cement mixtures) and ASTM D 560–2015 (standard test method for freezing and thawing of compacted soil–cement mixtures) specifications. As per the specifications, the weight loss for the specimens should not be more than 14% to call it as a durable material and the results are tabulated in Table 7.

Untreated soil specimens were failed in first cycle of wetting because of its inability to withstand moisture effect. The specimens after cement-treatment stood strong up to 12 cycles at all cement dosage variations. UCS test was carried out on the specimens after 12 cycles of wet-dry. The results showed that the UCS values were more than three times that of the 7 days moist cured specimens and more than two times that of the 60 days moist cured specimens. The cement-treated specimens showed increase in strength after alternate wet-dry cycles which make it suitable to use this combination in the weather conditions, wherein alternate wet and dry conditions are prominent.

From Table 8, it is evident that the UCS value of specimens after 12 cycles of freeze–thaw was found to be more than that of the UCS of 60 days moist cured specimens at all cement replacements except 3% replacement. Therefore, the soil–cement combination can be preferred even in case of countries with snow fall and colder weather.

4.2.5 Fatigue Test

The soil specimens were subjected to repeated loads or cyclic loads until failure and the resulting number gives the fatigue life of the specimens. The specimens with cement and soil combination were subjected to repeated loads under fatigue testing equipment in the laboratory. The cylindrical specimens having same size as that of UCS specimens were used in this experiment. Seven days cured soil specimens were taken for this study. 1/3rd, 1/2nd and 2/3rd of the loads with respect to the UCS values were applied on these specimens and the resulting number of repetitions up to failure are presented in Table 9.

It can be observed from Table 9 that, all the specimens were able to withstand minimum of 2 lakhs repetitions at all the three levels of loading. The soil mixed with all the cement replacement percentages showed higher endurance limit. Therefore, the treated soil can be efficiently used as a pavement material.

Table 7 Weight loss in percentage with respect to wet-dry cycles

No. of cycles	Percentage weight loss (%)			
	3% Cement		6% Cement	
	Wetting	Drying	Wetting	Drying
1	3.36	−8.29	2.11	−8.76
2	3.13	−8.74	0.97	−8.97
3	3.17	−8.44	2.36	−8.54
4	2.66	8.74	1.98	−8.95
5	2.52	9.14	1.9	−9.25
6	2.6	8.93	1.95	−8.98
7	2.42	9.16	0.28	−8.78
8	2.14	9.34	−0.01	−8.58
9	2	9.35	−0.14	−8.48
10	1.85	9.46	−0.28	9.07
11	1.75	9.56	−0.3	3.21
12	1.69	9.57	−0.38	9.3
UCS values (kPa)	1266		5206	
No. of cycles	9% cement		12% cement	
	Wetting	Drying	Wetting	Drying
1	2.45	−8.98	2.95	−9.23
2	2.59	−9.22	3.01	−9.39
3	2.7	−8.78	3.24	−8.8
4	2.33	−9.06	2.91	−9.16
5	2.35	−9.41	2.89	−9.6
6	2.38	−9.16	3	−9.29
7	2.23	−8.98	2.82	−9.1
8	2.31	−8.75	2.91	−8.76
9	2.43	−8.64	3.02	−8.74
10	2.15	−8.69	2.8	−9.12
11	2.1	−8.8	2.74	−9.24
12	2.1	−8.85	2.73	−9.34
UCS values (kPa)	6319		6470	

4.3 Phase 3: Tests on Lateritic Soil + Cement + Coir and Aggregates

In this phase, optimum percentage of cement was computed based on results from 7 days UCS strength. The cement content corresponding to strength of 1750 kPa is taken as design cement content and is considered to be adequate enough when

Table 8 Weight loss in percentage with respect to freeze–thaw cycles

No. of cycles	Percentage weight loss (%)			
	3% Cement		6% Cement	
	Freeze	Thaw	Freeze	Thaw
1	−0.48	−0.03	−0.5	0.11
2	−0.47	−0.21	−0.33	0.07
3	−0.22	−0.19	0.26	0.55
4	−0.1	0.37	0.35	1.03
5	0.07	0.45	0.65	1.09
6	−0.06	0.62	0.59	1.59
7	0.46	1.49	1.67	1.99
8	1.04	2.39	1.64	2.24
9	1.4	1.88	1.64	1.97
10	0.9	1.4	1.17	1.7
11	0.28	1.37	1.45	1.57
12	0.01	1.18	1.18	1.2
UCS values (kPa)	258		2160	
No. of cycles	9% Cement		12% Cement	
	Freeze	Thaw	Freeze	Thaw
1	−0.46	0.03	−0.32	0.18
2	−0.28	0.15	−0.17	19
3	0.47	0.67	0.28	0.47
4	0.37	0.71	0.14	0.59
5	0.49	0.72	0.34	0.68
6	0.28	1.15	0.31	0.93
7	1.15	1.51	0.85	1.68
8	1.14	1.76	1.28	2.19
9	1.44	1.64	1.93	2.2
10	1.2	1.57	1.78	1.78
11	1.08	1.48	1.67	1.74
12	1.06	1.29	1.62	1.73
UCS values (kPa)	3365		4673	

soil–cement mix has to be used as base course for pavement construction with light to medium traffic flow (Khanna and Justo 2011). From Table 5, it was found that 6% cement was optimum and the same is considered in phase 3.

Table 9 Fatigue test results of soil treated with various cement dosages

	1/3rd of UCS		1/2nd of UCS		2/3rd of UCS	
	Load (kg)	Cycles	Load (kg)	Cycles	Load (kg)	Cycles
Soil + 3% cement	35	200,000	53	200,000	71	200,000
Soil + 6% cement	70	200,000	105	200,000	140	200,000
Soil + 9% cement	78	200,000	118	200,000	157	200,000
Soil + 12% cement	82	200,000	123	200,000	164	200,000

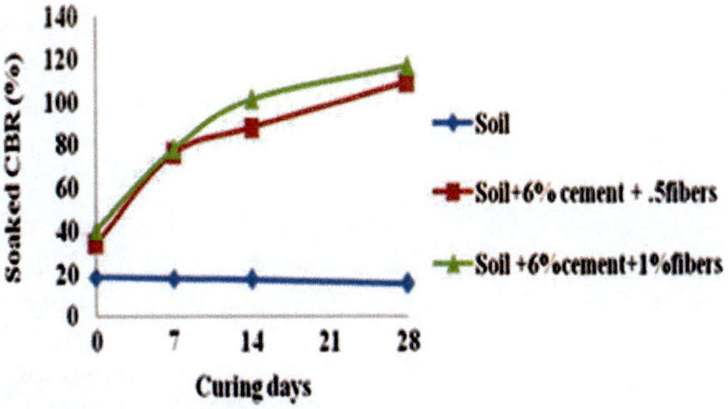

Fig. 1 Soaked CBR values for various curing periods

4.4 *Tests on Laterite Soil + Optimum Cement (6%) + Fibers*

Initially, weight of the soil needed for a particular test was determined. Then, 6% of that weight was replaced by cement. Coconut coir of length 20 mm was added in this phase. Coconut coir was used at dosage of 0.5% and 1% of weight of soil. In this phase, only compaction and CBR tests were carried out since UCS samples could not be molded as the soil could not reach molding water content at OMC conditions and the test results are presented in Fig. 1. It is observed that, CBR values for both the percentages of fibers increased tremendously as the curing period increased and it can be seen that values are very high compared to that of untreated soil.

4.5 *Tests on Laterite Soil + Optimum Cement (6%) + Aggregates*

In this case, the weight of the soil needed for a particular test was determined and 6% of that weight was replaced by cement. In this phase, three dosages of 10 mm down size aggregates (20, 25 and 30% of the soil weight) were mixed with soil. Due

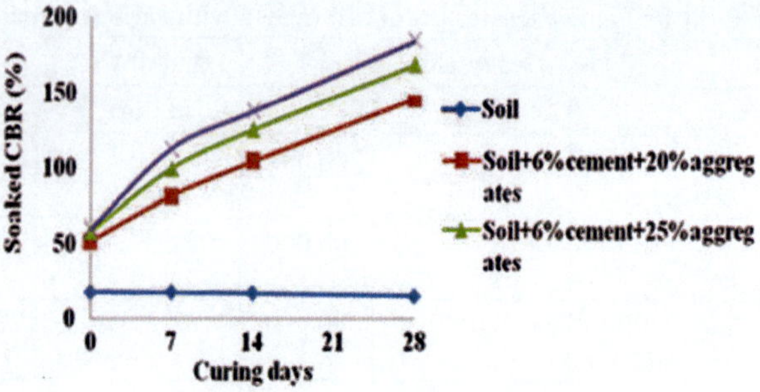

Fig. 2 Soaked CBR values for various curing periods

to the presence of aggregates, it was not possible to mold the UCS specimens, and hence only compaction and CBR tests were carried out. CBR test results at various curing periods are presented in Fig. 2. It is observed that, CBR values for all the three percentages of aggregates increased as the curing period increases and it can be seen that values are very high compared to that of untreated soil. These values were higher compared to that of the mix with soil and cement and the mix with soil, cement and fibers.

The soil specimens treated with cement, coir and aggregates were found to be having higher strength and can be recommended for construction of base course leading to an innovation in designing the pavement structure.

5 Conclusions

The study aimed at arriving with an innovative composition of base course layer for pavement construction, as well as modified subgrade with higher strength needed for highway construction in the places with lateritic soil.

For lateritic soil treated with 0, 3, 6, 9 and 12% cement:

- Up to 6% cement replacement, the value of MDD increases with increase in cement content at both standard and modified Proctor densities.
- In the similar context, the OMC value decreases up to 6% cement content and then increases for further increase in cement content.
- With increase in curing period, the UCS value of untreated soil decreases and that of treated soil increases.
- CBR of untreated soil specimens increases with number of curing days in case of treated specimens and it is constant for untreated specimens up to 14 days of curing and decreases beyond that.
- Durability test and fatigue test results prove that the specimens after treatment are stronger and durable and can sustain load repetitions of more than 2 lakhs. These factors make it clear that cement stabilization provides completely positive impact on strength, durability and fatigue behavior of treated lateritic soil.

For Lateritic Soil + Optimum Cement (6%) + Coconut Coir:

- For both standard and modified Proctor tests, MDD was higher in both the cases compared to that for untreated soil while OMC was lower than that for untreated soil.
- The CBR values for both the percentages of fibers increased tremendously as the curing period increases and the values are very high compared to that for untreated soil.

For Lateritic Soil + Optimum Cement (6%) + Aggregates:

- For both standard and modified Proctor tests, MDD was higher in all the cases compared to that of untreated soil, while OMC was lower than that for the untreated soil.
- The CBR values for all the three percentages of aggregates increased as the curing period increases and the values are very high compared to that of CBR values of untreated soil. These values are higher compared to that of soil and cement and even soil, cement and fibers mix.

References

Dempsey BJ, Thompson MR (1968) Durability properties of lime soil mixtures. Highway Res Rec 235:61–75

IS: 2720 Part I (1983) Preparation of dry soil samples for various tests. Bureau of Indian Standards, Manak Bhavan, New Delhi

IS: 2720 Part III (1980) Determination of specific gravity. Bureau of Indian Standards, Manak Bhavan, New Delhi

IS: 2720 Part IV (1980) Determination of grain size. Bureau of Indian Standards, Manak Bhavan, New Delhi

IS: 2720 Part V (1985) Determination of plastic and liquid limit. Bureau of Indian Standards, Manak Bhavan, New Delhi

IS: 2720 Part VII (1980) Determination of moisture content and dry density using light compaction. Bureau of Indian Standards, Manak Bhavan, New Delhi

IS: 2720 Part VIII (1983) Determination of moisture content and dry density using heavy compaction. Bureau of Indian Standards, Manak Bhavan, New Delhi

IS: 2720 Part X (1973) Determination of unconfined compressive strength. Bureau of Indian Standards, Manak Bhavan, New Delhi

IS: 2720 Part XI (1993) Determination of shear strength parameters of a specimen tested in unconsolidated undrained triaxial compressing without measurement of porewater pressure. Bureau of Indian Standards, Manak Bhavan, New Delhi

IS: 2720 Part 16 (1987) Laboratory determination of CBR. Bureau of Indian Standards, Manak Bhavan, New Delhi

Khanna SK, Justo CEG (2011) Highway engineering. Nem Chand and Bros, Roorkee, Uttarkhand

McConnell T (2009) Cement stabilization applied to highway subgrade. In: T N concrete pavement conference, Nashville, Portland Cement Association (PCA)

Model Studies to Restrain Swelling of Expansive Soil by Using Geostrip Reinforced Lime Fly Ash Columns

Vikrant Jain and B. V. S. Viswanadham

Abstract Estimation of swell pressure prior to any construction activity in expansive soil is an indispensable step. The purpose of this paper is to check the efficacy of mechanical and chemical means in restraining the swelling of expansive soil through a series of swell consolidation tests by installing geostrip reinforced fly ash-lime columns in compacted soil specimen. To begin with, several swell consolidation tests were performed in conventional oedometer cell to examine the influence of dry unit weight, initial water content, and sand drains on swell parameters of compacted soil specimen. Later, tests were conducted on custom made large-scale swell consolidation apparatus to check the efficacy of the proposed treatment method. The study reveals reinforcing the soil specimen with seven geostrip reinforced fly ash-lime columns at three times the diameter spacing results in minimal swelling. The proposed method can turn out to be an in situ method in order to restrain swelling of expansive subgrade.

Keywords Expansive soil · Swell consolidation tests · Swell potential · Swell pressure · Geostrip · Fly ash · Subgrade

1 Introduction

Expansive soils are termed as "problematic soils" in engineering prospective as they have intrinsic tendency to undergo volumetric changes with variation in moisture content. They swell as they imbibe water and consequently exert an upward force on overlying structure and shrink with evaporation of water resulting in desiccation cracks into the soil mass. The moisture variation can be attributed to precipitation,

V. Jain · B. V. S. Viswanadham (✉)
IIT Bombay, Mumbai, Maharashtra 400076, India
e-mail: viswam@civil.iitb.ac.in

V. Jain
e-mail: vikrantjain101993@gmail.com

M. Latha Gali and R. R. P. (eds.), *Problematic Soils and Geoenvironmental Concerns*, Lecture Notes in Civil Engineering 88,
https://doi.org/10.1007/978-981-15-6237-2_62

fluctuation of groundwater table, leakage from underground water pipeline, temperature, etc. This twofold nature of expansive soils can cause severe distress to lightly loaded civil engineering structures resting over them like low storey buildings, canal lining, pavements, sidewalks, basement wall, and slabs leading to structural instability (Chen 1988; Nelson and Miller 1992). In past many years, amount of damage caused by expansive soils is startling. Therefore, minimizing the swell potential of such soils is a great challenge.

Significant progress has been made in the past years to unravel the nature of expansive soils. Evaluation of swell pressure of expansive soil is a vital criterion to design any structure resting over it. Several ways to determine swell pressure have been documented in the literature (Sridharan et al. 1986; Basma et al. 1995; Nagaraj et al. 2009). Owing to expansive soils dual nature, counter heave measures need to be embodied for lightly loaded structures at design phase itself to minimize the risk of structural instability. Many researchers over a number of years tried several ways to mitigate this problem. Mechanical and physical means have been tried which involves mixing of non-swelling granular fraction with expansive soil to reduce its swell potential (Satyanarayana 1995; Katti 1978; Sivapullaiah et al. 1996; Ikizler et al. 2009) and mixing natural and synthetic fibres (Maher and Ho 1994; Viswanadham et al. 2009). Many chemical additives have been proposed by several researchers, which may bring change in the chemical environment around the soil particles and thus restrain swelling. Addition of lime was found to be quite effective in reducing the plasticity of soil mass (Bell 1988; Dash and Hussain 2011; Jha and Sivapullaiah 2015). Fly ash (by-product of coal industry) have also been utilized as a potential chemical reagent to arrest the swelling in expansive soils by the virtue of its capability to furnish divalent and trivalent cation which when ionized promotes flocculation of clay particles (Cokca 2001; Kolay and Ramesh 2016). Recently, granular pile anchor technique has been suggested where resistance to uplift of footing on expansive soil is provided by the anchor rod adhered to the bottom of the column to an anchor plate (Muthukumar and Shukla 2018). These techniques/admixtures have their own advantage and disadvantage in controlling the swelling characteristics of expansive soils. Some serious issues in aforementioned techniques can be listed as cost, material availability, environmental concerns, feasibility in field and durability.

With increasing population and traffic volume, the need is to widen the roads and increase number of lanes without disturbing the existing structure. If pavement happens to be on expansive soils, geotechnical performance of expansive soils needs to be ameliorated so that they can meet the serviceability requirements. Implementing anti-heave measures in expansive soil during design phase itself, therefore, can minimize to a greater extent the destruction to the overlying structure. Hence, the motivation of the study is to examine the efficacy of combined chemical and mechanical technique to restrain swelling of expansive soils which involves minimal excavation and replacement and can be extended in situ.

2 Material Properties

2.1 Soil

The soil used for the present study was blackish in colour, namely black cotton soil and was collected from a depth of 1 m at Nanded region near Pune city, Maharashtra, India. Free swell index of the soil was found to be 183.33%. Based on the unified soil classification system, soil was classified as CH.

2.2 Fly Ash and Lime

For the present study, fly ash was brought in from Dirk India, Nashik Plant and was visibly grey in colour. Fly ash was kept in the thermostatically controlled oven within a temperature range of 60–65 °C before using it for the experimental investigation. Table 1 shows basic properties of soil and fly ash used for the present study.

Table 1 Properties of soil and fly ash under assessment

Property	Soil	Fly ash
Specific gravity	2.61	2.22
Particle size distribution		
Sand (%)	19	2
Silt (%)	28	79
Clay (%)	58	19
Atterberg limits		
Liquid limit (%)	85.39	_a
Plastic limit (%)	38.61	_a
Plasticity index (%)	46.78	NP
USCS classification	CH	NP
*Compaction characteristics**		
Maximum dry unit weight (KN/m^3)	12.65	_a
Optimum moisture content (%)	34.5	_a
Free swell index (%)	183.33	Non-swelling

_a: Not relevant/ not carried out, * Standard Proctor compaction, NP: Non-plastic

2.3 *Mechanical Reinforcement*

2.3.1 Scaling Considerations

In small-scale 1 g physical modelling, geosynthetic material needs to be scaled down in order to comprehend accurate response of prototype structure. Taking into consideration geometric similitude conditions, the following relationship may be derived

$$\frac{b_m}{b_p} = \frac{1}{N}; \frac{t_m}{t_p} = \frac{1}{N} \tag{1}$$

$$\frac{A_m}{A_p} = \frac{b_m t_m}{b_p t_p} = \frac{1}{N^2} \tag{2}$$

where A is the cross-sectional area, b is the width, t is the thickness, N is the scale factor, subscript m indicates model dimensions and p indicates prototype dimensions.

Also, the scale factor for tensile load carrying capacity of model and prototype material can be deduced as,

$$\frac{T_m}{T_p} = \frac{E\varepsilon A_m}{E\varepsilon A_p} = \frac{1}{N^2} \tag{3}$$

where T is tensile load, E is the modulus of elasticity of geosynthetic material, ε is the strain in the material.

2.3.2 Geostrip

Taking into consideration above scaling laws, a polyester strand, 1 mm thick and 3 mm wide, coated with polyvinyl chloride was used as mechanical restraint in the experimental investigations.

3 Experimental Investigations

3.1 *Tests Conducted and Quantities Determined*

Several one-dimensional swell consolidation tests were performed on remoulded soil in an oedometer cell to analyse the factors affecting swell parameters of the compacted expansive soil specimen and also to check the efficacy of geostrip reinforced fly ash-lime columns in restraining its swelling. The purpose of the test was to determine swell potential ($\delta\%$) and swelling pressure (σ s), which are in turn indicative of degree of expansiveness of soil mass. The swell potential of the compacted soil specimen is the ratio of change in thickness (Δh) to the original thickness (h) expressed in per cent (Eq. 4).

$$\delta(\%) = \frac{\Delta h}{h} \times 100 \quad (4)$$

Swelling pressure can be taken as the vertical pressure required to bring back the completely swollen sample to its original state in terms of void ratio or thickness.

3.2 Test Equipment

Preliminary investigations were made to analyse the factors affecting swell parameters of compacted expansive soil specimen. This included series of swell consolidation test in conventional oedometer cell with fixed ring assembly of diameter (d) 75 mm and height (h) 25 mm.

In order to check the efficacy of mechanical anchors in restraining the swell potential of expansive soil, experiments in conventional oedometer cell would fall redundant. Experimental investigations may include several columns of larger diameter and reinforcing them with geostrip through mechanical anchors. Such arrangement was not at all feasible in small-scale oedometer cell due to its physical constraint. Considering all these facts and the size of consolidation frame available in the laboratory, a large-scale laboratory set-up was developed to carry out swell consolidation tests (Fig. 1). It consisted of a base plate, 200 mm in diameter and 20 mm thick with two diametrically opposite inlet valve and drainage channels. Further, it had outer

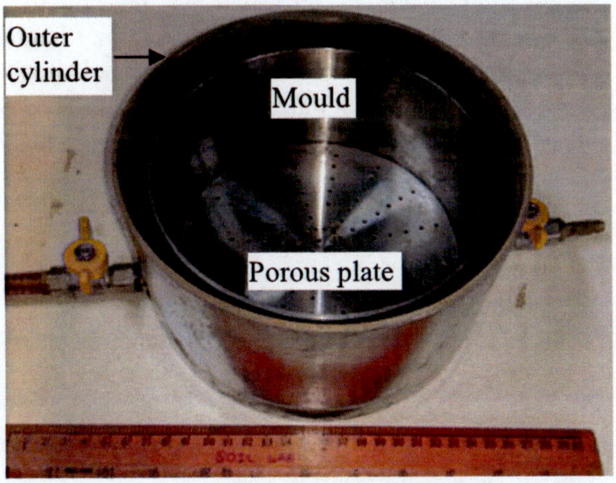

Fig. 1 Large swell consolidation cell

cylinder argon welded over base plate (to prevent water leakage) which was 90 mm in height and 3 mm thick with an external diameter of 200 mm.

Soil was compacted in the inner cylinder resembling soil mould which was 80 mm in height, 3 mm thick with an outer diameter of 170 mm. A solid steel plate of diameter 170 and 10 mm thick with pre-drilled holes was used at the top and bottom both as replacement for porous stone and loading plate. Taking into consideration, the increased diameter of soil sample, weights were accordingly fabricated to get the standard load intensities during loading phase.

3.3 Sample Preparation and Testing Procedure

To prepare test samples, the calculated amount of oven-dried soil passing 0.425 mm was hand mixed with required quantity of water to reach desired water content and kneaded to form a uniform mix. The soil sample was then wrapped in cling film sheet and kept in desiccator for 24 h for uniform distribution of water. In the case of sample from desiccator, it was compacted in the oedometer ring in three layers of equal thickness. In order to introduce vertical sand drains, compacted soil specimen was drilled at desired location throughout the depth with the help of a drill bit, 3 mm in diameter. Fine sand was then poured into the columns using paper cone. With all components in place, whole assembly was then mounted on loading frame and a seating load of 6.25 kPa (Sridharan et al. 1986) was kept on the loading yoke and water is allowed to enter into the soil specimen through water reservoir system placed at the specimen height. Cumulative heave was recorded through a dial gauge till the saturation heave occurs and subsequently the specimen is loaded in stages to get the swell pressure.

For conducting tests in developed large-scale swell consolidation apparatus, wet soil from the desiccator was compacted into the inner cylinder in three layers up to a height of 56.67 mm. In this way, it was possible to maintain the ratio of specimen's height and diameter as 1:3 and axial symmetric conditions as well. Custom made plywood with pre-punched holes at required spacing and orientation was used as a template to make columns into the compacted soil specimen with the help of a thin casing pipe of diameter 15 mm and filled with suitable column material. After placing non-woven geotextile on either side of the soil sample, geostrip was knotted at the bottom of the column along with the washer and anchored at the top of the column by switchboard screw assembly. Washer was supplied with a 3 mm hole in the centre for geostrip to pass through and 1 mm sized six holes all over the surface to facilitate the passage of water. This whole assembly was then placed inside the cell with a drainage layer of sand at top and bottom. In order to avoid stress concentration of loading plate over these switchboard screws, sand layer was laid at the top until the height of screws was matched. After laying the top sand layer, loading plate was placed over it and the whole assembly was mounted on loading frame for swell consolidation test.

Table 2 Swell consolidation test details (all tests were performed on black cotton soil compacted at OMC and MDD)

Test legend	No. of anchors	No. of columns	Column material	Column spacing	Nomenclature
T1	–	–	–	–	BCS
T2	7	7	Black cotton soil	3D	BCS + 7GRSC
T3	7	7	95% fly ash + 5% lime	3D	BCS + 7GRFC

BCS Black cotton soil, *GRSC* Geostrip reinforced soil column, *GRFC* Geostrip reinforced fly ash-lime column, *D* = Diameter of column installed

3.4 Testing Programme

The whole experimental investigation was a two-phase programme. In phase I, a series of conventional swell consolidation tests were performed to analyse the effect of dry unit weight, moulding water content and introducing vertical sand drains on swell parameters of compacted soil specimen. Soil sample was compacted into the ring at three different dry unit weights of 12.65, 11.77 and 11.28 kN/m^3 (1.29, 1.20 and 1.15 g/cc) at same water content of 34.5% to monitor effect of dry unit weight on swell parameters. Swell potential and swell pressure of dry soil powders were compared with that of soil compacted at OMC and MDD to examine the effect of moulding water content. Further, to achieve uniform water distribution across the depth of soil sample, nine sand drains of 3 mm diameter were introduced and consequently change in swell parameters were noted.

Now that the factors affecting the swell parameters of the soil have been evaluated, efforts were made in phase II to restrain the swell potential and consequently swell pressure of the compacted soil specimen by introducing geostrip reinforced fly ash-lime columns. All the tests were performed at OMC and MDD. Triangular arrangement for the columns was opted with column diameter (*D*) of 15 mm. Diameter of washer plate was kept as 30 mm (2D). For the chemical treatment, combination of 5% lime "(Dash and Hussain 2011; Jha and Sivapullaiah 2015)" and 95% fly ash was chosen as the column material. Table 2 presents the details of the swell consolidation tests performed.

4 Test Results and Discussion

4.1 Effect of Moulding Water Content

Figure 2 shows the significance of initial water content on swell potential of black cotton soil sample compacted at a dry unit weight of 12.65 kN/m^3. Final heave

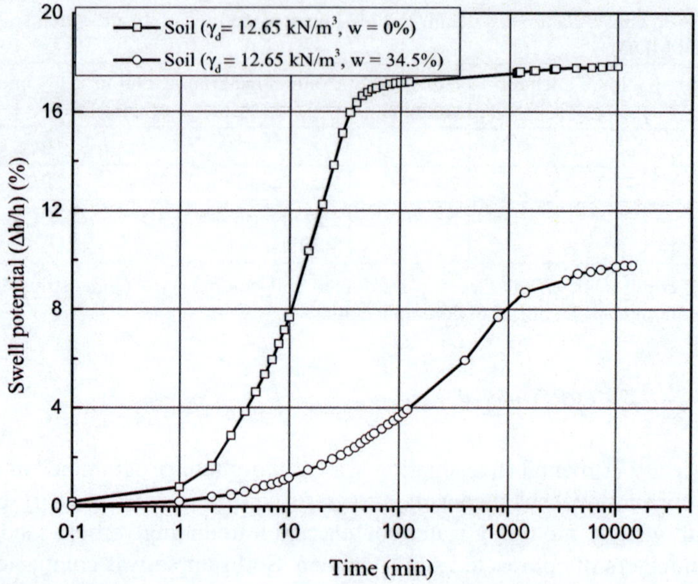

Fig. 2 Variation of swell potential versus log time with water content

decreased from 4.46 to 2.44 mm (45.29% reduction) with increase in the water content from 0 to 34.5%.

It can be also observed from Fig. 2 that swelling takes place in three phases, namely initial, primary and secondary phase. For dry soil sample, initial and primary phase do not last long and it attains saturation heave within 3 days as compared to the soil compacted at optimum moisture content whose saturation heave occurred after 7 days. Variation in swell pressure with change in water content can be seen in Fig. 3. Swelling pressure reduced from 285 to 260 kPa (8.7% reduction) as water content increased from 0 to 34.5%

4.2 *Effect of Dry Unit Weight*

Figures 4 and 5 compare swell potential and pressure of soil specimen compacted at different dry densities with same water content of 34.5%. Swell potential reduced from 9.76 to 7.2% as the dry unit weight was decreased from 12.65 to 11.28 kN/m^3.

Swelling pressure was also found to decrease with decrease in dry unit weight (Fig. 5). It came down from 260 to 195 kPa as the dry unit weight decreased from 12.65 to 11.28 kN/m^3. This could be attributed to the fact that at lower density, fewer number of soil particles occupy unit volume and consequently exerts less swell pressure on imbibition of water. Swelling pressure was also found to decrease with decrease in dry unit weight (Fig. 5). It came down from 260 to 195 kPa as the

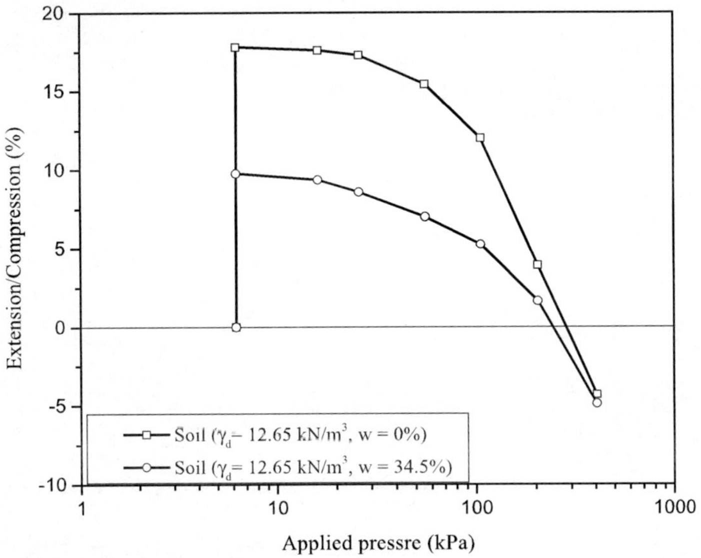

Fig. 3 δ-log *p* relationship for soil with water content

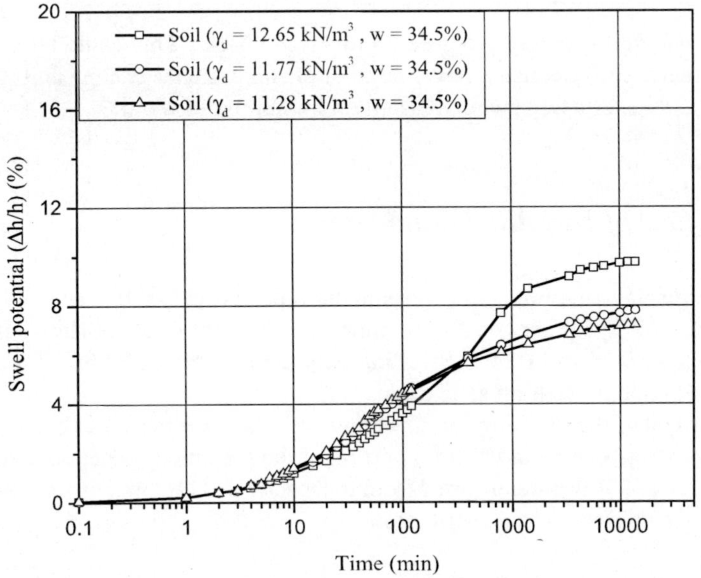

Fig. 4 Variation of swell potential versus log time with dry unit weight

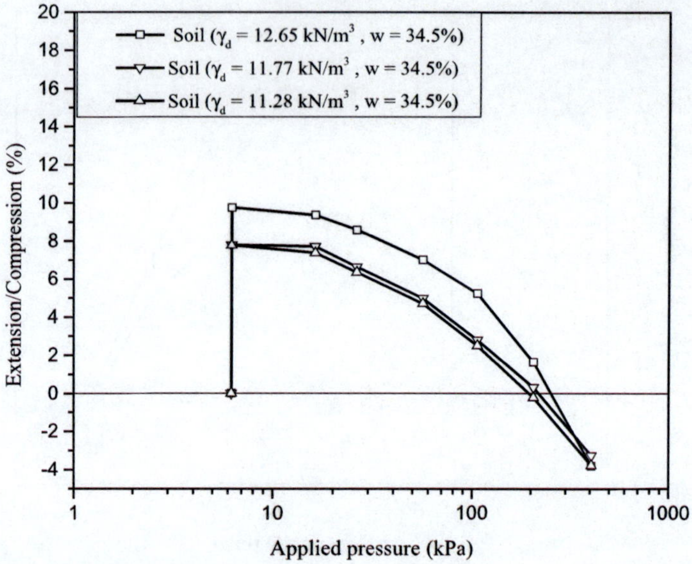

Fig. 5 δ-log *p* relationship for soil at different dry unit weight

dry unit weight decreased from 12.65 to 11.28 kN/m^3. This could be attributed to the fact that at lower density, fewer number of soil particles occupy unit volume and consequently exerts less swell pressure on imbibition of water.

4.3 Effect of Installing Sand Drains

Figure 6 displays the impact of introducing nine vertical sand drains in compacted expansive soil specimen. With the nine sand drains, swell potential marked an increase from 9.76 to 11.24%. Swelling pressure also increased by 17.3% (260 kPa to 305 kPa) upon installing sand drains (Fig. 7).

At the end of the test, three soil samples were kept for the water content determination from top, centre and bottom portion of the specimen. Water content of 42.12, 40.37 and 44.59% was determined from top, centre and bottom portion, respectively, indicating a uniform water distribution.

4.4 Effect of Combined Chemical–Mechanical Treatment

Table 3 summarizes the results of the large-scale swell consolidation tests carried out in order to restrain swelling of expansive soil. In order to evaluate the combined effect of chemical and mechanical restraint in arresting the swell potential of expansive soil

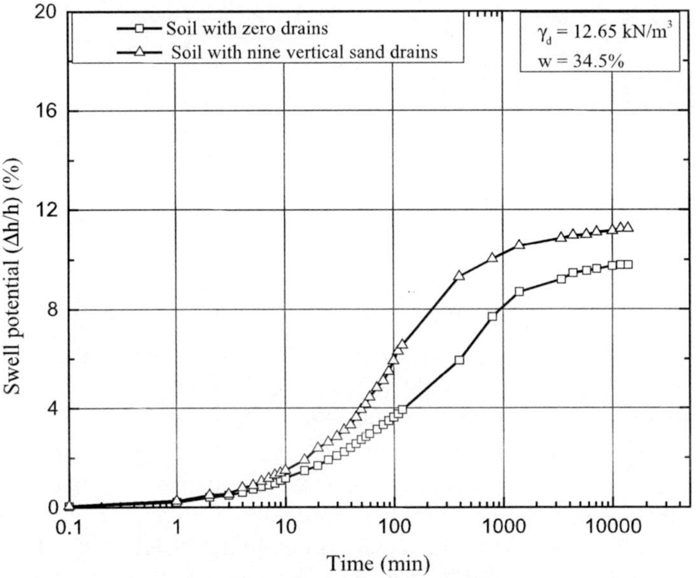

Fig. 6 Variation of percentage swell versus log time for soil with sand drains

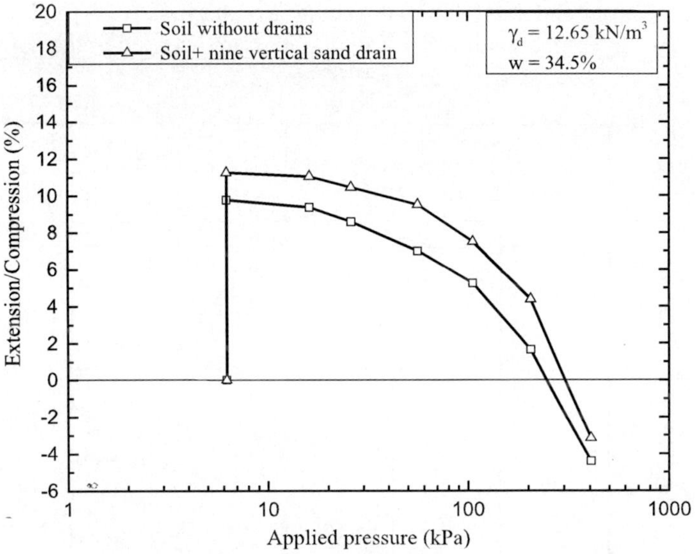

Fig. 7 δ-log p relationship for soil with sand drains

Table 3 Summary of large of large-scale swell consolidation test results

Legend	Test details	Column spacing	Swell potential (%)	% increase/decrease	Swell pressure (kPa)	% increase/decrease
T1	BCS	3D	11.35	Benchmark	265	Benchmark
T2	BCS + 7GRSC	3D	5.1	55.06 (–)	95	64.15 (–)
T3	BCS + 7GRFC	3D	2.41	78.70 (–)	50	81.13 (–)

sample, the installed columns were filled with 95% fly ash and 5% lime mix and were further anchored with geostrip. Swell potential declined from a value of 11.35–2.41% displaying an overall reduction of 78.70% (Fig. 8). Remarkable alleviation of 81.13% in swell pressure was also observed which dropped down from 265 to 50 kPa (Fig. 9)

This reduction in swell parameters can be attributed to the fact that as the swelling of the soil occurs, the polyester fibres in the geostrip are stretched and thus the geostrip arrests swelling of soil due to the tensile strength mobilized in it. Additionally, fly ash and lime brings appreciable changes in the chemical environment around soil particles.

Fly ash supply divalent and trivalent cations into the soil matrix in exchange for monovalent cations and thus leads to flocculation of dispersed clay particles. Further, as lime comes in contact with water, calcium ions are released in the pore fluid as a

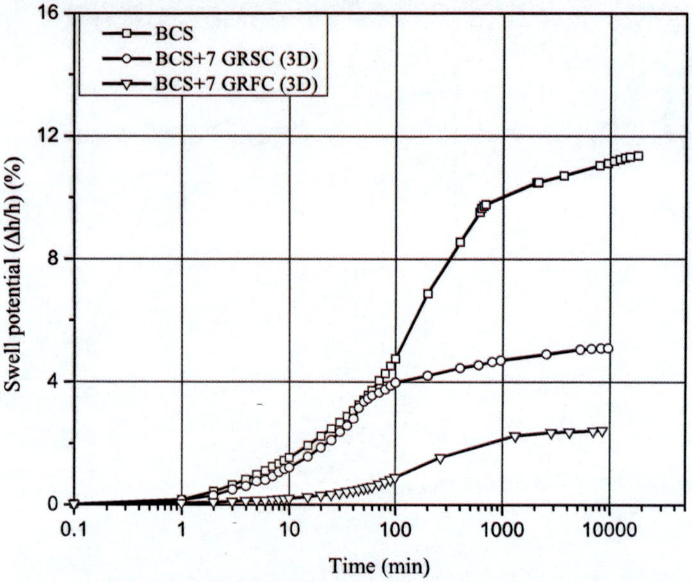

Fig. 8 Variation of percentage swell versus log time for soil with and without combined treatment

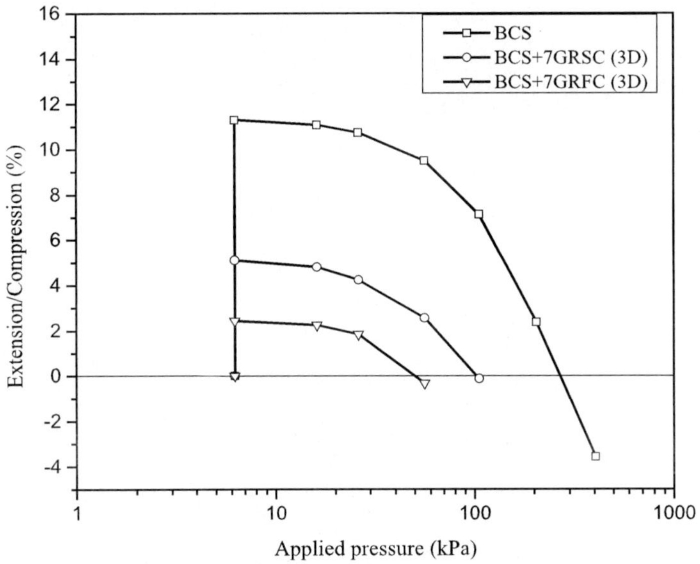

Fig. 9 δ-log p relationship for soil with and without combined treatment

result of which electrolytic concentration of pore fluid ameliorates, and thus thickness of diffuse double layer gets suppressed which in turn reduces the expansiveness of the soil.

5 Conclusions

From the experimental study performed, the following conclusions can be drawn,

1. For soil samples compacted at same dry unit weight, initial moisture content plays an important role in determining the swell parameters of soil sample. At same dry unit weight, swell potential increases with decrease in initial water content.
2. For soil sample compacted at same water content, dry unit weight at which it is compacted too plays a vital role in determining swelling behaviour of the soil sample. By increasing the dry unit weight, we allow a greater number of soil particle to fit in unit volume which in turn exerts large pressure upon water imbibition and thus swell potential and consequently swell pressure increases.
3. Introduction of vertical drains has a remarkable influence on swell parameters of soil. The percentage swell and swell pressure increased when soil sample was drilled with nine vertical sand drains. Swell potential increased by 15.16% and swell pressure marked an increase of 17.30% owing to the uniformity in water distribution throughout the sample.

4. With combined mechanical and chemical treatment, swell potential and pressure decreased by 78.70% and 81.13%, respectively, displaying the effectiveness of the technique. Thus, this could turn out to be an effective partial replacement technique to in situ restrain swelling of expansive subgrades.

Acknowledgements The authors would like to acknowledge the efforts and support provided by the staff of Geotechnical soil laboratory, Indian Institute of Technology Bombay, Mumbai, during the entire experimental study.

References

Basma AA, Al-Homoud AS, Husein A (1995) Laboratory assessment of swelling pressure of expansive soils. Appl Clay Sci 9(5):355–368

Bell FG (1988) Stabilization and treatment of clay soils with lime. Part II: some applications. Ground Eng 21(2):22–30

Chen FH (1988) Foundations on expansive soils. Elsevier Scientific Publishing Company, Amsterdam, pp 74–80

Cokca E (2001) Use of class C fly ashes for the stabilization of an expansive soil. J Geotech Geoenviron Eng 127(7):568–573

Dash SK, Hussain M (2011) Lime stabilization of soils: reappraisal. J Mater Civ Eng ASCE 24(6):707–714

Ikizler SB, Aytekin M, Vekli M (2009) Reductions in swelling pressure of expansive soil stabilized using EPS geofoam and sand. Geosynth Int 16(3):216–221

Jha AK, Sivapullaiah PV (2015) Mechanism of improvement in the strength and volume change behavior of lime stabilized soil. Eng Geol 198:53–64

Katti RK (1978) Search for solutions to problems in black cotton soils, 1st annual lecture. In: Proceedings of the Indian geotechnical conference, Indian Geotechnical Society, New Delhi V

Kolay PK, Ramesh KC (2016) Reduction of expansive index, swelling and compression behaviour of kaolinite and bentonite clay with sand and class C fly ash. Geotech Geol Eng 34(1):87–101

Maher MH, Ho YC (1994) Mechanical properties of kaolinite/fiber soil composite. J Geotech Eng 120(8):1381–1393

Muthukumar M, Shukla SK (2018) Swelling behaviour of expansive clay beds reinforced with encased granular pile anchors. Int J Geotech Eng 12(2):109–117

Nagaraj HB, Munnas MM, Sridharan A (2009) Critical evaluation of determining swelling pressure by swell-load method and constant volume method. Geotech Test J ASTM 32(4):305–314

Nelson DJ, Miller JD (1992) Expansive soils: problems and practice in foundation and pavement engineering. Wiley, New York

Satyanarayana B (1995) Swelling pressure and related mechanical properties of black cotton soils. Doctoral dissertation, Ph.D. thesis, Indian Institute of Science, Bangalore

Sivapullaiah PV, Sridharan A, Stalin VK (1996) Swelling behaviour of soil bentonite mixtures. Can Geotech J 33(5):808–814

Sridharan ASURI, Rao AS, Sivapullaiah PV (1986) Swelling pressure of clays. Geotech Test J ASTM 9(1):24–33

Viswanadham BVS, Phanikumar BR, Mukherjee RV (2009) Swelling behaviour of a geofiber-reinforced expansive soil. Geotext Geomembr 27(1):73–76

Effect of Bio-enzyme—Chemical Stabilizer Mixture on Improving the Subgrade Properties

C. M. Aswathy, Athira S. Raj, and M. K. Sayida

Abstract Conventional methods for strengthening the subgrade soil are time-consuming and are not economically feasible. In this paper, studies were performed to understand the effect of bio-enzyme, bio-enzyme–fly ash mixture and bio-enzyme–lime mixture on kaolinite clay collected from Thonnakkal, Thiruvananthapuram District, Kerala, India. In this study, a popular bio-enzyme known as terrazyme is used as stabilizer. Unconfined compressive strength tests and California bearing ratio tests were conducted on pure soil and soil mixed with terrazyme and its combinations with lime and fly ash. From the study, it was observed that there is a considerable increase in UCC and CBR value for the treated soil compared to untreated one. The optimum dosage of terrazyme obtained was 0.1 ml/kg and that of lime was 8%. With the increase in fly ash content on soil stabilized with terrazyme, an improvement in UCC value was observed, whereas there was not much improvement in CBR value up to 30% addition of fly ash compared to soil stabilized with terrazyme alone. Hence, the combination of terrazyme and lime is an effective method for stabilizing kaolinitic subgrade soil.

Keywords Kaolinite clay · Bio-enzyme · Terrazyme · Fly ash · Lime · Stabilization · CBR · UCC

1 Introduction

The importance of soil stabilization in road construction cannot be underestimated. Stabilized soil as road base has to bear weight of its future traffic and potential pavement. Numerous soil stabilization techniques have been used throughout history for ensuring a compact and reliable road. Common additives used in stabilizing soil are lime, cement, fly ash, etc. These chemical additives can improve the property of soil but their effect on the environment is damaging. In order to minimize these negative

C. M. Aswathy (✉) · A. S. Raj · M. K. Sayida
Department of Civil Engineering, College of Engineering Trivandrum, Thiruvananthapuram, Kerala, India
e-mail: aswa.cm@gmail.com

M. Latha Gali and R. R. P. (eds.), *Problematic Soils and Geoenvironmental Concerns*, Lecture Notes in Civil Engineering 88,
https://doi.org/10.1007/978-981-15-6237-2_63

effects, recently bio-enzymes have emerged as a new hope for soil stabilization. Bio-enzymes are natural, non-toxic, non-corrosive, liquid enzyme formulation fermented from vegetable extracts that improve the engineering properties of soil.

In their work, Puneet and Suneet (2014) used terrazyme to study its effect on the black cotton soil's unconfined compressive strength. The test resulted in a substantial increase in the unconfined compressive strength (UCS) of black cotton soil to 200%. The optimal terrazyme dosage for black cotton soil UCS improvement is 1 ml/per 5 kg of soil. Terrazyme soil treatment period has played a vital role in improving the quality and strength of the soil treated with terrazyme for 7 days.

Saini and Vaishnava (2015) conducted experiments with four separate terrazyme ratios on local soil treated. Terrazyme interacts with absorbed clay particle layer of water, thus increasing the thickness around soil surface. This results in the reduction of voids between soil particles, resulting in a closer orientation of soil particles with low compaction. Experimental analyzes are conducted taking into account dose and healing time.

Ramesh and Sagar (2015) examined the drying effect on the strength and other properties of terrazyme-stabilized expansive and non-expansive soils. Terrazyme stabilization showed good results in both black cotton soil and red earth engineering properties. Optimum dosage of terrazyme for black cotton soil is 200 ml/2.0m^3 and 200 ml/3.0m^3 terrazyme for red earth. Panchal et al. (2017) performed various tests on local soil sample with and without enzyme. Terrazyme improves the CBR values in road construction. Tests were repeated with mixing of local soil sample with different dosages of terrazyme with different curing periods. Period of treatment of bio-enzymatic soil played an important role in the improvement of the strength of soil. Ogundipe and Moses (2013) conducted tests on clayey soil with lime. 2–10% lime were added and tested. Optimum lime content is obtained as 8%.

Dimitris and Xiaoguang (2003) conducted experiments on the utilization of fly ash for stabilization/solidification of heavy metal contaminated soils. For the artificial soil mixes, two different types of clay were used: kaolinite and montmorillonite. The addition of fly ash during the quick lime–sulphate treatment of heavy metal contaminated soils is mainly responsible for their effective immobilization. Pandian et al. (2002) investigated the impact of two fly ash types Raichur fly ash (Class F) and Neyveli fly ash (Class C) on the black cotton soil's California bearing ratio (CBR). The amount of fly ash has been raised from 0 to 100%. Due to the dominance of clay fraction, the low CBR of black cotton soil is due to the inherent low strength. The addition of Raichur fly ash to soil increases the mix's CBR, resulting in a decrease of up to 60% to the first optimum level and the addition of fly ash beyond the optimum level. There is also an increase in strength in Neyveli fly ash with the rise in fly ash content, with additional pozzolonic reaction forming cemented compounds, resulting in strong binding between black cotton soil and fly ash particles.

From the data obtained from numerous literature reviews, it can be evaluated that a combination of bio-enzyme and chemical stabilizer has not conducted any studies. The aim of this paper is to study the effect of bio-enzyme with chemical stabilizers such as lime and fly ash. This will reduce costs and is also eco-friendly by combining bio-enzyme with chemical stabilizers.

Table 1 Characteristics of soil used for the study

Properties	Values
Specific gravity	2.4
Liquid limit (%)	51
Plastic limit (%)	27
Plasticity index (%)	24
Maximum dry density (g/cc)	1.52
Optimum moisture content (%)	28
Sand (%)	13
Silt (%)	50
Clay (%)	37
Unconfined compressive strength (N/mm^2)	131.7

2 Materials Used

2.1 Soil

The soil is collected from Thonnakkal Region, Thiruvananthapuram, which belongs to kaolinite category. Various laboratory tests were done to find soil properties such as specific gravity, Atterberg limits, optimum moisture content, maximum dry density, unconfined compressive strength and California bearing ratio. The soil is classified as MH based on plasticity chart. The physical and compaction characteristics of soil are summarized in Table1.

2.2 Terrazyme

Terrazyme is a natural, non-toxic, non-corrosive and non-inflammable liquid, produced by formulating vegetable extracts. They are perfectly soluble in water, brown in colour with smell of molasses. The application of terrazyme enhances weather resistance and increases load-bearing capacity of soil. This decreases the swelling capacity of the soil particles and reduces permeability. It increases the CBR values and density and decreases the OMC and plasticity index of the soil. Terrazyme changes the hydrophilic nature of clay materials to hydrophobic. Chemical composition of terrazyme is shown in Table 2.

Table 2 Chemical composition of terrazyme

Properties	Values
Specific gravity	1.05
pH value	3.50
Total dissolved salts	19.7 ppm
Hazardous content	None
Boiling point	212 °F
Evaporation rate	Same as water
Solubility in water	Complete
Reactivity data	Stable

2.3 *Fly Ash*

The fly ash used in this study was collected from Thermal Power Plant, Tuticorin. The specific gravity was obtained as 1.9. The fly ash is of class F category. The constituents of fly ash are given in Table 3.

3 Methodology

First aim is to find out the optimum terrazyme content. The index properties of the material are found out initially. The different dosages of terrazyme are given in Table 4.

An optimum percentage of terrazyme content was found out as 0.1 ml/kg of soil. Then in addition to the optimum content of terrazyme, lime is added in steps of 6, 8 and 10% and fly ash in steps of 10, 20 and 30% to soil to improve the strength properties. Light compaction test (IS: 2720, part 7, 1980), unconfined compressive strength test (IS: 2720, part 10, 1991) and California bearing ratio test (IS: 2720,

Table 3 Constituents of fly ash

Constituents	Quantity (%)
Sand	23
Silt	70
Clay	7

Table 4 Dosage rates of terrazyme

Dose	ml/kg of soil
Dosage 1	0.05
Dosage 2	0.1
Dosage 3	0.15

part 16, 1979) were the different tests performed in the laboratory. All the tests were done with different percentages of stabilizers. Specimens were prepared near the optimum water content and maximum dry density of different mixes. No curing period is allowed.

4 Results and Discussions

4.1 Optimum Terrazyme Content

To find out the optimum terrazyme content, CBR tests were carried out with different dosages of terrazyme. The load penetration curves for the samples at three different terrazyme dosages are given in Fig. 1. Variation of CBR values with different dosages of terrazyme is shown in Table 5.

With the addition of different dosage of terrazyme, the CBR value was observed to increase. It was seen that the maximum CBR value is obtained when the stabilizer dosage is 0.1 ml/kg. The CBR value increases by 8.5 times for the sample stabilized with 0.1 ml/kg terrazyme than the untreated sample.

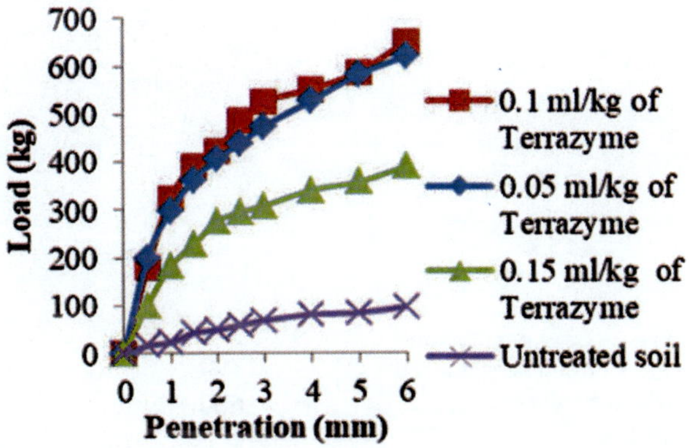

Fig. 1 Load penetration curve of sample stabilized with different dosages of terrazyme

Table 5 CBR values with different dosages of terrazyme

Terrazyme dosage(ml/kg)	CBR value (%)
0.0	4.1
0.05	31.99
0.1	35.55
0.15	21.56

4.2 California Bearing Ratio Test

To find out the optimum percentage of lime in lime–terrazyme mixture and fly ash–terrazyme mixture, CBR tests were carried out with different dosages of lime and fly ash. Variation of CBR value with respect to percentage of lime and fly ash is given in Table 6. The load penetration curves for the samples at three different percentages of lime and fly ash are given in Figs. 2 and 3, respectively.

With the addition of different percentages of lime with optimum terrazyme content, the CBR value was observed to increase. It was seen that the maximum CBR value is obtained when the percentage of lime is 8%. The CBR value increases by 22 times for the sample stabilized with 0.1 ml/kg terrazyme and 8% lime than the untreated sample. But with the addition of different percentages of fly ash to the optimum terrazyme content, CBR value shows an increase in trend than unstabilized

Table 6 Variation of CBR value with respect to percentage of lime and fly ash

Dosage of terrazyme (ml/kg)	Chemical stabilizers	Percentage (%)	CBR value (%)
0.1	Untreated soil	–	4.1
	Terrazyme treated soil	–	35.5
	Lime	6	73.466
		8	90.29
		10	71.09
	Fly ash	10	10.74
		20	21.48
		30	28.91

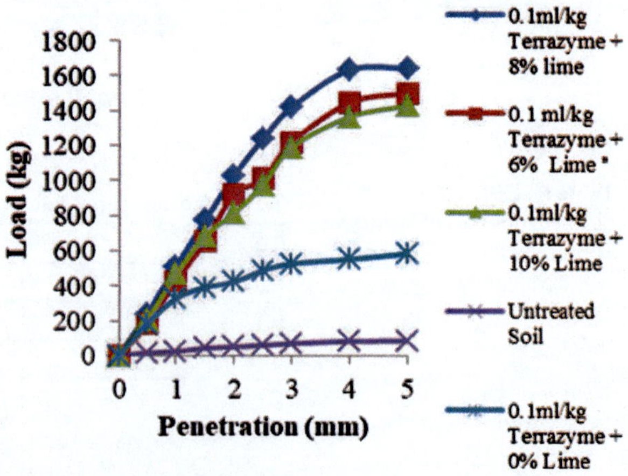

Fig. 2 Load penetration curve of sample stabilized with different dosages of lime

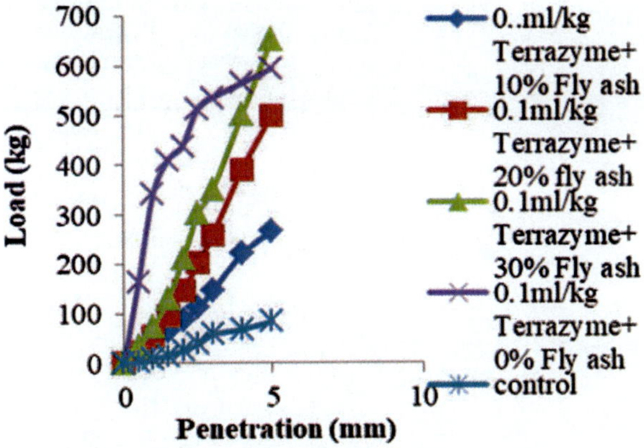

Fig. 3 Load penetration curve for different % fly ash and 0.1 ml/kg of terrazyme

soil, but not more than terrazyme stabilized soil. From the graph, it is clear that as the percentage of fly ash increases the CBR value also increases.

In the study of fly ash with terrazyme when mixed with kaolinite clay, the CBR value shows an increase from 4.1% to 28.91% (for 30% fly ash) but less than the CBR value 35.55%, obtained by adding 0.1 ml/kg of terrazyme alone with soil. From the data obtained from the literature review, it reveals that fly ash of C category has extra lime content, so it can impart more strength than class F fly ash. When soil blended with terrazyme alone, the strength of soil increased much higher than when it mixed with terrazyme and fly ash. Also, it can be observed that curing makes better results.

4.3 Unconfined Compressive Strength Test

Unconfined compressive strength (UCC) test was conducted with different percentages of lime and fly ash. Variation of UCC strength values with different dosages of lime and fly ash is shown in Table 7. The stress–strain curve of the samples at different percentages of lime and fly ash is shown in Figs. 4 and 5, respectively.

With the addition of different percentages of lime with optimum terrazyme content, the UCC strength was observed to increase. It was seen that the maximum strength is obtained when the percentage of lime is 8%. At 10% value tends to decrease. Increase in CBR value and UCC strength is due to the gradual formation of cementitious compounds in the soil as a result of the pozzolanic reaction between lime and clayey soil.

But with the addition of different percentages of fly ash with optimum terrazyme content, the UCC strength was observed to increase. It was seen that the strength is increased as the percentage of fly ash increases.

Table 7 Variation of UCC strength with respect to percentage of lime

Dosage of terrazyme (ml/kg)	Chemical stabilizers	Percentage (%)	UCC strength (N/mm^2)
0.1	Untreated soil	–	131.70
	Terrazyme treated soil	–	361.42
	Lime	6	478.26
		8	507.43
		10	445.3
	Fly ash	10	423.88
		20	457.17
		30	483.84

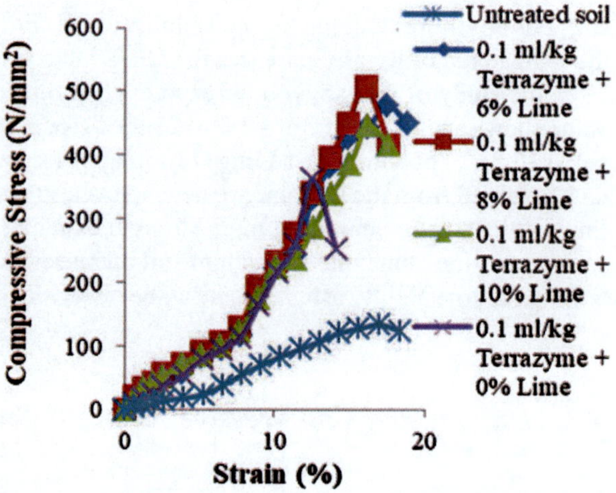

Fig. 4 Stress–strain curve of sample stabilized with different dosages of lime

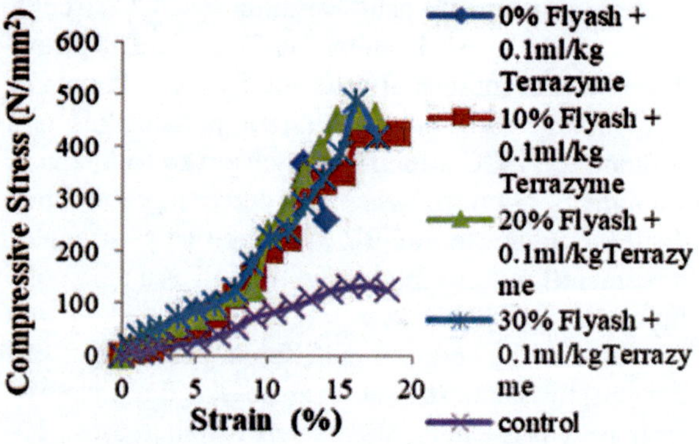

Fig. 5 Load penetration curve of sample stabilized with different dosages of fly ash–terrazyme (0.1 ml/kg) mixture

5 Conclusion

The optimum terrazyme content is obtained as 0.1 ml/kg. UCC strength for terrazyme treated clay is 361.42 N/mm^2, and CBR value is 35.55%. Optimum percentage of lime in lime–terrazyme mixture is 8%. After 8%, CBR value and UCC strength tend to decrease. The CBR value for the optimum is 90.29%, and UCC strength is increased to 507.43 N/mm^2. Fly ash–terrazyme mixture shows an increasing CBR value and UCC strength with increase in percentage. But CBR value of terrazyme treated soil is 35.55%, which is more than that of soil treated with initial percentage of fly ash–terrazyme mixture. For 30% fly ash–terrazyme mixture, UCC value obtained is 483.84 N/mm^2. As the percentage of fly ash increases, UCC value for fly ash–terrazyme treated soil also increases. Terrazyme-treated soil can act as a very good subgrade material without adding any other chemical stabilizer.

References

Dermatas D, Meng X (2003) Utilization of flyash for stabiliza tion/solidification of heavy metal contaminated soil. Eng Geol 70(3):377–391

Ogundipe, Moses O (2013) An investigation into the lime-stabilised clay as subgrade material. Int J Sci Technol Res 2(10):82–86

Pandian NS, Krishna KC, Leelavathamma B (2002) Effect of fly ash on the CBR behaviour of soils. In: Indian geotechnical conference, Allahabad, vol I, pp 183–186

Panchal S, Khan M, Sharma A (2017) Stabilization of soil using bio-enzyme. Int J Civ Eng Technol 8(1):234–237

Punnet A, Suneet K (2014) Effect of bio-enzyme stabilization on unconfined compressive strength of expansive soil. Int J Res Eng Technol 3(5):30–33

Ramesh HN, Sagar SR (2015) Effect of drying on the strength properties of terrazyme treated expansive and non-expansive soil. In: 50th Indian geotechnical conference, vol 14, no 7, pp 452–461

Saini V, Vaishnava P (2015) Soil stabilization using terrazyme. Int J Adv Eng Technol

Strength Properties of Laterite Soil Stabilized with Rice Husk Ash and Geopolymer

Sahana T. Swamy, K. H. Mamatha, S. V. Dinesh, and A. Chandrashekar

Abstract India is an agricultural country producing plenty of rice husks which is mostly used as fuel in the boilers for processing paddy, producing energy through direct combustion or by gasification. In India, about 122 million tonnes of paddy is produced annually and about 20–22% rice husk is generated from paddy and 20–25% of the total husk becomes as rice husk ash. After burning, each ton of paddy produces about 40 kg of rice husk ash. The rice husk creates great environment threat causing severe damage to the land and the atmosphere. Therefore, in this study, rice husk ash was used to stabilize the laterite soil which covers large area in India and elsewhere. Laterite soil was collected from Kodakani, Shimoga district, and was stabilized with varying percentages of rice husk ash, geopolymer, and rice husk-based geopolymer. A series of unconfined compressive strength tests was carried out on the above-specified sample conditions with varied dosages of the stabilizer and curing period. Based on the test results, it was found that the strength of the soil is increased by 2, 3, and 5 times, respectively, with geopolymer, RHA and RHA-based geopolymer as stabilizer with a curing period of 7 days. RHA-based geopolymer can be effectively used as a stabilizer for subgrade stabilization and this technique leads to an eco-friendly sustainable pavement.

Keywords Laterite soil · Stabilization · Rice husk ash · Geopolymer · Strength

1 Introduction

Soils are formed by weathering of rocks and in civil engineering, soil is widely used as construction material for which its strength plays a vital role. The structures

S. T. Swamy · A. Chandrashekar
Department of Civil Engineering, KVG College of Engineering, Sullia, Karnataka, India

K. H. Mamatha · S. V. Dinesh (✉)
Department of Civil Engineering, Siddaganga Institute of Technology, Tumakuru, Karnataka, India
e-mail: dineshsv204@gmail.com

M. Latha Gali and R. R. P. (eds.), *Problematic Soils and Geoenvironmental Concerns*, Lecture Notes in Civil Engineering 88,
https://doi.org/10.1007/978-981-15-6237-2_64

such as buildings, pavements, bridges, dams, etc., are built on soil and the soil has to withstand the load transferred by them. In day-to-day life, pavements are the basic networking agents and the design and construction of pavements with good materials and standards result in longer service life. In pavements, if the subgrade possesses good strength, the material requirement for the construction of successive layers is reduced as the subgrade with higher strength results in thinner pavement section. On the other hand, the pavement associated with weak subgrade is thick which requires increased volume of materials for the construction of successive layers. These days, good quality materials are under scarcity and are not available everywhere. In addition to this, pavements built over weak subgrade are prone to pre-mature failures in the form rutting and fatigue. Also, these days, the disposal of industrial waste products like fly ash, rice husk ash, etc., is of prime concern owing to rapid industrialization. Therefore, by considering these pre-mature failures and scarcity for good quality construction materials, the strength of subgrade needs to be improved which results in stronger, thinner, and economic pavement section. Also, the waste materials should be utilized effectively during stabilization of the soil which results in sustainable environment. Several techniques are available for improving the strength of subgrade, namely stabilization, reinforcement, etc. Under stabilization, the selected soil is mixed either with other selected soil or with binders thereby building a stronger pavement which can support heavy wheel loads and/or higher traffic. Among various methods of stabilization, chemical stabilization is very effective as it produces better strength and durability of the soil compared to other stabilization methods.

The husk produced by rice milling is used as fuel in the rice mills to generate steam for parboiling process and about 25% of the husk is converted into ash which is known as rice husk ash (RHA). As India being a major rice producing country, RHA is available in large quantity.

Geopolymer is an alkali-activated aluminosilicate binder which is formed by reacting raw solids which are rich in silica and alumina with a solution of alkali or alkali salts (Aziz and Mukri 2016). When these two materials are mixed, it will produce a mixture of gels and crystalline compounds which will eventually harden into a new strong matrix (Feng et al. 2004). Polycondensation reaction occurs in a high alkaline environment which reorganizes alumina and silica in a more stable Si–O–Al-type structure resulting in materials with increased mechanical strength (Wang et al. 2005; Rios et al. 2016). Alkaline-activated materials showed better performance as durability and stability can be increased, improvement from mechanical aspect compared to cement, and also, improved bond between the soil particles and binder (Torgal et al. 2012).

The selection of raw material is very important and the use of raw materials such as fly ash, bottom ash, volcanic ash, rice husk ash, palm oil ash, etc., will generally give better results with the geopolymer owing to the silica content present in these raw materials (Grytan et al. 2012).

There are many studies available for the effect of fly ash-based geopolymer on the physical and mechanical properties of soil (Swanepoel and Strydom 2002; Van

Jaarsveld et al. 2002; Bakharev 2005; Chindaprasirt et al. 2007; Detphan and Chindaprasiri 2009). But there are limited studies available on the influence of rice husk ash-based geopolymer on the soil properties (Della et al. 2002; Detphan and Chindaprasiri 2009). Therefore, in the present study, laterite soil was stabilized with rice husk ash-based geopolymer and the effectiveness of the stabilizer in improving the strength of the selected soil was investigated through a set of laboratory experiments. The effect of rice husk ash, geopolymer, and the rice husk ash-based geopolymer was analyzed and is reported.

2 Materials

2.1 Soil

Reddish light brown laterite soil was collected from Kodakani, Soraba Taluk, Shimoga District, Karnataka. The soil was tested for its engineering properties such as index properties, compaction characteristics, and strength characteristics and all the tests were carried out as per relevant IS codes. The engineering properties of the soil are tabulated in Table 1. Figure 1 shows the grain size distribution curve of the soil. Figure 2 shows the compaction curves of the soil under standard and modified proctor conditions. Based on the grain size distribution and plasticity characteristics, the soil is classified as A-2-4 as per HRB classification system and silty sand with group symbol SM as per IS classification system.

Table 1 Engineering properties of the soil

Properties	Laterite soil
Specific gravity	2.73
Atterberg's limits (%)	
Liquid limit	31
Plastic limit	NP
Compaction characteristics	
Standard proctor test	
Maximum dry unit weight (kN/m^3)	19.6
OMC (%)	13.0
Modified proctor test	
Maximum dry unit weight (kN/m^3)	20.0
OMC (%)	11
Unconfined compressive strength (kPa)	
Soaked	82
CBR (%)	
Soaked	6

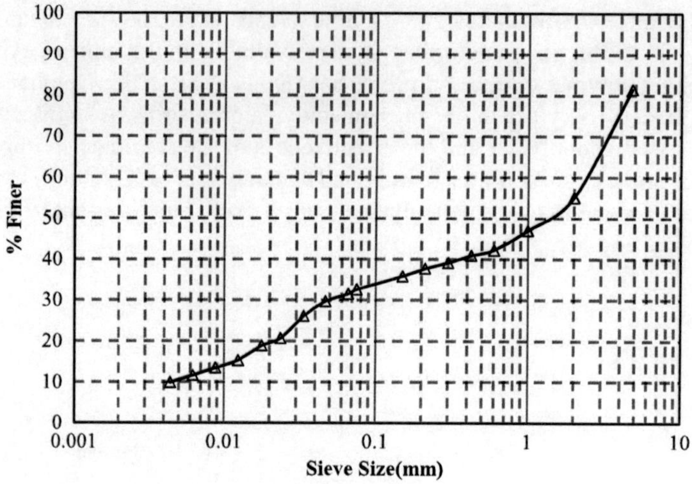

Fig. 1 Grain size distribution curve of the soil

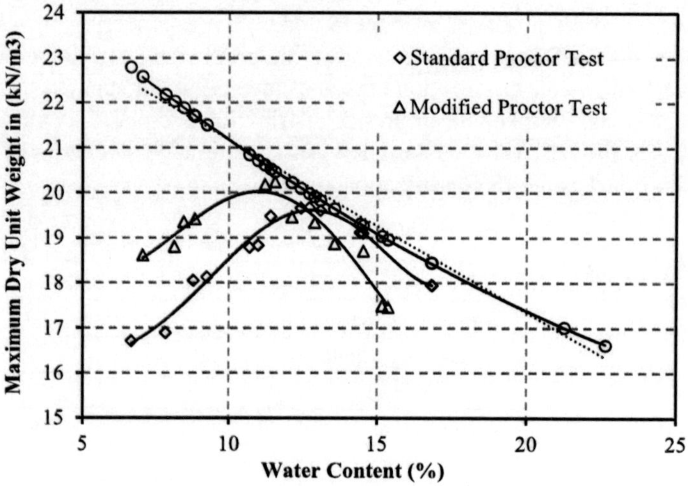

Fig. 2 Compaction curves of the soil

2.2 *Rice Husk Ash*

The rice husk ash (RHA) was collected from the Anthrasanahalli Industrial Area, Tumkur. The temperature at which the husk was burnt decides the quality of the RHA. In this study, the husk was burnt at a temperature of 600 °C and then the RHA was collected. In this study, the RHA was added in varying percentages, viz. 5, 7, 9 and 11% by weight of the soil.

2.3 Geopolymer

The geopolymer consists of Al–O–Si framework. The size of the cluster of geopolymer is of 5–10 nm. The combination of the alumina and silica will make the materials to bind firmly. In this study, geopolymer was prepared by using sodium hydroxide (NaOH) and sodium silicate (Na_2SiO_3).

The calculated amount of NaOH pellets was dissolved in distilled water to prepare 7 M concentration solution and Na_2SiO_3 was also dissolved in water. The ratio of sodium hydroxide to sodium silicate was maintained as 2:1. Both the solutions were prepared separately. In this study, geopolymer was added to the soil in varying percentages, viz. 10, 15, 20, and 25%.

3 Method

In this study, three different combinations were considered for investigation viz., soil + RHA, soil + geopolymer, and soil + RHA + geopolymer. In preparing the RHA stabilized soil, the desired amount of RHA was added to dry soil and mixed thoroughly and then desired amount of water was added followed by mixing to have a homogeneous mixture. In preparing geopolymer stabilized soil, the desired amount of geopolymer was first added to the soil and mixed thoroughly. Then desired amount of water is added followed by mixing to have a homogeneous mixture. In preparing RHA-based geopolymer, the desired amount of RHA was first added to the soil and mixed thoroughly. Then, the desired amount of geopolymer was added and mixed. Desired amount of water was then added to the above mixture and mixed properly to have a homogeneous mixture.

In this study, the Atterberg's limits and compaction characteristics of RHA stabilized soil were determined. Further, unconfined compressive strength of the soil was determined under all the stabilizing conditions. CBR of the soil under optimum stabilized condition was determined. In case of Atterberg's limits and compaction tests, the soil soon after adding the stabilizer was used for the respective tests. In case of UCS, the samples of 3.8 cm diameter and 7.6 cm height were prepared and cured for a period of 7, 14, and 28 days. In preparing the samples with varied percentages of RHA, the respective optimum moisture contents were considered. The unconfined compressive strength test was carried out under soaked condition. For soaking, the sample was covered by a rubber membrane with porous stone at the top and bottom. Then, the whole assembly was placed in a water bath and the water level was maintained as constant for 24 h duration. With this arrangement, water can only enter through the bottom and the samples get soaked through capillary action. At the end of 24 h, the sample along with membrane and porous stones was taken out of the water bath and kept in air for the removal of excess water. Then, the samples were subjected to unconfined compressive strength test. In case of CBR test, the

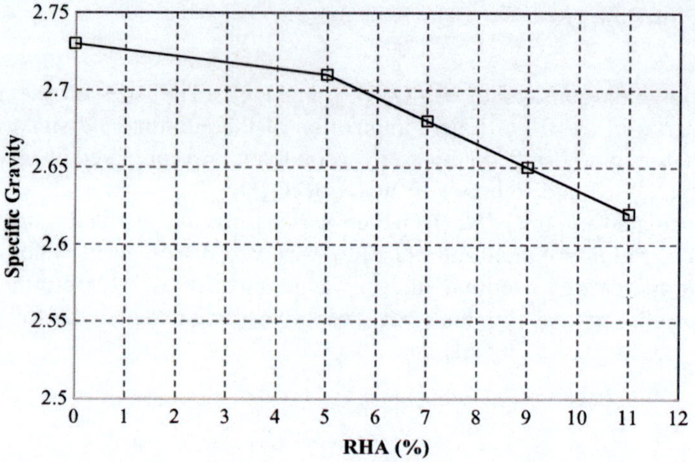

Fig. 3 Variation of specific gravity with percentage of RHA

CBR specimens were prepared, cured for a period of 7 days, and soaked in water for 4 days followed by air drying and testing.

4 Results and Discussions

4.1 Specific Gravity

Figure 3 shows the variation of specific gravity with the percentage of RHA. It is observed that, the specific gravity was reduced with the increased percentage of RHA. The specific gravity of the unstabilized soil is 2.73 and with the addition of 11% RHA, the value reduced to 2.62.

4.2 Atterberg's Limits

Figure 4 shows the variation of liquid limit with percentage of RHA. It is observed that the liquid limit increased with increase in RHA content. The value increased from 33% (unstabilized) to 40% (11% RHA). The increased liquid limit is attributed to increase fines content with the addition of RHA and increased specific surface area of the in stabilized soil.

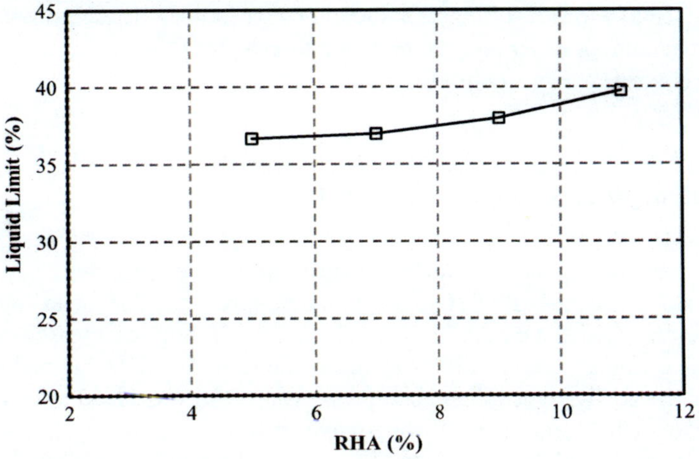

Fig. 4 Variation of liquid limit with percentage of RHA

4.3 *Compaction Characteristics*

Figure 5 shows the compaction curves of RHA stabilized soil under standard proctor condition. The soil was stabilized with 5, 7, 9, and 11% of RHA. It was observed that the maximum dry unit weight decreased and the optimum moisture content increased with the addition of RHA. The reduced dry unit weight is attributed to flocculation process which involves the electrostatic attraction between the positively charged and

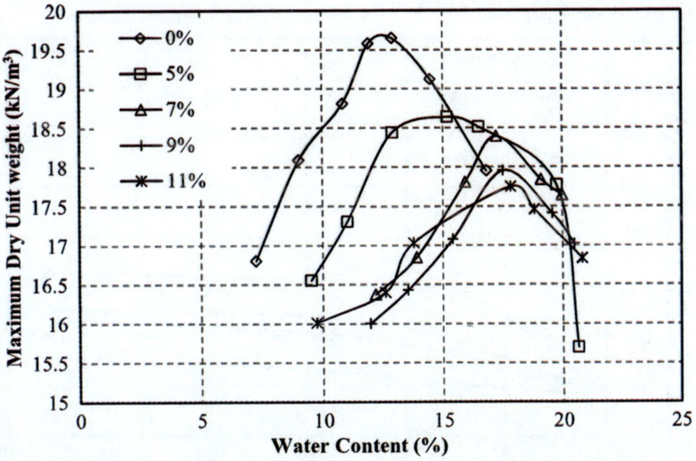

Fig. 5. Compaction curves of RHA stabilized soil under standard proctor condition

negatively charge ions resulting in attraction between fines. These floccules require increased amount of water for its lubrication thereby resulting in increased optimum moisture content.

4.4 Unconfined Compressive Strength

Figure 6 shows the variation of unconfined compressive strength with curing period for RHA stabilized soil with RHA content ranging from 5 to 11% with an increment of 2%. It is observed that the addition of RHA to the soil showed increased unconfined compressive strength and it further increased up to 9% of RHA beyond which the strength reduced. The unconfined compressive strength was found to increase with an increase in the curing period irrespective of RHA content. Up to 9% of RHA addition, it acts as binder and helps in holding the soil particles in dense state. Further addition of RHA results in flocculation thereby the strength reduces. It is found that the maximum strength was attained for curing period of 7 days irrespective of the RHA content and beyond 7 days of curing period, the strength development is marginal. With the addition of 9% of RHA and curing period of 7 days, the unconfined compressive strength increased by 3.5 times that of unstabilized soil. Based on the test results, it is concluded that 9% RHA is the optimum dosage and 7 days is the optimum curing period for stabilizing the selected soil.

Figure 7 shows the variation of unconfined compressive strength with curing period for geopolymer stabilized soil with geopolymer content ranging from 10 to 25% with an increment of 5%. With the addition of geopolymer, the strength of the soil increased and it further increased with an increase in the geopolymer content.

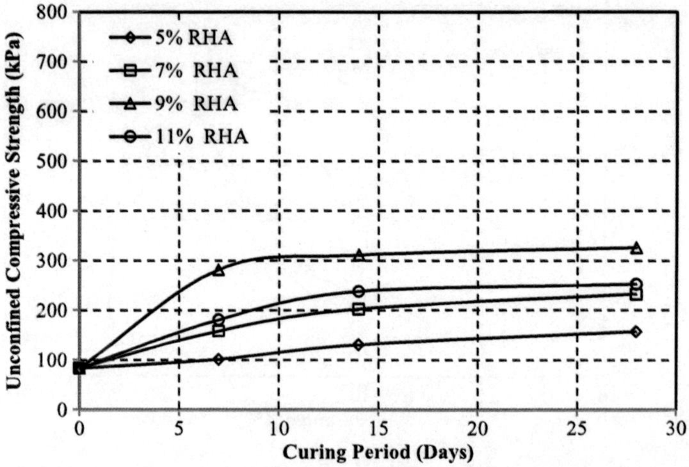

Fig. 6. Variation of unconfined compressive strength with curing period for RHA stabilized soil

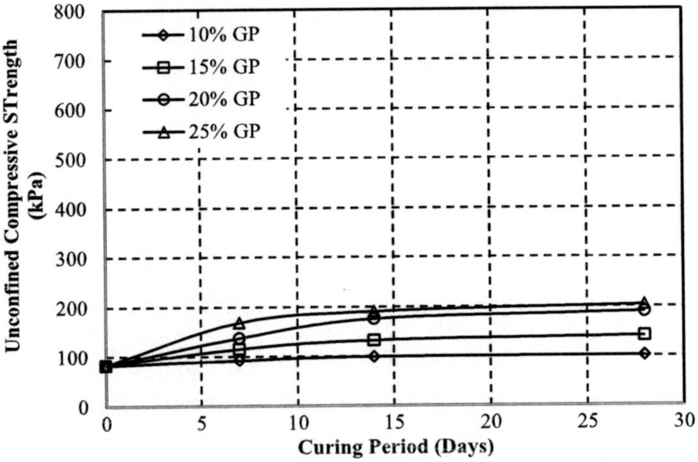

Fig. 7 Variation of unconfined compressive strength with curing period for geopolymer stabilized soil

The maximum strength was attained with a curing period of 7 days irrespective of geopolymer content and beyond 7 days of curing period, the strength development is marginal. With the addition of 25% geopolymer, the strength increased by 2 times when compared to that of unstabilized soil and the increased strength is attributed to the binding action of geopolymer. However, RHA stabilization is effective when compared with geopolymer stabilization.

Figure 8 shows the variation of unconfined compressive strength with curing period for RHA-based geopolymer stabilized soil with varied RHA and geopolymer contents. The addition of RHA-based geopolymer is found to increase the strength of the soil. The soil showed maximum strength with 10% geopolymer addition irrespective of the RHA content and further increase in geopolymer content showed reduced strength. The addition of 9% RHA showed maximum strength irrespective of geopolymer content and curing period. The addition of RHA-based geopolymer to the soil leads to geopolymerization where the small clusters of soil + RHA + geopolymer gather together to form a new covalent bond-framed structure. The addition of 9% RHA with 10% geopolymer increased the unconfined compressive strength of the soil by 5, 8, and 9 times when compared with that of unstabilized soil with a curing period of 7, 14, and 28 days, respectively. Based on the test results, is it concluded that the addition of 9% RHA with 10% geopolymer is effective for soil stabilization. In general, the RHA-based geopolymer soil stabilization is found to be effective when compared with that of RHA stabilization and geopolymer stabilization.

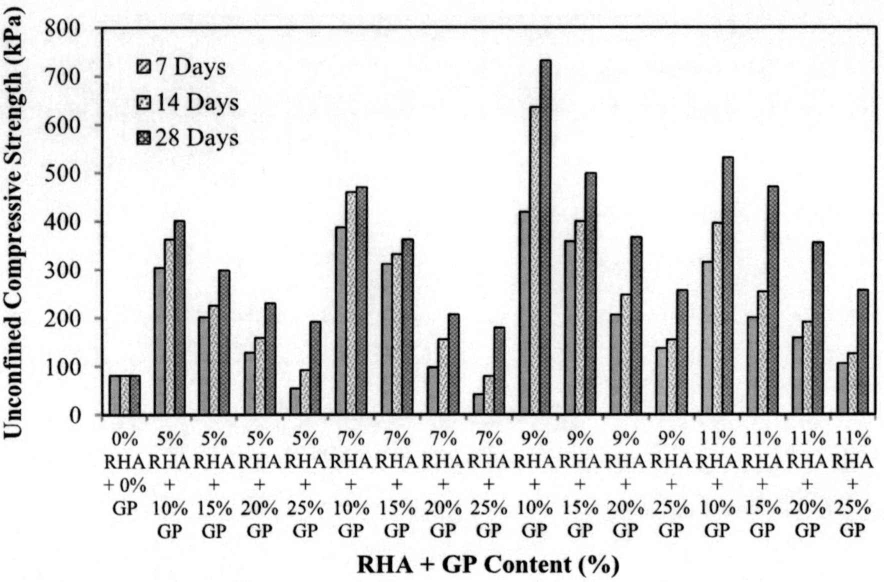

Fig. 8 Variation of unconfined compressive strength with curing period for RHA-based geopolymer stabilized soil

4.5 *California Bearing Ratio (CBR)*

CBR tests were carried out for the stabilized soil under soaked condition. The samples were stabilized with optimum stabilizer content (9% RHA + 10% geopolymer) and cured for a period of 7 days. The CBR of the soil was found to increase from 6% (unstabilized) to 13% (9% RHA + 10% GP stabilized). Based on the test results, it is concluded that the RHA-based geopolymer stabilization can be effectively used for subgrade stabilization for eco-friendly sustainable pavements.

5 Conclusions

A series of laboratory tests were carried out to evaluate the effectiveness of soil stabilization using RHA, geopolymer, and RHA-based geopolymer as stabilizer. Based on the test results, the following conclusions are drawn.

- The addition of RHA, geopolymer, and RHA-based geopolymer to the soil increases the strength of the soil. However, RHA-based geopolymer is a very effective stabilizer when compared to RHA and geopolymer alone.
- The strength of the soil was increased with an increase in the curing period irrespective of the stabilizer type. However, maximum strength is attained with a curing period of 7 days and beyond 7 days, the strength development is marginal.

- With 25% of geopolymer and a curing period of 7 days, the strength of the soil is increased by 2 times when compared with that of unstabilized soil.
- With 9% of RHA and a curing period of 7 days, the strength of the soil is increased by 3 times when compared with that of unstabilized soil.
- With 9% of RHA and 10% of geopolymer, the strength of the soil is increased by 5–9 times when compared with that of unstabilized soil with a curing period of 7 to 28 days.
- RHA-based geopolymer stabilization can be effectively used for subgrade stabilization and this technique leads to an eco-friendly sustainable pavements.

References

Aziz NNSNAB, Mukri M (2016) The effect of geopolymer to the compaction parameter of laterite soil. Middle East J Sci Res 24(5):1588–1593

Bakharev T (2005) Geopolymeric materials prepared using class F fly ash and elevated temperature curing. Cement Concr Compos 35(6):1224–1232

Chindaprasirt P, Chareerat T, Sirivivatnanon V (2007) Workability and strength of coarse high calcium fly ash geopolymer. Cement Concr Compos 29(3):224–229

Detphan S, Chindaprasirt P (2009) Preparation of fly ash and rice husk ash geopolymer. Int J Mater Metall Mater 16(6):720–726

Della VP, Kuhn L, Hotza D (2002) Rice husk ash as an alternative source for active silica production. Mater Lett 57(4):818–821

Feng D, Tan H, Ven Deventer J (2004) Ultrasound enhanced geopolymerization. J Mater Sci 39(2):571–580

Rios S, Ramos C, Antonio V, Cruz N, Rodrigues C (2016) Columbian soil stabilized with geopolymers for low cost roads. In: International conference on transportation geotechnics, vol 143, pp 1392–1400

Sarkar G, Islam Md R, Alamgir M, Rolonuzzaman Md (2012) Interpretation of rice husk ash on geotechnical properties of cohesive soil. Global J Res Eng Civ Struct Eng 12(2):1–7

Swanepoel JC, Strydon CA (2002) Utilisation of fly ash in a geopolymeric material. Appl Geochem 17(8):1143–1148

Torgal FP, Adollagnejad Z, Camoes AS, Jamshidi M, Ding Y (2012) Durability of alkali activated binders: a clear advantage over Portland cement or an unproven issue? Constr Build Mater 30:400–405

Van Jaarsveld JGS, Van Deventer JSJ, Lukey GC (2002) The effect of composition and temperature on the properties of fly ash and kaolinite based geopolymer. Chem Eng J 89(1–3):63–73

Wang H, Li H, Yan F (2005) Synthesis and mechanical properties of metakolinite-based geopolymer. Colloids Surf A 68(1):1–6

Bearing Capacity of Soft Clays Improved by Stone Columns: A Parametric Analysis

Suresh Prasad Singh, Indraneel Sengupta, and Mrinal Bhaumik

Abstract A detailed numerical analysis has been performed on a unit cell stone column by varying parameters such as shear strength of clay, angle of internal friction of the column material, slenderness ratio (*L*/*d*) of the column, area replacement ratio, and the modular ratio, Poisson's ratio of clay and column material. A parametric study is conducted for floating stone column by using the finite element package PLAXIS. A drained analysis was performed using Mohr–Coulomb criterion for both the materials. Validation was done by taking experimental data for a single unit cell stone column. From the analysis, it is observed that the bearing stress increases with increase in the aspect ratio (*L*/*d*) up to a particular range beyond which there is negligible change in the bearing capacity. Bulging is the main cause of failure of a single stone column when loaded alone. An increase in the area ratio caused an increase in the bearing value of the stone column-reinforced soil due to higher relative stiffness. The most influential parameters for the design of stone column-reinforced soil are the angle of internal friction of the column material and area replacement ratio whereas the cohesion of surrounding material, slenderness ratio and Poisson's ratio contribute to a lesser magnitude.

Keywords Unit cell stone column · Parametric analysis · Bearing improvement · Modular ratio · Slenderness ratio

S. P. Singh (✉) · I. Sengupta · M. Bhaumik
Department of Civil Engineering, National Institute of Technology, Rourkela, Odisha, India
e-mail: spsingh@nitrkl.ac.in

I. Sengupta
e-mail: indra3011@gmail.com

M. Bhaumik
e-mail: 217ce1025@nitrkl.ac.in

M. Latha Gali and R. R. P. (eds.), *Problematic Soils and Geoenvironmental Concerns*, Lecture Notes in Civil Engineering 88,
https://doi.org/10.1007/978-981-15-6237-2_65

1 Introduction

Soft clay deposits are mainly found in the marine environment and in India, they are found mainly along the Eastern and Western coastal regions. Soft clay possesses very low bearing capacity, undergoes large settlement, is highly compressible and has very low shear strength. Stone columns are inserted to improve the bearing capacity of the soft clays and to reduce the settlement. They act as a stiffer medium and also accelerate the rate of consolidation, thereby, increasing the shear strength of the surrounding clay. Various researchers have studied the behaviour of stone columns. Ambily and Gandhi (2007), Das and Pal (2013) have conducted model tests on stone columns to understand its behaviour based on the unit cell concept. Hanna et al. (2013) conducted numerical study to identify the mode of failure of a single column and a group of stone columns based on the unit cell concept. Hanna et al. (2014) also developed an analytical model to predict the bearing capacity of reinforced soft soil with stone column based on the limit equilibrium method. The present paper discusses the improvement of soft clay with the introduction of stone columns and the most influential parameters governing the bearing capacity of stone columns. A detailed numerical analysis has been carried out on a single unit cell stone column under a circular rigid raft considering the triangular pattern of stone columns. A parametric study is conducted by using the finite element package PLAXIS. The parameters varied are the shear strength of clay, the angle of internal friction of the column material, slenderness ratio (L/d) of the column, area replacement ratio, and the modular ratio, Poisson's ratio of clay and column material. The diameter of the column was taken as 1 m and the length varied as 2 m, 4 m, 5 m, 10 m, 15 m and 20 m. The spacing is varied as 1.74 m, 2.13 m, 3 m and 4.26 m to have area replacement ratios of 30%, 20%, 10% and 5%, respectively, whereas the undrained shear strength was varied as 10 kPa, 20 kPa, 30 kPa and 40 kPa. Friction angle of the column was varied as 40°, 45° and 50°. Failure pattern of a single column when loaded alone is also assessed and reported in this paper.

2 Validation

For checking the accuracy of the finite element analysis, validation was done on a unit cell model taken from published work done by Ambily and Gandhi (2007). Figure 1 shows the model for the unit cell stone column and Fig. 2 shows the comparison between bearing stress versus settlement curve for $S/d = 2$ and $C_u = 30$ kPa. The properties of the material used for the validation is shown in Table 1.

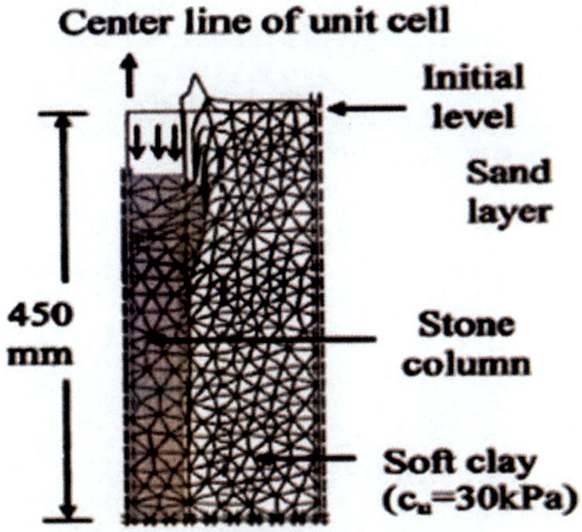

Fig. 1 Finite element model by Ambily and Gandhi (2007)

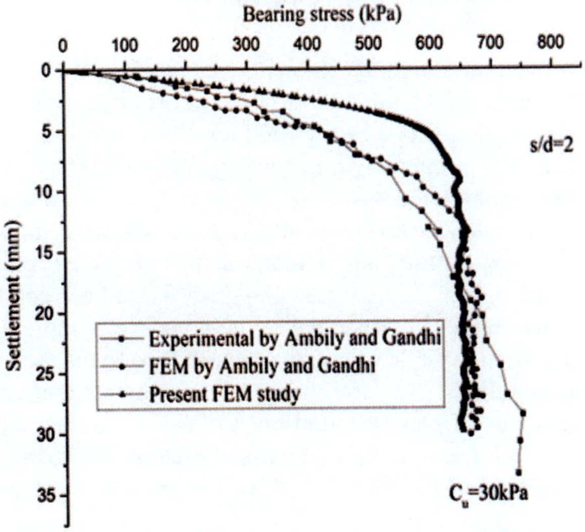

Fig. 2 Validation of PLAXIS by plotting bearing stress settlement profile

Table 1 Properties of materials used

Materials	C_u (kPa)	E (kPa)	μ	ϒ (kN/m^3)	φ (°)	Ψ (°)
Soft clay	30	5500	0.42	15.56	0	0
	14	3100	0.45	14.60		
	7	2150	0.47	13.60		
Stones	0	55,000	0.30	16.62	43	10

Table 2 Properties of materials used for analysis

Materials	C_u (kPa)	E (kPa)	μ	ϒ (kN/m^3)	φ (°)
Soft clay	10 20 30 40	2637 4018 5400 6781	0.46 0.44 0.42 0.40	14.12 14.97 15.52 16.78	0
Stones	0	25,257 50,633 101,505	0.30	15.32 16.73 18.15	40 45 50

3 Parametric Study

3.1 Analysis of Stone Column

Axisymmetric finite element analysis has been carried out considering Mohr–Coulomb criterion. Columns used were of floating nature because the depth of the soft clay deposit was very high and installing end bearing columns would be uneconomical. The circular rigid raft was modelled as plate element. A drained analysis was used for both the materials. The bearing stress was computed by varying parameters such as slenderness ratio, area replacement ratio, modular ratio, Poisson's ratio, friction angle of the column material and undrained cohesion of clay. From the various experimental data, a regression analysis has been conducted to develop a relationship between the undrained cohesion of clay (C_u) and modulus of elasticity of clay (E_c) and friction angle of column material (φ) to the modulus of elasticity of the column material (E_s). The correlation between the undrained cohesion of clay and Young's modulus of clay is found out to be $E_c = 138.15C_u + 1255$ and the relationship between friction angle of column material to Young's modulus of column material is given as $E_s = 96.821e^{0.1391}\varphi$. The properties of the material are as given in Table. 2.

3.2 Effect of Slenderness Ratio (L/d)

It is observed that with an increase in the slenderness ratio (L/d), the bearing stress increases up to a particular extent beyond which there is a negligible increase in the bearing stress. The slenderness ratio was varied for a particular value of the angle of internal friction, the area replacement ratio and the undrained cohesion of clay. The variation of slenderness ratio with bearing stress for varying modular ratio with constant S/d and angle of internal friction is as shown in Fig. 3. It is observed that there is a hump in the graph plotted between L/d and bearing stress. The hump determines the fact that after attaining a particular length of the stone column, with

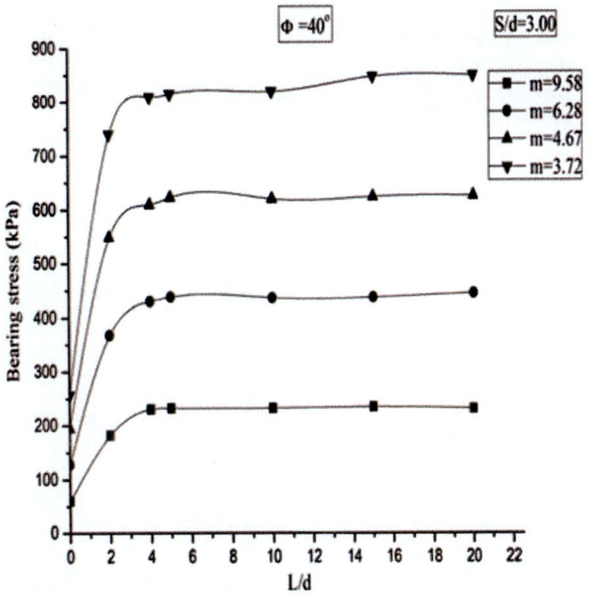

Fig. 3 Variation of L/d with bearing stress for $S/d = 3.00$ and $\varphi = 40°$

further increase in the length, there is a negligible increase in the bearing stress. The point at which negligible change in the bearing stress is found can be taken as the optimal length of the stone column. The optimal length varies for different modular ratios and area replacement ratios. Figures 4 and 5 show the variation of slenderness ratio with bearing stress for S/d values of 3 and 4.26, respectively. Comparing both the figures, it is observed that the bearing stress is higher for S/d ratio of 3 as compared to that of 4.26. This is due to the fact that with increase in the S/d, there is increase in the relative stiffness due to which the confinement in the reinforced soil increases and the soil can take higher load. Higher S/d ratio indicates lower area replacement ratio, which is defined as the area of stone column to the area of the unit cell.

3.3 Effect of Poisson's Ratio

Poisson's ratio is defined as the ratio of lateral strain to that of the longitudinal strain. Bearing stress decreases linearly with increase in the Poisson's ratio. This is attributed to the reason that with relatively higher Poisson's ratio, clay becomes more compressible in the lateral direction accommodating more deformation produced by the columns. Figure 6 shows the variation of bearing stress with Poisson's ratio for varying slenderness ratio, area replacement ratio of 5% and friction angle of 50°. Similar variations are observed for different area replacement ratios and friction angle of the column material. It is observed that clay having Poisson's ratio of 0.46 shows much lower bearing stress value as compared to a clay having Poisson's ratio of 0.40.

Fig. 4 Variation of L/d with bearing stress for $S/d = 4.26$ and $\varphi = 40°$

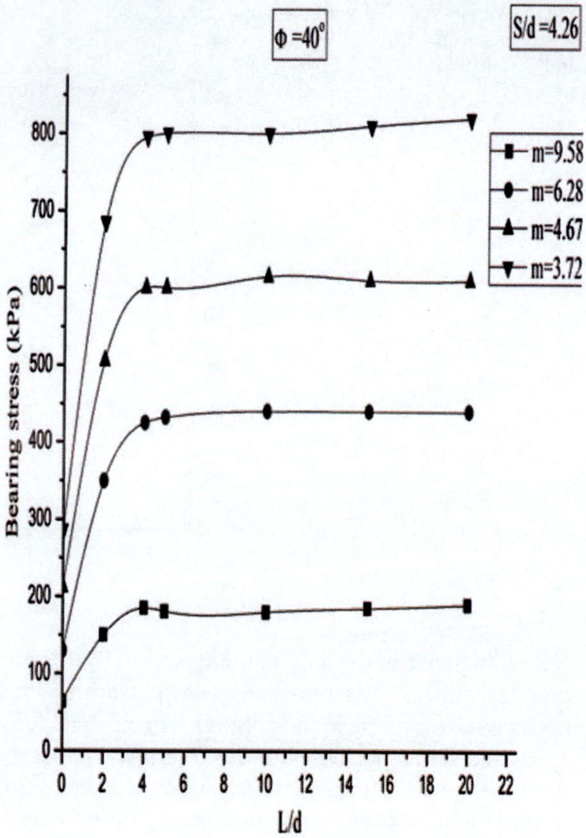

Clay having a lower value of Poisson's ratio would indicate a stiffer clayey material due to the increased shear strength and the lateral deformation for such soils is less as compared to the soils having lesser shear strength. There is not much difference in the bearing stress after L/d of 5 which can be assumed to be the optimum length as shown in Fig. 6. The variation of bearing stress with Poisson's ratio for varying slenderness ratio, $A_r = 5\%$ and $\varphi = 45°$ is presented in Fig. 7. From Figs. 6 and 7, it is observed that area replacement ratio remaining the same, the column material having higher friction angle will have higher bearing stress as is evident from Figs. 6 and 7, respectively.

3.4 *Effect of Modular Ratio*

To study the effect of modular ratio on the bearing stress, a graph has been plotted between modular ratio and the bearing stress as shown in Fig. 8 which illustrates

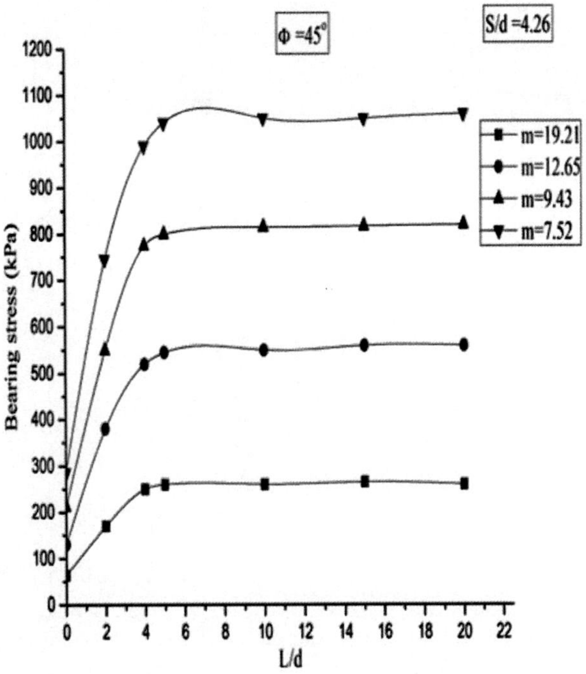

Fig. 5 Variation of L/d with bearing stress for $S/d = 4.26$ and $\varphi = 45^{\circ}$

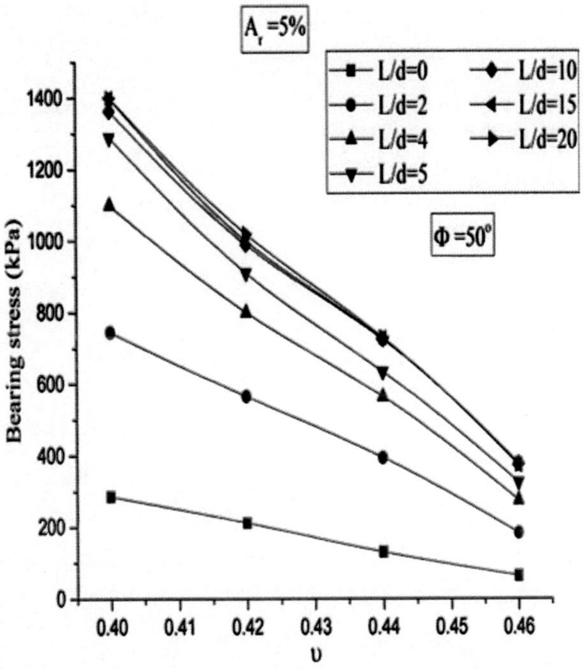

Fig. 6 Variation of bearing stress with Poisson's ratio for varying slenderness ratio, $A_r = 5\%$ and $\varphi = 50^{\circ}$

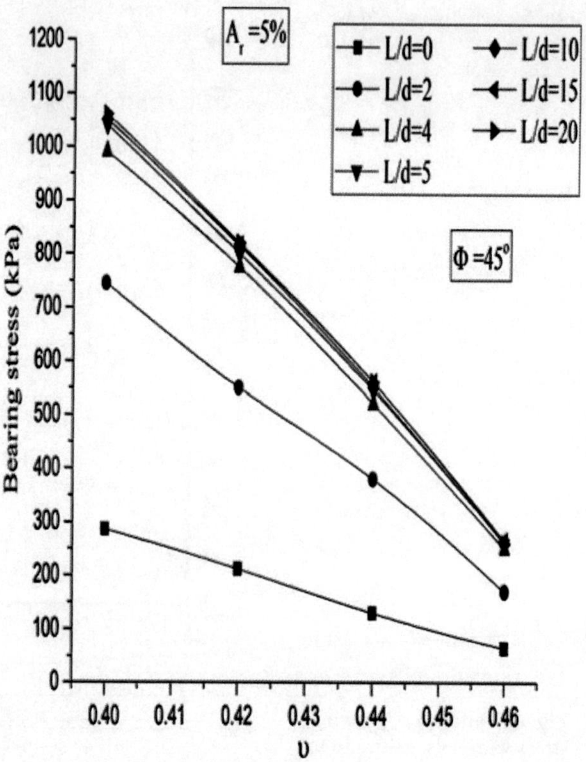

Fig. 7 Variation of bearing stress with Poisson's ratio for varying slenderness ratio, $A_r = 5\%$ and $\varphi = 45°$

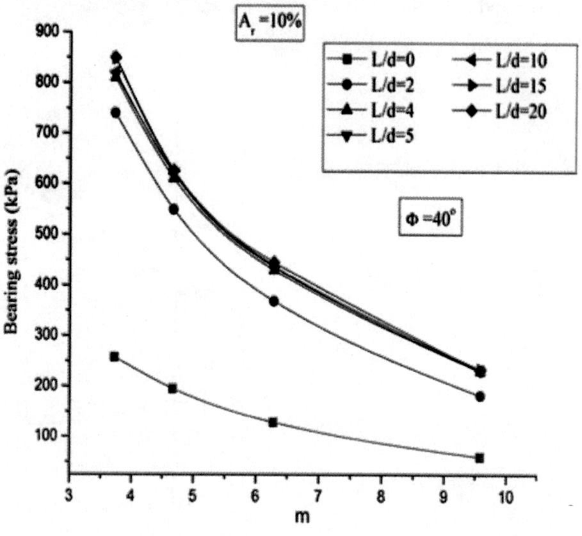

Fig. 8 Variation of modular ratio with bearing stress for $A_r = 10\%$ and $\varphi = 40°$ for varying L/d

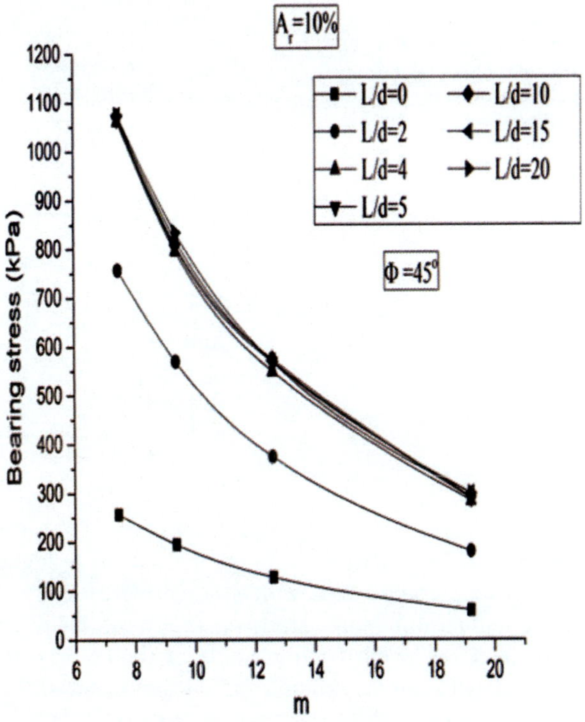

Fig. 9 Variation of modular ratio with bearing stress for $A_r = 10\%$ and $\varphi = 45°$ for varying L/d

that higher modular ratio causes a reduction in the bearing capacity for a constant friction angle as a higher modular ratio indicates low stiffness of the surrounding clay. This is attributed to the reason that low stiffness of clay would have a less confining effect. In the present study, the modular ratio has been varied by varying the stiffness of clay and the stiffness of the column material. Also, the clay having less stiffness would indicate a relatively higher Poisson's ratio which indicates that the clay becomes more compressible in the lateral direction accommodating more deformation produced by the stone columns, which causes bulging and subsequently causes a reduction in the bearing capacity (Fig. 9).

3.5 *Effect of Friction Angle*

It is observed that with an increase in friction angle of the column material, the stiffness of column increases with an exponential manner. With higher friction angle, the relative stiffness of the reinforced soil increases drastically. The increase in the bearing improvement is about 6.63 times compared to that of unreinforced soil for area replacement ratio of 10%, the undrained shear strength of 10 kPa and L/d of 10. Figure 10 shows the variation of friction angle with bearing stress for S/d of 3

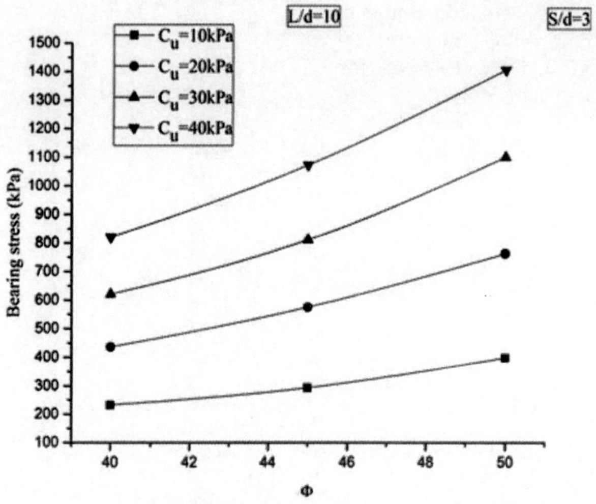

Fig. 10 Bearing stress variation with friction angle for varying undrained cohesion

having *L*/*d* of 10. Higher friction angle leads to higher stiffness due to friction and interlocking components which increase drastically. A column with higher friction angle will require more load to cause failure. Therefore, the bearing stress increases with increase in the friction angle of the column material due to higher relative stiffness between the column and the soil material (Fig. 11).

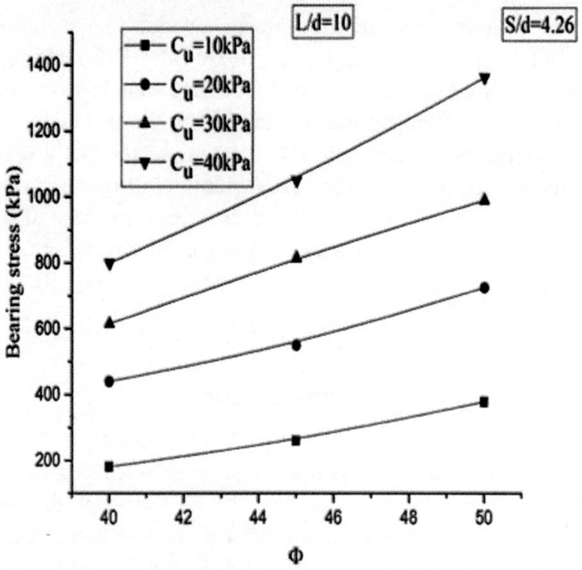

Fig. 11 Bearing stress variation with friction angle for varying undrained cohesion

Fig. 12 Plot of bearing improvement ratio with A_r for varying slenderness ratio and $\varphi = 40°$

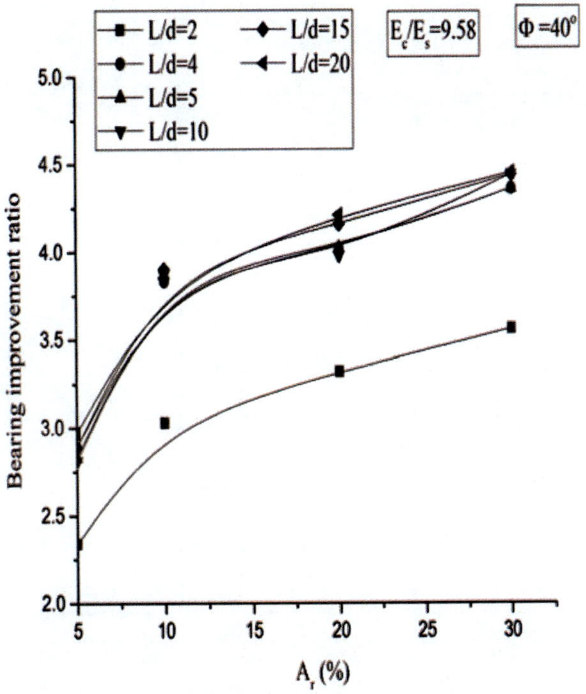

3.6 *Effect of Area Replacement Ratio (A_r)*

Area replacement ratio is defined as the column occupied area to the area of the surrounding soil. Higher area replacement ratio indicates larger area occupied by the columns, which causes an increase in the relative stiffness of the composite foundation system. The depth of failure wedge increases with increasing A_r, as more load is required to cause failure. Therefore, the bearing stress increases with increase in the area replacement ratio as shown in Fig. 12. The depth of bulging increases with higher area replacement ratio due to the increase in the relative stiffness, which causes the failure zone to extend at a greater depth. The improvement in the bearing value is about 4.3 times to that of the unreinforced soil which shows the effectiveness of stone column.

3.7 *Effect of Undrained Cohesion (C_u)*

The bearing stress increases with increase in the undrained cohesion of clay. This is because with an increase in the undrained cohesion, the shear strength of the surrounding soil increases. With the increase in the shear strength, the column will require more load to cause lateral deformation as the soil will provide higher lateral

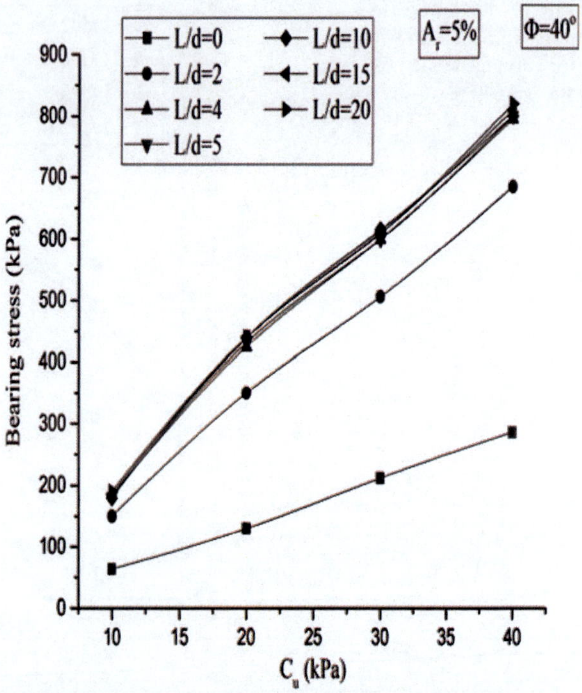

Fig. 13 Plot of bearing stress with C_u for varying slenderness ratio, $\varphi = 40°$ and $A_r = 5\%$

resistance. It is observed that the bearing stress increases in a linear fashion with the increase in the undrained cohesion. Figures 13 and 14 show the variation of bearing stress with cohesion for varying slenderness ratio. The slenderness ratio was varied as 0, 2, 4, 5, 10, 15 and 20 whereas the undrained cohesion was varied as 10 kPa, 20 kPa, 30 kPa and 40 kPa, respectively. For higher values of cohesion and higher values of *L/d*, the bearing stress is highest which signifies that the increase in the shear strength of the soil causes an increase in the bearing capacity of the composite structure.

4 Failure Modes of Stone Column

A single unit cell stone column when loaded alone fails due to bulging whereas a group of the stone column may fail due to bulging or due to shearing of the entire soil mass. From the analysis, it is observed that columns having a length greater than 4D, where D is the diameter of the stone column, fail due to bulging and columns having a length less than 4D fail due to punching. The typical failure modes are shown in Figs. 15 and 16.

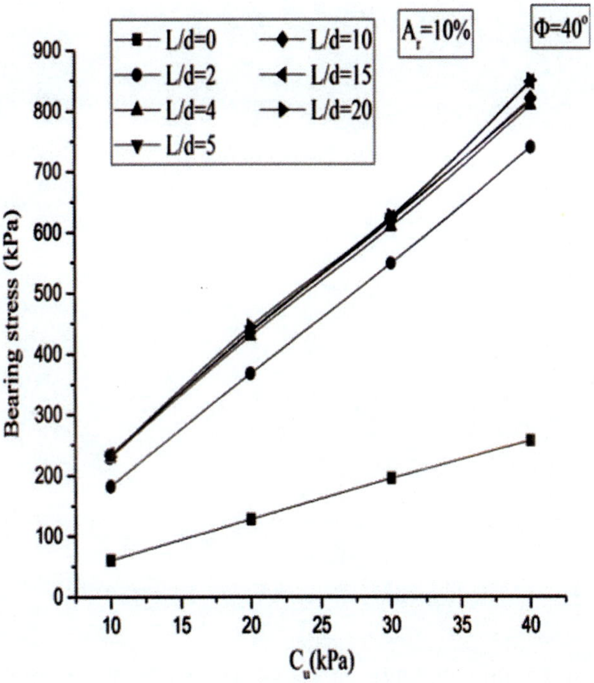

Fig. 14 Plot of bearing stress with C_u for varying slenderness ratio, $\varphi = 40°$ and $A_r = 10\%$

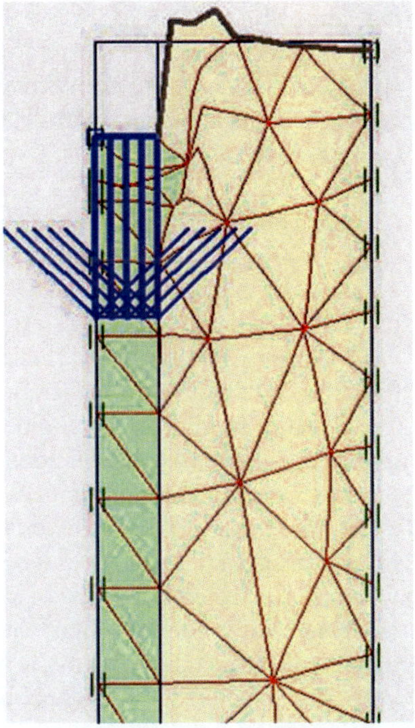

Fig. 15 Bulging mode of failure for *L*/*d* ratio of 10

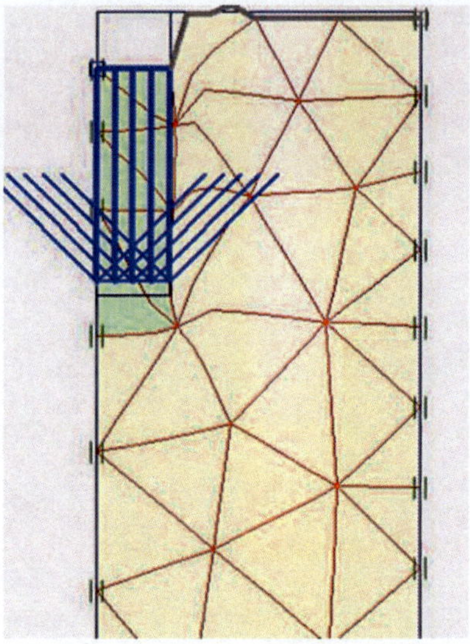

Fig. 16 Punching mode of failure for $L/d = 2$

5 Conclusion

In the present study, a parametric analysis has been conducted and each parameter was isolated and analysed for studying the nature of change of bearing capacity. From the parametric study, it is observed that with an increase in the slenderness ratio, the bearing stress increases up to an optimum value beyond which there is negligible change in the bearing stress. With relatively higher Poisson's ratio, the bearing stress decreases. This is because with relatively higher Poisson's ratio, clay becomes more compressible in the lateral direction accommodating more deformation produced by the columns. The Poisson's effect shows decreasing trend in bearing stress due to more lateral deformation and when the column is loaded, the surrounding soil does not provide much lateral resistance due to the low value of shear strength. A Poisson's ratio of 0.46 shows the very low value of bearing capacity as compared to that of a clay having Poisson's ratio of 0.40. A higher modular ratio shows a lower value of bearing stress. This is because the modular ratio is inversely proportional to the stiffness of clay and a higher modular ratio would indicate lower stiffness of clay. The effect of friction angle with bearing value shows increasing trend. The increase in the bearing improvement is about 6.63 times compared to that of unreinforced soil for area replacement ratio of 10%, the undrained shear strength of 10 kPa and L/d of 10. The bearing stress increases with increase in the area replacement ratio. The improvement in the bearing value is about 4.3 times to that of the unreinforced soil which shows the effectiveness of stone column. Bulging is the main cause of failure of a single stone column when loaded alone. An increase in the area ratio caused an increase in the bearing value of the stone column-reinforced soil due to higher relative stiffness. The most influential parameters for the design of stone column-reinforced soil are the angle of internal friction of the column material and area replacement ratio whereas the cohesion of surrounding material, slenderness ratio and Poisson's ratio contribute to a lesser magnitude. The depth of bulging increases with higher area replacement ratio due to the increase in the relative stiffness. A single unit cell stone column when loaded alone fails due to bulging. A column having the length greater than 4D, where D is the diameter of the stone column, fails due to bulging and columns having the length less than 4D fail due to punching.

References

Alonso JA, Jimenez R (2011) Reliability analysis of stone columns for ground improvement. In: Geo-risk 2011: risk assessment and management, pp 493–500

Ambily AP, Gandhi SR (2007) Behaviour of stone columns based on experimental and FEM analysis. J Geotech Geoenviron Eng 133(4):405–415

Asaoka A, Kodaka T, Nozu M (1994) Undrained shear strength of clay improved with sand compaction piles. Soils Found 34(4):23–32

Castro J (2014) Numerical modelling of stone columns beneath a rigid footing. Comput Geotech 60:77–87

Das P, Pal SK (2013) A study of the behavior of stone column in local soft and loose layered soil. EJGE, 18

Etezad M, Hanna AM, Ayadat T (2014) Bearing capacity of a group of stone columns in soft soil. Int J Geomech 15(2):04014043

Fattah MY, Zabar BS, Hassan HA (2016) Experimental analysis of embankment on ordinary and encased stone columns. Int J Geomech 16(4):04015102

Hanna AM, Etezad M, Ayadat T (2013) Mode of failure of a group of stone columns in soft soil. Int J Geomech 13(1):87–96

Comparative Assessment of Surface Soil Contamination Around Bellandur and Kengeri Lakes

M. T. Prathap Kumar, D. Jeevan Kumar, Ashutosh Kumar, Nikhil Jayaramulu Siregere, and T. V. Venu

Abstract The type and quantity of effluents and wastewater disposed of into highly polluted *Bellandur* and *Kengeri lakes* are different because of the nature and type of zone under the purview of two lakes. Study on the concentration of heavy metals in the lakebed sediments around both the lakes helps to identify the source of contamination and such a comparative study is almost non-existent. The present study comparatively assesses heavy metal contamination of surface soil around these lakes through grab sampling along with physical properties of soil to identify the presence of organic contents. Three locations of Kengeri Lake and four locations of Bellandur Lake were selected depending on the waste discharge locations. Samples of both the lakebed sediments indicate the presence of organic content. Both the lake sediments indicated heavy dosage of iron and chromium, in addition to nickel and zinc beyond the prescribed limits of FAO and WHO standards. The presence of mercury is also confirmed in both the contaminated lakebed sediments of both the lakes. However, the concentration of Fe is more compared to all heavy metals and the concentration of other heavy metals was found to be lower than permissible limits set by FAO as there was no defined source of heavy metal origin.

Keywords Heavy metals · Lakebed · Heavy metals

1 Introduction

Due to rapid urbanization and migration of people to cities, many lake bodies in and around these cities are being encroached upon, thereby subjected to contamination of these water bodies and lakebed sediments. Industrial wastes, atmospheric fall-outs and domestic wastes are among the major sources of heavy metals in urban sewage. Due to increased industrialization and urbanization, soils around these lakes trap

M. T. Prathap Kumar (✉) · D. Jeevan Kumar · A. Kumar · N. J. Siregere · T. V. Venu
Department of Civil Engineering, RNS Institute of Technology, Bengaluru, Karnataka, India
e-mail: drmtprathap@gmail.com

M. Latha Gali and R. R. P. (eds.), *Problematic Soils and Geoenvironmental Concerns*, Lecture Notes in Civil Engineering 88,
https://doi.org/10.1007/978-981-15-6237-2_66

heavy metals to a varying degree depending on heavy metal concentration in water and frequency of irrigation (Lokeshwari and Chandrappa 2006). It is a fact that heavy concentration of heavy metals in soil, water and air may pose health risk to humans endangering eco-system.

Bellandur and Kengeri lakes are some of the water bodies of Bengaluru City connected to the city's drainage system. Untreated and partially treated domestic sewerage and industrial wastewater from the catchment area of these lakes are let into these water bodies. Surrounding the catchment area of these lakes presents some of the important industries that include tech companies, public sector units, small-scale units (like plating and smelting industries), garment factories, distilleries, etc. These industries are found to release wastewater into storm water and sewerage drains that find their way into these lakes. Varalakshmi and Ganeshmurthy (2012) conducted a study in peri-urban Bengaluru to assess heavy metal contamination in water, soil and vegetable samples. Analyses revealed that concentration of cadmium (Cd) and chromium (Cr) in waters from all the tanks exceeded recommended levels of 0.01 and 0.1 mg/l, respectively, while content of lead (Pb) and nickel (Ni) is found to be in safe limits. It was found that the concentration of Cd was highest in the water of Bellandur tank (0.039 mg/l) and of Cr was highest in the water of Byramangala tank (0.311 mg/l). However, review of these studies indicates that samples used are concentrated away from source water bodies and samples sourced in these studies are from water samples that are being used for irrigation. One of the important aspects revealed in these studies is that the Cd concentration in all the vegetables grown around Varthur and Bellandur tanks exceeded the safe limit prescribed under the Prevention of Food Adulteration Act (PFA 1954). An assessment of water samples along with soil and crop plants using atomic absorption spectrophotometry by Lokeshwari and Chandrappa (2006) for seven heavy metals, viz. Fe, Zn, Cu, Ni, Cr, Pb and Cd. Revealed the fact that some of the heavy metals were present in rice and vegetables, beyond the limits of Indian standards. Evaluation of the degree of heavy metal contamination in lakes of Bangalore City and the extent of sediment concentration were conducted by Jumbe and Nandini (2009). The study indicated a more pronounced presence of Cu followed by Pb and Cd. Comparative assessment of Cd, Cu, Zn, Fe, Pb, Mn, Ni, and Co content in water and sediment from two stations along Red Sea coast by Al-Wesabi et al. (2015) concluded that untreated sewage shall not be let into the water that has significant impact on marine animals such as fisheries. Studies by Chiroma et al. (2014) on the accumulation of heavy metals in soil, water and plants revealed that soil samples have recorded heavy metal in excess of the permissible levels that ranged from severe pollution to slight contamination. Most of the heavy metals in the soil seem to originate sources that included atmospheric deposition, sewage irrigation, improper stacking of the industrial solid waste, mining activities. American Public health association published methods (2005) were referenced to examine waste water contamination. Investigations by Aslan et al. (2004) have indicated antropogenic impact factor that shall be considered prior to and during the recovery of water quality. Bagde et al. (1982) indicated the influence of coliform bacteria in a closed lake. Helen (2008), Huang et al. (2007) and Martin (2004) studied on the influence of heavy metals and their impact on the quality of lake waters. Further, Prajapati and Meravi (2014), Raju et al. (2013), Rao and Rao (1994) and Sulekh et al. (2012) investigated by assessing water quality of different lakes especially with regard to heavy metal concentration. Further the use of pesticides and fertilizers in soils around heavily contaminated lakes also posed significant effect on soil retention of heavy metals. Most of these studies are thus concentrated away from actual lakes around which the contaminated water is drained into the lake bodies. Very few studies thus have assessed the contamination of lakebed soils.

The objective of the present study is to assess levels of contamination of lakebed soils along with lake water contamination due to major toxic heavy metals, viz. Fe, As, Zn, Hg, Cd, Pb, Cr and Ni in both soil and water in both Bellandur and Kengeri lakes. The geotechnical properties of surface soils of both lakes were determined in order to identify the source responsible for such contamination since the type of effluents that is being discharged into both the lakes are different as they are strategically situated in different parts of Bengaluru. Further, the size of both the lakes is significantly different—Bellandur being the largest water body located in Bengaluru urban.

2 Sampling of Sediments

Grab samples were taken from along the banks of the lakebed. All samples were taken from the top 5 and 10 cm layer to a depth of over 30 cm. The results presented are mean of such samples tested in the laboratory. Further, the coordinates of these sampling points are mapped using GPS. Sampling tools were washed and dried with water before the next samples were collected. The collected samples were stored in polythene plastic containers. Samples were air-dried in the laboratory at room temperature, before being sieved under BIS:425-micron mesh. The samples were then stored in a polythene container. Figures 1 and 2 show sampling points located strategically around Bellandur Lake and Kengeri Lake, respectively.

The maximum allowable limits of heavy metals in soil, water and irrigation water have been established by standard regulatory bodies such as WHO guidelines. The heavy metals obtained in lake sediments of Kengeri Lake and Bellandur Lake are thus comparatively assessed using these permissible limits. The geotechnical properties of all the soil samples were evaluated in the laboratory using both air-dried and oven-dried samples. Table 1 shows sample number along with GPS coordinates and description of locations of soil sampling points.

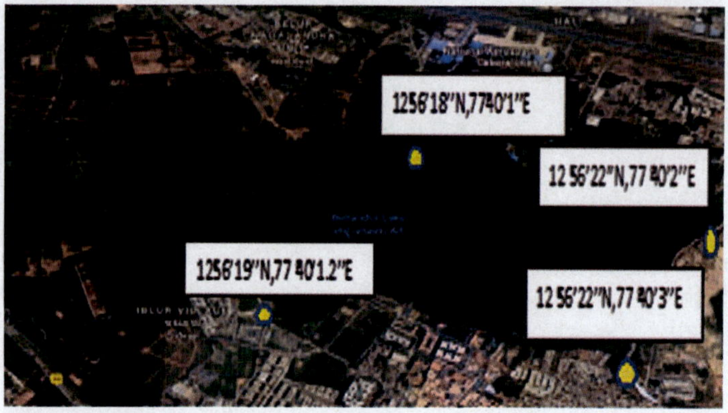

Fig. 1 Sampling points around Bellandur Lakebed

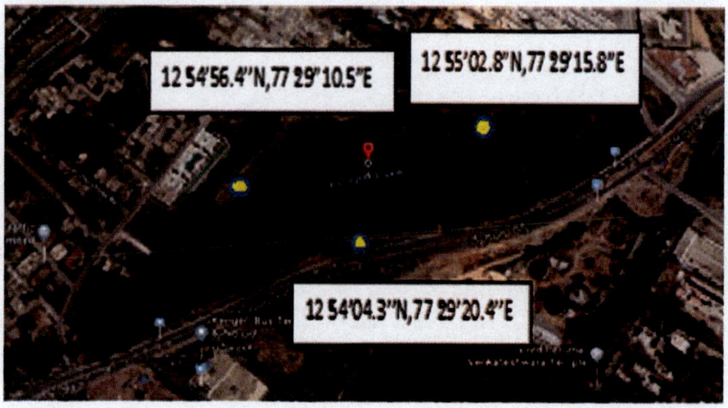

Fig. 2 Sampling points around Kengeri Lakebed

3 Results and Discussions

Both undisturbed and disturbed samples were collected in polyethylene covers and were brought to the laboratory and are air- and oven-dried for further testing, and the properties such as specific gravity, field density, dry density, natural water content were determined as per relevant IS standards. In both the lakes, the field density of all the samples was in the range of 12.4–16.4kN/m^3 with Kengeri like sediments recording higher range of field density. The in-situ water content was in the range of 22–80% for both the lake sediments. All the samples indicated loose to medium dense soils and soft in consistency.

Table 1 Sample locations and coordinates

Sample No.	Description	Coordinates
Bellandur Lake Sample Locations		
Sample-1(**SB1**)	Industrial and domestic discharge point	12° 56′ 18″ N, 77° 40′ 1″ E
Sample-2(**SB2**)	Intermediate point	12° 56′ 19″ N, 77° 40′ 1.2″ E
Sample-3(**SB3**)	Near road side	12° 56′ 22″ N, 77° 40′ 2″ E
Sample-4(**SB4**)	Outlet (waste weir)	12° 56′ 22″ N, 77° 40′ 3″ E
Kengeri Lake Sample Locations		
Sample-1(**SK1**)	Domestic waste discharge point	12° 54′ 56.4″ N, 77° 29′ 10.5″ E
Sample-2(**SK2**)	Intermediate point	12° 55′ 02.8″ N, 77° 29′ 15.8″ E
Sample-3(**SK3**)	Outlet (waste weir)	12° 54′ 04.3″ N, 77° 29′ 20.4″ E

3.1 Plasticity Characteristics of Lake Sediments

The plasticity properties were evaluated for both the lake sediments in terms of liquid limit and plastic limit for both the air-dried and oven-dried samples. Table 2 shows variations in plasticity for both the lake sediments along with liquid limit ratio. Liquid limit ratio (LLR) was determined as the ratio of liquid limit of oven-dried sample to the liquid limit of air-dried sample. Sample exhibiting LLR <0.75 indicates the presence of organic content. All the samples from Bellandur Lake sediment indicated the presence of organic content except sample SB4, which was collected

Table 2 LLR of Kengeri and Bellandur samples

LLR of Kengeri Lake Sediment				
Sample No.	Liquid limit (%)		LLR	Remarks
	Air-dried	Oven-dried		
SK1	43.15	31.93	0.74	Organic soil
SK2	40.05	36.25	0.91	Inorganic soil
SK3	35.32	33.20	0.94	Inorganic soil
LLR of Bellandur Lake Sediment				
SB1	48.8	32.20	0.66	Organic soil
SB2	33.01	23.10	0.70	Organic soil
SB3	27.5	19.80	0.72	Organic soil
SB4	24.05	22.126	0.92	Inorganic soil

near wastewater weir/outlet at Bellandur Lake. High velocity of water at the lake outlet discharging excess water seems to have washed and removed organic matters present in the sample SB4. For the case of Kengeri Lake sample SK1, which was taken at the domestic discharge point, indicated the presence of organic matter and as per its location, and the major effluent being discharged was domestic sewage which has contributed to the soil contamination. For samples SK2 and SK3 of Kengeri Lake, no such waste effluent discharge was noticed, and soil colour was reddish-brown indicating no contamination of lake sediments at these locations.

The presence of organic matter was assessed by conducting "*loss on ignition test (LOI)*" for all the lakebed sediments. The samples SK1 showed percentage of organic matter on LOI at around 12.8%, and both the other samples (SK2 and SK3) showed the percentage of organic matter on LOI less than 7.3. Similarly, all the samples (SB1, SB2 and SB3) showed the percentage of organic matter in excess of 15–16%, except that for sample SB4 that recorded at around 5.8%. Grain size distribution of all the samples of both the lakes indicated the samples are silty-clay, which had the potential to trap the organic matter, if effluents containing organic content are being discharged into the lake water. The trend thus confirms the presence of organic content only in one sample of Kengeri Lake and three samples of Bellandur Lake, indicating higher level of pollution of Bellandur Lakebed sediment. All the samples that indicated organic content in Bellandur Lake and their locations also confirmed that a large quantity of domestic and industrial wastes are being discharged at these locations.

3.2 Heavy Metal Concentration of Lakebed Sediments

Figures 3 and 4 show heavy metal concentration in lakebed sediments of Kengeri and Bellandur lakes. It can be clearly seen that the heavy metal concentration of lakebed sediment obtained from Kengeri Lake is smaller than those obtained for the samples of Bellandur Lake, indicating maximum pollution of Bellandur Lake.

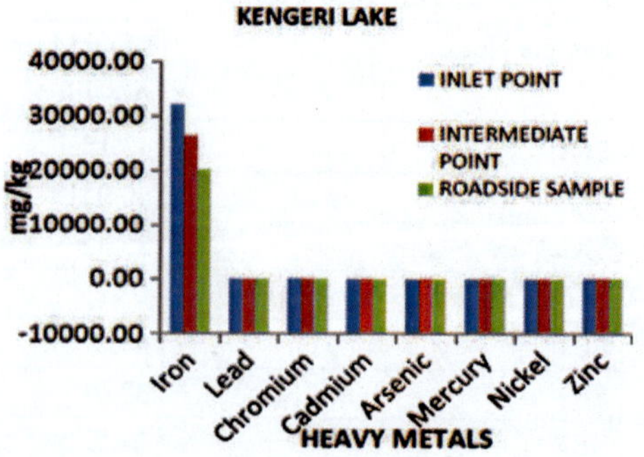

Fig. 3 Heavy metal concentration of Kengeri Lake sediment

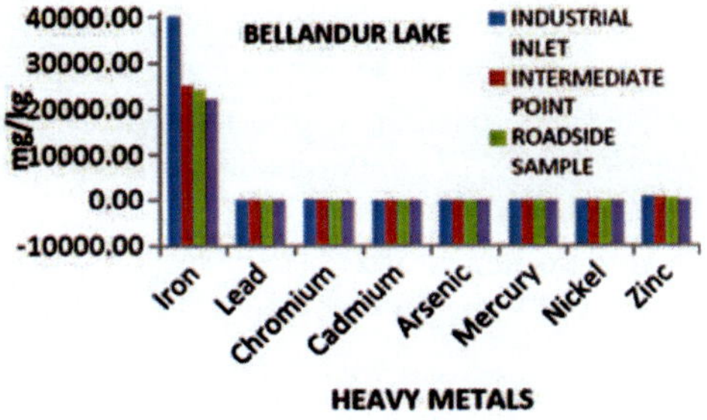

Fig. 4 Heavy metal concentration of Bellandur Lake sediment

Iron concentration of samples obtained from Kengeri Lake is found to be maximum though within permissible limit, for the samples where domestic sewage is being discharged. Similarly, in the case of samples of Bellandur Lake maximum iron concentrations occur where domestic/industrial effluent is being discharged. A minimum concentration is found near the outlet where water washes the lakebed sediments due to the high velocity of excess flowing water being discharged from the outlet. As far as lead content is concerned, both the lakebed sediments indicated concentration within permissible limits of 100 mg/kg. Regarding chromium concentration, samples of Kengeri Lake where domestic waste is being discharged indicated above permissible limit of 100 mg/l and almost all the samples of Bellandur Lake indicated higher concentrations than permissible limit (except near the outlet or waste weir) which is a cause of concern. Similarly, the cadmium and arsenic concentrations in both the lakebed sediments have shown less than permissible limits for all the samples obtained from different locations. However, the concentration of mercury obtained from the lake sediment extracted near domestic/industrial effluent discharged showed slightly higher concentrations though less than the permissible limit but greater than that prescribed for drinking water. The presence of mercury is a major cause for concern about the pollution of groundwater around both the lakes. The nickel concentration of lake sediments obtained around Kengeri Lake is within permissible limits, whereas that obtained for sample located near domestic/industrial effluent discharge point of Bellandur Lake indicated higher concentration than permissible limit. Similarly, it is the case for zinc concentration wherein samples around Kengeri Lake showed within permissible limits and samples obtained around Bellandur Lake showed greater than the permissible limits (except the sample near outlet).

Thus, the comparative assessment of heavy metal contamination of both the lake sediments conclusively proves that Bellandur Lake is polluted to greater extent. Being the largest lake body in Bangalore urban, it seems uncontrolled and untreated disposal of domestic as well as industrial waste has made this water body almost to be brought to the brink of dead lake unless, and until the discharge of sewage water is completely prevented. The water from the Bellandur Lake is being pre-treated and let into the lake at the location represented by SB2 which also indicating the presence of heavy metals beyond permissible limits.

The study thus proves that lake sediments trap the heavy metals present in water and hence is a cause for concern with regard to groundwater pollution. Since the wastewater discharged into the lake is not pre-treated, flows unhindered and is a major source of contamination of both the lakes. Hence, it is a prerequisite to pre-treat the waste effluent before it is being let into the lake water, or else, it is going to lead to "water disaster" (both surface and groundwater) in the surrounding areas.

4 Conclusions

The studies on the magnitude and extent of the heavy metals' deposition in the lakebed sediments along with the presence of organic matters suggest that the lakebed sediments in Bellandur Lake are trapped with large quantity of heavy metals than those in the Kengeri Lakebed sediments. This study therefore indicates the increasing levels of various heavy metals deposited in the sediment deposits of the lakebeds of the urban wetlands. If this trend is allowed to continue unabated, it is mostly likely that the local food web complexes in these fragile wetlands might be at the highest risk of induced heavy metals contamination. Based on the present study, the following major conclusions have been drawn.

i. Samples of both the lakebed sediments indicate the presence of organic content. Samples obtained from locations where sewage/waste domestic water being discharged indicated organic contamination of lakebed sediments.
ii. The samples were analysed for seven heavy metals (Zn, Fe, Cr, Cd, Pb, Hg and Ni) using standard procedures. Both the lake sediments indicated heavy dosage of iron and chromium, in addition to nickel and zinc beyond the prescribed limits of FAO and WHO standards.
iii. Contaminated samples of Kengeri Lake sediments showed iron content at around 32,100 mg/kg and that for Bellandur Lake contaminated sediment indicated at around 40,200 mg/kg. Chromium contamination was at around 150 mg/kg for Kengeri Lake sediment and that for Bellandur sediment it ranged up to 180 mg/kg. The presence of mercury is also confirmed in both the contaminated lakebed sediments of both the lakes.
iv. However, the concentration of Fe is more compared to all heavy metals that the values of other heavy metals concentration were found to be lower than permissible limits set by FAO as there was no defined source of heavy metal origin.

References

Al-Waesabi OA, Zindah A, Zari TA, Al-Hasawi ZM (2015) Int J Microbio Appl Sci 4(8):840–855
American Public Health Association (APHA) (2005) Standard Methods for the Examination of Water and Wastewater. 21st edn. Published by the American Public Health Association (APHA), the American Water Works Association (AWWA), and the Water Environment Federation (WEF)
Aslan-Yilmaz A, Okus E, Ovez S (2004) Bacteriological indicators of anthropogenic impact prior to and during the recovery of water quality in an extremely polluted estuary, Golden Horn, Turkey. Mar Pollut Bull 49:951–958
Bagde US, Varma AK (1982) Influence of physicochemical factors on the coliform bacteria in a closed lake water system. Int J Environ Stud (England) 18:237–241
Chiroma TM, Ebewele RO Hymore A (2014) Comparative assessment of heavy metal levels in soil vegetables and urban grey waste water used for irrigation in Yola and Kano. Int Ref J Eng Sci 3(2):01–09
Helen R, Paneerselvam (2008) Physico chemical analysis and role of Phytoplanktons in Bellandur Lake. In: Proceedings of Taal 2007: the 12th world lake conference, pp 1729–1736
Huang SS, Liao QL, Hua M et al (2007) Survey of heavy metal contamination and assessment of agricultural soils in Yangzhong district, Jiangsu Province, China. Chemosphere 67:2148–2155
Jumbe AS, Nandini N (2009) Heavy metals analysis and sediment quality values in urban lakes. Am J Environ Stud 5(6):678–687
Lokeshwari H, Chandrappa GT (2006) Impact of heavy metal contamination of Bellandur lake on soil and cultivated vegetation. Curr Sci 91(5):622–627
Martin CW (2004) Heavy metal storage in near channel sediments of the Lahn River, Germany. Geomorphology 61:275–285
Prajapati SK, Meravi N (2014) Heavy metal speciation of soil and Calotropis procera from thermal power plant area. Proc Int Acad Ecol Environ Sci 4(2):68–71
Prevention of Food Adulteration Act and Rules (1954) Ministry of Health, Government of India, pp 374–375
Raju KV, Somashekar RK, Prakash KL (2013) Spatio-temporal variation of heavy metals in Cauvery River basin. Proc Int Acad Ecol Environ Sci 3(1):59–75
Rao SM, Rao KSS (1994) Ground heaving from caustic soda solution spillage—a case study. Soils Found Jap Soc Soil Mech Found Eng 34:13–18
Sulekh C, Arendra S, Praveen K (2012) T, Assessment of water quality values in Porur Lake Chennai, Hussain Sagar Hyderabad and Vihar Lake Mumbai, India. Chem Sci Trans 1(3):508–515
Tripathi RM, Raghunath R, Mahapatra S, Sadasivan S (2001) Blood lead and its effect on Cd, Cu, Zn, Fe and hemoglobin levels of children. Sci Total Environ 2001(277):161–168
Varalakshmi LR, Ganeshamurthy AN (2012) Heavy metal contamination of water bodies, soils and vegetables in peri-urban areas: a case study in Bengaluru. J Hortic Sci 7(1):62–67

Micro-level Exploration of KOH-Contaminated Kaolinitic Clays Under Different Experimental Conditions

P. Lakshmi Sruthi and P. Hari Prasad Reddy

Abstract Contamination of soils due to caustic alkali has significant effect on the volume change behavior of soils, which in turn can have direct bearing on their geotechnical properties and can affect the stability of structures built on them. The existing literature strongly highlights the fact that failure of structures along with alterations in mineralogy and morphology occurs due to NaOH contamination. Efforts were made to simulate long-term effects of NaOH on mineralogical and morphological alteration by varying the experimental conditions in laboratory, so that preventive measures can be taken at a much faster rate. On the other hand, KOH is another strong alkali which is extensively used for commercial purposes. However, no studies were reported in literature to highlight the adverse effects of KOH. Thus, to understand the long-term effect of KOH on mineralogical and morphological alterations, a preliminary investigation is carried out in the present study by considering different experimental conditions (field contamination, long-term interactions, and temperature effects at 4 N KOH). Two types of kaolinitic clays, namely red earth and kaolin with varying mineral content were selected for the study. Micro-level investigations (XRD and SEM) have been carried out to noticeably understand impact of varying experimental conditions at particle-level interaction. Test results indicated that neogenic formations varied with type of experimental conditions along with variation in the morphology of soils. Further, it is observed that, long-term effects of KOH can be simulated within short period of time by conducting the experiments at elevated temperature.

Keywords Alkali · Kaolinitic · Neogenic formations · XRD · SEM

P. Lakshmi Sruthi (✉) · P. Hari Prasad Reddy
National Institute of Technology Warangal, Warangal, Telangana 506004, India
e-mail: plakshmisruthi@gmail.com

P. Hari Prasad Reddy
e-mail: ponnapuhari@gmail.com

M. Latha Gali and R. R. P. (eds.), *Problematic Soils and Geoenvironmental Concerns*, Lecture Notes in Civil Engineering 88,
https://doi.org/10.1007/978-981-15-6237-2_67

1 Introduction

Soil contamination due to accidental spills or leakages of industrial effluents, heavy metals, and pesticides from industrial wastes is being increasingly recognized in the literature. Chemical contamination of foundation soils, more specifically alkali contamination and consequent distress to the industrial structures (Kabanov et al. 1977; Chunikhin et al. 1998; Sivapullaiah et al. 2004), has been reported. The industries that may experience alkali contamination of foundation soils are paper and pulp, dye, ceramics, etc. When alkali solutions come in contact with clay minerals, dissolution of existing minerals and precipitation of new minerals may occur. Cuadros and Linares (1996), Boussen et al. (2015) investigated the influence of alkaline solutions on the transformation of clay minerals. Kawano and Tomita (1997) investigated the consequence of temperature, alkali concentration, and type of alkali on zeolites formation. Batch experiments were conducted by Bauer and Berger (1998) to evaluate the dissolution rate of kaolinite and smectite in KOH solution at 35 and 80 °C. Transformation of kaolinite mineral in KOH solutions was later continued by Bauer et al. (1998) at 35 and 80 °C. The mineral transformations depend on the nature and composition of the alkaline solutions and the nature of the reacting mineral. Studies on variation in experimental conditions (simulating the field contamination, long-term interactions and temperature effect) on the transformation of clay minerals when interacted with NaOH were reported by Sruthi and Reddy (2018) whereas no such studies were reported in literature regarding adverse effects of KOH, being another strong alkali, on kaolinitic soils. Thus, to understand the long-term effect of KOH on mineralogical and morphological alterations, a preliminary investigation is carried out in the present study by considering different experimental conditions (field contamination, long-term interactions, and temperature effects at 4 N KOH) on kaolinitic soils.

2 Materials and Methods

2.1 Soils Used

The natural red earth (R) soil sample was collected by open excavation, from a depth of one meter from the ground level from Warangal (Latitude: 17 98′ N and Longitude: 79 53′ E), India. Commercially, Kaolin (K) was obtained from Godavari Mines and Minerals, Visakhapatnam (Latitude: 17.6868° N, Longitude: 83.2185° E), India. The physical properties of soils are presented in Table 1. Soils passing through 425 μm sieve were used in the study.

Table 1 Physical properties of soils

Property	Red earth	Kaolin
Liquid limit, LL (%)	38	42.90.
Plastic limit, PL (%)	22.64	29.72
Plasticity Index, PI (%)	15.36	13.18
Specific gravity, G	2.62	2.70
Optimum Moisture content, (%)	20.74	27.05
Max. dry unit weight, (g/cc)	1.80	1.7

2.2 Chemicals Used

In this study, potassium hydroxide (4 N) was used. 4 N KOH solution was prepared by mixing 224.44 g of potassium hydroxide pellets in 1 L of distilled water to make 1 L solution. The solution was cooled by bringing it in contact with distilled water.

2.3 Methodology

Different experimental conditions were taken into consideration to identify its effect on the mineralogy and morphology of soil samples.

Condition 1: One-dimensional free swell test was carried out on Samples R1 and K1 inundated with 4 N KOH. The test procedure adopted was mentioned in Puppala et al. (2005). After attaining equilibrium swell, samples were terminated and oven dried.

Condition 2: Samples R2 and K2 were exposed to 4 N KOH with solid-to-liquid ratio of 1:10 similar to free swell index test and cured for a period of 100 days.

Condition 3: Soils were mixed with 4 N KOH solution in different volumes (liquid limit times), i.e., 7 LL for red earth and 10 LL for kaolin. The main reason for selecting these volumes is that the complete kaolinite mineral dissolution takes place in the respective soil samples (Sruthi and Reddy 2017). Liquid limit of the soil was taken as a measure to vary volume of interacting solution. After thoroughly mixing the soil samples with different volume of interacting solution, the samples were sealed in air tight container for a period of 24 h. After 24 h of interaction,

- Samples R3 and K3 mixed with alkali solutions at 7 LL and 10 LL, respectively, were oven dried at 110 °C
- Samples R4 and K4 mixed with alkali solutions at 7 LL and 10 LL, respectively, were filtered and oven dried at 110 °C
- Samples R5 and K5 mixed with alkali solutions at 7 LL and 10 LL, respectively, were kept under continuous agitation as soon as mixed with 4 N KOH solution at

the rate of 1000 rpm in magnetic stirrer for one day, then filtered, and oven dried at 110 °C.

All the samples were ground in mortar and pestle to fine powder. Representative samples in powder form were taken for XRD and SEM studies to analyze mineralogical and morphological changes due to soil-alkali interactions.

2.4 *X-ray Diffraction Studies*

PAN analytical diffractometer was used to assess the mineralogical variations in all the samples. The samples were scanned between 6° and 70° with a step size of 0.01°.

2.5 *Scanning Electron Microscopy Studies*

Morphological studies on soil samples were carried out using TESCAN VEGA 3LMU microscope. The samples were coated with gold using a sputter coater prior to scanning.

3 Materials and Methods

3.1 *Mineralogical Changes in Red Earth Due to KOH Interaction*

Figure 1 shows the XRD patterns of red earth interacted with 4 N KOH at different experiment conditions. XRD pattern of actual red earth comprises of very intense peaks of *quartz* (4.24, 3.340 and 1.821 and 1.369 Å) and less intense peaks related to *kaolinite* (7.14, 4.45, 3.59 and 2.56 Å) and *hematite* (2.69 and 2.51 Å). Field contamination condition was studied in the laboratory by performing free swell test. XRD patterns of Sample R1 showed new peaks pertaining to *lithosite* (4.21, 3.45, and 2.82 Å) and *natrolite* (potassium sodium aluminum silicate hydrate) at 6.90, 4.49, and 2.98 Å. In addition to lithosite and natrolite, new peaks pertaining to *muscovite* (3.00 and 2.56 Å) were observed in Sample R2. Complete dissolution of kaolinite mineral was observed from the XRD patterns of both Samples R1 and R2. Duration of interaction played a major role in complete dissolution of kaolinite mineral. Higher the duration of interaction higher is the mineralogical changes. Volume of interacting solution and the exposure period influence the nature and magnitude of new minerals formed. When red earth interaction with 4 N KOH is experimented by varying the experimental conditions and exposing the samples for 110 °C, mineralogical changes

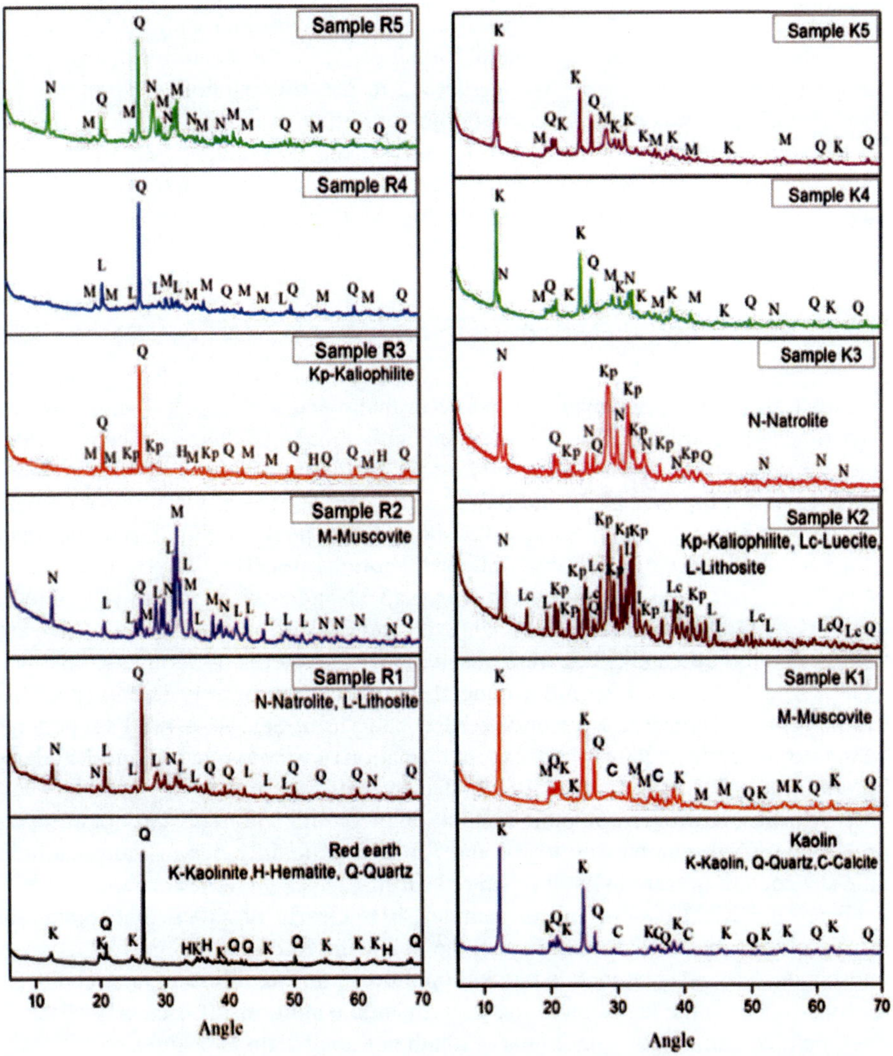

Fig. 1 XRD patterns of red earth interacted with 4 N KOH. (i) Sample R1, (ii) Sample R2, (iii) Sample R3, (iv) Sample R4, (v) Sample R5, (vi) Sample K1, (vii) Sample K2, (viii) Sample K3, (ix) Sample K4, (x) Sample K5

were observed. (i) XRD patterns of Sample R3 indicate complete mineral dissolution (kaolinite) and development of little traces of *kaliophilite* at 3.09, 2.59, and 2.13 Å and traces of *muscovite* were noticed. In this sample, complete dissolution of kaolinite mineral is mainly due to the higher volume of interacting solution. (ii) Surprisingly, XRD patterns of Sample R4 indicated less variation in the new mineral formations, i.e., very less intense peaks related to *lithosite* at 4.21 and 3.45 Å were observed. Filtering the interacting solution led to less soil–alkali interaction pertaining to less mineralogical changes. Therefore, volume of interacting solution plays a crucial role in mineral dissolution and new mineral formations. (iii) XRD patterns of Sample R5 indicated new mineral formations pertaining to natrolite. Complete dissolution of kaolinite mineral was noticed under continuous agitation and to exposure of 110 °C of the soil sample. The main reason for prominent mineralogical changes in this sample is continuous agitation which enables the soil–alkali mixtures to react at higher rate and thus higher are the mineralogical changes.

3.2 Mineralogy Alterations in Kaolin Due to KOH Interaction

Figure 1 shows the XRD patterns of kaolin interacted with 4 N KOH at different experimental conditions. The soil predominantly contains kaolinite mineral, which showed very intense peaks of *kaolinite* at 7.15, 4.36, 3.57, and 2.34 Å and less intense peaks of *quartz* (4.34 and 3.34 Å) and *calcite*, which is calcium carbonate at 3.03 Å. XRD patterns of Sample K1 showed new peaks pertaining to *muscovite* (4.46 and 2.85 Å). Complete dissolution of kaolinite mineral was not observed. In Sample K2, new peaks pertaining to *leucite* (5.31 and 3.25 Å), along with intense peaks pertaining to kaliophilite at 3.09, 2.59, and 2.13 Å, lithosite at 4.52, 3.16, and 2.82 Å and natrolite at 6.91 Å were observed. XRD patterns of Sample K3 indicate complete dissolution of kaolinite mineral and formation of very intense peaks of *kaliophilite and natrolite* was noticed. XRD patterns of Sample K4 did not undergo many variations in its mineralogy except formation of very less intense peaks related to *natrolite*. XRD patterns of K5 indicated formation of muscovite with very less intense peaks. However, complete dissolution of kaolinite mineral was not observed in K4 and K5 in comparison to R4 and R5. This is mainly due to more kaolinite mineral content present in kaolin than red earth.

From the XRD patterns of all samples, it is clearly observed that interaction of dissolved silica and alumina from the soil with KOH led to the formation of muscovite. Secondary minerals like kaliophilite, natrolite, lithosite, and leucite are formed based on the variation in the experimental conditions. It is clearly observed that, field contamination conditions of alkali and long-term exposures to alkali can be simulated at an early stage by exposing them to temperature. Thus, temperature along with volume of interaction plays a crucial role in simulating and understanding the long-term mineralogical changes in the contaminated samples at a very less interaction period.

3.3 Morphology Alterations in Red Earth Due to KOH Interaction

The SEM micrographs of red earth interacted with alkali solutions are shown in Fig. 2. Figure 2a shows the scanning electron microscope image of red earth which exhibited a compacted and aggregated form of morphology. Pseudo-tetragonal prisms were observed in Sample R1 (Fig. 2b) due to the formation of natrolite, which was also mentioned by IZA commission on zeolites. Morphology of Sample R2 (Fig. 2c) exhibited cubic crystalline zeolitic structure. In Sample R3 and R4, weathering with no severe morphological changes was observed (Fig. 2d, e) supporting XRD analysis. Whereas morphology of Sample R5 (Fig. 2f) resembled irregular non-radiating crystals indicating natrolite formation (Ibrahim 2004). Thus, all the soil samples exhibited different types of morphology under different experimental conditions and interaction periods. Red earth initially with compacted/aggregated texture transformed to pseudo-tetragonal prisms, cubic crystalline type, and non-radiating crystals.

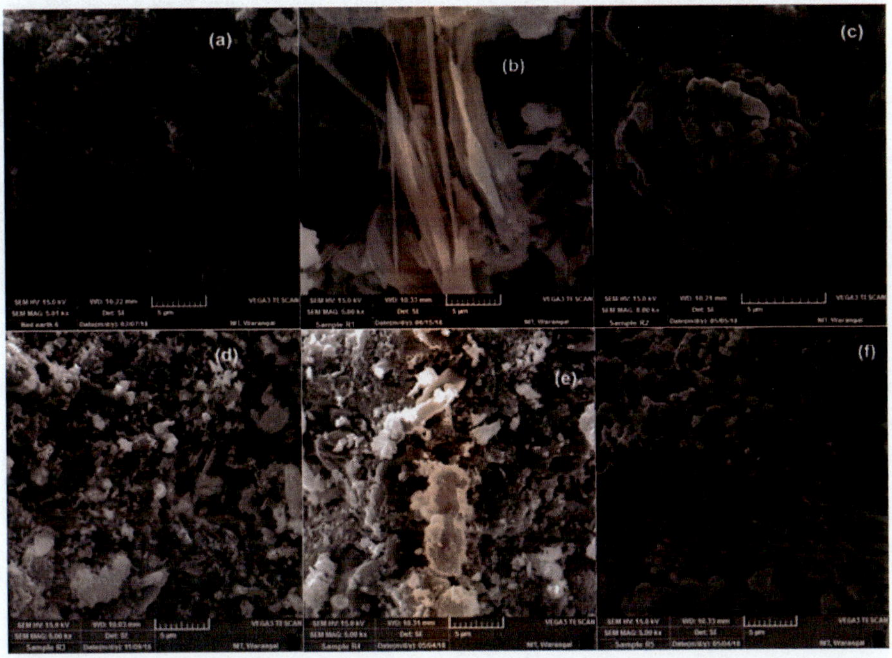

Fig. 2 SEM images of red earth interacted with 4 N KOH. **a** Water. **b** Sample R1. **c** Sample R2. **d** Sample R3. **e** Sample R4. **f** Sample R5

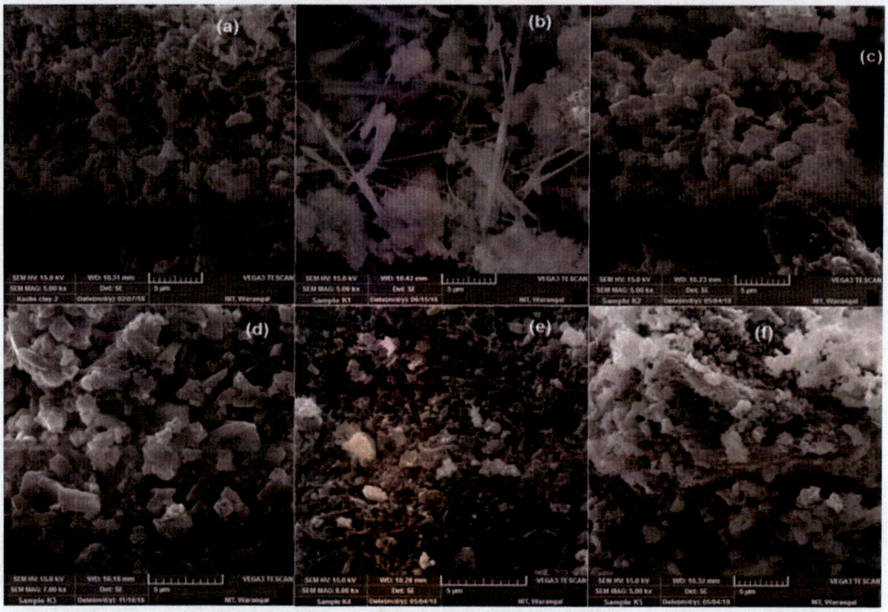

Fig. 3 SEM images of kaolin interacted with 4 N NaOH. **a** Water. **b** Sample K1. **c** Sample K2. **d** Sample K3. **e** Sample K4. **f** Sample K5

3.4 Morphology Alterations in Kaolin Due to KOH Interaction

The SEM micrographs of kaolin interacted with potassium hydroxide solution are shown in Fig. 3. Figure 3a reveals that kaolin used in the present study indicates highly loose fibrous nature with needle morphology. Morphology of Sample K1 exhibited disintegrated form with thin flaky particles (Fig. 3b). Though much variation in the mineralogy was not noticed, morphology exhibits quite a bigger variation. Sample K2 exhibited pseudo-cubic crystal morphology (Fig. 3c), which indicates the formation of *leucite* with more intensity. Formation of *kaliophilite* and *lithosite* along with *leucite* altered the texture of kaolin completely. Sample K3 exhibited irregular non-radiating crystals highlighting natrolite formation (Fig. 3d). Morphology of Sample K3 is completely different from untreated kaolin highlighting the complete dissolution of kaolin mineral supporting new mineral formations from XRD analysis. Sample K4 exhibited highly disintegrated and flaky morphology (Fig. 3e) whereas Sample K5 exhibited closely packed plate-shaped morphology highlighting the formation of muscovite (Fig. 3f). As complete dissolution of kaolin mineral did not take place in the Sample K1, K3, and K4, presence of kaolin mineral was highlighted by needle morphology and the new mineral formations were not clearly seen. Clear morphological changes were noticed in all the soil samples with respect to different field and laboratory conditions. Kaolin initially with needle form slowly disintegrated and transformed to crystalline, flaky, and plate-shaped morphology.

4 Conclusions

On the basis of results obtained, following conclusions can be drawn:

Neogenic formations were diversed with mineral content and experimental conditions.

Muscovites are the common minerals formed in almost all the samples. However, the intensity with which the muscovites are formed varied with respect to experimental conditions.

Exposure to more volume of interaction at elevated temperatures helps in simulating the long-term effects at an early stage.

Mineral content, volume of interacting solution, duration of interaction, and exposure to elevated temperatures have got significant importance in inducing mineralogical and morphological in all the samples studied.

References

Bauer A, Berger G (1998) Kaolinite and smectite dissolution rate in high molar KOH solutions at 35° and 80 °C. Appl Geochem 13:905–916

Bauer A, Velde B, Berger G (1998) Kaolinite transformation in high molar KOH solutions. Appl Geochem 13(5):619–629

Boussen S, Sghaier D, Chaabani F, Jamoussi B, Messaoud SB, Bennour A (2015) The rheological, mineralogical and chemical characteristic of the original and the Na_2CO_3—activated Tunisian swelling clay (Aleg formation) and their utilization as drilling mud. Appl Clay Sci 118:344–353

Chunikhin VG, Mavrodi VK, Kramarenko OA, Dobromil'skay NG (1998) Effect of leakage of industrial alkali solutions on the construction properties of soils. Soil Mech Found Eng 25(6):559–561

Cuadros J, Linares J (1996) Experimental kinetic study of the smectite-to-illite transformation. Geochim Cosmochim Acta 60:439–453

Ibrahim K (2004) Mineralogy and chemistry of natrolite from Jordan. Clay Miner 39:47–55

Kabanov VM, Lebedeva GA, Finke'shtein LI, Tkachenko GP, Shenin OS (1977) Swelling of soils due to wetting with solutions. Fundamenty Mekh Gruntov 5:31–32

Kawano M, Tomita K (1997) Experimental study on the formation of zeolites from Obsidian by interaction with NaOH and KOH solutions at 150 °C and 200 °C. Clay Clay Miner 45(3):365–377

Puppala AJ, Napat I, Rajan KV (2005) Experimental studies on ettringite- induced heaving in soils. J Geotech Geoenviron Eng 131(3):325–337

Sivapullaiah PV, Allam MM, Sankara G (2004) Structural distortion due to heaving of foundation soil induced by alkali contamination. In: Proceedings of the international conference on structural and foundation failures, Singapore, vol 1, pp 601–611

Sruthi PL, Reddy HPP (2017) Characterization of kaolinitic clays subjected to alkali contamination. Appl Clay Sci 146:535–547

Sruthi PL, Reddy HPP (2018) Mineralogical and morphological transformations of alkali contaminated kaolinitic clays under different experimental conditions. In: Proceedings of the international conference on advances in concrete, structural, & geotechnical engineering, February, BITS Pilani, India

Effect of Clay-Embedded Zeolite as Landfill Liner

P. A. Amalu and Ajitha B. Bhaskar

Abstract This study investigates the characteristics of natural zeolite amended kaolin clay, to be used as an impervious liner in the landfill. A landfill liner or composite liner is a low permeable barrier, which is laid down under engineered landfill sites. The ingredient zeolite was chosen due to its high absorption capacity for heavy metals as well as pozzolanic properties. A *B*/*Z* ratio of 0.5 was found to be an ideal landfill liner material considering its low hydraulic conductivity. Hydraulic conductivity tests on zeolite-embedded clay liner systems show that the hydraulic conductivity of all the mixtures prepared with varying proportions of zeolite, meet the common regulatory requirements. The micro-porous structure of natural zeolite results in a very high absorption rate, which in turn reduces the hydraulic conductivity. In order to estimate the optimum dosage of zeolite to the clay mixture, an extensive study on the strength and compaction characteristics was done.

Keywords Zeolite · Bentonite · Landfill liner

1 Introduction

Landfill liners are barriers, made of materials with very low hydraulic conductivity. Liner functions as a separator between the waste and outside soil, protecting the ground water from being contaminated. In practice, the clay liners are widely adopted due to its lower permeability and flexibility. Among the criteria required for design of landfill liner, one of the most important constrains is the hydraulic conductivity. In Kerala, a lot of environmental issues are occurring due to the absence of proper liner systems in landfills. Developing a liner system with locally available materials could resolve the issue of finding the cost-effective liner system in engineered landfills. This study focuses on the usage of locally available kaolin clay as landfill liner. Experimental analysis on the kaolin clay pointed out that, the hydraulic conductivity

P. A. Amalu (✉) · A. B. Bhaskar
College of Engineering Trivandrum, APJ Abdul Kalalm Technological University, Trivandrum, India
e-mail: amalupa12@gmail.com

M. Latha Gali and R. R. P. (eds.), *Problematic Soils and Geoenvironmental Concerns*, Lecture Notes in Civil Engineering 88,
https://doi.org/10.1007/978-981-15-6237-2_68

of the material is higher than the required values. Therefore, additives were added into kaolin clay for modifying its engineering properties to meet the liner criteria. The selected external ingredients were bentonite and zeolite. These ingredients have higher specific surface areas which in turn results in more absorption of liquid.

Bentonite is the type of clay which is abundant with montmorillonite clay minerals. Based upon the dominant element present within bentonite, different types of bentonites are available. Among these various types, calcium bentonite was used in this study. Calcium bentonite could absorb water several times its dry weight and hence widely used as an absorbent. Through chemical reactions, calcium bentonite can be converted into sodium bentonite. This process is called as sodium beneficiation. As the predominant mineral in calcium bentonite is montmorillonite, it is widely used as a filler as well as absorbent. Due to these properties, a wide spectrum of industries is using calcium bentonite for various purposes such as for the production of cosmetics, pellets of materials, and many more, a percentage of bentonite was added to the locally available clay.

Zeolite is advantageous with natural clay as landfill liner mixture because of its high adsorption capacity. Zeolite is commonly known for its industrial application as absorbents. These are naturally occurring minerals with considerably larger specific surface area.

Thus, clay-embedded zeolite with different zeolite content was developed for possible use as a barrier material. Zeolites are micro-porous, alumino-silicate minerals that can act as a filter in case of leachate through liner. Furthermore, the pozzolanicity characteristic of zeolite provides remarkable increase in compaction characteristics. Bentonite is also an effective absorbent consisting mostly of montmorillonite in a landfill liner mixture. Na-bentonite expands when absorbing as much as several times its dry mass in water.

In this study, bentonite and zeolite were mixed with clay in various proportions in order to find out the optimum mixture as landfill liner.

Most of the research work has been focused in understanding the change in behavior of zeolite-embedded clay liner system on various mix proportions. Limited knowledge is available on the influence of zeolite, on the properties of locally available Kaolin clay. In addition, the optimum percentage of zeolite dosage was not studied generally and thereby compressive strength as well as compaction characteristic was evaluated for various mix proportions. Also, the influence of zeolite on locally available Kaolin clay has rarely been considered by researchers. Hence, the motivation behind the present study is to understand the effect of zeolite in the clay available from nearby locality and its implementation as landfill liners in order to propose an economic and efficient solution for landfill liner systems.

Table 1 Properties of Kaolin clay

Particulars	Values
Specific gravity	2.37
Liquid limit (%)	53
Plastic limit (%)	33
Plasticity index (%)	20
Maximum dry density (g/cc)	1.5
Optimum moisture content (%)	27.8
Sand (%)	17
Silt (%)	48
Clay (%)	37
Compressive strength (kPa)	117

1.1 Soil

In this study, kaolin clay was purchased from English India Clay Ltd. Thiruvananthapuram. The properties of this clay are summarized in Table 1.

1.2 Bentonite

Bentonite is a montmorillonite rich clay mineral which possesses high swelling and absorption capacities. There are different types of bentonite, namely sodium bentonite, potassium bentonite, and calcium bentonite. This study considered the effects of calcium bentonite mixed with kaolin clay and zeolite.

1.3 Natural Zeolite Powder

Natural zeolite occurs on the earth crust in different forms of deposits. These deposits are identified and mined using heavy machineries and blasting equipment. The extracted zeolite ore is then pulverized into finer forms for commercial purpose. The basic properties of natural zeolite purchased from Intercity Chemicals, Chennai, are tabulated in Table 2.

Table 2 Properties of natural zeolite

Particulars	Values
Liquid limit (%)	44
Plastic limit (%)	24
Plasticity index (%)	17
Specific gravity	2.23

2 Methods

Various experiments were conducted on the sample mixture of liner system. The sample mixture consists of Kaolin clay, a fixed 10% calcium bentonite and varying percentages of natural zeolite powder. The test results were aimed to converge at the optimum mixture of a zeolite-embedded clay liner system that could provide the highest strength and lowest permeability characteristics.

2.1 Consistency Limits

The consistency limits, invented by Albert Atterberg, classified soils based on the extend of water it can contain at each stages. When added with water, soil shifts from solid state to semi-solid and liquid states. Precisely, the shift in each state occurs at a specific water content that depends on the type of soil. These water contents are experimentally calculated and are termed as shrinkage limit, plastic limit, and liquid limit. The moisture content above which the soil will behave as a liquid of feeble shear strength is termed as liquid limit. This limit could be determined by using a standard penetrometer. The plastic limit test was performed under fixed conditions. The water content that separates the plastic and semi-solid states of soil is the plastic limit, whereas while reducing the moisture content in a soil, there will be a point beyond which the reduction in water content will not change the volume of soil. This limit is defined as shrinkage limit of the soil. The tests were conducted under fixed conditions compiling to IS 2720-part 5 and part 6.

2.2 Specific Gravity

Specific gravity indicates how much heavier a material is when compared with water of equal volume. It is property related to that of soil formation and is independent of particle size. The tests were conducted under fixed conditions compiling to IS 2720-part 3.

2.3 Grain Size Analysis Using Hydrometer

Particle size distribution of soil particles, having sizes less than 75-μm, is often determined by a sedimentary process using a hydrometer, based on the analysis by Stoke's law. The fraction of clay and silt plays a vital role in determining the properties of fine-grained soil. This test is mainly carried out to obtain the percentage silt and clay size particles in the given soil sample. This method could not be adopted if less than 10% of material passes through 75-μm sieve. The tests were conducted under fixed conditions compiling to IS 2720-part 4.

2.4 Compaction Test

Engineering properties of the soil, such as its strength, stiffness, resistance to shrinkage, and imperviousness of soil, etc., can be improved by increasing the soil density. The terminology "*compaction*" defines the process of reduction in volume of soil by the reduction of air void in it under sudden impact loading. The compaction test is used to determine the right quantity of water to be used, during field compaction and the resulting degree of denseness. The test is performed as per IS 2720-part 7-1980.

2.5 Unconfined Compressive Strength Test

The test determines the maximum stress a cylindrical soil specimen can bear without failure, when kept unconfined and subjected to axial loading. The rate of strain was maintained at 1 mm/min. The results obtained from the unconfined compressive strength test could be used to determine the unconsolidated undrained shear strength of the clayey soil. The experiments were conducted as per IS 2720-part 10-1986.

2.6 Permeability

In this study, the permeability is calculated from consolidation test. The permeability characteristics of a soil sample are determined by using an oedometer. The main parameter obtained from this test will be coefficient of consolidation which will be used for further determination of permeability. The test is performed based on IS 2720-part 5-1986.

Table 3 Variation of dry density with zeolite content

Sample specification	Maximum dry density (g/cc)	Moisture content (%)
Clay	1.50	27.80
Clay + 10% B	1.47	28.88
Clay + 10% B + 10% Z	1.39	34.64
Clay + 10% B + 20% Z	1.44	35.75
Clay + 10% B + 30% Z	1.42	33.52
Clay + 10% B + 40% Z	1.39	36.48
Clay + 10% B + 50% Z	1.36	34.95

3 Results and Discussions

3.1 *Variation of Dry Density with Zeolite Content*

On the clay sample mixed with 10% of bentonite, the natural zeolite content was varied from 10 to 50%. Compaction tests were conducted as per standards on each of the mixtures. The value of maximum dry density obtained had a slight improvement with the increase in natural zeolite content up to 20%, after which there was a considerable reduction in dry density. This might be due to the increase in proportion of natural zeolite which adsorbs more moisture and thereby reduces the dry density. The compaction test result showed that clay bentonite mix with 20% of natural zeolite content has got a maximum dry density. The details of compaction test are summarized in Table 3. The graphical representation of variation in maximum dry density with change in moisture content and change in percentage of zeolite is shown in Figs. 1 and 2, respectively.

3.2 *Influence of Zeolite Content on Atterberg Limits*

The values of consistency limit obtained from the study by changing the proportion of natural zeolite were observed. It could be concluded from the observation that a liquid limit, plastic limit as well as plasticity index properties of the prepared sample tend to increase with the increment in zeolite percentage. The details of variation in these consistency limits are tabulated in Table 4. The graphical representations of these variations in Atterberg's limit are shown in Figs. 3 and 4.

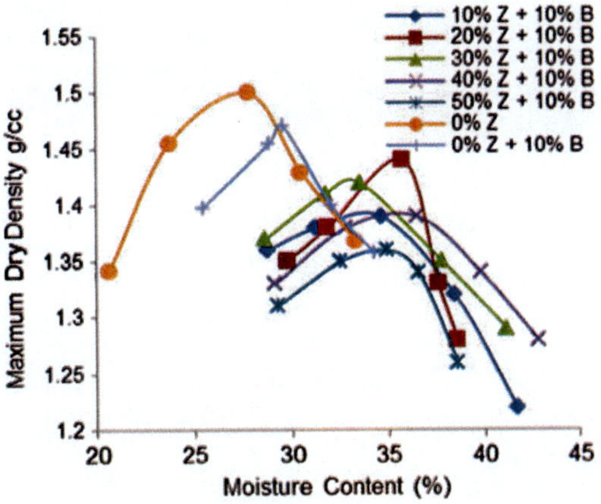

Fig. 1 Compaction characteristics of Kaolin clay, bentonite mixture with different zeolite content

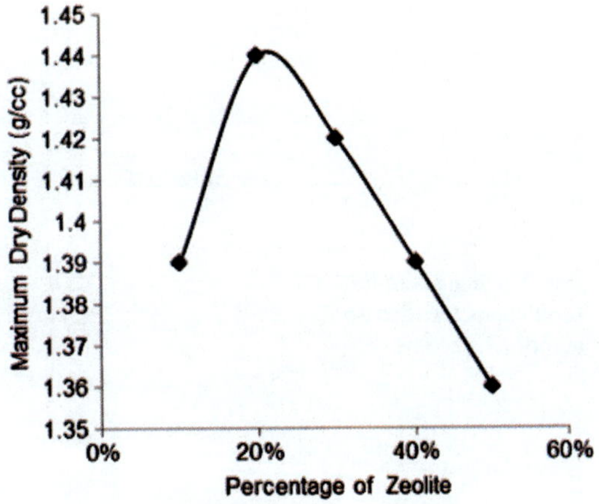

Fig. 2 Variation of dry density with zeolite content

3.3 *Influence of Zeolite Content on Unconfined Compressive Strength*

The increase in the percentage of zeolite content in the bentonite clay mix resulted in the improvement of unconfined compressive strength values. It could be inferred from the results obtained that the maximum strength was 280 kPa. This was obtained when 20% of natural zeolite was added to the bentonite clay mixture. The variations in compressive strength characteristics have been tabulated in Table 5. The stress–strain curves for various natural zeolite contents are graphically illustrated in Fig. 5.

The results obtained for changes in unconfined compressive strength indicated that beyond 20% of zeolite, there was a slight reduction in the compressive strength values. It is also evident that all the samples have met the minimum compressive strength requirement of liner criteria.

Table 4 Variation of consistency limits with zeolite content

Sample specification	Liquid limit (%)	Plastic limit (%)	Plasticity index
Clay + 10% Bentonite + 10% Zeolite	55.63	45.32	22.31
clay + 10% Bentonite + 20% Zeolite	59.00	36.50	22.50
clay + 10% Bentonite + 30% Zeolite	61.50	37.80	23.70
clay + 10% Bentonite + 40% Zeolite	62.92	38.64	24.28
clay + 10% Bentonite + 50% Zeolite	64.32	39.87	24.45

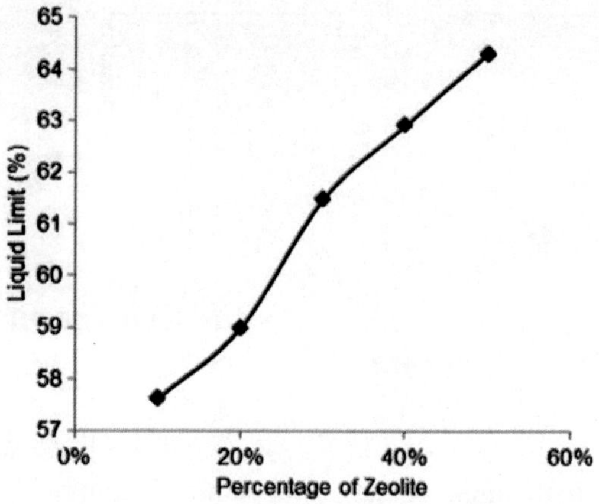

Fig. 3 Changes in liquid limit characteristics on addition of zeolite

3.4 *Variation of Hydraulic Conductivity with Zeolite Content*

The hydraulic conductivity values obtained for zeolite-embedded clay liner mixtures of various proportions denoted that the increase in zeolite percentage decreases the permeability of the liner system. This reduction in conductivity characteristics is mainly due to the porous molecular structure of the zeolite. Thereby, the zeolite entraps water molecule as well as other metal contaminants of sizes in coherence

Fig. 4 Changes in plastic limit characteristics on addition of zeolite

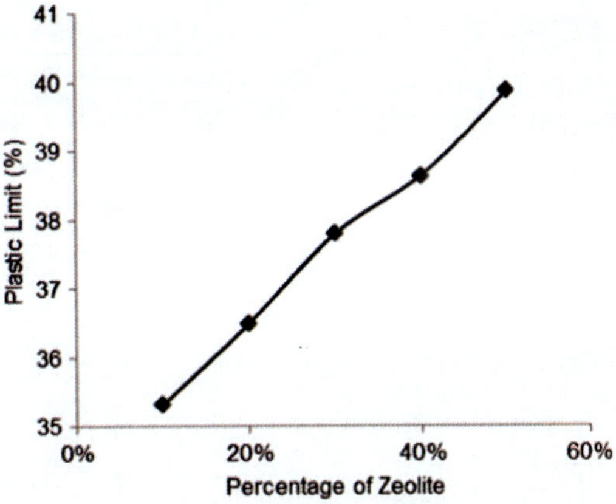

Table 5 Variation of unconfined compressive strength with zeolite content

Sample specification	Unconfined compressive strength value (kPa)
Clay + 10% Bentonite + 10% Zeolite	265
Clay + 10% Bentonite + 20% Zeolite	280
Clay + 10% Bentonite + 30% Zeolite	260
Clay + 10% Bentonite + 40% Zeolite	242
Clay + 10% Bentonite + 50% Zeolite	230

Fig. 5 Stress–strain behavior of different clay samples treated with zeolite

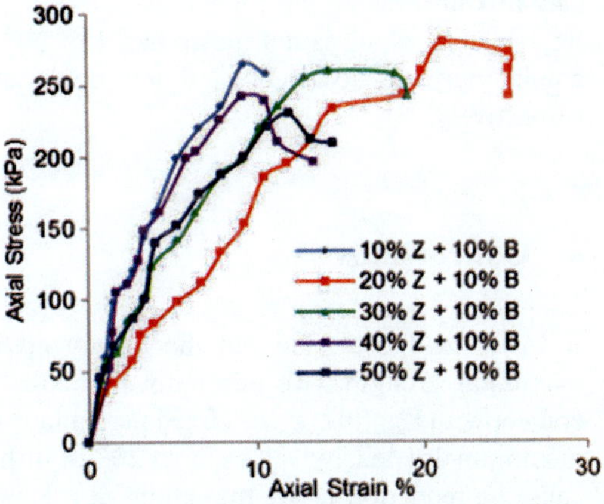

Table 6 Variation of hydraulic conductivity with zeolite content

Sample specification	Hydraulic conductivity (m/s)
Clay + 10% B + 10% Z	3.65×10^{-11}
Clay + 10% B + 20% Z	2.30×10^{-11}
Clay + 10% B + 30% Z	2.15×10^{-11}
Clay + 10% B + 40% Z	1.95×10^{-11}
Clay + 10% B + 50% Z	1.80×10^{-11}

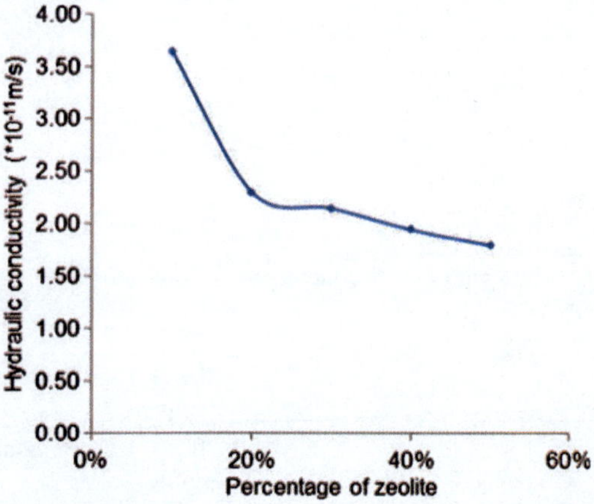

Fig. 6 Variation of hydraulic conductivity at different zeolite content in clay mixture

with inter-molecular pore spaces. This absorption of more water causes a reduction in hydraulic conductivity. In order to draw a conclusive statement regarding the optimum percentage of zeolite addition, the effect of 20% zeolite in the bentonite clay mixture was giving satisfactory results regarding the regulatory values for the liner system which is not more than 1×10^{-7} cm/s. These variations in hydraulic conductivity has been tabulated and graphically represented in Table 6 and Fig. 6, respectively.

4 Conclusions

It was intended to find out the optimized sample mixture which could render maximum strength with minimum hydraulic conductivity. The increase in zeolite content with Kaolin clay and fixed percentage of bentonite imparts an increase in the maximum dry density values up to 20%. Further increment in the zeolite percentage caused a reduction in the maximum dry density values which was an effect of the high moisture absorption characteristics of zeolite. The graphical representations of the unconfined compressive strength values point out that, the changes of strength characteristics resemble the similar trend in compaction properties. Thus the peak compressive strength was obtained for 20% zeolite content in the bentonite clay mix. From the results of consistency limit values, the increase in zeolite content causes a proportional increment in the consistency limits. The increment in consistency limit is the impact of the high absorption capacity per unit volume of natural zeolite. It could be inferred from the hydraulic conductivity values that when zeolite was added the hydraulic conductivity decreased significantly. But the addition of zeolite showed the reduction in hydraulic conductivity is due to the molecular sieve-like structure of zeolite that helps in the absorption considerable volume of water within its subatomic structure. Therefore, considering both strength as well as hydraulic conductivity criteria, the sample with zeolite content 20% mixed with 10% bentonite and 70% Kaolin clay can be taken as an optimum mixture for the liner system.

Bibliography

Hong C, Shackelford C (2011) Consolidation and hydraulic conductivity of zeolite-amended soil-bentonite backfills. J Geotech Geoenviron Eng 138(1):15–25

Joseph S, Varghese M (2017) Study on amended landfill liner using bentonite and zeolite mixtures. Int Res J Eng Technol 04(04):1130–1133

Kaya A, Durukan S (2004) Utilization of bentonite-embedded zeolite as clay liner. Appl Clay Sci 25(1):83–91

Kaya A, Durukan S, Oren A (2006) Determine the engineering properties of bentonite zeolite mixture. Tech J Turk Chamber Civ Eng 17(3):1075–1088

Kayabali K, Kezer H (1998) Testing the ability of bentonite amended natural zeolite (Clinoptilolite) to remove heavy metals from liquid waste. Environ Geol 34:5–102

Oncu S, Bilsel H (2017) Effect of zeolite utilization on volume change and strength properties of expansive soil as landfill barrier. Can Geotech J 54(9):1–44

Oren A, Ozdamar T (2013) Hydraulic conductivity of compacted zeolites. Waste Manage Res 31(6):634–640

Tuncan A, Tuncan M (2003) Use of natural zeolites as a landfill liner. Waste Manage Res 21(1):54–61

Yukselen Y, Aksoy V (2010) Characterization of two natural zeolites for geotechnical and geoenvironmental applications. Appl Clay Sci 50(1):130–136